Essential Numerical Computer Methods

Reliable Lab Solutions

Essential Numerical Computer Methods

Reliable Lab Solutions

Edited by

Michael L. Johnson

Departments of Pharmacology and Medicine
University of Virginia Health System
Charlottesville, VA, USA

ELSEVIER

AMSTERDAM • BOSTON • HEIDELBERG • LONDON
NEW YORK • OXFORD • PARIS • SAN DIEGO
SAN FRANCISCO • SINGAPORE • SYDNEY • TOKYO
Academic Press is an imprint of Elsevier

Academic Press is an imprint of Elsevier
30 Corporate Drive, Suite 400, Burlington, MA 01803, USA
525 B Street, Suite 1900, San Diego, CA 92101-4495, USA
32 Jamestown Road, London NW1 7BY, UK
Linacre House, Jordan Hill, Oxford OX2 8DP, UK

Material in the work originally appeared in Volumes 210, 240, 321, 383, 384, 454,
and 467 of *Methods in Enzymology* (1992, 1994, 2000, 2004, 2009, Elsevier Inc.)

Notice
No responsibility is assumed by the publisher for any injury and/or damage to persons
or property as a matter of products liability, negligence or otherwise, or from any use
or operation of any methods, products, instructions or ideas contained in the material
herein. Because of rapid advances in the medical sciences, in particular, independent
verification of diagnoses and drug dosages should be made

ISBN: 978-0-12-384997-7

For information on all Academic Press publications
visit our website at elsevierdirect.com

Transferred to Digital Printing in 2013

CONTENTS

CONTRIBUTORS

Numbers in parentheses indicate the pages on which the authors' contributions begin.

Robert D. Abbott (67), Division of Biostatistics, Department of Internal Medicine, University of Virginia School of Medicine, Charlottesville, Virginia, USA

G. Bard Ermentrout (293), Department of Mathematics, University of Pittsburgh, Pittsburgh, Pennsylvania, USA

James C. Boyd (561), Department of Pathology, University of Virginia Health System, Charlottesville, Virginia, USA

Marc Breton (461), University of Virginia Health System, Charlottesville, Virginia, USA

Emery N. Brown (187, 199), Statistics Research Laboratory, Department of Anesthesia and Critical Care, Massachusetts General Hospital, Harvard Medical School/MIT Division of Health Sciences and Technology, Boston, Massachusetts, USA

David E. Bruns (561), Department of Pathology, University of Virginia Health System, Charlottesville, Virginia, USA

William Clarke (461), University of Virginia Health System, Charlottesville, Virginia, USA

David F. Dinges (225), Unit for Experimental Psychiatry, University of Pennsylvania School of Medicine, Philadelphia, Pennsylvania, USA

Harold B. Dowse (479), School of Biology and Ecology, University of Maine, Orono, Maine, USA; Department of Mathematics and Statistics, University of Maine, Orono, Maine, USA

Leon S. Farhy (309, 585), Department of Medicine, Center for Biomathematical Technology, University of Virginia, Charlottesville, Virginia, USA

Howard P. Gutgesell (67), Division of Pediatric Cardiology, Department of Pediatrics, University of Virginia School of Medicine, Charlottesville, Virginia, USA

E. R. Henry (81), Laboratory of Chemical Physics, NIDDK, National Institutes of Health, Bethesda MD, USA

J. Hofrichter (81), Laboratory of Chemical Physics, NIDDK, National Institutes of Health, Bethesda MD, USA

Abdul Salam Jarrah (529), Virginia Bioinformatics Institute at Virginia Tech, Blacksburg, Virginia, USA; Department of Mathematics and Statistics, American University of Sharjah, Sharjah, United Arab Emirates

Michael L. Johnson (1, 37, 55, 173), University of Virginia Health System, Charlottesville, VA, USA

Boris Kovatchev (461), University of Virginia Health System, Charlottesville, Virginia, USA

Michelle Lampl (173), Emory University, Atlanta, GA, USA

Reinhard Laubenbacher (529), Virginia Bioinformatics Institute at Virginia Tech, Blacksburg, Virginia, USA

Leslie M. Loew (399), R. D. Berlin Center for Cell Analysis and Modeling, University of Connecticut Health Center, Farmington, CT, USA

Greg Maislin (225), Biomedical Statistical Consulting, Wynnewood, Pennsylvania, USA

Anthony L. McCall (585), Department of Medicine, Center for Biomathematical Technology, University of Virginia, Charlottesville, Virginia, USA

Jay I. Myung (511), Department of Psychology, Ohio State University, Columbus, Ohio, USA

Erik Olofsen (225), Department of Anesthesiology, P5Q, Leiden University Medical Center, 2300 RC, Leiden, The Netherlands

Steven M. Pincus (141), Independent Mathematician, Guilford, CT, USA

Mark A. Pitt (511), Department of Psychology, Ohio State University, Columbus, Ohio, USA

Douglas Poland (257), Department of Chemistry, The Johns Hopkins University, Baltimore, Maryland, USA

James C. Schaff (399), R. D. Berlin Center for Cell Analysis and Modeling, University of Connecticut Health Center, Farmington, CT, USA

Christopher H. Schmid (187, 199), Biostatistics Research Center, Division of Clinical Care Research, New England Medical Center, Tufts University, Boston, Massachusetts, USA

Boris M. Slepchenko (399), R. D. Berlin Center for Cell Analysis and Modeling, University of Connecticut Health Center, Farmington, CT, USA

Walter F. Stafford III (337), Analytical Centrifugation Research Laboratory, Boston Biomedical Research Institute, Boston, MA, USA; Department of Neurology, Harvard Medical School, Boston, MA, USA

Martin Straume (37, 55, 173), COBRA, Inc.

Yun Tang (511), Department of Psychology, Ohio State University, Columbus, Ohio, USA

Joel Tellinghuisen (361), Department of Chemistry, Vanderbilt University, Nashville, Tennessee, USA

Hans P. A. Van Dongen (225), Unit for Experimental Psychiatry, University of Pennsylvania School of Medicine, Philadelphia, Pennsylvania, USA

Donald G. Watts (23), Department of Mathematics and Statistics, Queen's University, Kingston, Ontario, Canada K7L 3N6

F. Eugene Yates (425), The John Douglas French Alzheimer's Foundation, 11620 Wilshire Blvd, Suite 270, Los Angeles, CA, USA

PREFACE

Many of the chapters within this volume were first published almost two decades ago. Since then, these basic algorithms have not changed. However, what has changed is the huge increase in computer speed and ease of use along with the corresponding decrease in the costs. The increase in computer speed has made the use of some algorithms common that were almost never used in biochemistry laboratories two decades ago. During the past two decades, the training of the majority of senior M.D.s and Ph.D.s in clinical or basic disciplines at academic research and medical centers has not kept pace with advanced coursework in mathematics, numerical analysis, statistics, or computer science.

Nevertheless, the use of computers and computational methods has become ubiquitous in biological and biomedical research. One primary reason is the emphasis being placed on computers and computational methods within the National Institutes of Health (NIH) Roadmap. Another factor is the increased level of mathematical and computational sophistication among researchers, particularly among junior scientists, students, journal reviewers and NIH Study Section members. Yet another is the rapid advances in computer hardware and software that make these methods far more accessible to the rank-and-file members of the research community.

There exists a general perception that the applications of computers and computer methods in biological and biomedical research are either basic statistical analysis or the searching of DNA sequence data bases. While these are important applications, they only scratch the surface of the current and potential applications of computers and computer methods in biomedical research. The various chapters within this volume include a wide variety of applications that extend far beyond this limited perception. The chapters within this volume are basically in chronological order of original publication in Methods in Enzymology volumes 210, 240, 321, 383, 384, 454, and 467. This chronological order also provides a general progression from basic numerical methods to more specific biochemical and biomedical applications.

CHAPTER 1

Use of Least-Squares Techniques in Biochemistry[1]

Michael L. Johnson

University of Virginia Health System
Charlottesville, VA, USA

I. Update

This chapter was originally published (Johnson, 1992) under the title "Why, When, and How Biochemists Should Use Least-Squares" and this descriptive title clearly states the purpose of this chapter. This chapter emphasizes the underlying assumptions and philosophy of least-squares fitting of nonlinear equations to experimental data.

In the last two decades, the basic algorithms have not changed. Most are based on the Gauss-Newton algorithm. However, what has changed is the huge increase

---

[1] This article was originally published as "Why, When, and How Biochemists Should Use Least Squares" in *Analytical Biochemistry*, Volume 206 (1992). Reprinted with permission from Academic Press.

ESSENTIAL NUMERICAL COMPUTER METHODS
                                    1   DOI: 10.1016/B978-0-12-384997-7.00001-7

in computer speed and ease of use along with the corresponding orders of magnitude decrease in cost. These factors have combined to make the least-squares fitting of nonlinear equations to experimental data a common task in all modern research laboratories. Many of the available software packages will perform the required least-squares fits, but fail to address subjects such as goodness-of-fit, joint confidence intervals, and correlation within the estimated parameters. This chapter provides an introduction to these subjects.

The increase in computer speed has made some algorithms possible that were almost never used in biochemistry laboratories two decades ago. One such algorithm is the use of bootstraps (Chernick, 1999) for the determination of parameter confidence intervals. If this chapter were being written today (Johnson, 2008), I would have stressed their use.

II. Introduction

There are relatively few methods available for the analysis of experimental data in the biochemical laboratory. Graphical methods and least-squares (regression) methods are by far the most common. Unfortunately, both classes of analysis methods are commonly misused. The purpose of this chapter is to explain why, when, and how a biochemist should use least-squares techniques and what confidence can be assigned to the resulting estimated parameters.

One classic group of biochemical experiments involves measuring the response of a system to an external perturbation. Temperature-jump experiments perturb the chemical equilibrium of a solution by rapidly increasing the temperature of the solution and subsequently monitoring an observable, like absorbance, as a function of time. Here, the absorbance is the observable (i.e., the variable) that is dependent on the experiment, and time is the variable that can be independently controlled by the experimental protocol.

Another example of this general class is the ligand-binding titration experiment. The investigator measures the amount of a ligand bound (the dependent variable) by fluorescence, absorbance, or radioactive counting. To do so, the investigator titrates the ligand concentration (the independent variable). Note that the ligand concentrations might be either the total or the free ligand concentration, depending on the experimental protocol.

In these examples, and all others of this class, the investigator has measured a response caused by a perturbation of the system. The next step is to obtain the parameters of the system that characterize the chemical processes by "analyzing" the data. In the above examples, these parameters, the desired answers, might be the relaxation half-lives or macroscopic binding constants. Alternatively, the desired parameters might be the microscopic forward and reverse reaction rates of the biochemical system.

Analysis of these data requires that the biochemist assume a mathematical relationship between the observed quantities, the dependent variables, and the independent variables. This relationship is the fitting function. In the past, analysis of relaxation experiments, such as temperature jump, assumed that the mathematical relationship was a single exponential decay. Based on this assumption, the investigator would commonly perform a logarithmic transformation of the dependent variable and create a graph of, for example, the logarithm of absorbance as a function of time. If the original assumption of a single exponential process is correct, then the graph will be a straight line with a slope related to the relaxation rate of the chemical process. A single class of binding sites is a common assumption for ligand binding experiments. This, in turn, implied that the mathematical relationship for the amount bound as a function of free, or unbound, ligand was a rectangular hyperbola. A consequence of this mathematical relationship is that various transformations of the data, such as a Scatchard plot, will yield a straight line with a slope related to the binding affinity. It was quickly realized that the assumption of a single biochemical process was generally not valid. Generalizations of these graphical procedures for consideration of multiple processes were attempted but with generally poor results.

The desired result of the analysis of any experimental data is to obtain the set of parameters of the biochemical reaction with the maximum likelihood (ML), highest probability, of being correct. This is the most critical lesson of this review. We do not care what the slope of a log plot is; we want the relaxation rate constants with the ML of being correct. We do not care what the slope of a Scatchard plot is; we want the ligand binding constants with the highest probability of being correct.

Does a Scatchard plot, or a logarithmic plot, yield parameter values with the ML of being correct? Generally, they do not (Johnson and Frasier, 2010). These methods are mathematically correct if the experimental data contain no experimental uncertainties. They fail because they do not correctly consider the experimental uncertainties present in all experimental data. Why then were these graphical methods developed and commonly reported? The evaluation of the parameters with the ML of being correct requires a high-speed digital computer to perform the calculations, but the development of the graphical methods occurred before high-speed digital computers were commonly available to the biochemical researcher. At that stage graphical methods were the only practical ones for the analysis of the experimental data. Should these methods still be used? They may aid the investigator in visualizing the data, but the methods should not be used for determining parameter values.

The most common alternative to graphical analysis in use in the biochemical laboratory today is nonlinear least-squares (NLLS). To use a NLLS method, an investigator must assume a functional form for the mathematical relationship between the dependent and independent variables of the experiments in terms of a series of desired parameters. This functional form is not restricted to a form that can be transformed into a straight line, as with the graphical procedures. NLLS is a process of "fitting" the experimental data to almost any functional form by evaluating an optimal set of parameters for the fitting function.

Does a NLLS method yield parameter values with the highest probability of being correct? It may, if the NLLS analysis procedure is correctly formulated and correctly used (Johnson and Faunt, 2010; Johnson and Frasier, 2010; Straume *et al.*, 1992).

III. Nonlinear Least–Squares

NLLS refers to a group of different mathematical algorithms that perform a "best-fit" of a fitting function to a set of experimental data. The objective of this best-fit operation is to obtain a set of "optimal" parameters for the fitting function such that the fitting function will correctly describe the original data and average out the experimental uncertainties. NLLS is a special case of a more general class of parameter estimation procedures known as ML techniques. For linear and NLLS procedures, the definition of best-fit is that the weighted sum of the squares of the difference between the dependent variables and the fitting function (WSSR) is a minimum when evaluated at the optimal parameter values and the independent variables. These differences are the deviations and/or the residuals:

$$\mathrm{WSSR}(\alpha) = \sum_{i=1}^{n} \left(\frac{Y_i - F(X_i, \alpha)}{\sigma_i} \right)^2 = \sum_{i=1}^{n} \left(\frac{r_i}{\sigma_i} \right)^2 \tag{1}$$

where the weighted sum of the squares of the residuals, WSSR, is a function of the parameters, represented here as the vector α, and the n data point, $X_i Y_i$. The σ_i refers to the statistical weight of the particular data point. This statistical weight is

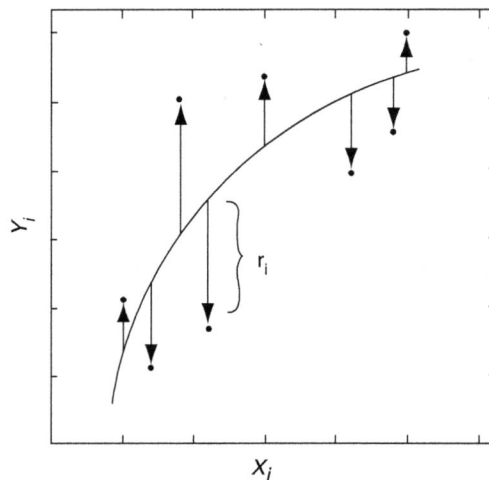

Fig. 1 Graphical representation of the residuals, r_i, of a least-squares parameter estimation procedure. It is the weighted sum of the squares of these residues, WSSR, that is minimized by the least-squares process. Note that the residuals are perpendicular to the X axis, not perpendicular to the fitted curve or the Y axis. From Johnson and Frasier (2010) with permission.

the standard error of the particular Y_i observation. For an unweighted analysis all the σ_i values are identical and usually set to 1. The r_i values in Eq. (1) are graphically depicted in Fig. 1. It is the weighted sum of the squares of the vertical distances that are minimized, not of the horizontal or perpendicular distances.

The temperature-jump experiment mentioned above is a useful example for defining some terms used in the discussion of NLLS. In the classic Eigen temperature-jump apparatus, a sample solution is rapidly heated while the absorbance of the solution is recorded as a function of time. Here, the absorbance (A) is the dependent variable Y_i. Time is the independent variable X_i. In the example presented in Fig. 1, there are only a single dependent and a single independent variable. NLLS is equally valid for experiments with multiple dependent and/or independent variables. The analysis of these data requires a specific form for the mathematical relationship that is used to predict the absorbance as a function of time for a set of parameters that are to be determined by the analysis procedure. For the simplest temperature-jump experiments, this mathematical relationship, the fitting function, is of the form

$$A = (A_0 - A_\infty)e^{-K \cdot \text{time}} + A_\infty \qquad (2)$$

where the parameters to be estimated by NLLS are the chemical relaxation rate K, the initial absorbance (after the temperature jump but before any relaxation has occurred) A_0, and the absorbance after the relaxation process is complete (i.e., infinite time) A_∞. More complex relaxation processes can be analyzed with a fitting function that is a summation of several exponential terms.

There are many different NLLS algorithms: the Nelder-Mead (Caceci and Cacheris, 1984; Nelder and Mead, 1965), Gauss-Newton (Johnson and Faunt, 2010; Johnson and Frasier, 2010; Straume *et al.*, 1992), Marquardt-Levenberg (Marquardt, 1963), and steepest-descent (Bevington, 1969; Johnson and Frasier, 2010), among others. The actual mathematical details of these algorithms are discussed elsewhere and are not repeated here (Bevington, 1969; Caceci and Cacheris, 1984; Johnson and Faunt, 2010; Johnson and Frasier, 2010; Marquardt, 1963; Nelder and Mead, 1965; Straume *et al.*, 1992). For some fitting problems, a particular one of these algorithms may be preferable, whereas other problems may call for a different algorithm (Bevington, 1969; Johnson and Faunt, 2010; Johnson and Frasier, 2010; Straume *et al.*, 1992). For most parameter estimation problems, these algorithms have many common features. All the algorithms will find a set of parameters α that minimize the weighted sum of the squares of the deviations between the fitting function and the data [WSSR(α) in Eq. (1)]. When correctly used, all the algorithms will yield the same optimal parameter values. All the algorithms require the user to provide initial estimates of the parameter values, and they all work by an iterative process of using the initial estimate of the parameters to provide a better estimate of the parameters. The algorithms then iteratively use the better estimate of the parameters as the initial estimate and return an even better estimate, until the parameter values do not change within some specified limit. The validity and usefulness of all the algorithms are based on the same set of assumptions. Does this "least-squares best-fit" provide parameter values with the ML of being correct? Only sometimes will the parameters estimated by NLLS correspond to the desired ML estimates.

Linear least-squares (LLS) is a special case of NLLS. Technically, for LLS the second, and higher, derivatives of the fitting function with respect to the parameters are all zero, whereas for NLLS these derivatives are not zero. An example of a linear fitting function is a simple polynomial equation like $Y = A + BX$. The practical difference between LLS and NLLS is that if the second, and higher, derivatives are all zero then the Gauss-Newton algorithm will require only a single iteration for any initial "guesses" of the fitting parameter values. This, in turn, means that for LLS, the required initial values of the fitting parameters can all be zero. The polynomial least-squares equations found in almost every textbook on this subject (Chemical Rubber Co., 2010) can be derived from the Gauss-Newton NLLS method by assuming that the initial parameter values are zero and performing only a single iteration. Consequently, the restrictions and limitations of NLLS all apply to LLS, and NLLS can always be used instead of LLS. Therefore, only NLLS is discussed here.

IV. Why Use NLLS Analysis Procedures?

There is only one valid reason for using a NLLS analysis procedure: when correctly applied, NLLS will yield parameter values with the highest probability, the ML, of being correct. When NLLS cannot be correctly applied, it should not be used.

Some have claimed that least-squares are always valid because least-squares methods will always provide a set of parameters that correspond to a minimum in the variance-of-fit.[2] Why would we want a minimum variance-of-fit, that is, a minimum WSSR? We desire the parameters with the highest probability of being correct. The parameter values corresponding to the minimum variance-of-fit are not necessarily the parameter values with the highest probability of being correct. The next section discusses the assumptions required for the parameters corresponding to a minimum variance-of-fit to have the highest probability of being correct. The assumptions outlined are *sufficient* to ensure that a least-squares procedure will yield parameter values with the highest probability of being correct. For an arbitrary fitting function, these assumptions are also *necessary* to

---

[2] Variance-of-fit is the average of the weighted squares of the differences between the data points and the fitting function, as shown in Fig. 1. The variance-of-fit is calculated as the *WSSR*, from Eq. (1), divided by the number of data points. Thus, a minimum variance-of-fit corresponds to a minimum *WSSR*, that is, a least-squares minimum. The variance-of-fit is a commonly used, and abused, measure of the quality of a fit. It is generally, and sometimes incorrectly, assumed that the lower the variance-of-fit, the better is the fit of the data.

demonstrate the relationship between ML methods and least-squares methods. However, for a few specific fitting functions, it can be demonstrated that one or more of these assumptions are not required.

V. When to Use NLLS Analysis Procedures

Again, there is only one valid reason to use a NLLS analysis procedure: only if NLLS can be correctly applied to the data. The algebraic demonstration that NLLS will yield a set of estimated parameters that have the ML of being correct for an arbitrary fitting function requires a series of assumptions about the characteristics of the experimental data (Bates and Watts, 1988; Johnson and Faunt, 2010; Johnson and Frasier, 2010; Straume *et al.*, 1992). Specifically, it is the characteristics of the experimental uncertainties contained in the experimental data that must be assumed. Therefore, if these assumptions are valid, then NLLS should be used. Conversely, if these assumptions are invalid, then NLLS should generally not be used. The remainder of this section concentrates on these assumptions and their consequences. Several assumptions listed below are interrelated and are corollaries of other assumptions. Most of the assumptions apply to NLLS and to almost every other method of data analysis.

A. Assumption 1: No Experimental Uncertainty

The demonstration that NLLS is a ML method requires the assumption that the independent variables contain no experimental uncertainty. In practical terms, this assumption means that the precision of the independent variables is much better than the precision of the dependent variables. It is this assumption that allows NLLS to minimize a function of the vertical deviations shown in Fig. 1. For the temperature-jump experiment, this assumption is that the time measurement is significantly more precise than the absorbance measurement. Here, the experimental protocol can clearly be designed such that this assumption is reasonable. Note that a Scatchard analysis generally violates this assumption (Johnson and Faunt, 2010; Johnson and Frasier, 2010; Straume *et al.*, 1992). For Scatchard plots, the experimental uncertainties have been transformed such that they are no longer vertical. Consequently, if an investigator represents ligand binding data as a Scatchard plot, then it is usually not valid to apply LLS to calculate the best slope of the plot. ML methods other than NLLS that can be used for the analysis of experimental data with uncertainties in the independent variables are described elsewhere (Acton, 1959; Bajzer and Prendergast, 2010; Bard, 1974; Bates and Watts, 1988; Johnson, 1985).

B. Assumption 2: Gaussian Uncertainties

The demonstration that NLLS is a ML method also requires the assumption that the experimental uncertainties of the dependent variable must follow a Gaussian (i.e., a random or bell-shaped) distribution with a mean of zero. This means that if the experiment is performed thousands of times, the distributions of values of the individual data points are Gaussian distributions.

This assumption is usually reasonable for the experimental data as collected by the experimenter. In biochemistry, only two types of experimental uncertainty distributions are usually observed: Gaussian and Poisson distributions (Bevington, 1969). Radioactive, photon counting, and similar experiments yield Poisson uncertainty distributions. If the number of counts is high, these Poisson distributions can be closely approximated by Gaussian distributions (Bevington, 1969; Johnson and Faunt, 2010; Johnson and Frasier, 2010; Straume et al., 1992). Almost every other source of uncertainty in biochemical work will yield a Gaussian distribution. Sample handling and preparation uncertainties such as pipetting, weighing, and dilution will yield a Gaussian distribution.

The investigator should not perform any nonlinear transformations of the dependent variables, the Y axis, that will alter the distribution of uncertainties between the collection of the data and the analysis of the data (Johnson and Faunt, 2010; Johnson and Frasier, 2010; Johnson, 1985; Straume et al., 1992). A nonlinear transformation refers to a transformation of the variable other than a simple addition or multiplication. Logarithms, exponentials, powers, and inverses are examples of nonlinear transformations. In the previously described temperature-jump experiment, the original data probably contain a Gaussian distribution of experimental uncertainties in the absorbance and comparatively little uncertainty in the time values. Owing to the complexity of the NLLS fitting process, an investigator might prefer to create a plot of the logarithm of the absorbance as a function of time and then subsequently evaluate the slope of the resulting straight line by LLS. There are several reasons why this is not a statistically valid procedure, but at this point consider the distribution of uncertainties in the dependent variable. The logarithmic transformation of the dependent variable changes the form of the distribution of experimental uncertainties on the absorbance. The logarithmic transformation of a Gaussian is not a Gaussian (Johnson and Faunt, 2010; Johnson and Frasier, 2010; Straume et al., 1992). If the experimental uncertainty distribution is not a Gaussian then LLS cannot be used to evaluate the parameters of the straight line. This problem cannot be corrected by "appropriate weighting factors" (Johnson and Faunt, 2010; Johnson and Frasier, 2010; Straume et al., 1992). Consequently, the logarithmic transformation of the data has created a fitting equation of a significantly simpler form but, in the process, has precluded the use of LLS for the analysis.

The commonly used reciprocal plots, such as the Lineweaver-Burk plot, also violate the assumption of a Gaussian distribution of experimental uncertainties. The original enzyme velocities probably follow a Gaussian distribution, but the inverse of the velocities used in the Lineweaver-Burk plot generally does not contain a Gaussian distribution of experimental uncertainties. Consequently, the reciprocal plots generate a fitting equation of a simpler form and create a distribution of uncertainties that precludes the use of least-squares as an analysis method.

An investigator should not perform nonlinear transformations of the dependent variables before proceeding with the analysis of the data (Johnson and Faunt, 2010; Johnson and Frasier, 2010; Johnson, 1985; Straume et al., 1992) if the original data contain Gaussian uncertainties. However, transformations of the dependent variable are valid if the transformations are performed to convert a non-Gaussian distribution of experimental uncertainties to a Gaussian distribution of experimental uncertainties. This is the only statistically valid reason to perform nonlinear transformations of the dependent variables. The reverse hemolytic plaque assay (Leong et al., 1985) is an example of the type of experiment where the original distribution of uncertainties is a skewed distribution that is approximately an exponential of a Gaussian. For this experimental protocol, it is best to perform a logarithmic transformation of the dependent variable, namely, the plaque size. For the reverse hemolytic plaque assay, this nonlinear transformation will transform the distribution of experimental uncertainties such that they are approximately Gaussian.

The LLS and NLLS techniques allow transformations of the independent variables, that is, the X axis (Johnson, 1985). This is because NLLS assumes that no experimental uncertainty exists in the independent variables.

It is also possible to convert an experimental protocol that yields experimental uncertainty distributions that are not Gaussian to a protocol that yields a Gaussian distribution by replicate measurements of the experimental data points. The central limit theorem of calculus states that the mean of a group of numbers will have a Gaussian uncertainty distribution even if the individual replicates have uncertainty distributions that are not Gaussian (Mathews and Walker, 1970). Therefore, the mean of a group of replicate measurements will have a more Gaussian-like distribution than the individual replicates. Consider a standard radioactively labeled hormone binding experiment. Usually the amount of bound hormone is determined by radioactive counting with relatively low numbers of counts. Therefore, the distribution of experimental uncertainties in the amount bound should follow a Poisson distribution. These experiments are usually performed as a series of replicate experiments at each hormone concentration, with the means used for the analysis of the data. According to the central limit theorem these mean values will tend to have a Gaussian distribution, rather than a Poisson distribution. Therefore, NLLS can be used to estimate parameters from the mean values of an experiment of this type.

This does not mean that hormone binding experiments should be performed as a series of replicates. Given a choice between 10 data points measured in triplicate and 30 individual data points, it is better to measure the 30 individual data points at different hormone concentrations and count the radioactivity of each data point long enough that the Poisson distribution of uncertainties can be approximated as a Gaussian distribution. Some experimenters feel that having the triplicates will allow an obvious bad point to be eliminated. Although this is true, having 30 singlet observations of hormone binding would also allow an obvious bad point to be eliminated as the observations must consistently follow a smooth binding isotherm. What is gained by more singlet observations is the ability to evaluate how well the calculated curve actually describes the data, that is, the ability to evaluate the "goodness-of-fit" and test the hypothesis that the fitting equation is consistent with the experimental data.

If the experimental protocol cannot be altered, or the data manipulated, to create a Gaussian distribution of experimental uncertainties, then the NLLS method should generally not be used. The reader is referred to the more general ML methods that can be formulated without the assumption of a Gaussian distribution (Acton, 1959; Bajzer and Prendergast, 2010; Bard, 1974; Bates and Watts, 1988; Johnson and Faunt, 2010; Johnson and Frasier, 2010; Straume et al., 1992).

It is assumed that no systematic uncertainties exist within the data. Any type of systematic uncertainty would require either a non-Gaussian distribution of uncertainties or a nonzero mean of the uncertainties. Thus, this assumption is a corollary of Assumption 2 which states that the experimental uncertainties are Gaussian with a mean of zero. However, it is treated separately here because of its consequences. Consider the logarithmic plot of the temperature-jump experiment. For this plot, it is the logarithm of the difference between the absorbance and the final absorbance $(A_i - A_\infty)$ that is plotted. Here, the value of A_∞ must be estimated first, the logarithms of the differences calculated, and then the slope determined. Small errors in the determination of A_∞ will create systematic uncertainties in the values of the logarithms and will be reflected as a systematic error in the evaluation of the slope. Thus systematic errors will appear in the evaluation of the relaxation rate constants (Johnson and Faunt, 2010; Johnson and Frasier, 2010; Straume et al., 1992). Table I and Figs. 2 and 3 present an example of this problem. Figure 2 presents a synthetic data set. Table I presents the results of three NLLS analyses of this data with different assumed values for A_∞. Case 4 in Table I is an analysis of these data with A_∞ as an additional estimated variable. Figure 3 presents the corresponding logarithmic plots with two different assumed values for A_∞. There is no method by which the least-squares process can detect systematic uncertainties, or be modified to consider systematic uncertainties, like those shown in Fig. 3. These systematic errors

Table I
Least–Squares Analysis of Data in Fig. 2[a]

Case	$A_0 - A_\infty$	K	A_∞
1	10.0 (9.6, 10.4)	1.00 (0.94, 1.06)	0.50[b]
2	9.7 (9.3, 10.2)	1.14 (1.07, 1.22)	1.00[b]
3	10.3 (9.9, 10.7)	0.89 (0.84, 0.95)	0.00[b]
4	9.9 (9.4, 10.3)	1.08 (0.90, 1.25)	0.78 (0.10, 1.42)

[a]Values in parentheses are the ± 1 SD joint confidence intervals for these parameters. See text for details.

[b]These values were assumed for the analysis of this case.

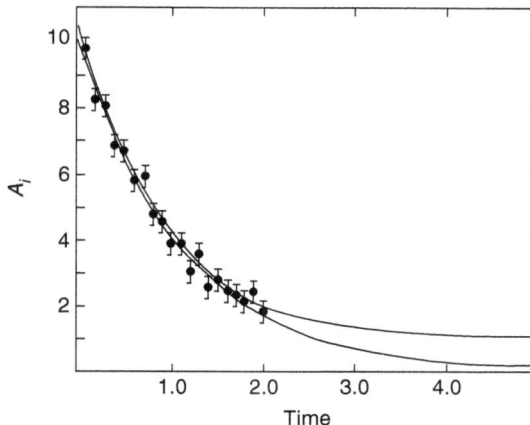

Fig. 2 Synthetic data set with pseudorandom experimental uncertainty added. The data were generated with $A_0 = 10.5$, $A_\infty = 0.5$, and a decay rate $K = 1.0$. Definitions are the same as those for Eq. (2). Pseudorandom noise was added with a standard deviation of 0.26. The solid lines correspond to cases 2 and 3 in Table I. From Johnson and Frasier (2010) with permission.

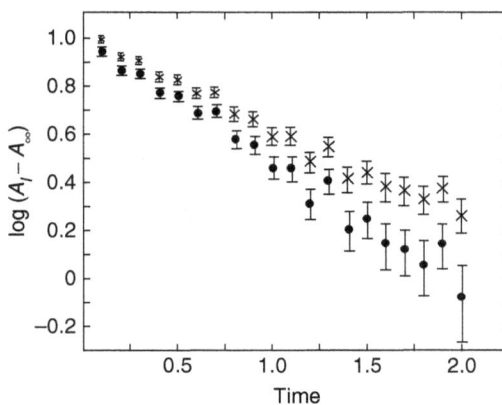

Fig. 3 Logarithmic plots of the data shown in Fig. 2. The lower set of data points was generated by assuming A_∞ is equal to 1.0 (case 2 in Table I), and the upper set of points was generated by assuming that A_∞ is equal to 0.0 (case 3 in Table I). Note that the resulting slopes are distinctly different. From Johnson and Frasier (2010) with permission.

cannot be corrected by appropriate weighting factors. As far as the least-squares parameter estimation procedure is concerned, systematic uncertainties simply do not exist. Systematic uncertainties should be eliminated by changing the data collection protocol.

This general type of problem occurs whenever a parameter is estimated and then assumed to be correct for subsequent analysis steps. The subsequent analysis does not include any possibility of considering the consequences of the uncertainties of the previous steps. The entire analysis should be performed by a single-step multiple-parameter estimation process like NLLS that considers the joint uncertainties of the parameters simultaneously.

The validity of the use of NLLS is also dependent on the assumption that the fitting function [e.g., Eq. (2)] is the correct mathematical description of the nonrandom processes contained within the data. Stated another way, NLLS assumes that the dependent variables of the data can be described as the sum of the random (Gaussian) experimental noise and the fitting function evaluated at the corresponding independent variables and the optimal parameter values. The second assumption and both of its corollaries are simply different statements of the requirement that the residuals be a measure of the random uncertainties of the data.

A self-evident, but commonly overlooked, consequence of this assumption is that an incorrect fitting equation will result in the estimated parameters having no physical meaning. For example, consider the binding of oxygen to human hemoglobin. Human hemoglobin exists in solution as an $\alpha\beta$ dimer that self-associates to form a $\alpha_2\beta_2$ tetramer (Johnson and Ackers, 1977; Johnson and Lassiter, 1990). The dimer binds two oxygens, and the tetramer binds four oxygens. Until relatively recently, the most common fitting equation for the analysis of oxygen binding to human hemoglobin contained only four parameters, namely, the four Adair binding constants of the tetrameric hemoglobin (Johnson and Ackers, 1977; Johnson and Lassiter, 1990). The assumption was that the hemoglobin concentration was high enough to preclude the formation of the dimeric species. Thus, the consequences of the dissociation of the tetramer into dimers and the binding of oxygen by the dimers were neglected. It has been shown that a fitting equation that neglects the dimeric species will yield incorrect answers for the Adair binding constants even at the hemoglobin concentrations found within red blood cells (Johnson and Ackers, 1977; Johnson and Lassiter, 1990). For other examples of this type, the reader is referred to Johnson and Frasier (2010). The lesson is that parameters estimated by any curve fitting procedure are dependent on the assumed form of the fitting equation. The presentation, or publication, of parameters determined by these methods should always include a statement about the assumed molecular mechanism.

One further comment about the nature of the fitting function is in order. All least-squares algorithms require that the fitting function be continuous at each of the data points. Furthermore, most least-squares algorithms also require that the first derivatives of the fitting function with respect to all of the parameters being estimated be continuous at each of the data points. The fitting functions can have discontinuities as long as they do not coincide with the experimental data.

C. Assumption 3: Independent Observations

For a NLLS procedure to produce parameter values with the highest probability of being correct, the individual data points must be independent observations. This is a standard statistical assumption for almost every type of statistical and mathematical analysis. A common source of data points with uncertainties that are not independent occurs wherever data are collected by an automated data acquisition system with a time response that is slow compared to the time between the data points. When data are collected in this manner, the instrument response will cause an apparent serial correlation between successive data points; in other words, if one data point contains a random uncertainty, then the subsequent data points will have a tendency to have an uncertainty in the same direction.

The best method to approach the analysis of data that has been perturbed by the response characteristics of an instrument is to include the instrument response function in the analysis procedure. An approach of this type is used for the analysis of time-correlated single-photon counting (TCSPC) fluorescence lifetime measurements. If the fluorescent molecules are instantaneously excited, then the fluorescence intensity as a function of time is the intensity decay law, $I(t)$. The form of $I(t)$ can differ depending on the particular mechanism for fluorescence emission; the exact form is not important for this discussion. The problem is that the fluorescent molecules cannot be instantaneously excited by the instrument. The flash lamp or laser pulse has a finite width. The data collected by the instrument contain information about the intensity decay, $I(t)$, and about the time dependence of the intensity of the excitation lamp, $L(t)$. The correct fitting function for TCSPC data is a combination of the lamp intensity function and the intensity decay function. The correct combination of these two functions is the convolution integral of the two functions, $L(t) \otimes I(t)$. By fitting to the convolution integral, the systematic uncertainties, introduced because of the finite pulse width of the excitation lamp, are included in the fitting function. The convolution integral correctly describes the experimental data and, thus, allows the use NLLS for the analysis of the data.

D. Assumption 4: Large Number of Data Points

There must be sufficient data points to provide a good random sampling of the random experimental uncertainties. This assumption is not actually required to demonstrate that least-squares provide a ML estimate of the parameter values. This assumption is, however, required for the assignment of realistic measures of the accuracy/precision of the estimated parameters. The theoretical minimum number of data points is equal to the number of parameters being simultaneously estimated. Because each data point contains experimental uncertainty, significantly more data points than the minimum are required. The system is "overdetermined" when more than the minimum number of data points is used. Unfortunately, there is no method to access the actual number of data points required to provide a good random sampling of the experimental uncertainties.

Experimental data should never be smoothed. Data smoothing is commonly, and incorrectly, used to improve the quality of experimental data. However, once the experimental data have been smoothed, it is impossible to obtain a good random sampling of the random experimental uncertainties of the original data. Furthermore, all smoothing algorithms will perturb the information within the data as well as remove noise from the data. Improving the quality of the experimental data is equivalent to increasing the information content of the data. However, the process of smoothing experimental data does not add information to the experimental data. The smoothed experimental data actually contain less information than the original data because of the perturbations caused by the smoothing process. The only method to increase the information content of an experimental data set is to collect more experimental data.

When the experimental uncertainties contained within a data set are consistent with the above assumptions, it is appropriate to use a NLLS procedure for the analysis of the data. Conversely, NLLS should probably not be used if these assumptions are not satisfied. A ML method probably can be formulated for experimental data with almost any distribution of uncertainties in the dependent and independent variables (Johnson, 1985).

VI. What Confidence Can Be Assigned to Results of NLLS Analysis?

There are two steps required for the analysis of experimental data by NLLS or any other method. The first is to find the set of parameters with the ML of being correct, and the second is to find realistic measures of the accuracy of those parameters. When we determined that the relative mass ratio, M_r, of a protein is 90,000, what does this number mean? If the accuracy of the determination is $\pm 80,000$, then we know relatively little. However, if the accuracy is ± 1000, we might be able to use the M_r to increase our understanding of the protein. A functional measure of the accuracy of the determined values is actually more important than the optimal values. If we knew that the M_r was probably between 89,000 and 91,000, would we care if the value with the highest probability of being correct was 90,000 or 90,001? An investigator should always provide a realistic measure of the precision of the determined values when such values are reported.

This section discusses the determination of confidence intervals for parameters determined by NLLS methods. The confidence intervals are a measure of the precision to which a group of parameters can simultaneously be determined from a limited set of data. Confidence intervals are measures of the precision of the measurement based on a single set of data. If the assumptions required are valid, these confidence intervals will also provide a good measure of the absolute accuracy of the determined parameters.

It should be clear that there is no exact theory for the evaluation of confidence intervals for nonlinear fitting equations. All the methods are extensions of the methods developed for LLS and, therefore, require a linear fitting equation. These methods all assume that the fitting equation can be approximated as a first-order series expansion in the estimated parameters. This assumption is always valid for linear fitting equations. For nonlinear fitting equations, this assumption is usually reasonable for small perturbations of the parameter values from the corresponding minimum least-squares values.

There are several approximate methods for the evaluation of the confidence intervals of simultaneously estimated parameters that can be used with NLLS methods. The most commonly used one, the "asymptotic standard errors" (Johnson and Faunt, 2010; Johnson and Frasier, 2010), is both the easiest to calculate and by far the least accurate for most applications. Asymptotic standard errors nearly always provide underestimates of the actual confidence limits of the determined parameters. It is the use of the asymptotic standard errors that is responsible for the perception among many investigators that the confidence intervals reported by NLLS procedures are so inaccurate that they cannot be used for any practical purpose. This perception is correct because almost every commonly available least-squares analysis program either reports no measure of the precision of the determined parameters or reports the values of the asymptotic standard errors. There are many other published methods that provide realistic estimates of the confidence intervals of parameters determined by NLLS methods. The reason that these other methods are rarely used is that they are significantly more complex and require significantly more computer time for evaluation. They also require a considerably more complex computer program.

Most NLLS procedures require, or provide for, the evaluation of the "information matrix." The information matrix is the basis for most methods commonly used for the evaluation of the confidence intervals, or the precision, of determined parameters. This information matrix is also called the Hessian matrix, H. The individual j, k elements of the matrix are defined as

$$H_{j,k} = \sum_{i=1}^{n} \frac{1}{\sigma_{i^2}} \left[\frac{\partial F(X_i, \alpha)}{\partial \alpha_j} \cdot \frac{\partial F(X_i, \alpha)}{\partial \alpha_k} \right] \tag{3}$$

where the summation is over the n data points. $F(X_i, \alpha)$ is the fitting function evaluated at a particular independent variable X_i and optimal estimate of the fitting parameters α. The J and k subscripts refer to particular fitting parameters, that is, particular elements of the α vector and the H matrix.

The variance-covariance matrix is evaluated by multiplying the inverse of the Hessian matrix by the variance of the random uncertainties of the experimental data. Usually the variance of the residuals (variance-of-fit) is assumed to be a reliable estimate of the true variance of random experimental uncertainties of the data. This is Assumption 4 from the previous section. This is true only in the asymptote as the number of data points approaches infinity. In this context, infinity is simply enough data points to provide a good random sampling of the experimental uncertainties of the data. The inverse of H times the variance-of-fit is the asymptotic variance-covariance matrix, AVC. The diagonal elements of the AVC matrix are the squares of the asymptotic standard errors of the corresponding simultaneously estimated parameters. The off-diagonal elements of AVC are the covariances of the parameters.

Most NLLS procedures report the asymptotic standard errors of the parameters as the measure of the confidence, the precision, of the estimated parameters. Three assumptions were made to obtain these confidence estimates: we assumed that the fitting equation was linear; that the number of data points is near infinite; and that the covariance terms can be neglected. The first is probably a reasonable assumption. The second may be a reasonable assumption. The third assumption is usually unreasonable. When parameters are simultaneously determined by NLLS, they will usually have a significant covariance. The consequence of neglecting the covariances is that the confidence intervals will significantly underestimate the actual range of the confidence intervals for the simultaneously determined parameters. Consequently, the resulting measures of the precision of the determined parameters that neglect the covariance are not reasonable. The reader should question the validity of computer programs that report the asymptotic standard errors of determined parameters without also reporting the corresponding covariances.

The assumption that the covariance terms can be neglected is equivalent to assuming that the fitting parameters are all orthogonal. Parameters are mathematically orthogonal if the corresponding off-diagonal elements of the inverse of the Hessian matrix are zero; that is, if the cross-correlation and covariance of the parameters are zero. Operationally, if the parameters are orthogonal, then the evaluation of the parameters does not depend on the values of the other parameters. This means that the values of the parameters can be evaluated separately, and a simultaneous NLLS procedure is not required. Note that the orthogonality of fitting parameters is dependent on both the actual form of the fitting equation and the individual data points being fit. For example, a Fourier series is an orthogonal equation, but the Fourier coefficients are orthogonal only if there are $2m + 1$ equally spaced data points per primary period of the sine and cosine function.

A Fourier analysis is one of the few cases that a biochemist is likely to encounter in which these assumptions are valid. A Fourier analysis is equivalent to a least-squares fit of the experimental data to the function

$$Y_i = \sum_{l=0}^{m} \left[a_l \text{ cosine}\left(\frac{2\pi X_i l}{\text{period}}\right) + b_l \text{ sine}\left(\frac{2\pi X_i l}{\text{period}}\right) \right] \qquad (4)$$

where the parameters to be estimated are the coefficients of the sine and cosine terms, a_l and b_l, and b_0 is fixed at zero. There are $2m + 1$ parameters estimated in a Fourier analysis. Because all the second derivatives of Y with respect to the parameters to be estimated are zero, this is a linear fitting problem. If the data points are equally spaced in the independent variable X, if the number of data points is equal to the number of estimated parameters $n = 2m + 1$, and if the period is equal to $(n + 1)/n$ times the difference between the largest and smallest independent variable, then the off-diagonal elements of the inverse Hessian matrix H are zero and the parameters are orthogonal. If these assumptions about the spacing of the data points are not met, then the coefficients from a Fourier analysis will not be orthogonal even though the basis functions are orthogonal. If a classic Fourier series analysis is performed without these assumptions being met, then it will yield incorrect estimates of the Fourier coefficients. For almost every other fitting equation (including a simple straight line like $Y = A + BX$), the parameters will not be orthogonal.[3]

If the fitting equation is not orthogonal in the parameters, the covariance terms will be nonzero and cannot be neglected for the estimation of the uncertainties of the estimated parameters. If the covariances cannot be neglected, then the asymptotic standard errors do not provide the investigator with reasonable estimates of the uncertainties of the fitted parameters. The consequence of neglecting the covariances is that the confidence intervals for the determined parameters will be significantly underestimated. This underestimate can commonly be a factor of two or three. Thus the investigator might significantly underestimate the standard errors of the determined parameters and reach incorrect conclusions about the significance of the results. Asymptotic standard errors should not be used as an estimate of the confidence of parameters determined by either LLS or NLLS.

What, then, can be used to evaluate confidence intervals of simultaneously determined parameters? Monte Carlo methods are the very best, but they require a tremendous amount of computer time (Straume and Johnson, 2010a) and, therefore, are usually impractical and are not discussed here. One could create a large grid of all combinations of the fitting parameters and then search for where the increase of the variance is statistically significant. These grid search methods will usually provide good estimates of the regions of the parameter grid where the parameters are not significantly different. These regions are the joint confidence regions for the parameters and can usually be approximated as a multidimensional ellipse. Figure 4 presents a typical elliptically shaped confidence region obtained by a grid search. Because grid search methods also require a large amount of computer time, they are generally not used.

---

[3] Note that orthogonal forms of simple polynomial equations like $Y = A + BX$ can be created by a transformation of the independent variables, X (Bevington, 1969; Acton, 1959). The transformed equation is of the form $Y = a + B(X - \beta)$, where $\beta = (\sum X_i)/N$ and $a \neq A$. Note that it is β that makes the equation orthogonal and that the value of β is not determined by the form of the equation; β is determined by the distribution of values of the independent variable, X_j. This is similar to a Fourier series in that the parameters are orthogonal only if the distribution of the independent variable is correct, even though the sine and cosine basic functions are orthogonal!

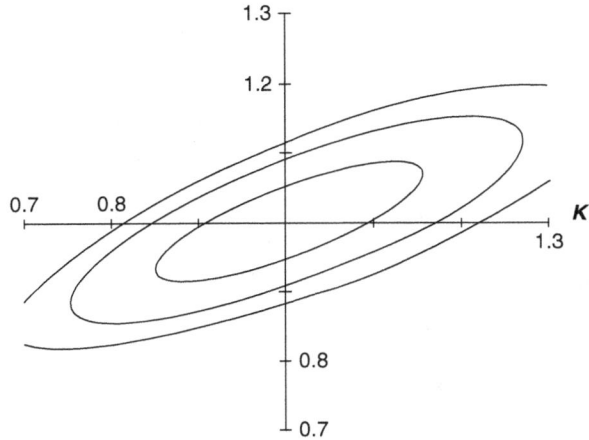

Fig. 4 Examples of 68% (± 1 SE), 95% (± 2 SE), and 99% (± 3 SE) confidence contours. This is the result of a two-parameter estimation problem taken from Johnson and Faunt (2010) with permission. In this least-squares problem the two parameters were K and $A_0 - A_\infty$ with definitions as for Eq. (2). Note that these confidence regions are almost elliptically shaped and that the axes of the ellipse do not correspond to the parameter axes. The rotations of the ellipse axes are a measure of the covariance between the fitted parameters.

The question of what increase in the variance (or WSSR) is statistically significant provides the groundwork for the following discussion. The standard definition of statistical significance in this context is

$$\frac{\text{WSSR}(\alpha')}{\text{WSSR}(\alpha)} = 1 + \frac{p}{n-p} F(p, n-p, 1-\text{PROB}) \tag{5}$$

where p is the number of parameters being simultaneously estimated, n is the number of data points, and F is the upper $1-\text{PROB}$ quantile for Fisher's F distribution with p and $n-p$ degrees of freedom (Bates and Watts, 1988). Equation (5) can be used to compare the probability, PROB, that any set of parameters á is statistically different from the optimal parameters α. The validity of Eq. (5) is based on two assumptions. It assumes that the observations are independent and, therefore, that the number of degrees of freedom for the problem is $n-p$. It also assumes a linear fitting equation. The derivation of the right-hand side of Eq. (5) requires that the WSSR at any point á be the sum of the WSSR at the point α and a WSSR arising from the change of the parameters from α to á. This separation of the WSSR into component parts is valid only for linear equations. However, the assumption that the fitting equation is approximately linear for small changes in α is usually reasonable.

The functional form of the elliptically shaped joint confidence region is available for linear equations (Bard, 1974; Bates and Watts, 1988; Box, 1960; Johnson and

Faunt, 2010; Johnson and Frasier, 2010; Straume *et al.*, 1992). The joint confidence region for a particular PROB is the ellipsoid á

$$(\alpha' - \alpha)^{\mathrm{T}} H^{\mathrm{T}} H (\alpha' - \alpha) \leq ps^2 F(p, n - p, 1 - \text{PROB}) \tag{6}$$

where

$$s^2 = \frac{\text{WSSR}(\alpha)}{n - p} \tag{7}$$

and the other variables are as previously defined. The derivation of Eq. (6) makes the assumption that the parameters are not orthogonal. Equation (6) models the variance as a quadratic shaped space near the point of minimum variance where the parameters have the ML of being correct. Therefore, the joint confidence intervals derived from Eq. (6) only make the assumption that the fitting equation is linear. This assumption is usually reasonable for small perturbations of the estimated parameters. The use of Eqs. (6) and (7) for the evaluation of the joint confidence intervals provides a significantly better estimate of the precision of the determined parameters than the asymptotic standard errors.

Equations (6) and (7) predict the quadratic shape of the variance space from the Hessian matrix evaluated at α. This is possible because of the assumption of a linear fitting equation. My preference is to use Eq. (5) for an actual search for all parameters á corresponding to any desired probability (Johnson and Faunt, 2010; Johnson and Frasier, 2010; Straume *et al.*, 1992). This search can be limited to specific directions from the optimal values α to save computer time. If p parameters are being estimated, the search can be limited to $4p$ direction. First, each α_i is searched in both directions, while holding the remaining α_j terms, $i \neq j$, at their optimal values. Second, Eqs. (6) and (7) are used to evaluate the directions of the axes of the multidimensional ellipse of the joint confidence intervals. The ellipse axes are also searched in both directions for values of á that are different at the same probability levels. The evaluation of these directions simply involves the rotation of the coordinate system such that the off-diagonal elements of the inverse of the Hessian matrix in the new coordinate system are all zero (Johnson and Faunt, 2010; Johnson and Frasier, 2010). In this new coordinate system, the new parameters are orthogonal. The joint confidence regions are the extreme values of the statistically acceptable parameters found by the search. The search for statistically significant sets of parameters á eliminates some, but not all, of the consequences of the assumption that the fitting equation is linear. Therefore, this search procedure will provide joint confidence intervals that are more precise than the joint confidence intervals predicted by Eqs. (6) and (7).

Another more accurate and more time consuming procedure is to search each of the parameters in both directions, much like the search described in the previous paragraph. The primary difference is that at each point within each of the searches an additional least squares procedure is performed to evaluate all of the other model parameters (Bates and Watts, 1988). For example, consider the case of the fit of Eq. (2) to a set of data. The primary parameter estimation will determine the optimal

(ML) values of A_0, A_∞, and K. Then to determine the confidence region for each of these parameters, each is searched in both directions for a significant increase in the weighted sum-of-squared residuals, WSSR, as given by Eq. (8). Specifically, for the search along the K axis, the value of the WSSR is evaluated by fixing K at a series of specific values and then performing additional parameter estimations to determine A_0 and A_∞ that are conditional on the specific value of K. Note that for each of these searches the WSSR reflects the consequence of the variation of a specific single parameter. Thus, Eq. (8) is analogous to Eq. (5) except for p being equal to 1.

$$\frac{\text{WSSR}(\alpha')}{\text{WSSR}(\alpha)} = 1 + \frac{1}{n-1} F(1, n-1, 1 - \text{PROB}) \tag{8}$$

It is interesting that the joint confidence intervals for nonlinear problems are not symmetrical. Suppose that we have determined a free energy change for some biochemical process. Further, suppose that we have evaluated the joint confidence region for this free energy change and that it is symmetrical. We can then express the value of the free energy change as some value plus or minus a value of the uncertainty. If we want to express the value of the corresponding equilibrium constant also, we can perform the appropriate nonlinear transformation. However, when we attempt to transform the joint confidence interval of the free energy change into a joint confidence interval for the equilibrium constant, we find that the interval is no longer symmetrical and cannot be expressed as plus or minus a single value. A careful examination of Fig. 4 and Table I shows that the elliptically shaped confidence region is not quite symmetrical and/or not centered at the optimal values α. Therefore, the reader should question the validity of any NLLS computer program that provides a symmetrical estimate of the confidence intervals of the determined parameters.

VII. Conclusions

Our choice of methods for the analysis of experimental data is extremely limited. The methods that are available always make assumptions about the nature of the experimental data being analyzed. An investigator needs to be aware of the requirements placed on the data by these assumptions before collecting the data. It is while the experiment is being designed that the data collection protocol can most readily be altered to be compatible with the available data analysis methods.

When publishing results, a realistic measure of the precision of the determined values should accompany the published values. These are essential for the reader to evaluate the significance of the values reported. Asymptotic standard errors should not be used as an estimate of the confidence of parameters simultaneously determined by either LLS or NLLS. Joint confidence intervals are preferred as they are more accurate than asymptotic standard errors.

Some investigators consider the results of a computer analysis as gospel. Computers are not oracles, however, and computer programmers sometimes make inappropriate assumptions. Programmers commonly use approximations to speed either the programming or the time of execution of programs, and they do make mistakes. Some computer programs are correct for one application, but when used for different applications the methods no longer apply. It is necessary to be aware of the assumptions made by the programmer about the nature of the experimental data being analyzed, and one must be aware of the basic assumptions of the method of analysis. The investigator must always question the applicability of any method of analysis for each particular problem. After results are obtained from a computer the next question should be, "Does this result have any physical meaning?" Do not assume that the values are correct because they come from a computer analysis of the data.

Whenever possible, investigators should devise methods to "test" their analysis programs. The tests might be with real sets of data that have known answers, for example, measuring the fluorescence lifetime of a compound with a known lifetime. These tests also might involve simulated experiments with realistic amounts of pseudorandom experimental uncertainties added (Johnson and Faunt, 2010; Johnson and Frasier, 2010; Straume and Johnson, 2010a; Straume et al., 1992). The need to include realistic experimental uncertainties in simulated data cannot be overemphasized. Many analysis methods work well for test cases without experimental noise and fail with even small amounts of experimental noise present.

This chapter has attempted to present the basic ideas and assumptions of linear and NLLS analysis methods. It does not include the rigorous mathematical descriptions of the methods, nor does it include a discussion of topics like the propagation of errors based on joint confidence intervals (Johnson and Faunt, 2010; Johnson and Frasier, 2010; Straume et al., 1992), analysis of the randomness of residuals (Straume and Johnson, 2010b; Straume et al., 1992), goodness-of-fit criteria (Straume and Johnson, 2010b; Straume et al., 1992), global analysis (Johnson and Faunt, 2010; Johnson and Frasier, 2010; Straume et al., 1992), weighting functions (Di Cera, 2010), and the advantages of alternate sets of fitting parameters (Johnson and Faunt, 2010; Johnson and Frasier, 2010; Straume et al., 1992). For the actual methods, the reader is referred to other articles (Caceci and Cacheris, 1984; Johnson and Faunt, 2010; Johnson and Frasier, 2010; Johnson, 1985; Leong et al., 1985; Marquardt, 1963; Nelder and Mead, 1965; Straume and Johnson, 2010a,b; Straume et al., 1992). More complete general discussions of these topics are available for the beginner (Acton, 1959; Bevington, 1969; Caceci and Cacheris, 1984; Johnson and Faunt, 2010; Johnson and Frasier, 2010; Johnson, 1985; Leong et al., 1985; Straume and Johnson, 2010a,b; Straume et al., 1992; Di Cera, 2010) and for the mathematician (Bard, 1974; Bates and Watts, 1988).

Acknowledgments

This work was supported, in part, by the University of Virginia Diabetes Endocrinology Research Center Grant USPHS DK-38942, the University of Virginia National Science Foundation Science and Technology Center for Biological Timing, and National Institutes of Health Grants GM-28928 and GM-35154.

I acknowledge and thank Dr. Ludwig Brand, Dr. D. Wayne Bolen, Dr. William Jakoby, and Dr. Dima Toptygin for comments on this review. Software for NLLS analysis is available on written request from the author (Michael L. Johnson, Departments of Pharmacology and Internal Medicine, Box 448, University of Virginia Health Sciences Center, Charlottesville, VA 22908).

References

Acton, F. S. (1959). Analysis of Straight Line Data. Wiley, New York.

Bajzer, Z., and Prendergast, F. G. (1992). Maximum Likelihood Analysis of Fluorescence Data. *Methods Enzymol.* **210,** 200–237.

Bard, Y. (1974). *In* "Nonlinear Parameter Estimation" p. 67. Academic Press, New York.

Bates, D. M., and Watts, D. G. (1988). Nonlinear Regression Analysis and Its Applications. Wiley, New York.

Bevington, P. R. (1969). Data Reduction and Error Analysis for the Physical Sciences. McGraw-Hill, New York.

Box, G. E. P. (1960). *Anal. NY Acad. Sci.* **86,** 792.

Caceci, M. S., and Cacheris, W. P. (1984). *Byte* **9**(5), 340.

Chemical Rubber Co. (1964). *In* "Standard Mathematical Tables," 13th edn., p. 425. Chemical Rubber Co., Cleveland, OH.

Chernick, M. R. (1999). Bootstrap Methods, A Practitioner's Guide. Wiley Interscience, New York.

Di Cera, E. (1992). Use of Weighting Functions in Data Fitting. *Methods Enzymol.* **210,** 68–87.

Johnson, M. L. (1985). The Analysis of Ligand Binding Data with Experimental Uncertainties in the Independent Variables. *Anal. Biochem.* **148,** 471–478.

Johnson, M. L. (2008). Nonlinear least-squares fitting methods. *Methods Cell Biol.* **84,** 781–805.

Johnson, M. L., and Ackers, G. K. (1977). Resolvability of Adair Constants from Oxygenation Curves Measured at Low Hemoglobin Concentrations. *Biophys. Chem.* **7,** 77–80.

Johnson, M. L., and Faunt, L. M. (1992). Parameter Estimation by Least-Squares. *Methods Enzymol.* **210,** 1–37.

Johnson, M. L., and Frasier, S. G. (1985). Nonlinear Least-Squares Analysis. *Methods Enzymol.* **117,** 301–342.

Johnson, M. L., and Lassiter, A. E. (1990). Consequences of Neglecting the Hemoglobin Concentration on the Determination of Adair Binding Constants. *Biophys. Chem.* **37,** 231–238.

Johnson, M. L. (1992). Why, when, and how biochemists should use least-squares. *Anal. Biochem.* **206,** 215–225.

Leong, D. A., Lau, S. K., Sinha, Y. N., Kaiser, D. L., and Thorner, M. O. (1985). *Endocrinology (Baltimore)* **116,** 1371.

Marquardt, D. W. (1963). *SIAM J. Appl. Math.* **14,** 1176.

Mathews, J., and Walker, R. L. (1970). *In* "Mathematical Models of Physics," 2nd edn., p. 383. Benjamin/Cummings, Menlo Park, CA.

Nelder, J. A., and Mead, R. (1965). *Comput. J.* **7,** 308.

Straume, M., and Johnson, M. L. (1992a). A Monte Carlo Method for Determining Complete Probability Distributions of Estimated Model Parameters. *Methods Enzymol.* **210,** 117–129.

Straume, M., and Johnson, M. L. (1992b). Analysis of Residuals: Criteria for Determining Goodness-of-Fit. *Methods Enzymol.* **210,** 87–106.

Straume, M., Frasier-Cadoret, S. G., and Johnson, M. L. (1992). Least-Squares Analysis of Fluorescence Data. *In* "Topics in Fluorescence Spectroscopy," (J. R. Lakowicz, ed.), p. 177. Plenum, New York.

CHAPTER 2

Parameter Estimates from Nonlinear Models

Donald G. Watts

Department of Mathematics and Statistics
Queen's University, Kingston
Ontario, Canada K7L 3N6

Parameter estimates from linear models enjoy all sorts of important useful properties. For example, if the response is assumed to be well described by a model of the form

$$f(\mathbf{x}, \boldsymbol{\beta}) = \beta_0 + \beta_1 x_1 + \beta_2 x_2$$

where β_0, β_1, and β_2 are parameters and x_1 and x_2 are factors, then one can derive exact expressions for the unique least-squares estimates of the parameters and for regions of plausibility, such as joint and marginal confidence regions. These expressions are helpful in assessing how well we "know" a parameter and whether a model may be simplified, for example, by removing one or more parameters/factors, or whether a parameter can be assumed to have a particular value, and so on.

For models in which the parameters appear nonlinearly, for example, a compartment model,

$$f(t, \boldsymbol{\theta}) = \theta_1 \exp(-\theta_2 t) + \theta_3 \exp(-\theta_4 t)$$

ESSENTIAL NUMERICAL COMPUTER METHODS
Copyright © 1994 by Elsevier Inc. All rights reserved.

DOI: 10.1016/B978-0-12-384997-7.00002-9

none of the properties enjoyed by linear models pertain; it is not even possible to derive expressions for the least-squares estimates of the parameters, let alone exact regions of plausibility. The most common approach to stating plausibility regions for parameters and for deciding whether a nonlinear model can be simplified is to use linear approximation confidence regions. Unfortunately, linear approximation regions can be extremely misleading.

In this chapter, we present improved procedures for assessing how well we know nonlinear parameters. The methods require some extra computing after the model has been fitted to a data set, but the computing is efficient and easily accomplished.

I. Introduction

Fitting nonlinear models to data relies heavily on procedures used to fit linear models. Accordingly, we begin with a brief review of fitting linear models, including how to assess the quality of parameter estimates for such fits. We then discuss fitting nonlinear models and application of linear model methods for assessing the quality of parameter estimates for nonlinear models. Finally, we discuss a more accurate and valid procedure for characterizing the behavior of estimates of parameters in nonlinear models, illustrating the approach and the insights it gives using a model for frontal elution affinity chromatography and a compartment model.

A. Linear Regression: An ELISA Example

The activity of a monoclonal antibody was determined as a function of antibody concentration using an enzyme-linked immunosorbent assay (ELISA). The assay response (optical density) and antibody concentration (ng/ml) are listed in Table I and plotted in Fig. 1. As can be seen from the plot, the response shows some tendency to increasing variance at higher doses, but we ignore this in development of the model. For these data, the straight-line model

$$f(x_n, \boldsymbol{\beta}) = \beta_1 + \beta_2 x_n \tag{1}$$

is seen to be appropriate, where $f(x_n, \boldsymbol{\beta})$ is the absorbance (optical density) and x is the antibody concentration (ng/ml).

A formal statistical treatment for fitting a linear model is to assume a set of data consisting of N values of P factors, x_{np}, $n = 1, 2, \ldots, N$, $p = 1, 2, \ldots, P$, and the corresponding values of a response, y_n. The model can be written in matrix form as

$$\mathbf{Y} = \mathbf{X}\boldsymbol{\beta} + \mathbf{Z} \tag{2}$$

where $\mathbf{Y}$ is the $N \times 1$ vector of random variables representing the responses, Y_n, $\mathbf{X}$ is the $N \times P$ derivative matrix, $\boldsymbol{\beta}$ is the $P \times 1$ vector of unknown parameter values, and $\mathbf{Z}$ is the vector of random variables representing the noise infecting the data. The noise for each case is assumed to be normally distributed with mean 0 and variance σ^2, and independent from case to case. The quantity $\mathbf{X}\boldsymbol{\beta}$ is called the

Table I
Absorbance and Antibody Concentration for Elisa Study

Dose (ng/ml)	Absorbance	Dose (ng/ml)	Absorbance
10	0.041	80	0.340
10	0.043	80	0.347
20	0.081	100	0.402
20	0.087	100	0.407
40	0.185	150	0.671
40	0.187	150	0.690
60	0.267	200	0.853
60	0.269	200	0.878

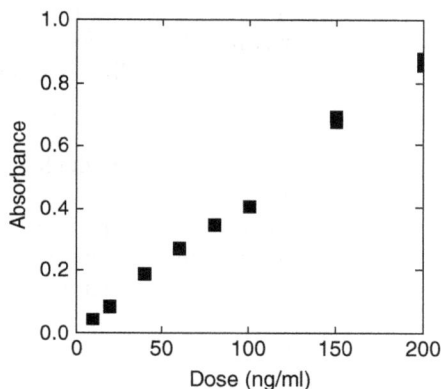

Fig. 1 Plot of absorbance versus antibody concentration for ELISA study.

expectation function, and the model is termed linear because the derivative of the expectation function with respect to any parameter does not depend on any of the parameters (Bates and Watts, 1988). The straight-line model [Eq. (1)] is easily seen to be linear because $\partial f/\partial\beta_1 = 1$ and $\partial f/\partial\beta_2 = x_n$, and neither of these involves any parameters.

Classical statistical analysis (Draper and Smith, 1981) shows that for a linear model, the least squares estimates of $\boldsymbol{\beta}$, given data $\mathbf{y}$, are

$$\hat{\boldsymbol{\beta}} = (\mathbf{X}^T\mathbf{X})^{-1}\mathbf{X}^T\mathbf{y} \tag{3}$$

where $\hat{\boldsymbol{\beta}} = (\hat{\beta}_1, \hat{\beta}_2, \ldots, \hat{\beta}_P)^T$. The least squares estimator can also be shown to be normally distributed with expected value $\boldsymbol{\beta}$ and variance-covariance matrix $(\mathbf{X}^T\mathbf{X})^{-1}\sigma^2$. It follows that parameter β_p has estimated standard error

$$se(\hat{\beta}_p) = s[\{(\mathbf{X}^T\mathbf{X})^{-1}\}_{pp}]^{1/2} \tag{4}$$

Table II
Parameter Summary for ELISA Model

Parameter	Estimate	Standard error	Correlation
$\beta_1 \times 10^{-4}$	−9.04	74.7	
$\beta_2 \times 10^{-3}$	4.37	0.73	−0.80

where $s^2 = S(\boldsymbol{\beta})/(N - P)$ is the variance estimate given by the minimum sum of squares divided by the degrees of freedom, $N - P$, and so a $1 - \alpha$ confidence interval is

$$\hat{\beta}_p \pm t(N - P; \alpha/2)se(\hat{\beta}_p) \tag{5}$$

where $t(N - P; \alpha/2)$ is the value that isolates an area $\alpha/2$ under the right tail of the Student"s t distribution with $N - P$ degrees of freedom. Furthermore, a $(1 - \alpha)$ joint parameter inference region for all the parameters is given by

$$(\beta - \hat{\beta})^T \mathbf{X}^T \mathbf{X}(\boldsymbol{\beta} - \hat{\boldsymbol{\beta}}) \leq Ps^2 F(P, N - P; \alpha) \tag{6}$$

where $F(P, N - P; \alpha)$ is the value which isolates an area α under the right tail of Fisher"s F distribution with P and $N - P$ degrees of freedom.

For the absorbance data, the residual variance is $s^2 = 3.18 \times 10^{-4}$ with 14 degrees of freedom. Parameter summary statistics are given in Table II, and joint confidence regions are ellipses, as shown in Fig. 2.

B. Nonlinear Regression

Data on the elution volume of the human immunodeficiency virus (HIV) protein p24gag as a function of the soluble protein concentration were presented in Rosé *et al.* (1992) The analytical affinity chromatography data are listed in Table III and plotted in Fig. 3.

The model proposed for the elution volume as a function of the soluble p24gag concentration is

$$f(x, \boldsymbol{\theta}) = \frac{\theta_1}{\{1 + 4/[(1 + 8x/\theta_2)^{1/2} - 1]\}x} \tag{7}$$

where x is the soluble p24gag concentration and θ_1 and θ_2 are the unknown parameters. The model is nonlinear because at least one of the derivatives with respect to the parameters involves at least one of the parameters (Bates and Watts, 1988), for example,

$$\frac{\partial f}{\partial \theta_1} = \frac{\theta_1}{\{1 + 4/[(1 + 8x/\theta_2)^{1/2} - 1]\}x}$$

For a nonlinear model with expectation function $f(x_n, \boldsymbol{\theta})$, a formal statistical analysis involves writing the model for the nth case as

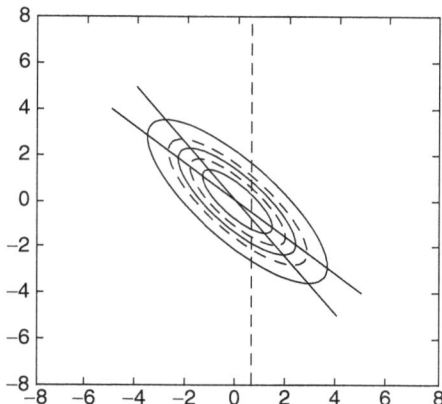

Fig. 2 Joint 60%, 80%, 90%, 95%, and 99% inference regions for β_1 and β_2 for the absorbance data. The straight lines are profile trace plots, which intersect the inference regions where the tangents to the curves are vertical and horizontal.

Table III
Net Retarded Elution Volume and Soluble p24gag Concentration

p24gag concentration (μM)	Elution volume ($\times 10^{-4}$ l)	p24gag concentration (μM)	Elution volume ($\times 10^{-4}$ l)
0.141	7.30	8.27	4.59
0.282	7.12	16.8	3.55
0.652	6.47	36.5	2.30
1.54	6.05	91.0	0.167
3.20	5.63		

$$Y_n = f(x_n, \boldsymbol{\theta}) + Z_n \qquad (8)$$

where $\boldsymbol{\theta} = (\theta_1, \ldots, \theta_P)^T$ is a $P \times 1$ parameter vector. As for the linear model [Eq. (2)], the disturbances Z_n are assumed to be normally distributed with mean 0, constant variance, σ^2, and independent from case to case.

Unlike the linear model [Eq. (2)], no analytical results exist for the estimates and their distributions—there is not even an explicit solution for the least squares estimates. Instead, we must resort to iterative techniques to achieve convergence to $\hat{\boldsymbol{\theta}}$. Once convergence has been achieved, the properties of the estimates are usually assumed to be well represented by linear approximations evaluated at the least squares estimates $\hat{\boldsymbol{\theta}}$. For example, the linear approximation variance-covariance matrix is taken to be $(\mathbf{V}^T\mathbf{V})^{-1}s^2$, where $\mathbf{V} = \partial\boldsymbol{\eta}/\partial\boldsymbol{\theta}^T$ is the derivative matrix with rows $\partial f(x_n, \boldsymbol{\theta})/\partial\boldsymbol{\theta}^T$ evaluated at $\hat{\boldsymbol{\theta}}$, and $s^2 = S(\hat{\theta})/(N - P)$ is the variance estimate.

The linear approximation standard error for the parameter θ_p is, by analogy with Eq. (4),

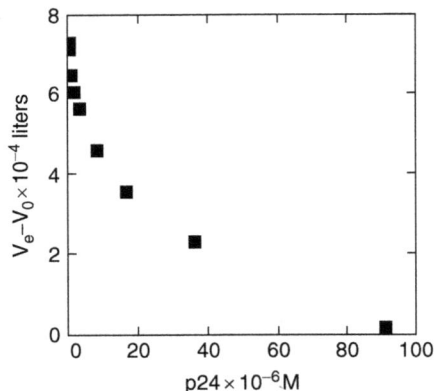

Fig. 3 Plot of net retarded elution volume versus soluble p24gag concentration.

$$se(\hat{\theta}_p) = s[\{(\mathbf{V}^T\mathbf{V})^{-1}\}_{pp}]^{1/2} \tag{9}$$

and a linear approximation $(1 - \alpha)$ marginal confidence interval is, by analogy with Eq. (5),

$$\hat{\theta}_p \pm t(N - P; \alpha/2)se(\hat{\theta}_p). \tag{10}$$

Finally, a linear approximation $(1 - \alpha)$ joint parameter inference region for the parameters is taken to be

$$(\boldsymbol{\theta} - \hat{\boldsymbol{\theta}})^T\mathbf{V}^T\mathbf{V}(\boldsymbol{\theta} - \hat{\boldsymbol{\theta}}) \leq Ps^2F(P, N - P; \alpha) \tag{11}$$

which corresponds to Eq. (6).

For the elution data, convergence to the least squares estimates was obtained with a residual variance of $s^2 = 2.08 \times 10^{-9}$ with 7 degrees of freedom. The least squares estimates and some linear approximation summary statistics are given in Table IV. The parameter correlation estimate is 0.99.

C. Profile Plots and Profile Traces

The methods presented so far rely on statistical theory which is too dependent on the linear model. By removing this restriction, we are able to recognize that, in the situation where the noise is normally distributed with constant variance, all the important information about the parameters is embodied in the sum of squares function

Table IV
Linear Approximation Parameter Summary for Elution Example

| | | | Linear approximation | | Likelihood: 99% region | |
| | | | 99% Region | | | |
Parameter	Estimate	Standard error	Lower	Upper	Lower	Upper
$\theta_1 \times 10^{-8}$	2.16	0.39	0.80	3.52	1.18	4.07
$\theta_2 \times 10^{-5}$	3.00	0.61	0.88	5.12	1.51	6.11

$$S(\boldsymbol{\theta}) = \sum_{n=1}^{N}[y_n - f(\mathbf{x}_n, \boldsymbol{\theta})]^2 \qquad (12)$$

which, for a given data set, depends only on the parameters. Consequently an informative and meaningful description of the sum of squares surface provides all the inferential information about the parameters.

The important features of a sum of squares function are (1) the location of the minimum, given by the least squares estimates $\hat{\boldsymbol{\theta}}$, (2) the value of the minimum, $S(\hat{\boldsymbol{\theta}})$, and (3) the behavior near the minimum. For the special case of a linear model, these quantities can all be specified analytically: (1) the location of the minimum is $\hat{\boldsymbol{\beta}} = (\mathbf{X}^T\mathbf{X})^{-1}\mathbf{X}^T\mathbf{y}$, (2) the value of the minimum is $S(\hat{\boldsymbol{\beta}}) = \mathbf{y}^T\mathbf{y} - \hat{\boldsymbol{\beta}}^T\mathbf{X}^T\mathbf{X}\hat{\boldsymbol{\beta}}$, and (3) the behavior is $S(\boldsymbol{\beta}) = S(\hat{\boldsymbol{\beta}}) + (\boldsymbol{\beta} - \hat{\boldsymbol{\beta}})^T\mathbf{X}^T\mathbf{X}(\boldsymbol{\beta} - \hat{\boldsymbol{\beta}})$. Consequently, parameter inference regions are concentric ellipsoids corresponding to specific levels of the sum of squares (e.g., see Fig. 2). In particular, a joint $(1 - \alpha)$ inference region corresponds to the contour specified by $S(\boldsymbol{\theta}) = S_F$, where

$$S_F = S(\hat{\boldsymbol{\theta}})\left[1 + \frac{P}{N - P}F(P, N - P; \alpha)\right]. \qquad (13)$$

A marginal $(1 - \alpha)$ inference interval for parameter θ_p can also be determined from the sum of squares surface, because the end points correspond to two special points on the contour specified by $S(\boldsymbol{\theta}) = S_t$, where

$$S_t = S(\hat{\boldsymbol{\theta}})\left[1 + \frac{t^2(N - P; \alpha/2)}{N - P}\right]. \qquad (14)$$

At these special points, the other parameters are at their conditional minimum values, say, $\tilde{\boldsymbol{\theta}}(\theta_p)$.

As discussed in the introduction, we cannot write an explicit expression for the least squares estimates for a nonlinear model, but we can determine the location of the minimum, $\hat{\boldsymbol{\theta}}$, and the value at the minimum, $S(\hat{\boldsymbol{\theta}})$, using an iterative procedure. The remaining two tasks are to describe the behavior near the minimum and to express the behavior in terms of parameter joint and marginal inference regions. The first of these tasks can be done very efficiently by profiling the sum of squares surface (Bates and Watts, 1988); the second simply requires determining the values of the parameters at which the profile sum of squares correspond to the critical values of S_t and S_F.

D. Profiling

1. Calculations

Profiling a sum of squares surface involves the following calculations: (1) Select the profile parameter θ_p. Specify the increment $\Delta = 0.1 \times se(\hat{\theta}_p)$. (2) Initialize $\theta_p = \hat{\theta}_p$ and $\tilde{\boldsymbol{\theta}}(\theta_p) = \hat{\boldsymbol{\theta}}$. (3) Increment $\theta_p = \theta_p + \Delta$. Use previous $\tilde{\boldsymbol{\theta}}(\theta_p)$ as starting value. Converge to $\tilde{\boldsymbol{\theta}}(\theta_p)$, the profile trace vector. Store θ_p, $\tilde{\boldsymbol{\theta}}(\theta_p)$, and $\tilde{S}(\theta_p)$, the profile sum of squares. Repeat (3) as "necessary." (4) Set $\Delta = -\Delta$. Go to (2). When finished with parameter θ_p, go to (1) and repeat until all parameters are profiled.

For two parameters, the calculations involve incrementing from $\hat{\theta}_1$ a small positive amount, converging to $\tilde{\theta}_2$, storing θ_1, $\tilde{\theta}_2$, and $\tilde{S}$, incrementing θ_1 again, converging to $\tilde{\theta}_2$, storing θ_1, $\tilde{\theta}_2$, and $\tilde{S}$, and so on, until enough information has been obtained to allow calculation of the 99% likelihood interval upper end point. Then return to $\hat{\theta}_1$ and increment from there in small negative amounts, repeating the calculations until enough information has been obtained to allow calculation of the 99% likelihood interval lower end point. Then return to $\hat{\boldsymbol{\theta}}$ and profile on θ_2, first using positive increments and then negative increments.

2. Converting to Likelihood Regions

Expressing the behavior of the sum of squares surface in terms of parameter inference regions involves finding the values of θ_p which produce a profile sum of squares equal to the critical value S_t defined in Eq. (14). This could be done by plotting $\tilde{S}(\theta_p)$ versus θ_p and then finding the values θ_p where $\tilde{S}(\theta_p) = S_t$, but a more informative approach is as follows:

1. Convert parameter values θ_p to studentized values:

$$\delta(\theta_p) = (\theta_p - \hat{\theta}_p)/se(\hat{\theta}_p) \qquad (15)$$

2. Convert values of the profile sum of squares to profile t values equal to the square root of the "relative excess" sum of squares:

$$\tau(\theta_p) = sgn(\theta_p - \hat{\theta}_p)\{[\tilde{S}(\theta_p) - S(\hat{\boldsymbol{\theta}})]/s^2\}^{1/2} \qquad (16)$$

3. Plot the profile t values $\tau(\theta_p)$ versus $\delta(\theta_p)$.
4. The points defining a $(1 - \alpha)$ marginal interval correspond to the points where

$$\tau(\theta_p) = \pm t(N - P; \alpha/2) \qquad (17)$$

Therefore, an exact likelihood interval is obtained by refracting the critical value $\pm t(N-P; \alpha/2)$ on the vertical scale through the profile t curve onto the horizontal scale.

5. Finally, convert from δ to θ to express the likelihood interval end points in terms of the original parameters.

The transformations in steps 1 and 2 are advantageous because it is then easy to compare parameters from the same model and data set or from different models and different data sets. It is also easy to see how nonlinear the behavior of a parameter is because, if we used profiling to investigate a model which is linear in the parameters, the profile t plot would be a straight line at 45° through the origin. Therefore, departures of a profile t plot from the 45° reference line reveal how nonlinear that parameter is. More important, however, is the fact that exact marginal likelihood intervals can be calculated—there is no need to rely on linear approximations.

E. Profile t Plots for Elution Data

Profile t plots for the parameters θ_1 and θ_2 for the elution data and model are given in the figures on the diagonal of Fig. 4. The tau (τ) curves lie below the linear reference lines, indicating that the sum of squares surface falls steeply as the parameter value approaches the least-squares estimate from below and rises slowly as the parameter value increases above the least-squares estimate.

Exact likelihood intervals can be obtained by refracting the value of the confidence coefficient, $t(N - P, \alpha/2)$, on the vertical (τ) scale through the point on the profile t curve onto the horizontal (δ) scale, then converting from δ to θ. For this example with $N - P = 7$, to obtain a 99% likelihood interval we find $t(7, 0.005) = 3.50$, and so refraction of ± 3.50 onto the horizontal axis through the $\tau(\theta_1)$ curve gives $\delta(\theta_1) = (-2.28, +5.35)$. These convert to end points $(1.18, 4.07)$, which are very different from the (symmetric) linear approximation interval $(0.80, 3.52)$. Similarly, for θ_2, the exact interval is $(1.51, 6.11)$, which is very different from the linear approximation interval $(0.88, 5.12)$.

F. Profile Trace Plots

Further useful meaningful information can be obtained from pairwise plots of the components of the trace vector $\tilde{\boldsymbol{\theta}}(\theta_p)$ versus the profile parameter θ_p, that is, a plot of $\tilde{\theta}_q(\theta_p)$ versus θ_p and of $\tilde{\theta}_p(\theta_q)$ versus θ_q on the same figure. For a linear model, a trace plot of $\tilde{\beta}_q(\beta_p)$ versus β_p will be a straight line through the origin with slope given by the correlation between the parameters [derived from the appropriate element of the matrix $(\mathbf{X}^T\mathbf{X})^{-1}$; see, e.g., Fig. 2]. Note that the profile trace values correspond to points where the sum of squares contours have vertical or horizontal tangents.

For a nonlinear model, the traces will be curved but will still intersect parameter joint likelihood contours at points of vertical and horizontal tangency. This

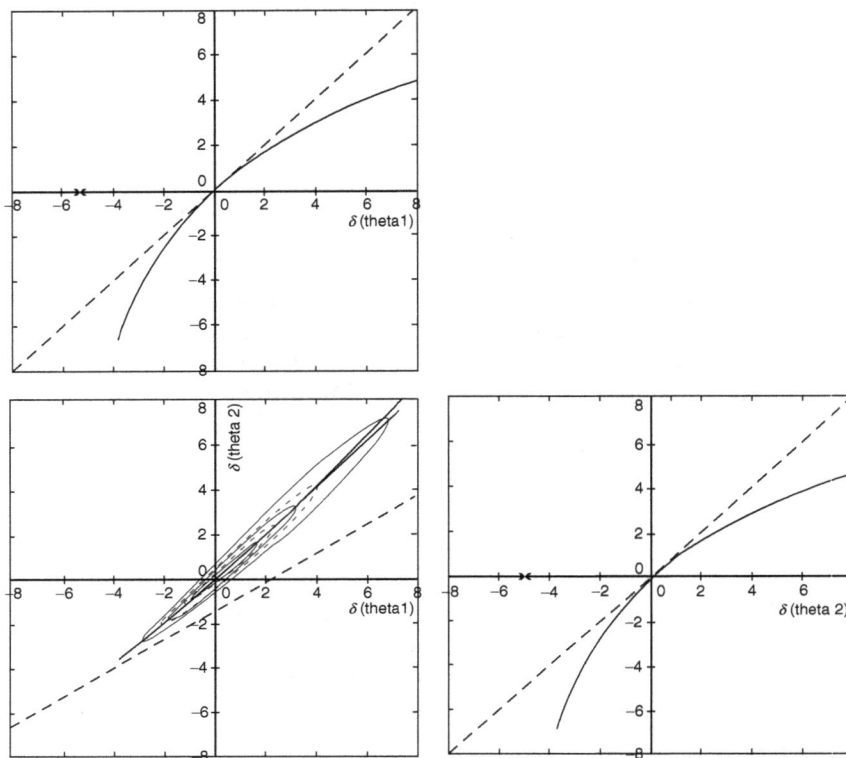

Fig. 4 Profile t and trace plots for the parameters in the elution volume model. On the diagonal plots, the solid line is the profile t function and the dashed line is the linear reference. On the (off-diagonal) profile trace plot, the solid and dashed closed curves denote the 60%, 80%, 90%, 95%, and 99% joint likelihood regions.

information, together with information from the profile t plots, can be used to obtain accurate sketches of the joint regions, as described by Bates and Watts (1988) The traces and sketches reveal useful information about interdependence of the parameter estimates caused by the form of the expectation function, the experimental design used in the investigation, and the actual data values obtained. Such information can provide valuable insights for inference and for model building, as shown below.

A profile trace plot for the parameters θ_1 and θ_2 for the elution data and model is given in the off-diagonal plot of Fig. 4. The solid intersecting curves are the profile traces, and the solid and dashed closed curves correspond to 60, 80, 90, 95, and 99% joint likelihood regions. The profile traces are quite straight, but they lie on top of one another for $\theta < \hat{\theta}$ and are very close together for $\theta > \hat{\theta}$. The closeness of the profile traces is expected, because of the large linear approximation correlation coefficient. The joint regions are fairly elliptical but are nonsymmetric about the least squares point.

For this model and data set, the parameters show little nonlinear interaction, because the joint regions are essentially ellipses which are squashed in for $\theta < \hat{\theta}$ and stretched out for $\theta > \hat{\theta}$. In other examples (see, e.g., Bates and Watts, 1988) the joint regions are curved as well as differentially extended, so the interdependence between the parameters changes as the parameters deviate from the least squares values. In other words, the sum of squares surface is curved as well as nonparabolic.

G. Tetracycline Metabolism

Data on the metabolism of tetracycline were presented by Wagner (1967) In this experiment, a tetracycline compound was administered orally to a subject, and the concentration of tetracycline hydrochloride in the serum (μg/ml) was measured over a period of 16 h. The data are plotted in Fig. 5.

A two-compartment model with delay was fitted in the form

$$f(t, \boldsymbol{\theta}) = \frac{\theta_3 \theta_1 \{\exp[-\theta_1(t - t_0)] - \exp[-\theta_2(t - t_0)]\}}{\theta_2 - \theta_1}. \tag{18}$$

The parameters θ_1 and θ_2 are transfer coefficients, and the parameter θ_3 corresponds to an initial concentration. To ensure positive values for these parameters, we let $\phi_p = \ln(\theta_p)$, with $p = 1, 2$. Summary statistics for the ϕ parameters are given in Table V. The residuals for this model were well behaved, indicating a good fit. Profile plots for the parameters are given in Fig. 6.

Because there are only 5 degrees of freedom for the residual variance, the critical values for $t(N - P, \alpha/2)$ and $F(2, N - P; \alpha)$ are very large for $\alpha = 0.10, 0.05$, and 0.01, and so it is not possible to determine the 90%, 95%, and 99% joint inference regions. The profile t plots show that the parameters behave fairly linearly up to about the 95% level $[t(5, 0.025) = 2.57]$. The upper end point of the 99% likelihood interval for ϕ_1 is

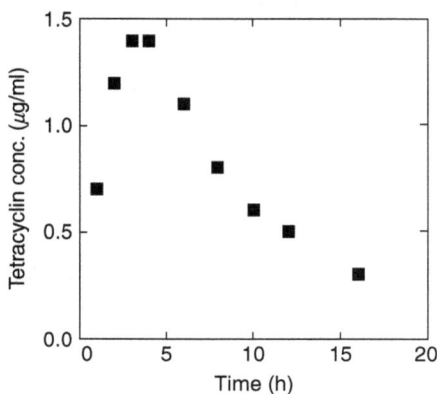

Fig. 5 Plot of tetracycline hydrochloride concentration vesus time.

Table V
Linear Approximation Parameter Summary for Tetracycline Model

Parameter	Estimate	Standard error	Correlation
ϕ_1	−1.91	0.097	
ϕ_2	−0.334	0.176	−0.86
ϕ_3	2.31	0.198	−0.92, 0.99
t_0	0.412	0.095	−0.54, 0.81, 0.77

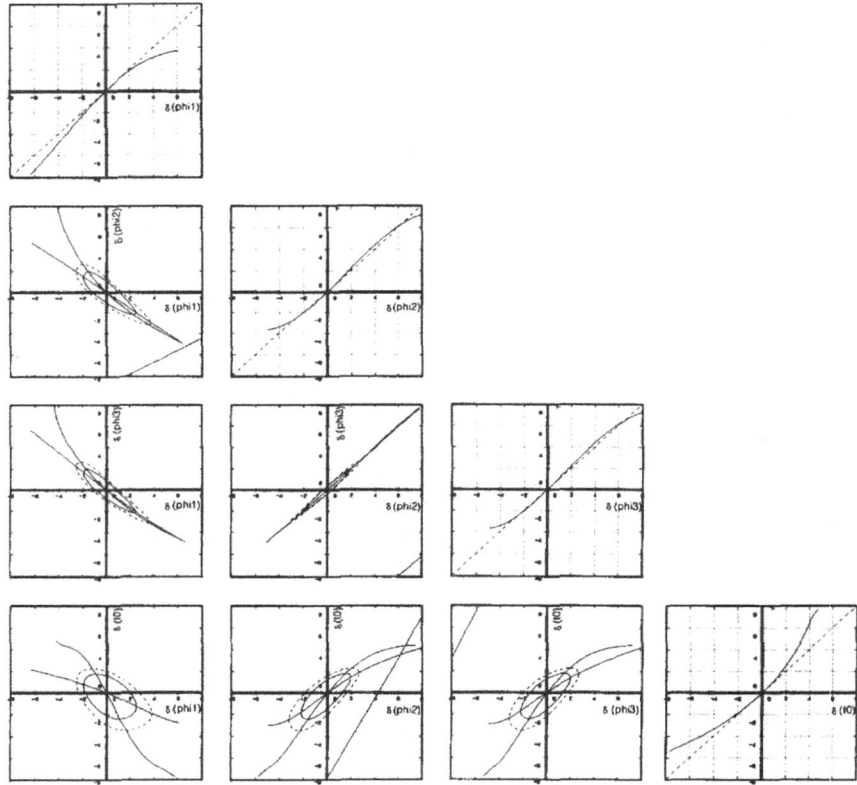

Fig. 6 Profile plots for the parameters in the tetracycline compartment model. On the diagonal plots the solid line is the profile t function and the dashed line is the linear reference. On the off-diagonal profile trace plots, the solid and dashed closed curves denote the 60% and 80% joint likelihood regions.

not defined, nor are the lower 99% end points for ϕ_2 and ϕ_3. The profile trace plots show fairly linear association between the parameters, and there is very strong association between the (ln) transfer coefficient ϕ_2 and the (ln) initial concentration ϕ_3.

II. Discussion

Profile plots can be extremely useful in nonlinear model building as they remove the gross dangers involved when using linear approximation standard errors and confidence regions. Computing the values for the profile t and profile trace plots is extremely efficient because excellent starting estimates are available (the values from the previous convergence), and because the dimension of the nonlinear problem is reduced $(P - 1)$. The plots are especially informative when laid out in the form of a matrix.

The plots provide important meaningful information about the estimation situation, in particular the extent to which linear approximation statistics can be relied on. If the profile t plots and the profile trace plots are nearly straight, then one can summarize the situation nicely with the linear approximation statistics; if not, then perhaps uniparameter transformations can be used such that the new parameters exhibit near linear behavior and, again, a few numbers can be used to summarize the situation for the new parameters. For example, for the tetracycline model, the ln(rate constants) and ln(initial concentration) parameters (ϕ) are quite well behaved, whereas the original $(\theta = \exp(\phi))$ parameters are very badly behaved.

If the parameter estimates are highly nonlinear, then it is best to use the profile t and trace plots to summarize the estimation situation, possibly after a linearizing reformulation. The profile plots will always provide accurate marginal and pairwise likelihood regions for the parameters to the extent allowed by the model, the data, and the experimental design.

Profiling also provides insights into the estimation situation by revealing how the experimental design could be improved. For example, the design for the tetracycline experiment would be improved by simply increasing the number of observation points to increase the degrees of freedom for residuals. Increasing N from 9 to 14 in such a study would probably not require much effort, but the degrees of freedom for residuals would be doubled, thereby reducing the critical values for $t(N - P, 0.005)$ and $F(2, N - P; 0.01)$ from 4.03 and 13.3 to 3.17 and 7.56 (down about 20% and 40%), respectively. This would dramatically improve the precision with which the parameters could be specified.

Finally, profiling is not limited to the uniresponse situation with normal (Gaussian) noise because the general form for a profile t function is

$$\tau^2 = -2\left(\frac{N - P}{N}\right)(\log \text{ likelihood ratio}) \tag{19}$$

and therefore, it can be used in any situation for which a likelihood function can be derived for the parameters, for example, time series analysis (Lam and Watts, 1991), and for logistic regression in which the response is proportion surviving. Other applications of profiling are given by Bates and Watts (1990) and Watts (1994).

Acknowledgments

I am grateful to Robert Gagnon and Preston Hensley of Smith Kline Beecham for providing data for the ELISA and the elution examples. Support for this research was provided by the Natural Sciences and Engineering Research Council of Canada.

References

Bates, D. M., and Watts, D. G. (1988). Nonlinear Regression Analysis and Its Applications. Wiley, New York.

Bates, D. M., and Watts, D. G. (1990). *Chemom. Intell. Lab. Syst.* **10,** 107.

Draper, N. R., and Smith, H. (1981). Applied Regression Analysis. 2nd edn. Wiley, New York.

Lam, R. L. H., and Watts, D. G. (1991). *J. Time Ser. Anal.* **12**(3), 225.

Rosé, S., Hensley, P., Shannessy, D. J. O., Culp, J., Debouck, C., and Chaiken, I. (1992). *Proteins Struct. Funct. Genet.* **13,** 112.

Wagner, J. G. (1967). *Clin. Pharmacol. Ther.* **8,** 201.

Watts, D. G. (1994). *Can. J. Chem. Eng.* **72**.

CHAPTER 3

Analysis of Residuals: Criteria for Determining Goodness-of-Fit

Martin Straume★ and Michael L. Johnson†

★COBRA, Inc.

†University of Virginia Health System
Charlottesville, VA, USA

I. Update

Many currently available software packages provide investigators who have a limited mathematics, numerical analysis, and computer skills with the ability to perform least-squares fits of nonlinear equations to experimental data. However, many of these do not provide the user with simple methods for addressing the question: Do the equation and estimated parameters provide a good description of the experimental observations?

Many investigators assume that if the computer fits the equation to the data, then it must be a good fit! But consider the simple example of fitting a straight line to data which are actually a sine wave. In this case, the computer will return an intercept and slope of zero, and will provide a terrible description of the underlying sine wave. This chapter addresses several standard approaches for the analysis of goodness-of-fit.

It is the goodness-of-fit methods which allow the user to perform hypothesis testing. If hypothesis based fitting equations and parameters can provide a good description of the experimental data, then the hypothesis is consistent with the data. Conversely, if hypothesis based fitting equations and parameters cannot provide a good description of the experimental data, then either the hypothesis is wrong or the data are bad (Johnson, 2010, 1992).

II. Introduction

Parameter-estimation procedures provide quantitation of experimental data in terms of model parameters characteristic of some mathematical description of the relationship between an observable (the dependent variable) and experimental variables [the independent variable(s)]. Processes such as least-squares minimization procedures (Johnson and Faunt, 1992; Johnson and Frasier, 1985) will produce the maximum likelihood model parameter values based on minimization of the sum of squared residuals (i.e., the sum of the squares of the differences between the observed values and the corresponding theoretical values calculated by the model employed to analyze the data). There are assumptions regarding the properties of experimental uncertainty distributions contained in the data that are implicit to the validity of the least-squares method of parameter estimation, and the reader is referred to Johnson and Faunt (1992) and Johnson and Frasier (1985) for a more detailed discussion. The widespread availability of computer hardware and software (particularly that implementing parameter-estimation algorithms such as least-squares) translates into commonplace implementation of parameter-estimation algorithms and, on occasion, perhaps a not-close-enough look at the appropriateness of particular mathematical models as applied to some experimental data.

Of course, just how critical a determination of the appropriateness of fit of a model is required will vary depending on the significance of the data, the phenomenon, and the interpretation being considered. When looking at simple, routine analytical applications (e.g., linear or polynomial empirical fits of protein assay standard curves, or perhaps analysis for single-exponential decay in kinetic enzyme assays for first-order rate constant estimates to use for defining specific activities during steps of purification procedures), it may not be particularly important to examine carefully the quality of fit produced by the model used to analyze the data. An empirical or "lower-order" estimate of the behavior of some system property in these cases is fully sufficient to achieve the goals of the analysis. However, when quantitatively modeling detailed aspects of biomolecular properties, particularly when asking more advanced theoretical models to account for experimental data of ever increasing quality (i.e., more highly determined data), many sophisticated numerical methods and complex mathematical modeling techniques are often implemented. In these cases, a careful eye must be directed toward consideration of the ability of the model to characterize the available experimentally determined system properties reliably, sometimes to quite exquisite levels of determination.

To perform these types of detailed analyses (and, in principle, for any analysis), data must be generated by experimental protocols that provide data (1) possessing experimental uncertainties that are randomly distributed and (2) free of systematic behavior not fully accounted for by the mathematical model employed in analysis. A mathematical model must be defined to describe the dependence of the observable on the independent variable(s) under experimental control. The definition of an appropriate mathematical model involves considerations of how to transform the experimentally observed system behavior into a mathematical description that permits physical interpretation of the model parameter values. In this way, information about the biomolecular phenomena underlying the system response is quantitatively defined. Such modeling efforts can become quite specific when addressing molecular level interpretations of the functional, structural, and thermodynamic properties of biological systems.

Ongoing biochemical and biophysical studies to elucidate the molecular and thermodynamic foundations of macromolecular structure-function relationships have been producing data from experiments designed to test, to ever finer levels of detail, behavior predicted or theorized to exist as based on modeling efforts. All complementary experimental information available about a particular system must be incorporated into comprehensive mathematical models to account fully for all the known properties of a system. Therefore, data regarding structural properties, functional properties, influences of experimental conditions (e.g., ionic strength, pH, and ligand concentration), and any other specifically relevant system variables must, in principle, all be consistent with a common model descriptive of the system under study to be comprehensively valid. Approximations in data analysis applications such as these are therefore no longer tolerable so as to achieve an accurate and precise characterization of biochemical or biophysical properties. Neither are approximations necessary given the recent increases in computational capacity in terms of hardware capabilities as well as software availability and theoretical advancements. Analyses of better determined experimental data sometimes indicate deficiencies in current interpretative models, thereby prompting a closer look at the system and how it is best modeled mathematically. The consideration of residuals (the differences between observed and calculated dependent variable values) becomes a very important element in the overall data analysis process in cases where attempts to model detailed molecular system properties mathematically are being pursued.

The significance of subtle behavior in residuals may suggest the presence of a significant system property that is overlooked by the current mathematical model. But a more fundamental role served by examination of residuals is in providing information on which to base a judgment about the appropriateness of a particular mathematical description of system behavior as a function of some independent, experimental variable(s). If an examination of the residuals obtained from a parameter-estimation procedure on some experimental data yields the conclusion that the data are reliably characterized by the mathematical model (i.e., that a good fit to the data is obtained possessing no unaccounted for residual systematic behavior in the data), this is not to say that this represents justification for necessarily accepting the model as correct. Rather, it indicates that the model employed is sufficient to characterize the behavior of the experimental data. This is the same as saying that the data considered provide no reason to reject the current model as unacceptable. The residuals for a case in which a "good fit" is obtained then, in principle, represent the experimental uncertainty distribution for the data set. However, if an examination of the residuals indicates inconsistencies between the data and the behavior predicted by the analysis, then the current model may correctly be rejected and considered unacceptable (unless some other source for the residual systematic behavior is identified).

When considering residuals, a qualitative approach is often the most revealing and informative. For example, generating plots to represent visually trends and correlations provides a direct and often unambiguous basis for a judgment on the validity of a fit. Of course, quantitative methods to test more rigorously particular properties of residuals sometimes must be considered in order to quantitate the statistical significance of conclusions drawn as a result of data analysis. Some of the available methods for considering residuals will be discussed below with the aid of illustrative examples.

III. Scatter Diagram Residual Plots

Visualizing residuals is commonly performed by generating scatter diagrams (Armitage, 1977a). Residuals may be plotted as a function of various experimental variables to permit convenient identification of trends that may not have been accounted for by the analytical model. Residuals are most commonly plotted as a function of either the values of the independent variable(s) (e.g., time in kinetic experiments or ligand concentration in ligand binding experiments) or the calculated values of the dependent variable (i.e., the values of the experimental observable calculated from the model). However, residual plots versus some other functional relationship of the independent variable(s) or some other potentially significant variable that was not explicitly considered in the original model may also provide information about important relationships that have not been previously identified.

In Fig. 1 is presented a simulated scatter diagram to illustrate the type of information potentially provided by visual inspection of residual plots. The circles represent pseudo-Gaussian distributed residuals with a standard deviation of 1.0 and a mean of 0.0. The points denoted by crosses represent similar bandwidth noise as seen in the pseudo-Gaussian distributed points but possessing higher-order structure superimposed on them. Visual inspection of such plots permits ready identification of deficiencies in the ability of an analytical model to describe adequately the behavior of experimental data if nonrandom residuals are obviously present (e.g., as in the residuals represented by the crosses in Fig. 1). This type of

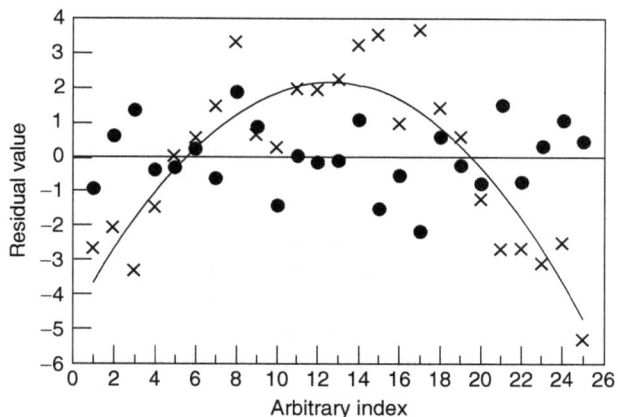

Fig. 1 Scatter diagrams for two sets of synthetic residuals generated to demonstrate a normally distributed set of residuals and another that exhibits a trend in the behavior of the residual values as a function of residual number (presented as an arbitrary index in this example).

observation would suggest that there exists some systematic behavior in the data (as a function of the variable against which the residuals were plotted) that was not well accounted for by the model employed in analysis.

An examination of the trend in residuals as a function of some particular variable space may even provide information about the type of quantitative relationship that must be accommodated by the analytical model that currently is not. However, correctly accounting for any newly incorporated relationships into currently existing analytical models requires re-evaluation of the data set(s) originally considered. This is necessary so as to simultaneously estimate values for all the model parameters characteristic of the model, both previously existing and newly incorporated. Quantifying phenomena originally omitted from consideration by a model must not be attempted by analyzing the resulting residuals. Correlation among parameters must be accommodated during a parameter-estimation procedure so as to produce the true best-fit parameter values that accurately characterize the interdependence between parameters of the model and between these parameters and the properties of the data being analyzed (the dependence of the experimental observable on the independent experimental variables as well as on the distribution of experimental uncertainty in the data).

IV. Cumulative Probability Distributions of Residuals

Another visual method for examining residuals involves generating a cumulative frequency plot (Bard, 1974). The information provided by this form of consideration of residuals is related to the randomness of the distribution of residual values. The process requires that the residuals be ordered and numbered sequentially such that

$$r_1 < r_2 < r_3 < \ldots < r_n$$

where r_i is the ith residual value. A quantity P_i is then defined such that

$$P_i = (i - 0.5)/n.$$

Here, P_i, the cumulative probability, represents a statistical estimate of the theoretical probability of finding the ith residual (out of n total residuals) with a value of r_i if they are distributed randomly (i.e., Gaussian or normally distributed residuals). A graph of the standard normal deviate, or Z-value (which represents the number of standard deviations from the mean), corresponding to the cumulative probability P_i versus the values of the ordered residuals will then produce a straight line of points, all of which will be very near the theoretical cumulative probability line if the residuals are distributed randomly. The Z-values corresponding to particular levels of probability may be obtained from tabulations in statistics books or calculated directly by appropriate integration of the function defining Gaussian distributed probability.

The cumulative probability plots corresponding to the two sets of simulated residuals presented in Fig. 1 are shown in Fig. 2. The points for the pseudorandom residuals (circles) form a linear array with all points in close proximity to the theoretical line. The slope of this line is 1.0 in the manner plotted in this graph, corresponding to a standard deviation of 1.0 for this particular distribution of residuals. The points for the distribution exhibiting residual structure in the scatter diagram of Fig. 1 (crossed points) can be seen to generally follow along their theoretical line (with a slope of 2.5 corresponding to an apparent standard deviation of 2.5); however, they show systematic behavior and occasionally deviate

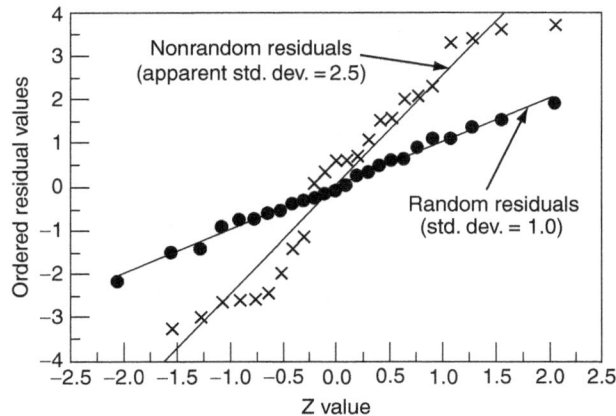

Fig. 2 Cumulative frequency plots for the two sets of residuals presented in Fig. 1. The ordered residual values are plotted relative to the Z-value (corresponding to the number of estimated standard deviations from the mean, in this case zero) characteristic of Gaussian distributed residuals. The estimated standard deviations are 1.0 and 2.5 for the Gaussian and nonrandom residuals, respectively, as reflected in the slopes of the theoretical lines.

considerably from the line (relative to the near superposition for the pseudoran-domly distributed residuals). This level of deviation in a cumulative probability plot suggests that the data are not well characterized by the model used to describe their behavior because the resulting residuals clearly exhibit nonrandom behavior.

V. χ^2 Statistic: Quantifying Observed Versus Expected Frequencies of Residual Values

To assess the properties of distributions of residuals more quantitatively, one may generate a discrete, theoretical residual probability distribution (based on an assumption of randomly distributed residuals) and compare the distribution of observed residual values with these expected frequencies (Armitage, 1977b; Daniel, 1978). A histogram is in effect created in which the range of residual values is divided into a number of intervals such that at least one residual (out the n total residuals being considered) is expected to exist in each interval. The expected frequencies are then compared to the observed frequencies by the relationship

$$\chi^2 = \Sigma[(O_i - E_i)^2/E_i]$$

which is summed over each interval considered. Here, O_i represents the observed number of residuals possessing values within the range defined by interval i. Analogously, E_i is the expected number of residuals in this interval if the residuals are randomly distributed. The value of this calculated parameter will be distributed approximately as the χ^2 statistic (for the desired level of confidence and the number of degrees of freedom). The significance of the χ^2 statistic is that it represents a method for quantifying the probability that the distribution of residuals being considered is not random.

When the χ^2 statistic is applied to the residual distributions presented in Fig. 1 (see Table I), we find that the pseudo-Gaussian residual distribution produces a χ^2 value of 1.326, whereas that which possessed residual structure had a χ^2 value of 7.344. In this case, nine intervals were considered, the central seven being finite with a width of one-half the (apparent, in the case of the nonrandom residuals) standard deviation, with the two extreme intervals considering the remaining probability out to $\pm\infty$. For a total of 25 residuals, this choice of interval width and number produced expected frequencies of at least one for each of the intervals considered (1.0025 being the lowest, for the two end intervals). When considering this type of analysis of a residual distribution, small expected frequencies must be dealt with so as to produce intervals with at least one for an expected frequency. With small numbers of residuals, this may become a necessary concern.

A χ^2 value of 1.326 means that there is between a 1% and 2.5% chance that the pseudo-Gaussian residuals in Fig. 1 are not randomly distributed. This is the derived level of confidence indicated by this χ^2 value with 7 degrees of freedom [in this case, the number of degrees of freedom is 9 (the number of intervals)

Table I
Residual Probability Distribution per Interval[a]

	Observed		
Z-Interval	Pseudo-Gaussian	Structured	Expected
$-\infty, -1.75$	1	1	1.0025
$-1.75, -1.25$	2	1	1.6375
$-1.25, -0.75$	2	6	3.0250
$-0.75, -0.25$	6	6	4.3675
$-0.25, 0.25$	4	4	4.9350
$0.25, 0.75$	4	3	4.3675
$0.75, 1.25$	3	4	3.0250
$1.25, 1.75$	2	0	1.6375
$1.75, +\infty$	1	0	1.0025

[a] For $\chi^2 = \Sigma[(O_i - E_i)^2/E_i]$, where O_i is the observed number of residuals with values in interval i and E_i is the expected number of residuals with values in interval i. χ^2 (pseudo-Gaussian) = 1.326; χ^2 (structured) = 7.344. The probabilities associated with these values of χ^2 (for seven degrees of freedom, nine intervals minus the two constraints for estimating the mean and variance of the distributions) verify that the Gaussian distributed residuals are correctly identified as being Gaussian distributed [χ^2 (pseudo-Gaussian) = 1.326], whereas the nonrandom residuals are confirmed to be non-Gaussian [χ^2 (structured) = 7.344].

minus 1 (for the requirement that $\sum O_i = \sum E_i$) minus 1 (for the estimation of an apparent standard deviation) equals 7]. The considerably larger χ^2 value of 7.344 for the structured residuals of Fig. 1 indicates a significantly higher probability that the residuals are indeed not randomly distributed, supporting the conclusion drawn by inspection of the scatter diagrams in Fig. 1.

VI. Kolmogorov–Smirnov Test: An Alternative to the χ^2 Statistic

As an alternative to the χ^2 method for determining whether the residuals generated from an analyis of data by a mathematical model are randomly distributed, one may apply the Kolmogorov-Smirnov test (Daniel, 1978). The Kolmogorov-Smirnov test has a number of advantages over the χ^2 treatment. Whereas the χ^2 approach requires compartmentalization of residuals into discrete intervals, the Kolmogorov-Smirnov test has no such requirement. This relaxes the constraint of possessing a sufficient number of residuals so as to significantly populate each of the intervals being considered in the χ^2 analysis. And to provide a closer approximation to a continuous distribution, the χ^2 approach requires consideration of a large number of intervals. The Kolmogorov-Smirnov approach requires no discrete approximations but rather provides a quantitative basis for making a statistical comparison between the cumulative distribution of a set of residuals and any theoretical cumulative probability distribution (i.e., not limited to only a Gaussian probability distribution).

The statistic used in the Kolmogorov-Smirnov test, D, is the magnitude of the greatest deviation between the observed residual values at their associated cumulative probabilities and the particular cumulative probability distribution function with which the residuals are being compared. To determine this quantity, one must consider the discrete values of the observed residuals, r_i, at the cumulative probability associated with each particular residual, P_i,

$$P_i = [(i - 0.5)/n]; 1 \le i \le n$$

relative to the continuous theoretical cumulative probability function to which the distribution of residuals is being compared (e.g., that of a Gaussian distribution possessing the calculated standard deviation, as visually represented in the cumulative probability plots of Fig. 2). The continuous nature of the theoretical cumulative probability function requires that both end points of each interval defined by the discrete points corresponding to the residuals be considered explicitly. The parameter D is therefore defined as

$$D = \max_{1 \le i \le n}\{\max[|r_i(P_i) - r_{\text{theory}}(P_i)|, |r_{i-1}(P_{i-1}) - r_{\text{theory}}(P_i)|]\}.$$

The deviations between the observed and theoretical values are thus considered for each end of each interval defined by the observed values of $r_i(P_i)$ and $r_{i-1}(P_{i-1})$. The value of this statistic is then compared with tabulations of significance levels for the appropriate number of residuals (i.e., sample size), a too-large value of D justifying rejection of the particular theoretical cumulative probability distribution function as incorrectly describing the distribution of residuals (at some specified level of confidence).

VII. Runs Test: Quantifying Trends in Residuals

The existence of trends in residuals with respect to either the independent (i.e., experimental) or dependent (i.e., the experimental observable) variables suggests that some systematic behavior is present in the data that is not accounted for by the analytical model. Trends in residuals will often manifest themselves as causing too few runs (consecutive residual values of the same sign) or, in cases where negative serial correlation occurs, causing too many runs. A convenient way to assess quantitatively this quality of a distribution of residuals is to perform a runs test (Bard, 1974). The method involves calculating the expected number of runs, given the total number of residuals as well as an estimate of variance in this expected number of runs.

The expected number of runs, R, may be calculated from the total number of positive and negative valued residuals, n_p and n_n, as

$$R = \{[2n_p n_n/(n_p + n_n)] + 1\}.$$

The variance in the expected number of runs, σ_R^2, is then calculated as

$$\sigma_R^2 = \{[2n_p n_n (2n_p n_n - n_p - n_n)]/[(n_p + n_n)^2 (n_p + n_n - 1)]\}.$$

A quantitative comparison is then made between the expected number of runs, R, and the observed number of runs, n_R, by calculating an estimate for the standard normal deviate as

$$Z = |(n_R - R + 0.5)/\sigma_R|.$$

When n_p and n_n are both greater than 10, Z will be distributed approximately as a standard normal deviate. In other words, the calculated value of Z is the number of standard deviations that the observed number of runs is, from the expected number of runs for a randomly distributed set of residuals of the number being considered. The value of 0.5 is a continuity correction to account for biases introduced by approximating a discrete distribution with a continuous one. This correction is $+0.5$ (as above) when testing for too few runs and is -0.5 when testing for too many runs. The test is therefore estimating the probability that the number of runs observed is different from that expected from randomly distributed residuals. The greater the value of Z, the greater the likelihood that there exists some form of correlation in the residuals relative to the particular variable being considered.

In Table II is presented an application of the runs test to the residuals of Fig. 1. The results clearly indicate that the number of runs expected and observed for the pseudo-Gaussian distributed residuals agree quite well (Z values of 0.83 and 1.24), whereas the agreement between expected and observed numbers of runs with the structured residuals is very different (Z values of 4.05 and 4.48). The probability that the distributions of residuals exhibit the "correct" number of runs is therefore statistically acceptable in the former case (less than 0.83 standard deviations from expected) and statistically unacceptable in the latter (more than 4 standard deviations from expected). A cutoff value for acceptability may be considered to be

Table II
Runs Test Applied to Residuals[a]

| Parameter | Residuals | |
	Pseudo-Gaussian	Structured
n_p	12	15
n_n	13	10
R	13.48	13
σ_R^2	5.97	5.5
n_R	16	3
Z_{tf}	1.24	4.05
Z_{tm}	0.83	4.48

[a] The Z values derived from this analysis suggest that the expected number of runs is encountered in the Gaussian distributed residuals ($Z_{tf} = 1.24$ and $Z_{tm} = 0.83$), a situation that is not the case for the nonrandom residual distribution ($Z_{tf} = 4.05$ and $Z_{tm} = 4.48$). Subscripts refer to testing for too few or too many runs.

2.5–3 standard deviations from the expected value (corresponding to probabilities of approximately 1-0.25%). The values calculated in this illustrative example fall well to either side of this cutoff range of Z values.

VIII. Serial Lag$_n$ Plots: Identifying Serial Correlation

A situation in which too few runs are present in the residual distribution indicates positive serial correlation. Too many runs, on the other hand, is a situation characteristic of negative serial correlation in residuals. Serial correlation suggests systematic behavior with time and, in fact, is often considered in this context. However, this sort of serial dependence may also be of significance when considering parameter spaces other than time.

Visualization of this phenomenon is best achieved by lag$_n$ serial correlation plots (Draper and Smith, 1981). The residual values are plotted against each other, each value plotted versus the one occurring n units before the other. As demonstrated in Fig. 3 for the two residual distributions presented in Fig. 1, considerable correlation is suggested in the structured residuals in the lag$_1$, lag$_2$, and lag$_3$ serial plots but is no longer obvious in the lag$_4$ plot, whereas the pseudo-Gaussian residual distribution produces no obvious trend in any of the serial lag$_n$ plots. The presence of positive serial correlation (i.e., adjacent residuals with the same sign) is evidenced by positive sloping point distributions (as demonstrated by the distribution exhibiting residual structure in Fig. 3), whereas negative slopes characterize negative serial correlation (i.e., adjacent residuals with opposite sign). In the absence of correlation, the points will be randomly clustered about the origin of the plot (as demonstrated by the pseudo-Gaussian distributed residuals in Fig. 3).

IX. Durbin–Watson Test: Quantitative Testing for Serial Correlation

The Durbin-Watson test provides a quantitative, statistical basis on which to judge whether serial correlation exists in residuals (Draper and Smith, 1981). The test permits estimation of the probability that serial correlation exists by attempting to account for effects of correlation in the residuals by the following formula

$$r_i = \rho r_{i-1} + Z_i.$$

Here, r_i and r_{i-1} correspond to the ith and $(i - 1)$th residuals, and ρ and Z_i provide for quantitation of any effects of serial correlation. This approach is based on the null hypothesis that the residuals are distributed in a manner free of any serial correlation and possessing some constant variance. The above expression relating adjacent residuals is a means to quantify deviation from the hypothesized

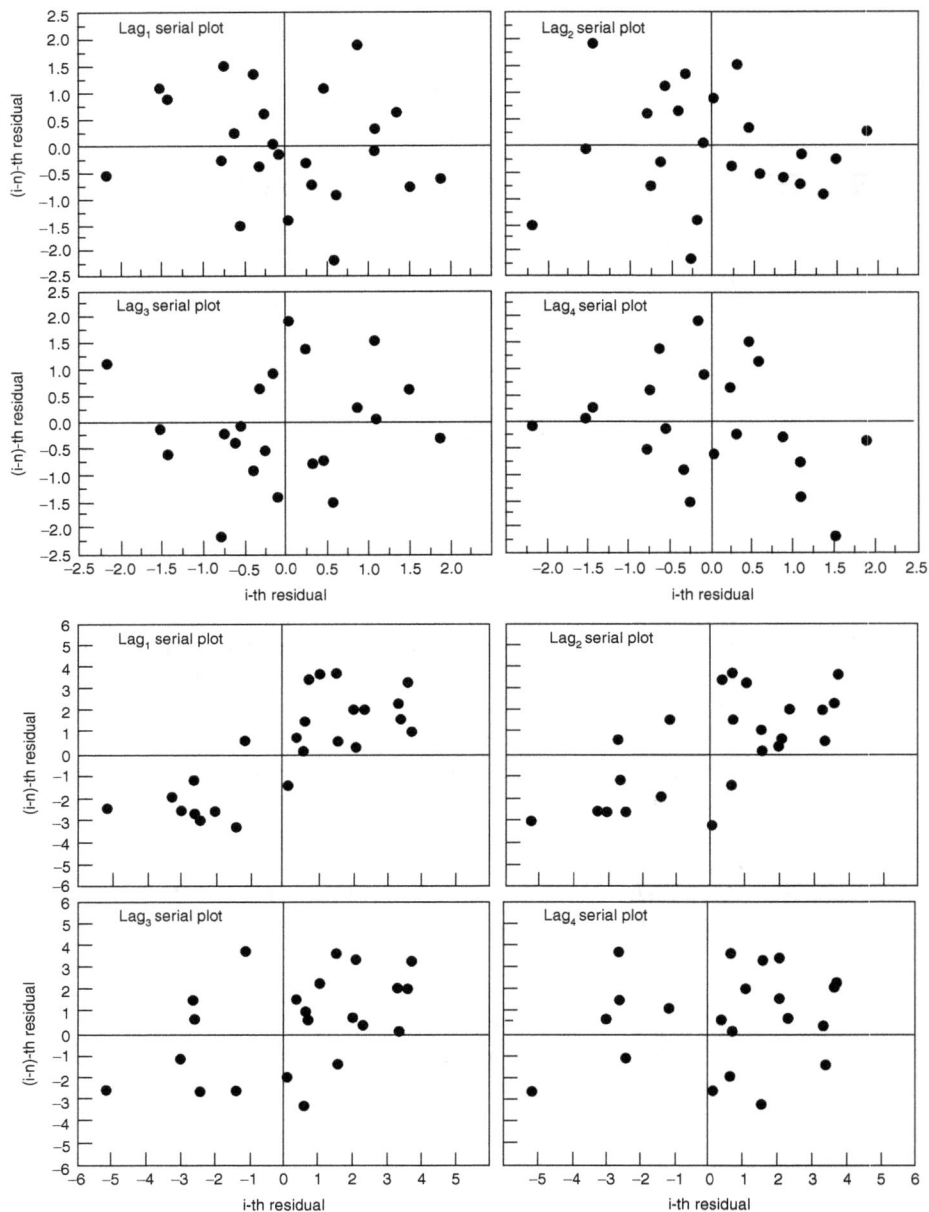

Fig. 3 Lag$_n$ serial plots for the residual distributions presented in Fig. 1 are displayed for lags of 1, 2, 3, and 4. It is apparent that the nonrandom residuals of Fig. 1 (bottom plots) possess some positive serial correlation as demonstrated by positive slopes in the point distributions for the lag$_1$, lag$_2$, and lag$_3$ cases but disappearing by the lag$_4$ plot. The Gaussian distributed residuals (top plots), on the other hand, exhibit no indication of serial correlation in any of the four serial lag$_n$ plots.

ideal case. The underlying assumptions are that the residual values, the r_i's, as well as the values for the Z_i's in the above equation each possess constant variances, which for the case of the residuals is given by

$$\sigma^2 = \sigma_r^2/(1 - \rho^2).$$

The case in which no serial correlation is indicated is given by $\rho = 0$. The variance then reduces to that estimated from the original distribution (and the null hypothesis is accepted).

A parameter is calculated to provide a statistical characterization of satisfying the null hypothesis and therefore addressing whether any correlation is suggested in the residuals being considered. This parameter is given by

$$d = \sum_{i=2}^{n}(r_i - r_{i-1})^2 / \sum_{i=1}^{n} r_i^2$$

In assigning a probability to the likelihood of serial correlation existing in the considered residuals, two critical values of d (a lower and an upper value, d_1 and d_u) are specified, thus defining a range of values associated with a specified probability and the appropriate number of degrees freedom. Tables of these critical values at various levels of confidence may be consulted (Draper and Smith, 1981) and testing of the following three conditions may be permitted: $\rho > 0$, $\rho < 0$, or $\rho \neq 0$. When considering the first case ($\rho > 0$), $d < d_1$ is significant (at the confidence level specified in the table used), and serial correlation with $\rho > 0$ is accepted. A value of $d > d_u$ indicates that the case $\rho > 0$ may be rejected, that is, that one is *not* justified in assigning any positive serial correlation to the distribution of residuals being considered. Intermediate values of d (between d_1 and d_u) produce an "inconclusive" test result. The case of $\rho < 0$ is considered in an analogous manner except that the value of $(4 - d)$ is used in comparisons with tabulated values of d_1 and d_u. The same process as outlined above for the first case applies here as well. The test for $\rho \neq 0$ is performed by seeing whether $d < d_1 \, or \, (4 - d) < d_1$. If so, then $\rho \neq 0$ at twice the specified level of confidence (it is now a two-sided test). If $d > d_u \, and \, (4 - d) > d_u$, then $\rho = 0$ at twice the specified level of confidence. Otherwise, the test is considered "inconclusive." To resolve this "inconclusive" occurrence, one may assume the conservative approach and reject once the more stringent criterion of the two is exceeded.

X. Autocorrelation: Detecting Serial Correlation in Time Series Experiments

Experimental data collected as a time series typically exhibit serial correlations. These serial correlations arise when the random uncertainties superimposed on the experimental data tend to have values related to the uncertainties of other data points that are close temporally. For example, if one is measuring the weight of a test animal once a month and the data are expressed as a weight gain per month, negative serial correlation may be expected. This negative serial correlation is expected because a positive experimental error in an estimated weight gain for one month (i.e., an overestimate) would cause the weight gain for the next month to be underestimated.

A basic assumption of parameter-estimation procedures is that the experimental data points are independent observations. Therefore, if the weighted differences between experimental data points and the fitted function (the residuals) exhibit such a serial correlation, then either the observations are not independent or the mathematical model did not correctly describe the experimental data. Thus, the serial correlation of the residuals for adjacent and nearby points provides a measure of the quality of the fit.

The autocorrelation function provides a simple method to present this serial correlation for a series of different lags, k (Box and Jenkins, 1976). The lag refers to the number of data points between the observations for a particular autocorrelation. For a series of N observations, Y_t, with a mean value of μ, the autocorrelation function is defined as

$$\beta_k = \hat{\sigma}_k / \hat{\sigma}_0$$

for $k = 0, 1, 2, \ldots, K$, where the autocovariance function is

$$\hat{\sigma}_k = \frac{1}{n} \sum_{t=1}^{n-k} (Y_t - \mu)(Y_{t+k} - \mu)$$

for $k = 0, 1, 2, \ldots, K$. In these equations, K is a maximal lag less than n. The autocorrelation function has a value between -1 and $+1$. Note that the autocorrelation function for a zero lag is equal to 1 by definition.

The expected variance (Moran, 1947) of the autocorrelation coefficient of a random process with independent, identically distributed random (normal) errors is

$$\text{var}(\beta_k) = \frac{n - k}{n(n + 2)}$$

where μ is assumed to be zero.

Autocorrelations are presented graphically as a function of k. This allows an investigator to compare easily the autocorrelation at a large series of lags k with the corresponding associated standard errors (square root of the variance) to decide if any significant autocorrelations exist.

XI. χ^2 Test: Quantitation of Goodness–of–Fit

After verification that the residuals resulting from a model parameter-estimation process to a set of data are indeed free of any systematic trends relative to any variables of significance (i.e., dependent or independent variables), a quantitative estimate of the adequacy of the particular model in describing the data is possible. Calculation of the χ^2 statistic is a common quantitative test employed to provide a statistical estimate of the quality of fit of a theoretical, mathematical description of the behavior of a system to that measured experimentally (Bevington, 1969). The value of the χ^2 statistic varies approximately as the number of degrees of freedom in situations where the mathematical description is correct and only random fluctuations (i.e., experimental noise) contribute to deviations between calculated and observed dependent values. The χ^2 statistic is defined as

$$\chi^2 = \sum_{i=1}^{n} \left[\frac{Y_{\text{obs},i} - Y_{\text{calc},i}}{\sigma_i} \right]^2$$

that is, as the sum over all n data points of the squared, normalized differences between each observed and calculated value of the dependent variable ($Y_{\text{obs},i} - Y_{\text{calc},i}$), normalized with respect to the error estimate for that particular point (σ_i).

The required knowledge of an accurate estimate for the uncertainty associated with each observed value makes it challenging sometimes to implement this test. It is just these estimated uncertainties that give the χ^2 test its statistical significance, by appropriately normalizing the residuals. By dividing this calculated value of χ^2 by the number of degrees of freedom, the reduced χ^2 value is obtained. The number of degrees of freedom is defined as the number of data points (n) minus the number of parameters estimated during analysis (p) minus 1 (i.e., $NDF = n - p - 1$). The value of the reduced χ^2 value will quite nearly approximate 1 if both (1) the estimated uncertainties, σ_i, are accurate and (2) the mathematical model used in analysis accurately describes the data. With accurate knowledge of the experimental uncertainty, it is possible to define statistically the probability that a given model is an accurate description of the observed behavior.

XII. Outliers: Identifying Bad Points

In any experimental measurement, occasionally values may be observed that produce an unusually large residual value after an analysis of the data is performed. The existence of an outlier (or "bad point") suggests that some aberration may have occurred with the measurement of the point. The presence of such a point in the data set being analyzed may influence the derived model parameter values significantly relative to those that would be obtained from an analysis without the apparent outliers. It is therefore important to identify such "bad points" and perhaps reconsider the data set(s) being analyzed without these suspect points.

Visual inspection of residual scatter diagrams often reveals the presence of obvious outliers. Cumulative frequency plots will also indicate the presence of outliers, although perhaps in a less direct manner. Visualization methods may suggest the presence of such points, but what method should be used to decide whether a point is an outlier or just a point with a low probability of being valid? A method to provide a quantitative basis for making this decision derives from estimating the apparent standard deviation of the points after analysis. This is calculated as the square root of the variance of fit obtained from analysis of an unweighted data set. The variance of fit is defined as the sum of the squared residuals divided by the number of degrees of freedom (the number of data points minus the number of parameters being estimated). In the case that the model employed is capable of reliably characterizing the data (i.e., capable of giving a "good fit"), the distribution of residuals will, in principle, represent the distribution of experimental uncertainty. Any residuals possessing values that are more than approximately 2.5–3 standard deviations from the mean have only a 1–0.25% chance of being valid. When considering relatively large data sets (of the order of hundreds of points or more), the statistical probability of a residual possessing a value 3 standard deviations from the mean suggests that such a point should be expected about once in every 400 data points.

XIII. Identifying Influential Observations

The presence of outliers (as discussed in the previous section) may produce derived model parameter values that are biased as a result of the influence of outliers. Methods to test for influential observations may be applied to determine the influence of particular data points or regions of independent variable space on the parameters of the analytical model (Draper and Smith, 1981). The influence a potential bad point may have on the resulting model parameter values will be dependent on whether there exist other data points in the immediate vicinity of the suspect point (i.e., in an area of high data density) or whether the point is relatively isolated from others. And if there are regions of low data density, influential observations may not be made apparent by looking for outliers. That is because the relatively few points defining part of an independent parameter space may be largely responsible for determination of one or a few particular model parameters but have very little influence on other model parameters. These points will then represent a particularly influential region of independent parameter space that may strongly affect the outcome of an analysis but may at the same time be difficult to identify as being bad points.

One approach is to omit suspected influential regions of data from consideration during analysis to see if any portion of the complete data set can be identified as being inconsistent with results suggested by consideration of other regions of independent parameter space. A difficulty that may be encountered is that particular regions of independent parameter space may be almost exclusively responsible for determining particular model parameters. Omitting such regions of data from analysis may not permit a complete determination of all the parameters characteristic of the model. If such a situation is encountered, it indicates that a higher level of determination is necessary in this region of independent parameter space and that the experimental protocol during acquisition of data should be modified to permit more data to be accumulated in this "influential window."

The various quantitative methods that have been developed to address influential observations (Draper and Smith, 1981) generally involve reconsideration of multiple modified data sets in which some points have been omitted from consideration. The variation in the values of the derived model parameters arising from considering multiple such modified data sets then indicates the degree to which particular regions of data influence various model parameters. If an influential region of independent parameter space is identified, a relatively easy fix to the dilemma is to change the data acquisition protocol to take more experimental measurements over the influential region of independent parameter space.

XIV. Conclusions

Qualitative and quantitative examination of residuals resulting from analysis of a set (or sets) of experimental data provides information on which a judgment can be made regarding the validity of particular mathematical formulations for reliably characterizing the considered experimental data. With the advances in biochemical and biophysical instrumentation as well as computer hardware and software seen in recent years (and the anticipated advances from ongoing development), quantitative descriptions of biological system properties are continuously being better determined. Deficiencies in current models characteristic of system behavior are often recognized when more highly determined experimental data become available for analysis. Accommodation of these recognized deficiencies then requires evolution of the particular mathematical description to more advanced levels. In so doing, a more comprehensive understanding of the biochemical or biophysical properties of the system often results.

An interpretation of derived model parameter values implicitly relies on the statistical validity of a particular mathematical model as accurately describing observed experimental system behavior. The concepts and approaches outlined in the present chapter provide a survey of methods available for qualitatively and quantitatively considering residuals generated from data analysis procedures. In those cases where very precise interpretation of experimental observations is required, a thorough, quantitative consideration of residuals may be necessary in order to address the statistical validity of particularly detailed mathematical models designed to account for the biochemical or biophysical properties of any experimental system of interest.

Acknowledgments

This work was supported in part by National Institutes of Health Grants RR-04328, GM-28928, and DK-38942, National Science Foundation Grant DIR-8920162, the National Science Foundation Science and Technology Center for Biological Timing of the University of Virginia, the Diabetes Endocrinology Research Center of the University of Virginia, and the Biodynamics Institute of the University of Virginia.

References

Armitage, P. (1977a). "Statistical Methods in Medical Research," 4th edn., p. 316. Blackwell, Oxford.

Armitage, P. (1977b). "Statistical Methods in Medical Research," 4th edn., p. 319. Blackwell, Oxford.

Bard, Y. (1974). "Nonlinear Parameter Estimation." p. 201. Academic Press, New York.

Bevington, P. R. (1969). "Data Reduction and Error Analysis for the Physical Sciences" p. 187. McGraw-Hill, New York.

Box, G. E. P., and Jenkins, G. M. (1976). "Time Series Analysis Forecasting and Control," p. 33. Holden-Day, Oakland, CA.

Daniel, W. W. (1978). "Biostatistics: A Foundation for Analysis in the Health Sciences," 2nd edn., Wiley, New York.

Draper, N. R., and Smith, R. (1981). "Applied Regression Analysis," 2nd edn., p. 153. Wiley, New York.

Johnson, M. L. (1992). Why, when, and how biochemists should use least-squares. *Anal. Bioche.* **206,** 215–225.

Johnson, M. L. (2010). Use of Least-Squares Techniques in Biochemistry. Chapter 1. of this volume.

Johnson, M. L., and Faunt, L. M. (1992). Parameter estimation by least-squares. *Methods Enzymol.* **210,** 1–37.

Johnson, M. L., and Frasier, S. G. (1985). Nonlinear least-squares analysis. *Methods Enzymol.* **117,** 301–342.

Moran, P. A. P. (1947). *Biometrika* **34,** 281.

CHAPTER 4

Monte Carlo Method for Determining Complete Confidence Probability Distributions of Estimated Model Parameters

Martin Straume★ and Michael L. Johnson†

★COBRA, Inc.

†University of Virginia Health System
Charlottesville, VA, USA

I. Update

The fitting of equations to experimental data involves numerical procedures which estimate the numerical values of the parameters of the equations that have the highest probability of being correct (Johnson, 1992, 2010, 2008). Commonly the next step will be a statistical comparison of these parameter values with some *a priori* values, such as zero. If the parameter is not different from zero then the term is not significant, etc.

This is not as simple a task as one might envision. For linear equations, like orthogonal polynomials, a simple equation for the evaluation of the parameter standard errors exists, tha is, asymptotic standard errors. For non-orthogonal

DOI: 10.1016/B978-0-12-384997-7.00004-2

linear equations, such as ordinary polynomials, these commonly used asymptotic standard errors fail to include the contribution due to the co-variance between the estimated parameters, that is, they typically will significantly underestimate the actual errors in the estimated parameters and will thus overestimate the significance of the results (Johnson, 1992, 2010, 2008). For nonlinear fitting equations, no analytical solution exists, only approximations exist.

This chapter provides an example of a Monte-Carlo approach for the evaluation of confidence intervals of estimated parameters that can be used for orthogonal equations that are either linear or nonlinear. The reader is encouraged to also examine Bootstrap Methods (Chernick, 1999).

II. Introduction

The quantitative analysis of experimental data generally involves some numerical process to provide estimates for values of model parameters (least-squares, Johnson and Faunt, 1992; method of moments, Small, 1992; maximum entropy, Press *et al.*, 1986a; Laplace transforms, Ameloot, 1992; etc.). The derived parameter values are, in turn, interpreted to provide information about the observed properties of the experimental system being considered. This fundamental process applies for the simplest of analyses (e.g., protein determinations employing standard curves) as well as for the highly sophisticated modeling algorithms in use today for interpretation of a broad spectrum of complex biomolecular phenomena.

The primary objective of a quantitative analysis is derivation of the values corresponding to the best estimates for the parameters of the model employed to characterize the experimental observations. System properties may then be inferred by a physical interpretation of the significance of the model parameter values. However, the level of confidence one can have in the interpretation of derived parameter values depends strongly on the nature and magnitude of the confidence probability distribution of the parameter values about their most probable (or best-fit) values.

Determination of reliable estimates of confidence intervals associated with model parameters may be critical in discerning between alternative interpretations of some biomolecular phenomena (e.g., the statistical justification for existence of quaternary enhancement in human hemoglobin oxygen-binding behavior (Straume and Johnson, 1989). In a case such as this, the most probable derived value is significant, but the shape and breadth of the distribution of expected parameter values, given the experimental uncertainties associated with the data sets being analyzed, are also of critical importance with regard to arriving at a statistically significant conclusion. Knowledge of complete confidence probability distributions as well as the correlation that exists among parameters or between parameters and the experimental independent variable(s) is also of value for identifying influential regions of independent parameter space (e.g., extent of binding saturation in a ligand-binding experiment) as well as for pointing out the relative behavior of parameters between different models used to interpret the same data (e.g., models that explicitly account for ligand-linked cooperative binding versus those allowing nonintegral binding stoichiometries to accommodate effects arising from cooperativity; Correia et al., 1991).

The determination of confidence intervals for parameters estimated by numerical techniques can be a challenging endeavor for all but the simplest of models. Methods for estimation of parameter confidence intervals vary in the level of sophistication necessarily employed to obtain reliable estimates (Beecham, 1992). Implementation of parameter spaces that minimize statistical correlation among the parameters being determined may permit extraction of moderately accurate estimates of confidence intervals with relative ease. However, the great majority of parameter estimation procedures employed in interpretation of biophysical data are cast in terms of complex mathematical expressions and processes that require evaluation of nonorthogonal, correlated model parameters.

Accommodation of statistical thermodynamic equations like those describing multiple, linked equilibria (e.g., as in the case of oxygen-linked dimer-tetramer association in human hemoglobin as a function of protein concentration) or processes such as iterative interpolation or numerical integration involves solving complex mathematical relationships using nontrivial numerical methods. Additionally, comprehensive modeling of multidimensional dependencies of system properties (e.g., as a function of temperature, pH, ionic strength, and ligand concentration) often requires relatively large numbers of parameters to provide a full description of system properties. Mathematical formulations such as these therefore often involve mathematical relationships and processes sufficiently complex as to obscure any obvious correlations among model parameters as well as between the parameters and data (e.g., through effects of regions of influential observations; Correia et al., 1991; Straume and Johnson, 1992, 2010). It therefore becomes difficult to identify conveniently parameter spaces that minimize correlation, creating a potentially more challenging situation with regard to confidence interval determination.

The numerical procedures that have been developed for estimating confidence intervals all involve some approximations, particularly about the shape of the confidence probability distribution for estimated parameters (Beecham, 1992). Sometimes, these approximations may produce grossly incorrect estimates, particularly with more simplistic methods applied to situations exhibiting correlation. Errors in estimates of confidence intervals usually arise from the inability of the estimation procedure to account for high levels of sometimes complex, nonlinear correlation among the parameters being estimated. Improving the accuracy of confidence interval estimates therefore requires implementation of more thorough mathematical procedures designed to eliminate or reduce the influence of approximations regarding the shape of parameter variance space that reduce the reliability of lower-order methods.

III. Monte Carlo Method

Of course, the ultimate objective is to have available the entire joint confidence probability distributions for each of the parameters being estimated in an analysis. The Monte Carlo approach is unique in the sense that it is capable of determining confidence interval probability distributions, in principle, to any desired level of resolution and is conceptually extremely easy to implement (Bard, 1974; Press *et al.*, 1986b). The necessary information for application of a Monte Carlo method for estimating confidence intervals and probability distribution profiles is 2-fold: (1) an accurate estimate of the distribution of experimental uncertainties associated with the data being analyzed and (2) a mathematical model capable of accurately characterizing the experimental observations.

The Monte Carlo method is then applied by (1) analysis of the data for the most probable model parameter values, (2) generation of "perfect" data as calculated by the model, (3) superposition of a few hundred sets of simulated noise on the "perfect" data, (4) analysis of each of the noise-containing, simulated data with subsequent tabulation of each set of most probable parameter values, and finally (5) assimilation of the tabulated sets of most probable parameter values by generating histograms. These histograms represent discrete approximations of the model parameter confidence probability distributions as derived from the original data set and the distribution of experimental uncertainty contained therein.

The level of resolution attainable in determining confidence probability profiles by this method is dependent on the number of Monte Carlo "cycles" performed (i.e., the number of noise-containing, simulated data sets considered). The more cycles carried out, the more accurate will be the resolution of the probability distribution. In practice, this means that the amount of computer time needed to generate a probability distribution will be of the order of 100–1000 times that required for an individual parameter estimation (i.e., after ~100–1000 Monte Carlo cycles). This method must therefore be considered a "brute force" type of approach to the determination of parameter confidence intervals. Although the computational time required by the Monte Carlo method can be substantial, no other method is so easy to implement yet capable of providing information as complete about profiles of confidence probability distributions associated with estimated model parameters.

IV. Generating Confidence Probability Distributions for Estimated Parameters

Implementation of the Monte Carlo confidence probability determination method requires the initial estimation of the set of most probable parameter values that best characterize some set(s) of experimental observations according to a suitable mathematical model (i.e., one capable of reliably describing the data). [At this point, we will proceed under the assumption that the mathematical model being used to analyze the data is "valid." The reader is referred to discussions addressing concepts related to judging the validity of analytical models as descriptors of experimental data in terms of either statistical probability (Straume and Johnson, 1992) or theoretical prediction (Johnson and Faunt, 1992) (as opposed to simply empirical "fitting").] With this set of best-fit model parameter values in hand, a set of "noise-free" data is next generated to produce a data set made up of simulated "experimental points" calculated at exactly the same independent variable values as those occurring in the original data. For example, suppose that in a ligand-binding experiment measurements of some experimental observable are made as a function of ligand concentration at, say, 0.1, 0.2, 0.25, 0.3, 0.33, 0.37, and 0.4 μM ligand. After the data are analyzed by an applicable model for the most probable parameter values characteristic of this data set, theoretical values of the "expected" observable quantity are calculated from the model at 0.1, 0.2, 0.25, 0.3, 0.33, 0.37, and 0.4 μM ligand using the best-fit parameter values. The calculated dependent variable values (the simulated "experimental points") therefore correspond to those values produced by evaluating the analytical model at the same independent variable values encountered in the original data and employing the derived best-fit parameter values.

In performing an analysis of the experimental data (to obtain the most probable model parameter values), uniform, unit weighting of each experimental data point is usually employed (i.e., each data point possesses a weighting factor of 1). In cases where independent estimates of uncertainties are available for each of the observed experimental values, weighting of the data by their estimated standard deviation is desirable because a more statistically accurate parameter estimation will result (Johnson and Faunt, 1992). This provides a basis for directly calculating the variance of fit of the analytical model to the experimental data. The square root of this variance of fit represents the estimated standard deviation in the experimental data. In cases where variable weighting is employed, the square root of the variance becomes a relative indicator of the quality of fit (relative to the absolute values of the uncertainties used in weighting the data during the analysis). The assumptions underlying this assignment are (1) that the model employed in analysis is capable of accurately describing the data, (2) that the experimental uncertainty in the data is randomly distributed, and (3) that there is no systematic behavior in the data that is not accounted for by the analytical model. When these three conditions are satisfied, this estimate of the standard deviation of the experimental data permits realistic approximations of the actual experimental uncertainty to be synthesized and superimposed on the noise-free, simulated dependent variable values.

Pseudorandom noise (Forsythe *et al.*, 1977), with a distribution width defined by the estimated standard deviation, is generated to be consistent with the actual experimental uncertainty encountered in the original data. This pseudorandom noise is added to the noise-free data set to produce simulated data possessing a distribution of experimental uncertainty throughout the data. With variably weighted data, the magnitude of the pseudorandom noise that is added for a particular data point is proportional to the estimated uncertainty associated with the data point. A data set such as this corresponds to one possible distribution of noise on the simulated, noise-free data and accounts for both average system properties as well as experimental uncertainties. A few hundred such simulated, noise-containing data sets are generated and subsequently analyzed in the same manner and by the same analytical model as was done with the original experimental data. The most probable model parameter values derived from the analysis of these simulated, noise-containing data sets are then recorded as a group for each case considered.

An alternative way to generate synthetic noise sets is to rely on the residuals actually produced as a result of the parameter estimation. With this approach, the residuals obtained from an analysis are "reshuffled" to redistribute them among the independent parameter values encountered in the original data. Again, uniform, unit weighting is straightforward and direct, whereas variably weighted data must take into account the variable relative uncertainties associated with data obtained at different values of independent parameter space. This approach may in some sense be viewed as "more correct" in that the actual noise distribution obtained from analysis of the data is used—it is just redistributed among the available independent variable values. No assumptions about the shape of the actual probability distribution function are involved.

At this point exists a tabulation of a few hundred sets of most probable model parameter values obtained from analysis of a spectrum of simulated data sets. The properties of this group of data sets are meant to represent statistically what would be expected had this many actual experiments been done. This information may be assimilated in terms of probability distributions by generating histograms of relative probability of occurrence as a function of parameter value (as in Figs. 1 and 2). These examples involved determinations of 500 simulated data sets, the results of which were distributed into 51-element histograms to produce readily discernible confidence probability distributions (Straume and Johnson, 1989). The resolution of the determined probability distribution is dependent on the number of simulated data sets considered and may be improved by analyzing a greater number. In the example presented herein, 51-element histograms were employed because they were judged as providing sufficient resolution as well as providing intervals sufficiently populated to offer a statistically significant sample size.

V. Implementation and Interpretation

Knowledge of the full confidence probability distribution of model parameters provides a most rigorous way to address questions regarding resolvability of parameters characteristic of a mathematical model. The distribution of parameter confidence probability is dependent on the scatter or noise present in the experimental data as well as on the correlation between parameters of the model. The mathematical linkage of these coupled properties of the data and the analytical model parameters must be accounted for when estimating parameter confidence intervals and when propagating uncertainties between parameter spaces.

Consider the example of propagating uncertainty for the case of a difference between two derived free energy changes, as in the case for oxygen binding to

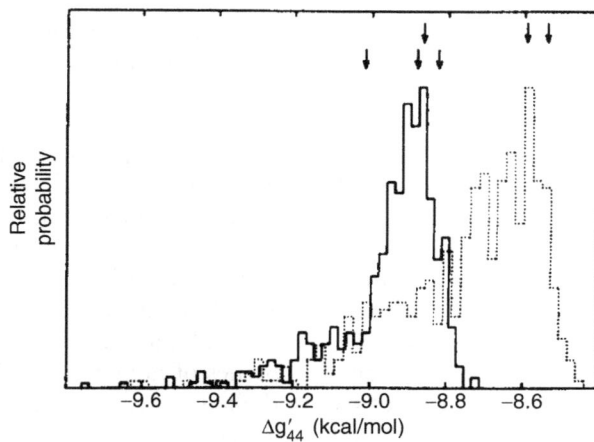

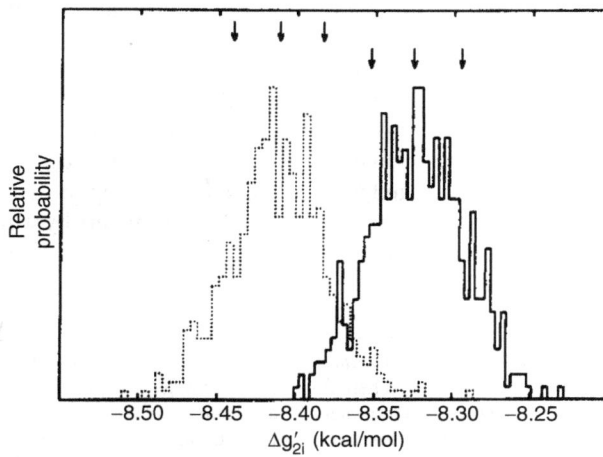

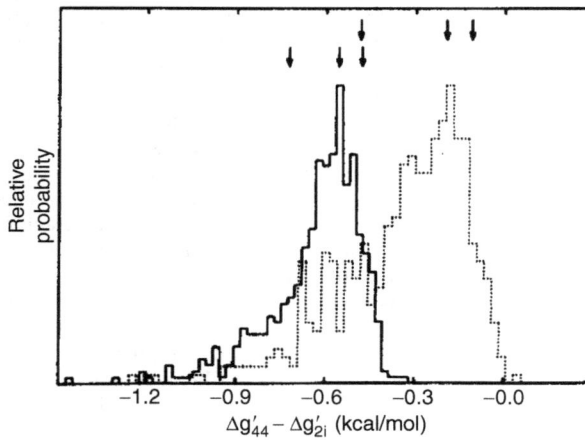

human hemoglobin (Straume and Johnson, 1989). The quaternary enhancement effect in human hemoglobin (as quantified by the quaternary enhancement free energy change, Δg_{QE}) may be defined as the difference between the free energy changes associated with oxygenation of the last available site of hemoglobin tetramers (Δg_{44}) and that for binding oxygen to the last available site in dissociated hemoglobin dimers (Δg_{22}, or Δg_{2i} for the case of noncooperative oxygen binding by hemoglobin dimers). The quaternary enhancement free energy difference is therefore $\Delta g_{QE} = \Delta g_{44} - \Delta g_{2i}$. The significance of this parameter at the molecular level is that it quantifies the cooperative oxygen-binding free energy gained by the macromolecular association of hemoglobin dimers to triply ligated tetramers.

The equilibrium for the molecular association of dimers to tetramers is coupled to the oxygen binding properties of human hemoglobin. Mathematical modeling of the behavior of oxygen-linked dimer-tetramer association involves estimating parameters characteristic of the thermodynamic linkage scheme for this system (Straume and Johnson, 1989). Oxygen-binding isotherms obtained over a range of protein concentrations represent the two-dimensional data considered. When analyzed, six model parameters require estimation. The actual parameter spaces employed were those empirically judged to provide the most robust parameter estimation (of those examined; Johnson *et al.*, 1976; Straume and Johnson, 1988).

The analysis provides the most probable values for the oxygen-binding free energy changes associated with binding at each step in the thermodynamic linkage scheme. Two of these are Δg_{44} and Δg_{2i}, the parameters by which the quaternary enhancement effect is most obviously defined. Estimates of joint confidence intervals for these

---

Fig. 1 Derived confidence probability distributions obtained from application of the Monte Carlo method are presented here for three free energy change parameters characteristic of oxygen binding by human hemoglobin tetramers. The parameter Δg_{44} is the intrinsic free energy change for addition of the last (i.e., fourth) oxygen to hemoglobin tetramers, whereas Δg_{2i} is that for oxygenation of dimer binding sites. Because oxygen binding by dimers has been experimentally shown to be noncooperative, both free energy changes Δg_{21} (for binding of the first oxygen) and Δg_{22} (for binding to the second site) are equal and therefore identified as Δg_{2i}. The quaternary enhancement effect (see text for further details) is quantified by the difference $\Delta g_{44} - \Delta g_{2i}$. The quaternary enhancement free energy change is therefore a composite parameter that requires evaluation of the difference between the values of the two constituent parameters by which it is defined, Δg_{44} and Δg_{2i}. The confidence probability distribution for the quaternary enhancement free energy change is demonstrated by these results to reside exclusively in negative free energy space. This leads to the conclusion that, given the experimental data sets considered, quaternary enhancement is indeed indicated to exist under the conditions of the experimental observations. The two distributions presented in each graph correspond to the results obtained by considering two independent variable protein concentration oxygen-binding data sets [the solid lines are derived from the data of Chu *et al.* (1984) (four binding isotherms at four protein concentrations for a total of 283 data points), and the dotted lines are for the data of Mills *et al.* (1976) (five binding isotherms at four protein concentrations for a total of 236 data points)]. The arrows in the upper parts of the graphs correspond to estimates of the most probable and the upper and lower 67% confidence limits for the distributions from the data of Chu *et al.* (lower set of arrows) and Mills *et al.* (upper set of arrows). [Reproduced from Straume and Johnson, 1989, by copyright permission of the Biophysical Society.]

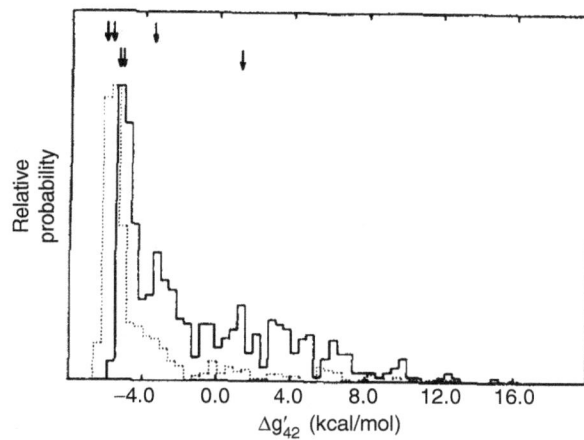

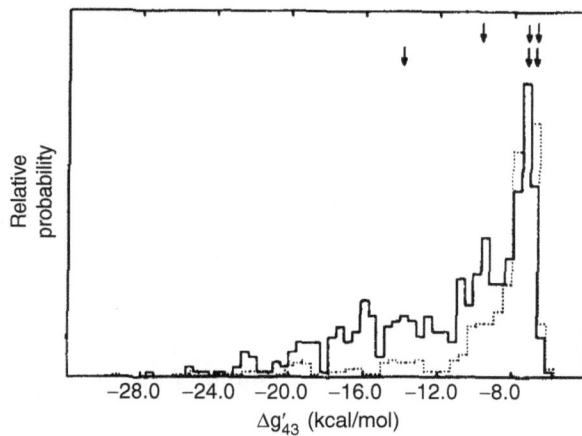

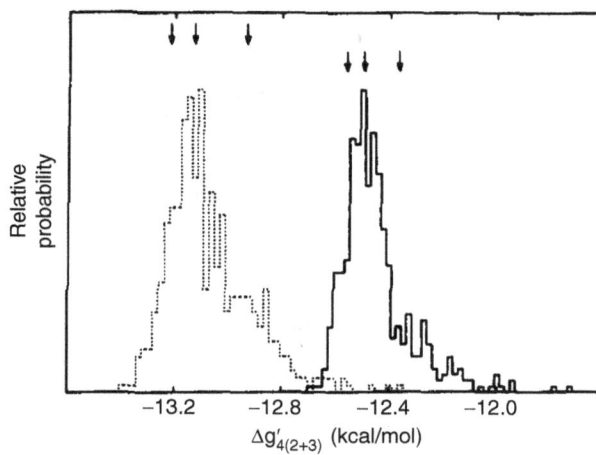

derived model parameters are also possible; however, they are difficult to obtain reliably using numerical methods that search the analytical variance space. An estimate of Δg_{QE} now requires subtracting $\Delta g_{44} - \Delta g_{2i}$. But what about the confidence interval associated with this best-estimate value of Δg_{QE}? If confidence intervals for Δg_{44} and Δg_{2i} are determined, a propagation of these uncertainties to that of Δg_{QE} is possible. To account for correlation, however, rigorous methods to map the variance spaces of Δg_{44} and Δg_{2i} to that of Δg_{QE} would have to be performed. This can be a quite challenging task with an analytical model as involved as the thermodynamic linkage scheme considered here for human hemoglobin.

In Fig. 1, we see the confidence probability distributions for Δg_{44} and Δg_{2i} as determined by application of the Monte Carlo method to two different sets of oxygen-binding isotherms (Straume and Johnson, 1989). The distributions for Δg_{2i} are seemingly symmetric, whereas those for Δg_{44} are noticeably skewed toward negative free energy space. The distributions obtained for Δg_{QE} ($\Delta g_{44} - \Delta g_{2i}$, see Fig. 1) are also (not surprisingly) noticeably skewed toward negative free energy space. The significant point here is that the confidence probability distributions of Δg_{QE} for either data set remain entirely in the negative free energy domain. This result supports the conclusion that association of hemoglobin dimers to form triligated tetramers is accompanied by a cooperative free energy change for oxygen binding. In this case, the molecular structural changes experienced by hemoglobin tetramers (relative to dissociated dimers) are responsible for the enhanced average oxygen affinity of triligated tetramers. This conclusion about thermodynamic properties, in turn, provides information that contributes to elucidating the molecular mechanisms for transduction of information which ultimately modifies a functionally significant biological property of this system.

Although the confidence probability distributions for Δg_{44}, Δg_{2i}, and Δg_{QE} do not exhibit strong effects of parameter correlation or evidence of highly asymmetric

---

Fig. 2 Derived confidence probability distributions for the intermediate tetramer oxygen-binding free energy changes Δg_{42}, Δg_{43}, and $\Delta g_{4(2+3)}$ are presented. Binding of the second oxygen to singly ligated hemoglobin tetramers is characterized by Δg_{42}, and binding of the third oxygen to doubly ligated tetramers is determined by Δg_{43}. These two free energy change parameters exhibit very broad and highly asymmetric confidence probability distributions. The distributions for the free energy change associated with binding of two oxygens to singly ligated tetramers to produce triply ligated tetramers, $\Delta g_{4(2+3)}$, however, is only moderately asymmetric and spans a much narrower range of free energy space. This property of the parameter confidence probability distributions leads to the conclusion that the free energy change for adding two oxygens to singly ligated hemoglobin tetramers may be confidently determined, whereas the partitioning of this free energy change between the two steps (singly-to-doubly ligated and doubly-to-triply ligated) is very poorly resolvable (from the experimental data considered in this analysis). Propagation of the highly correlated and asymmetric uncertainties of Δg_{42} and Δg_{43} to estimate those of $\Delta g_{4(2+3)}$ would require performing a sophisticated mapping of the three variance spaces relative to each other to provide reliable uncertainty estimates for $\Delta g_{4(2+3)}$. By using the Monte Carlo method, propagation of uncertainties is quite straightforward because the method implicitly accounts for all parameter correlation effects. Possessing the tabulated results from a Monte Carlo confidence probability determination therefore permits generation of complete probability profiles for any other parameter space of interest, as long as it may be obtained from the parameters for which distributions have been determined.

variance spaces, the same is not the case for Δg_{42}, Δg_{43}, and $\Delta g_{4(2+3)}$ (see Fig. 2). Here, Δg_{42} is the average free energy change for adding the second oxygen to hemoglobin tetramers, Δg_{43} is that for adding the third, and $\Delta g_{4(2+3)}$ is that for adding the second and third oxygens (i.e., for proceeding from singly ligated tetramers to triligated ones). As clearly shown, Δg_{42} and Δg_{43} show very broad and highly asymmetric (in opposite directions) confidence probability distributions. However, the probability distributions for $\Delta g_{4(2+3)}$ (the sum of Δg_{42} and Δg_{43}) are symmetric and span much narrower ranges of free energy space than does either Δg_{42} or Δg_{43}. Here is a case where effects of both strong correlation and highly asymmetric parameter variance spaces are demonstrated. The conclusion from the standpoint of a physical interpretation is that it is possible to quantify with considerable confidence the free energy change associated with going from singly to triply ligated tetramers but not how this free energy change is partitioned between adding the second and adding the third oxygens (at least from the particular data being considered in this analysis).

VI. Conclusion

The application of ever more sophisticated analytical protocols to interpretation of experimental data has been made possible largely from ongoing advances in computer technology, both in terms of computational power and speed as well as affordability. Biological scientists thus now have convenient access to analytical capabilities superior in many ways to that available in the past. Continued developments in both computer hardware and software will undoubtedly lead to more widespread use of sophisticated parameter-estimation algorithms that may, in principle, be applied to any analytical situation.

The estimation of most probable (or best-fit) model parameter values is, of course, the primary objective of the great majority of analytical procedures. However, the statistical validity of an interpretation of system properties (based on the most probable derived parameter values) may be critically dependent on the nature of the confidence probability distributions associated with these parameters. In those cases where detailed knowledge of entire confidence probability distributions is needed, the Monte Carlo method is capable of providing the necessary information while minimizing the number of assumptions that are implicit (to varying degrees) in other confidence interval estimation protocols.

The total computer time needed to carry out a Monte Carlo confidence probability determination is directly proportional to the number of Monte Carlo "cycles" needed to produce the desired level of resolution in the probability profile (typically in the range of ~ 500 estimations). Therefore, although other, more approximate methods will produce estimates of parameter confidence intervals using considerably less computer time, the Monte Carlo approach described here circumvents the approximations implicit in these methods and produces the most accurate, experimentally based and numerically derived profiles of entire confidence probability distributions associated with estimated parameters of any analytical model as applied to any particular data set(s).

The Monte Carlo method also implicitly fully accounts for all correlations among model parameters. After the original most probable parameter values obtained from a Monte Carlo analysis are tabulated, it is possible to generate directly complete confidence probability distributions for any composite parameters (e.g., Δg_{QE} or $\Delta g_{4(2+3)}$) from knowledge of the distributions of and correlations between constituent parameters (i.e., Δg_{44} and Δg_{2i} or Δg_{42} and Δg_{43} for Δg_{QE} and $\Delta g_{4(2+3)}$, respectively). Propagating uncertainties in this way requires no assumptions about the correlation among parameters and obviates the need for complex mapping of variance spaces to convert from one parameter space to another.

Acknowledgments

This work was supported in part by National Institutes of Health Grants RR-04328, GM-28928, and DK-38942, National Science Foundation Grant DIR-8920162, the National Science Foundation Science and Technology Center for Biological Timing of the University of Virginia, the Diabetes Endocrinology Research Center of the University of Virginia, and the Biodynamics Institute of the University of Virginia.

References

Ameloot, M. (1992). Laplace Deconvolution of Fluorescence Decay Surfaces. *Methods Enzymol.* **210**, 279–305.

Bard, Y. (1974). "Nonlinear Parameter Estimation." p. 46. Academic Press, New York.

Beechem, J. (1992). Global Analysis of Biochemical Data. *Methods Enzymol.* **210**, 37–54.

Chernick, M. R. (1999). "Bootstrap Methods, A Practitioner's Guide." Wiley Interscience, New York.

Chu, A. H., Turner, B. W., and Ackers, G. K. (1984). *Biochemistry* **23**, 604.

Correia, J. J., Britt, M., and Chaires, J. B. (1991). *Biopolymers* (in press).

Forsythe, G. E., Malcolm, M. A., and Molter, C. B. (1977). "Computer Methods for Mathematical Computations." p. 240. Prentice-Hall, Englewood Cliffs, New Jersey.

Johnson, M. L. (1992). Why, when, and how biochemists should use least-squares. *Anal. Biochem.* **206**, 215–225.

Johnson, M. L. (2008). Nonlinear least-squares fitting methods. *Methods Cell Biol.* **84**, 781–805.

Johnson, M. L. (2010). Chapter 1 of this volume.

Johnson, M. L., and Faunt, L.M (1992). this volume [1].

Johnson, M. L., Halvorson, H. R., and Ackers, G. K. (1976). *Biochemistry* **15**, 5363.

Mills, F. C., Johnson, M. L., and Ackers, G. K. (1976). *Biochemistry* **15**, 5350.

Press, W. H., Flannery, B. P., Teukolsky, S. A., and Vetterling, W. T. (1986a). "Numerical Recipes: The Art of Scientific Computing." p. 430. Cambridge Univ. Press, Cambridge.

Press, W. H., Flannery, B. P., Teukolsky, S. A., and Vetterling, W. T. (1986b). "Numerical Recipes: The Art of Scientific Computing." p. 529. Cambridge Univ. Press, Cambridge.

Small, E.W. (1992). Method of Moments and Treatment of Nonrandom Error. *Methods Enzymol.* **210**, 237–279.

Straume, M., and Johnson, M. L. (1988). *Biochemistry* **27**, 1302.

Straume, M., and Johnson, M. L. (1989). *Biophys. J.* **56**, 15.

Straume, M., and Johnson, M. L. (2010). this volume [3].

Straume, M., and Johnson, M. L. (1992). Analysis of Residuals: Criteria for Determining Goodness-of-Fit. *Methods Enzymol.* **210**, 87–106.

CHAPTER 5

Effects of Heteroscedasticity and Skewness on Prediction in Regression: Modeling Growth of the Human Heart

Robert D. Abbott★ and Howard P. Gutgesell[†]

★Division of Biostatistics
Department of Internal Medicine
University of Virginia School of Medicine
Charlottesville, Virginia, USA

[†]Division of Pediatric Cardiology
Department of Pediatrics
University of Virginia School of Medicine
Charlottesville, Virginia, USA

I. Introduction

Two of the most common characteristics of data include heteroscedasticity (heterogeneity of variance) and skewness. Unfortunately, these are features that are often ignored or improperly considered for their impact on inference and estimation in a statistical model of biological relationships. As is often the case, however, it can be extremely time consuming and computationally messy to

DOI: 10.1016/B978-0-12-384997-7.00005-4

consider skewness and heteroscedasticity. Although assumptions of homogeneity of variance and symmetry will frequently lead to reasonable conclusions, there are many occasions, in terms of efficiency and prediction, where ignoring important distributional properties can have serious consequences.

This chapter examines the effect of heteroscedasticity and skewness as it affects inference and prediction in regression. An example is presented in pediatric cardiology where interest centers on developing models of growth of the normal human heart as a function of increasing body size. Such models are useful in helping identify growth that is both normal and abnormal. The data presented represent one of the most common types of statistical problems that are encountered in modeling biological relationships. The selected example serves as a useful means for describing the general effects of heteroscedasticity and skewness on traditional analytical procedures which commonly require assumptions of constant variance and symmetry for valid statistical testing to be undertaken. We illustrate the consequences of ignoring heteroscedasticity and skewness based on three methods of estimation.

II. Example from Modeling Growth of the Human Heart

The data presented here were gathered by the Division of Pediatric Cardiology in the Department of Pediatrics at the University of Virginia Health Sciences Center. The data are derived from 69 normal children ranging in age from infancy to 18 years. Among the many indices of cardiac development, aortic valve area (AVA) is used in this chapter to reflect size of the human heart (Gutgesell and Rembold, 1990). The index used for body size will be body surface area (BSA), a quantity derived from height and weight that has been widely used as a proxy for growth and physical development (Gutgesell and Rembold, 1990).

In this chapter, interest is in developing a regression model that can be used to predict a range of normal values of AVA for a given BSA. In particular, we wish to establish boundaries which discriminate between normal and abnormal values of AVA. We are also more interested in the lower boundary than the higher boundary of AVA prediction because low AVA is more likely to be a marker of abnormal cardiac development (aortic stenosis). Although the prognostic significance of low AVA values is not addressed, low AVA is a precursor of various cardiac conditions that are observed later in adult life, including left ventricular hypertrophy and cardiac failure.

The data from the 69 normal children are displayed in Fig. 1. The lower range of BSA levels (<0.5 m^2) correspond to infants, and the higher values (>1.5 m^2) correspond to teenagers. There is considerable heteroscedasticity in the data; in other words, as BSA increases, the variability in AVA increases. Although skewness in the data is less apparent, it will be shown that its effect on prediction in this example is at least as important as the effect of heteroscedasticity.

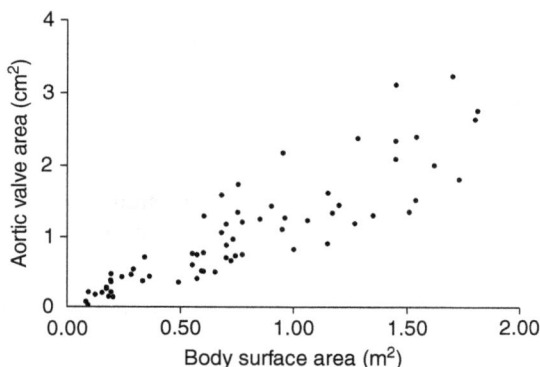

Fig. 1   Observed levels of aortic valve area and body surface area in 69 normal children.

The first step in modeling the relationship between AVA and BSA might be to fit a simple straight line through the data. In fact, there is no evidence from the data to suggest that more complicated models would be better. Similar relationships have been observed in adults (Davidson *et al.*, 1991). The model we use will have the form

$$Y_i = \alpha + \beta X_i + \varepsilon_i, \tag{1}$$

for $i = 1, \ldots, N$, where for the ith child, $Y_i = \mathrm{AVA}_i$, $X_i = \mathrm{BSA}_i$, and ε_i represents the random error around the regression line. The unknown regression coefficients are α and β. For notational simplicity, we suppress the i subscript.

Once the regression model has been estimated, we next create boundaries which encompass 95% of the normal values for a given BSA. The selection of 95%, although arbitrary, is not an uncommon level of confidence that is selected in medicine for the purpose of classification and prediction. It can always be easily changed to meet specific needs.

In terms of modeling growth of the human heart, the boundaries can help define a range of values which are characteristic of normal growth in AVA with increasing BSA. Although the original purpose of collecting the data set was to describe the physiology of cardiac development through the early years of life, one could also imagine that the boundaries have clinical use as well. For example, a child who falls outside the boundaries might be thought of as unusual or atypical. Although the child might still have normal cardiac development, falling outside the boundaries which encompass 95% of normal heart growth suggests that the child is more like children who are abnormal. Misclassification here would be a type I error. Further monitoring of the child by the cardiologist may be warranted if the child has an AVA which is unusually low. For a given BSA, a child who falls within the 95% boundary is more likely to resemble normal children and should probably be classified as normal (a mistake would be a type II error).

III. Methods of Estimation

A. Method I: Ordinary Least Squares

From ordinary least squares, (Neter and Wasserman, 1974) based on a sample of size N, we choose estimates a and b to estimate α and β in Eq. (1), respectively, where a and b minimize the error sum of squares between the observed value of Y and the expected value from the estimated equation $a + bX$; that is, we choose a and b to minimize

$$\Sigma[Y - (a + bX)]^2 = \Sigma e^2. \tag{2}$$

In ordinary least squares estimation, we assume that the errors (ε) in Eq. (1) are normally distributed around the regression line. In particular, it is assumed that the variance of the errors in Eq. (1) is constant (homoscedastic) with changing levels in X and that the distribution of the errors is symmetric.

Once the regression line is estimated, then a 95% prediction interval for a new observation is given as

$$95\% \text{prediction interval} = a + bX \pm z_{0.975}(s_p), \tag{3}$$

where

$$s_p = [s^2 + \text{Var}(a) + X^2 \text{Var}(b) + 2X \text{Cov}(a, b)]^{1/2} \tag{4}$$

and $z_{0.975} = 1.96$ is the 97.5th percentile from a normal distribution. With smaller sample sizes, z is usually replaced with the 97.5th percentile from a t distribution with $n - 2$ degrees of freedom. In Eq. (4), Var(a) and Var(b) are the estimated variances of a and b, respectively. Cov(a, b) is an estimate of the covariance between a and b, and $s^2 = \Sigma e^2/(n - 2)$.

Estimation of Eq. (1) and the 95% prediction interval in Eq. (3) based on ordinary least squares yields the result displayed in Fig. 2. There are two major deficiencies in the estimation procedure, although the simple linear regression model appears to fit the data well. The first deficiency is the result of assuming that the variance of errors is constant across the regression line. This assumption results in the prediction that all infants will have normal cardiac development. In fact, for ranges of BSA less than 0.5 m^2, a low AVA in infants will always be within the range of normal growth (it is impossible to have a negative AVA). Constant variance also assumes that approximately the same prediction interval width can be used for infants and teenagers. Clearly, it is easier to identify unusual cardiac development in the older age ranges based on this assumption, with the opposite effect occurring in infants. The circled observations in Fig. 2 illustrate this point.

The second deficiency in ordinary least squares estimation as it is used for the cardiac data is now more apparent. In Fig. 2 the data are skewed to the right, that is, there are greater deviations in AVA that fall above the regression line than below. As a result, it would be more unusual to find data that fall below the lower prediction boundary as compared to that falling above the upper boundary. This

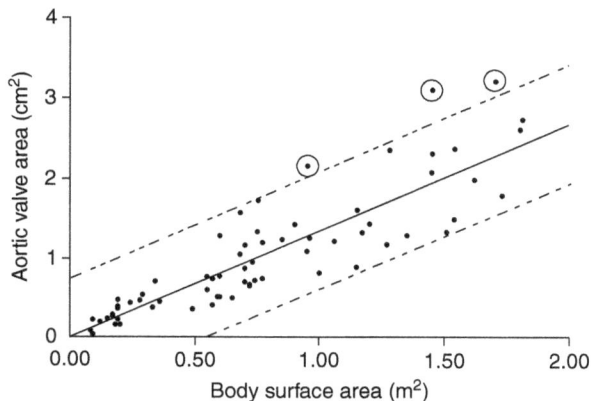

Fig. 2 The 95% prediction intervals for aortic valve area based on a prediction from body surface area using ordinary least squares estimation.

becomes more serious as younger children are studied. For the cardiac data, this is a serious problem as abnormal cardiac growth is associated with low AVA. If the lower boundary is placed too low, however, then abnormalities will be harder to identify. It is more common, as confirmed in Fig. 2, to find unusually high values of AVA, but there is no evidence suggesting that high AVA has any adverse consequences—it may even be associated with cardiovascular benefits.

B. Method II: Weighted Least Squares

If we ignore, for the moment, the problem imposed by skewness, we should at least consider alternatives to ordinary least squares estimation, which will account for the heteroscedasticity in the data. Although several alternatives exist, the best known is weighted least squares estimation (Neter and Wasserman, 1974).

In weighted least squares, we again are interested in estimating the parameters in Eq. (1), but now we relax our assumption that the variance of the errors is constant. Figure 3 shows how the variability in levels of AVA changes with BSA. Here, the estimated standard deviation of AVA is calculated within specific ranges of BSA. The linear increase in the estimated standard deviations suggests that the variance of the errors, $\mathrm{Var}(\varepsilon)$, might be modeled as follows:

$$\mathrm{Var}(\varepsilon) = \varphi^2(\gamma + \delta\mathbf{X})^2. \tag{5}$$

Here, γ and δ are regression coefficients associated with the linear increase in the standard deviation of the errors as a function of BSA. As can be seen below, ϕ^2 is a constant variance term for the errors ε^* that appear in the following reparameterized model:

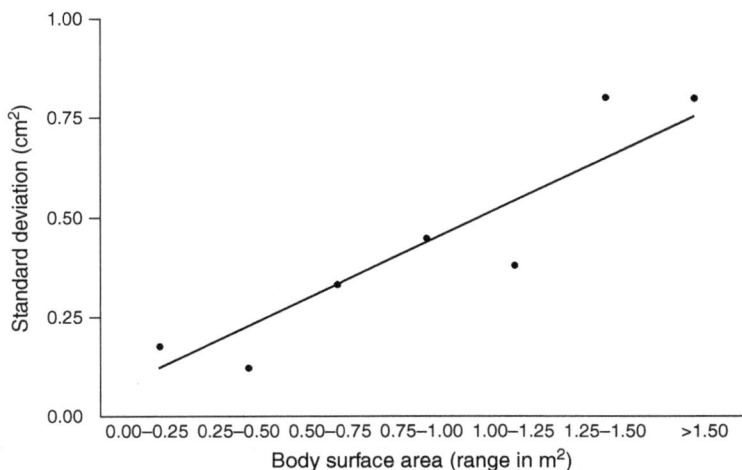

Fig. 3   Standard deviation of aortic valve area for various ranges of body surface area.

$$Y/(\gamma + \delta X) = \alpha/(\gamma + \delta X) + \beta X/(\gamma + \delta X) + \varepsilon/(\gamma + \delta X) \qquad (6)$$

which is equivalent to the linear regression model

$$Y^* = \alpha Z + \beta X^* + \varepsilon^*. \qquad (7)$$

Notice that the parameters α and β are the same as in Eq. (1), but now the variance (ϕ^2) of the errors (ε^*) in Eq. (7) is a constant. The idea was to reparameterize Eq. (1) with heteroscedastic errors and create a new Eq. (7) where the errors can be assumed to be homoscedastic. If we choose to estimate γ and δ in Eq. (5) using ordinary least squares, then we can proceed to estimate the parameters in Eq. (7) and approximate 95% prediction intervals also using ordinary least squares. In our example, estimates of γ and δ are from the estimated regression line in Fig. 3. The 95% prediction interval is then approximated by Eq. (3), but s_p is replaced with

$$s_p = [h^2(g + dX)^2 + \mathrm{Var}(a) + X^2\,\mathrm{Var}(b) + 2X\,\mathrm{Cov}(a, b)]^{1/2}. \qquad (8)$$

Here, g and d are ordinary least squares estimates of γ and δ in Eq. (5), respectively, and $h^2 = \Sigma e^{*2}/(n-2)$ is an estimate of ϕ^2. Notice that use of Eq. (8) ignores the error in estimating γ, δ, and ϕ^2.

Figure 4 presents the results from the weighted least squares approach. Although the problem of heteroscedasticity has been addressed, the problem of skewness continues to be an issue. For the upper 95% boundary, two additional individuals are identified in the younger age range with abnormally high levels of AVA, and one older subject identified as having an abnormally high AVA based on ordinary least squares estimation (see Fig. 2) is now within the normal range of cardiac development. Unfortunately, except for this last subject, not accounting for skewness has resulted in an upper boundary which may be too low for younger

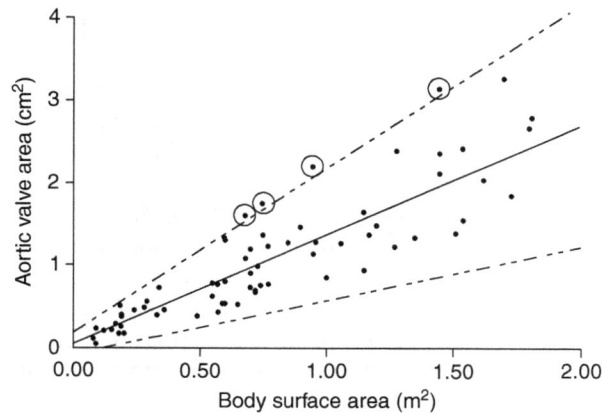

Fig. 4 The 95% prediction intervals for aortic valve area based on a prediction from body surface area using weighted least squares estimation.

children. Normal cardiac development has a greater chance of being identified as being abnormally high. Although misclassification in this region has consequences which are probably unimportant, the misclassification that occurs can be largely attributed to ignoring the long tails that occur in skewed data.

Although the prospects for identifying abnormalities in infants have improved, it does not occur for low values of AVA where abnormal cardiac development is more likely to occur. In fact, across all values of BSA, none of the AVA values is close to the lower 95% boundary where abnormal cardiac development is of greatest interest. Not accounting for skewness has created a lower boundary which has increased the range of normal values for AVA in which abnormal cardiac development has a greater chance of being classified as normal. Misclassification in this region is most serious.

C. Method III: Transform Both Sides

Although other methods exist for modeling heteroscedasticity and skewness, the most common or logical alternative to ordinary least squares and weighting is based on transformations. One approach includes a modified power transformation (Box and Cox, 1964). The idea is to produce a single expression for a transformation that includes the most common types of transformations that are used to induce constant variance and approximate normality. If Y is a dependent variable as in Eq. (1), then we let

$$h(Y, \lambda) \quad = (Y^\lambda - 1)/\lambda \text{ if } \lambda \neq 0 \\ = \log(Y) \text{ if } \lambda = 0, \tag{9}$$

where λ is a new power transformation parameter. Choosing $\lambda = 0$ means taking the log transformation, choosing $\lambda = 0.5$ means taking the square root transformation, choosing $\lambda < 0$ includes inverse transformations, and choosing $\lambda = 1.0$ means taking no transformation.

Such transformations can have several effects on inducing symmetric errors. If the data are skewed to the left, then $\lambda > 1$. An example of such data includes gestational time. When $\lambda > 1$, $h(Y, \lambda)$ is a convex function in Y. This means that for data which are left skewed values in the long left tail will be pushed to the right and values in the blunted right tail will be pulled apart. The effect is to make the data more symmetric.

If data are skewed to the right, as in the heart growth data, then $\lambda < 1$. Here, $h(Y, \lambda)$ is a concave function in Y, which has a similar but opposite effect on data than when $\lambda > 1$. When data are symmetric, $\lambda = 1$. Notice that when λ tends asymptotically to 0, $(Y^\lambda - 1)/\lambda$ tends asymptotically to $\log(Y)$. The log is probably the most common type of transformation used in medical research for correcting for right skewed data. Its use falls among the class of transformations ($\lambda < 1$) for data with long right tails.

Because selection of the log transformation in medical research is often arbitrary, the advantage of the transformation defined by Eq. (9) is that it can provide a means for systematically searching for better transformations. Although the log transformation is one of many possible transformations included in Eq. (9), alternatives for right skewed data which may be better than the log can be considered. Transformations for left skewed data are also included as well. In addition, corrections to heteroscedasticity can occur (Carroll and Ruppert, 1988). One single transformation, however, does not always work. Even after estimation of λ, one still has to question model adequacy, heteroscedasticity, and skewness, as should be the case in ordinary and weighted least-squares estimation.

The problem with the transformation in Eq. (9) is that now the heart growth model becomes

$$h(Y, \lambda) = \alpha + \beta X + \varepsilon. \tag{10}$$

This is not the simple model we began with that describes a straight line relationship between AVA and BSA. In fact, the relationship is destroyed under Eq. (10). To preserve the simple linear relationship that is supported by Fig. 1 and similar data from adults, instead of transforming Y we transform both sides of Eq. (1) as follows:

$$h(Y, \lambda) = h(\alpha + \beta X, \lambda) + \varepsilon$$

or

$$(Y^\lambda - 1)/\lambda = [(\alpha + \beta X)^\lambda - 1]/\lambda + \varepsilon. \tag{11}$$

Equation (11) is referred to as the transform both sides (TBS) model (Carroll and Ruppert, 1984). The relationship between Y and X is preserved, but now ε represents the error in prediction of $h(Y, \lambda)$ from $h(\alpha + \beta X, \lambda)$. Of course, the error in Eq. (11) subjects it to the usual distributional concerns of heteroscedasticity and skewness, but it is hoped that the TBS model has altered these deficiencies in a way that makes prediction reasonable.

Although one can always generalize Eq. (11) further, it is sufficient in the heart growth example to assume that the TBS model has resulted in symmetric errors and homoscedasticity. Carroll and Ruppert (1988) consider the case when the error variance is not constant. Implementing both weighting and transformation is a possible option. One could also consider different power transformations for the right-hand and left-hand sides of Eq. (11). The major disadvantage of Eq. (11), and more complicated alternatives, is that we now have a nonlinear model, although a linear relationship between Y and X remains preserved inside Eq. (11). Although computationally more messy, it may be worth the effort to achieve symmetry and homoscedasticity for making proper predictions.

Fortunately, for the heart growth data, the TBS model results in errors that are nearly symmetric and homoscedastic. As a result, we assume that the errors (ε) are independent and identically distributed with mean $\mu = 0$ and variance $\sigma^2 = 1$. An assumption of a normal distribution can also be imposed on the errors. If normality is assumed, then based on the conditional density of Y given X, the likelihood of the data is maximized for fixed α, β, and λ when σ^2 is estimated by

$$s^2 = \Sigma[h(Y, \lambda) - h(\alpha + \beta X, \lambda)]^2/N. \tag{12}$$

Maximizing the likelihood is then equivalent to choosing a, b, and k to estimate α, β, and λ, respectively, that minimize

$$\Sigma\{[h(Y, k) - h(a + bX, k)]/G^k\}^2, \tag{13}$$

where $G = (\Pi Y)^{1/N}$ is the geometric mean of Y. Minimizing Eq. (13) seems like a reasonable goal to achieve even in the presence of nonnormality.

The problem with estimating α, β, and λ based on minimizing Eq. (13) is that most nonlinear regression computer routines do not allow the response to depend on unknown parameters, in this case λ. One solution is to create a "pseudomodel" (Carroll and Ruppert, 1988). Here, we fit the model

$$0 = [h(Y, \lambda) - h(\alpha + \beta X, \lambda)]/G^\lambda + \varepsilon, \tag{14}$$

that is, we regress a "pseudoresponse" of 0 onto the right-hand side of Eq. (14). Note that the least-squares estimates from the pseudomodel minimize Eq. (13) and are therefore maximum likelihood estimates of α, β, and λ in Eq. (11).

To approximate the 95% prediction intervals for Y, we first transform Eq. (11) back into the original units of Y as follows:

$$Y = [(\alpha + \beta X)^\lambda + \lambda\varepsilon]^{1/\lambda}. \tag{15}$$

If we ignore estimation errors in α, β, λ, and σ^2, then the pth quantile of Y given X is

$$q_\mathrm{p}(Y|X) = [(\alpha + \beta X)^\lambda + \lambda F^{-1}(p)]^{1/\lambda}, \tag{16}$$

where ε has distribution F. If the errors are normally distributed, then we replace $F^{-1}(p)$ with $s\Phi^{-1}(p)$. An approximate 95% confidence interval becomes

$$95\%\text{Confidence interval} \cong [q_{0.025}(Y|X), q_{0.975}(Y|X)]. \tag{17}$$

For large samples, ignoring the errors in α, β, λ, and σ^2 has few consequences. Because of the computational complexity involved, the errors in estimating these parameters are often ignored in practice (Carroll and Ruppert, 1988). In our example, ignoring the errors is probably not important as there is some mitigating effect arising from the negative covariance between estimates of α and β. More refined estimation methods which include simulation and resampling techniques are described elsewhere (Carroll and Ruppert, 1991). As will be seen, the simpler formulation based on Eqs. (16) and (17) and assumption of normality of the errors or estimation of the empirical distribution function produce results that are unlikely to be improved in the heart growth example. With large enough samples, the prediction intervals would be expected to include an average of 95% of the observations across the entire range of X. Large departures from this percentage may indicate that ignoring the errors in α, β, λ, and σ^2 is inappropriate.

The fitted regression line from the TBS Eq. (11) and 95% prediction intervals based on Eqs. (16) and (17) for the heart growth data are displayed in Fig. 5. The dots represent the observed levels of $Y = \text{AVA}$ after transformation, $(Y^k - 1)/k$, and the solid line is given by $[(a + bX)^k - 1]/k$ where $X = \text{BSA}$. The dashed lines represent the 95% prediction boundaries where the distribution of the errors is derived from the empirical distribution function. Assume that normality has little effect on the placement of the boundaries. The distribution of the errors around the regression line now appears homogeneous across values of BSA. Figure 6 displays the standard deviations for the errors in the TBS model for various ranges of BSA. The estimated standard deviations appear to be more constant. Skewness is also

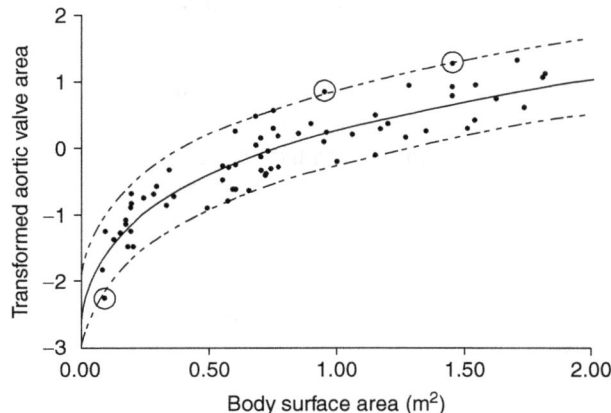

Fig. 5 The 95% prediction intervals for transformed aortic valve area based on a prediction from body surface area using transform both sides estimation.

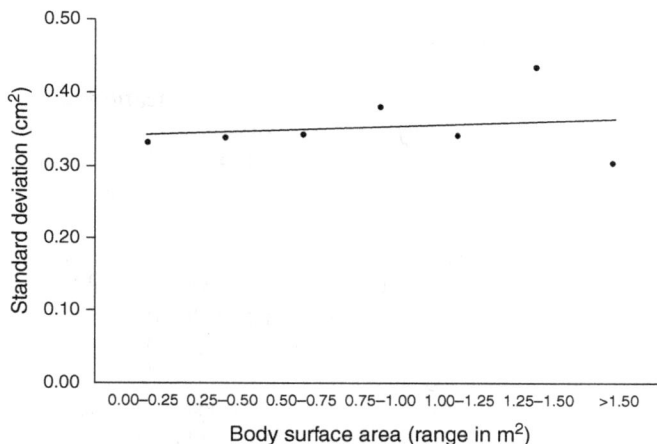

Fig. 6 Standard deviation of transformed aortic valve area for various ranges of body surface area.

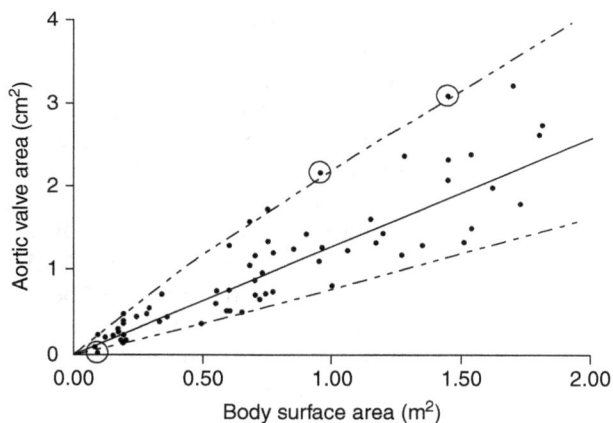

Fig. 7 The 95% prediction intervals for untransformed aortic valve area based on a prediction from body surface area using transform both sides estimation.

less apparent as there are now some low values of AVA which fall below or are closer to the lower prediction boundary.

Transforming back to the original units yields Fig. 7. The effect on prediction by using the TBS model is more apparent here. The 95% prediction intervals account for the heteroscedasticity in the errors, as does the weighted least squares approach in Fig. 4. However, the lower prediction boundary has now been raised, correcting for the problem imposed by right skewness. There are now clear opportunities for abnormal cardiac development to be identified in infants as well as older children.

IV. Discussion

Correct prediction and proper classification of diseased conditions is essential to the practice of medicine. It is often the case that defining an abnormal condition is determined by what is normal. Identifying individuals who fail to conform to normal standards of physical development is important in the diagnosis of disease and in determining prognosis.

In this chapter, we are interested in developing a regression model which can be used to predict a range of normal values of AVA in children for a given BSA. In particular, we are interested in establishing boundaries which discriminate between normal and abnormally low values of AVA on an individual basis. Although these boundaries may have some clinical use, our focus is on the effects of heteroscedasticity and skewness in helping to estimate the boundaries. We believe that these effects are extremely common in medicine where minimal levels of certain elements are common in biological systems (zero is always the best lower bound) but excesses can have wide variation. An example of such data that are right skewed is serum triglyceride. The opposite occurs with gestational time.

In an investigation of growth, it seems natural to expect that increases in anatomy will promote greater diversity of structures that can comprise an area. Some structures develop more quickly than others as increased heterogeneity evolves between individuals. Increased heterogeneity with growth can also be expected as exposure to diversity in diet, pollution, and social conditions increases.

Assuming that a statistical model is correct and the errors are independent, heteroscedasticity and skewness can have a major effect on classic regression analysis for estimation and prediction. As can be seen in Fig. 2, assuming that the errors have constant variance across all levels of BSA results in frequent misclassifications of infants with abnormal AVA as being normal. Older children with normal AVA have a greater chance of being misclassified as abnormal.

Assuming that the errors are symmetrically distributed around the regression line makes it difficult to identify abnormal cardiac development at any age. This is apparent from Fig. 4 where the lower prediction boundary is set too low, particularly in infants. The TBS model, however, accounts for skewness in the data, the key factor which is most likely to influence misclassification of abnormally low levels of AVA as being normal.

For our example, we believe that ignoring the errors in α, β, λ, and σ^2 has negligible effects on the placement of the prediction boundaries. The distribution of points near the prediction boundaries is rather similar for older and younger children. Exceptions may occur near BSA levels of 2.0 m (Davidson *et al.*, 1991), but this could be due to the limited sample size in this range of BSA. The scatter of points relative to the prediction boundaries is consistent with what would be expected from creating 95% prediction intervals. Of course, improved placement of the boundaries can be accomplished by resampling techniques and other methods (Carroll and Ruppert, 1991), but generalizing these ideas to a broad range of linear and nonlinear regression models is difficult.

Table I

Regression Coefficient Estimates Based on Ordinary and Weighted Least Squares and Transform Both Sides Methods

Method of estimation	Coefficient	Estimate	Standard error	p Value
Ordinary least squares	α	0.0128	0.0819	0.876
	β	1.3356	0.0869	<0.001
Weighted least squares	α	0.0434	0.0338	0.204
	β	1.2915	0.0761	<0.001
Transform both sides	α	0.0279	0.0292	0.339
	β	1.2656	0.0715	<0.001
	λ	0.2090	0.0955	0.029

Use of the TBS method has also resulted in regression coefficients that are similar to those estimated by ordinary and weighted least squares (see Table I). In all estimation approaches, the estimates of α are similar and not significantly different from zero. The estimates of β are also similar but statistically significant, suggesting that AVA has a true relationship to BSA. Often, the TBS method results in more efficient parameter estimates. The standard errors for the estimates of α and β are smaller than those for the other estimation procedures. From the TBS model, the standard error of the estimate of λ is determined numerically from the negative of the inverse of the second derivative of the likelihood with respect to λ (Carroll and Ruppert, 1988). The estimate of λ in Table I is significantly different from zero, which suggests that the log transformation may not be suitable. It is also significantly different from one, which suggests that some correction is in order.

In general, we believe that heterogeneity of variance and skewness are common characteristics in the distribution of data. The effects of these features can be small, but for data similar to the heart growth example, the effect on prediction can be large. Weighting alone may not be sufficient, although the TBS method may not always work either. Nevertheless, awareness of the effects of heteroscedasticity and skewness on prediction is important, as is the need to identify methods which can correct for their influence. Transformation and weighting offer many possibilities.

Acknowledgments

Supported by National Heart, Lung, and Blood Institute Contract NO1-HC-05102 (to the Honolulu Heart Program) and by National Institutes of Health Grant RR-00847 (to the University of Virginia General Clinical Research Center).

References

Box, G. E. P., and Cox, D. R. (1964). *J. R. Stat. Soc. Ser. B* **26**, 211.
Carroll, R. J., and Ruppert, D. (1984). *J. Am. Stat. Assoc.* **79**, 321.

Carroll, R. J., and Ruppert, D. (1988). "Transformation and Weighting in Regression." Chapman & Hall, New York.

Carroll, R. J., and Ruppert, D. (1991). *Technometrics* **33,** 197.

Davidson, W. R., Jr., Pasquale, M. J., and Fanelli, C. (1991). *Am. J. Cardiol.* **67,** 547.

Gutgesell, H. P., and Rembold, C. M. (1990). *Am. J. Cardiol.* **65,** 662.

Neter, J., and Wasserman, W. (1974). "Applied Linear Statistical Models." Richard D. Irwin, Inc., Homewood, IL.

CHAPTER 6

Singular Value Decomposition: Application to Analysis of Experimental Data

E. R. Henry and J. Hofrichter

Laboratory of Chemical Physics
NIDDK, National Institutes of Health
Bethesda, MD, USA

ESSENTIAL NUMERICAL COMPUTER METHODS

DOI: 10.1016/B978-0-12-384997-7.00006-6

I. Update

Since the publication of this article, the expansion in the scope of application of SVD-based analysis techniques to experimental data has been driven largely by the evolution of the computing field itself. On the one hand, the orders-of-magnitude increase in both CPU speeds and memory capacities has enabled the treatment of data sets far larger than could have been contemplated even a decade ago, extending, for example, to SVD analysis of sets of electron-density maps in X-ray crystallography (e.g., "Analysis of experimental time-resolved crystallographic data by singular value decomposition," Rajagopal *et al.*, 2004). On the other hand, the availability of the SVD algorithm itself in off-the-shelf analysis programs such as MATLAB and Mathematica, as well as its incorporation into what has become the *de facto* standard of matrix analysis libraries, the LAPACK library, have placed the necessary facilities in the hands of researchers and programmers everywhere.

The compact representation of the essential information content of a data set provided by the SVD was touted in the article as an important advantage of this approach over so-called global analysis, especially in the face of limited computing resources. The dramatic increase in available computational power noted above has to some extent rendered this specific point moot. However, the unique "cross-sectional" view of a data set provided by proper use and interpretation of the SVD makes it an extremely powerful diagnostic and analytical tool, independently of any considerations of computational efficiency. In a subsequent article (Henry, 1997), we have presented a unified view of modeling of matrix data sets applicable to both SVD-based and global analysis procedures; we have often integrated both methodologies into our own work in order to best exploit their respective strengths.

Tips and Caveats Concerning the Use of the Technique:

We believe that this technique has a place in any toolkit for the numerical analysis of experimental data. However, like most analytical tools, its effective use benefits from a deeper understanding of its specific powers and limitations. In this article we have attempted to bring out the various types of information that SVD and associated processing algorithms are able to reveal about a set of measurements. First and foremost, it must be emphasized that it is a purely mathematical procedure, the workings of which are in no way a reflection of the physical reality underlying the measurements. This analysis provides a convenient way to estimate the effective rank of a set of measurements—that is, within measurement errors the number of linearly independent components present in varying proportions in the data—but the actual such components provided by the analysis are simply a basis set defined entirely by the mathematical properties of the data set and should not be assigned *a priori* any deeper physical significance. To use this basis set as more than a noise-filtered representation of the original data set requires some sort of model framework by which a description of physically meaningful species in terms of this basis set may be produced.

The need for an appreciation of the mathematical aspects of the analysis extends to some of the possible post-processing steps that were described. One caveat of this sort was discussed in the article, but bears repeating. Because the rotation algorithm can be very effective in further separating signal from noise in the SVD output, there might be a temptation to incorporate a larger fraction of this output (i.e., a larger number of component vectors) in the procedure. However, it may easily be demonstrated that, in the limit in which the entire SVD output is included in the rotation procedure, the number of mathematical constraints introduced is sufficient to fix the final rotated vectors to a form that is completely *independent* of the initial input. That is, the output of the rotation of the entire set of V vectors from the SVD of a measured data matrix is identical to the output of the rotation of the V vectors from the SVD of any *random* matrix of the same size. This observation adds considerable weight to the caution in the article against using more than a small fraction of the SVD output in the rotation procedure.

In the article we have gone to great lengths to characterize the statistical properties of the SVD procedure in the presence of measurement noise, using idealized simulations of experimental data sets. For most applications, these basic observations, as well as semi-quantitative estimates such as that in Eq. 33, should provide a good starting point for a sound statistical treatment of the results. However, there is no true substitute for a detailed understanding of the sources and statistical properties of the measurement errors in a specific experiment, which can provide the basis for simulations to guide the interpretation of the results of the SVD analysis.

II. Introduction

The proliferation of one- and two-dimensional array detectors and rapid scanning monochromators during the 1980s has made it relatively straightforward to characterize chemical and biochemical systems by measuring large numbers of spectra (e.g., absorption or emission spectra) as a function of various condition parameters (e.g., time, voltage, or ligand concentration). An example of such a data set is shown in Fig. 1. These data were obtained by measuring absorption difference spectra as a function of time after photodissociation of bound carbon monoxide from modified hemoglobin. The difference spectra are calculated with respect to the CO-liganded equilibrium state. We will use this data set as an illustrative example at several points in the following discussion. As such experiments have become easier to carry out; two alternative approaches, one based on singular value decomposition (SVD) (Golub and VanLoan, 1996; Horn and Johnson, 1990,1994; Lawson and Hanson, 1974) and the other called global analysis (Golub and Pereyra, 1973; Johnson *et al.*, 1981; Knutson *et al.*, 1983; Nagel *et al.*, 1982), have emerged as the most general

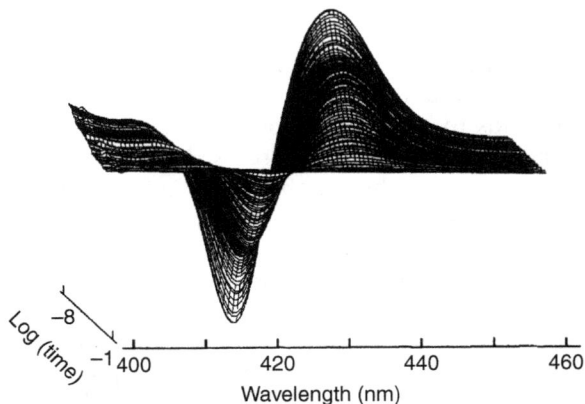

Fig. 1 Time-resolved absorption-difference spectra measured after photodissociation of $\alpha_2(\text{Co})\beta_2(\text{FeCO})$hemoglobin by 10 ns, 532 nm laser pulses. The original data consisted of 91 sets of intensities measured for both photodissociated and reference (equilibrium sample) portions of the same sample at 480 channels (wavelengths) using an optical multichannel analyzer (OMA) and vidicon detector (Hofrichter *et al.*, 1983, 1985; Murray *et al.*, 1988). Background counts from the vidicon measured in the absence of the measuring flash and baseline intensities measured in the absence of the photodissociating flash were also collected. The spectra were calculated by subtraction of the backgrounds from each set of measured intensities and calculation of the absorbance-difference spectra as the logarithm of the ratio of the corrected intensities. These spectra were then corrected for the appropriate baseline. The resulting spectra were averaged using a Gaussian filter having the spectral bandwidth of the spectrograph (4 pixels) and then truncated to 101 wavelength points at approximately 0.8 nm intervals to produce the results shown. Positive signals arise from deoxy photoproducts and negative signals from the CO-liganded reference state(Data courtesy of Colleen M. Jones.).

approaches to the quantitative analysis of the resulting data. Before beginning a detailed discussion of SVD, it is worthwhile to compare briefly these two alternative approaches.

Suppose that we have collected a set of time-resolved spectra (e.g., the data in Fig. 1) measured at n_λ wavelengths and n_t times which we wish to analyze in terms of sums of exponential relaxations. That is, we wish to represent the measured data matrix in the form

$$A_{ij} = A(\lambda_i, t_j) = \sum_{n=1}^{n_k} a_n(\lambda_i)e^{-k_n t_j}, \tag{1}$$

for each λ_i. An obvious approach to solving this problem is to use global analysis, in which all of the n_λ vectors of time-dependent amplitudes (i.e., all of the columns of the data matrix) are simultaneously fitted using the same set of n_k relaxation rates $\{k\}$ (Knutson *et al.*, 1983; Nagel *et al.*, 1982). The total number of parameters which must be varied in carrying out this fit is $(n_\lambda + 1) \times n_k$. Such a fit to the unsmoothed data represented in Fig. 1 would require fitting $91(480) = 43,680$ data points to a total of $(480 + 1)5 = 2405$ parameters; reducing the data by averaging over the spectral bandwidth, pruning of regions where the signals are relatively small, and sampling at 101 wavelengths reduces this to fitting $91(101) = 9191$ data points using $(101 + 1)5 = 510$ parameters.

To determine the number of relaxations necessary to fit the data, some statistical criterion of goodness-of-fit must be used to compare the fits obtained for different assumed values for n_k, the number of relaxations. The fitting of data will be discussed in more detail in Section V.E on the application of physical models and is also discussed at length elsewhere in this volume. The value of n_k determined from the fitting procedure provides a lower limit for the number of kinetic intermediates which are present in the system under study.[1] Another piece of information which is useful in the analysis of such data is the number of spectrally distinguishable molecular species (n_s) which are required to describe the data set. It becomes difficult to determine this number from inspection of real experimental data when it exceeds two, in which case isosbestic points cannot be used as a criterion. In the case of global analysis, the only method for estimating n_s is indirectly (and ambiguously) from the number of relaxations, n_k.

We now turn to the SVD-based analysis of the same data. If the system under observation contains n_s species which are spectrally distinguishable, then Beer's law requires that the measured spectrum at time t_j can be described as a linear combination of the spectra of these species:

$$A_{ij} = A(\lambda_i, t_j) = \sum_{n=1}^{n_s} f_n(\lambda_i)c_n(t_j), \qquad (2)$$

where A_{ij} is the element of measured spectrum A_j (the spectrum measured at time t_j) sampled at wavelength λ_i, $f_n(\lambda_i)$ is the molar absorbance of species n at wavelength λ_i multiplied by the sample pathlength, and $c_n(t_j)$ is the concentration of species n at time t_j. This result does not depend on the number of species present in the system or the size of the data matrix (i.e., how many spectra are measured and the number of wavelengths on which the spectra are sampled). One of the most useful and remarkable properties of an analysis based on SVD is that it provides a

---

[1] The simplest kinetic model for a system which contains n_s species is one in which species interconvert only via first-order reactions. Such a system may be described by an $n_s \times n_s$ matrix containing the elementary first-order rates. The kinetics of such a system may, in most cases, be completely described in terms of a set of exponential relaxations with rates given by the eigenvalues of the rate matrix. If the system comes to equilibrium, one of these eigenvalues is zero, leaving $n_s - 1$ nonzero relaxation rates. If the eigenvalues of the rate matrix are nondegenerate, all relaxations are resolved in the kinetic measurement, and all of the species in the system are spectrally distinguishable, then the number of relaxations is one less than the number of species, $n_k = n_s - 1$. Because the spectra of all of the kinetic intermediates may not be distinguishable, the number of relaxations often equals or exceeds the number of distinguishable spectra, that is, $n_k \geq n_s$. Under conditions where two or more species exchange so rapidly that the equilibration cannot be resolved by the experiment, both the number of relaxations and the number of observed species will be reduced.

determination of n_s which is independent of any kinetic analysis. In the absence of measurement errors this number is the rank of the data matrix (Golub and VanLoan, 1996; Horn and Johnson, 1990; Lawson and Hanson, 1974). For real data, SVD provides information which can be used to determine the effective rank of the data matrix (i.e., the number of species which are distinguishable given the uncertainty of the data) which provides a lower limit for n_s. This determination is discussed in more detail below in Sections V.C and V.D which describe the analysis of SVD output and the rotation procedure.

When SVD is used to process the data matrix prior to carrying out the fit, the output is a reduced representation of the data matrix in terms of a set of n_s basis spectra and an associated set of n_s time-dependent amplitude vectors. A second important property of SVD is that if the set of output components (pairs of basis spectra and amplitude vectors) is ordered by decreasing size, each subset consisting of the first n components provides the best n-component approximation to the data matrix in the least-squares sense (Golub and VanLoan, 1996; Horn and Johnson, 1990; Lawson and Hanson, 1974). It is therefore usually possible to select a subset containing only n_s of the output components which describe the data matrix $\mathbf{A}$ to within experimental precision. Once n_s has been determined, fitting the data requires modeling the amplitudes for only n_s time-dependent amplitude vectors instead of the n_λ vectors required by global analysis. The total number of parameters which must be varied in carrying out the fit is, therefore, $(n_s + 1) \times n_k$. The determination of the number of relaxations required to best fit the data is accomplished using a weighted fitting procedure which is directly comparable to that used for the global analysis of the data, except that it requires fitting of a much smaller set of time-dependent amplitude vectors.

The effectiveness of this procedure is illustrated by the SVD of the data in Fig. 1, the first six components of which are presented in Fig. 2. The spectra and time-dependent amplitude vectors which describe the first two components clearly exhibit signals which are present in the data. Note, however, the progressive decrease in the singular values, s_i, and the signal-to-noise ratios of the subsequent amplitude vectors. Given this result, if n_s were chosen based on a visual inspection of Fig. 2, one might estimate n_s to be only 2; that is, nearly all of the information in the data can be described in terms of only the first two basis spectra and their associated amplitudes. Fitting the first two amplitude vectors from the SVD to five exponential relaxations would require fitting only $91(2) = 182$ data points using only $(2 + 1)5 = 15$ parameters, as compared with the 9191 data points and 510 parameters required by global analysis of the data in Fig. 1.

This brief discussion and the example point out the advantages of using SVD in carrying out such an analysis when the number of wavelengths on which the data are sampled is large (i.e., $n_\lambda \gg n_s$). The use of SVD as an intermediate filter of the data matrix not only provides a rigorous and model-independent determination of n_s,

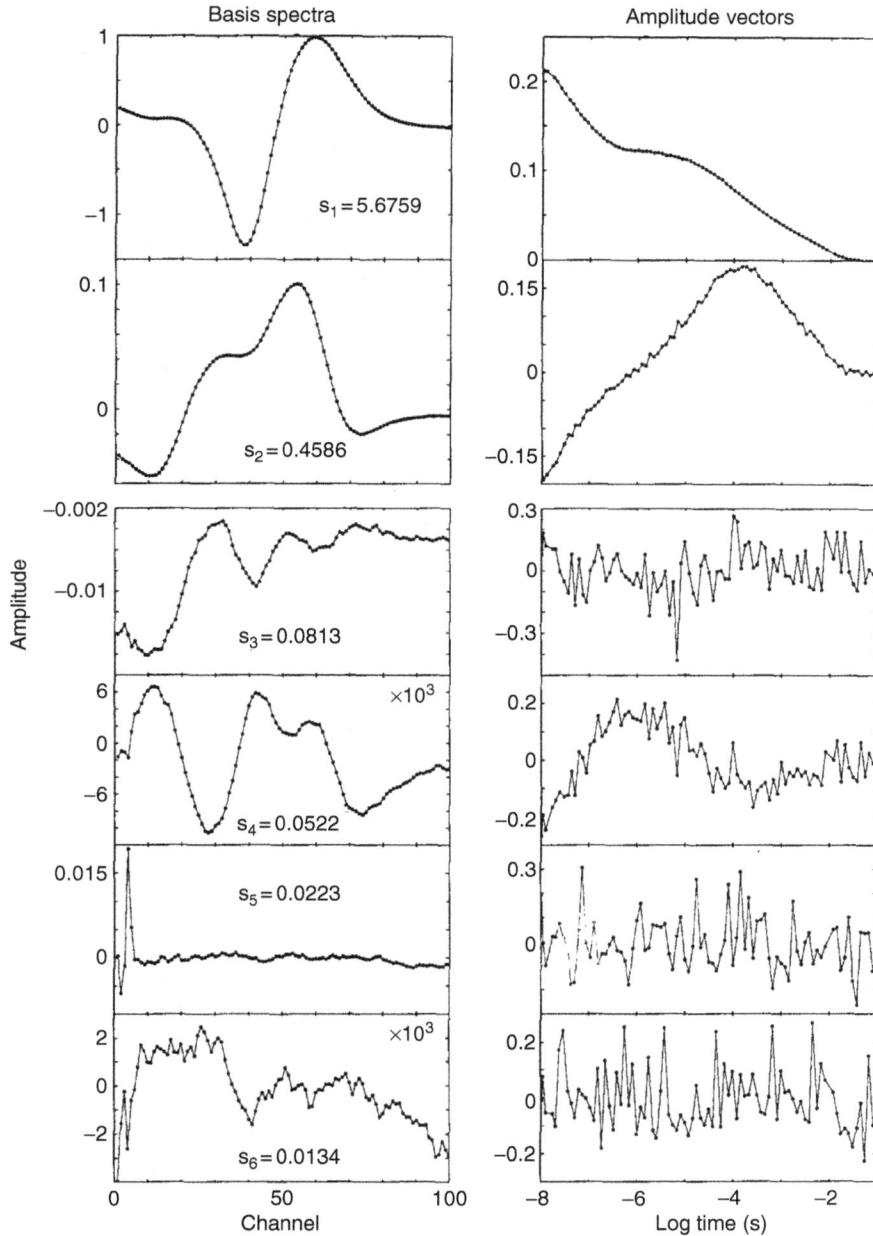

Fig. 2 Singular value decomposition of the data in Fig. 1. The basis spectra (columns of $\mathbf{U} \cdot \mathbf{S}$) are plotted on the left, and the corresponding time-dependent amplitudes (columns of $\mathbf{V}$) are plotted on the right. The first 10 singular values were as follows: $s_1 = 5.68$; $s_2 = 0.459$; $s_3 = 0.0813$; $s_4 = 0.0522$; $s_5 = 0.0223$; $s_6 = 0.0134$;

but also enormously simplifies the fitting problem. If the data set includes experiments at only a small number of wavelengths, so that the number of wavelengths is smaller than the number of species in the system which exhibit distinguishable spectra, then $n_s \cong n_\lambda$ and SVD offers no clear advantage in the analysis. This brief discussion also points out why the use of SVD proliferated in the 1980s. Earlier experiments usually consisted of measuring time traces at a small set of selected wavelengths. Only the availability of array detectors and efficient data acquisition computers has made it possible to analyze sets of data sampled on a sufficiently dense array of wavelengths to demand the increases in processing efficiency which result from the use of SVD.

The matrix of data can be derived from a wide variety of experiments. Examples include sets of time-resolved optical spectra, obtained using either a rapid-scanning stopped-flow spectrometer (Cochran *et al.*, 1980) or a pulse-probe laser spectrometer (Hofrichter *et al.*, 1983, 1985; Milder *et al.*, 1991; Murray *et al.*, 1988), and equilibrium spectra obtained during potentiometric (Hendler *et al.*, 1986; Subba Reddy *et al.*, 1986) or pH (Frans and Harris, 1985) titrations. This analysis has also been applied to other types of spectra, such as circular dichroism (Hennessey and Johnson, 1981; Johnson, 1988) and optical rotatory dispersion spectra (McMullen *et al.*, 1965). The only constraint imposed by the analysis presented here is that the measured signal be linear in the concentrations of the chemical species. The data matrix can then be described by an expression analogous to Eq. (2). In general, the index j runs over the set of experimental conditions which are varied in measuring the spectra. In the case of time-resolved spectroscopy, this index includes, but is not necessarily limited to, the time variable, whereas in pH or potentiometric titrations it would include the solution pH or voltage, respectively.

If all of the spectra, $f_n(\lambda)$, are known with sufficient accuracy, then the problem of determining the sample composition from the spectra is easily solved by linear regression. More often, however, the spectra of only a subset of the species are known, or the accuracy with which the reference spectra are known is insufficient to permit the analysis of the data to be carried out to within instrumental precision. Under these conditions one is interested in determining both the number and the shapes of a minimal set of basis spectra which describe all of the spectra in the data matrix. Because the information contained in the data matrix almost always over-determines the set of basis spectra, the algorithm must be robust when faced with

---

$s_7 = 0.0109$; $s_8 = 0.0072$; $s_9 = 0.0047$; $s_{10} = 0.0043$. The data produce two significant basis spectra for which the time-dependent amplitudes have large signal-to-noise ratios. The first, which has a signal-to-noise ratio of about 250, results primarily from a decrease in the amplitude of the deoxy-CO difference spectrum, and its amplitude monitors the extent of ligand rebinding. The second, which has a signal-to-noise ratio of about 30, arises from changes in the spectra of the deoxy photoproduct and hence reflects changes in the structure of the molecule in the vicinity of the heme chromophore. The amplitudes of the SVD components are plotted as the points connected by solid lines.

rank-deficient matrices. SVD is optimally suited to this purpose. Two alternative procedures can be used to calculate the decomposition. One is to calculate it directly using an algorithm which is also called singular value decomposition (SVD), and the other is to use a procedure called principal component analysis (PCA; Anderson, 1963; Cochran and Horne, 1980; Shrager and Hendler, 1982; Shrager, 1986).[2] PCA was used in most of the early applications of rank-reduction algorithms to experimental data (Cochran and Horne, 1977, 1980; Kankare, 1970; Sylvestre *et al.*, 1974). The output of the decomposition provides a set of basis spectra in terms of which all of the spectra in the data set can be represented to within any prescribed accuracy. These spectra are not the spectra of molecular species, but are determined by the mathematical properties of the SVD itself, most significantly by the least-squares property mentioned above. These spectra and their corresponding amplitudes can be used in a variety of ways to extend the analysis and thereby obtain the spectra of the molecular species. This problem is discussed in detail in Section V.E. A historical summary of the approaches which have been brought to bear on this problem has been presented by Shrager (1986).

Practical applications of SVD to data analysis followed only after the development of an efficient computer algorithm for computing the SVD (Golub and Kahan, 1965; Golub and Reinsch, 1970) and the experimental advances discussed above. Much of the existing literature which addresses the application of SVD to spectroscopic data has focused on describing specific algorithms for extracting the number of spectral components which are necessary to describe the data and for determining the concentrations of molecular intermediates from the basis spectra (Cochran and Horne, 1977, 1980; Shrager, 1984; Sylvestre *et al.*, 1974). Since beginning to collect this type of data almost a decade ago, we have made extensive use of SVD in the analysis of time-resolved spectroscopic data. In addition to the utility of SVD in the quantitative analysis of data, we have found that a truncated SVD representation of the data also provides an ideal "chart paper" for array spectroscopy, in that it allows data to be compared both qualitatively and quantitatively at a range of levels of precision and also to be stored in a compact and uniquely calculable format. This application of SVD is extremely important to the experimental spectroscopist, since it is very difficult to compare directly raw data sets which may contain as many as several hundred thousand data points. Moreover, because no assumptions are required to carry out the SVD portion of the analysis, it provides a simple intermediate screen of the relative quality of "identical" data sets which permits the selection of both representative and optimal data for further analysis.

We begin this chapter with a brief summary of the properties of the singular value decomposition which are relevant to data analysis. We then describe how the SVD of a noise-free data set for which the spectra, f, and concentration, c, vectors

---

[2] In this chapter we use the abbreviation SVD to refer both to the decomposition itself and to the SVD algorithm and the abbreviation PCA to refer specifically to the calculation of the SVD by the eigenvalue-eigenvector algorithm (see below).

[Eq. (2)] are known can be calculated from consideration of the integrated over-laps[3] of these components. Because data analysis necessarily begins with matrices which are "noisy" at some level of precision, we next consider some of the properties of the SVD of matrices which contain noise. This section begins with a brief description of the SVD of random matrices (i.e., matrices which contain *only* noise). We then use perturbation theory to explore how the random amplitudes are distributed in the SVD output when noise is added to a data matrix which has a rank of one, a simple example which enables a quantitative analysis of the noise-averaging properties of SVD. The discussion of noisy matrices continues by describing an asymptotic treatment which permits the best estimate of the noise-free matrix to be calculated in the presence of noise, and concludes with a brief discussion of a special case in which the noise amplitudes are not random over all of the data matrix, but are highly correlated along either the rows or columns of **A**.

With this theoretical background, we proceed to a step-by-step description of how SVD-based analysis is carried out on real data. The steps include preparation and preprocessing of the data, the calculation of the SVD itself, and a discussion of how the SVD output is analyzed to determine the effective rank of the data matrix. This discussion includes the description of a "rotation" procedure which can be used to distinguish condition-correlated amplitude information from randomly varying amplitudes of nonrandom noise sources in the data matrix, the mathematical treatment of which is presented in the Appendix. The analysis of real data necessarily includes the use of molecular models as a means of obtaining from the data information about the system under study. We next describe how the output of the SVD procedures is used as input data for fitting to models and the weighting of the SVD output which optimizes the accuracy with which the fit describes the original data. In Section VI, we present simulations of the SVD-based analysis of sets of time-resolved spectra for the kinetic system $A \to B \to C$. These simulations address in some detail the effects of both random and nonrandom noise on data where the information content is known *a priori,* and they explore the range of noise amplitudes for which the rotation algorithm results in useful improvement of the retained SVD components.

---

[3] The integrated overlaps of two continuous spectra, $f_1(\lambda)$ and $f_2(\lambda)$, and of two sets of concentrations defined as continuous functions of conditions x, $c_1(x)$ and $c_2(x)$, are defined, respectively, as

$$\int_0^\infty f_1(\lambda)f_2(\lambda)\mathrm{d}\lambda; \int_0^\infty c_1(x)c_2(x)\mathrm{d}x$$

If f_1 and f_2 are vectors which represent the spectra $f_1(\lambda)$ and $f_2(\lambda)$ sampled on a discrete set of wavelengths $\{\lambda_1\}$, and c_1 and c_2 are vectors which consist of the concentrations $c_1(x)$ and $c_2(x)$ sampled on a discrete set of x values $\{x_i\}$, then the overlaps defined above are closely approximated by either $f_1 \cdot f_2$ or $c_1 \cdot c_2$ multiplied by the size of the appropriate sampling interval. We will conventionally ignore the sampling interval, which appears as a scale factor when comparing the overlaps of vectors sampled on the same points, and use the dot product as the definition of the "overlap" between two vectors.

III. Definition and Properties

The existence of the SVD of a general rectangular matrix has been known for over 50 years (Eckhart and Young, 1939). For an $m \times n$ matrix $\mathbf{A}$ of real elements $(m \geq n)$ the SVD is defined by

$$\mathbf{A} = \mathbf{U}\mathbf{S}\mathbf{V}^\mathrm{T}, \tag{3}$$

where $\mathbf{U}$ is an $m \times n$ matrix having the property that $\mathbf{U}^\mathrm{T}\mathbf{U} = \mathbf{I}_n$, where $\mathbf{I}_n$ is the $n \times n$ identity matrix, $\mathbf{V}$ is an $n \times n$ matrix such that $\mathbf{V}^\mathrm{T}\mathbf{V} = \mathbf{I}_n$, and $\mathbf{S}$ is a diagonal $n \times n$ matrix of nonnegative elements[4]. The diagonal elements of $\mathbf{S}$ are called the singular values of $\mathbf{A}$ and will be denoted by s_k, $k \in \{1, \ldots, n\}$. The columns of $\mathbf{U}$ and $\mathbf{V}$ are called the left and right singular vectors of $\mathbf{A}$, respectively (Golub and VanLoan, 1996; Horn and Johnson, 1990; Lawson and Hanson, 1974). The singular values may be ordered (along with the corresponding columns of $\mathbf{U}$ and $\mathbf{V}$) so that $s_1 \geq s_2 \geq \ldots \geq s_n \geq 0$. With this ordering, the largest index r such that $s_r > 0$ is the rank of $\mathbf{A}$, and the first r columns of $\mathbf{U}$ comprise an orthonormal basis of the space spanned by the columns of $\mathbf{A}$. An important property of the SVD is that for all $k \leq r$, the first k columns of $\mathbf{U}$, along with the corresponding columns of $\mathbf{V}$ and rows and columns of $\mathbf{S}$, provide the best least-squares approximation to the matrix $\mathbf{A}$ having a rank of k. More precisely, among all $m \times n$ matrices $\mathbf{B}$ having rank k, the matrix $\mathbf{B} \equiv \mathbf{A}_k \equiv \mathbf{U}_k\mathbf{S}_k\mathbf{V}_k^\mathrm{T}$, where $\mathbf{U}_k$ and $\mathbf{V}_k$ consist of the first k columns of $\mathbf{U}$ and $\mathbf{V}$, respectively, and $\mathbf{S}_k$ consists of the first k rows and columns of $\mathbf{S}$, yields the smallest value of $\|\mathbf{A} - \mathbf{B}\|$.[5] Furthermore, the magnitude of the difference $\|\mathbf{A} - \mathbf{A}_k\| = (s_{k+1}^2 + \cdots + s_n^2)^{1/2}$ (Golub and VanLoan, 1996; Horn and Johnson, 1990; Lawson and Hanson, 1974).

The relationship between SVD and principal component analysis (PCA)[2] may be seen in the following way. Given the matrix $\mathbf{A}$ with the decomposition shown in Eq. (3), the matrix product $\mathbf{A}^\mathrm{T}\mathbf{A}$ may be expressed as

$$\begin{aligned} \mathbf{A}^\mathrm{T}\mathbf{A} &= (\mathbf{U}\mathbf{S}\mathbf{V}^\mathrm{T})^\mathrm{T}\mathbf{U}\mathbf{S}\mathbf{V}^\mathrm{T} \\ &= \mathbf{V}\mathbf{S}\mathbf{U}^\mathrm{T}\mathbf{U}\mathbf{S}\mathbf{V}^\mathrm{T} \\ &= \mathbf{V}\mathbf{S}^2\mathbf{V}^\mathrm{T}. \end{aligned} \tag{4}$$

---

[4] There is some variability in the precise representation of the SVD. The definition given by Lawson and Hanson,[3] for example, differs from that given here in that both $\mathbf{U}$ and $\mathbf{V}$ are square matrices ($m \times m$ and $n \times n$, respectively), and $\mathbf{S}$ is defined to be $m \times n$, with the lower $(m - n) \times n$ block identically zero. The definition given here has advantages in terms of storage required to hold the matrices $\mathbf{U}$ and $\mathbf{S}$. The SVD is similarly defined for an arbitrary matrix of complex numbers. We assume, without loss of generality, that all of the matrices appearing in this chapter consist of real numbers.

[5] The matrix norm used here is the so-called Frobenius norm, defined for an $m \times n$ matrix $\mathbf{M}$ as $\|\mathbf{M}\| = \left(\sum_{i=1}^m \sum_{j=1}^n M_{ij}^2\right)^{1/2}$.

The diagonal elements of S^2 (i.e., the squares of the singular values of A) are the eigenvalues, and the columns of V are the corresponding eigenvectors, of the matrix $A^T A$. A principal component analysis of a data matrix A has traditionally derived the singular values and the columns of V from an eigenvalue-eigenvector analysis of the real symmetric matrix $A^T A$, and the columns of U either from the eigenvectors corresponding to the n largest eigenvalues of the reverse product AA^T [$= US^2U^T$, by a derivation similar to that shown in Eq. (4)], or by calculating $U = AVS^{-1}$. Although obtaining the matrices U, S, and V via this procedure is mathematically equivalent to using the direct SVD algorithm (Golub and Reinsch, 1970), the latter procedure is more robust and numerically stable and is preferred in most practical situations (Golub and Reinsch, 1970; Shrager, 1986).

A. Singular Value Decomposition of Known Data Matrix

To understand how SVD sorts the information contained in a noise-free data matrix it is instructive to consider the SVD of matrices having the form of Eq. (2). To generalize Eq. (2), the $m \times n$ matrix A may be written

$$A = FC^T, \qquad (5)$$

where the $m \times r$ matrix F consists of a set of r column vectors $\{F_i\}$ which are the spectra of r individual species and the $n \times r$ matrix C is a set of corresponding amplitude vectors $\{C_i\}$, describing the condition-dependent concentrations of these r species. The sets of vectors $\{F_i\}$ and $\{C_i\}$ are both assumed to be linearly independent. The matrix A will then have rank r. We now consider the $r \times r$ matrices $F^T F$ and $C^T C$ which consist of the overlaps of all possible pairs of vectors in $\{F_i\}$ and $\{C_i\}$, respectively:

$$\begin{aligned}(F^T F)_{ij} &= F_i \cdot F_j \\ (C^T C)_{ij} &= C_i \cdot C_j.\end{aligned} \qquad (6)$$

The eigenvalues and eigenvectors of the $r \times r$ product of these two matrices, $F^T F C^T C$, have a simple relationship to the SVD of A. If v is an eigenvector of this matrix with eigenvalue λ, then

$$F^T F C^T C v = \lambda v. \qquad (7)$$

Premultiplying Eq. (7) by C yields

$$\begin{aligned}CF^T F C^T C v &= C\lambda v, \\ (CF^T F C^T)Cv &= \lambda(Cv), \\ A^T A(Cv) &= \lambda(Cv).\end{aligned} \qquad (8)$$

The vector Cv is therefore an eigenvector of the matrix $A^T A$ with the same eigenvalue.

Because the eigenvalues of $A^T A$ are the squared singular values of A, and the normalized eigenvectors are the columns of the matrix V in Eq. (4), it follows that the r eigenvalues of $F^T F C^T C$ are the squares of the r nonzero singular values of A.

Multiplying each corresponding eigenvector by $\mathbf{C}$ yields (to within a normalization factor) the corresponding column of $\mathbf{V}$. Note that the transpose of the overlap product matrix $(\mathbf{F}^T\mathbf{F}\mathbf{C}^T\mathbf{C})^T = \mathbf{C}^T\mathbf{C}\mathbf{F}^T\mathbf{F}$ has the same set of eigenvalues but a different set of eigenvectors. If ω is the eigenvector corresponding to eigenvalue λ, then by a derivation similar to the above:

$$\mathbf{C}^T\mathbf{C}\mathbf{F}^T\mathbf{F}\omega = \lambda\omega,$$
$$\mathbf{F}\mathbf{C}^T\mathbf{C}\mathbf{F}^T\mathbf{F}\omega = \lambda(\mathbf{F}\omega), \tag{9}$$
$$\mathbf{A}\mathbf{A}^T(\mathbf{F}\omega) = \lambda(\mathbf{F}\omega).$$

$\mathbf{F}\omega$ is therefore an eigenvector of the matrix $\mathbf{A}\mathbf{A}^T$. Normalization of $\mathbf{F}\omega$ yields the column of the matrix $\mathbf{U}$ corresponding to the singular value given by $\lambda^{1/2}$. The remaining $n - r$ columns of $\mathbf{U}$ and $\mathbf{V}$, corresponding to singular values which are equal to zero, may be made up of arbitrary orthonormal sets of vectors which are also orthogonal to the first r column vectors constructed as described here.

A useful result of this analysis is that, because the columns of $\mathbf{V}$ and $\mathbf{U}$ may be formed simply by normalizing the sets of vectors $\{\mathbf{C}v\}$ and $\{\mathbf{F}\omega\}$, respectively, the individual elements of the eigenvectors v and ω are the coefficients with which the various columns of $\mathbf{C}$ and $\mathbf{F}$ are mixed to produce the columns of $\mathbf{V}$ and $\mathbf{U}$. This analysis of the overlap product matrix thus allows us to understand quantitatively how, in the absence of noise, SVD constructs the output matrices from the spectra and concentrations of the species which generate the input matrix.

IV. Singular Value Decomposition of Matrices Which Contain Noise

To this point we have discussed the SVD of hypothetical data constructed from spectra and concentrations of a set of species which are known with arbitrary accuracy. In the analysis of experimental data, one is confronted with matrices which contain noise in addition to the desired information. One objective of the experimental spectroscopist is to extract the data from the noise using the smallest possible number of *a priori* assumptions. To take full advantage of SVD in accomplishing this task it is important to understand how SVD deals with matrices in which the individual elements include random as well as nonrandom contributions. Although some insight into this problem can be obtained from algebraic analysis, the problems encountered in the analysis of data are generally too complex to solve analytically, and simulations are required. In this section, we use both algebraic analysis and simulations in treating some relatively simple examples which we have selected to illustrate the general principles involved in dealing with noisy matrices.

We begin with a description of the SVD of random matrices. We have carried out simulations to obtain distributions of singular values for a set of $m \times n$ matrices and for square matrices of finite size, and we compare these results with the known analytical result in the asymptotic limit of infinite matrix size. Next, we illustrate the noise-averaging properties of SVD by asking how random noise is partitioned among the SVD components in the case where the noise-free data matrix has a rank of one. We then present a procedure which generalizes an earlier treatment by Cochran and Horne (1980, 1977) which specifies a weighting of the data matrix which can be used to obtain the best estimate of the noise-free data matrix from noisy data if the matrix of variances of the noise component is known. Finally, we consider the problem of noise which can be described as the outer product of two vectors (i.e., the noise amplitude matrix has a rank of 1).

A. Random Matrices

To explore the effects of noise in the data matrix, A, on the SVD of A we begin by considering matrices which contain only random elements. Figure 3A depicts the distributions of the singular values for matrices of dimensions 10×10, 100×100, and 1000×1000. The distributions were calculated from simulations in which a total of $2(10^5)$ singular values were generated from the SVD of matrices having the specified size, each element of which was a normally distributed random variable having mean value zero and variance σ^2. Note that the rank of the $n \times n$ noise matrix is always close to n. This result can be readily understood, since it is not generally possible to write any one random vector of length n as a linear combination of the remaining $n - 1$ random vectors. This distribution is described in the limit as $n \to \infty$ by the so-called quarter-circle law (Trotter, 1984; Wigner, 1967).

$$P(x) = \frac{1}{\pi}(4 - x^2)^{1/2}, \quad x \equiv s/\sigma n^{1/2}. \tag{10}$$

The distribution function describes the quarter-circle on the interval [0,2], also shown in Fig. 3A. The simulations show that the distribution of singular values closely approximates the quarter-circle distribution, even for relatively small matrices. Characteristic distortions, which are largest when n is small (10×10), are present in the regions of the maximum and minimum singular values but the asymptotic limit becomes a very good first-order description of the distribution of singular values for matrices larger than 100×100, a size which is often exceeded in the collection of real experimental data.

There is no analytical theory to describe the distribution of singular values for an $m \times n$ matrix where $m \neq n$. If $m > n$, then it is almost always possible to write $m - n$ of the rows of the matrix as linear combinations of a subset of n rows which are linearly independent. If the singular values of an $m \times n$ matrix are compared with those of an $n \times n$ matrix, where both are composed of random elements having the same variance, one intuitively expects that each singular value of the $m \times n$ matrix will, on the average, be larger than the corresponding singular value of the $n \times n$ matrix. This expectation is confirmed by the results of simulations which were carried out to determine the distribution of singular values for matrices varying from 200×200 to 1000×200 which are presented in Fig. 3B. The results show that the entire distribution of singular values shifts to higher values, with the magnitude of the shift being correlated with $m - n$. The results in Fig. 3B suggest that the largest singular value from the distribution increases roughly as $m^{0.3}$. It is

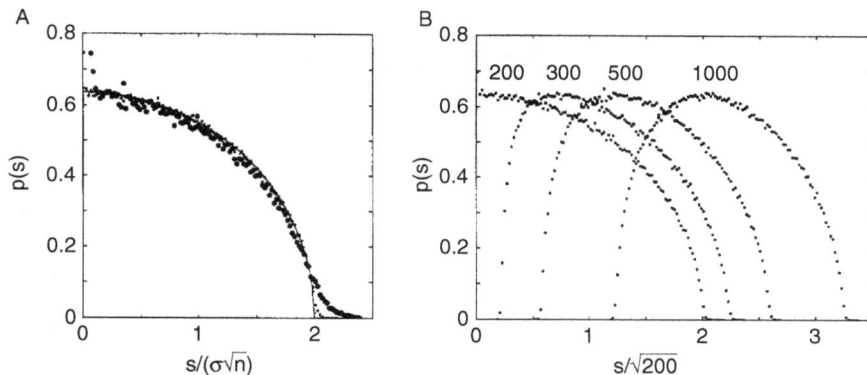

Fig. 3 Distributions of singular values of matrices of normally distributed random numbers having zero mean. (A) Calculated and asymptotic distributions for square matrices. The distribution predicted in the limit of infinite matrix size, described by the quarter circle law [Eq. (10)], is shown as the solid line. The average distributions obtained from calculation of a total of $2(10^5)$ singular values for matrices of the following sizes are shown for comparison: ($\bullet$) 10×10; ($\cdot$) 100×100; ($\cdot$) 1000×1000. (B) Calculated distributions for $m \times n$ matrices where $m \geq n$. The average distributions obtained by calculating a total of $2(10^5)$ singular values for matrices of the following sizes are shown: 200×200; 300×200; 500×200; 1000×200. The number of rows is indicated above each distribution.

important to note that when $m > n$, the entire set of singular values is effectively bounded away from zero, so a random matrix which is not square can be confidently assumed to have full rank (i.e., rank = $\min\{m,n\}$). In the simulations in Section VI.A, we shall see that this conclusion can be generalized to matrices which contain nonrandom as well as random amplitudes.

B. Noise-Averaging by Singular Value Decomposition

As discussed above, the first component of the SVD of the matrix **A** provides the best one-component least-squares approximation to **A**. For a data set which consists of n spectra that are identical except for the admixture of random noise, the first singular vector (U_1) is, to within a scale factor, very nearly identical to the average of all of the spectra in the data matrix. This example illustrates the averaging properties of SVD. In this section, we use perturbation theory to examine these properties in more detail for a particularly simple case. We consider a data matrix $\mathbf{A}_0$ which has a rank of 1 (i.e., $\mathbf{A}_0$ can be described as the outer product of two column vectors a and b, $\mathbf{A}_0 = ab^{\mathrm{T}}$). We now add to $\mathbf{A}_0$ a random noise matrix, $\boldsymbol{\varepsilon}$, each element of which is a normally distributed random variable having a mean value zero and variance σ^2, that is,

$$\mathbf{A} = \mathbf{A}_0 + \boldsymbol{\varepsilon} = ab^{\mathrm{T}} + \boldsymbol{\varepsilon}. \tag{11}$$

One would like to know how the noise represented by $\boldsymbol{\varepsilon}$ alters the singular values and vectors of the matrix **A**. If we consider for the moment the error-free data matrix $\mathbf{A}_0$, we can write

$$\mathbf{A}_0 = s_0 U_0 V_0^{\mathrm{T}}, \tag{12}$$

where $U_0 = a/\|a\|$, $V_0 = b/\|b\|$, and the singular value $s_0 = \|a\|\|b\|$. Second-order perturbation theory can be used to determine how these quantities are modified by the addition of the random matrix $\boldsymbol{\varepsilon}$.

We begin by calculating the eigenvalues and eigenvectors of the matrices $\mathbf{AA}^{\mathrm{T}}$ and $\mathbf{A}^{\mathrm{T}}\mathbf{A}$.

$$\begin{aligned} \mathbf{AA}^{\mathrm{T}} &= \mathbf{A}_0\mathbf{A}_0^{\mathrm{T}} + \mathbf{A}_0\boldsymbol{\varepsilon}^{\mathrm{T}} + \boldsymbol{\varepsilon}\mathbf{A}_0^{\mathrm{T}} + \boldsymbol{\varepsilon}\boldsymbol{\varepsilon}^{\mathrm{T}} \\ &= \mathbf{A}_0\mathbf{A}_0^{\mathrm{T}} + \mathbf{W}. \end{aligned} \tag{13}$$

The perturbed values of the largest eigenvalue, s^2, and the corresponding eigenvector, U, of this matrix can then be written

$$s^2 = s_0^2 + U_0^{\mathrm{T}}\mathbf{W}U_0 + \sum_{n \neq 0} \frac{(U_0^{\mathrm{T}}\mathbf{W}U_n)^2}{s_0^2}$$

$$U = U_0 + \sum_{n \neq 0} \frac{(U_0^{\mathrm{T}}\mathbf{W}U_n)}{s_0^2} U_n, \tag{14}$$

where the $\{U_n\}$ are a set of normalized basis spectra which are orthogonal to U_0. We proceed by calculating the matrix elements in Eq. (14) and then calculating the expected values and variances of the resulting expressions for s^2 and U. The results, in which only terms that are first order in σ^2 have been retained, may be summarized by

$$s^2 \cong s_0^2 + (m+n-1)\sigma^2 + 2s_0\varepsilon_s,$$

$$s \cong s_0 + \frac{(m+n-1)\sigma^2}{2s_0} + \varepsilon_s, \, [\mathrm{var}(\varepsilon_s) = \sigma^2],$$

$$U_i \cong U_{0i} + \frac{\varepsilon_{U_i}}{s_0}, \, [\mathrm{var}(\varepsilon_{U_i}) = (1 - U_{0i}^2)\sigma^2], \tag{15}$$

where ε_s and the ε_{U_i} are random variables having zero mean and the indicated variances.[6] A similar calculation for the matrix $\mathbf{A}^{\mathrm{T}}\mathbf{A}$ yields the result

---

[6] We have used the following properties of random variables in this derivation and in the discussion which follows. First, if X is any fixed vector and Y is a vector of random variables of mean zero and variance σ_Y^2, then

$$\langle X{\cdot}Y \rangle = 0, \, \mathrm{Var}(X{\cdot}Y) = |X|^2\sigma_Y^2.$$

Furthermore, if the individual elements of Y are independent and normally distributed, then X $\cdot$ Y is also normally distributed. Second, if Z is also a vector of random variables of variance σ_Z^2, which are independent of those in Y, then

$$\langle Y{\cdot}Z \rangle = \sum_{i-1}^{n} \langle Y_i \rangle \langle Z_i \rangle, \, \mathrm{Var}(Y{\cdot}Z) = \sum_{i-1}^{n} (\sigma_Y^2 \langle Z_i \rangle + \sigma_Z^2 \langle Y_i \rangle + \sigma_Y^2 \sigma_Z^2).$$

$$V_j = V_{0j} + \frac{\varepsilon_{V_j}}{s_0}, [\text{var}(\varepsilon_{V_j}) = (1 - V_{0j}^2)\sigma^2]. \tag{16}$$

The results in Eqs. (15) and (16) show that, while each element of the input matrix, $\mathbf{A}$, has variance σ^2, each element of the U and V vectors of the perturbed data matrix has a variance which is somewhat less than $(\sigma/s_0)^2$, and the variance of the singular value s is simply σ^2. As the matrix, $\mathbf{A}$, becomes large the squares of the individual elements of the normalized vectors U_0 and V_0 will, in most cases, become small compared to 1, and the variance of each of the individual elements of U and V will approach $(\sigma/s_0)^2$.

We expect *a priori* that the averaging of multiple determinations of a variable, each of which is characterized by random error of variance σ^2, decreases the error in the average value by a factor of $d^{1/2}$, where d is the number of determinations. It is interesting to consider an example for which the above results may be easily compared with this expectation. If we choose the matrix $\mathbf{A}_0$ to be an $m \times n$ matrix of ones, U_0 and V_0 are constant vectors having values of $1/m^{1/2}$ and $1/n^{1/2}$, respectively, and $s_0 = (mn)^{1/2}$. Equations (15) and (16) then become

$$s \cong (mn)^{1/2} + \frac{(m + n - 1)\sigma^2}{2(mn)^{1/2}} + \varepsilon_s, [\text{var}(\varepsilon_s) = \sigma^2]$$

$$U_i \cong \frac{1}{m^{1/2}} + \frac{\varepsilon_{U_i}}{(mn)^{1/2}}, \left[\text{var}(\varepsilon_{U_i}) = \left(1 - \frac{1}{m}\right)\sigma^2 \right] \tag{17}$$

$$V_j \cong \frac{1}{n^{1/2}} + \frac{\varepsilon_{V_j}}{(mn)^{1/2}}, \left[\text{var}(\varepsilon_{V_j}) = \left(1 - \frac{1}{n}\right)\sigma^2 \right].$$

If the elements of the basis spectrum, U, were obtained by simply fitting the noisy data, $\mathbf{A}$, with the V_0 vector from the noise-free data which has n identical elements, one would expect a relative error in the fitted "U" vector of $\sigma/n^{1/2}$. Use of the corresponding procedure to obtain the amplitude vector, V, should produce a relative error of $\sigma/m^{1/2}$. The predictions of Eq. (17) are very close to these expected results: the variances of both the U and the V vectors are slightly less than would be obtained from the fits. This can be rationalized by the fact that one degree of freedom, that is, variations in the sum of the squares of the entries of the data matrix, is incorporated into the variations of s. Each element of the filtered matrix, reconstructed from the first singular value and vectors, sUV^T, can now be calculated

$$A_{ij} \cong 1 + \frac{(m + n - 1)}{mn}\sigma^2 + \frac{\varepsilon_{U_i}}{n^{1/2}} + \frac{\varepsilon_{V_j}}{m^{1/2}} + \frac{\varepsilon_s}{(mn)^{1/2}}, n, m \gg 1, \tag{18}$$

where terms of higher order than $1/(mn)^{1/2}$ have been neglected. The amplitude of the noise in the reconstructed matrix is thus also significantly reduced from that of the input matrix if both n and m are large. This reduction results from discarding the amplitudes of the higher SVD components which are derived almost exclusively from the random amplitudes of ε.

These results point out a number of useful features in designing experiments to maximize signal-to-noise in the SVD-reduced representation of the data. Increasing the size of the data matrix in either dimension improves the signal-to-noise ratio in the singular vectors if it increases the magnitude of s_0. For the additional data to contribute to s_0, the added points must contain meaningful amplitude information and hence cannot include regions in which there is little or no absorbance by the sample. Increasing the size of the data matrix also does not help if it can only be accomplished by simultaneously increasing the standard deviation in the measurement for each data point. In most cases, the size of the data matrix must be determined by compromises. For example, increasing the value of m (i.e., increasing the wavelength resolution of the experiment) reduces the number of photons detected per resolution element of the detector. At the point where the noise in the measured parameter is dominated by statistical fluctuations in the number of photons detected (shot noise), further increasing the resolution will increase σ as $m^{1/2}$, so no improvement in signal-to-noise in the SVD output will result from accumulating more densely spaced data. Increasing the size of the data set by using a greater number of conditions necessarily increases the time required for data acquisition. In this case, reduction in the quality of the data matrix, perhaps by spectrometer drift or long-term laser drifts, may offset the improvements expected from increasing the number of conditions sampled.

C. Statistical Treatment of Noise in Singular Value Decomposition Analysis

We have seen in Section IV.A that a matrix which includes random noise nearly always has full rank (i.e., rank $= \min\{m,n\}$). The presence of measurement noise in a data matrix thus complicates not only the best estimate of the error-free data contained therein but even the determination of the effective rank of the matrix. Two attempts have been made to treat quantitatively the statistical problems of measurement errors in the principal component analysis of data matrices. Based on a series of simulations using sets of synthetic optical absorption spectra having a rank of 2 in the presence of noise of uniform variance, Sylvestre et al. (1974) proposed that an unbiased estimate of the variance could be obtained by dividing the sum of squared residuals obtained after subtraction of the rank r representation of a $p \times n$ data matrix by the quantity $(n - r)(p - r)$. This result is useful as a criterion in determining the rank of a matrix if the noise is uniform (see Section V.C). This analysis was generalized by Cochran and Horne (1977) to the case where the matrix of variances σ_{ij}^2 of the elements of the data matrix, A_{ij}, is any matrix having a rank of 1, rather than a constant matrix. Cochran and Horne (1977) also introduced a scheme for statistical weighting of the data matrix prior to PCA so that the effective rank, r, is more easily determined and the rank-r representation of the data is optimized. In this section, we discuss this analysis and its extension to the case where the matrix of variances, σ_{ij}^2, is arbitrary and establish a connection between such a weighting scheme and SVD-based analysis.

Consider a single set of measurements, arranged as an $m \times n$ data matrix $\mathbf{A}$. Successive determinations of $\mathbf{A}$ will differ because of measurement errors owing to noise and other factors. If multiple determinations of **A** were carried out, its expected value, $\langle\mathbf{A}\rangle$, could be calculated by averaging the individual elements. In the limit of a very large number of determinations, the matrix $\langle\mathbf{A}\rangle$ will become the best estimate of the error-free matrix. In the following discussion we make constant use of the fact that the SVD of **A** is closely related to the eigenvector-eigenvalue analyses of the matrices $\mathbf{A}\mathbf{A}^T$ and $\mathbf{A}^T\mathbf{A}$. We consider the expected values $\langle\mathbf{A}\mathbf{A}^T\rangle$ and $\langle\mathbf{A}^T\mathbf{A}\rangle$ that would be generated by making an infinite number of determinations of **A** and accumulating the averages of the resulting two product matrices. If we assume that individual elements of **A** may be treated as independent variables uncorrelated with other elements, the components of the average matrix $\langle\mathbf{A}\mathbf{A}^T\rangle$ may be written

$$
\begin{aligned}
\langle\mathbf{A}\mathbf{A}^T\rangle_{ij} &= \left\langle \sum_k A_{ik}A_{jk} \right\rangle \\
&= \sum_k \langle A_{ik}A_{jk}\rangle \\
&= \sum_k (\langle A_{ik}\rangle\langle A_{jk}\rangle(1-\delta_{ij}) + \langle A_{ik}^2\rangle\delta_{ij}),
\end{aligned}
\tag{19}
$$

where δ_{ij} is the Kronecker delta. If we now define the elements of the variance matrix as

$$
\sigma_{ij}^2 = \langle A_{ij}^2\rangle - \langle A_{ij}\rangle^2,
\tag{20}
$$

Eq (19) can be rewritten as

$$
\langle\mathbf{A}\mathbf{A}^T\rangle_{ij} = \sum_k (\langle A_{ik}\rangle\langle A_{jk}\rangle + \sigma_{ik}^2\delta_{ij}).
\tag{21}
$$

Similarly, the elements of the average matrix $\langle\mathbf{A}^T\mathbf{A}\rangle$ may be written

$$
\langle\mathbf{A}^T\mathbf{A}\rangle_{ij} = \sum_k (\langle A_{ki}\rangle\langle A_{kj}\rangle + \sigma_{ki}^2\delta_{ij}).
\tag{22}
$$

These two results may be recast in matrix notation as

$$
\begin{aligned}
\langle\mathbf{A}\mathbf{A}^T\rangle &= \langle\mathbf{A}\rangle\langle\mathbf{A}\rangle^T + \mathbf{X}, \\
\langle\mathbf{A}^T\mathbf{A}\rangle &= \langle\mathbf{A}\rangle^T\langle\mathbf{A}\rangle + \mathbf{Y},
\end{aligned}
\tag{23}
$$

X and **Y** are diagonal matrices, whose diagonal elements consist of sums of rows and columns of the matrix of variances, respectively, that is,

$$
\begin{aligned}
X_{ij} &= \left(\sum_k \sigma_{ik}^2\right)\delta_{ij}, \\
Y_{ij} &= \left(\sum_k \sigma_{ki}^2\right)\delta_{ij}.
\end{aligned}
\tag{24}
$$

In Eq. (23) the effects of measurement errors on the expectation values of $\mathbf{AA^T}$ and $\mathbf{A^TA}$ have been isolated in the matrices $\mathbf{X}$ and $\mathbf{Y}$, respectively. In general, these matrices are not simple multiples of the identity matrices of the appropriate size, so there is no simple relationship between the eigenvectors of $\langle \mathbf{AA^T} \rangle$ and those of $\langle \mathbf{A} \rangle \langle \mathbf{A} \rangle^T$ or between the eigenvectors of $\langle \mathbf{A^TA} \rangle$ and those of $\langle \mathbf{A} \rangle^T \langle \mathbf{A} \rangle$. In the special case in which the matrix of variances σ_{ij}^2 has a rank of 1, Cochran and Horne (1977) showed that it is possible to obtain diagonal matrices $\mathbf{L}$ and $\mathbf{T}$ such that the weighted or transformed matrix $\mathbf{A_w} = \mathbf{LAT}$ produces an expected value of the first product matrix of the form

$$\langle \mathbf{A_w A_w^T} \rangle = \langle \mathbf{A_w} \rangle \langle \mathbf{A_w} \rangle^T + c\mathbf{I_m}, \tag{25}$$

where c is an arbitrary constant and $\mathbf{I_m}$ is the $m \times m$ identity matrix. Although not discussed by Cochran and Horne (1977), it may also be shown that the same choices of $\mathbf{L}$ and $\mathbf{T}$ produce an expected value of the reverse product matrix $\langle \mathbf{A_w^T A_w} \rangle$ which has a similar form. Equation (25) is significant because it shows that the eigenvectors of the "noise-free" product $\langle \mathbf{A_w} \rangle \langle \mathbf{A_w} \rangle^T$ are now identical to those of the average of "noisy" matrices $\mathbf{A_w A_w^T}$, with eigenvalues offset by the constant c; a similar description holds for the reverse products.

It may be shown that for an arbitrary matrix of variances σ_{ij}^2 it is possible to construct diagonal matrices $\mathbf{L}$ and $\mathbf{T}$ such that the transformed matrix $\mathbf{A_W} = \mathbf{LAT}$ satisfies the following conditions:

$$\begin{aligned} \langle \mathbf{A_W A_W^T} \rangle &= \langle \mathbf{A_W} \rangle \langle \mathbf{A_W} \rangle^T + a\mathbf{I_m}, \\ \langle \mathbf{A_W^T A_W} \rangle &= \langle \mathbf{A_W} \rangle^T \langle \mathbf{A_W} \rangle + b\mathbf{I_n}, \end{aligned} \tag{26}$$

where a and b are constants such that $a/b = n/m$. This analysis shows that, by using the matrices $\mathbf{L}$ and $\mathbf{T}$, which can be determined from the matrix of variances σ_{ij}^2, it is possible to produce indirectly the singular value decomposition of the weighted noise-free matrix $\langle \mathbf{A_W} \rangle$ from the averages of the noisy products $\mathbf{A_W A_W^T}$ and $\mathbf{A_W^T A_W}$. It should be emphasized that this result is only rigorous in the limit of a large number of determinations of the data matrix $\mathbf{A}$ (and hence $\mathbf{A_W}$). The efficacy of such weighting schemes in improving the estimate of the noise-free data obtained from the analysis of a single determination of $\mathbf{A}$ can only be established by numerical simulations which incorporate the known characteristics of both the signal and the noise. Because the noise distribution in our experiments (e.g., Fig. 1) is nearly uniform, our experience with such schemes is severely limited. For this reason we will not discuss this issue in any detail in this chapter. One can argue intuitively, however, that the utility of such procedures for individual data matrices should depend both on the size of the data matrix and on the detailed distribution of the noise. That is, as the data matrix becomes large, a single data set should be able to sample accurately the noise distribution if the distribution of the variances is smoothly varying. On the other hand, the noise distribution might never be accurately sampled if the variances are large for only a very small number of elements of the data matrix.

Implementation of this general statistical weighting scheme requires solving a system of $m + n$ simultaneous nonlinear equations (Henry, unpublished) and using the resulting diagonal matrices $\mathbf{L}$ and $\mathbf{T}$ to calculate the weighted data matrix $\mathbf{A_w} = \mathbf{LAT}$. The SVD of this matrix is then calculated, and the output screened and/or postprocessed by any of the methods discussed in this chapter (see below), yielding a set of r basis spectra $\mathbf{U_w}'$ and amplitudes $\mathbf{V_w}'$ for which $\mathbf{A_w} \cong \mathbf{U_w'V_w'T}$. A set of basis spectra and amplitudes of the unweighted matrix $\mathbf{A}$ which are consistent with those of $\mathbf{A_w}$ may then be constructed by simply "undoing" the weighting separately on $\mathbf{U_w}'$ (using $\mathbf{L}^{-1}$) and on $\mathbf{V_w}'$ (using $\mathbf{T}^{-1}$), that is,

$$\mathbf{A} \simeq \mathbf{U'V^T},$$
$$\mathbf{U'} = \mathbf{L}^{-1}\mathbf{U_w'}, \qquad (27)$$
$$\mathbf{V'} = \mathbf{T}^{-1}\mathbf{V_w'}.$$

It is important to note that the final basis spectra and amplitudes are generally neither normalized nor orthogonal, but these mathematical properties are not usually critical for the subsequent steps in data analysis (see below). As discussed by Cochran and Horne (1980, 1977) one of the advantages of producing a weighted matrix satisfying Eq. (26) is that, if $\mathbf{A_w}$ has rank r, then the last $m - r$ eigenvalues of $\langle \mathbf{A_w A_w^T} \rangle$ will equal a. This is equivalent to having only the first m singular values of $\langle \mathbf{A_w} \rangle$ nonzero. This suggests that one measure of the success in applying the procedure to a finite data set might be the extent to which it pushes one set of singular values toward zero and away from the remaining set.

D. Singular Value Decomposition of Matrices Containing Rank-1 Noise

In addition to the random noise which we have discussed to this point, data may contain signals which have random amplitudes when examined along either the rows or the columns of the data matrix, but nonrandom amplitudes when examined along the other set of variables. One example of a situation in which noise has these characteristics arises in single-beam optical spectroscopy using array detectors, where changes in the output energy of the source or in the sensitivity of the detector result in constant offsets across the entire measured spectrum. The amplitude of these offsets is highly correlated along the wavelength direction of the data matrix, but uncorrelated along the conditions dimension. Another example arises in the measurement of condition-dependent absorbances at a single wavelength, such as kinetic traces or titration curves, where the limits of the absorbance changes can often only be obtained by extrapolation of the data to conditions where precise measurement is not possible (e.g., infinite or zero time; complete saturation with substrate). Uncertainty in the value of the extrapolated absorbance can generate errors which are present with equal amplitude in all of the data measured at a single wavelength, but which vary from wavelength to wavelength.

The influence of this type of noise on the SVD output may be addressed using the formalism developed in Section III.A. We consider a noise-free $m \times n$ data matrix which can be written as the sum of the outer products of a small set of basis m-vectors $\{F_{0i}\}$ and corresponding amplitude n-vectors $\{C_{0i}\}$, namely, $\mathbf{A_0} = \mathbf{F_0 C_0^T}$. We consider the situation in which the noise **N** may also be written as the product of two vectors: X, an m-vector which describes the noise amplitudes as a function of the isolated variable and Y, an n-vector which describes the noise amplitudes as a function of the remaining variables. In other words, $\mathbf{N} = XY^{\mathrm{T}}$ is rank-1.[7] We can then write the full data matrix as $\mathbf{A} = \mathbf{A_0} + \mathbf{N} = \mathbf{F}\mathbf{C}^{\mathrm{T}}$, where **F** and **C** are formed by simply appending the column vectors X and Y to the matrices $\mathbf{F_0}$ and $\mathbf{C_0}$, respectively. As discussed in Section III.A, the SVD of a matrix of this form is completely determined by eigenvalue-eigenvector analyses of the overlap product matrix $\mathbf{F}^{\mathrm{T}}\mathbf{F}\mathbf{C}^{\mathrm{T}}\mathbf{C}$ and its transpose.

Either the vector X or the vector Y may contain the random amplitudes. For purposes of discussion, we will assume that the randomness appears only in the amplitude vector Y, which we assume to be a set of independent, normally distributed random variables. We will also assume for simplicity that the "noise-free" matrices $\mathbf{F_0}$ and $\mathbf{C_0}$ each have a single column; these column vectors will also be called F_0 and C_0, respectively. The analyses for situations in which X is the random vector, and in which the noise-free data matrix consists of more than one component, proceed in a similar fashion. We will assume further that both X and F_0 are normalized vectors, so that the overall amplitude information is contained in the vectors Y and C_0. Then the overlap matrix $\mathbf{F}^{\mathrm{T}}\mathbf{F}$ may be written simply as

$$\mathbf{F}^{\mathrm{T}}\mathbf{F} = \begin{bmatrix} 1 & \Delta \\ \Delta & 1 \end{bmatrix}, \tag{28}$$

where Δ is the overlap of the normalized vectors F_0 and X.

The statistical properties of the overlap product matrix $\mathbf{F}^{\mathrm{T}}\mathbf{F}\mathbf{C}^{\mathrm{T}}\mathbf{C}$ and its transpose are now determined by the statistical properties of the random overlap matrix $\mathbf{C}^{\mathrm{T}}\mathbf{C}$. Using the results of Note 6, the expected value and variance of $C_0 \cdot Y$ become

$$\langle C_0 \cdot Y \rangle = 0$$

$$\mathrm{Var}(C_0 \cdot Y) = \sum_i \sigma_{yi}^2 \langle (C_0)_i \rangle^2 \tag{29}$$

$$= \sigma_Y^2 |C_0|^2$$

and $C_0 \cdot Y$ is normally distributed.

The expected value and variance of $Y \cdot Y$ may be determined in a similar fashion. The results, quoted here without proof, are

$$\langle Y \cdot Y \rangle = n\sigma_Y^2,$$

$$\mathrm{Var}(Y \cdot Y) = 2n(\sigma_Y^2)^2. \tag{30}$$

However, in this case the resulting values of $Y \cdot Y$ are not normally distributed, but are characterized by the skewed distribution

---

[7] This situation in which the noise amplitude matrix is rank-1 must be distinguished from the case in which the matrix of variances of the noise is rank-1, which was discussed in a different context in Section IV.C.

$$p(x)\mathrm{d}x = \frac{2^{n/2}\sigma_Y^n}{\Gamma(n/2)} x^{n/2-1}\mathrm{e}^{-x/(2\sigma_Y^2)}\mathrm{d}x, \tag{31}$$

where $x = Y \cdot Y$, $\Gamma(\ldots)$ is the gamma function, and n is the number of elements in Y.

With these results, the overlap matrix $\mathbf{C^T C}$ in the case of two vectors may be written

$$\mathbf{C^T C} = \begin{bmatrix} |C_0|^2 & 0 \pm \sigma_Y \\ 0 \pm \sigma_Y & n\sigma_Y \pm (2n)^{1/2}\sigma_Y^2 \end{bmatrix}, \tag{32}$$

where the notation $a \pm b$ denotes a random variable with expected value a and variance b^2. Because every instance of $\mathbf{C^T C}$ in an ensemble is symmetric, the two off-diagonal elements are in fact the same normally distributed random variable (derived from the inner product of the random vector Y with C_0). This simplifying feature is offset by the fact that the lower right diagonal element in Eq. (32) is a random variable (derived from the inner product of the random vector Y with itself) which is *not* normally distributed and is neither independent of nor representable in terms of the off-diagonal elements. If the variance of this element (which is second order in $\sigma_Y^2/|C_0|^2$) is neglected, the analysis simplifies to the diagonalization of overlap product matrices which are functions of a single normally distributed random variable. Even in this approximation analytical expressions for the distributions of eigenvalues and eigenvectors of such matrices are unmanageably complex.

It is, however, possible to determine the statistical properties of the SVD of the perturbed data matrix by explicit simulation. The aim of such simulations is to produce an ensemble of noisy data sets, the mean of which corresponds to some prescribed, noise-free data set, and use this ensemble to calculate the statistical properties (means and variances) of the singular values and vectors. In most cases it is necessary to explicitly calculate the SVD of a large number of matrices synthesized by adding random noise having specified characteristics to the elements of the noise-free data matrix. In the present situation, however, the simulation procedure is greatly simplified because it is only necessary to create ensembles of overlap product matrices $\mathbf{F^T F C^T C}$. Because $\mathbf{F^T F}$ is determined by the overlaps of the (normalized) basis vectors, it can be specified by simply specifying the magnitude of the off-diagonal elements, Δ. An ensemble of the elements of $\mathbf{C^T C}$ which involve the random amplitudes can then be constructed by calculating the overlaps of an ensemble of random amplitude vectors with the various fixed amplitude vectors and with themselves. The results of a set of such simulations are presented in Fig. 4.

Figure 4 shows the extent of mixing of the random amplitudes, Y, with C_0 and the spectrum, X, with F_0 as a function of the spectral overlap, Δ, at a number of different values of the root-mean-square (RMS) noise amplitude, σ_Y. Let us start by examining the results in Fig. 4B, which describe the mixing of the basis spectrum X with F_0. When the RMS amplitude of the "noise" spectrum is significantly

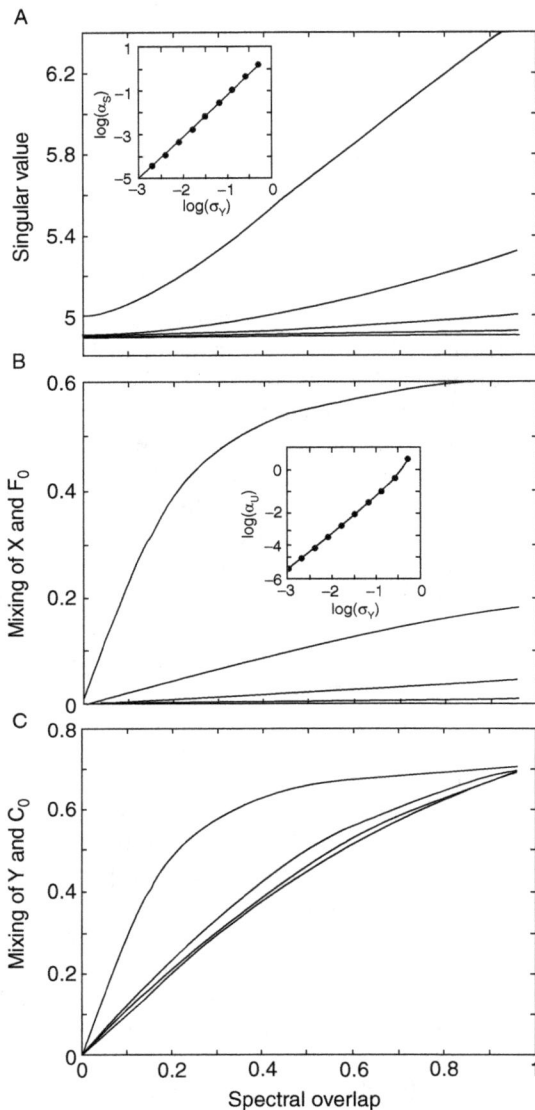

Fig. 4 Mixing between data and noise when both can be described by matrices having a rank of 1. Simulations were carried out at each of 10 noise amplitudes, σ_Y, spaced by factors of 2 from 0.001 to 0.512 and at each of 20 values of the overlap, Δ, ranging from 0 to 0.96 in increments of 0.04. Each simulation was performed as follows. The matrix $\mathbf{F}$ was first formed as in Eq. (28) for a prescribed value of the overlap Δ. A noise-free amplitude vector, C_0, having elements $(C_0)_i = 0.5[\exp(-k_1 t_i) + \exp(-k_2 t_i)]$, for a set of 71 values of t_i, uniformly spaced on a logarithmic grid from 10^{-8} to 10^{-1} s, was first calculated using $k_1 = 10^6\ \mathrm{s}^{-1}$ and $k_2 = 10^3\ \mathrm{s}^{-1}$. An ensemble of 10^4 vectors, Y, each consisting of 71 normally distributed random numbers with variance σ_Y^2 and a mean value of zero was used to construct an ensemble of overlap product matrices $\mathbf{F}^{\mathrm{T}}\mathbf{F}\mathbf{C}^{\mathrm{T}}\mathbf{C}$ where $\mathbf{C}^{\mathrm{T}}\mathbf{C}$ has the form $\begin{bmatrix} C_0 \cdot C_0 & C_0 \cdot Y \\ C_0 \cdot Y & Y \cdot Y \end{bmatrix}$.

smaller than the noise-free data, SVD effectively discriminates against the "noise" spectrum. Figure 4B shows that the effectiveness of this discrimination depends on both Δ and the noise amplitude, σ_Y. For low to moderate noise levels the mixing coefficient increases linearly with Δ, and the inset to Fig. 4B shows that the initial slope of the curves increases roughly as σ_Y^2. In interpreting these results, we must remember that the amplitude of the "noise" spectrum is a random variable, so SVD is able to average and thereby substantially reduce these amplitudes in producing the output signal spectrum. This cancellation, however, can only be as effective as the ability of SVD to distinguish between the "signal" and "noise" spectra. The extent to which these spectra are distinguishable is determined by Δ. When Δ is zero, there is no mixing of noise with signal at any noise amplitude. When Δ is nonzero, SVD mixes the signal and noise spectra to produce orthogonal basis spectra, and the extent of mixing increases roughly as $\Delta \sigma_Y^2$. At high noise amplitudes the curves become nonlinear, and appear to saturate at a value of $2^{-1/2}$.

We now examine the mixing of the random amplitudes, Y, with the unperturbed composition vector, C_0, described by Fig. 4C. The extent of the mixing is essentially independent of the noise amplitude when the noise is small. As Δ increases, the mixing coefficient increases monotonically to a value of $2^{-1/2}$. Recall that, by design, *all* of the information on the amplitude of the perturbation is contained in the norm of the vector Y. As a result, the amplitude-independent mixing coefficient in Fig. 4C actually reflects the fact that the noise content of the first amplitude vector increases in *direct* proportion to the amplitude of the perturbation. The only operative discrimination against the random amplitudes in deriving the first amplitude vector is the overlap. At the largest noise amplitudes the mixing coefficient approaches its saturating value at smaller values of Δ.

The contribution of the random amplitudes to the first amplitude vector is proportional to σ_Y, and their contribution to the first spectrum is proportional to σ_Y^2. This is a direct consequence of the averaging of this contribution over the

---

The eigenvalues and eigenvectors of these matrices and their transposes were then used to construct an ensemble of singular values and mixing coefficients as described in the text. (A) Averaged singular values, s_1. The inset in (A) describes the dependence of α_s on σ_Y (the square root of the variance of Y) where α_s is determined by fitting the initial portion of the curve to $\alpha_s \Delta^2$. (B) Mixing coefficients which describe the singular vectors U_1. These coefficients describe the mean amplitude of X, the normalized spectrum associated with the random amplitudes which is mixed with F_0 to generate U_1 under each set of conditions. The coefficients for U_1 depend on both the overlap, Δ, and the noise amplitude, σ_Y. The dependence of the mixing coefficient on the overlap is approximately linear for values of the mixing coefficient less than about 0.2. The inset in (B) describes the dependence of the initial slope of the curves in (B), α_U, on σ_Y. The second derivative of the curves in (A), α_s, and the slopes in (B), α_U, can both be approximately represented by the relation $\alpha = A\sigma_Y^2$ for $\sigma_Y < 0.1$. (C) Mixing coefficients which describe the singular vectors V_1. These coefficients describe the mean amplitude of the random amplitude vector, Y, which are mixed with C_0 to generate V_1. These coefficients depend primarily on the overlap, Δ, and are nearly independent of the noise amplitude, σ_Y. In each graph, the uppermost line represents the results of the calculations for $\sigma_Y = 0.512$, and each lower line represents the results for a value of σ_Y which is successively smaller by a factor of 2.

random amplitudes by SVD. Given these results it becomes possible to rationalize the dependence of the singular values on the noise amplitudes shown in Fig. 4A. At low noise amplitudes, the singular value increases quadratically with increasing Δ. The second derivative at $\Delta = 0$ increases in proportion to σ_Y^2 as shown in the inset to Fig. 4A. We have seen in Section IV.B that addition of a small random noise amplitude to a nonrandom matrix increases the singular value in proportion to its variance, σ^2. The observed results from the simulations parallel this behavior. The first-order effect of adding the noise at low noise amplitudes is to increase the random component in the amplitude vector in direct proportion to $\Delta\sigma_Y$. Without the normalization imposed by SVD, these random amplitudes would be expected to increase the norm of this vector by an amount proportional to $(\Delta\sigma_Y)^2$. This increase then appears in the singular value when the amplitude vector is normalized. No contribution to the singular value is expected from the mixing of F_0 and X, since both spectra are normalized prior to the mixing. At the highest noise amplitudes, σ_Y becomes comparable to the mean value of C_0, and an additional small contribution of Y to the singular value can be perceived as an offset in the value of the singular value s at $\Delta = 0$. This probably results from the fact that the random amplitudes have, at this point, become comparable to the "signal" amplitudes, and the "noise" component can no longer be treated as a perturbation.

These simulations provide considerable insight into the performance of SVD for data sets which contain one or more component spectra together with noise described by a well-defined spectrum having random amplitudes. The results show that when a perturbation having these characteristics is present in a data set, it will have a much larger effect on the amplitude vectors than on the spectra. Our observation that the degree of mixing with the signal spectrum increases as σ_Y^2 suggests that any steps taken to minimize such contributions will be particularly helpful in improving the quality of the resulting component spectra. The noise contribution to the amplitude vectors increases only in direct proportion to σ_Y, so reduction of the noise amplitude will be less effective in improving these output vectors. There are other analytical methods which can be used to supplement the ability of SVD to discriminate against such contributions. One such method, the so-called rotation algorithm, is discussed in Section V.D. Because the mixing of the random amplitudes, Y, with the "signal" component, C_0, is directly proportional to the overlap between the spectra associated with these amplitudes, Δ, the results further argue that, in some cases, it may be advantageous to select a form for the data which minimizes this overlap. If, for example, the "noise" arises primarily from baseline offsets mentioned above, then the overlap can be minimized by arranging the collection and preprocessing of the data so that the spectra which are analyzed by SVD are different spectra rather than absolute spectra. The spectral signature of such random components in a specific experiment (corresponding to X) can usually be determined by analysis of a data set which contains no "signal" but only experimentally random contributions. We shall return to this point when discussing the simulations presented in Section VI.B below in which random noise comparable to that discussed in Sections IV.A and IV.B has also been included in the data matrix.

V. Application of Singular Value Decomposition to Analysis of Experimental Data

Having considered some of the properties of the SVD of noise-free and noisy matrices, we now turn to the problem of applying SVD to the analysis of experimental data. The actual calculation of the SVD of a data matrix is only one of a series of steps required to reduce and interpret a large data set. For the purposes of this discussion we shall break the procedure into four steps. The first step is the organization of the experimental measurements into the matrix form required by SVD. In addition to the processing of the measured signals to produce the relevant experimental parameter (e.g., absorbance, linear dichroism, corrected fluorescence intensity) this step might include some preconditioning (i.e., truncation or weighting) of the data. The second step is the calculation of the SVD of the data matrix. The third step is the selection of a subset of the singular values and vectors produced by the SVD that are judged sufficient to represent the original data to within experimental error (i.e., the determination of the effective rank of $\mathbf{A}$). In some cases this step may be preceded or followed by some additional processing of the matrices produced by the SVD. We describe one such procedure, a rotation of subsets of the left and right singular vectors which optimizes the signal-to-noise ratio of the retained components. The effects of this rotation procedure are explored in more detail by the simulations described below. The final step is the description of the reduced representation of the original data set that is produced in the first three steps in terms of a physical model. This step most often involves least-squares fitting.

A. Preparation of Data Matrix

To carry out the SVD of a set of data, the data must be assembled as a matrix $\mathbf{A}$ which is arranged so that each column contains a set of measurements for which a single isolated variable ranges over the same set of values for each column, the values of all of the other variables remaining fixed.[8] Different columns of $\mathbf{A}$ then correspond to different sets of values for the remaining variables. For example, the data in Fig. 1 consist of optical absorption-difference spectra (i.e., a difference in

---

[8] The first step in any analysis is the reduction of the raw data to produce values for the desired experimental parameter. This operation usually includes adjustment of the measured data for offsets, instrument response, and instrument background, as well as correction for baselines and other experimental characteristics. We assume that all such calculations which are specific to a given experimental technique and instrument have been carried out and tested by appropriate control experiments which demonstrate, for example, the applicability of Eq. (2) to data collected and analyzed by these procedures.

optical densities between the photoproduct and the unphotolyzed sample measured as a function of wavelength) obtained at different times after photo-dissociation. To reduce these data using SVD, we create a data matrix **A**, each column of which contains a single spectrum (i.e., varies only with wavelength). The matrix **A** is then built up from such column vectors (spectra) measured under different conditions [in this case, times as described by Eq. (2)]. In a properly constructed matrix each row then corresponds to a single wavelength.

Three types of preprocessing of the data matrix, **A**, might be contemplated prior to calculation of the SVD. We shall refer to them as truncation, smoothing, and weighting. Truncation refers to the reduction of the size of the data matrix by selection of some subset of its elements; smoothing refers to any procedure in which noise is reduced by averaging of adjacent points; weighting refers to scaling of the data matrix to alter systematically the relative values of selected elements. Truncation of the data set, the first of these operations, may always be carried out. The effect of truncation is to reduce the size of the data matrix and thereby delimit the range of the experimental conditions. Truncation is clearly desirable if some artifact, such as leakage of light from a laser source into the spectrograph, preferentially reduces the quality of data on a subset of the data points. Smoothing of the data could, in principle, be performed either with respect to the isolated variable (i.e., "down the columns" of **A**) or with respect to the remaining variables (i.e., "across the rows" of **A**). As we have seen in the discussion of the noise-averaging properties presented in the previous section, SVD itself acts as an efficient filter to suppress random measurement noise in the most significant components. A data set reconstructed from the SVD components is therefore effectively noise-filtered without the artifacts that may arise when some of the more popular smoothing algorithms are used. For this reason, there is no clear advantage to pre-smoothing a data set across either variable, unless such an operation is to take place in conjunction with a sampling operation in order to reduce the data matrix to a size determined by limits on either computational speed or computer memory.

The statistical discussion of noise in Section IV.C suggests that it would be advantageous to weight the data matrix in accordance with the measured variances of its individual elements. A detailed discussion of the desirability of and strategies for statistical weighting is beyond the scope of this chapter. It would appear, however, from the discussion of Cochran and Horne (1980, 1977) that a weighting procedure should probably be incorporated into the analysis both in cases where the variances of the data set have been very well characterized and in cases where the variances range over values which differ by a significant factor. In the latter cases, it is likely that any reasonable weighting scheme will produce better results than no weighting at all. It is difficult to judge *a priori* whether weighting will significantly improve the accuracy of the SVD analysis. The only unambiguous method for determining the effects of weighting for a given type of data appears to be to carry out statistical simulations that incorporate the known properties of the data as well as the variances characteristic of the measurement system.

B. Calculation of Singular Value Decomposition

The computation of the SVD of a data matrix is the most clear-cut of all the analytical steps in the treatment of experimental data. The input matrix can be either $\mathbf{A}$ or $\mathbf{A_w}$, depending on whether the weighting procedure has been used. When the SVD of the data matrix $\mathbf{A}$, assembled as described above, is calculated [Eq. (3)], the left singular vectors of $\mathbf{A}$ (the columns of $\mathbf{U}$) are an orthonormal set of basis spectra which describe the wavelength dependencies of the data, and the corresponding right singular vectors (the columns of $\mathbf{V}$) are normalized sets of time-dependent amplitudes for each basis spectrum (see Fig. 2). The singular values, $\{s_i\}$, are the normalization factors for each basis spectrum U_i and amplitude vector V_i. Thoroughly tested FORTRAN subroutines for computing the SVD, based on the work of Golub and co-workers (Golub and Kahan, 1965; Golub and Reinsch, 1970), are generally available as part of the LAPACK (Anderson *et al.*, 1999) and *Numerical Recipes* (Press *et al.*, 2007) subroutine packages. The reader is referred to the original work for a discussion of the computational details of the SVD algorithm, which are outside the scope of this chapter (Golub and Kahan, 1965; Golub and Reinsch, 1970; Golub and VanLoan, 1996; Press *et al.*, 2007).

C. Analysis of Singular Value Decomposition Output

The SVD provides a complete representation of the matrix $\mathbf{A}$ as the product of three matrices $\mathbf{U}$, $\mathbf{S}$, and $\mathbf{V}$ having well-defined mathematical properties. Equation (3) represents the $m \times n$ elements of $\mathbf{A}$ in terms of $m \times n$ (elements of $\mathbf{U}$) $+ n \times n$ (elements of $\mathbf{V}$) $+ n$ (diagonal elements of $\mathbf{S}$) $= (m + n + 1)n$ numbers.[9] The effective use of SVD as a data reduction tool therefore requires some method for selecting subsets of the columns of $\mathbf{U}$ and $\mathbf{V}$ and corresponding singular values which provide an essentially complete representation of the data set. This selection then specifies an "effective rank" of the matrix $\mathbf{A}$. In practice, a reasonable selection procedure produces an effective rank which is much less than the actual number of columns of $\mathbf{A}$, effecting a drastic reduction in the number of parameters required to describe the original data set.

A first criterion for the selection of usable components is the magnitude of the singular values, since the ordered singular values provide a quantitative measure of the accuracy of the representation of the original data matrix $\mathbf{A}$ in terms of any subset of the columns of $\mathbf{U}$ and $\mathbf{V}$. In the absence of measurement noise and other perturbations, the number of nonzero singular values is the number of linearly independent component spectra required to describe the data set. In experimental

---

[9] The number of independent parameters required to specify $\mathbf{U}$, $\mathbf{S}$, and $\mathbf{V}$ is reduced because these numbers are constrained by the mathematical properties of the matrices $\mathbf{U}$ and $\mathbf{V}$. A total of $n(n + 1)/2$ constraints arise from the orthonormality conditions on the columns of each matrix, giving a total of $n(n + 1)$ constraints. The total number of independent parameters is therefore $(m + n + 1)n - n(n + 1) = mn$, which is the number of independent parameters in the matrix $\mathbf{A}$.

data, however, the presence of noise results in all nonzero singular values (see Section IV.A). Despite this complexity, it is still possible to use the singular values, together with an estimate of the measurement uncertainties, to determine how many component spectra are sufficient to describe the data set to within experimental error. If the data have not been weighted, and the variance, σ^2, is identical for all elements of the data matrix, it is reasonable to argue that a component $k + 1$ may be considered negligible if the condition

$$\|\mathbf{A} - \mathbf{U}_k \mathbf{S}_k \mathbf{V}_k^{\mathrm{T}}\| = \sum_{i=k+1}^{n} s_i^2 \leq \mu v \sigma^2 \qquad (33)$$

is satisfied. $\mathbf{U}_k$, $\mathbf{S}_k$, and $\mathbf{V}_k$ are the representation of $\mathbf{A}$ in terms of k basis vectors and their corresponding amplitudes, as defined in Section III.A, and μ and v are related to the size of the data matrix. This expression simply states that the neglect of this and all subsequent components should yield a reconstructed data matrix that differs from the original by an amount that is less than the noise. The choice of μ and v rests on the determination of the number of degrees of freedom for the representation which remain after the selection of k basis vectors. Shrager has suggested that $\mu = m$ and $v = n$ (Shrager and Hendler, 1982; Shrager, 1984, 1986). The results of Sylvestre *et al.* (1974) mentioned in Section IV.C suggest that a better choice may be $\mu = m - k$ and $v = n - k$. The index r of the least significant component that does not satisfy this condition is then an estimate of the effective rank of $\mathbf{A}$, and the first through the rth components are retained for further consideration.

Some guidance in selecting significant components from the SVD of a weighted data matrix is obtained from the work of Cochran and Horne (1977, 1980). Weighting of the data using the procedure described in Section IV.C produces a weighted matrix $\mathbf{A}_{\mathbf{w}}$ such that, if $\mathbf{A}_{\mathbf{w}}$ has rank r, then the last $m - r$ eigenvalues of $\langle \mathbf{A}_{\mathbf{w}} \mathbf{A}_{\mathbf{w}}^{\mathrm{T}} \rangle$ will have the same value, a [Eq. (26)]. This is equivalent to having only the first m singular values of $\langle \mathbf{A}_{\mathbf{w}} \rangle$ nonzero. Successful application of the weighting algorithm thus produces a set of singular values which are pushed toward zero away from the remaining set. If such a bifurcation is found, the point at which the singular values separate can be used to estimate of the rank of the matrix.

Another reasonable criterion for the selection of usable components from the SVD is the signal-to-noise ratio of the left and right singular vectors (columns of $\mathbf{U}$ and $\mathbf{V}$). Under some experimental conditions, particularly when noise is present which is random only along one dimension of the data matrix (see Sections IV.D and VI.B), selection of usable components from the SVD based on singular values alone can produce a representation of the data matrix that approximates the original to within experimental error, but in which some of the selected components do not contain enough signal to lend themselves to further analysis (e.g., by least-squares fitting with a physical model). An example of such behavior is seen in the SVD presented in Fig. 2, where the amplitude of the third basis spectrum exhibits almost no time-correlated "signal," but the fourth component, which is only about half as large, clearly does. Under these circumstances additional criteria may be required to select those components for which the signal-to-noise ratios are sufficiently large to be candidates for further processing.

A useful measure of the signal-to-noise ratio of given columns of $\mathbf{U}$ (U_i) and $\mathbf{V}$ (V_i), introduced by Shrager and co-workers (Shrager and Hendler, 1982; Shrager, 1984, 1986), are their autocorrelations defined by

$$C(U_i) = \sum_{j=1}^{m-1} U_{j,i} U_{j+1,i}, \tag{34}$$

$$C(V_i) = \sum_{j=1}^{n-1} V_{j,i} V_{j+1,i}, \tag{35}$$

where $U_{j,i}$ and $V_{j,i}$ represent the jth elements of the ith columns of $\mathbf{U}$ and $\mathbf{V}$, respectively. Because the column vectors are all normalized to unity, those vectors which exhibit slow variations from row to row ("signal") will have values of the autocorrelation that are close to but less than one. Rapid row-to-row variations ("noise") will result in autocorrelations which are much less than one, and possibly negative. (The minimum possible value is -1.) For column vectors with many elements (>100 rows) that are subjectively "smooth," autocorrelation values may exceed 0.99, whereas values less than about 0.8 indicate signal-to-noise ratios approaching 1. Components which have been selected based on singular value can be further screened by evaluating the autocorrelations of the corresponding columns of $\mathbf{U}$ and $\mathbf{V}$ and rejecting the component if either autocorrelation falls below some threshold value. A proper choice of this threshold depends on the number of elements in the columns being considered and other experimental details.

D. Rotation Procedure

The presence of measurement noise and other random components in the data matrix decreases the effectiveness with which SVD extracts useful information into the rank-ordered singular values and vectors. As we have seen in Section IV.D, when the magnitudes of signal and noise components of the data become comparable, they may be mixed in the SVD. The signal amplitude is "spread" by this mixing over two or more of the singular values and vectors. In some cases, the columns of $\mathbf{U}$ and $\mathbf{V}$ ordered by decreasing singular value do not exhibit monotonically decreasing signal-to-noise ratios (see Fig. 2). A component which is primarily "noise" may actually be sufficiently large to supersede a signal component in the hierarchy. If this problem is addressed by simply discarding the "noise" component from the data, one effectively introduces "holes" in the set of retained components where components having large amplitudes are ignored and those having small amplitudes are retained. In other cases one encounters a set of components which satisfy the condition in Eq. (33) and contain some signal, but are not individually of sufficient quality to pass a signal-to-noise test such as the autocorrelation criterion just described. Because such small signals are almost always of interest, some procedure for "concentrating" the signal content from a number of such noisy components into one or a very small number of vectors to be retained for further analysis can be extremely useful.

One such optimization procedure transforms a selected set of such noisy components by finding normalized linear combinations for which the autocorrelations [Eqs. (34) and (35)] are maximized. The autocorrelations may be optimized either for the columns of $\mathbf{U}$ [Eq. (34)] or for the columns of $\mathbf{V}$ [Eq. (35)]. The choice depends on whether the signal-to-noise ratio of the determinations as a function of the isolated variable (e.g., wavelength), or as a function of the remaining variables (e.g., time, pH), is considered more important. For purposes of discussion, the transformations will be applied to a set of p columns of $\mathbf{V}$ to be denoted by $\{V_k\}$, where the indices k are taken from the set $\{k1, k2, \ldots, kp\}$. Clearly, blocks of consecutive columns of either matrix are the most obvious candidates for transformation, because they correspond to blocks of consecutively ordered singular values, but this choice is not required by the algorithm. It is only necessary that the processing of the columns of one matrix be accompanied by the compensatory processing of the corresponding columns of the other matrix so that the contribution of the product of the two matrices to the decomposition in Eq. (3) is preserved. The problem then is to determine coefficients $\{r_i\}$, for $i = 1, \ldots, p$, such that the autocorrelation of the normalized vector

$$\mathbf{V}' = r_1\mathbf{V}_{k1} + \cdots + r_p V_{kp} \tag{36}$$

is a maximum. Because the set of vectors $\{V_k\}$ is an orthonormal set, the requirement that $\mathbf{V}'$ be normalized is enforced by the constraint $r_1^2 + \cdots + r_p^2 = 1$. The solution of this problem is described in the Appendix. The procedure yields p distinct sets of coefficients $\{r_i\}$ for which the autocorrelations of the transformed vectors given by Eq. (36) have zero derivatives (yielding some maxima, some minima, and some saddle points) with respect to the coefficients. The transformed vectors with the largest autocorrelations may then be inspected individually to determine whether they should be retained for subsequent analysis.

To represent the effect of this transformation on the entire matrix $\mathbf{V}$, the p sets of coefficients $\{r_i\}$ provided by the transformation procedure may be arrayed as columns of an orthogonal matrix $\mathbf{R}_{\{k\}}$ (see Appendix). This matrix may be viewed as describing a rotation of the ordered set of orthonormal vectors $\{V_k\}$ onto a transformed set of orthonormal vectors $\{V_k'\}$. We can define an $n \times n$ matrix $\mathbf{R}$ by

$$\begin{aligned} R_{ij} &= \delta_{ij} \quad \text{if } i \text{ or } j \notin \{k\}, \\ R_{ki,kj} &= (R_{\{k\}})_{ij}, \end{aligned} \tag{37}$$

that is, by embedding the matrix $\mathbf{R}_{\{k\}}$ into an identity matrix using the indices $\{k\}$. It is easily verified that $\mathbf{R}$ is also an orthogonal matrix. We can then define a transformed matrix $\mathbf{V}^{\mathbf{R}}$ in terms of the entire original matrix $\mathbf{V}$ by

$$\mathbf{V^R} = \mathbf{VR}. \tag{38}$$

The columns of V that are in the set $\{V_k\}$ are transformed in $\mathbf{V^R}$ to the corresponding vectors in the set $\{V'_k\}$, and columns of V that are not in $\{V_k\}$ are carried over to $\mathbf{V^R}$ unchanged. If we define a transformed $\mathbf{U^R}$ matrix by

$$\mathbf{U^R} = \mathbf{USR} \tag{39}$$

then the decomposition in Eq. (3) may be written

$$\begin{aligned}
\mathbf{A} &= \mathbf{USV^T} \\
&= \mathbf{USRR^TV^T} \\
&= (\mathbf{USR})(\mathbf{VR})^T \\
&= \mathbf{U^R}(\mathbf{V^R})^T,
\end{aligned} \tag{40}$$

where we have exploited the orthogonality of $\mathbf{R}$ (i.e., $\mathbf{RR^T} = \mathbf{I}_n$) on the second line. The matrices $\mathbf{U^R}$ and $\mathbf{V^R}$ contain new "basis vectors" and amplitudes, respectively, in terms of which the data matrix $\mathbf{A}$ may be represented.[10] It is important to point out that, while the columns of $\mathbf{V^R}$ still comprise an orthonormal set of vectors, the columns of $\mathbf{U^R}$ are neither normalized nor orthogonal. Furthermore, the mixing of different components results in the loss of the optimal least-squares property (see Section III) when the data matrix is described in terms of any but the complete set of transformed vectors produced by this procedure.

The set of column vectors produced by the rotation procedure (columns of $\mathbf{V^R}$) are mutually "uncorrelated" (in the sense that the symmetrized cross-correlation matrix defined in the Appendix is now diagonal). One consequence of this fact is that variations which are correlated in the original columns of V (the "signal" distributed by the SVD over several components) will tend to be isolated in single vectors after the rotation. Another consequence is that columns of V which are uncorrelated will not be mixed by the rotation procedure. Therefore, one anticipates that components having totally random amplitudes (i.e., those which result from random noise in the data matrix) which are included in the rotation will not be significantly altered by the rotation procedure and will subsequently be eliminated on screening of the transformed vectors based on the autocorrelation criterion. Extension of this line of reasoning suggests that including in the set of rotated vectors additional vectors beyond those that might be expected to contain usable signal will not significantly alter the characteristics of the transformed vectors which contain signal and will be retained after rotation.

The question of which components to include in the rotation procedure has no simple answer. It is clear that even two components which have very high signal-to-

---

[10] In practice, of course, only those columns of U and V whose indices are in the set $\{k\}$ need be transformed by postmultiplication by $\mathbf{R}_{\{k\}}$ the remaining columns of V are simply carried over unchanged, and the remaining columns of U are multiplied by the corresponding singular values to produce the transformed basis vectors.

noise ratios (i.e., autocorrelations which are close to 1) may be mixed in the transformation if their variations are correlated in the sense defined above. As a result, any component of the SVD output which is interesting or useful in its present form, either for mathematical reasons or based on experience, should be excluded from the rotation procedure (Hofrichter *et al.*, 1983, 1985). Furthermore, although the discussion in the previous paragraph suggests that it is "safe" to include more components than are clearly required, the set of included components should be kept to some small fraction of the total set.[11] A procedure that we have used with some success with data matrices of about 100 columns is to select as candidates roughly 10% of the components which have the largest singular values, exclude any of these which either should not be mixed with other components for some reason or will not be significantly improved by such mixing, and apply the rotation procedure to the rest (Hofrichter *et al.*, 1985, 1991; Murray *et al.*, 1988).

An example which demonstrates the effectiveness of rotation in reordering and improving the autocorrelations of the amplitude vectors is shown in Fig. 5. Columns 3 through 10 of the SVD shown in Fig. 2 were included in the rotation, which was carried out with the expectation of removing random contributions to the small signal observed in V_4 of the SVD. Columns 1 and 2 were excluded because their singular values were, respectively, 70 and 5.6 times larger than that of the "noise" component 3 and the signal-to-noise ratios of these components were already about 250 and 30, respectively. The first effect of rotation is that which was anticipated: the signal-to-noise ratio in the third amplitude vector is improved from about 2 to more than 7 and the derivative-shaped "signal" in channels 1 through 40 is concentrated in the third basis spectrum.[12] The autocorrelation of the rotated V_3^R is 0.933, whereas that of V_4 is only 0.742. The second effect is to suppress the random offset amplitudes represented by the third component in the original SVD (Fig. 2) to the point that they do not even appear in the first six components after rotation. The bulk of the offset amplitude actually appears as component 8 of the rotated SVD, and the autocorrelation of V_8^R is slightly less than zero (-0.12).

---

[11] Additional constraints placed on the elements of the transformed vectors by the rotation procedure tend to determine the individual elements as the size of the included set approaches that of the complete set. Specifically, if p vectors out of a total of n columns of **V** are included, the $p \times n$ elements of the resulting transformed vectors are subject to p^2 constraints—$p(p-1)/2$ from the fact that the symmetrized cross-correlation matrix (see Appendix) has all off-diagonal elements equal to zero, $p(p-1)/2$ from the orthogonality of all the transformed vectors, and p from the normalization of each vector. As p approaches n, these constraints obscure any relationship between the shapes of the untransformed and the transformed vectors, and the set of vectors required to represent the signal content of the original data matrix will actually increase rather than decrease.

[12] This "signal" arises from the perturbation of the absorption spectra of the cobalt porphyrins in the α chains of the hybrid hemoglobin tetramer. The time course for this spectral change is distinctively different from that of the second component (Hofrichter *et al.*, 1985).

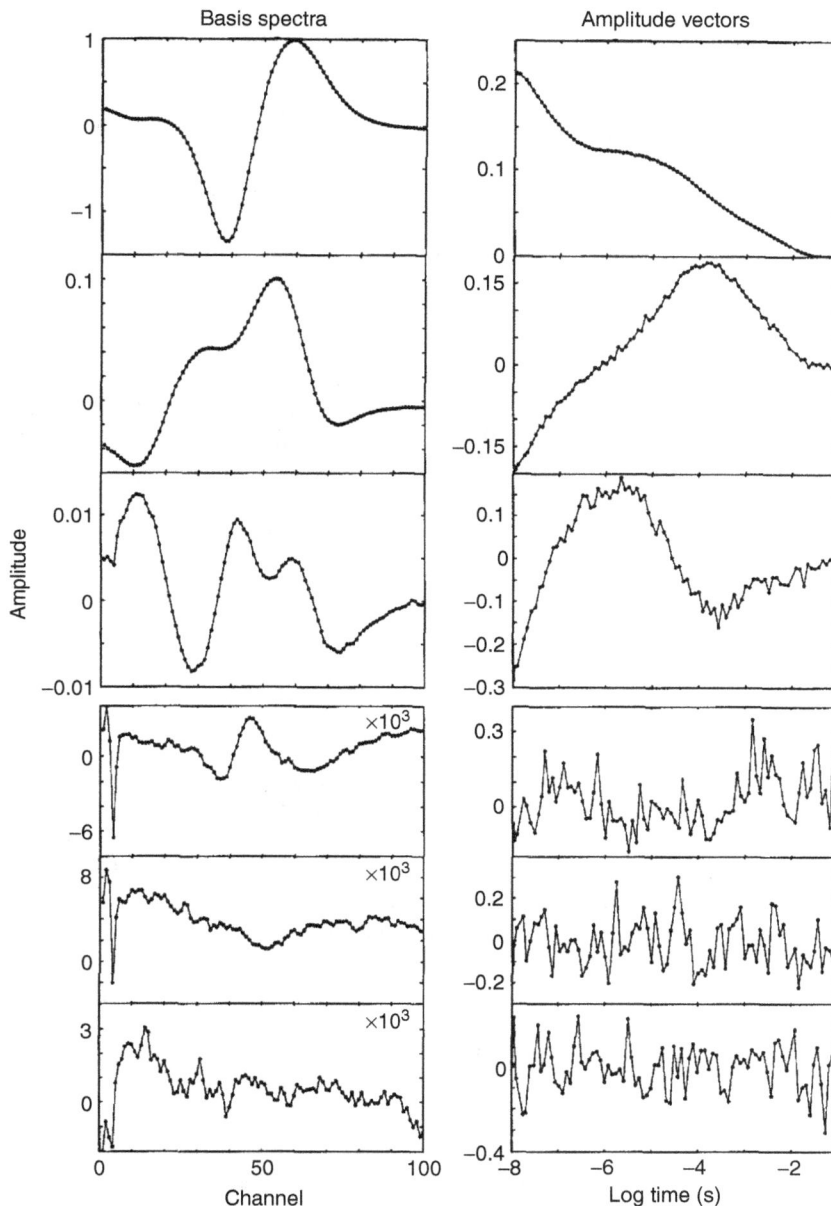

Fig. 5 Rotated SVD of the data in Fig. 1. Components 3 through 10 of the SVD for which the first six components are shown in Fig. 2 were rotated using the algorithm discussed in the text and derived in the Appendix. The autocorrelations of the components included in the transformation (3–10) were 0.149, 0.746, 0.062, −0.089, 0.337, −0.010, 0.031, and 0.099 before rotation. The signal-to-noise ratio for the component with the highest autocorrelation (V_4) evaluated by comparing a smoothed version of this

E. Application of Physical Models to Processed Singular Value Decomposition Output

The discussion to this point has dealt with the purely mathematical problem of using SVD and associated processing to produce a minimal but faithful representation of a data matrix in terms of basis vectors and amplitudes. The next step in the analysis of the data is to describe this representation of the data matrix in terms of the concentrations and spectra of molecular species. This step requires that some physical model be invoked to describe the system and an optimization procedure be carried out to adjust the parameters of the model so that the differences between the data and the model description are minimized. Several assumptions are inherent in such a procedure. First, a set of pure states or species which are accessible to the system must be enumerated. The measured spectra are then assumed to be linear combinations of the spectra of the various pure species, weighted by their populations under each set of conditions [e.g., Eq. (2)]. The dependence of the populations of these species [the $\{c_n\}$ of Eq. (2)] on the conditions is further assumed to be quantitatively described by a kinetic or thermodynamic model. If the model provides for r distinct species, then the first two of these assumptions permit the (m wavelengths) $\times$ (n conditions) matrix $\mathbf{A}$ to be written in the form of Eq. (5), that is,

$$\mathbf{A} = \mathbf{FC}^{\mathrm{T}}, \tag{41}$$

where the columns of the $m \times r$ matrix $\mathbf{F}$ contain the spectra of the individual species, and the corresponding columns of the $n \times r$ matrix $\mathbf{C}$ contain the populations of the species as a function of conditions.

The most common means for reducing a representation of a set of experimental data to a description in terms of a physical model is through the use of least-squares fitting. Using this approach, the amplitudes of all of the vectors which describe the data matrix would be simultaneously fitted to the model to produce a set of coefficients which describe the spectra of each of the species in the model as well as the dependence of the species concentrations on experimental conditions (Shrager, 1986). A common alternative to using molecular or physical models to directly fit the data is to assume functional forms which result from analysis of generalized or simplified models of the system and to use these forms to fit the data. For example, if the kinetics of a system can be described by a set of first-order or pseudo-first-order processes, then the kinetics of the changes in system composition can be described by sums of exponentials, with relaxation rates which are the eigenvalues of the first-order rate matrix.[1] Under these circumstances, the time-dependent vectors which describe the changes in the spectra can be empirically described by sums of exponential relaxations, and fitting can be carried out using functions of this form.

---

component with the original is approximately 2. The autocorrelations of transformed components 3 through 10 were 0.932, 0.473, 0.277, 0.191, 0.001, −0.115, −0.165, and −0.268, and their normalized amplitudes were 0.058, 0.015, 0.041, 0.011, 0.023, 0.057, 0.014, and 0.029. The signal-to-noise ratio for the most highly correlated component (V_3^{R}) is about 7.

Similarly, pH titration curves can be assumed to be sums of simple Henderson-Hasselbach curves describing the uncoupled titration of individual groups, and the measured dependence of the spectra on pH can be fitted to sums of these curves (Frans and Harris, 1985; Shrager, 1986). Because use of this approach requires the assumption of some functional form, it is therefore less rigorous than the use of an explicit model. It also does not permit direct determination of the spectra of the species in the model. As pointed out in Section II, the advantage of using the output of SVD in any fitting procedure is that the number of basis spectra required to describe the data matrix, and hence the number of amplitudes which must be fitted, is minimized by the rank reduction which has been accomplished by SVD.

Suppose that a population matrix $\mathbf{C}'$ is derived from a specific set of model parameters. If $\mathbf{C}'$ has rank r so that $(\mathbf{C}'^{\mathrm{T}}\mathbf{C}')^{-1}$ exists, the generalized inverse of $\mathbf{C}'^{\mathrm{T}}$ can be written as $\mathbf{C}'(\mathbf{C}'^{\mathrm{T}}\mathbf{C}')^{-1}$ (Lawson and Hanson, 1974), and the corresponding matrix $\mathbf{F}'$ of species spectra which minimizes the difference $\|\mathbf{A} - \mathbf{F}'\mathbf{C}'^{\mathrm{T}}\|$ may be written (Cochran and Horne, 1980; Lawson and Hanson, 1974)

$$\mathbf{F}' = \mathbf{A}\mathbf{C}'(\mathbf{C}'^{\mathrm{T}}\mathbf{C}')^{-1}. \tag{42}$$

Least-squares fitting of the matrix $\mathbf{A}$ with the model then requires varying the parameters of the model in some systematic way so that the population matrix $\mathbf{C}'$ calculated from the parameters, and the matrix $\mathbf{F}'$ of spectra calculated using Eq. (42), result in the smallest possible value of the difference $\|\mathbf{A} - \mathbf{F}'\mathbf{C}'^{\mathrm{T}}\|$. The suitability of the model as a description of the measurements would then be assessed on the basis of how well the final matrices $\mathbf{F}'$ and $\mathbf{C}'$ describe the original data.

This approach of least-squares fitting the entire data matrix, commonly referred to as global analysis, has been applied in a large number of studies. Examples include the analysis of sets of spectra obtained from pH titrations of multicomponent mixtures (Frans and Harris, 1984), analysis of fluorescence decay curves (Knutson et al., 1983), and analysis of flash photolysis data on the bacteriorhodopsin photocycle (Mauer et al., 1987a,b; Nagel et al., 1982; Xie et al., 1987). In principle it provides the most complete possible description of a data matrix in terms of a postulated model; however, it has certain features that make it difficult to use in many cases. The most obvious difficulties are associated with the matrix $\mathbf{F}'$, which specifies the spectra of the species in the model. If the number of wavelengths (m) sampled in collecting the data matrix is large, this matrix, which contains the extinction coefficient of each of the r species at each of m wavelengths, is also large, containing a total of $m \times r$ adjustable parameters. The fitting procedure then tends to become computationally cumbersome, in that every iteration of a search algorithm in parameter space requires at least one recalculation of $\mathbf{F}'$ using Eq. (42) or the equivalent. It should be noted that in most of the applications cited above the number of wavelengths included in the analysis was 15 or less. Furthermore, numerical instabilities may arise in the direct application of Eq. (42) if $\mathbf{C}'$ is rank-deficient, or nearly so, because calculation of the inverse of $\mathbf{C}'^{\mathrm{T}}\mathbf{C}$ then becomes problematic.

SVD provides a reduced representation of a data matrix that is especially convenient for a simplified least-squares fitting process. In the most general terms, after SVD and postprocessing have been performed, an essentially complete representation of the data matrix $\mathbf{A}$ in terms of k components may be written

$$\mathbf{A} \cong \mathbf{U}'\mathbf{V}'^{\mathbf{T}}, \tag{43}$$

where $\mathbf{U}'$ is a matrix of k basis spectra, and $\mathbf{V}'$ contains the amplitudes of the basis spectra as a function of conditions. If only the SVD has been performed, then $\mathbf{U}'$ consists of the k most significant columns of $\mathbf{US}$, and $\mathbf{V}'$ the corresponding columns of $\mathbf{V}$; if a rotation or similar procedure has been performed as well, then $\mathbf{U}'$ consists of the k most significant columns of the matrix $\mathbf{U}^{\mathbf{R}}$ [Eq. (39)] and $\mathbf{V}'$ the corresponding columns of the matrix $\mathbf{V}^{\mathbf{R}}$ [Eq. (38)]. If the data have been weighted prior to SVD, then $\mathbf{U}'$ consists of the k most significant columns of $\mathbf{U}'$ and $\mathbf{V}'$ the corresponding columns of $\mathbf{V}'$ as calculated from Eq. (27). The assumed completeness of the representations of $\mathbf{A}$ in Eqs. (41) and (43) suggests the *ansatz* that the columns of any matrix $\mathbf{C}'$ of condition-dependent model populations may be written as linear combinations of the columns of $\mathbf{V}'$. This linear relationship between $\mathbf{C}'$ and $\mathbf{V}'$ may be inverted, at least in the generalized or least-squares sense, so that we can write formally[13]

$$\mathbf{V}' \cong \mathbf{C}'\mathbf{P}. \tag{44}$$

In the least-squares fit, the model parameters used to calculate $\mathbf{C}'$ and the set of linear parameters $\mathbf{P}$ are varied to produce a population matrix $\mathbf{C}' = \hat{\mathbf{C}}$ and a parameter matrix $\mathbf{P} = \hat{\mathbf{P}}$ such that the difference $\|\mathbf{V}' - \mathbf{C}'\mathbf{P}\|$ is minimized. The optimal approximation to $\mathbf{V}'$ will be denoted $\hat{\mathbf{V}}(\equiv \hat{\mathbf{C}}\hat{\mathbf{P}})$. This then yields the further approximation

$$\mathbf{A}' = \mathbf{U}'\mathbf{V}'^{\mathbf{T}} \cong \mathbf{U}'\hat{\mathbf{V}}^{\mathbf{T}} = \mathbf{U}'\hat{\mathbf{P}}^{\mathbf{T}}\hat{\mathbf{C}}^{\mathbf{T}} \equiv \hat{\mathbf{F}}\hat{\mathbf{C}}^{\mathbf{T}}, \tag{45}$$

where the matrix $\hat{\mathbf{F}}$ is the set of corresponding "least-squares" species spectra. Equation (45) permits the identification of $\hat{\mathbf{F}}$ in terms of the basis spectra:

$$\hat{\mathbf{F}} = \mathbf{U}'\hat{\mathbf{P}}^{\mathbf{T}}. \tag{46}$$

It is important to note that, because all of the species spectra must be represented in terms of the set of basis spectra which comprise $\mathbf{U}'$, the matrix $\hat{\mathbf{P}}$ is much smaller than the matrix $\hat{\mathbf{F}}$. Accordingly, the number of adjustable parameters which must be specified in fitting the SVD representation of the data is significantly reduced relative to the number required to fit the original data matrix.

---

[13] The formal inversion [Eq. (44)] may optionally be used to facilitate the fitting procedure (Golub and Pereyra, 1973). When the model parameters which produce the population matrix $\mathbf{C}'$ are varied in each step of the optimization, the generalized inverse may be used to produce the matrix of linear parameters, $\mathbf{P}$, which produces the best approximation to $\mathbf{V}'$ corresponding to the specified set of model parameters.

This somewhat formal discussion may be made clearer by considering an example from the field of time-resolved optical absorption spectroscopy, which is similar to the example described in Section II. Recall that the data consist of a set of absorption spectra measured at various time delays following photodissociation of bound ligands from a heme protein by laser pulses (Hofrichter et al., 1983, 1985). Each column of the data matrix $\mathbf{A}$ describes the absorbances at each wavelength measured at a given time delay. After the SVD and postprocessing, we are left with a minimal set of basis spectra $\mathbf{U}'$ and time-dependent amplitudes $\mathbf{V}'$ (see Fig. 5). Suppose that we now postulate a "model" which states that the system contains r "species," the populations of which each decay exponentially with time with a characteristic rate:

$$C_{ij} = e^{-\kappa_j t_i}, j = 1, \ldots, r \text{ and } i = 1, \ldots, m, \tag{47}$$

where the set $\{t_i\}$ represents the times at which the spectra (columns of $\mathbf{A}$) are measured and the set $\{\kappa_j\}$ represents the characteristic decay rates of the populations of the various "species." The fit in Eq. (45) optimizes the relation

$$V'_{ij} \cong \sum_{q=1}^{r} P_{qj} e^{-k_q t_i}. \tag{48}$$

Producing an optimal least-squares approximation to $\mathbf{V}'(\equiv \hat{\mathbf{V}})$ clearly involves simultaneously fitting all the columns of $\mathbf{V}'$ using linear combinations of exponential decays, with the same set of rates $\{\hat{\kappa}_q\}$, but with distinct sets of coefficients $\{P_{qj}\}$, for each column j. The resulting best-fit rates $\{\hat{\kappa}_q\}$ then produce a best-fit set of "model" populations $\hat{\mathbf{C}}$ and best-fit coefficients $\{\hat{\mathbf{P}}_{qj}\}$ [Eq. (48)]. The set of "species" spectra $\hat{\mathbf{F}}$ which produce a best fit to the matrix $\mathbf{U}'$ are then obtained from Eq. (46), that is,

$$\hat{F}_{iq} = \sum_{j=1}^{k} \hat{\mathbf{P}}_{qj} U'_{ij}. \tag{49}$$

Although this "model" is admittedly highly contrived, in that descriptive kinetic models involving interconverting species will not produce species populations which all decay to zero as simple exponentials, it illustrates the general fitting problem.

Least-squares fitting the columns of $\mathbf{V}'$ obtained from SVD, when the residuals from each column of $\mathbf{V}$ are correctly weighted, is mathematically equivalent to least-squares fitting the entire data matrix using the global analysis procedure described in Section II. Shrager (1986) has shown that for SVD alone (no postprocessing) the two procedures in fact yield the same square deviations for any set of parameters if the sum of squared residuals from each of the columns of $\mathbf{V}$ is weighted by the square of the respective singular value. In other words, the function to be minimized in the simultaneous fit to all the columns of $\mathbf{V}$ should be

$$\phi^2 = \sum_{i=1}^{m} s_i^2 \| V'_i - (\mathbf{C}'\mathbf{P})_i \|^2, \tag{50}$$

where V_i' is the ith column vector of $\mathbf{V}'$. It is shown in the Appendix that, if a rotation has been performed as described in Eqs. (36)–(40), approximately the same squared deviations will be obtained if all the columns of $\mathbf{V^R}$ are fit with the ith column weighted by the ith diagonal element W_{ii} of the matrix $\mathbf{W} = \mathbf{R^T S^2 R}$.

In practice the SVD is truncated to generate $\mathbf{V}'$, and only a small fraction of the columns of $\mathbf{V}$ are included in the fitting procedure. This is equivalent to setting the weighting factors of the remaining columns to zero. If the truncation is well designed, then the columns of $\mathbf{V}$ which are discarded either have small weighting factors, s_i^2, or have autocorrelations which are small enough to suggest that they contain minimal condition-correlated signal content. If a rotation procedure described in Eqs. (36)–(40) has been performed prior to selecting $\mathbf{V}'$, then singular values of very different magnitudes may be mixed in producing the retained and discarded columns of $\mathbf{V^R}$ and their corresponding weighting factors (see Appendix). Because the rotation procedure is designed to accumulate the condition-correlated amplitudes into the retained components, the discarded components, while not necessarily small, also have little or no signal content. In both cases the neglected components clearly contribute to the sum of squared residuals, ϕ^2. Because their condition-correlated amplitudes are small, however, their contribution to ϕ^2 should be nearly independent of the choice of fitting parameters. To the extent that this is true, parameters optimized with respect to either truncated representation of the data should closely approximate those which would have been obtained from fitting to the entire data set.

In summary, an SVD-based analysis almost always simplifies the process of least-squares fitting a data matrix with a physical model by reducing the problem to that of fitting a few selected columns of $\mathbf{V}'$. Reducing the rank of the data matrix also minimizes the number of parameters which must be varied to describe the absorption spectra of the molecular species [the elements of the matrix $\hat{\mathbf{P}}$ in Eq. (46)]. Attention must be paid to the proper choice of weighting factors in order to produce a result which faithfully minimizes the deviations between the fit and the full data matrix, but the increase in the efficiency of fitting afforded by this approach argues strongly for its use under all conditions where the rank of the data matrix is significantly smaller than the number of rows and/or columns (i.e., rank$\ll \min\{m, n\}$).

VI. Simulations for a Simple Example: The Reaction A → B→ C

To explore in more detail the effects on the SVD output of introducing noise into data sets we have carried out simulations of noisy data for the simple kinetic system

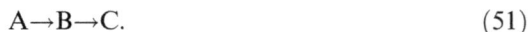

$$A \rightarrow B \rightarrow C. \tag{51}$$

This model was used to generate sets of data consisting of sample absorption spectra and difference spectra (with difference spectra calculated as sample spectrum − C) using rates $k_{AB} = 10^6 \, s^{-1}$ and $k_{BC} = 10^3 \, s^{-1}$. The spectra of A, B, and C were represented as peaks having Gaussian bandshapes centered about wavelengths $\lambda_A = 455$ nm, $\lambda_B = 450$ nm, and $\lambda_C = 445$ nm. The bandwidths (half-widths at $1/e$ of maximum) and peak absorbances for the three species were chosen to be $\Delta_A = 20$ nm, $\Delta_B = 18$ nm, and $\Delta_C = 16$ nm and $\varepsilon_A c_t l = 0.9$ OD, $\varepsilon_B c_t l = 1.0$ OD, and $\varepsilon_C c_t l = 1.1$ OD ($c_t l$ is the product of the total sample concentration and path length). These spectra were selected so that the ordered nonzero singular values of the noise-free data successively decreased by a factor of between 5 and 10. These data thus represent many of the problems encountered in the processing of real data in which some processes may produce changes in absorbance as large as 1 OD, whereas other processes produce changes as small as a few thousandths of an optical density unit. To derive reliable kinetic information under such unfavorable circumstances, careful consideration must be given to the effects of measurement noise on the data analysis.

Two different types of noise were added to the data. The first noise component, which we refer to as random noise, was selected independently for each element of the data matrix from a Gaussian distribution having an expectation value of zero and variance σ_r^2. Random noise simulates shot noise or instrumental noise in the determination of each experimental absorbance. The assumption that the amplitude of the random noise component is constant over the entire data matrix is certainly an oversimplification for real data: shot noise results from random deviations in the number of photons measured for each data point, and therefore depends on a number of factors, including the intensity of the source and the optical density of the sample. Moreover, the photon and electronic noise actually appears in the measured intensities, not in the absorbance, which is calculated from the logarithm of the ratio of two intensities. The second noise source consists of a spectrum having an identical absorbance at each wavelength, but having a different amplitude for every measured spectrum, selected from a Gaussian distribution with mean value zero and variance σ_λ^2. We shall refer to noise having these characteristics as wavelength-correlated noise. This noise approximates changes in the DC baseline of the spectrometer. In single-beam spectroscopy, such noise can arise from changes in the output energy of the lamp or changes in the sensitivity of the detector. In double-beam spectroscopy, it can result from electronic drift. In both cases, however, σ_λ can be significantly larger than the error inherent in the determination of the dependence of a given spectrum on wavelength, which is characterized by σ_r.

We have already seen that independent addition of these two kinds of noise to noise-free data has qualitatively different effects on the SVD. Based on the results of Sections IV.A and IV.B, random noise is expected to introduce a spectrum of singular values similar in magnitude to those obtained for a random matrix and to perturb the singular values and vectors of the noise-free data as discussed in Section IV.B. The effects of adding wavelength-correlated noise have been explored in Fig. 4 for the case where the noise-free data matrix is rank-1. As shown there, the SVD contains only a single extra component, which arises primarily from the constant spectrum assumed as the noise source.

To examine the statistical properties of the SVD of data sets having specified amplitudes for random (σ_r) and wavelength-correlated (σ_λ) noise, the SVD of each of 5000 independently generated data matrices was calculated. Each matrix contained the identical "signal," which consisted of absorbances at 101 wavelengths and 71 time points evenly spaced on a logarithmic time grid, as well as randomly selected noise. For each set of 5000 trials, the means and standard deviations of the individual singular values and of the individual elements of the appropriate singular vectors were calculated. In calculating the statistical properties of the SVD, one is confronted with the problem of choosing the sign of each of the SVD components. Because SVD only determines unambiguously the sign of each of the products $U_i \cdot V_i$, some independent criterion must be used to choose the sign of each of the U_i or V_i. The algorithm chosen in these simulations was to require that the inner product of the left singular vector with the corresponding left singular vector obtained from the SVD of the noise-free data matrix be positive. To present the results of the simulations in a compact form we have chosen to display the singular values, together with the square root of the mean of the variances of the relevant singular vectors (noise amplitudes). The singular values facilitate comparison of the magnitude of the noise contributions with those of the signals which result from the noise-free data. The noise amplitude provides a compact characterization of the signal-to-noise ratio for a given parameter or vector.

A. Effects of Random Noise

The first set of simulations was carried out to explore the consequences of adding random noise of arbitrary amplitude to the data. Based on the discussion of noise presented earlier, addition of random noise to a noise-free data matrix would be expected to have two effects on the SVD output. First, the random amplitudes will generate a set of additional singular values having amplitudes comparable to those of a random matrix having the same size and noise amplitude [see Eq. (10) and Fig. 3] in addition to those which derive from the noise-free data. When the noise amplitude is small, all of these values should be significantly smaller than the smallest singular value of the noise-free data, and the noise should not interfere with the ability of SVD to extract the spectral information from the data. Second, the noise should perturb the singular values and vectors which derive from the noise-free data by the addition of random amplitudes as shown in Eqs. (15) and (16) for the case in which the noise-free data matrix has a rank of 1. One objective of the simulations was to extend this analysis to explore both data sets which had a rank higher than 1 and the effects of larger noise amplitudes. In particular, we were interested in determining the random noise amplitudes at which signals became unrecoverable. It is intuitively expected that the noise amplitudes must become large compared to the signal for this to occur, so information on this point cannot be obtained by treating the noise as a perturbation.

An example of the input data at a relatively low noise amplitude is shown in Fig. 6B, and the results of the simulations are summarized in Fig. 7. The averages of the first three singular values are shown in Fig. 7A, and the square roots of the variances of the first three singular values and singular vectors are plotted as a function of σ_r in Fig. 7B–D. The results in Fig. 7A show that, for small σ_r ($<3 \times 10^{-2}$), the average values of s_1, s_2, and s_3 are essentially unperturbed from the values obtained from the noise-free data. Figure 7B shows that the presence of the noise in the data matrix at these noise amplitudes is observable as an increase in the variances of the first three singular values and vectors. The square roots of the variances of s_1, s_2, and s_3 are each very nearly equal to σ_r. Figure 7B and C show that the square roots of the variances of the singular vectors, σ_U and σ_V, are very nearly equal to σ_r/s_j, where s_j is the relevant singular value from the noise-free data. Because these noise amplitudes are small compared to all of the SVD components of the noise-free data, these results can be compared directly to the results predicted by the perturbation treatment in Section IV.B.

The observed variances suggest that it may be possible to generalize Eqs. (15) and (16) which state that, for a data matrix having a rank of 1, the square root of the variance of s_1 is equal to σ_r, whereas the square roots of the variances of the vectors U_1 and V_1 approximate σ_r/s_1 for large matrices. The proposed generalization of these equations would predict that the jth SVD component is described by

$$s_j \cong s_{j0} + \frac{(m+n-1)\sigma^2}{2s_{j0}} + \varepsilon_{s_j}, \, [\mathrm{var}(\varepsilon_{s_j}) = \sigma_r^2],$$

$$U_{ji} \cong U_{ji0} + \frac{\varepsilon_{U_{ji}}}{s_{j0}}, \, [\mathrm{var}(\varepsilon_{U_{ji}}) = (1 - U_{ji0}^2)\sigma_r^2], \quad\quad (52)$$

$$V_{ji} \cong V_{ji0} + \frac{\varepsilon_{V_{ji}}}{s_{j0}}, \, [\mathrm{var}(\varepsilon_{V_{ji}}) = (1 + V_{ji0}^2)\sigma_r^2].$$

A more careful examination of the variances of the singular values and vectors obtained from these simulations shows that these approximations are quite accurate, describing the variance of the first SVD component to within 1%, and the second and third components to within 2-3%. We can conclude from this analysis that if the noise in the data is small and purely random, then the variances of the singular values and vectors should be related as described by Eq. (52). Failure of the variances to meet this criterion argues strongly for the presence of other sources of noise in the data.

As σ_r rises above 0.03 OD, s_3 begins to increase. The noise amplitudes for all the singular values and vectors continue to increase in direct proportion to σ_r. Once σ_r exceeds 0.1 OD, the square roots of the variances of the third singular vectors, U_3 and V_3, saturate at the values expected from completely random vectors ($m^{-1/2}$ and $n^{-1/2}$, respectively), showing that the elements of these vectors have become

A

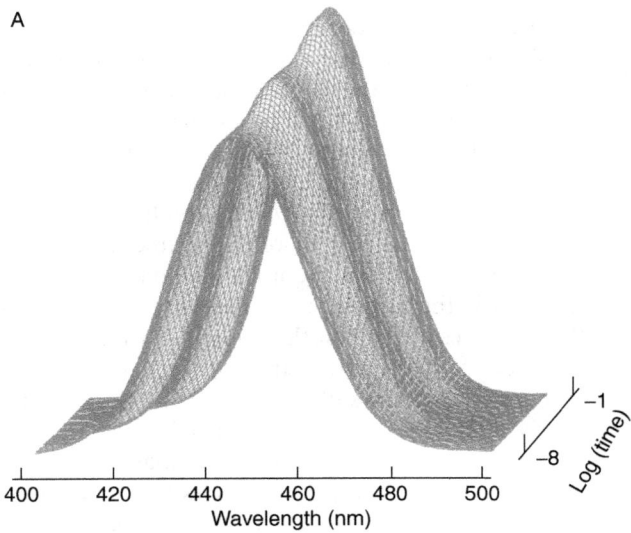

B

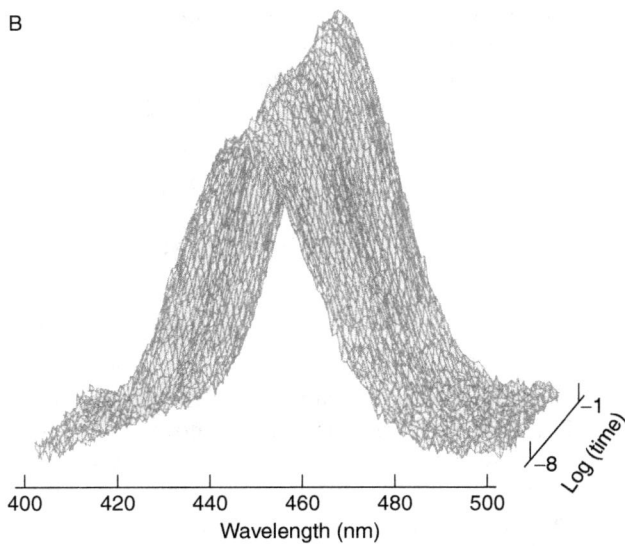

C

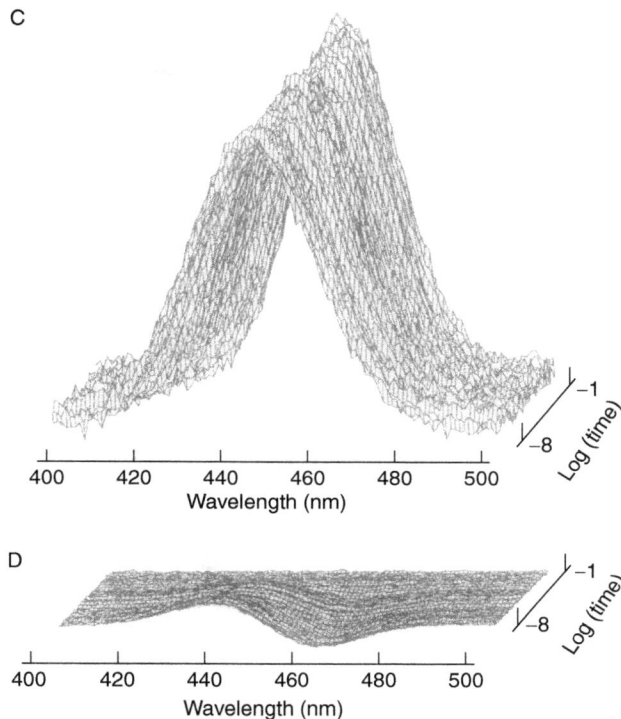

D

Fig. 6 Sample data sets from simulations of the effects of random and wavelength-correlated noise on singular values and vectors. (A) The noise-free data. The spectra are calculated as the sums of the spectra of the three species A, B, and C times the concentrations calculated for a population which is 100% A at $t = 0$, using the rates $k_{AB} = 10^6 \text{ s}^{-1}$ and $k_{BC} = 10^3 \text{ s}^{-1}$. The spectra are described in the text. The time points were chosen on a logarithmic grid with 10 points per decade, beginning at 10^{-8} and ending at 10^{-1} s. Random and wavelength-correlated noise having amplitudes σ_r and σ_λ, respectively, were added to the noise-free data. (B) Spectra with random noise only: $\sigma_r = 0.016$ OD; $\sigma_\lambda = 0$. (C) Spectra with both random and wavelength-correlated noise: $\sigma_r = 0.016$ OD; $\sigma_\lambda = 0.016$ OD. (D) Difference spectra with both random and wavelength-correlated noise. The difference spectra were calculated by subtracting the spectrum of pure C from all of the calculated spectra. The noise amplitudes in (D) are identical to those in (C).

completely uncorrelated. This result implies that the third SVD component of the noise-free data has been replaced by a component which is generated by the random noise. Further increasing σ_r produces proportional increases in s_3, the singular value associated with this pair of random basis vectors. This behavior is analogous to that observed for random matrices, where the U and V vectors are random and the variance of the random variable appears only as a scale factor multiplying the singular values [Fig. 3; Eq. (10)]. At the maximum values of σ_r,

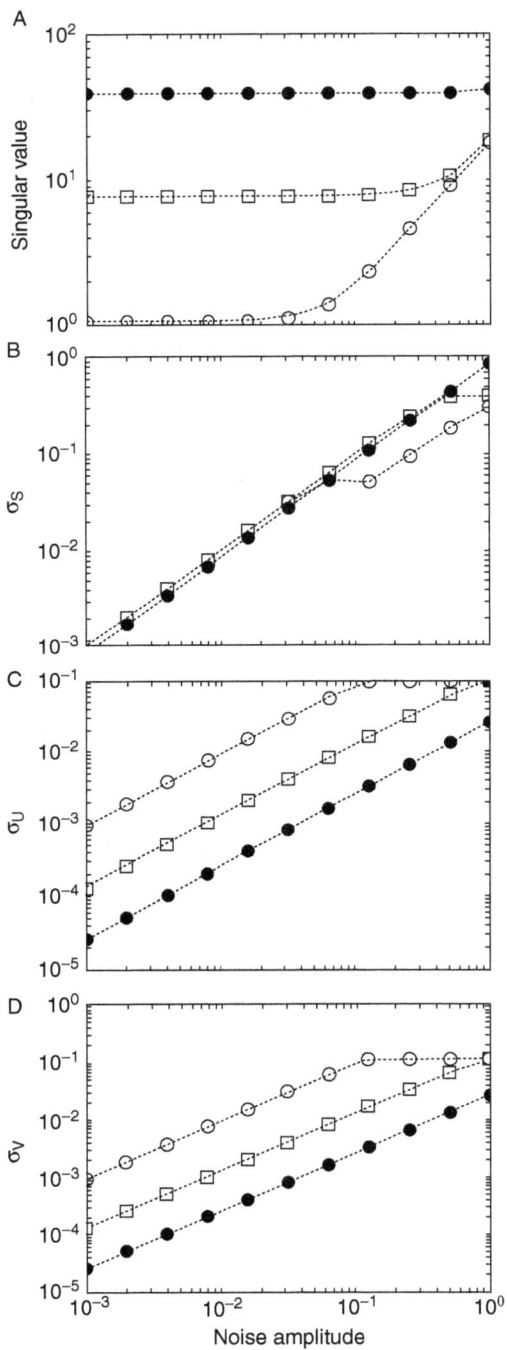

similar behavior is observed in the second singular value and vectors as the noise becomes sufficiently large that its contribution to the spectrum of singular values begins to dominate s_2.

The dependence of the standard deviations of the singular values and vectors suggests that a smooth connection can be made between the perturbation treatment for low noise amplitudes and the results obtained from the simulations of random matrices described in Fig. 3 at higher noise amplitudes. To quantitatively compare the results of the simulations with those in Fig. 3 we first calculate from Eq. (10) that, for an $n \times n$ square matrix, the largest singular value resulting from the random amplitudes is equal to $s_{\max} = 2\sigma_r n^{1/2}$. Inspection of the results in Fig. 3B for $m/n < 2$ suggests that the corresponding first approximation for an $m \times n$ matrix is $s_{\max} \cong 2\sigma_r(mn)^{1/4}$. The second and third singular vectors in Fig. 7, which have singular values $s_2 = 7.7$ and $s_3 = 1.04$, become dominated by random noise when the noise amplitudes exceed 1.0 and 0.12 OD, respectively. The corresponding values of $s_{\max}$ calculated from the noise amplitudes are 9.2 and 1.1, very close to s_2 and s_3, respectively. If this result can be generalized, it suggests that the random noise dominates a signal component k when $s_{\max} \geq s_k$. To restate this conclusion, the signal described by a given singular value, s_j, and its associated singular vectors, U_j and V_j, becomes totally obscured by noise when the random noise amplitude, σ_r, exceeds $s_j/2(mn)^{1/4}$. Attempts to improve the quality of signal components in the presence of random noise by rotation or other postprocessing algorithms were uniformly unsuccessful (a result which is expected because the noise components generated by SVD are almost completely uncorrelated).

B. Combined Effects of Random and Wavelength-Correlated Noise

The rotation procedure described in Section V.D and derived in the Appendix was designed to discriminate between signals which are correlated in one dimension of the data matrix and those which are not. The simulations just described show that this procedure has little or no effect when confronted with random noise, which produces SVD components which are uncorrelated in both dimensions of the data matrix. We anticipate, however, that this algorithm will be more

Fig. 7 Effects of the amplitude of random noise on the SVD of simulated data for the reaction A → B → C. The spectra were calculated as described in the text. (A) The first three singular values; (B) square root of the variance of the first three singular values; (C, D) square root of the average variance of the first three left (C) and right (D) singular vectors. For each noise amplitude, σ_r, the SVD of each of 5000 independently generated data matrices was calculated. For each set of 5000 trials the means and variances of each of the singular values, s_1, s_2, and s_3, and of each element of the three most significant singular vectors (columns 1–3 of the matrices **U** and **V**) were calculated. The value plotted for each singular vector and noise amplitude in (C) and (D) is the square root of the mean value of the variances calculated for all of the elements of that singular vector. (Preliminary calculations in which the dispersions of the variances were examined showed that variations were too small to be analyzed from the results of these simulations.)

successful in discriminating against mixing of wavelength-correlated noise with the noise-free data. To explore the effectiveness of this procedure we performed two sets of simulations in which both wavelength-correlated noise and random noise were added to the data. In the first set of simulations the spectra were calculated as in Section VI.A above, and in the second set the same spectra were used but the spectrum of pure C was subtracted from the sample spectrum at each time point to produce a difference spectrum. Sample data sets at the same noise amplitudes are shown in Fig. 6C and D. Recall that in the absence of random noise, addition of wavelength-correlated noise increases the rank of the matrix of the absolute spectra from 3 to 4 and the rank of the matrix of difference spectra from 2 to 3.

At each of three amplitudes for the random noise ($\sigma_r = 0.016$, 0.032, and 0.064 OD), σ_λ was increased systematically from a point where the noise amplitudes were much smaller than the smallest SVD component of the noise-free data to a point where the contribution of this component to the resulting data matrix was larger than that of the second component. At each set of noise amplitudes 5000 data sets were generated and analyzed by SVD. For the absorption spectra components 3 through 10 were then rotated as described above to optimize the autocorrelation of the retained V_3^R component. For the difference spectra components 2 through 10 were rotated to optimize the retained V_2^R component.

Figure 8 presents the first three singular values and the noise amplitudes of the smallest singular vectors (U_3 and V_3) of the noise-free data in the first case where the data are absorption spectra. The random amplitudes of the wavelength-correlated noise would produce a singular value, $s_\lambda = (mn\sigma_\lambda^2)^{1/2}$ if the noise were isolated in a single SVD component. The noise amplitudes thus become comparable to the signal represented by the third SVD component of the noise-free data when σ_λ is slightly larger than 0.01 OD. Figure 8A permits us to track the magnitude of this component once it exceeds s_3 from the noise-free data. Note that s_3 doubles when the noise amplitude reaches a value of about 0.03 OD and then climbs steadily until it reaches a value of about 0.1 OD, at which point it levels off. This is the region in which we are primarily interested.

If we now examine the results in Fig. 8C, we find that the square root of the variance of V_3, $\sigma(V_3)$, begins to increase detectably at values of σ_λ as small as 0.004 OD. When σ_λ becomes about 10 times larger ($\sim$0.05 OD), $\sigma(V_3)$ approaches the value expected for a random variable $n_t^{-1/2} \cong 0.119$, strongly suggesting that the third SVD component results almost exclusively from the wavelength-correlated noise at these noise amplitudes. This conclusion is supported by examining U_3 of the individual SVDs under these conditions, which shows that the U_3 of the noise-free data has been replaced by a nearly constant spectrum arising from the offsets. The variance of V_3 increases monotonically as σ_λ increases between these two values. This result argues that most of the information originally contained in component 3 of the noise-free data has been displaced into component 4 by the wavelength-correlated noise when σ_λ is larger than 0.05 OD.

These results show that the sorting of wavelength-correlated noise from the data by SVD is moderately efficient as long as s_λ is about a factor of 3 smaller than the

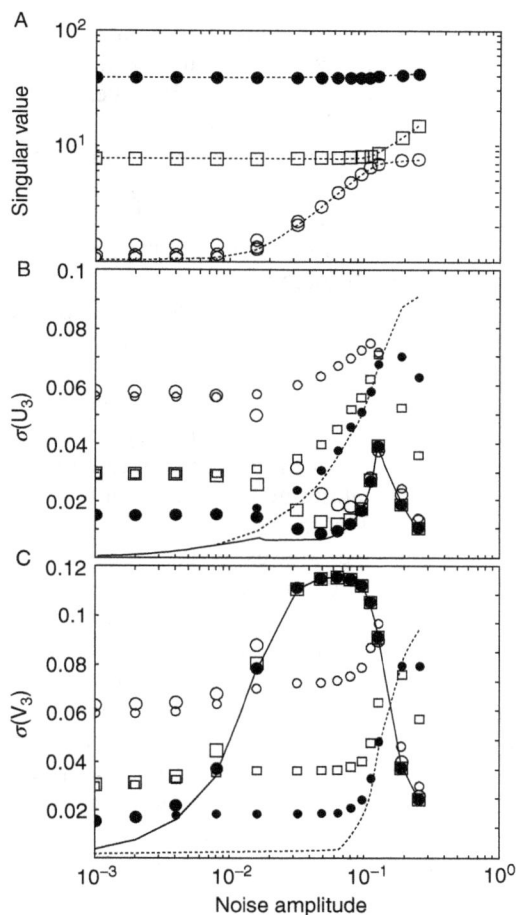

Fig. 8 Effects of the RMS amplitude of an offset spectrum on the singular values and variances of the third singular vectors of simulated absorption spectra for the reaction A → B → C in the presence of random noise. Simulations were carried out using the spectra described in the text and analysis procedures identical to those described in the legend to Fig. 7, both in the absence of random noise and at random noise amplitudes (σ_r) of 0.016, 0.032, and 0.064 OD. (A) The first three singular values. The dashed lines are the result in the absence of random noise; the results for all values of σ_r are plotted using the identical symbol: s_1, filled circles; s_2, open squares; s_3, open circles. (B, C) Square root of the average variance of the third singular vectors, U_3 (b) and V_3 (C). The results are shown both before (large symbols and solid lines) and after (small symbols and dashed lines) rotation to optimize the autocorrelation of the retained right singular vector, V_3^R.

singular value of a given component ($\sigma_\lambda < 0.004$ OD). What is even more interesting is that SVD is able to sort efficiently when s_λ is more than a factor of 3 *larger* than this singular value ($\sigma_\lambda > 0.04$ OD). There is a "mixing zone," delimited by the requirement that the amplitude of the wavelength-correlated noise, measured by s_λ, be within a factor of about 3 of that of the signal, measured by s_3, in which significant mixing occurs. At all values of σ_λ, rotation of the SVD output has a dramatic effect on $\sigma(V_3)$, reducing it to a value which is almost identical to that obtained in the presence of only the random noise component. This improvement persists until σ_λ exceeds 0.1, where the wavelength-correlated noise is almost an order of magnitude larger than the third component of the noise-free data.

If we now examine the noise amplitude for U_3 shown in Fig. 8B we find that it remains small throughout this mixing zone and even decreases as σ_λ becomes large enough to dominate the third SVD component. This result is consistent with the results of the simulations shown in Fig. 7, which showed that the noise amplitude in each of the left singular vectors produced by mixing of random noise amplitudes with the data is inversely proportional to its singular value. Because U_3 of the noise-free data and the spectral signature associated with the wavelength-correlated noise are smooth, the contribution of random noise to the noise amplitude of U_3 should be determined by the magnitude of s_3. When the SVD output is rotated to optimize $\sigma(V_3)$, the variance of U_3^R increases systematically as σ_λ increases, but never reaches the value expected if the spectrum were completely uncorrelated. It is also interesting to note that the random noise component acts almost only as a "background" to the wavelength-correlated noise in all of the output vectors. The noise amplitude contributed by the wavelength-correlated noise appears to be simply superimposed on this background.

A second set of simulations was carried out to explore to what extent the mixing of wavelength-correlated noise was dependent on the preprocessing of the data. The noise-free data used in these simulations were identical to those used above except the spectrum of pure C was subtracted at each time point. The data are therefore representative of data processed to produce difference absorption spectra at a given time point with the equilibrium sample used as a reference. The sample data set in Fig. 6D shows that the calculation of difference spectra removes much of the signal from the data matrix. From the point of view of SVD, the major consequence of this change is effectively to remove the first SVD component, which corresponds to the average absorption spectrum of the sample, from the analysis. The first SVD component now corresponds to the average *difference* spectrum observed in the simulated experiment.

The first two singular values and the noise amplitudes obtained for the second singular vectors (U_2 and V_2) are shown in Fig. 9. Note that the values of s_1 and s_2 of the noise-free data in these simulations are only slightly larger than s_2 and s_3 in Fig. 8. The dependence of the variance of V_2 on σ_λ, shown in Fig. 9C, is qualitatively similar to that found for V_3 in Fig. 8C. The ability of the rotation algorithm to reduce the variance of V_2 is also qualitatively similar to that shown in Fig. 8C, but the reduction in $\sigma(V_2)$ which results from rotation in Fig. 9C is even larger than that for $\sigma(V_3)$ in Fig. 8C. As in the previous simulation, rotation is able to "rescue" the original signal-to-noise ratio in V_2^R up to the point where the noise becomes as large as the first component of the noise-free data. At this point, significant noise

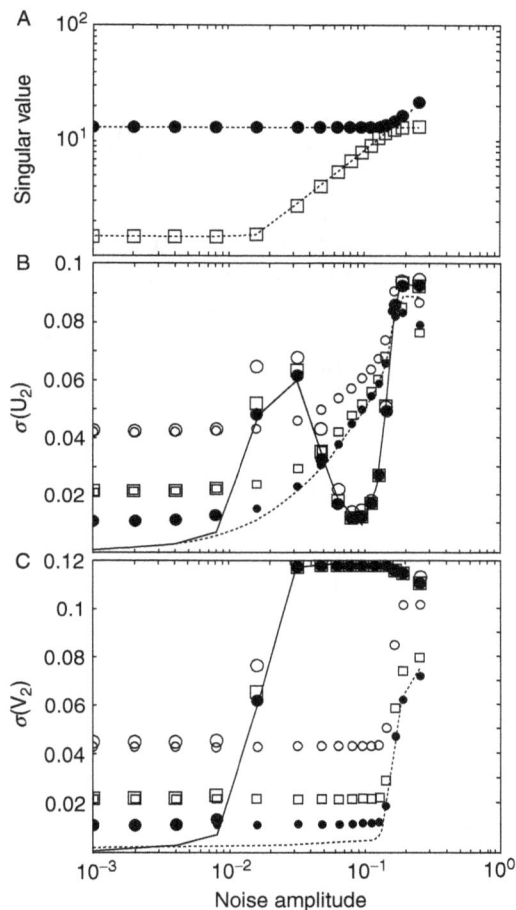

Fig. 9 Effects of the RMS amplitude of an offset spectrum on the singular values and variances of the second singular vectors of simulated difference spectra for the reaction A → B → C in the presence of random noise. Simulations were carried out using the spectra described in the text except that the spectrum of pure C was subtracted from the calculated sample spectrum at each time point and using analysis procedures identical to those described in the legend to Fig. 7. (A) The first two singular values. The dashed lines are the result in the absence of random noise, and the results for all values of σ_r are plotted using the same symbol: s_1, filled circles; s_2, open squares. (B, C) Square root of the average variance of the second singular vectors, U_2 (b) and V_2 (C). The results are shown both before (large symbols and solid lines) and after (small symbols and dashed lines) rotation to optimize the autocorrelation of the retained right singular vector, V_2^R. It should be noted that in the absence of random noise the SVD consists of only three singular values and vectors, so rotation only involves the mixing of V_2 and V_3.

amplitudes begin to mix into V_1. Comparable simulations which included rotation of V_1 show that the original signals in both V_1 and V_2 can be recovered even at these very large noise amplitudes.

Closer comparison of the results in Figs. 8 and 9 shows that there are significant differences between the analyses of the absorption spectra and the difference spectra. One major difference is the width of the "mixing zone," or transition region in which the relevant component (V_3 in Fig. 8 or V_2 in Fig. 9) is a combination of signal and wavelength-correlated noise. For the difference spectra (Fig. 9C) this zone is significantly narrower, covering about a factor of 3 change in σ_λ. (An intermediate value of $\sigma(V_2)$ is obtained for only a single value of σ_λ.) In Fig. 8C, on the other hand, this transition zone extends over at least a factor of 10 in σ_λ. The RMS deviations of V_3 in Fig. 8C are also somewhat greater than the corresponding deviations in V_2 in Fig. 9C at low noise amplitudes ($\sigma_\lambda < 0.02$). These results suggest that SVD alone sorts the wavelength-correlated noise from the signal more efficiently when the input data are in the form of difference spectra.

Another significant difference in the two simulations is that the variance of U_2 from the difference spectra also becomes very large in the transition region in Fig. 9B. There is essentially no evidence of such an effect in Fig. 8B. Because the offset spectrum has no wavelength dependence, this cannot arise from high-frequency contributions to U_2 in the SVD of a single data set. Rather, it must result from variations in the sign and magnitude of the offsets which are mixed with the U_2 of the noise-free data in different simulated data sets. Because the relative sign of the signal and offset contributions to U_2 is presumably determined by statistical fluctuations in the offset amplitudes, it is not unreasonable that the offset contribution varies both in magnitude and sign from data set to data set. The major difference between the two simulations is the absence of the largest SVD component in the difference spectra simulations. Examination of individual SVD outputs shows mixing of significant offset amplitudes with this component. Our tentative interpretation of this difference, then, is that significant mixing of the offsets with the average absorption spectrum can occur when the absorption spectra are used as data (Fig. 8B), but that this mixing cannot occur when difference spectra are used (Fig. 9B).[14]

The results of these simulations can be summarized as follows. As the amplitude of the wavelength-correlated noise increases, it first mixes with the components of the noise-free data, systematically increasing the random component of $V_3(V_2)$ and altering $U_3(U_2)$ (components in parentheses refer to the simulations of the difference spectra). As the noise amplitude increases further, the third (second) SVD component obtained from the noisy data results primarily from the wavelength-correlated offset, and its singular value, s_3 (s_2), increases in proportion to σ_λ. At these noise amplitudes $V_3(V_2)$ is essentialy a random vector. For the absorption spectra, the increased value of s_3 reduces $\sigma(U_3)$, suggesting that nearly all of the noise in U_3 results from random noise contributions, but more complex behavior is observed for the U_2 obtained from the difference spectra. Under these conditions most of the information contained in the third

[14] The possibility cannot be ruled out that a remaining ambiguity in the determination of the sign of U_2 as calculated in these simulations contributes to the variance of U_2 at higher noise levels.

(second) component of the SVD of the noise-free data has been displaced into the fourth (third) SVD component.

The rotation procedure is extremely effective in reclaiming the signals represented by V_3 (V_2) under all conditions where σ_λ is less than about 0.1. Processing of components 3 through 10 (2 through 10) by this algorithm dramatically reduces the noise of V_3 (V_2). The cost of this decrease in the noise of the rotated V_3 is a significant increase in the noise amplitude of the rotated U_3 (U_2). The noise amplitude of the rotated V_3(V_2) only increases at the point where the wavelength-correlated noise becomes comparable in magnitude to the second (first) SVD component of the noise-free data. At this point the wavelength-correlated noise begins to mix significantly with the second (first) SVD component, and a different processing algorithm (specifically, a rotation which includes V_2(V_1) in the rotated set) would be required to extract the signal from the noise. The ineffectiveness of the rotation algorithm when confronted with random noise amplitudes, noted earlier, is strongly reinforced by these simulations. We therefore conclude that the rotation procedure can be of significant benefit in extracting correlated noise contributions to the data matrix which are comparable in magnitude to a given signal component. When the signal is significantly larger than the noise, rotations appear to be of questionable benefit, sometimes resulting in a less efficient extraction of the signal by reducing the difference in the norms of the mixed components.

VII. Summary

In writing this chapter we have attempted to point out the significant advantages which result from the application of SVD and complementary processing techniques to the analysis of large sets of experimental data. Although SVD performs no "magic," it does efficiently extract the information contained in such data sets with a minimum number of input assumptions. Moreover, SVD is the *right* approach to use for such a reduction. The least-squares property which we have described both in the introduction (Section II) and in Section III, together with the theorem which shows that fitting of appropriately weighted SVD output is equivalent to fitting of the entire data set, argues that there are few, if any, additional risks which accompany the use of SVD as part of the analysis procedure.

The simple examples of the SVD of noisy matrices (Section IV) together with the simulations in Section VI teach several lessons in regard to the choice of an algorithm for processing data. First, when random noise is mixed with data the results are totally predictable. SVD very effectively averages over random contributions to the data matrix. However, when the random noise amplitude becomes sufficiently large to completely submerge the averaged signal amplitude from a given SVD component, that component becomes unresolvable. This result is a straightforward extension of conventional signal-averaging wisdom. The only possible approaches to extracting additional information when all of the higher SVD components are uncorrelated (i.e., they result from random noise) are either to increase the size of the data matrix and hence increase the number of observations which are averaged or to improve the quality of the data matrix by decreasing the noise level in the experiment. Second, when the data contain random amplitudes of a correlated noise source (e.g., the wavelength-correlated noise in our simulations) the mixing of the "signal," represented by a given SVD component, k, with the noise becomes significant when σ_λ becomes sufficiently large that the contribution of the noise component to the data matrix, $s_\lambda = (mn\sigma_\lambda^2)^{1/2}$, becomes comparable in magnitude to the "signal" represented by that SVD component, s_k. When the noise is small, it does not seriously degrade the component, but is efficiently sorted by SVD into a separate component, appearing as a larger than random contribution to the spectrum of singular values which result from the noise. When the noise becomes significantly larger than a given SVD component, the ability of SVD to sort by signal magnitude again separates such noise from the desired signal relatively cleanly, but the efficiency with which the two components are separated appears to depend on the specific data being analyzed. When such noise is mixed with the signal component in the SVD, dramatic improvement in signal-to-noise ratios can be produced by rotation of the mixed and highly cross-correlated SVD components to improve the autocorrelation of the retained component(s).

In conclusion, it must be noted that in the examples and simulations which we have presented we know the "signal" *a priori* and are also using noise having well-defined characteristics. As a result, it is considerably easier to understand the behavior observed in these cases than when one is confronted with real data. The results of the SVD analysis either with or without the rotation procedure depend on both the spectra which are used as input and on the detailed characteristics of the noise which is added to the data. Furthermore, the distribution of correlated noise in the SVD output can be altered by preprocessing as simple as the calculation of difference spectra. Extrapolation of these results would suggest that, in order to optimize the processing algorithm for any specific data, the experiment and its processing should be tested by simulations which use spectra that closely match those measured and noise that closely approximates that measured for data sets in which there is no signal. The only obvious alternative to using simulations appears to be to process the same experimental data using a variety of different procedures and to compare the results. Either of these approaches may pay dividends in the analysis of real experimental data.

Acknowledgments

We are grateful to Attila Szabo for numerous discussions and suggestions, as well as for critical comments on the manuscript. We also thank our colleagues in using SVD, William A. Eaton, Anjum Ansari, and Colleen M. Jones, for helpful comments and suggestions. We especially thank Colleen M. Jones for providing the data used as the example.

Appendix: Transformation of SVD Vectors to Optimize Autocorrelations

Suppose that we have an orthonormal set of column vectors $\{V_i\}$, for example, the set of right singular vectors of an $m \times n$ matrix. The autocorrelation of column vector V_i is defined by

$$C(V_i) = \sum_{j=1}^{n-1} V_{ji} V_{j+1,i}, \qquad (A.1)$$

where V_{ji} is the jth element of the vector V_i and n is the number of elements in each vector. The rotation procedure produces a linear transformation $\mathbf{R}_{\{k\}}$ which takes a given subset consisting of the p basis vectors $\{V_k\}$ ($k \in \{k1, \ldots, kp\}$) to a new set of vectors $\{V_k'\}$ for which the autocorrelations as defined in Eq. (A.1) are optimized. The problem is to find coefficients r_{ji}, the elements of the matrix $\mathbf{R}_{\{k\}}$, such that the autocorrelations of the vectors

$$V_j' = \sum_{i \in \{k\}} r_{ji} V_i \qquad (A.2)$$

are optimized. From Eq. (A.1) we have

$$\begin{aligned}
C(V_i') &= \sum_{j=1}^{n-1} V_{ji}' V_{j+1,i}' \\
&= \sum_j \left(\sum_k r_{ik} V_{jk} \right) \left(\sum_p r_{ip} V_{j+1,p} \right) \qquad (A.3) \\
&= \sum_{k,p} r_{ik} r_{ip} \left(\sum_j V_{jk} V_{j+1,p} \right).
\end{aligned}$$

If we define the $p \times p$ cross-correlation matrix $\mathbf{X}$ by

$$X_{ij} = \sum_k V_{ki} V_{k+1,j} \qquad (A.4)$$

then Eq. (A.3) may be written in the compact form

$$C(V') = \sum_{i,p} r_i r_p X_{ip}, \qquad (A.5)$$

where we have also dropped the vector index j for brevity.

We now require that the coefficients r_i produce extremum values of the autocorrelation in Eq. (A.5), subject to the constraint that the normalized and orthogonal vectors V_i are transformed into normalized vectors V_j', that is, $\Sigma r_i^2 = 1$. These requirements are easily formulated using the method of undetermined multipliers:

$$\frac{\partial F}{\partial r_i} = 0, i = 1, \ldots, m, \tag{A.6}$$

where

$$F = C(V') + \lambda \sum_i r_i^2. \tag{A.7}$$

Using Eq. (A.5), Eq. (A.6) becomes

$$\sum_p (X_{ip} + X_{pi}) - 2\lambda r_i = 0. \tag{A.8}$$

This may be rewritten in matrix-vector notation as

$$\mathbf{X}^S r = \lambda r, \tag{A.9}$$

where the matrix $\mathbf{X}^S = (\mathbf{X} + \mathbf{X}^T)/2$ is the symmetrized cross-correlation matrix and r is the vector of coefficients r_i.

Equation (A.9) represents a simple eigenvalue problem for the real symmetric matrix $\mathbf{X}^S$. The individual eigenvectors r_j ($j = 1, \ldots, p$) of $\mathbf{X}^s$ are sets of coefficients r_{ji} that produce distinct transformed vectors V'_j from the starting vectors V_i according to Eq. (A.2). It is easily shown that the corresponding eigenvalues λ_j are in fact the autocorrelations of the transformed vectors V'_j. The matrix $\mathbf{R}_{\{k\}}$ of the linear transformation that we seek is just the matrix of column vectors r_j, and the transformation in Eq. (A.2) may be rewritten

$$\mathbf{V}^R = \mathbf{V}\mathbf{R}_{\{k\}}, \tag{A.10}$$

where $\mathbf{V}$ and $\mathbf{V}^R$ are the matrices of untransformed and transformed vectors, respectively. The matrix $\mathbf{R}_{\{k\}}$ is clearly orthogonal, that is, $\mathbf{R}_{\{k\}}^{-1} = \mathbf{R}_{\{k\}}^T$, allowing the transformation of Eq. (A. 10) to be identified as a rotation. The eigenvectors r_j that make up the matrix $\mathbf{R}_{\{k\}}$ are conventionally arranged in order of decreasing eigenvalues, so that the columns of $\mathbf{V}^R$ will be arranged in order of decreasing autocorrelations of the transformed vectors.

It should be pointed out that the conditions of Eq. (A.6) only ensure extremum values of the autocorrelation with respect to the coefficients r_i, not necessarily maximum values. In general, the procedure produces a set of p vectors V'_j some of which have autocorrelations that are maxima, some of which have autocorrelations that are minima, and some of which have autocorrelations that represent saddle points in the space of coefficients r_i. For this reason, the procedure may be viewed as producing a new set of autocorrelations, some of which are improved (with respect to the original set) at the expense of others. It should also be noted that this straightforward procedure for reducing the optimization problem to an eigenvalue problem for determining the transformation matrix $\mathbf{R}_{\{k\}}$ is easily generalized to optimization of a much broader class of autocorrelation and other bilinear functions than has been considered here.

One final topic of discussion is the proper choice of weighting factors to be used when performing least-squares fits using transformed columns of $\mathbf{V}$. As discussed in the text, fits using the untransformed columns of $\mathbf{V}$ require that the fit to each column be weighted by the square of the corresponding singular value. In this case, for any set of model parameters the weighted squared deviation between $\mathbf{V}$ and the matrix $\widetilde{\mathbf{V}}(\widetilde{\mathbf{V}} \equiv \mathbf{C}'\mathbf{P}$ in Section V.E) of fitted columns produced by explicit evaluation of the model may be written

$$\phi^2 = \|\mathbf{S}(\mathbf{V}^\mathbf{T} - \widetilde{\mathbf{V}}^\mathbf{T})\|^2. \tag{A.11}$$

Inserting the product $\mathbf{R}\mathbf{R}^\mathbf{T} = \mathbf{I}_n$, where $\mathbf{R}$ is the full orthogonal transformation matrix defined in Eq. (37), and using the definition of Eq. (A.10),

$$
\begin{aligned}
\phi^2 &= \|\mathbf{S}\mathbf{R}\mathbf{R}^\mathbf{T}(\mathbf{V}^\mathbf{T} - \widetilde{\mathbf{V}}^\mathbf{T})\|^2 \\
&= \|\mathbf{S}\mathbf{R}[(\mathbf{V}^\mathbf{R})^\mathbf{T} - (\widetilde{\mathbf{V}}^\mathbf{R})^\mathbf{T}]\|^2.
\end{aligned} \tag{A.12}
$$

We define the matrix Δ as

$$\Delta = \mathbf{V}^\mathbf{R} - \widetilde{\mathbf{V}}^\mathbf{R}. \tag{A.13}$$

Using the identity

$$\|\mathbf{M}\|^2 = Tr(\mathbf{M}^\mathbf{T}\mathbf{M}), \tag{A.14}$$

where $Tr\,(\ldots)$ signifies the matrix trace operation, we can write

$$
\begin{aligned}
\phi^2 &= Tr(\Delta\mathbf{R}^\mathbf{T}\mathbf{S}^2\mathbf{R}\Delta^\mathbf{T}) \\
&= \sum_i \sum_{kl} \Delta_{ik}(\mathbf{R}^\mathbf{T}\mathbf{S}^2\mathbf{R})_{kl}\Delta_{il}.
\end{aligned} \tag{A.15}
$$

If the model being used is adequate to describe the data matrix to within some tolerance, then within some neighborhood of the minimum of Eq. (A.11) in parameter space it is reasonable to expect that the deviations represented by the different columns of Δ will be uncorrelated. If this is the case, we can write

$$\sum_i \Delta_{ik}\Delta_{il} = \sum_i \delta_{kl}\Delta_{ik}^2, \tag{A.16}$$

where δ_{kl} is the Kronecker delta. This allows us to simplify Eq. (A.15) to

$$
\begin{aligned}
\phi^2 &\cong \sum_k (\mathbf{R}^\mathbf{T}\mathbf{S}^2\mathbf{R})_{kk} \sum_i \Delta_{ik}^2 \\
&= \sum_k W_{kk}\|\Delta_k\|^2,
\end{aligned} \tag{A.17}
$$

where the matrix $\mathbf{W} = \mathbf{R}^\mathbf{T}\mathbf{S}^2\mathbf{R}$ and Δ_k is the kth column of the matrix Δ. Thus, subject to caveats concerning the assumptions leading to Eq. (A.16), it is reasonable to choose the squares of the corresponding diagonal elements of the matrix $\mathbf{R}^\mathbf{T}\mathbf{S}^2\mathbf{R}$ as weighting factors in fits to columns of $\mathbf{V}^\mathbf{R}$.

Another way to estimate the weighting factors is via the amplitudes of the columns of $\mathbf{U}^\mathbf{R}$ corresponding to the normalized columns of $\mathbf{V}^\mathbf{R}$ being fit. In the absence of rotations, the weighting factors for fitting the columns of $\mathbf{V}$ are simply the squared amplitudes of the corresponding columns of $\mathbf{US}$. By analogy, the weighting factors for fitting the columns of $\mathbf{V}^{\mathbf{R}}$ could be chosen as the squared amplitudes of the corresponding columns of $\mathbf{U}^{\mathbf{R}}$. These squared amplitudes are given by the diagonal elements of the product $(\mathbf{U}^{\mathbf{R}})^{\mathbf{T}}\mathbf{U}^{\mathbf{R}}$. By Eq. (39) this product may be written

$$
\begin{aligned}
(\mathbf{U}^{\mathbf{R}})^{\mathbf{T}}\mathbf{U}^{\mathbf{R}} &= \mathbf{R}^{\mathbf{T}}\mathbf{S}\mathbf{U}^{\mathbf{T}}\mathbf{U}\mathbf{S}\mathbf{R} \\
&= \mathbf{R}^{\mathbf{T}}\mathbf{S}^2\mathbf{R}.
\end{aligned}
\tag{A.18}
$$

Thus, this approach to estimating the weighting factors yields the same result as was produced by the first method.

References

Anderson, T. W. (1963). *Ann. Math. Stat.* **34**, 122.

Anderson, E. *et al.* (eds.) (1999). "LAPACK Users' Guide," 3rd edn. SIAM, Philadelphia, PA.

Cochran, R. N., and Horne, F. H. (1977). *Anal. Chem.* **49**, 846.

Cochran, R. N., and Horne, F. H. (1980). *J. Phys. Chem.* **84**, 2561.

Cochran, R. N., Horne, F. H., Dye, J. L., Ceraso, J., and Suetler, C. H. (1980). *J. Phys. Chem.* **84**, 2567.

Eckhart, C., and Young, G. (1939). *Bull. Am. Math. Soc.* **45**, 118.

Frans, S. D., and Harris, J. M. (1984). *Anal. Chem.* **56**, 466.

Frans, S. D., and Harris, J. M. (1985). *Anal. Chem.* **57**, 1718.

Golub, G., and Kahan, W. (1965). *SIAM J. Numer. Anal. Ser. B* **2**, 205.

Golub, G. H., and Pereyra, V. (1973). *SIAM J. Numer. Anal.* **10**, 413.

Golub, G. H., and Reinsch, C. (1970). *Numer. Math.* **14**, 403.

Golub, G., and VanLoan, C. (1996). "Matrix Computations," 3rd edn. The Johns Hopkins University Press, Baltimore, MD.

Hendler, R. W., Subba Reddy, K. V., Shrager, R. I., and Caughey, W. S. (1986). *Biophys. J.* **49**, 717.

Hennessey, J. P. Jr., and Johnson, W. C. Jr. (1981). *Biochemistry* **20**, 1085.

Henry, E. R. (1997). The use of matrix methods in the modeling of spectroscopic data sets. *Biophys. J.* **72**, 652.

Hofrichter, J., Sommer, J. H., Henry, E. R., and Eaton, W. A. (1983). *Proc. Natl. Acad. Sci. USA* **80**, 2235.

Hofrichter, J., Henry, E. R., Sommer, J. H., and Eaton, W. A. (1985). *Biochemistry* **24**, 2667.

Hofrichter, J., Henry, E. R., Szabo, A., Murray, L. P., Ansari, A., Jones, C. M., Coletta, M., Falcioni, G., Brunori, M., and Eaton, W. A. (1991). *Biochemistry* **30**, 6583.

Horn, R. A., and Johnson, C. R. (1990). "Matrix Analysis," Cambridge Univ. Press, Cambridge.

Horn, R. A., and Johnson, C. R. (1994). "Topics in Matrix Analysis," Cambridge Univ. Press, Cambridge.

Johnson, W. C. Jr. (1988). *Annu. Rev. Biophys. Biophys. Chem.* **17**, 145.

Johnson, M. L., Correira, J. J., Yphantis, D. A., and Halvorson, H. R. (1981). *Biophys. J.* **36**, 575.

Kankare, J. J. (1970). *Anal. Chem.* **42**, 1322.

Knutson, J. R., Beechem, J. M., and Brand, L. (1983). *Chem. Phys. Lett.* **102**, 501.

Lawson, C. L., and Hanson, R. J. (1974). "Solving Least-Squares Problems," Prentice-Hall, Englewood Cliffs, New Jersey.

Mauer, R., Vogel, J., and Schneider, S. (1987a). *Photochem. Photobiol.* **46**, 247.

Mauer, R., Vogel, J., and Schneider, S. (1987b). *Photochem. Photobiol.* **46**, 255.

McMullen, D. W., Jaskunas, S. R., and Tinoco, I. Jr. (1965). *Biopolymers* **5**, 589.

Milder, S. J., Thorgeirsson, T. E., Miercke, L. J. W., Stroud, R. M., and Kliger, D. S. (1991). *Biochemistry* **30,** 1751.

Murray, L. P., Hofrichter, J., Henry, E. R., Ikeda-Saito, M., Kitagishi, K., Yonetani, T., and Eaton, W. A. (1988). *Proc. Natl. Acad. Sci. USA* **85,** 2151.

Nagel, J. F., Parodi, L. A., and Lozier, R. H. (1982). *Biophys. J.* **38,** 161.

Press, W. H., Flannery, B. P., Teukolsky, S. A., and Vetterling, W. T. (2007). "Numerical Recipes: The Art of Scientific Computing," 3rd ed. Cambridge Univ. Press, Cambridge.

Rajagopal, S., *et al.* (2004). *Acta Cryst* **D60,** 860.

Shrager, R. I. (1984). *SIAM J. Alg. Disc. Methods* **5,** 351.

Shrager, R. I. (1986). *Chemom. Intell. Lab. Syst.* **1,** 59.

Shrager, R. I., and Hendler, R. W. (1982). *Anal. Chem.* **54,** 1147.

Subba Reddy, K. V., Hendler, R. W., and Bunow, B. (1986). *Biophys. J.* **49,** 705.

Sylvestre, E. A., Lawton, W. H., and Maggio, M. S. (1974). *Technometrics* **16,** 353.

Trotter, H. F. (1984). *Adv. Math.* **54,** 67.

Wigner, E. (1967). *SIAM Rev.* **9,** 1.

Xie, A. H., Nagle, J. F., and Lozier, R. H. (1987). *Biophys. J.* **51,** 627.

CHAPTER 7

Irregularity and Asynchrony in Biologic Network Signals

Steven M. Pincus

Independent Mathematician, Guilford
CT, USA

I. Update

Since the publication of this article, both the primary methods espoused herein, approximate entropy (ApEn) and cross-ApEn, have been applied very frequently to an increasingly wide variety of settings within biology, medicine, and epidemiology. These include an upsurge in applications to the analysis of electroencephalogram (EEG) data; extensive studies of balance and motor control; further applications to heart rate and respiratory data as well as very broad continued application to the analysis of hormonal dynamics within endocrinology. Representatives of these studies are applications of ApEn to classify the depth of anesthesia with greater accuracy than previous benchmark standard methods (Bruhn *et al.*, 2000), and a means to assess clinically subtle post-concussion motor control in an evaluation of recovery status (Cavanaugh *et al.*, 2006).

Moreover, ApEn has been recently applied to several studies within psychiatry, primarily to the analysis of daily mood rating data of sadness and depression (Pincus *et al.*, 2008), and as well, has been demonstrated to be a strong predictor of the onset of menopause on the basis of disrupted patterns of menstrual bleeding, when applied to a seminal database of a long term, longitudinal study of womens' menstrual cycles (Weinstein *et al.*, 2003).

Finally, more theoretical mathematical aspects have been developed as part of an application to financial data (Pincus and Kalman, 2004), as well as in a theoretical probability study of the limiting nature of the ApEn distribution, which was shown to be chi-squared (Rukhin, 2000).

II. Introduction

Series of sequential data arise throughout biology, in multifaceted contexts. Examples include (1) hormonal secretory dynamics based on frequent, fixed-increment samples from serum, (2) heart rate rhythms, (3) EEGs, and (4) DNA sequences. Enhanced capabilities to quantify differences among such series would be extremely valuable as, in their respective contexts, these series reflect essential biological information. Although practitioners and researchers typically quantify mean levels, and oftentimes the extent of variability, it is recognized that in many instances, the persistence of certain patterns or shifts in an "apparent ensemble amount of randomness" provide the fundamental insight into subject status. Despite this recognition, formulas and algorithms to quantify an "extent of randomness" have not been developed and/or utilized in the above contexts, primarily as even within mathematics itself, such quantification technology was lacking until very recently. Thus except for the settings in which egregious changes in serial features presented themselves, which specialists are trained to detect visually, subtler changes in patterns would largely remain undetected, unquantified, and/or not acted on.

Recently, a new mathematical approach and formula, ApEn, has been introduced as a quantification of *regularity* of data, motivated by both the above application needs (Pincus, 1991) and by fundamental questions within mathematics (Pincus and Kalman, 1997; Pincus and Singer, 1996). This approach calibrates an ensemble extent of sequential interrelationships, quantifying a continuum that ranges from totally ordered to completely random. The central focus of this review is to discuss ApEn, and subsequently cross-ApEn (Pincus and Singer, 1996; Pincus *et al.*, 1996a), a measure of two-variable asynchrony that is thematically similar to ApEn.

Before presenting a detailed discussion of regularity, we consider two sets of time series (Figs. 1 and 2) to illustrate what we are trying to measure. In Fig. 1A-F, the data represent a time series of growth hormone (GH) levels from rats in six distinct physiologic settings, each taken as a 10-min sample during a 10-h lights-off ("dark") state (Gevers *et al.*, 1998). The end points (a) and (f) depict, respectively, intact male and intact female serum dynamics; (b) and (c) depict two types of neutered male rats; and (d) and (e) depict two classes of neutered female rats. It appears that the time series are becoming increasingly irregular as we proceed from (a) to (f), although specific feature differences among the sets are not easily pinpointed. In Fig. 2, the data represent the beat-to-beat heart rate, in beats per minute, at equally spaced time intervals. Figure 2A is from an infant who had an aborted SIDS (sudden infant death syndrome) episode 1 week prior to the recording, and Fig. 2B is from a healthy infant (Pincus *et al.*, 1993). The standard deviations (SD) of these two tracings are approximately equal, and while the aborted SIDS infant has a somewhat higher mean heart rate, both are well within the normal range. Yet tracing (A) appears to be more regular than tracing (B). In both of these instances, we ask these questions: (1) How do we quantify the apparent differences in regularity? (2) Do the regularity values significantly distinguish the data sets? (3) How do inherent limitations posed by moderate length time series, with noise and measurement inaccuracies present as shown in Figs. 1 and 2, affect statistical analyses? (4) Is there some general mechanistic hypothesis, applicable to diverse contexts, that might explain such regularity differences?

The development of ApEn evolved as follows. To quantify time series regularity (and randomness), we initially applied the Kolmogorov-Sinai (K-S) entropy (Kolmogorov, 1958) to clinically derived data sets. The application of a formula for K-S entropy (Grassberger and Procaccia, 1983a; Takens, 1983) yielded intuitively incorrect results. Closer inspection of the formula showed that the low magnitude noise present in the data greatly affected the calculation results. It also became apparent that to attempt to achieve convergence of this entropy measure, extremely long time series would be required (often 1,000,000 or more points), which even if available, would then place extraordinary time demands on the computational resources. The challenge was to determine a suitable formula to quantify the concept of regularity in moderate-length, somewhat noisy data sets, in a manner thematically similar to the approach given by the K-S entropy.

Historical context further frames this effort. The K-S entropy was developed for and is properly employed on truly chaotic processes (time series). Chaos refers to output from deterministic dynamic systems, where the output is bounded and aperiodic, thus appearing partially "random." Recently, there have been myriad

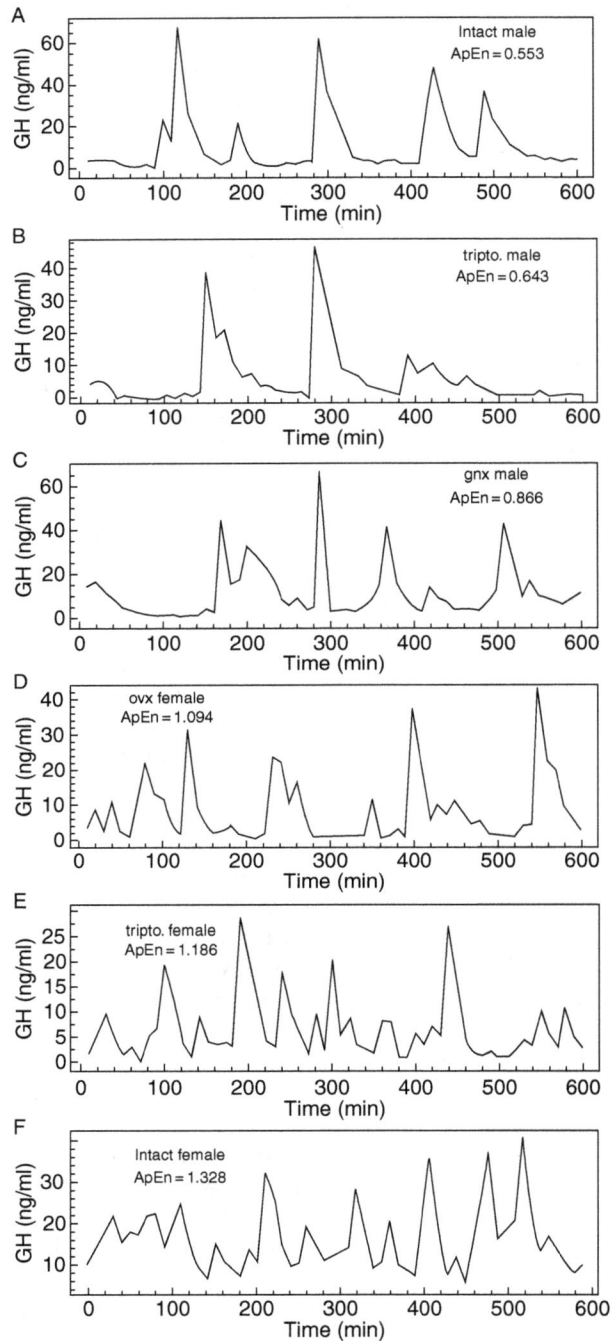

Fig. 1 Representative serum growth hormone (GH) concentration profiles, in ng/ml, measured at 10-min intervals for 10 h in the dark. In ascending order of ApEn values, and hence with increasing irregularity or disorderliness: (A) intact male, (B) triptorelin-treated male, (C) gonadectomized male, (D) ovariectomized female, (E) triptorelin-treated female, and (F) intact female rats.

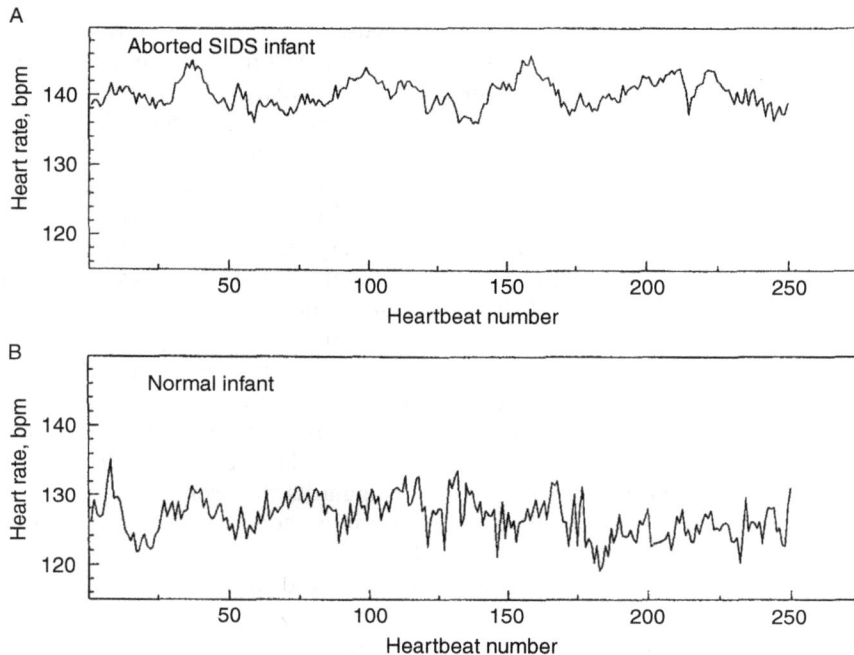

Fig. 2 Two infant quiet sleep heart rate tracings with similar variability, SD (A) aborted SIDS infant, SD = 2.49 beats per minute (bpm), ApEn(2, 0.15 SD, 1000) = 0.826; (B) normal infant, SD = 2.61 bpm, ApEn(2, 0.15 SD, 1000) = 1.463.

claims of chaos based on analysis of experimental time series data, in which correlation between successive measurements has been observed. Because chaotic systems represent only one of many paradigms that can produce serial correlation, it is generally inappropriate to infer chaos from the correlation alone. The mislabeling of correlated data as "chaotic" is a relatively benign offense. Of greater significance, complexity statistics that were developed for application to chaotic systems and are relatively limited in scope have been commonly misapplied to finite, noisy, and/or stochastically derived time series, frequently with confounding and nonreplicable results. This caveat is particularly germane to biologic signals, especially those taken *in vivo*, because such signals usually represent the output of a complicated network with both stochastic and deterministic components. We elaborate on these points in the later section titled Statistics Related to Chaos. With the development of ApEn, we can now successfully handle the noise, data length, and stochastic/composite model constraints in statistical applications.

As stated, we also discuss cross-ApEn (Pincus and Singer, 1996; Pincus *et al.*, 1996a), a quantification of asynchrony or conditional irregularity between two signals. Cross-ApEn is thematically and algorithmically quite similar to ApEn, yet with a critical difference in focus: it is applied to two time series, rather than a single series, and thus affords a distinct tool from which changes in the extent of synchrony in interconnected systems or networks can be directly determined. This quantification strategy is thus especially germane to many biological feedback and/or control systems and models for which cross-correlation and cross-spectral methods fail to fully highlight markedly changing features of the data sets under consideration.

Importantly, we observe a fundamental difference between regularity (and asynchrony) statistics, such as ApEn, and variability measures: most short- and long-term variability measures take raw data, preprocess the data, and then apply a calculation of SD (or a similar, nonparametric variation) to the processed data (Parer *et al.*, 1985). The means of preprocessing the raw data varies substantially with the different variability algorithms, giving rise to many distinct versions. However, once preprocessing of the raw data is completed, the processed data are input to an algorithm for which the order of the data is immaterial. For ApEn, the order of the data is the essential factor; discerning changes in order from apparently random to very regular is the primary focus of this statistic.

Finally, an absolutely paramount concern in any practical time series analysis is the presence of either artifacts or nonstationarities, particularly clear trending. If a time series is nonstationary or is riddled with artifacts, little can be inferred from moment, ApEn, or power spectral calculations, because these effects tend to dominate all other features. In practice, data with trends suggest a collection of heterogeneous epochs, as opposed to a single homogeneous state. From the statistical perspective, it is imperative that artifacts and trends first be removed before meaningful interpretation can be made from further statistical calculations.

III. Quantification of Regularity

A. Definition of ApEn

ApEn was introduced as a quantification of regularity in time series data, motivated by applications to relatively short, noisy data sets (Pincus, 1991). Mathematically, ApEn is part of a general development of approximating Markov chains to a process (Pincus, 1992); it is furthermore employed to refine the formulations of i.i.d. (independent, identically distributed) random variables, and normal numbers in number theory, via rates of convergence of a deficit from maximal irregularity (Pincus and Kalman, 1997; Pincus and Singer, 1996, 1998). Analytical properties for ApEn can be found in Pincus (1991), Pincus and Singer (1996), Pincus and Huang (1992), Pincus and Goldberger (1994); in addition, it provides a finite sequence formulation of randomness, via proximity to maximal irregularity (Pincus and Kalman, 1997; Pincus and Singer, 1996). Statistical evaluation is given in Pincus and Huang (1992), Pincus and Goldberger (1994).

ApEn assigns a nonnegative number to a sequence or time series, with larger values corresponding to greater apparent process randomness (serial irregularity) and smaller values to more instances of recognizable patterns or features in data.

ApEn measures the logarithmic likelihood that runs of patterns that are close for m observations remain close on next incremental comparisons. From a statistician's perspective, ApEn can often be regarded as an ensemble parameter of process autocorrelation: smaller ApEn values correspond to greater positive autocorrelation; larger ApEn values indicate greater independence. The opposing extremes are perfectly regular sequences (e.g., sinusoidal behavior, very low ApEn) and independent sequential processes (very large ApEn).

Formally, given N data points $u(1)$, $u(2)$,..., $u(N)$, two input parameters, m and r, must be fixed to compute ApEn [denoted precisely by ApEn(m, r, N)]. The parameter m is the "length" of compared runs, and r is effectively a filter. Next, form vector sequences $x(1)$ through $x(N - m + 1)$ from the $\{u(i)\}$, defined by $x(i) = [u(i), \ldots, u(i + m - 1)]$. These vectors represent m consecutive u values, commencing with the ith point. Define the distance $d[x(i), x(j)]$ between vectors $x(i)$ and $x(j)$ as the maximum difference in their respective scalar components. Use the sequence $x(1), x(2), \ldots, x(N - m + 1)$ to construct, for each $i \leq N - m + 1$,

$$C_i^m(r) = [\text{number of } x(j) \text{ such that } d[x(i), x(j)] \leq r]/(N - m + 1). \quad (1)$$

The $C_i^m(r)$ values measure within a tolerance r the regularity, or frequency, of patterns similar to a given pattern of window length, m. Next, define

$$\Phi^m(r) = (N - m + 1)^{-1} \sum_{i=1}^{N-m+1} \ln C_i^m(r), \quad (2)$$

where ln is the natural logarithm. We define ApEn by

$$\text{ApEn}(m, r, N) = \Phi^m(r) - \Phi^{m+1}(r). \quad (3)$$

Via some simple arithmetic manipulation, we deduce the important observation that

$$
\begin{aligned}
-\text{ApEn} = \ & \Phi^{m+1}(r) - \Phi^m(r) \\
= \ & \text{average over } i \text{ of } \ln[\text{conditional probability}] \\
& \text{that} |u(j + m) - u(i + m)| \leq r \\
& \text{given that } |u(j + k) - u(i + k)| \leq r \text{ for } k = 0, 1, \ldots, m - 1].
\end{aligned}
\quad (4)
$$

When $m = 2$, as is often employed, we interpret ApEn as a measure of the difference between the probability that runs of value of length 2 will recur within tolerance r, and the probability that runs of length 3 will recur to the same tolerance. A high degree of regularity in the data would imply that a given run of length 2 would often continue with nearly the same third value, producing a low value of ApEn.

ApEn evaluates both dominant and subordinate patterns in data; notably, it will detect changes in underlying episodic behavior that do not reflect in peak occurrences or amplitudes (Pincus and Keefe, 1992), a point that is particularly germane to numerous diverse applications. Additionally, ApEn provides a direct barometer of feedback system change in many coupled systems (Pincus and Keefe, 1992; Pincus, 1994).

ApEn is a relative measure of process regularity, and can show significant variation in its absolute numerical value with changing background noise characteristics. Because ApEn generally increases with increasing process noise, it is appropriate to compare data sets with similar noise characteristics, that is, from a common experimental protocol.

ApEn is typically calculated via a short computer code; a FORTRAN listing for such a code can be found in Pincus *et al.* (1991) ApEn is nearly unaffected by noise of magnitude below r, the *de facto* filter level, and it is robust or insensitive to artifacts or outliers: extremely large and small artifacts have little effect on the ApEn calculation, if they occur infrequently.

Finally, to develop a more intuitive, physiological understanding of the ApEn definition, a multistep description of its typical algorithmic implementation, with figures, is presented in Pincus and Goldberger (1994).

IV. Implementation and Interpretation

A. Choice of m, r, and N

The value of N, the number of input data points for ApEn computations, is typically between 50 and 5000. This constraint is usually imposed by experimental considerations, not algorithmic limitations, to ensure a single homogeneous epoch. On the basis of the calculations that included both theoretical analysis (Pincus and Huang, 1992; Pincus and Keefe, 1992; Pincus, 1991) and numerous clinical applications (Fleisher *et al.*, 1993; Kaplan *et al.*, 1991; Pincus and Viscarello, 1992; Pincus *et al.*, 1991; 1993) we have concluded that for both $m = 1$ and $m = 2$, and $50 \leq N \leq 5000$, values of r between 0.1 and 0.25 SD of the $u(i)$ data produce good statistical validity of ApEn(m, r, N). For such r values, we demonstrated (Pincus and Huang, 1992; Pincus and Keefe, 1992; Pincus, 1991) the theoretical utility of ApEn(1, r) and ApEn(2, r) to distinguish data on the basis of regularity for both deterministic and random processes, and the clinical utility in the aforementioned applications. These choices of m and r are made to ensure that the conditional frequencies defined in Eq. (4) are reasonably estimated from the N input data points. For smaller r values than those indicated, one usually achieves poor conditional probability estimates as well, while for larger r values, too much detailed system information is lost.

To ensure appropriate comparisons between data sets, it is strongly preferred that N be the same for each data set. This is because ApEn is a *biased* statistic; the expected value of ApEn(m, r, N) generally increases asymptotically with N to a well-defined, limit parameter denoted ApEn(m, r). Restated, if we had 3000 data points, and chose $m = 2$, $r = 0.2$ SD, we would expect that ApEn applied to the first 1000 points would be smaller than ApEn applied to the entire 3000-point time series. Biased statistics are quite commonly employed, with no loss of validity. As an aside, it can be shown that ApEn is asymptotically unbiased, an important theoretical property, but that is not so germane to "everyday" usage. This bias is discussed elsewhere (Pincus and Goldberger, 1994; Pincus and Huang, 1992), and techniques to reduce bias via a family of ε estimators are provided. However, note that ultimately, attempts to achieve bias reduction are model specific; thus, as stated earlier, it is cleanest to impose a (nearly) fixed data length mandate on all ApEn calculations.

B. Family of Statistics

Most importantly, despite algorithmic similarities, ApEn(m, r, N) is not intended as an approximate value of K-S entropy (Pincus and Huang, 1992; Pincus *et al.*, 1991; Pincus, 1991). It is imperative to consider ApEn(m, r, N) as a *family* of statistics; for a given application, system comparisons are intended with fixed m and r. For a given system, there usually is significant variation in ApEn(m, r, N) over the range of m and r (Pincus and Huang, 1992; Pincus *et al.*, 1991, 1993).

For fixed m and r, the conditional probabilities given by Eq. (4) are precisely defined probabilitistic quantities, marginal probabilities on a coarse partition, and contain a great deal of system information. Furthermore, these terms are finite, and thus allow process discrimination for many classes of processes that have infinite K-S entropy (see below). ApEn aggregates these probabilities, thus requiring relatively modest data input.

C. Normalized Regularity

ApEn decrease frequently correlates with SD decrease. This is not a "problem," as statistics often correlate with one another, but typically we desire an index of regularity decorrelated from SD. We can realize such an index, by specifying r in ApEn(m, r, N) as a fixed percentage of the sample SD of the *individual* subject data set (time series). We call this *normalized regularity*. Normalizing r in this manner gives ApEn a translation and scale invariance to absolute levels (Pincus *et al.*, 1993) in that it remains unchanged under uniform process magnification, reduction, or constant shift higher or lower. Choosing r via this procedure allows sensible regularity comparisons of processes with substantially different SDs. In most clinical applications, it is the normalized version of ApEn that has been employed, generally with $m = 1$ or $m = 2$ and $r = 20\%$ of the SD of the time series.

D. Relative Consistency

Earlier we commented that ApEn values for a given system can vary significantly with different m and r values. Indeed, it can be shown that for many processes, ApEn(m, r, N) grows with decreasing r like $\log(2r)$, thus exhibiting infinite variation with r (Pincus and Huang, 1992). We have also claimed that the utility of ApEn is as a relative measure; for *fixed* m and r, ApEn can provide useful information. We typically observe that for a given time series, ApEn(2,0.1) is quite different from ApEn(4, 0.01), so the question arises as to which parameter choices (m and r) to use. The guidelines above address this, but the most important requirement is consistency. For noiseless, theoretically defined deterministic dynamic systems, we have found that when K-S entropy(A) $\leq$ K-S entropy(B), then ApEn(m, r)(A) $\leq$ ApEn(m, r)(B) and conversely, for a wide range of m and r. Furthermore, for both theoretically described systems and those described by experimental data, we have found that when ApEn(m_1, r_1)(A) $\leq$ ApEn(m_1, r_1) (B), then ApEn(m_2, r_2)(A) $\leq$ ApEn(m_2, r_2)(B), and conversely. This latter property also generally holds for parameterized systems of stochastic (random) processes, in which K-S entropy is infinite. We call this ability of ApEn to preserve order a relative property. It is the key to the general and clinical utility of ApEn. We see no sensible comparisons of ApEn(m, r)(A) and ApEn(n, s)(B) for systems A and B unless $m = n$ and $r = s$.

From a more theoretical mathematical perspective, the interplay between meshes [(m, r) pair specifications] need not be nice, in general, in ascertaining which of two processes is "more" random. In general, we might like to ask this question: Given no noise and an infinite amount of data, can we say that process A is more regular than process B? The *flip-flop pair* of processes (Pincus and Huang, 1992) implies that the answer to this question is "not necessarily": in general, comparison of relative process randomness at a prescribed level is the best one can do. That is, processes may appear more random than processes on many choices of partitions, but not necessarily on all partitions of suitably small diameter (r). The *flip-flop pair* is two i.i.d. processes A and B with the property that for any integer m and any positive r, there exists $s < r$ such that ApEn(m, s)(A) $<$ ApEn (m, s)(B), and there exists $t < s$ such that ApEn(m, t)(B) $<$ ApEn(m, t)(A). At alternatingly small levels of refinement given by r, process B appears more random and less regular than process A followed by appearing *less* random and more regular than process A on a still smaller mesh (smaller r). In this construction, r can be made arbitrarily small, thus establishing the point that process regularity is a relative [to mesh, or (m, r) choice] notion.

Fortunately, for many processes A and B, we can assert more than relative regularity, even though both A and B will typically have infinite K-S entropy. For such pairs of processes, which have been denoted as *completely consistent pairs* (Pincus and Huang, 1992), whenever ApEn(m, r)(A) $<$ ApEn(m, r)(B) for any specific choice of m and r, it follows that ApEn(n, s)(A) $<$ ApEn(n, s)(B) for *all* choices of n and s. Any two elements of {MIX(p)} (defined below), for example, appear to be completely consistent. The importance of completely consistent pairs is that we can then assert that process B is more irregular (or random) than process A, without needing to indicate m and r. Visually, process B appears more random than process A at any level of view. We indicate elsewhere (Pincus and Goldberger, 1994) a conjecture that should be relatively straightforward to prove, that would provide a sufficient condition to ensure that A and B are a completely consistent pair, and would indicate the relationship to the autocorrelation function.

E. Model Independence

The physiologic modeling of many complex biological systems is often very difficult; one would expect accurate models of such systems to be complicated composites, with both deterministic and stochastic components, and interconnecting network features. The advantage of a *model-independent* statistic is that it can distinguish classes of systems for a wide variety of models. The mean, variability, and ApEn are all model-independent statistics in that they can distinguish many classes of systems, and all can be meaningfully applied to $N > 50$ data points. In applying ApEn, therefore, we are not testing for a particular model form, such as deterministic chaos; we are attempting to distinguish data sets on the basis of regularity. Such evolving regularity can be seen in both deterministic and stochastic models (Pincus and Huang, 1992; Pincus and Keefe, 1992; Pincus, 1991; Pincus, 1994).

F. Statistical Validity: Error Bars for General Processes

Ultimately, the utility of any statistic is based on its replicability. Specifically, if a fixed physical process generates serial data, we would expect statistics of the time series to be relatively constant over time; otherwise, we would have difficulty ensuring that two very different statistical values implied two different systems (distinction). Here, we thus want to ascertain ApEn variation for typical processes (models), so we can distinguish data sets with high probability when ApEn values are sufficiently far apart. This is mathematically addressed by SD calculations of ApEn, calculated for a variety of representative models; such calculations provide "error bars" to quantify probability of true distinction. Via extensive Monte Carlo calculations, we established the SD of ApEn (2, 0.2 SD, 1000) < 0.055 for a large class of candidate models (Pincus and Huang, 1992; Pincus and Keefe, 1992). It is this small SD of ApEn, applied to 1000 points from various models, that provides its utility to practical data analysis of moderate-length time series. For instance, applying this analysis, we deduce that ApEn values that are 0.15 apart represent nearly 3 ApEn SDs, indicating true distinction with error probability nearly $p = 0.001$. Similarly, the SD of ApEn(1, 0.2 SD, 100) < 0.06 for many diverse models (Pincus and Huang, 1992; Pincus and Keefe, 1992), thus providing good replicability of ApEn with $m = 1$ for the shorter data length applications.

G. Analytic Expressions

For many processes, we can provide analytic expressions for ApEn(m, r). Two such expressions are given by Theorems 1 and 2 (Pincus, 1991):

Theorem 1. Assume a stationary process $u(i)$ with continuous state space. Let $\mu(x, y)$ be the joint stationary probability measure on $\mathbf{R}^2$ for this process, and $\pi(x)$ be the equilibrium probability of x.

Then
ApEn(1, r)

$$= - \int \mu(x,y) \log \left[\int_{z=y-r}^{y+r} \int_{w=x-r}^{x+r} \mu(w,z) \mathrm{d}w \mathrm{d}z \Big/ \int_{w=x-r}^{x+r} \pi(w) \mathrm{d}w \right] \mathrm{d}x \mathrm{d}y.$$

Theorem 2. For an i.i.d. process with density function $\pi(x)$ for any $m \geq 1$,

$$\mathrm{ApEn}(m,r) = - \int \pi(y) \log \left[\int_{z=y-r}^{y+r} \pi(z) \mathrm{d}z \right] \mathrm{d}y.$$

Theorem 1 can be extended in straightforward fashion to derive an expression for ApEn(m, r) in terms of the joint [($m + 1$)-fold] probability distributions. Hence, we can calculate ApEn(m, r) for Gaussian processes, as we know the joint probability distribution in terms of the covariance matrix. This important class of processes (for which finite sums of discretely sampled variables have multivariate normal distributions) describes many stochastic models, including solutions to ARMA (autoregressive-moving average) models and to linear stochastic differential equations driven by white noise.

Moreover, from a different theoretical setting, ApEn is related to a parameter in information theory, *conditional entropy* (Blahut, 1987). Assume a finite state space, where the entropy of a random variable X, Prob($X = a_j$) $= p_j$, is $H(X) := -\sum p_j \log p_j$, and the entropy of a block of random variables $X_1, \ldots, X_n = H(X_1, \ldots, X_n) := - \sum \sum \cdots \sum p^n (a_{j1}, \ldots, a_{jn}) \log p^n (a_{j1}, \ldots, a_{jn})$. For two variables, the conditional entropy $H(Y \| X) = H(X, Y) - H(X)$; this extends naturally to n variables. Closely mimicking the proof of Theorem 3 of Pincus (1991), the following theorem is immediate: for $r < \min_{j \neq k} |a_j - a_k|$, ApEn($m$, r) $= H(X_{m+1} \| X_1, \ldots, X_m)$; thus in this setting, ApEn is a conditional entropy. Observe that we do not assume that the process is mth-order Markov, that is, that we fully describe the process; we aggregate the mth-order marginal probabilities. The rate of entropy $= \lim_{n \to \infty} H(X_n \| X_1, \ldots, X_{n-1})$ is the discrete state analog of the K-S entropy. However, we cannot go from discrete to continuous state naturally as a limit; most calculations give ∞. As for differential entropy, there is no fundamental physical interpretation of conditional entropy (and no invariance; see Blahut, 1987, p. 243) in continuous state.

V. Representative Biological Applications

ApEn has recently been applied to numerous settings both within and outside biology. In heart rate studies, ApEn has shown highly significant differences in settings in which moment (mean, SD) statistics did not show clear group distinctions (Fleisher *et al.*, 1993; Kaplan *et al.*, 1991; Pincus and Viscarello, 1992; Pincus *et al.*, 1991; 1993) including analysis of aborted SIDS infants and of fetal distress. Within neuromuscular control, for example, ApEn showed that there was greater control in the upper arm and hand than in the forearm and fingers (Morrison and

Newell, 1996). In applications to endocrine hormone secretion time series data based on as few as $N = 60$ points, ApEn has shown vivid distinctions ($P < 10^{-10}$; nearly 100% sensitivity and specificity in each study) between normal and tumor-bearing subjects for GH (Hartman et al., 1994), adrenocorticotropin (ACTH) and cortisol (van den Berg et al., 1997), and aldosterone (Siragy et al., 1995), with the tumorals markedly more irregular, a pronounced and consistent gender difference in GH irregularity in both human and rat (Pincus et al., 1996b), highly significant differences between follicle stimulating hormone (FSH) and luteinizing hormone (LH) both in sheep (Pincus et al., 1998) and in humans (Pincus et al., 1997), and a positive correlation between advancing age and each of greater irregularity of (1) GH (Veldhuis et al., 1995) and of (2) LH and testosterone (Pincus et al., 1996a). We next discuss briefly the gender difference findings in GH, to further develop intuition for ApEn in an application context.

A. Sample Application: Gender Differences in GH Serum Dynamics

In two distinct human subject studies (employing, respectively, immunoradiometric assays and immunofluorimetric assays), women exhibited significantly greater irregularity than their male counterparts, $P < 0.001$ in each setting, with almost complete gender segmentation via ApEn in each context (Pincus et al., 1996b). ApEn likewise vividly discriminates male and female GH profiles in the adult intact rat (Gevers et al., 1998; Pincus et al., 1996b), $P < 10^{-6}$, with nearly 100% sensitivity and specificity (Fig. 3A). More remarkably, in rats that had been castrated prior to puberty, the ApEn of GH profiles in later adulthood is able to separate genetically male and female animals (Gevers et al., 1998). Among intact animals and rats treated prepubertally either with a long-acting GnRH agonist or surgical castration, the following rank order of ApEn of GH release emerged, listed from maximally irregular to maximally regular: intact female, GnRH-agonist-treated female, ovariectomized female, orchidectomized male, GnRH-agonist-treated male, and intact male animal (Gevers et al., 1998), illustrated in Fig. 1. ApEn was highly significantly different between the pooled groups of neutered female and neutered male animals, $P < 10^{-4}$, confirmed visually in Fig. 3B.

More broadly, this application to the rat studies indicates the clinical utility of ApEn. ApEn agrees with intuition, confirming differences that are visually "obviously distinct," as in the comparisons in Figs. 1A and F, intact male versus female rats. Importantly, ApEn can also uncover and establish graded and oftentimes subtle distinctions, as in comparisons of Figs. 1B-E, the neutered subject time series. Furthermore, these analyses accommodated both a point-length restriction of $N = 60$ samples (10-h dark period, 10-min sampling protocol) and a typically noisy environment (due to assay inaccuracies and related factors), representative of the types of constraints that are usually present in clinical and laboratory settings.

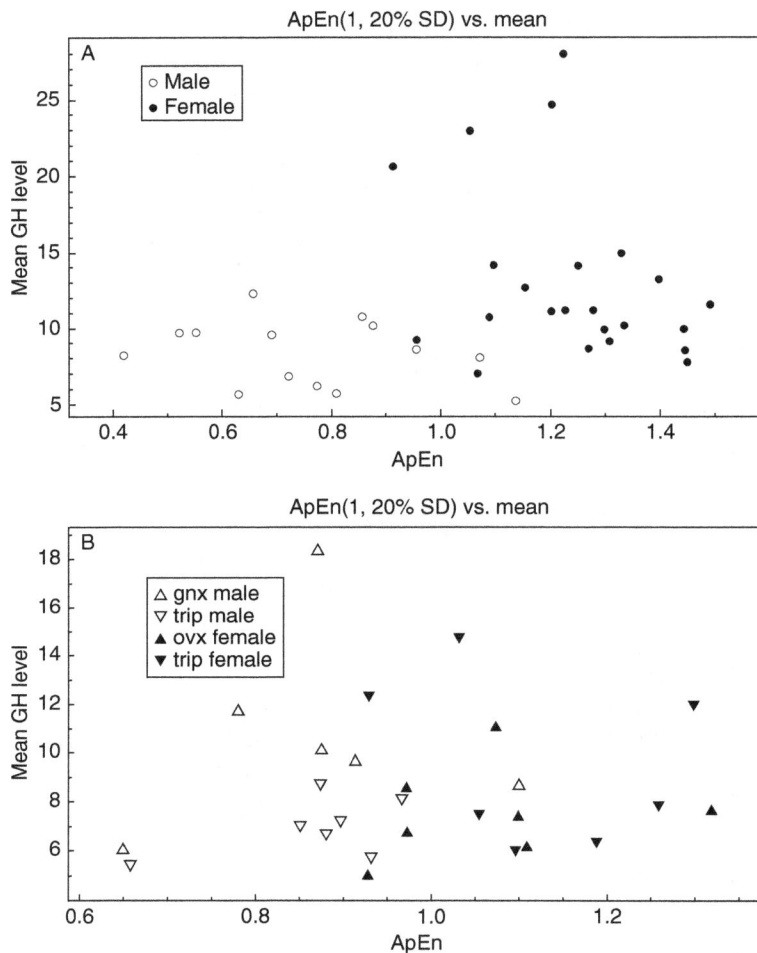

Fig. 3 Scatterplots of mean serum GH concentrations (ng/ml) versus ApEn(1, 20% SD) values in (A) individual intact male and female rats; (B) in surgically gonadectomized (gnx) or pharmocologically neutered (GnRH agonist triptorelin treatment) male and female rats.

VI. Relationship to Other Approaches

A. Feature Recognition Algorithms

The orientation of ApEn is to quantify the amount of regularity in time series data as a single parameter (number). This approach involves a different and complementary philosophy than do algorithms that search for particular pattern features in data. Representative of these latter algorithms are the pulse detection algorithms central to endocrine hormone analysis, which identify the number of peaks in pulsatile data and their locations (Urban *et al.*, 1988). When applied to clearly pulsatile data, such pulse detection algorithms have provided significant capability in the detection of abnormal hormone secretory patterns. However, such algorithms often ignore "secondary" features whose evolution may provide substantial information. For instance, ApEn will identify changes in underlying episodic behavior that do not reflect changes in peak occurrences or amplitudes, whereas the aforementioned pulse identification algorithms generally ignore such information. Also, ApEn can be applied to those signals in which the notion of a particular feature, such as a pulse is not at all clear, for example, an EEG time series. We recommend applying feature recognition algorithms in conjunction with ApEn when there is some physical basis to anticipate repetitive presence of the feature.

B. Statistics Related to Chaos

The historical development of mathematics to quantify regularity has centered around various types of entropy measures. Entropy is a concept addressing system randomness and predictability, with greater entropy often associated with more randomness and less system order. Unfortunately, there are numerous entropy formulations, and many entropy definitions cannot be related to one other (Pincus, 1991). K-S entropy, developed by Kolmogorov and expanded on by Sinai, allows one to classify *deterministic* dynamic systems by rates of information generation (Kolmogorov, 1958). It is this form of entropy that algorithms such as those given by Grassberger and Procaccia (1983a) and by Eckmann and Ruelle (1985) estimate. There has been keen interest in the development of these and related algorithms (Takens, 1983) in the last few years, as entropy has been shown to be a parameter that characterizes chaotic behavior (Shaw, 1981).

However, K-S entropy was not developed for statistical applications, and it has major debits in this regard. The original and primary motivation for K-S entropy was to handle a highly theoretical mathematics problem: determining when two Bernoulli shifts are isomorphic. In its proper context, this form of entropy is primarily applied by ergodic theorists to well-defined theoretical transformations, for which no noise and an infinite amount of "data" are standard mathematical assumptions. Attempts to utilize K-S entropy for practical data analysis represent out-of-context application, which often generates serious difficulties, as it does here. K-S entropy is badly compromised by steady, (even very) small amounts of noise, generally requires a vast amount of input data to achieve convergence (Ornstein and Weiss, 1990; Wolf *et al.*, 1985), and is usually infinite for stochastic (random) processes. Hence a "blind" application of the K-S entropy to practical time series will only evaluate system noise, not the underlying system properties. All of these disadvantages are key to the present context as most biological time series likely comprise both stochastic and deterministic components.

ApEn was constructed along lines thematically similar to those of K-S entropy, though with a different focus: to provide a widely applicable, statistically valid formula that will distinguish data sets by a measure of regularity (Pincus *et al.*, 1991; Pincus, 1991). The technical observation motivating ApEn is that if joint probability measures for reconstructed dynamics that describe each of two systems are different, then their marginal probability distributions on a fixed partition, given by conditional probabilities as in Eq. (4), are likely different. We typically need orders of magnitude fewer points to accurately estimate these marginal probabilities than to accurately reconstruct the "attractor" measure defining the process. ApEn has several technical advantages in comparison to K-S entropy for statistical usage. ApEn is nearly unaffected by noise of magnitude below r, the filter level; gives meaningful information with a reasonable number of data points; and is finite for both stochastic and deterministic processes. This last point gives ApEn the ability to distinguish versions of composite and stochastic processes from each other, whereas K-S entropy would be unable to do so.

Extensive literature exists about understanding (chaotic) deterministic dynamic systems through reconstructed dynamics. Parameters such as correlation dimension (Grassberger and Procaccia, 1983b), K-S entropy, and the Lyapunov spectrum have been much studied, as have been techniques to utilize related algorithms in the presence of noise and limited data (Broomhead and King, 1986; Fraser and Swinney, 1986; Mayer-Kress *et al.*, 1988). Even more recently, prediction (forecasting) techniques have been developed for chaotic systems (Casdagli, 1989; Farmer and Sidorowich, 1987; Sugihara and May, 1990). Most of these methods successfully employ embedding dimensions larger than $m = 2$, as is typically employed with ApEn. Thus in the purely *deterministic dynamic system* setting, for which these methods were developed, they reconstruct the probability structure of the space with greater detail than does ApEn. However, in the general (stochastic, especially correlated stochastic process) setting, the statistical accuracy of the aforementioned parameters and methods is typically poor; see Pincus (1991), Pincus and Singer (1996), Pincus (1995) for further elucidation of this operationally central point. Furthermore, the prediction techniques are no longer sensibly defined in the general context. Complex, correlated stochastic and composite processes are typically not evaluated, because they are not truly chaotic systems. The relevant point here is that because the dynamic mechanisms of most biological signals remain undefined, a suitable statistic of regularity for these signals must be more "cautious," to accommodate general classes of processes and their much more diffuse reconstructed dynamics.

Generally, changes in ApEn agree with changes in dimension and entropy algorithms for low-dimensional, deterministic systems. The essential points here, ensuring broad utility, are that (1) ApEn can potentially distinguish a wide variety of systems: low-dimensional deterministic systems, periodic and multiply periodic systems, high-dimensional chaotic systems, and stochastic and mixed (stochastic and deterministic) systems (Pincus and Keefe, 1992; Pincus, 1991); and (2) ApEn is applicable to noisy, medium-sized data sets, such as those typically encountered in biological time series analysis. Thus, ApEn can be applied to settings for which the

K-S entropy and correlation dimension are either undefined or infinite, with good replicability properties as discussed below. Evident, yet of paramount importance, is that the data length constraint is key to note; for example, hormone secretion time series lengths are quite limited by physical (maximal blood drawing) constraints, typically <300 points.

C. Power Spectra, Phase Space Plots

Generally, smaller ApEn and greater regularity correspond in the spectral domain to more total power concentrated in a narrow frequency range, in contrast to greater irregularity, which typically produces broader banded spectra with more power spread over a greater frequency range. The two opposing extremes are (1) periodic and linear deterministic models, which produce highly peaked, narrow-banded spectra, with low ApEn values; and (2) sequences of independent random variables, for which time series yield intuitively highly erratic behavior, and for which spectra are very broad banded, with high ApEn values. Intermediate to these extremes are autocorrelated processes, which can exhibit complicated spectral behavior. These autocorrelated aperiodic processes can be either stochastic or deterministic chaotic. In some instances, evaluation of the spectral domain will be insightful, when pronounced differences occur in a *particular* frequency band. In other instances, there is oftentimes more of an ensemble difference between the time series, both viewed in the time domain and in the frequency domain, and the need remains to encapsulate the ensemble information into a single value to distinguish the data sets.

Also, greater regularity (lower ApEn) generally corresponds to greater ensemble correlation in phase space diagrams. Such diagrams typically display plots of some system variable $x(t)$ versus $x(t - T)$, for a fixed "time lag" T. These plots are quite in vogue, in that they are often associated with claims that correlation, in conjunction with aperiodicity, implies chaos. A cautionary note is strongly indicated here. The labeling of bounded, aperiodic, yet correlated output as *deterministic* chaos has become a false cognate. This is incorrect; application of Theorem 6 in Pincus (1992) proves that any n-dimensional steady-state measure arising from a deterministic dynamic system model can be approximated to arbitrary accuracy by that from a *stochastic* Markov chain. This then implies that any given phase space plot could have been generated by a (possibly correlated) stochastic model. The correlation seen in such diagrams is typically real, as is the geometric change that reflects a shift in ensemble process autocorrelation in some comparisons. However, these observations are entirely distinct from any claims regarding underlying model form (chaos vs. stochastic process). Similarly, in power spectra, decreasing power with increasing frequency (oftentimes labeled $1/f$ decay) is also a property of process correlation, rather than underlying determinism or chaos (Pincus, 1994).

VII. Mechanistic Hypothesis for Altered Regularity

It seems important to determine a unifying theme suggesting greater signal regularity in a diverse range of complicated neuroendocrine systems. We would hardly expect a single mathematical model, or even a single family of models, to govern a wide range of systems; furthermore, we would expect that *in vivo*, each physiologic signal would usually represent the output of a complex, multinodal network with both stochastic and deterministic components. Our mechanistic hypothesis is that in a variety of systems, greater regularity (lower ApEn) corresponds to greater component and subsystem autonomy. This hypothesis has been mathematically established via analysis of several very different, representational (stochastic and deterministic), mathematical model forms, conferring a robustness to model form of the hypothesis (Pincus and Keefe, 1992; Pincus, 1994). Restated, ApEn typically increases with greater system coupling and feedback, and greater external influences, thus providing an explicit barometer of autonomy in many coupled, complicated systems.

Many endocrine hormone findings, including those indicated above, suggest that hormone secretion pathology usually corresponds to greater signal *irregularity*. Accordingly, a possible mechanistic understanding of such pathology, given this hypothesis, is that healthy, normal endocrine systems function best as relatively closed, autonomous systems (marked by regularity and low ApEn values), and that accelerated feedback and too many external influences (marked by irregularity and high ApEn values) corrupt proper endocrine system function.

It would be very interesting to attempt to experimentally verify this hypothesis in settings where some of the crucial network nodes and connections are known, via appropriate interventions to normal neuroendocrine (more generally, biological network) flow, coupled with signal analysis at one or more output sites.

VIII. Cross–ApEn

Cross-ApEn is a measure of asynchrony between two time series (Pincus and Singer, 1996; Pincus *et al.*, 1996). As for ApEn, it is a two-parameter family of statistics, with m and r taking the same meaning as in the ApEn setting, herein fixed for application to the paired time series $\{u(i)\}$, $\{v(i)\}$. Cross-ApEn measures, within tolerance r, the (conditional) regularity or frequency of v patterns similar to a given u pattern of window length m. It is typically applied to standardized u and v time series. Greater asynchrony indicates fewer instances of (sub)pattern matches, quantified by larger cross-ApEn values. Figure 4, taken from a recent study of paired ACTH-cortisol dynamics in Cushing's disease (Roelfsema *et al.*, 1988), illustrates the cross-apEn quantification, with greater ACTH-cortisol secretory asynchrony in the diseased subject, compared to the control.

Cross-ApEn is generally applied to compare sequences from two distinct yet intertwined variables in a network. Thus, we can directly assess network, and not

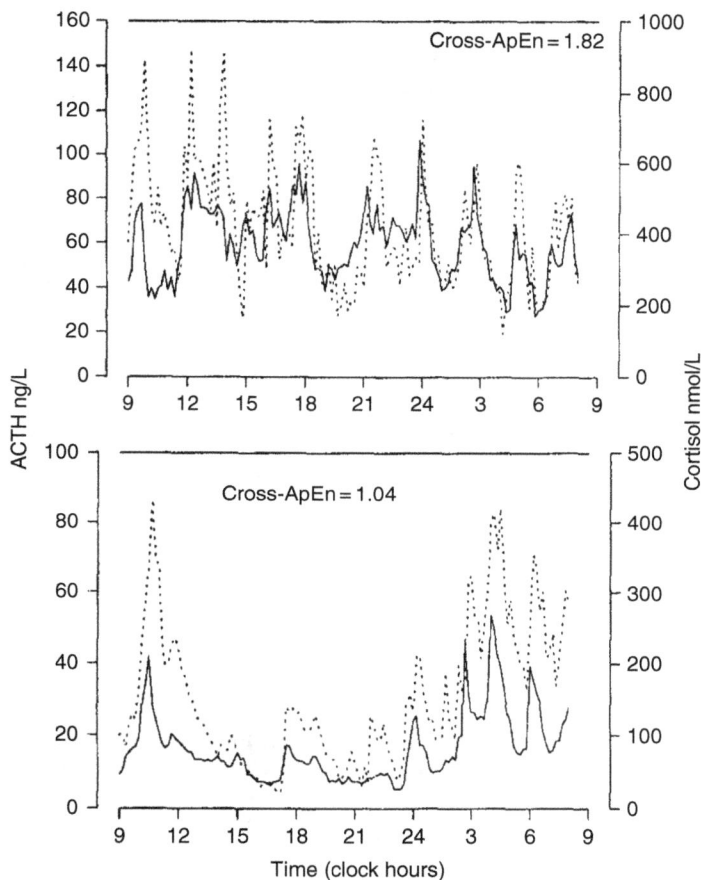

Fig. 4 Plasma concentrations of ACTH (dotted line) and cortisol (continuous line) in a female patient with Cushing's disease (upper panel) and a control subject (lower panel), each sampled at 10-min intervals for 24 h.

just nodal, evolution under different settings, for example, to evaluate uncoupling and/or changes in feedback and control. Hence, cross-ApEn facilitates analyses of output from myriad complicated networks, avoiding the requirement to model the underlying system. This is especially important as accurate modeling of (biological) networks is often nearly impossible—even a full description of all system nodes and pathways is typically unknown in most biologic systems, to say nothing of subsequent good mathematical approximations of the resultant internetwork dynamics. The key point, similarly for ApEn, is that full model specification is not required to realize an effective discrimination strategy. Furthermore, of course, there is a paucity of general *multi*variate time series statistical tools, as discussed further below.

In addition to the evident means to potentially discriminate network aspects of systems, cross-ApEn allows us to now address the following critical, yet generic, network issue: Are system changes primarily nodal (one-variable) or, rather, pathway or central control alterations (multivariate)? An answer to this question is not only essential to basic system understanding, but also a prime determinant in choosing, for example, therapy/intervention strategies to attempt to restore pathobiologic milieus to more normative settings. Also, given multiple-node networks, we can successively probe pairwise, via cross-ApEn, to determine the weakest or altered (paired) links in the system. Furthermore, in cross-ApEn applications, paired (*u*, *v*) signal inputs can be within known subnetworks, for example, the FSH-LH hormone secretory system, or alternatively, from less obviously related, broader networks, for example, the EEG-LH system. Analysis of this latter setting allows us to address obliquely central control changes, in instances in which direct evaluation of the same would be effectively impossible. Concomitantly, to generate the *u*, *v* paired time series, one can utilize (quite) distinct sampling frequencies for each series. The technical point is that because we are interested in discrimination, rather than full model specification of the joint measure, we only require a fixed (common) protocol applied throughout to all data sets in a study.

The precise definition, introduced in Pincus and Singer (1996), Definition 5, given next, is thematically similar to that for ApEn.

A. Definition of Cross-ApEn

Let $u = [u(1), u(2), \ldots, u(N)]$ and $v = [v(1), v(2), \ldots, v(N)]$ be two length - N sequences. Fix input parameters m and r. Form vector sequences $x(i) = [u(i), u(i+1), \ldots, u(i+m-1)]$ and $y(j) = [v(j), v(j+1), \ldots, v(j+m-1)]$ from u and v, respectively. For each $i \le N - m + 1$, set $C_i^m(r)(v\|u) =$ (number of $j \le N - m + 1$ such that $d[x(i), y(j)] \le r)/(N - m + 1)$, where $d[x(i), y(j)] = \max_{k=1,2,\ldots,m} [|u(i+k-1) - v(j+k-1)|]$, that is, the maximum difference in their respective scalar components. The $C_i^m(r)$ values measure within a tolerance r the regularity, or frequency, of (*v*–) patterns similar to a given (*u* –) pattern of window length m.

Then define $\Phi^m(r)(v\|u)$ as the average value of $\ln C_i^m(r)(v\|u)$, and, finally, define cross-ApEn$(m, r, N)(v\|u) = \Phi^m(r)(v\|u) - \Phi^{m+1}(r)(v\|u)$.

Typically, we apply cross-ApEn with $m = 1$ and $r = 0.2$ to *standardized u* and *v* time series data, that is, for each subject, we apply cross-ApEn(1, 0.2) to the $\{u^*(i), v^*(i)\}$ series, where $u^*(i) = [u(i) - \text{mean } u]/\text{SD } u$ and $v^*(i) = [v(i) - \text{mean } v]/\text{SD } v$. This standardization, in conjunction with the choice of m and r, ensures good replicability properties for cross-ApEn for the data lengths to be studied. To establish a theoretical statistical validity of cross-ApEn as so employed, we studied a range of two-variable vector AR(2) processes, and several types of coupled two-variable analogs of the "variable lag" process described below, for each of which we applied cross-ApEn(1, 0.2) to standardized time series (*x*, *y* pair) outputs, 50 replicates of $N = 150$-point data lengths per process. For each process studied, SD (cross-ApEn) was ≤ 0.06, the SD calculated from the cross-ApEn values from the

50 replicates; this imparts reasonable replicability properties similar to that for ApEn (Pincus and Huang, 1992; Pincus and Keefe, 1992; Pincus *et al.*, 1996a). This degree of reproducibility is not unexpected as, qualitatively, cross-ApEn is a parameter that aggregates low-order, two-variable joint distributions at a moderately coarse resolution (determined by *r*).

As a representative example of application of cross-ApEn to biological data, we now consider the following study.

B. LH-T Study, Men

In recent years, many studies have been concerned with LH and testosterone (T) serum concentration time series in both younger and older men, both to better understand the physiology of reproductive capacity and, clinically, to assess, for example, a loss of libido, or decreased reproductive performance. Furthermore, there is considerable interest in determining whether a hypothesized male climacteric (or so-called andropause) at least partially analogous to menopause in the woman exists and, if so, in what precise sense. While considerable insight has already been gained from many studies, nontrivial controversies remain concerning several classes of findings, including primary determinations of whether overall mean levels of LH and T decrease with increasing age.

A study was performed to determine possible secretory irregularity shifts with aging within the LH-T axis (Pincus *et al.*, 1996a). Serum concentrations were derived for LH and T in 14 young (21–34 yr) and 11 older (62–74 yr) healthy men. For each subject, blood samples were obtained at frequent (2.5-min) intervals during a sleep period, with an average sampling duration of 7 h. Although mean (and SD) of LH and T concentrations were indistinguishable in the two age groups, for each of LH and T, older men have consistently and highly significantly more irregular serum reproductive-hormone concentrations than younger men: for LH, aged subjects had greater ApEn values (1.525 ± 0.221) than younger individuals (1.207 ± 0.252), $P < 0.003$, while for testosterone, aged subjects had greater ApEn values (1.622 ± 0.120) than younger counterparts (1.384 ± 0.228), $P < 0.004$.

Probably a yet mechanistically more important finding in this study (Pincus *et al.*, 1996a) was seen via cross-ApEn analysis. Cross-ApEn was applied to the paired LH-T time series; statistically, even more vividly than for the irregularity (ApEn) analyses, older subjects exhibited greater cross-ApEn values (1.961 ± 0.121) compared to younger subjects (1.574 ± 0.249), $P < 10^{-4}$, with nearly 100% sensitivity and specificity, indicating greater LH-T asynchrony in the older group (Fig. 5). Moreover and notably, no significant LH-T linear correlation (Pearson R) differences were found between the younger and older cohorts, $P > 0.62$ (Fig. 5). Several possibilities for the source of the erosion of LH-testosterone synchrony are discussed (Pincus *et al.*, 1996a), although a clear determination of this source awaits future study. Mechanistically, the results implicate (LH-T) network uncoupling as marking male reproductive aging, for which we now have several quantifiable means to assess.

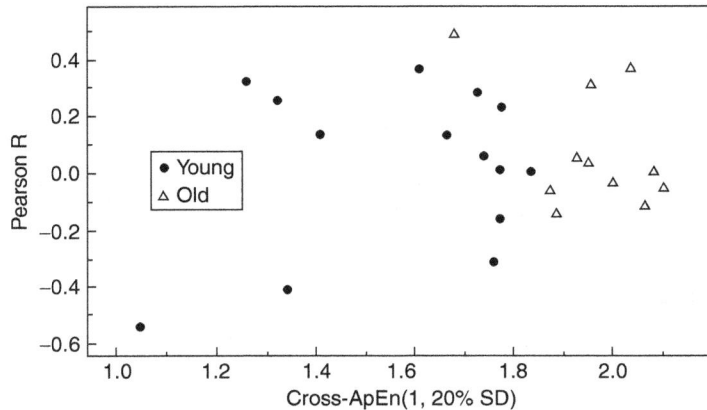

Fig. 5 Individual subject cross-ApEn values versus cross-correlation (Pearson R), applied to the joint LH-testosterone time series in healthy young (●) versus older men (Δ).

As another endocrinologic example of cross-ApEn utility, in a study of 20 Cushing's disease patients versus 29 controls (Roelfsema *et al.*, 1988), cross-ApEn of ACTH-cortisol was greater in patients (1.686 ± 0.051) than in controls (1.077 ± 0.039), $P < 10^{-15}$, with nearly 100% sensitivity and specificity, suggesting compromise of hormonal pathways and feedback control in diseased subjects, atop that previously seen for more localized nodal secretory dynamics of each hormone individually (van den Berg *et al.*, 1997). Figure 4 displays representative serum profiles from this study. Additionally, healthy men and women showed progressive erosion of bihormonal ACTH-cortisol synchrony with increased aging (Roelfsema *et al.*, 1988) via cross-ApEn, similar to the LH-T erosion of synchrony in men noted above, suggesting that increased cross-ApEn (greater asynchrony) of paired secretory dynamics is an ubiquitous phenomenon with advancing age.

C. Complementarity of ApEn and Cross-ApEn to Correlation and Spectral Analyses

Mathematically, the need for ApEn, and particularly for cross-ApEn, is clarified by considering alternative parameters that might address similar concepts. In comparing two distinct signals or variables (e.g., to assess a degree of synchrony), primary parameters that one might employ include the cross-correlation function (including Pearson R) and the cross-spectrum (Chatfield, 1989), with single-variable counterparts, the autocorrelation function and the power spectrum. Evaluation of these parameters often is insightful, but with relatively short length data sets, statistical estimation issues are nontrivial and, moreover, interpretation of the sample cross-correlation function is highly problematic, unless one employs a model-based prefiltering procedure (Chatfield, 1989, p. 139). Furthermore, "standard" spectral estimation methods such as the FFT (fast Fourier transform) can be shown to be inconsistent and/or so badly biased that findings may be qualitatively incorrect, especially in the presence of outliers and nonstationarities. This is vividly demonstrated by Thomson (1990) who recently developed a superior multiple-data-window technique with major advantages compared to other spectral estimation techniques (Kuo et al., 1990; Thomson, 1990). These difficulties are mirrored in the cross-spectrum, in addition to an often serious bias in estimation of coherency in short series.

Most importantly, the autocorrelation function and power spectrum, and their bivariate counterparts, are most illuminating in linear systems, for example, SARIMA (seasonal autoregressive integrated moving average) models, for which a rich theoretical development exists (Box and Jenkins, 1976). For many other classes of processes, these parameters are often relatively ineffective at highlighting certain model characteristics, even apart from statistical considerations. To illustrate this point, consider the following simple model, which we denote as a "variable lag" process: this consists of a series of quiescent periods, of variable length duration, interspersed with identical positive pulses of a fixed amplitude and frequency. Formally, we recursively define an integer time-valued process denoted VarLag whose ith epoch consists of (a quiescent period of) values $= 0$ at times $t_{i-1} + 1, t_{i-1} + 2, \ldots, t_{i-1} + \text{lag}_i$, immediately followed by the successive values $\sin(\pi/6)$, $\sin(2\pi/6), \sin(3\pi/6), \sin(4\pi/6), \sin(5\pi/6), \sin(6\pi/6) = 0$ at the next 6 time units, where lag_i is a random variable uniformly distributed on (randomly chosen between) the integers between 0 and 60, and t_{i-1} denotes the last time value of the $(i-1)$st sine-pulse. Figure 6A displays representative output from this process, with Fig. 6B giving a closer view of this output near time $t = 400$. The power spectrum and autocorrelation function calculations shown in Figs. 6C and E were calculated from a realization of length $N = 100,000$ points. (The somewhat coarse sampling of the pulse in the above process definition was chosen to approximate typical sampling resolution in actual clinical studies.)

Processes consisting of alternately quiescent and active periods would seem reasonable for biologists to consider, as they appear to model a wide variety of phenomena. However, within mathematics, such processes with a variable quiescent period are not commonly studied. To the biologist, output from the above model would be considered smoothly pulsatile, especially with the identical pulses; the variable lag process would be most readily distinguished from its constant lag counterpart (for which $\text{lag}_i = 30$ time units for all i) via a decidedly positive SD for the interpulse duration time series, in the variable lag setting, as opposed to $SD = 0$ (constant interpulse duration) in the constant lag setting. The essential point here, however, is that for VarLag, the power spectrum and autocorrelation function somewhat confound, as seen in Figs. 6C and E. On the basis of these figures alone, the pulsatile nature of the time series realizations is hardly evident, and for all $k \geq 6$, the autocorrelation coefficient r_k at lag k is insignificantly different from 0. In contrast, the power spectrum and autocorrelation function confirm the periodicity of the constant lag analog, shown in Figs. 6D and F, as expected. Significantly, the issues here are in the parameters, rather than in statistical

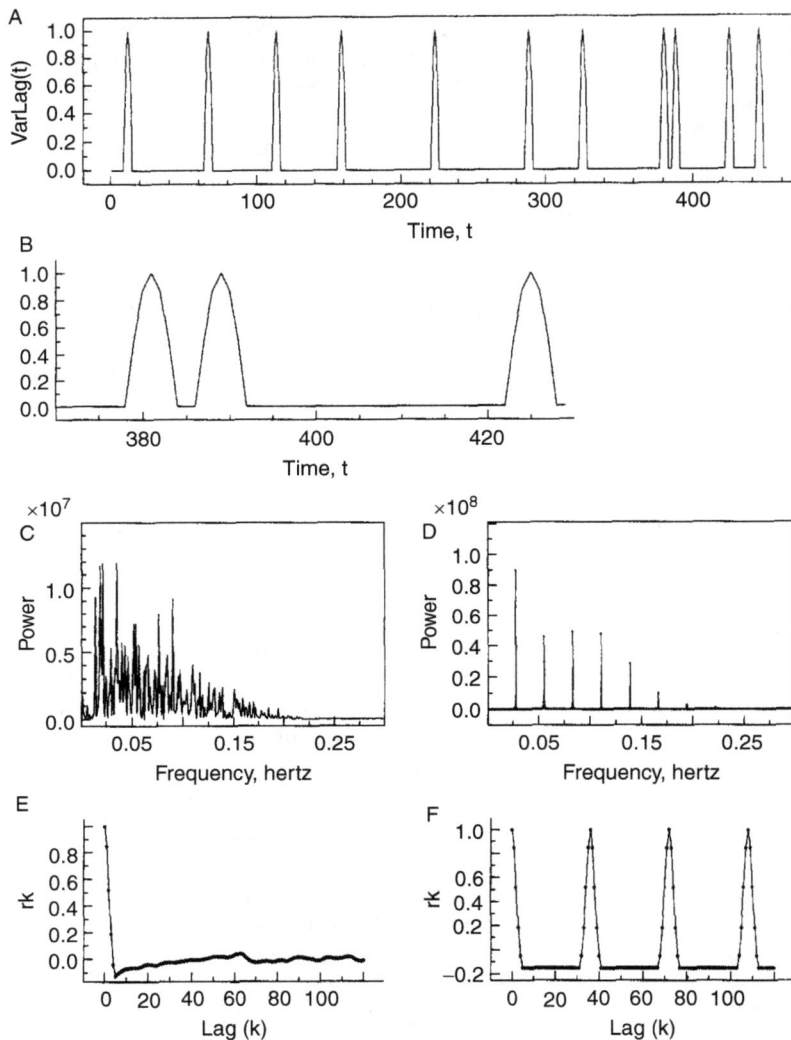

Fig. 6 (A) Representative time series for a "variable lag" sine-wave process denoted VarLag (see text for formal definition); (B) close-up view of (A), near time $t = 400$; (C) power spectrum for VarLag; (D) power spectrum for a constant (fixed) lag analog of VarLag; (E) autocorrelogram corresponding to (C); (F) autocorrelogram corresponding to (D). Parts (C)-(F) are all derived from time series of length $N = 100,000$ points.

inadequacies based on an insufficiently long output, or on artifacts (outliers), as Figs. 6C-F were derived from calculations based on 100,000 points from a purely theoretical model.

Similar limitations of the spectra and autocorrelation function are inherent to wide classes of processes. From a general mathematical framework, we can construct large classes of variable lag processes simply by considering point processes (Bremaud, 1981), in which we replace the "point" occurrence by a pulse occurrence, the pulse itself of either a fixed or variable form. The associated counting process could be of any character, and need not be so special as Poisson or renewal (as in the above example). Also, variable lags between events to be compared are the normative case in nonlinear (deterministic and stochastic) differential equations, in Poisson clumping models (Aldous, 1989), and in output variables in typical (adaptive) control theory models and queueing network models. Notably, for many two-dimensional analogs of variable lag processes, and indeed for many two-dimensional systems in which no small set of dominant frequencies encapsulates most of the total power, the cross-spectrum and the cross-correlation function often will similarly fail to highlight episodicities in the underlying model and data, and thus fail to highlight concomitant changes to such episodic components.

In contrast to the autocorrelation function and spectral differences between the above variable lag and constant lag processes, the respective ApEn(1, 20% SD) values for the two processes are in close agreement: mean ApEn = 0.195 for the variable lag process, while ApEn = 0.199 for the constant lag setting. This agreement in ApEn values manifests the primary requirement of matching (sub)patterns within data, while relaxing the requirement of a dominant set of frequencies at which these subpatterns occur. The two-variable analog of ApEn, given by cross-ApEn, similarly enables one to assess synchrony in many classes of models. It thus should not be surprising that in many studies, for example, the LH-T study (Pincus *et al.*, 1996a), cross-correlation (Pearson *R*) does not show significant group differences, whereas cross-ApEn does (as in Fig. 5).

It should be emphasized, nonetheless, that Figs. 6C-F neither invalidate spectral power and (lagged) autocorrelation calculations, nor do they violate a properly oriented intuition. The broad-banded spectrum in Fig. 6C, and the negligible lagged autocorrelation in Fig. 6E for lag $\geq$ 6 time units, primarily reflect the independent, identically distributed, relatively broad distribution of the variable lag$_i$. Visually this conforms to viewing Fig. 6A from afar, in effect (nearly) ignoring the nature of each pulse, and instead *de facto* primarily focusing on the "random" timing of the peaks as the process of interest. The viewpoint taken by ApEn is thus complementary to the spectrum and correlogram, more *de facto* focusing on (close-up) similarities between active pulses, for example, from the perspective given in Fig. 6B, while in effect nearly ignoring the nature of the quiescent epoch aspect of the process. The utility of ApEn and cross-ApEn to biologists is based on the recognition that in many settings, changes in the episodic character of the *active* periods within time series appear to mark physiologic and pathophysiologic changes—thus there is a concomitant need for quantitative methods that primarily address this perspective, for example, ApEn and cross-ApEn.

======   ## IX. Toward More Faithful Network Modeling

In modeling general classes of real biologic networks, we anticipate that any single homogeneous model form, such as deterministic dynamic systems, ARMA models, stochastic differential equations, or Markov chains or processes, is inadequate. At the least, we would expect faithful models in many settings to incorporate queueing network and (adaptive) control theory considerations, as well as the likelihood that different components in a network are defined by polymorphous and distinct mathematical model forms. Queueing models arise naturally in multinode network analysis with interconnections; control models arise from considering the brain (or some focal component) as an intelligent central processor, possibly altering system characteristics on the basis of, for example, a threshold response. Queueing theory has developed largely within communication (traffic) theory (Gross and Harris, 1985) and computer network analysis (Allen, 1990; Jackson, 1957), while control theory has developed toward optimizing performance in engineering systems (Fleming and Rishel, 1975). Notably, analytical developments from these fields may not be directly suitable to physiologic network modeling, not surprisingly as these fields were not driven by biological context. Two physiologically motivated problems within these fields that seem worthy of significant effort are to describe the behavior of (1) queueing networks in which some nodes are coupled oscillators, and (2) (adaptive) control systems in which there is a balking probability p with which the control strategy is not implemented. Problem (2) could model some diseases, in which messages may not reach the controller or the controller may be too overwhelmed to respond as indicated.

Several "decision theoretic" modeling features that fall under the umbrella of queueing theory seem especially appropriate (and timely) to impose on many biological networks, to achieve faithful characterizations of true network protocols, both qualitatively as well as quantitatively. These aspects of traffic theory include (1) "broadcast" signaling (of a central controller); (2) priority service; (3) alternative routing hierarchies; (4) "finite waiting areas" for delayed messages, incorporating the possibility (and consequences) of "dropped" or lost messages; and (5) half-duplex transmission, in which on a given pathway between two sources, only one source at a time may use the transmission pathway. In addition, we must always clarify the "network topology" or routing configuration among nodes in a network, that is, determine which pairs of nodes have (direct) pathways to one another, before beginning to address quantitative specifications of signal transmission along the putative pathways. All of these features can be described quantitatively, via decision-theoretic point processes and, typically, resultant network performance is then evaluated by large-scale numerical programs that "simulate" the stochastic environment (Law and Kelton, 1991). General versions of such programs require considerable expertise and time to write, and are commercially available from a few sources, though regrettably, these are quite expensive to procure, and are usually targeted to specialists. Furthermore, only in very specialized settings do the mathematical descriptions of the decision-theoretic constraints allow for purely analytic (as opposed to simulation, that is, so-called "numerical Monte Carlo" methods) solutions. This is quite possibly the reason why this essential yet specialized branch of applied probability theory is relatively unknown.

The above perspective strongly motivates the requirement that for effective and broadest utility, statistics developed for general network analysis be model independent, or at least provide robust qualitative inferences across a wide variety of network configurations. The observation that both ApEn and cross-ApEn are model independent, that is, functionals of the presented sequences (time series), and are not linked to a prescribed model form, fits squarely with this perspective.

X. Spatial (Vector) ApEn

A spatial (vector) version of ApEn was recently developed to quantify and grade the degrees of irregularity of planar (and higher dimensional) arrangements (Singer and Pincus, 1998). Spatial ApEn appears to have considerable potential, both theoretically and empirically, to discern and quantify the extent of changing patterns and the emergence and dissolution of traveling waves, throughout multiple contexts within both biology and chemistry. This is particularly germane to the detection of subtle or "insidious" structural differences among arrays, even where clear features or symmetries are far from evident.

One initial application of spatial ApEn will facilitate an understanding of both its potential utility and, simultaneously, of precisely what the quantification is doing. In Singer and Pincus (1998), we clarified and corrected a fundamental ambiguity (flaw) in R. A. Fisher's specification of experimental design (Fisher, 1925; 1935). Fisher implicitly assumed throughout his developments that all Latin squares (n row $\times$ n column arrangements of n distinct symbols where each symbol occurs once in each row and once in each column) were equally and maximally spatially random, and subsets of such Latin squares provided the underpinnings of experimental design. In the example below, even in the small sized 4×4 Latin square case, we already see that spatial ApEn quantifies differences among the candidate squares. (We then proposed an experimental design procedure based on *maximally irregular* Latin squares, eliminating the flaw (Singer and Pincus, 1998).

The precise definition of spatial ApEn is provided as Definition 1 in Singer and Pincus (1998). Thematically, again, it is similar to that for ApEn, both in the form of comparisons (determining the persistence of subpatterns to matching subpatterns), and in the input specification of window length m and *de facto* tolerance width r. The critical epistemologic novelty is that in the planar and spatial case, given a multidimensional array A and a function u on A (spatial time series), we specify a vector direction $\mathbf{v}$, and consider irregularity in A along the vector direction $\mathbf{v}$. We denote this as vector-ApEn$_\mathbf{v}$ $(m, r)(u)$; in instances in which the array values are discrete, for example, integers, as in the Latin square example below, we often set r to 0, thus monitoring precise subpattern matches, and suppress r in the vector-ApEn notation, with the resultant quantity denoted vector-ApEn$_v(m)$ (u). Descriptively, vector-ApEn$_v(m)(u)$ compares the logarithmic frequency of matches of blocks of length m (for $m \geq 1$) with the same quantity for blocks of length $m + 1$. Small values of vector-ApEn imply strong regularity, or persistence, of patterns in u in the vector direction v, with the converse interpretation for large values. The vector direction v designates arrangements of points on which the irregularity of u is specified, *a priori*, to be of particular importance. For example, if $v = (0, 1)$, then vector-ApEn measures irregularity along the rows of A, and disregards possible patterns, or the lack thereof, in other directions; $v = (1, 0)$ focuses on column irregularity; and $v = (1, 2)$ or $(2, 1)$ or $(-1, 2)$ emphasizes knight's move (as in chess) patterns. In typical applications, it is necessary to guarantee irregularity in two or more directions simultaneously. This requires evaluation of vector-ApEn for a set V of designated vectors. For example, simultaneous row, column, and diagonal irregularity assessment entails calculation of vector-ApEn for all elements v in $V = \{(1, 0); (0, 1); (1, -1); (1, 1)\}$.

Example 1 illustrates vector-ApEn for four Latin squares, and as noted above, also clarifies the remarks concerning Fisher's ambiguity in the specification of experimental design.

Example 1

Consider the following four 4×4 Latin squares.

A	B	C	D
1 2 3 4	1 2 3 4	1 2 3 4	1 2 3 4
2 3 4 1	3 4 1 2	4 3 2 1	3 1 4 2.
3 4 1 2	4 3 2 1	3 1 4 2	2 4 1 3
4 1 2 3	2 1 4 3	2 4 1 3	4 3 2 1

For A, vector-ApEn$_{(1,0)}(1)$ = vector-ApEn$_{(0,1)}(1)$ = 0; for B, vector-ApEn$_{(1,0)}(1)$ = vector-ApEn$_{(0,1)}(1)$ = 0.637; for C, vector-ApEn$_{(1,0)}(1)$ = 0.637, and vector-ApEn$_{(0,1)}(1)$ = 1.099; and for D, vector-ApEn$_{(1,0)}(1)$ = vector-ApEn$_{(0,1)}(1)$ = 1.099. These calculations manifest differing extent of feature replicability in the (1, 0) and (0, 1) directions, with A quite regular in both directions, B intermediately irregular in both directions, C maximally irregular in rows, yet intermediate in columns, and D maximally irregular in both rows and columns. Alternatively, in A, for example, in rows, there are three occurrences each of four pairs [(1, 2), (2, 3), (3, 4), and (4, 1)], and no occurrences of the other eight possible pairs. In B, in rows, four pairs occur twice [(1, 2), (2, 1), (3, 4), and (4, 3)], while four pairs occur once [(1, 4), (2, 3), (3, 2), and (4, 1)]. In D, in rows, each of the 12 pairs (i, j), $1 \leq i, j \leq 4$, $i \neq j$, occur precisely once. (Similar interpretation follows readily for columns.)

Several broad application areas illustrate the proposed utility of vector-ApEn to frequently considered settings within biology and chemistry. First, we anticipate that vector-ApEn will bear critically on image and pattern recognition determinations (Grenander, 1993), to assess the degree of repeatability of prescribed features. Sets of base atoms would be shapes of features of essential interest; moreover, these can be redefined (as indicated in Singer and Pincus, 1998) either on the same scale as the original atoms, or on a much larger scale, thus providing a more macroscopic assessment of spatial irregularity.

Also, many models within physics and physical chemistry are lattice-based systems, for example, the nearest neighbor Ising model and the classical Heisenberg model, which have been employed to model a magnet (via spin), a lattice gas, alloy structure, and elementary particle interactions (Israel, 1979). Determining relationships between changes in vector-ApEn in these models and physical correlates would seem highly worthwhile, either theoretically or experimentally. Also, within solid-state physical chemistry, we speculate that grading the extent of array disorder may prove useful in assessing or predicting (1) crystal and alloy strength and/or stability under stresses; (2) phase transitions, either liquid-to-gas, solid-to-liquid, or frigid-to-superconductive; and (3) performance characteristics of semiconductors.

Lastly, the analysis of traveling waves oftentimes requires a quantification of subtle changes, particularly as to the extent of formation and, conversely, the extent of dissolution or dissipation of wave fronts, above and beyond an identification of primary wave "pulses" and resultant statistical analyses. Although considerable signal-to-noise analysis methodology has been developed for and applied to this setting, to clarify wave fronts, in the ubiquitous instances where the extent of insidious or subordinate activity is the primary feature of interest, a critical and further assessment of the wave patterns is required, to which vector-ApEn should readily apply, both in two- and three-dimensional settings. This recognition may be particularly critical near the genesis of an upcoming event of presumed consequence. One representative, quite important application of this perspective is to (atrial) fibrillation and arrhythmia detection within cardiac physiology.

XI. Summary and Conclusion

The principal focus of this chapter has been the description of both ApEn, a quantification of serial irregularity, and of cross-ApEn, a thematically similar measure of two-variable asynchrony (conditional irregularity). Several properties of ApEn facilitate its utility for biological time series analysis: (1) ApEn is nearly unaffected by noise of magnitude below a *de facto* specified filter level; (2) ApEn is robust to outliers; (3) ApEn can be applied to time series of 50 or more points, with good reproducibility; (4) ApEn is finite for stochastic, noisy deterministic, and composite (mixed) processes, the last of which being likely models for complicated biological systems; (5) increasing ApEn corresponds to intuitively increasing process complexity in the settings of (4); and (6) changes in ApEn have been shown mathematically to correspond to mechanistic inferences concerning subsystem autonomy, feedback, and coupling, in diverse model settings. The applicability to medium-sized data sets and general stochastic processes is in marked contrast to capabilities of "chaos" algorithms such as the correlation dimension, which are properly applied to low-dimensional iterated deterministic dynamical systems. The potential uses of ApEn to provide new insights in biological settings are thus myriad, from a complementary perspective to that given by classical statistical methods.

ApEn is typically calculated by a computer program, with a FORTRAN listing for a "basic" code referenced above. It is imperative to view ApEn as a family of statistics, each of which is a relative measure of process regularity. For proper implementation, the two input parameters m (window length) and r (tolerance width, *de facto* filter) must remain fixed in all calculations, as must N, the data length, to ensure meaningful comparisons. Guidelines for m and r selection are indicated above. We have found normalized regularity to be especially useful, as in the GH studies discussed above; "r" is chosen as a fixed percentage (often 20%) of the subject's SD. This version of ApEn has the property that it is decorrelated from process SD—it remains unchanged under uniform process magnification, reduction, and translation (shift by a constant).

Cross-ApEn is generally applied to compare sequences from two distinct yet interwined variables in a network. Thus, we can directly assess network, and not just nodal evolution, under different settings—for example, to directly evaluate uncoupling and/or changes in feedback and control. Hence, cross-ApEn facilitates analyses of output from myriad complicated networks, avoiding the requirement to fully model the underlying system. This is especially important as accurate modeling of (biological) networks is often nearly impossible. Algorithmically and insofar as implementation and reproducibility properties are concerned, cross-ApEn is thematically similar to ApEn.

Furthermore, cross-ApEn is shown to be complementary to the two most prominent statistical means of assessing multivariate series, correlation and power spectral methodologies. In particular, we highlight, both theoretically and by case study examples, the many physiological feedback and/or control systems and models for which cross-ApEn can detect significant changes in bivariate asynchrony, yet for which cross-correlation and cross-spectral methods fail to clearly highlight markedly changing features of the data sets under consideration.

Finally, we introduce spatial ApEn, which appears to have considerable potential, both theoretically and empirically, in evaluating multidimensional lattice structures, to discern and quantify the extent of changing patterns, and for the emergence and dissolution of traveling waves, throughout multiple contexts within biology and chemistry.

References

Aldous, D. (1989). "Probability Approximations via the Poisson Clumping Heuristic," Springer-Verlag, Berlin.

Allen, A. O. (1990). "Probability, Statistics, and Queueing Theory: With Computer Science Applications," 2nd edn., Academic Press, San Diego.

Blahut, R. E. (1987). "Principles and Practice of Information Theory," pp. 55–64. Addison-Wesley, Reading, MA.

Box, G. E. P., and Jenkins, G. M. (1976). "Time Series Analysis: Forecasting and Control," Holden-Day, San Francisco, CA.

Bremaud, J. P. (1981). "Point Processes and Queues: Martingale Dynamics," Springer-Verlag, New York.

Broomhead, D. S., and King, G. P. (1986). *Physica D* **20**, 217.

Bruhn, J., Ropcke, H., Rehberg, B., Bouillon, T., and Hoeft, A. (2000). *Anesthesiology* **93**, 981.

Casdagli, M. (1989). *Physica D* **35**, 335.

Cavanaugh, J. T., Guskiewicz, K. M., Giuliani, C., Marshall, S., Mercer, V. S., and Stergiou, N. (2006). *J. Athl. Train.* **41**, 305.

Chatfield, C. (1989). "The Analysis of Time Series: An Introduction," 4th edn. Chapman and Hall, London.

Eckmann, J. P., and Ruelle, D. (1985). *Rev. Mod. Phys.* **57**, 617.

Farmer, J. D., and Sidorowich, J. J. (1987). *Phys. Rev. Lett.* **59**, 845.

Fisher, R. A. (1925). "Statistical Methods for Research Workers," Oliver & Boyd, Edinburgh, UK.

Fisher, R. A. (1935). "The Design of Experiments," Oliver & Boyd, Edinburgh, UK.

Fleisher, L. A., Pincus, S. M., and Rosenbaum, S. H. (1993). *Anesthesiology* **78**, 683.

Fleming, W. H., and Rishel, R. (1975). "Deterministic and Stochastic Optimal Control," Springer-Verlag, Berlin.

Fraser, A. M., and Swinney, H. L. (1986). *Phys. Rev. A* **33**, 1134.

Gevers, E., Pincus, S. M., Robinson, I. C. A. F., and Veldhuis, J. D. (1998). *Am. J. Physiol.* **274**, R437 (*Regul. Integrat.* 43).

Grassberger, P., and Procaccia, I. (1983a). *Phys. Rev. A* **28**, 2591.

Grassberger, P., and Procaccia, I. (1983b). *Physica D* **9**, 189.

Grenander, U. (1993). General Pattern Theory, Oxford University Press, Oxford, UK.

Gross, D., and Harris, C. M. (1985). Fundamentals of Queueing Theory, 2nd edn. John Wiley and Sons, New York.

Hartman, M. L., Pincus, S. M., Johnson, M. L., Mathews, D. H., Faunt, L. M., Vance, M. L., Thorner, M. O., and Veldhuis, J. D. (1994). *J. Clin. Invest.* **94**, 1277.

Israel, R. (1979). Convexity in The Theory of Lattice Gases, Princeton University Press, New Jersey.

Jackson, J. R. (1957). *Operations Res.* **5**, 518.

Kaplan, D. T., Furman, M. I., Pincus, S. M., Ryan, S. M., Lipsitz, L. A., and Goldberger, A. L. (1991). *Biophys. J.* **59**, 945.

Kolmogorov, A. N. (1958). *Dokl. Akad. Nauk SSSR* **119**, 861.

Kuo, C., Lindberg, C., and Thomson, D. J. (1990). *Nature* **343**, 709.

Law, A. M., and Kelton, W. D. (1991). "Simulation Modeling and Analysis," 2nd edn. McGraw-Hill, New York.

Mayer-Kress, G., Yates, F. E., Benton, L., Keidel, M., Tirsch, W., Poppl, S. J., and Geist, K. (1988). *Math. Biosci.* **90**, 155.

Morrison, S., and Newell, K. M. (1996). *Exp. Brain Res.* **110**, 455.

Ornstein, D. S., and Weiss, B. (1990). *Ann. Prob.* **18**, 905.

Parer, W. J., Parer, J. T., Holbrook, R. H., and Block, B. S. B. (1985). *Am. J. Obstet. Gynecol.* **153**, 402.

Pincus, S. M. (1991). *Proc. Natl. Acad. Sci. USA* **88**, 2297.

Pincus, S. M. (1992). *Proc. Natl. Acad. Sci. USA* **89**, 4432.

Pincus, S. M. (1994). *Math. Biosci.* **122**, 161.

Pincus, S. M. (1995). *Chaos* **5**, 110.

Pincus, S., and Kalman, R. E. (1997). *Proc. Natl. Acad. Sci. USA* **94**, 3513.

Pincus, S., and Kalman, R. E. (2004). *Proc. Natl. Acad. Sci. USA* **101**, 13709.

Pincus, S., and Singer, B. H. (1996). *Proc. Natl. Acad. Sci. USA* **93**, 2083.

Pincus, S., and Singer, B. H. (1998). *Proc. Natl. Acad. Sci. USA* **95**, 10367.

Pincus, S. M., and Goldberger, A. L. (1994). *Am. J. Physiol.* **266**, H1643, (*Heart Circ. Physiol.* 35).

Pincus, S. M., and Huang, W. M. (1992). *Commun. Stat. Theory Methods* **21,** 3061.

Pincus, S. M., and Keefe, D. L. (1992). *Am. J. Physiol.* **262,** E741, (*Endocrinol. Metab.* 25).

Pincus, S. M., and Viscarello, R. R. (1992). *Obstet. Gynecol.* **79,** 249.

Pincus, S. M., Gladstone, I. M., and Ehrenkranz, R. A. (1991). *J. Clin. Monit.* **7,** 335.

Pincus, S. M., Cummins, T. R., and Haddad, G. G. (1993). *Am. J. Physiol.* **264,** R638 (*Regul. Integrat.* 33).

Pincus, S. M., Mulligan, T., Iranmanesh, A., Gheorghiu, S., Godschalk, M., and Veldhuis, J. D. (1996a). *Proc. Natl. Acad. Sci. USA* **93,** 14100.

Pincus, S. M., Gevers, E., Robinson, I. C. A. F., van den Berg, G., Roelfsema, F., Hartman, M. L., and Veldhuis, J. D. (1996b). *Am. J. Physiol.* **270,** E107, (*Endocrinol. Metab.* 33).

Pincus, S. M., Veldhuis, J. D., Mulligan, T., Iranmanesh, A., and Evans, W. S. (1997). *Am. J. Physiol.* **273,** E989(*Endocrinol. Metab.* 36).

Pincus, S. M., Padmanabhan, V., Lemon, W., Randolph, J., and Midgley, A. R. (1998). *J. Clin. Invest.* **101,** 1318.

Pincus, S. M., Schmidt, P. J., Palladino-Negro, P., and Rubinow, D. R. (2008). *J. Psychohist. Res.* **42,** 337.

Roelfsema, F., Pincus, S. M., and Veldhuis, J. D. (1988). *J. Clin. Endocr. Metab.* **83,** 688.

Rukhin, A. L. (2000). *J. Appl. Prob.* **37,** 88.

Shaw, R. (1981). *Z. Naturforsch. A* **36,** 80.

Singer, B. H., and Pincus, S. (1998). *Proc. Natl. Acad. Sci. USA* **95,** 1363.

Siragy, H. M., Vieweg, W. V. R., Pincus, S. M., and Veldhuis, J. D. (1995). *J. Clin. Endocrinol. Metab.* **80,** 28.

Sugihara, G., and May, R. M. (1990). *Nature* **344,** 734.

Takens, F. (1983). "Atas do 13. Col. Brasiliero de Matematicas," Rio de Janerio, Brazil.

Thomson, D. J. (1990). *Phil. Trans. R. Soc. Lond. A* **330,** 601.

Urban, R. J., Evans, W. S., Rogol, A. D., Kaiser, D. L., Johnson, M. L., and Veldhuis, J. D. (1988). *Endocr. Rev.* **9,** 3.

van den Berg, G., Pincus, S. M., Veldhuis, J. D., Frolich, M., and Roelfsema, F. (1997). *Eur. J. Endocrinol.* **136,** 394.

Veldhuis, J. D., Liem, A. Y., South, S., Weltman, A., Weltman, J., Clemmons, D. A., Abbott, R., Mulligan, T., Johnson, M. L., Pincus, S., Straume, M., and Iranmanesh, A. (1995). *J. Clin. Endocrinol. Metab.* **80,** 3209.

Weinstein, M., Gorrindo, T., Riley, A., Mormino, J., Niedfeldt, J., Singer, B., Rodríguez, G., Simon, J., and Pincus, S. (2003). *Am. J. Epidemiol.* **158,** 782.

Wolf, A., Swift, J. B., Swinney, H. L., and Vastano, J. A. (1985). *Physica D* **16,** 285.

CHAPTER 8

Distinguishing Models of Growth with Approximate Entropy

Michael L. Johnson,* Michelle Lampl,[†] and Martin Straume[‡]

*University of Virginia Health System
Charlottesville, VA, USA

[†]Emory University
Atlanta, GA, USA

[‡]COBRA, Inc.

I. Update

A part of the basic scientific method is to postulate and compare conflicting hypotheses, which purport to describe specific experimental results. The present chapter describes a novel approach to distinguish models with Approximate Entropy (ApEn; Pincus, 2000, 2010). This chapter provides a comparison of two models of longitudinal growth in children; the Saltation and Stasis model (Johnson and Lampl, 1995; Johnson et al., 1996, 2001; Lampl and Johnson, 1993, 1997, 1998a, 2000; Lampl et al., 1992, 2001) versus the smooth continuous growth model. Nevertheless, the basic approach has broad applicability.

The Saltation and Stasis model (Lampl *et al.*, 1992) describes longitudinal growth as a series of very short duration growth events followed by a stasis or refractory period. Most other models of growth assume a slowly varying, continuous growth hypothesis. ApEn (Pincus, 2000, 2010) is a measure of the short-term temporal irregularity of a time-series and is thus perfect to distinguish models such as these. In the present case, a bootstrap approach was utilized to evaluate the distribution of the expected ApEn values for each of these two models. It was observed that these distributions did not overlap and that the ApEn of the experimental data was within the expected distribution for the Saltation and Stasis model.

II. Introduction

Researchers are commonly faced with having to differentiate among two or more hypotheses (i.e., theories, models, etc.) based on how well they describe the actual experimental data. Frequently, this is done by translating the mechanistic theory, or hypothesis, into a mathematical model and then "fitting" the model to the experimental data. This "fitting" process is commonly done by a least squares parameter estimation procedure (Bates and Watts, 1988; Johnson, 1994; Johnson and Faunt, 1992; Johnson and Frasier, 1985; Nelder and Mead, 1965).

When the quality of the 'fit' of the experimental data is significantly different between the models (i.e., theories), they can usually be distinguished by applying the goodness-of-fit criteria (Armitage, 1977; Bard, 1974; Bevington, 1969; Box and Jenkins, 1976; Daniel, 1978; Draper and Smith, 1981; Johnson and Straume, 2000; Straume and Johnson, 1992), such as a runs test. However, if these tests are not conclusive, how can the researcher distinguish between the theories?

This chapter presents a unique application of ApEn (Johnson and Straume, 2000; Pincus, 1991, 1992, 1994, 2000), ApEn, to distinguish between models of growth in children (Johnson, 1993, 1999; Johnson and Lampl, 1994, 1995; Johnson *et al.*, 1996; Lampl and Johnson, 1993, 1997, 1998a,b; Lampl *et al.*, 1992, 1995, 1997, 1998). While this specific example is only applicable to some types of experimental data, it does illustrate the broad applicability of ApEn.

Historically, growth (Johnson, 1993, 1999; Johnson and Lampl, 1994, 1995; Johnson *et al.*, 1996; Lampl and Johnson, 1993, 1997, 1998a,b; Lampl *et al.*, 1992, 1995, 1997, 1998) has been considered to be a smooth, continuous process that varies little from day to day. In this model, growth rates change gradually on a timescale of months or years, and not hours or days. However, when Lampl *et al.* measured the lengths and heights of infants and adolescents at daily intervals, it was observed that large changes in growth rates occured between some days and no growth occured between other days (Johnson, 1993, 1999; Johnson and Lampl, 1994, 1995; Johnson *et al.*, 1996; Lampl and Johnson, 1993, 1997, 1998a,b; Lampl *et al.*, 1992, 1995, 1997, 1998). These daily observations led to the development of the *saltation and stasis hypothesis* and the mathematical model of growth (Johnson,

1993, 1999; Johnson and Lampl, 1994, 1995; Johnson *et al.*, 1996; Lampl and Johnson, 1993, 1997, 1998a,b; Lampl *et al.*, 1992, 1995, 1997, 1998).

The saltation and stasis hypothesis states growth (i.e., saltation) occurs over a very short time and then the organism enters a refractory period of little or no growth (i.e., stasis). In this context, "a very short time" means less than the interval between the measurements and "little or no growth" means less than what could be measured.

Previously, we utilized the classical goodness of-fit criteria (Armitage, 1977; Bard, 1974; Bevington, 1969; Box and Jenkins, 1976; Daniel, 1978; Draper and Smith, 1981; Johnson and Straume, 2000; Straume and Johnson, 1992), such as autocorrelation, to demonstrate that the saltation and stasis model provided a better description of the experimental observations than is obtainable withthe more classical growth models. However, for some data sets, the goodness-of-fit tests do not provide a clear distinction between the models and hypotheses. With these data sets in mind, we developed a new method to distinguish models of growth that are based on a modified version of the ApEn, metric of experimental data.

The basic procedure for the use of ApEn to distinguish these models involves calculation of the ApEn value for the original data sequence and the expected ApEn values, with standard errors, for each of the growth models. The observed ApEn value is then compared with the distributions of expected ApEn values for each of the growth models being tested. ApEn quantifies the regular versus the irregular nature of a time series. This test of adequacy of the growth models is based on the fact that the degree of regularity of the experimental observations is quantifiable.

III. Definition and Calculation of ApEn

ApEn was formulated by Pincus to statistically discriminate a time series by quantifying the regularity of a time series (Pincus, 1991, 1992, 1994, 2000). Of particular significance is the ability of ApEn to reliably quantify the regularity of a finite length time series, even in the presence of noise and measurement inaccuracy. This is a property unique to ApEn and one that is not shared by other methods common to nonlinear dynamic systems theory (Pincus, 1994).

Specifically, ApEn measures the logarithmic likelihood of patterns of a run in a time series that are close for m consecutive observations and remain close even when considered as $m + 1$ consecutive observations. A higher probability of remaining close (i.e., greater regularity) yields smaller ApEn values, whereas, greater independence among sequential values of a time series yields larger ApEn values.

Calculation of ApEn requires prior definition of the two parameters m and r. The parameter m is the length of run to be compared (as alluded to above) and r is a filter (the magnitude that will discern "close" and "not close" as described below). ApEn values can only be validly compared when computed for the same m, r, and N values (Pincus, 1994), where N is the number of data points in the time series being considered. Thus, ApEn is specified as ApEn(m,r,N). For optimum statistical validity, ApEn is typically implemented using m values of 1 or 2 and r values of approximately 0.2 standard deviations of the series being considered (Pincus, 1994).

ApEn is calculated according to the following formula (Pincus, 1994): Given N data points in a time series, $u(1), u(2), \ldots, u(N)$, the set of $N - m + 1$ possible vectors, $x(i)$, are formed with m consecutive u values such that $x(i) = [u(i), \ldots, u(i + m - 1)]^T$, $i = 1, 2, \ldots, N - m + 1$. The distance between vectors $x(i)$, and $x(j)$, $d[x(i), x(j)]$, is defined as the maximum absolute difference between corresponding elements of the respective vectors. For each of the $N - m + 1$ vectors $x(i)$, a value for $C_i^m(r)$ is computed by comparing all $N - m + 1$ vectors, $x(j)$ to vector $x(i)$ such that

$$C_i^m(r) = \frac{\text{number of } x(j) \text{ for which } d[x(i), x(j)] \leq r}{N - m + 1}. \tag{1}$$

These $N - m + 1$ $C_i^m(r)$ values measure the frequency with which patterns that are similar to the pattern given by $x(i)$ of length m within tolerance r, were encountered. Note that for all i, $x(i)$ is always compared relative to $x(i)$ (i.e., to itself), so that all values of $C_i^m(r)$ are positive. Now, define

$$\Phi^m(r) = \frac{\sum_{i=1}^{N-m+1} \ln C_i^m(r)}{N - m + 1} \tag{2}$$

from which the ApEn(m, r, N) is given by

$$\text{ApEn}(m, r, N) = \Phi^m(r) - \Phi^{m+1}(r) \tag{3}$$

ApEn, defined in this way, can be interpreted (with $m = 1$, for example) as a measure of the difference between (1) the probability that runs of length 1 will recur within the tolerance r and (2) the probability that runs of length 2 will recur within the same tolerance (Pincus, 1994).

IV. Modifications of ApEn Calculation for this Application

Two modifications were made to the standard methods for calculation of ApEn for the present use. The ApEn calculation is normally performed on a stationary time series (i.e., a series of data where the mean of the data is not a function of time). Clearly, measures of growth such as height generally increase with time and consequently the ApEn values were calculated on these nonstationary time series. This is the first modification. As described above, ApEn(m, r, N) is a function of the run length size m; r the magnitude that will discern "close" and "not close"; and N the length of the time series. Normally, r is expressed in terms of the standard deviation of the stationary time series; r is expressed in terms of the experimental variability or uncertainty of the time series. When applied to a nonstationary time series (e.g., growth) it is more logical to express r in terms of the actual measurement uncertainties. Thus, the second modification of the ApEn calculation was to express r in terms of the known levels of error in measurement.

V. Growth Models

Numerous mathematical models are available for describing the growth process. The saltation and stasis model is unique in that it is based on a mechanistic hypothesis that describes how growth proceeds. This mechanistic hypothesis was translated into a mathematical form, where growth is described as a series of distinct instantaneous events of positive growth (i.e., saltations) that are separated by stasis or periods of no growth:

$$\text{Height} = \sum_{k=1}^{i} G_i, \tag{4}$$

where the summation is over each of the observations, and G_i is zero during a stasis interval but has a positive value for the measurement intervals where a saltation occurred.

This mathematical model does not require that the saltation events be instantaneous. It simply requires that the growth events occur in a lesser time period compared to the interval between the observations. Under these conditions, the experimental observations do not contain any information about the actual shape of the saltation event and thus can simply be approximated as a step function.

Virtually all other growth models (Gasser *et al.*, 1990; Karlberg, 1987; Preece and Baines, 1978) assume that growth is a smooth continuous process that varies little from day to day. These models assume that small changes in growth rates occur on a timescale of months or years, and not by hours or days. Furthermore, virtually all of these models are not based on a hypothesis or a theory. These models are simply empirical descriptions of growth. They are generally formulated to describe only a few observations per year. Thus, these models do not—and cannot—describe growth patterns that vary on a daily time frame.

It is impossible to test a set of experimental data against every possible slowly varying smooth, continuous mathematical form. Consequently, we decided to use an exponential rise

$$\text{Height} = A_0 - A_1 e^{-k\text{Time}}$$

and/or polynomials of order 1-6

$$\text{Height} = \sum_{i=1}^{6} A_i \text{Time}^i$$

as surrogates for the infinite number of possible slowly varying empirical mathematical forms that can be utilized to describe growth.

VI. Expected Model-Dependent Distribution of ApEn

The expected distributions of ApEn values for any growth model (e.g., saltation and stasis, polynomial, exponential) are evaluated by using a Monte Carlo procedure. This involves simulating a large number of growth patterns and observing the distribution of the resulting ApEn values.

The first step in this Monte Carlo procedure is to fit, using least squares, a particular growth model to a particular set of experimental observations. This provides an optimal model-dependent description of the growth pattern (i.e., the best calculated curve for the particular model) and a set of residuals. The residuals are the differences between the experimental observations and calculated optimal curve.

The next step is to simulate a large number (e.g., 1000) of calculated growth patterns for the particular growth model. These calculated growth patterns are the best calculated curves for the particular model with "pseudorandom noise" added to it. The expected model-dependent distribution of the ApEn values is then calculated from the ApEn values for each of the large number of simulated model-dependent growth patterns.

The "pseudorandom noise" can be generated by three possible methods. First, the noise can be calculated by generating a Gaussian distribution ofpseudorandom numbers with a variance equal to the variance of the residuals and a mean of zero. Second, the actual residuals can simply be shuffled (i.e., selected in a random order) and added back to the best calculated curve in a different order. Third, the actual residuals can be shuffled with replacement, as is done in the bootstrap procedures (Efron and Tibshirani, 1993). The "shuffled with replacement" choice means that for each of the simulated growth patterns ~37% of the residuals are randomly selected and not used while the same number of the remaining residuals are used twice to obtain the requisite number of residuals. A different 37% of the residuals are selected for each of the simulated growth patterns. The order of the selected residuals is randomized (i.e., shuffled) before they are added back to the calculated growth pattern.

The expected distribution of the ApEn values, as determined for each of the growth models, is compared with the observed ApEn value from the experimental observations. The observed ApEn value will be within either one or more distributions or may be within none. The conclusion of this test is that if the observed ApEn value is within an expected distribution for a particular mathematical model, then the data are consistent with that model. Conversely, if the observed ApEn value is not within an expected distribution for a particular mathematical model, then the data are inconsistent with that model. If the observed ApEn is within the expected distributions for more than one model, then these models cannot be distinguished by this test.

VII. Example of this Use of ApEn

Figure 1 presents a typical set of daily experimental observations of the height of an infant. The solid line, in Fig. 1, corresponds to the saltation and stasis analysis of these data. The intervals of saltation and stasis are visually obvious. The analysis of these data by the saltation and stasis model indicated that there were 13 statistically significant ($P < 0.05$) saltations at days 94, 106, 117, 119, 132, 144, 159, 168, 175, 183, 192, 201, and 216. These saltations are not at regular intervals; they are episodic and not periodic. The average saltation amplitude in this infant was 0.9 cm. The standard deviation of the differences (i.e., residuals) between the calculated optimal saltation and stasis model and the experimental observations is 0.3 cm.

The irregular nature of the distributions in Fig. 2, and those that follow, is due to the fact that only 1000 Monte Carlo cycles were used for their generation. These distributions become increasingly smooth as the number of Monte Carlo cycles used increases. However, 1000 cycles are usually sufficient to characterize the distributions and only require a few seconds on a 450-MHz Pentium II PC.

The various model-dependent distributions of ApEn(2, 0.4, 118) for the data shown in Fig. 1 are presented in Fig. 2. The panels, from top to bottom, correspond to the expected distribution for the saltation and stasis model, the expected distributions for first- through sixth-degree polynomials, the Karlberg infant model (Karlberg, 1987), and the exponential rise model. These distributions are the results of 1000 Monte Carlo cycles, with the noise being generated by shuffling the residuals. The nonstationary height series was not detrended. The value of r was set at 0.4 cm, and not at a fraction of the standard deviation of the data points. The square in each of the panels in Fig. 2 is the actual observed ApEn(2, 0.4, 118)

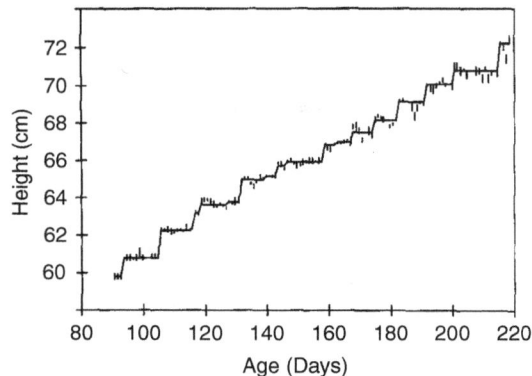

Fig. 1 A typical set of daily experimental observations of the height of an infant. The solid line corresponds to the saltation and stasis analysis of these data. The saltation and stasis model analysis indicated that there were 13 statistically significant ($P < 0.05$) saltations at 94, 106, 117, 119, 132, 144, 159, 168, 175, 183, 192, 201, and 216 days.

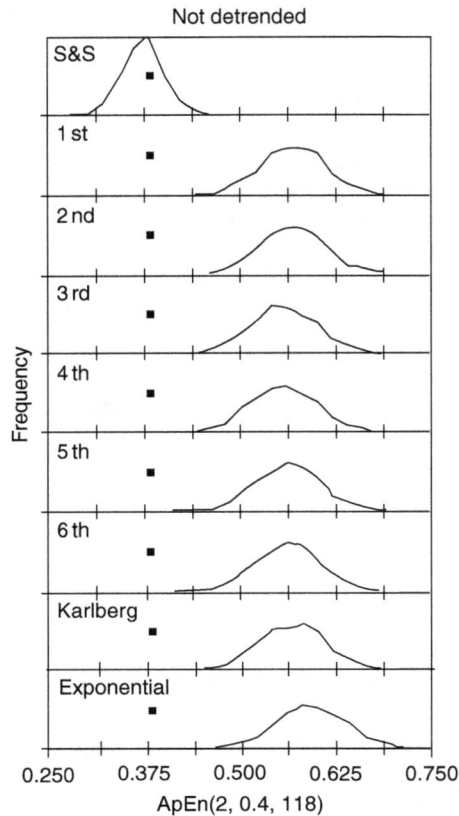

Fig. 2 The various model-dependent distributions of ApEn(2, 0.4, 118) for the data presented in Fig. 1. The panels from top to bottom correspond to the expected distribution for the saltation and stasis model; the expected distribution for first- through sixth-degree polynomials; the Karlberg infant model; and the exponential rise model. The square in each of the panels is the actual observed ApEn(2, 0.4, 118) value for the original data. These analyses are the results of 1000 Monte Carlo cycles with the noise being generated by shuffling the residuals. The nonstationary height series was not detrended.

value for the original data. It is clear (from the figure) that this value is consistent with the expected distribution for the saltation and stasis model and that it is inconsistent with the other models. Although not shown, the ApEn(1, 0.4, 118) values provide analogous results.

Figure 3 presents the same analysis as in Fig. 2, except that the nonstationary time series was detrended by subtracting the best least squares straight line before the analyses were performed. Clearly, while the results are numerically slightly different, the conclusions remain the same. Again, the ApEn(1, 0.4, 118) values provide analogous results (not shown).

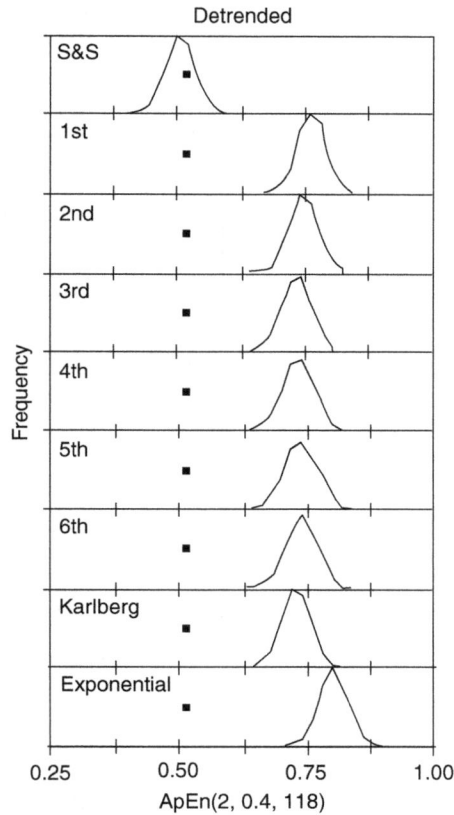

Fig. 3 The same analysis as presented in Fig. 2 except that the nonstationary time series was detrended by subtracting the best least squares straight line before the analyses were performed.

When the analysis, as depicted in Fig. 2, was repeated with the pseudorandom noise being created by either shuffling with replacement (i.e., a bootstrap) or by a Gaussian distribution, the results were virtually identical to those shown in Fig. 2. Consequently, the plot of these results is not repeated.

Figure 4 presents the same analysis as is presented in Fig. 2 except that the r value was set to 0.2 cm. Note that while the figure appears somewhat different, the conclusions remain the same. When r is increased to 0.6 cm, the results are virtually identical to those shown in Fig. 2. It is interesting to note that the ApEn(1, 0.2, 118) distributions (not shown) do not exhibit a reversal of magnitude that the ApEn(2, 0.2, 118) distributions show.

From a comparison of Figs. 2 and 4, it appears that at some intermediate value of r (e.g., 0.27) the ApEn(2, 0.27, 118) distributions coincide, but the models cannot be distinguished by that metric (not shown). However, as shown in

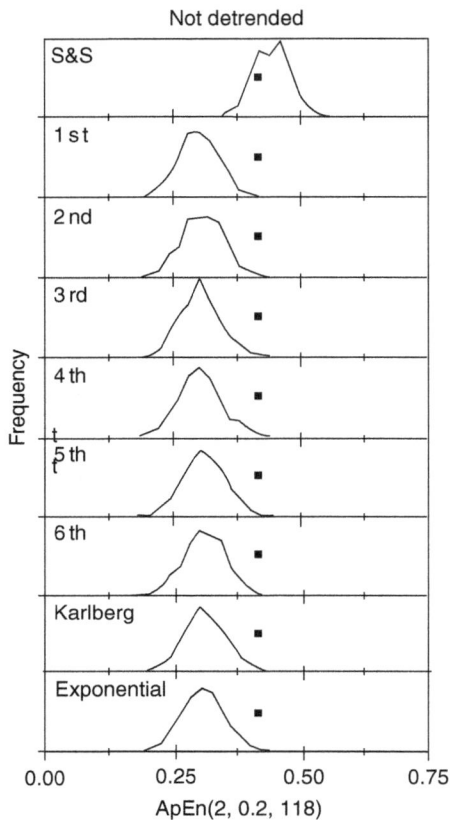

Fig. 4 The same analysis as presented in Fig. 2 except that the value of *r* was decreased to 0.2 cm. The nonstationary height series was not detrended.

Fig. 5, the ApEn(1, 0.27, 118) distributions are clearly distinguishable. The ApEn (2, 0.27, 118) distributions are clearly distinguishable if the time series is detrended. This is shown in Fig. 6.

VIII. Conclusion

This chapter presents examples of how ApEn, can be used to distinguish between mathematical models and their underlying mechanistic hypotheses that purport to describe the same experimental observations. If the expected distributions of ApEn for the different models do not overlap, then it is expected that the ApEn can be utilized to distinguish these models and hypotheses. However, if the distributions overlap significantly, then no conclusion can be drawn.

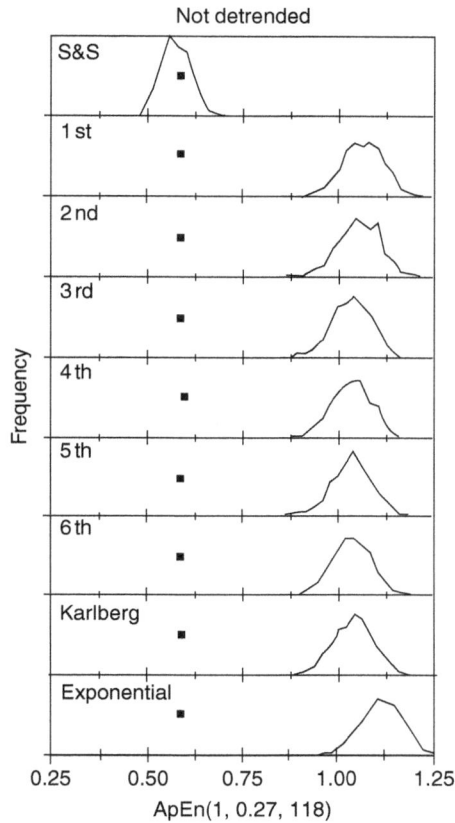

Fig. 5   The ApEn(1, 0.27, 118) distributions for the data in Fig. 1. The nonstationary height series was not detrended.

From the various figures shown above, it is obvious that the ApEn distributions are fairly robust to variations in the values of m and r. The ApEn distributions do not appear to be sensitive, in these examples, to the method of random noise generation (i.e., pseudorandom number generator, shuffling, or boot-strapping) for the Monte Carlo process.

Furthermore, the ApEn distributions appear to be relatively insensitive to the stationary nature of the data when the value of r is expressed as an absolute quantity that is dependent on known experimental measurement errors. However, as shown in Fig. 6, there are cases where the results are better if the data are detrended. Using a straight line as a detrending function is both sufficient and preferable.

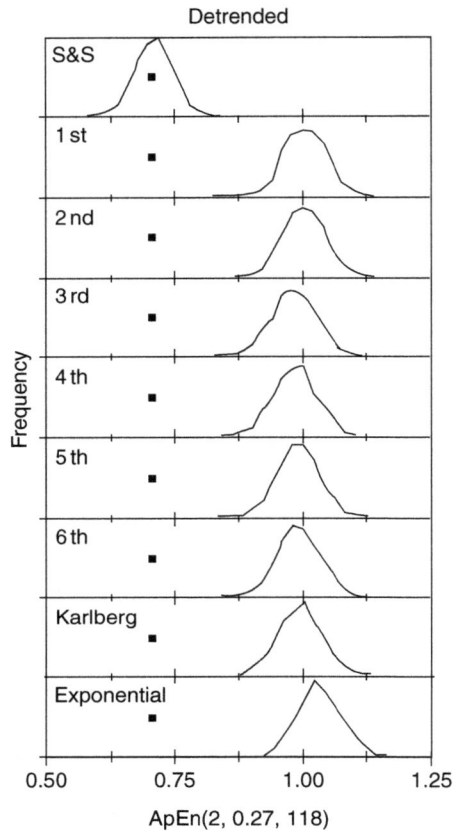

Fig. 6 The ApEn(2, 0.27, 118) distributions for the data in Fig. 1. The nonstationary time series was detrended by subtracting the best least squares straight line before the analyses were performed.

Acknowledgments

The authors acknowledge the support of the National Science Foundation Science and Technology Center for Biological Timing at the University of Virginia (NSF DIR-8920162), the General Clinical Research Center at the University of Virginia (NIH RR-00847), and the University of Maryland at Baltimore Center for Fluorescence Spectroscopy (NIH RR-08119).

References

Armitage, P. (1977). "Statistical Methods in Medical Research," 4th edn., p. 391. Blackwell, Oxford.
Bard, Y. (1974). "Nonlinear Parameter Estimation," p. 201. Academic Press, New York.
Bates, D. M., and Watts, D. G. (1988). Nonlinear Regression Analysis and Its Applications. John Wiley and Sons, New York.

Bevington, P. R. (1969). "Data Reduction and Error Analysis in the Physical Sciences," p. 187. McGraw-Hill, New York.

Box, G. E. P., and Jenkins, G. M. (1976). "Time Series Analysis Forecasting and Control," p. 33. Holden-Day, Oakland, California.

Daniel, W. W. (1978). "Biostatistics: A Foundation for Analysis in the Health Sciences," 2nd edn. John Wiley and Sons, New York.

Draper, N. R., and Smith, R. (1981). "Applied Regression Analysis," 2nd edn., p. 153. John Wiley and Sons, New York.

Efron, B., and Tibshirani, R. J. (1993). "An Introduction to the Bootstrap." Chapman and Hall, New York.

Gasser, T., Kneip, A., Ziegler, P., Largo, R., and Prader, A. (1990). *Ann. Hum. Biol.* **13**, 129.

Johnson, M. L. (1993). Analysis of Serial Growth Data. *Am. J. Hum. Biol.* **5**, 633–640.

Johnson, M. L. (1994). Use of Least-Squares Techniques in Biochemistry. *Methods Enzymol.* **240**, 1–23.

Johnson, M. L. (1999). Methods for the analysis of saltation and stasis in human growth data. "Saltation and Stasis in Human Growth and Development," pp. 101–120. Smith-Gordon, London.

Johnson, M. L., and Faunt, L. M. (1992). Parameter Estimation by Least-Squares Methods. *Methods Enzymol.* **210**, 1.

Johnson, M. L., and Frasier, S. G. (1985). Nonlinear Least-Squares Analysis. *Methods Enzymol.* **117**, 301.

Johnson, M. L., and Lampl, M. (1994). Artifacts of Fourier Series Analysis. *Methods Enzymol.* **240**, 51–68.

Johnson, M. L., and Lampl, M. (1995). Methods for the evaluation of saltatory growth in infants. *Methods Neurosci.* **28**, 364–387.

Johnson, M. L., and Straume, M. (2000). Approximate Entropy as Indication of Goodness of Fit. *Methods Enzymol.* **321**, 207–216.

Johnson, M. L., Veldhuis, J. D., and Lampl, M. (1996). Is growth saltatory? The usefulness and limitations of distribution functions in analyzing pulsatile data. *Endocrinology* **137**, 5197–5204.

Johnson, M. L., Straume, M., and Lampl, M. (2001). The use of regularity as estimated by approximate entropy to distinguish saltatory growth. *Ann. Hum. Biol.* **28**, 491–504.

Lampl, M., and Johnson, M. L. (1993). A case study of daily growth during adolescence: A single spurt or changes in the dynamics of saltatory growth? *Ann. Hum. Biol.* **20**, 595–603.

Lampl, M., and Johnson, M. L. (1997). Identifying saltatory growth patterns in infancy: a comparison of results bases on measurement protocol. *Am. J. Hum. Biol.* **9**, 343–355.

Lampl, M., and Johnson, M. L. (1998a). Normal human growth as saltatory: Adaptation through irregularity. *In* "Applications of Nonlinear Dynamics to Developmental Process Modeling," (K. M. Newell, and P. C. Molenaar, eds.), pp. 15–38. Lawrence Erlbaum Associates, New York.

Lampl, M., and Johnson, M. L. (1998b). Wrinkles Induced by the Use of Smoothing Procedures Applied to Serial Growth Data. *Ann. Hum. Biol.* **25**, 187–202.

Lampl, M., and Johnson, M. L. (2000). Distribution methods for the analysis of nonlinear longitudinal data. *Methods Enzymol.* **321**, 182–195.

Lampl, M., Veldhuis, J. D., and Johnson, M. L. (1992). Saltation and stasis: A model of human growth. *Science* **258**, 801–803.

Lampl, M., Cameron, N., Veldhuis, J. D., and Johnson, M. L. (1995). Patterns of Human Growth – Reply. *Science* **268**, 445–447.

Lampl, M., Johnson, M. L., and Frongillo, E. A., Jr (1997). Stasis Without Saltations? A Comment on "No Evidence for Saltation in Human Growth". *Ann. Hum. Biol.* **24**, 65–68.

Lampl, M., Ashizawa, K., Kawabata, M., and Johnson, M. L. (1998). An Example of Variation and Pattern in Saltation and Stasis Growth Dynamics. *Ann. Hum. Biol.* **25**, 203–219.

Lampl, M., Johnson, M. L., and Frongillo, E. A. (2001). Mixed distribution analysis identifies saltation and stasis growth. *Ann. Hum. Biol.* **28**, 403–411.

Straume, M., and Johnson, M. L. (1992). Analysis of Residuals: Criteria for Determining Goodness-of-Fit. *Methods Enzymol.* **210**, 87–105.

Karlberg, J. (1987). *Stat Med* **6,** 185.

Nelder, J. A., and Mead, R. (1965). *Comput. J.* **7,** 308.

Pincus, S. (2010). this volume.

Pincus, S. M. (1991). *Proc. Natl. Acad. Sci. USA* **88,** 2297.

Pincus, S. M. (1992). *Proc. Natl. Acad. Sci. USA* **89,** 4432.

Pincus, S. M. (1994). *Methods Enzymol.* **240,** 68.

Pincus, S. M. (2000). Irregularity and Asynchrony in Biological Network Signals. *Methods in Enzymol.* **321,** 149–182.

Preece, M. A., and Baines, M. J. (1978). *Ann. Hum. Biol.* **5,** 1.

CHAPTER 9

Application of the Kalman Filter to Computational Problems in Statistics

Emery N. Brown* and Christopher H. Schmid[†]

*Statistics Research Laboratory
Department of Anesthesia
Massachusetts General Hospital
Boston, Massachusetts, USA

[†]Biostatistics Research Center
Division of Clinical Care Research
New England Medical Center
Tufts University
Boston, Massachusetts, USA

I. Introduction

The Kalman filter is a linear filtering algorithm that was developed in the early 1960s to solve estimation and control problems in engineering such as monitoring the position and velocity of a satellite orbiting the earth using signals received at ground tracking stations (Kalman, 1960; Meinhold and Singpurwalla, 1983). The algorithm significantly reduced the amount of information that had to be stored and the computational costs required to analyze such problems in two ways; one by assuming that the probability densities of the noise and error in the systems are well approximated by their first two moments and, two, by taking advantage of the inherent Markov structure of the problem. The Markov property means that information about the system that is required to describe its state at time t depends

DOI: 10.1016/B978-0-12-384997-7.00009-1

only on the history that was summarized at the last time, say $t - 1$. A complete description of the history of the system did not have to be retained. In the case of an orbiting earth satellite, the state of the system may contain, among other information, the position and velocity of the satellite.

To define the Kalman filter algorithm, we first define the observation equation

$$Y_t = F_t\theta_t + z_t \tag{1}$$

and the system equation

$$\theta_t = G_t\theta_{t-1} + w_t, \tag{2}$$

where θ_t is the state vector at time t, F_t is a known transformation matrix, Y_t is the observed data, z_t is a random variable with mean zero, and the covariance matrix Z_t, G_t describes the expected change in position and velocity of the satellite over time, and w_t is a random perturbation with mean zero and with covariance matrix W_t. The random variables z_t and w_t are assumed to be uncorrelated. Let $Y_{t-1}^* = (Y_1, \ldots, Y_{t-1})$. Starting with $\hat{\theta}_{t-1|t-1}$ and $\Sigma_{t-1|t-1}$ the estimates, respectively, of the state vector and covariance matrix at time $t - 1$, the Kalman filter algorithm is defined by the following set of equations:

$$\hat{\theta}_{t|t-1} = G_t\hat{\theta}_{t-1|t-1} \tag{3a}$$

$$R_{t|t-1} = G_t\sum\nolimits_{t-1|t-1}G_t^T + W_t \tag{3b}$$

$$K_t = R_{t|t-1}F_t^T[F_tR_{t|t-1}F_t^T + Z_t]^{-1} \tag{3c}$$

$$\hat{\theta}_{t|t} = \hat{\theta}_{t|t-1} + K_t[Y_t - F_t\hat{\theta}_{t|t-1}] \tag{3e}$$

$$\sum\nolimits_{t|t} = [I - K_tF_t]R_{t|t-1} \tag{3d}$$

given the initial conditions θ_0 and Σ_0, Eq. (3a) is the prior mean and Eq. (3b) is the prior variance. The latter describes the error in the prediction of θ_t, given $\hat{\theta}_{t|t-1}$ and Y^*_{t-1}. The matrix K_t is the Kalman gain, $\hat{\theta}_{t|t}$ is the Kalman filter estimate of the state vector given Y^*_t, and $\Sigma_{t|t}$ is the posterior variance of θ_t, given Y^{*t}.

Schweppe (1965) showed that, under the assumption that z_t and w_t are Gaussian random variables, important computational savings could be made by using the Kalman filter to evaluate the Gaussian likelihood functions for models with the Markov structure. Statisticians later realized that, in addition to this application, the Kalman filter greatly facilitated the computation of posterior probability densities in Bayesian linear models with Gaussian errors and the estimation of model parameters when data were missing in Gaussian models with a Markov structure. In this chapter, we illustrate some of the computational problems to which the Kalman filter has been successfully applied in statistics. Section II illustrates its use to evaluate a Gaussian likelihood, where the observational error process is a Gaussian, serially correlated, noise. In Section III, we describe how the Kalman filter can be used to compute posterior densities for Bayesian statistical models. Section IV demonstrates the use of the Kalman filter in evaluating a Gaussian likelihood as part of the expectation and maximization (EM) algorithm. In Section V, we mention some extensions.

II. Evaluating Gaussian Likelihood Using the Kalman Filter

To illustrate how the Kalman filter may be used to evaluate a Gaussian likelihood function, we consider the problem of modeling human biological rhythm data. Core temperature data are an often-studied biological rhythm, used to estimate the properties of the human biological clock (Czeisler et $al.$, 1989, 1990). Brown and Czeisler (1992) showed that a reasonable statistical description of core temperature data, collected on the constant routine protocol, is given by the two harmonic regression and first-order serial correlated noise model. To define the model, we assume that a sequence of temperature measurements $y_1, \ldots, y_N$ is made on a human subject and that these data obey Eq. (4)

$$y_t = s_t + v_t \tag{4}$$

for $t = 1, \ldots, N$ where

$$s_t = \mu + \sum_{r=1}^{2} A_r \cos\left(\frac{2\pi r}{\tau} t\right) + B_r \sin\left(\frac{2\pi r}{\tau} t\right) \tag{5}$$

$$v_t = \alpha v_{t-1} + \varepsilon_t$$

and it is assumed that $|\alpha| < 1$ and the ε_t values are distributed as Gaussian random variables with mean zero and variance σ_ε^2. Let $y = (y_1, \ldots, y_N)^T)$, $B = (\mu, A_1, B_1, A_2, B_2)^T$, and $v = (v_1, \ldots, v_N)^T$ and we define

$$X(\tau) = \begin{bmatrix} 1 & \cos\left(\frac{2\pi}{\tau}\right) & \sin\left(\frac{2\pi}{\tau}\right) & \cos\left(\frac{4\pi}{\tau}\right) & \sin\left(\frac{4\pi}{\tau}\right) \\ \vdots & \vdots & \vdots & \vdots & \vdots \\ 1 & \cos\left(\frac{2\pi}{\tau}N\right) & \sin\left(\frac{2\pi}{\tau}N\right) & \cos\left(\frac{4\pi}{\tau}N\right) & \sin\left(\frac{4\pi}{\tau}N\right) \end{bmatrix}$$

to be the $N \times 5$ harmonic regression design matrix. Equation (4) may then be rewritten in the matrix notation as

$$y = X(\tau)B + v. \tag{6}$$

From the assumption that the ε_t values are Gaussian random variables, it follows that v is a multivariate Gaussian random vector with zero mean and an $N \times N$ covariance matrix which we denote as Γ. Given B, τ, α, and σ_ε^2, the joint probability density of the core temperature data is

$$f(y|B, \tau, \alpha, \sigma_\varepsilon^2) = \left(\frac{1}{2\pi\sigma_\varepsilon^2}\right)^{N/2} |\Gamma|^{-1/2} \exp\left\{-\frac{1}{2\sigma_\varepsilon^2} S_N\right\}, \qquad (7)$$

where $|\Gamma|$ is the determinant of Γ and

$$
\begin{aligned}
S_N &= [y - X(\tau)B]^T \Gamma^{-1} [y - X(\tau)B] \\
&= \{L[y - X(\tau)B]\}^T \{L[y - X(\tau)B]\},
\end{aligned}
\qquad (8)
$$

where $\Gamma^{-1} = L^T L$ is the inverse of Γ, and L is the Cholesky factor of Γ^{-1}. An objective of the core temperature data analysis is to estimate the model parameters B, τ, and α, using maximum likelihood, so that the phase and amplitude of the circadian rhythm of the subject may be determined (Brown and Czeisler, 1992). As v obeys a first-order autoregression, the model has a Markov structure (Jones, 1980). This, combined with the fact that the ε_t terms are Gaussian in nature, suggests that the Kalman filter may be used to evaluate the likelihood function, which can then be numerically maximized to estimate the model parameters. As minimizing the -2 log likelihood is equivalent to and more tractable numerically than maximizing the likelihood, the computational algorithm is developed in terms of the -2 log likelihood function. Taking logarithms of both sides of Eq. (7) yields

$$\log f(B, \tau, \alpha, \sigma_\varepsilon^2 | y) = -\frac{N}{2} \log(\sigma_\varepsilon^2) - \frac{S_N}{2\sigma_\varepsilon^2} - \frac{1}{2} \log|\Gamma| - \frac{N}{2} \log(2\pi), \qquad (9)$$

where we write $f(B, \tau, \alpha, \sigma_\varepsilon^2 | y)$ to indicate that we are viewing Eq. (7) as a function of the parameters B, τ, α, and σ_ε^2 for the fixed, observed data vector, y.

Differentiating Eq. (9) with respect to σ_ε^2 yields

$$\frac{d \log f(B, \tau, \alpha, \sigma_\varepsilon^2 | y)}{d\sigma_\varepsilon^2} = -\frac{N}{2\sigma_\varepsilon^2} + \frac{S_N}{2(\sigma_\varepsilon^2)^2}. \qquad (10)$$

Setting the right-hand side equal to 0 and solving for σ_ε^2 gives its maximum-likelihood estimate,

$$\hat{\sigma}_\varepsilon^2 = \frac{S_N}{N}. \qquad (11)$$

Substituting Eq. (11) into Eq. (9), neglecting the constants, and multiplying both sides by -2 gives

$$-2 \log f(B, \tau, \alpha | y) = \log(S_N) + \log|\Gamma|, \qquad (12)$$

which is the -2 log likelihood. In analyses where it is important to distinguish between Eqs. (9) and (12), -2 times of the left-hand side of Eq. (9) is called the -2 log likelihood, and Eq. (12) is the concentrated -2 log likelihood, since its dependence on σ_ε^2 has been removed. As this distinction is not important in our problem, we refer to Eq. (12) as the -2 log likelihood. The model parameters that minimize $-2 \log f(B, \tau, \alpha | y)$ are the maximum-likelihood estimates (Priestly, 1981). We denote them as $\hat{B}$, $\hat{\tau}$, and $\hat{\alpha}$. To minimize Eq. (12), we note first that the minimization problem is only two-dimensional because, given τ and α, the maximum-likelihood estimate of B is the generalized least-squares estimate which is defined as

$$\hat{B}(\tau, \alpha) = [X^*(\tau)^T X^*(\tau)]^{-1} X^*(\tau)^T y^*,$$

where $X^*(\tau) = LX(\tau)$ and $y^* = Ly$. If we substitute $\hat{B}(\tau, \alpha)$ for B in Eq. (12), the -2 log likelihood becomes

$$-2 \log f(\tau, \alpha | y) = \log(\hat{S}_N) + \log|\Gamma|, \qquad (13)$$

where

$$\hat{S}_N = [y^* - X^*(\tau)\hat{B}(\tau, \alpha)]^T [y^* - X^*(\tau)\hat{B}(\tau, \alpha)].$$

The technical problem that needs to be solved in order to minimize Eq. (13) is to efficiently compute Ly, $LX(t)$, and $|\Gamma|$ at each step of a numerical minimization algorithm. Computing L is equivalent to performing a Gram-Schmidt orthogonalization on the vectors that define the column space of Γ (Wecker and Ansley, 1983). Given the Markov structure and the Gaussian error properties of the problem, these tasks can be efficiently accomplished by using the Kalman filter. Let X_j be the jth column of $X(\tau)$, X^*_j be the jth column of $X^*(\tau)$, and $\theta_{t,j}$ be a 1×1 dimensional state vector for $j = 1, \ldots, 5$. It follows from Wecker and Ansley (1983) that the Kalman filter yields the following algorithm for computing the t, jth element of LX:

$$\theta_{t|t-1j} = G_t \theta_{t-1|t-1,j} \qquad (14a)$$

$$R_{t|t-1} = G_t \sum_{t-1|t-1} G_t + 1 \qquad (14b)$$

$$d_t = F_t R_{t|t-1} F_t^T \qquad (14c)$$

$$X^*_{t,j} = X_{t,j} - F_t \theta_{t|t-1,j}] d_t^{-1/2} \qquad (14d)$$

$$\theta_{t|t,j} = \theta_{t|t-1,j} + R_{t|t-1} F_t^T [X_{t,j} - F_t \theta_{t|t-1,j}] d_t^{-1} \qquad (14e)$$

$$\sum_{t|t} = R_{t|t-1}[I - F_t^T F_t R_{t|t-1}] d_t^{-1} \qquad (14f)$$

for $t = 1, \ldots, N$ and $j = 1, \ldots, 5$, where the implicit observation equation is $X_{t,j} = F_t \theta_{t,j}$, $F_t = 1$, and $G_t = \alpha$. The determinant of Γ is

$$|\Gamma| = \prod_{t=1}^{N} d_t.$$

Equation (14d) is the step added to the standard Kalman filter algorithm to carry out the Gram-Schmidt procedure. The Kalman filtering procedure is repeated five times, once for each column of X. The vector Ly is computed similarly by substituting Y_t for $X_{t,j}$ in Eq. (14d), with the implicit observation equation $Y_t = F_t \theta_t$. It follows from Jones (1980) that the initial conditions are $\theta_0 = 0$ and $\Sigma_0 = 1$. Estimation of τ and α is carried out using nonlinear minimization procedures to minimize Eq. (13). The advantage of using the Kalman filter for this model is that Γ^{-1} and $|\Gamma|$ are computed in a highly efficient manner; this involves the computing of only the reciprocals and needs no submatrix inversions. The first-order autoregression model is a special case of the general autoregressive moving average process of orders p and q [ARMA(p,q)] (Priestly, 1981). For this more general error model, the order of the largest submatrix which must be inverted with the Kalman filter algorithm is max $(p, q + 1)$, instead of possibly N, for an arbitrary, unpatterned covariance matrix. The Cholesky factor algorithm, described in Eqs. (14a)–(14f), extends easily to the case in which v is any ARMA(p, q) process. The initial conditions for the algorithm are given in Jones (1980).

III. Computing Posterior Densities for Bayesian Inference Using the Kalman Filter

Bayesian inference is an approach to statistical analysis based on the Bayes' rule in which the probability density of a model parameter θ is determined conditional to the observed experimental data, Y. This conditional probability density of θ given Y is defined as

$$f(\theta|Y) = \frac{f(\theta)f(Y|\theta)}{f(Y)},$$

where $f(\theta)$ is the prior probability density of θ, which summarizes the knowledge of θ before observing Y, $f(Y|\theta)$ is the likelihood of Y given θ, and

$$f(Y) = \int f(\theta)f(Y|\theta)\mathrm{d}\theta$$

is a normalizing constant. Duncan and Horn (1972) and Meinhold and Singpurwalla (1983) demonstrated that the Kalman filter could be used to compute the posterior densities for certain Gaussian linear models in Bayesian statistics. For the satellite tracking problem mentioned earlier, we might wish to characterize the uncertainty in our determination of the state of the satellite at time t given the data at time Y_t^*. This may be expressed formally, in terms of the Bayes' rule, as

$$f(\theta_t|Y_t^*) = \frac{f(\theta_t|Y_{t-1}^*)}{f(Y_t|\theta_t, Y_{t-1}^*)}f(Y_t^*). \tag{15}$$

In this case $f(\theta_t|Y_{t-1}^*)$ summarizes the uncertainty in the knowledge of the state of the satellite at time t, given that the observations made up to time t - 1, and $f(Y_t|\theta_t, Y_{t-1}^*)$ describe the likelihood of the observations at time t, and given the expected state of the system at time t and the observations made up to time t - 1. From the specifications of the model stated in Section I and the Gaussian assumptions on w_t and v_t, it can be shown that the density $f(\theta_t | Y_{t-1}^*)$ is Gaussian with mean vector $\hat{\theta}_{t|t-1}$ and covariance matrix $R_{t|t-1}$ [Eq. (3b)], and that the density $f(\theta_t | Y_t^*)$ [Eq. (3c)] and covariance matrix $\Sigma_{t|t}$ [Eq. (3d)] (Meinhold and Singpurwalla, 1983). The manipulations required to compute the probability density of $f(\theta_t | Y_t^*)$ are entirely defined by the Kalman filter because θ_t depends linearly on θ_{t-1} and also because any Gaussian density is completely described by its mean and its covariance matrix. This suggests that in Bayesian statistical models having similar structure, the Kalman filter can be used to compute posterior probability densities.

For example, consider the regression model

$$Y = XB + \varepsilon,$$

where Y is an $N \times 1$ vector of observations, X is an $N \times p$ design matrix, B is a $p \times 1$ vector of regression coefficients, and ε is an $N \times 1$ Gaussian error vector with zero mean and a known covariance matrix Γ. To complete the Bayesian formulation of the problem, we assume that the prior probability density of B is Gaussian with mean μ and variance Σ. We are interested in computing the posterior probability density of B when given Y. Each observation gives information about the regression coefficient, so if we define the state vector as $B_t = B$, then for $t = 1, \ldots, N$ we have the following observation and state equations

$$Y_t = X_t^T B_t + \varepsilon_t,$$
$$B_t = G_t B_{t-1},$$

where X_t is the tth row of X and G_t is a $p \times p$ identity matrix. Applying the Kalman filter, as defined in Eqs. (3a)–(3e), for $t = 1, \ldots, N$ with $F_t = X_t^T$, $W_t = 0$, $Z_t = \Gamma$ and initial conditions $\theta_0 = B_0$ and $\Sigma_0 = \Sigma$, yields the posterior mean and covariance matrix of B, when Y is given. In a closed form, the mean and covariance matrix were given by Lindley and Smith (1972) as

$$E(B|Y) = (X^T \Gamma^{-1} X + \Sigma^{-1})^{-1} (X^T \Gamma^{-1} Y + \Sigma^{-1} B_0),$$
$$\mathrm{Var}(B|Y) = (X^T \Gamma^{-1} X + \Sigma^{-1})^{-1}.$$

IV. Missing Data Problems and the Kalman Filter

The sequential manner in which the Kalman filter can be used to evaluate Gaussian likelihoods has led to the development of an efficient technique for handling missing data in a time series estimation problem. If we intend to collect data from time $t = 1, \ldots, N$ and some of the observations Y_t are missing, then the fact that these data are not observed must be accounted for in the parameter estimation. As many statistical time series models are linear, have a Markov structure, and assume Gaussian errors, three approaches to evaluating the likelihood function with missing data are possible. The first approach fills in the missing observations with a simple estimate such as the mean. Of the three possibilities, this is the least desirable because it fails to account for the random variation in the distribution of the missing data and also ignores the structure of the model in the problem. The second approach is to evaluate the likelihood, as illustrated in Eqs. (9) through (14). As the observations are not assumed to be evenly spaced, the Kalman filter algorithm proceeds directly from Y_t to Y_{t+k}, assuming k missing observations. The likelihood is then maximized, as described in Section II. Jones (1980) has described this method for fitting ARMA models with missing observations.

The third approach is to perform the maximum-likelihood estimation with the EM algorithm (Dempster *et al.*, 1977). Under the assumptions that the missing data are missing at random and that the sampling density of the data belongs to the exponential family, the EM algorithm takes in to account the unobserved information by replacing the missing components of the sufficient statistics of the data with their expected values which are conditional on the observed data. Missing at random means there is no relation between the value of the missing observation and the probability that it is not observed. To define the EM algorithm, we follow the discussion in Little and Rubin (1987) and let Y_{obs} denote the observed data, Y_{miss} denote the missing data, and β denote the model parameter to be estimated. The log likelihood for the observed data may then be written as

$$\log f(\beta|Y_{\text{obs}}) = \log f(\beta|Y_{\text{obs}}, Y_{\text{miss}}) - \log f(Y_{\text{miss}}|Y_{\text{obs}}, \beta) \qquad (16)$$

Assuming that an estimate of β, say $\beta^{(\ell)}$, is given and taking expectations of both sides of Eq. (16) with respect to $f(Y_{\text{miss}} \mid Y_{\text{obs}}, \beta)$ we obtain

$$\log f(\beta|Y_{\text{obs}}) = Q(\beta|\beta^{(\ell)}) - H(\beta|\beta^{(\ell)}), \qquad (17)$$

where

$$Q(\beta|\beta^{(\ell)}) = \int \log f(\beta|Y_{\text{obs}}, Y_{\text{miss}}) f(Y_{\text{miss}}|Y_{\text{obs}}, \beta^{(\ell)}) d Y_{\text{miss}} \qquad (18)$$

and

$$H(\beta|\beta^{(\ell)}) = \int \log f(Y_{\text{miss}}|Y_{\text{obs}}, \beta) f(Y_{\text{miss}}|Y_{\text{obs}}, \beta^{(\ell)}) d Y_{\text{miss}}.$$

The EM algorithm maximizes Eq. (17) by iterating them between the expectation (E) step [Eq. (18)], in which $Q(\beta|\beta^{(\ell)})$ is determined given $\beta^{(\ell)}$, and the maximization (M) step, given below, in which a $\beta^{(\ell+1)}$ is found such that

$$Q(\beta^{(\ell+1)}|\beta^{(\ell)}) \geq Q(\beta^{(\ell)}|\beta^{(\ell)})$$

It can be shown that the term $H(\beta|\beta^{(\ell)})$ need not be considered in order to maximize the log likelihood of the observed data in Eq. (17) (Little and Rubin, 1987). For a linear time series model with a Markov structure and Gaussian errors, evaluating $Q(\beta|\beta^{(\ell)})$ entails computing $E(Y_{\text{miss}}|Y_{\text{obs}}, \beta^{(\ell)})$ and $\text{Cov}(Y_{\text{miss}}|Y_{\text{obs}}, \beta^{(\ell)})$ because, as noted in Section III, any Gaussian density is completely described by its mean and covariance matrix. The parameter $\beta^{(\ell+1)}$ is estimated by maximizing $Q(\beta^{(\ell+1)}|\beta^{(\ell)})$.

For many probability densities belonging to the exponential family, $\beta^{(\ell+1)}$ can be analytically derived by equating the estimated complete sufficient statistics to their expected values and solving the same. For more general probability densities, it is found numerically. Iteration between the E and M steps is continued until $\beta^{(\ell)}$ converges.

Shumway and Stoffer (1982) have applied the EM algorithm for estimating and predicting total physician expenditures using time series data from the Social Security Administration (1949–1973) and the Health Care Financing Administration (1965–1976). They used a state-space model with Gaussian observation errors to account for the differences in the two estimates of expenditure. They showed that the expectation and covariance of the unknown true expenditures, which were conditional on the observed data, and the current estimates of the model parameters could be efficiently computed by combining the Kalman filter with the fixed-interval smoothing algorithm (Ansley and Kohn, 1982) and the state-space covariance algorithms. For the Kalman filter algorithm defined in Eqs. (1)–(3), the associated fixed-interval smoothing algorithm is

$$\hat{\theta}_{t|N} = \hat{\theta}_{t|t} + A_t(\hat{\theta}_{t+1|N} - G_{t+1}\hat{\theta}_{t|t}) \tag{19a}$$

$$A_t = \Sigma_{t|t} G_{t+1}^T R_{t+1|t}^{-1} (t = N - 1, \ldots, 1) \tag{19b}$$

$$\Sigma_{t|N} = \Sigma_{t|t} + A_t(\Sigma_{t+1|N} - R_{t+1|t})A_t^T \tag{19c}$$

and the state-space covariance algorithm is

$$\Sigma_{t,u|N} = A_t \Sigma_{t+1,u|N} \quad (t < u \leq N) \tag{20}$$

for $t = N - 1, \ldots, 1$, where $\hat{\theta}_{t|N}$ is the expected value of the state vector given Y_N^*, $\Sigma_{t|N}$ is the covariance matrix of the state vector when given Y_N^*, and $\Sigma_{t,u|N}$ is the covariance between the state vectors θ_t and θ_u when given Y_N^*. The form of the state-space covariance algorithm given in Eq. (20) is according to DeJong and MacKinnon (1988). The $\hat{\theta}_{t|N}$ terms represent $E(Y_{\text{miss}}|Y_{\text{obs}}, \beta^{(\ell)})$, whereas, the $\text{Cov}(Y_{\text{miss}}|Y_{\text{obs}}, \beta^{(\ell)})$ terms are determined from $\Sigma_{t|N}$ and $\Sigma_{t,u|N}$. Embedding the Kalman filter, the fixed-interval smoothing algorithm, and the state-space covariance algorithm into the E step efficiently exploits the Markov structure and linear Gaussian error features;characteristics of many statistical time series models.

The Kalman filter and the EM algorithm approaches to the time series missing data problem may also be used when covariates are introduced. They yield the same parameter estimates even if the covariates are completely observed. If there are missing observations in both the covariates and in the time series, then the EM algorithm is the preferred method of analysis, as shown by Schmid (2010) in an investigation that studied the relation between children's lung function and their age, height, gender, and airway responsiveness.

The fixed-interval smoothing algorithm in Eqs. (19a)–(19c) was developed originally as an extension to the Kalman filter for sequential estimation problems in which the observation interval is fixed; and it is important that the state vector estimate at each time point be dependent on all the data (Sage and Melsa, 1971). The algorithm has a clear interpretation in terms of the satellite tracking problem. The state vector $\hat{\theta}_{t|N}$ represents a corrected estimate of the position and velocity of the satellite, after observing all the ground tracking data from time 1 to N. We see from Eq. (19a) that the estimate $\hat{\theta}_{t|N}$ is arrived at by taking a linear combination of $\hat{\theta}_{t|t}$ and $\hat{\theta}_{t+1|N}$.

V. Extensions of the Kalman Filter Algorithm

Many extensions of the Kalman filter algorithm are possible. First, the linear assumption in the state-space model may be relaxed, so that both the transition equation and the observation equation become nonlinear functions of the state vectors. Examples of these models and the computational algorithms to estimate them are described by Sage and Melsa (1971), Kitagawa (1987), and Carlin *et al.* (1992) Second, the Gaussian assumption may be replaced with more general error models. Kitagawa (1987), West and Harrison (1989), and Carlin *et al.* (1992), discuss some of these extensions. Third, the combined Kalman filter and fixed-interval smoothing algorithms may be viewed as a special case of a graphical model in which the N nodes are arranged in a line, and in which the model information first propagates across the nodes from left to right and then propagates from right to left. The information from the observed data enters at each node, as the algorithm makes its first pass from left to right. In a general graphical model, the nodes may be aligned in any spatial configuration, the information may propagate in any direction between the nodes, and the errors are not necessarily Gaussian (Lauritzen and Spiegelhalter, 1988). These models have tremendous flexibility and are gaining greater use since their computational complexity is becoming less of an issue (Spiegelhalter *et al.*, 1993). Finally, Dempster (2010) has developed, in terms of belief functions, a general extension of the Kalman filter principles that is appropriate for computation with both probabilistic and nonprobabilistic information. Meehan (1993) has successfully applied these methods to solve the problems of estimating the posterior probability densities of the parameters in a stochastic differential equation model of diurnal corstisol patterns.

References

Ansley, C. F., and Kohn, R. (1982). *Biometrika* **69,** 486.
Brown, E. N., and Czeisler, C. A. (1992). *J. Biol. Rhythms* **7,** 177.
Carlin, B. P., Polson, N. G., and Stoffer, D. S. (1992). *J. Am. Stat. Assoc.* **87,** 493.
Czeisler, C. A., Kronauer, R. E., Allan, J. S., Duffy, J. F., Jewett, M. E., Brown, E. N., and Ronda, J. M. (1989). *Science* **244,** 1382.
Czeisler, C. A., Johnson, M. P., Duffy, J. F., Brown, E. N., Ronda, J. M., and Kronauer, R. E. (1990). *N. Engl. J. Med.* **322,** 1253.
DeJong, P., and MacKinnon, M. J. (1988). *Biometrika* **75,** 601.
Dempster, A. P. (2010). "Normal Belief Functions and the Kalman Filter," Research Report S-137. Department of Statistics, Harvard University, Cambridge, MA.

Dempster, A. P., Laird, N. M., and Rubin, D. B. (1977). *J. R. Stat. Soc. B* **39**, 1.

Duncan, D. B., and Horn, S. D. (1972). *J. Am. Stat. Assoc.* **67**, 815.

Jones, R. H. (1980). *Technometrics* **22**, 389.

Kalman, R. E. (1960). *J. Basic Eng.* **82**, 34.

Kitagawa, G. (1987). *J. Am. Stat. Assoc.* **82**, 1032.

Lauritzen, S. L., and Spiegelhalter, D. J. (1988). *J. R. Stat. Soc. B* **50**, 157.

Lindley, D. V., and Smith, A. F. M. (1972). *J. R. Stat. Soc. B* **34**, 1.

Little, R. J. A., and Rubin, D. B. (1987). "Statistical Analysis with Missing Data." Wiley, New York.

Meehan, P. M. (1993). "A Bayesian Analysis of Diurnal Cortisol Series" Ph. D. dissertation. Department of Statistics, Harvard University, Cambridge, MA.

Meinhold, R. J., and Singpurwalla, N. D. (1983). *Am. Stat.* **37**, 123.

Priestly, M. B. (1981). "Spectral Analysis and Time Series," Academic Press, London.

Sage, A. P., and Melsa, J. L. (1971). "Estimation Theory with Applications to Communications and Control," McGraw-Hill, New York.

Schmid, C. H. (2010). *J. Am. Stat. Assoc.* invited revision.

Schweppe, F. C. (1965). *IEEE Trans. Inf. Theory* **IT-11**, 61.

Shumway, R. H., and Stoffer, D. S. (1982). *Time Ser. Anal.* **3**, 253.

Spiegelhalter, D. J., Dawid, A. P., Lauritzen, S. L., and Cowell, R. G. (1993). *Stat. Sci.* **8**, 219.

Wecker, W. E., and Ansley, C. F. (1983). *J. Am. Stat. Assoc.* **78**, 81.

West, M., and Harrison, J. (1989). "Bayesian Forecasting and Dynamic Models", Springer-Verlag, New York.

CHAPTER 10

Bayesian Hierarchical Models

Christopher H. Schmid* and Emery N. Brown[†]

*Biostatistics Research Center
Division of Clinical Care Research
New England Medical Center
Boston, Massachusetts, USA

[†]Statistics Research Laboratory
Department of Anesthesia and Critical Care
Massachusetts General Hospital
Harvard Medical School/MIT Division of Health Sciences and Technology
Boston, Massachusetts, USA

I. Introduction

Many random processes generating data for which statistical analyses are required involve multiple sources of variation. These sources represent randomness introduced at different levels of a nested data structure. For example, in a

DOI: 10.1016/B978-0-12-384997-7.00010-8

multicenter study, responses vary within individuals, between individuals, within sites, and between sites. When the multiple measurements taken on individuals are intended to be replicates, the variation is called measurement error.

For example, in a study of growth that we will explore in detail later, measurements of height were taken daily for 3 months on a young child. Despite careful attention to the measurement process, multiple observed readings taken consecutively in a 10-min time span were not identical. In addition to this measurement variability, each child measured varied with respect to the amount of growth and the time at which growth occurred. We will see how a hierarchical model can address these issues.

The standard statistical approach for describing processes with multiple components of variation uses mixed models of fixed and random effects. The fixed effects are quantities about which inference is to be made directly, whereas the random effects are quantities sampled from a population about which inference is desired. Typically, the variance components of the process represent the variances of these populations. An alternative method of description consists of a hierarchy of simpler models, each of which describes one component of variation with a single random effect.

Meta-analysis, a technique used to pool data from different studies in order to estimate some common parameter such as a treatment effect in medical studies, is a particular example of this type (Carlin, 1992; DuMouchel, 1990; Morris and Normand, 1992; Smith *et al.*, 1995). Assume that y_t represents the observed treatment effect from the tth study. The objective of the analysis is to synthesize the results of all of the studies, either summarizing them in a single number as an average treatment effect applicable to all of the studies or as a set of effects describing different subgroups of individuals or studies. Let θ_t be the parameters describing the tth study and let these study-level parameters be expressed in terms of a set of parameters ζ common to all the studies. The hierarchy of models can then be expressed as

$$y_t|\theta_t \sim f(y_t|\theta_t) \quad t = 1, 2, \ldots, T \tag{1}$$

$$\theta_t|\zeta \sim g(\theta_t|\zeta) \quad t = 1, 2, \ldots, T, \tag{2}$$

where the notation indicates that f is the probability distribution of y_t given θ_t and g is the distribution of θ_t given ζ. If inference is desired only about ζ, the two expressions can be combined to describe a distribution for y_t in terms of ζ, but the hierarchical structure permits inference also to be made about the study-specific parameters θ_t.

This hierarchical structure defines the dependence between the many system parameters in such a way that the model can have enough parameters to fit the data while not being overfit. Nonhierarchical models are not as effective because they must either model only the population structure or be overparameterized to describe the individual structure. To the extent that these studies are exchangeable *a priori,* the hierarchical structure also permits the use of information from one study to give a more accurate estimate of the outcome of another study.

The different levels of variation may be further characterized by ascribing causes to them on the basis of covariates. Thus, in Eq. (2), ζ may consist of regression parameters that describe the effect of covariates, creating systematic differences between studies. Model checking may be accomplished by extending the model to incorporate nonlinearity and test for robustness to assumptions.

Maximum likelihood computation of mixed models, a standard technique for fitting hierarchical structures such as those given by Eqs. (1) and (2), assumes normality of the parameter estimates. Although adequate for the large amount of data usually available to estimate the common parameters ζ, asymptotic methods fail to give good estimates for the study-specific parameters θ_t when data are sparse in the *t*th study, especially if the underlying processes are not Gaussian. In such instances, external information about model parameters may be needed to inform the current data. This external information can be incorporated by setting up a Bayesian model.

Bayesian inference is an approach to statistical analysis based on Bayes rule in which model parameters θ are considered as random variables in the sense that knowledge of them is incomplete. Prior beliefs about model parameters θ are represented by a probability distribution $\pi(\theta)$ describing the degree of uncertainty with which these parameters are known. Combining this prior distribution with the data in the form of a likelihood for these parameters leads to a posterior distribution $\pi(\theta|Y)$ of the belief about the location of the unknown parameters θ conditional on the data Y. This posterior probability density of θ given Y is defined as

$$f(\theta|Y) = \frac{f(\theta)f(Y|\theta)}{f(Y)},$$

where $f(\theta)$ is the prior probability density of $\theta, f(Y|\theta)$ which is the likelihood of Y given θ, and

$$f(Y) = \int f(\theta)f(Y|\theta)\mathrm{d}\theta$$

is the marginal distribution of the data and serves as a normalizing constant so that the posterior density integrates to one as required by the rules of probability densities. Statements of the probability of scientific hypotheses and quantification of the process parameters follow directly from this posterior distribution.

A Bayesian version of the hierarchical model can be set up by simply placing a probability distribution on the population parameter ζ in the second stage. Thus, together with Eqs. (1) and (2), we have a third stage

$$\zeta \sim h(\zeta), \tag{3}$$

where the function h represents the prior distribution of ζ. Of course, the Bayesian model is not restricted to only three stages, although in practice this number is often sufficient.

The use of Bayesian methodology has several advantages. It provides a formal framework for incorporating information gained from previous studies into the current analysis. As the data collected accrue and are analyzed, conclusions can be revised to incorporate the new information. The information from new data is combined with knowledge based on previously processed information (possibly data) to reach a new state of knowledge. The Bayesian analysis also provides estimates of the model parameters and their uncertainties in terms of complete posterior probability densities. Furthermore, inference is not restricted just to model parameters. Posterior densities of any functions of the model parameters may be simply computed by resampling methods to be discussed later.

Since its original description by Lindley and Smith (1972). the Bayesian hierarchical model has become a standard tool in Bayesian applications applied to problems in such diverse areas as corporate investment (Smith, 1973), growth curves (Fearn, 1975; Strenio et al., 1983), and educational testing (Rubin, 1981). A text of case studies in biometrical applications of Bayesian methods devoted an entire section to different hierarchical models (Berry and Stang, 1996) and a panel of the National Academy of Sciences advocated the hierarchical model as a general paradigm for combining information from different sources (Graver et al., 1992).

Before describing tools for computing hierarchical models in Section III, we discuss the Gaussian model in the next section. Not only is this the most commonly employed hierarchical structure, but it also permits an illuminating interpretation of model parameters as averages between quantities in different levels of the structure. We will illustrate this interpretation using a state-space representation and computation of the Gaussian form of the Bayesian hierarchical model by forward and backward Kalman filter algorithms (Brown and Schmid, 1994).

The connection between Bayesian linear models and the Kalman filter has been previously made in several contexts. Duncan and Horn (1972) and Meinhold and Singpurwalla (1983) illustrated how the Kalman filter could be used to estimate the means and marginal covariance matrices of the random variables in the nonhierarchical Bayesian linear model with Gaussian error. As part of the E-step in an EM algorithm, Shumway and Stoffer (1982) combined the Kalman filter and the fixed-interval smoothing algorithms (Ansley and Kohn, 1982) to compute the posterior densities of both the state vectors and the missing data in a Gaussian linear state-space model with measurement error. Schmid extended their model to a general first-order autoregressive model for continuous longitudinal data (Schmid, 1996). Wecker and Ansley (1983) also combined the Kalman filter and fixed-interval smoothing algorithms to compute sequentially approximate posterior densities of the state-space vectors in a polynomial spline model.

We illustrate the application of these algorithms to meta-analysis with an example in Section IV. Section V describes a hierarchical model for saltatory growth and Section VI outlines Monte Carlo methods needed to compute non-Gaussian hierarchical models. Section VII presents some methods for model checking and we close with a brief conclusion in Section VIII.

II. The Gaussian Model

Assuming Gaussian probability distributions, we can express the three-stage hierarchical model as follows. The first stage describes a linear model relating the response $\mathbf{y}_t$ to a set of predictors $\mathbf{X}_t$ as

$$\mathbf{y}_t = \mathbf{X}_t\boldsymbol{\beta}_t + \boldsymbol{\varepsilon}_t \quad \boldsymbol{\varepsilon}_t \sim \mathbf{N}(0, \mathbf{V}_t) \tag{4}$$

for $t = 1, \ldots, T$ where $\mathbf{y}_t$ is an N_t-dimensional vector of data on the tth experimental unit, $\mathbf{X}_t$ is a known $N_t \times p$ design matrix, $\boldsymbol{\beta}_t$ is a p-dimensional vector of regression coefficients, and $\boldsymbol{\varepsilon}_t$ is an N_t-dimensional error vector with covariance matrix $\mathbf{V}_t$. We assume that $\boldsymbol{\varepsilon}_t$ and $\boldsymbol{\varepsilon}_u$ are uncorrelated for $t \neq u$.

The second stage describes the prior distribution of each $\boldsymbol{\beta}_t$ as

$$\boldsymbol{\beta}_t = \mathbf{Z}_t\boldsymbol{\gamma} + \boldsymbol{\omega}_t \quad \boldsymbol{\omega}_t \sim N(0, \mathbf{W}_1), \tag{5}$$

where $\mathbf{Z}_t$ is a known $p \times q$ design matrix, $\boldsymbol{\gamma}$ is a q-dimensional vector of hyperparameters, $\boldsymbol{\omega}_t$ is distributed as a p-dimensional Gaussian random vector with covariance matrix $\mathbf{W}_1$, and the random errors $\boldsymbol{\omega}_t$ and $\boldsymbol{\omega}_u$ are uncorrelated for $t \neq u$.

The third stage defines the prior distribution for $\boldsymbol{\gamma}$ as

$$\boldsymbol{\gamma} = \mathbf{G}_0\boldsymbol{\gamma}_0 + \boldsymbol{\nu} \quad \boldsymbol{\nu} \sim N(0, \mathbf{W}_2) \tag{6}$$

in terms of a known $q \times r$ matrix $\mathbf{G}_0$, r hyperparameters $\boldsymbol{\gamma}_0$, and a q-dimensional vector $\boldsymbol{\nu}$ distributed as Gaussian with covariance matrix $\mathbf{W}_2$. Prior knowledge of the hyperparameters in the third stage is often imprecise and a noninformative prior distribution such as one defined by $\mathbf{W}_2^{-1} = 0$ is frequently used.

The objective of the Bayesian analysis for the hierarchical model is to compute the marginal and joint posterior densities of the random variables $(\boldsymbol{\beta}_t | \mathbf{y}_t^*, \boldsymbol{\gamma}_0, \mathbf{W}_1, \mathbf{W}_2)$ for $t = 1, \ldots, T$ and $(\boldsymbol{\gamma} | \mathbf{y}_t^*, \boldsymbol{\gamma}_0, \mathbf{W}_1, \mathbf{W}_2)$ where $\mathbf{y}_t^* = (\mathbf{y}_1, \ldots, \mathbf{y}_t)$. Because of the linear structure of the model and its Gaussian error assumptions, it suffices to determine the first two moments of these densities.

For the present, assume the covariance matrices $\mathbf{V}_t$, $\mathbf{W}_1$, and $\mathbf{W}_2$ to be known. This is not a serious limitation because the algorithms are easily fit into an iterative scheme, such as Gibbs sampling (Gilks *et al.*, 1996), for simultaneously estimating the mean and variance parameters.

It might appear that this three-stage model has too many parameters for the amount of data collected. But note that by collapsing Eqs. (4)–(6), we can express $\mathbf{y}_t$ as a normal distribution with mean $\mathbf{X}_t\mathbf{Z}_t\mathbf{G}_0\boldsymbol{\gamma}_0$ and variance $\mathbf{V}_t + \mathbf{X}_t\mathbf{W}_1\mathbf{X}_t' + \mathbf{X}_t\mathbf{Z}_t\mathbf{W}_2\mathbf{Z}_t'\mathbf{X}_t'$. The parameters describing the data are simply $\boldsymbol{\gamma}_0$, $\mathbf{V}_t$, $\mathbf{W}_1$, and $\mathbf{W}_2$. If $N_t = 1$, the model reduces to a regression with parameters $\boldsymbol{\gamma}_0$ conditional on the variance parameters.

A. State-Space Formulation

A state-space model typically provides a Markov representation of a system that evolves over time or space. It is defined by two equations: the state equation, which describes the temporal or spatial evolution of the system, and the observation equation, which defines the relationship between the observed data and the true state of the system. For the three-stage model in Eqs. (4)–(6), a state equation is written as

$$\boldsymbol{\theta}_t = \mathbf{F}_t \boldsymbol{\theta}_{t-1} + \boldsymbol{\omega}_t^* \tag{7}$$

for $t = 1, \ldots, T$ where $\boldsymbol{\theta}_t = \begin{pmatrix} \boldsymbol{\beta}_t \\ \boldsymbol{\gamma} \end{pmatrix}$, $\boldsymbol{\omega}_t^* = \begin{pmatrix} \boldsymbol{\omega}_t \\ 0 \end{pmatrix}$ is a $(p + q)$ vector with zero mean and covariance matrix $\mathbf{W}_1^* = \begin{pmatrix} \mathbf{W}_1 & 0 \\ 0 & 0 \end{pmatrix}$, $\mathbf{F}_t = \begin{pmatrix} 0 & \mathbf{Z}_t \\ 0 & \mathbf{I}_q \end{pmatrix}$ is the $(p + q) \times (p + q)$ transition matrix, and $\mathbf{I}_q$ is the $q \times q$ identity matrix. An observation equation is written as

$$\mathbf{y}_t = \mathbf{X}_t^* \boldsymbol{\theta}_t + \boldsymbol{\varepsilon}_t, \tag{8}$$

where $\mathbf{X}_t^*$ is the $N_t \times 2p$ matrix defined as $\mathbf{X}_t^* = (\mathbf{X}_t 0)$.

The state-space model divides $\boldsymbol{\gamma}$ and the $\boldsymbol{\beta}_t$ among the state vectors $\boldsymbol{\theta}_t$ so that they may be estimated sequentially by the Kalman filter, fixed-interval smoothing, and state-space covariance algorithms working sequentially forward and then backward in time as each piece of data $\mathbf{y}_t$ is incorporated. Starting with initial estimates $\hat{\boldsymbol{\theta}}_0$ and $\mathbf{S}_0$ of the state vector and its covariance matrix, the forward pass uses the Kalman filter to compute the expectation and covariance of $\boldsymbol{\theta}_t$ given $\mathbf{y}_t^*$ for $t = 1, \ldots, T$. The final forward step gives the correct expectation and covariance for $\boldsymbol{\theta}_T$, but the estimated moments of $\boldsymbol{\theta}_t$ for $t < T$ are incompletely updated, using only the data up to time t. Therefore, starting with the completely updated $\boldsymbol{\theta}_T$, the fixed-interval smoothing and covariance algorithms work back from time T to time 0, updating $\hat{\boldsymbol{\theta}}_t$ and $\mathbf{S}_t$ for $\mathbf{y}_u^*$ when $u > t$, finally obtaining the fully updated moments $\hat{\boldsymbol{\theta}}_{t|T}$ and $\mathbf{S}_{t|T}$. The details of one step for each of these algorithms follow.

B. Kalman Filter

Starting with $\hat{\boldsymbol{\theta}}_{t-1|t-1}$, the estimate of $\boldsymbol{\theta}_{t-1}$ given $\mathbf{y}_{t-1}^*$, and its covariance matrix $\mathbf{S}_{t-1|t-1}$, the Kalman filter algorithm for this state-space model is

$$\hat{\boldsymbol{\theta}}_{t|t-1} = \mathbf{F}_t \hat{\boldsymbol{\theta}}_{t-1|t-1} \tag{9}$$

$$\mathbf{S}_{t|t-1} = \mathbf{F}_t \mathbf{S}_{t-1|t-1} \mathbf{F}_t^T + \mathbf{W}_1^* \tag{10}$$

$$\mathbf{K}_t = \mathbf{S}_{t|t-1} \mathbf{X}_t^{*T} (\mathbf{X}_t^* \mathbf{S}_{t|t-1} \mathbf{X}_t^{*T} + \mathbf{V}_t)^{-1} \tag{11}$$

$$\hat{\boldsymbol{\theta}}_{t|t} = \hat{\boldsymbol{\theta}}_{t|t-1} + \mathbf{K}_t (\mathbf{y}_t - \mathbf{X}_t^* \hat{\boldsymbol{\theta}}_{t|t-1}) \tag{12}$$

$$\mathbf{S}_{t|t} = (\mathbf{I}_{p+q} - \mathbf{K}_t \mathbf{X}_t^*) \mathbf{S}_{t|t-1} \tag{13}$$

for $t = 1, \ldots, T$ where $\hat{\boldsymbol{\theta}}_0 = \begin{pmatrix} \mathbf{Z}_t \mathbf{G}_0 \boldsymbol{\gamma}_0 \\ \mathbf{G}_0 \boldsymbol{\gamma}_0 \end{pmatrix}$ and $\mathbf{S}_0 = \begin{pmatrix} \mathbf{W}_1 + \mathbf{Z}_t \mathbf{W}_2 \mathbf{Z}_t^T & \mathbf{Z}_t \mathbf{W}_2 \\ \mathbf{W}_2 \mathbf{Z}_t^T & \mathbf{W}_2 \end{pmatrix}$ are the initial mean and covariance matrices, respectively.

C. Fixed–Interval Smoothing Algorithm

The associated fixed-interval smoothing algorithm is

$$\hat{\boldsymbol{\theta}}_{t|T} = \hat{\boldsymbol{\theta}}_{t|t} + \mathbf{A}_t(\hat{\boldsymbol{\theta}}_{t+1|T} - \hat{\boldsymbol{\theta}}_{t+1|t}) \tag{14}$$

$$\mathbf{A}_t = \mathbf{S}_{t|t} \mathbf{F}_{t+1}^T \mathbf{S}_{t+1|t}^{-1} \tag{15}$$

$$\mathbf{S}_{t|T} = \mathbf{S}_{t|t} + \mathbf{A}_t(\mathbf{S}_{t+1|T} - \mathbf{S}_{t+1|t})\mathbf{A}_t^T \tag{16}$$

recursively computed from $t = T-1, \ldots, 1$ where $\hat{\boldsymbol{\theta}}_{t|T}$ and $\mathbf{S}_{t|T}$ are, respectively, the estimate of $\boldsymbol{\theta}_t$ and its covariance matrix given $\mathbf{Y}_T^*$. The initial conditions, $\hat{\boldsymbol{\theta}}_{T|T}$ and $\mathbf{S}_{T|T}$, are obtained from the last step of the Kalman filter. Thus, computation of $\hat{\boldsymbol{\theta}}_{t|T}$ and $\mathbf{S}_{t|T}$ requires only the output of $\hat{\boldsymbol{\theta}}_{t+1|T}$ and $\mathbf{S}_{t|t}$ from the tth forward step and the output of $\hat{\boldsymbol{\theta}}_{t+1|T}$ and $\mathbf{S}_{t+1|T}$ from the previous backward step.

D. State-Space Covariance Algorithm

The state-space covariance algorithm (DeJong and Mackinnon, 1988) gives the covariance between $\hat{\boldsymbol{\theta}}_{t|T}$ and $\hat{\boldsymbol{\theta}}_{u|T}$ as

$$\mathbf{S}_{t,u|T} = \mathbf{A}_t \mathbf{S}_{t+1,u|T} \quad t < u \leq T. \tag{17}$$

By the definitions of the Kalman filter and the fixed-interval smoothing algorithm, the probability density of $(\boldsymbol{\beta}_t | \mathbf{y}_T^*, \boldsymbol{\gamma}_0, \mathbf{W}_1, \mathbf{W}_2)$ is the Gaussian density whose mean vector is the first p components of $\hat{\boldsymbol{\theta}}_{t|T}$ and whose covariance matrix is the left upper $p \times p$ submatrix of $\mathbf{S}_{t|T}$. The probability density of $(\boldsymbol{\gamma} | \mathbf{y}_T^*, \boldsymbol{\gamma}_0, \mathbf{W}_1, \mathbf{W}_2)$ is the Gaussian density whose mean is the second q components of any $\hat{\boldsymbol{\theta}}_{t|T}$ and whose covariance matrix is the right lower $q \times q$ submatrix of any $\mathbf{S}_{t|T}$. The posterior covariance between $\boldsymbol{\beta}_t$ and $\boldsymbol{\gamma}$ is given by the off-diagonal blocks of $\mathbf{S}_{t|T}$ for $t = 1, \ldots, T$, whereas the posterior covariance between $\boldsymbol{\beta}_t$ and $\boldsymbol{\beta}_u$ for $t \neq u$ is given by the left upper $p \times p$ submatrix of $\mathbf{S}_{t,u|T}$. The posterior covariance between $\boldsymbol{\beta}_t$ and $\boldsymbol{\gamma}$ can be used to assess the proximity of the tth experimental unit to the mean of the second stage of the hierarchy.

When no order dependence is assumed *a priori* among the $\mathbf{y}_t$ values, the estimates of the posterior densities are independent of the sequence in which the data enter the algorithm. The sequence chosen determines only the point in the algorithm at which a particular posterior density is estimated. As we show next, the

Kalman filter updating makes the effect on the posterior densities of adding new data simple to understand and simple and quick to compute.

E. Quantifying Contribution of New Information

Consider the simplest form of the model where each unit has one response (i.e., $N_t = 1$) and a simple mean is estimated at each stage so that

$$y_t = \beta_t + \varepsilon_t \tag{18}$$

$$\beta_t = \gamma + \omega_t \tag{19}$$

$$\gamma = \gamma_0 + v. \tag{20}$$

Because only the Kalman (forward) filter is needed to compute the posterior of γ, we only need to examine Eqs. (9)–(13) to understand how the population mean γ is computed.

Let us write $\gamma^{(t-1)}$ for the expectation $E(\gamma | \mathbf{y}_{t-1}^*)$ of γ given $\mathbf{y}_{t-1}^*$. Application of Eqs. (11)–(13) gives the expectation of γ after the new datum y_t is observed as the weighted average

$$\gamma^{(t)} = (1 - \omega_1)\gamma^{(t-1)} + \omega_1 y_t \tag{21}$$

with posterior variance

$$V(\gamma | \mathbf{y}_t^*) = V_\gamma^{(t)} = (1 - \omega_1) V_{\gamma^{(t-1)}}, \tag{22}$$

where the weight $\omega_1 = V_{\gamma^{(t-1)}}/(V_{\gamma^{(t-1)}} + W_1 + V_t)$.

The updated posterior mean for γ is a compromise between the previous estimate of the mean, $\gamma^{(t-1)}$, and the new data, y_t. If the new observation, y_t, is larger than expected, then the expectation of γ will increase; if it is smaller than expected, the expected value of γ will decrease. Large measurement variability of the new data (large V_t), large between-unit variation (large W_1), or precise prior information about γ (small $V_{\gamma^{(t-1)}}$) leads to small changes in γ. Conversely, substantial change in the posterior of γ will occur when y_t is precisely measured, experimental units are homogeneous, or γ is not well estimated.

The posterior variance is reduced by an amount related to the previous estimate of the variance $V_{\gamma^{(t-1)}}$, the variance of the new data V_t, and the between-unit variance W_1. Reduction is greatest when V_t and W_1 are small relative to $V_{\gamma^{(t-1)}}$ because then the tth unit is providing a lot of information relative to that provided by previous units.

These conditions are intuitively reasonable. If the new data are imprecisely measured, we should not have much confidence in them and therefore would not want them to affect our model estimates substantially. Likewise, large between-unit variation implies only a loose connection between the units and so a new observation will have less effect on the existing structure. Finally, new data should be less influential if we already have a good estimate of the population mean.

To consider information about β_t, we note that from Eq. (9), the best estimate of β_t before seeing y_t is the population mean $\gamma^{(t-1)}$ and the estimated precision is $V_{\gamma^{(t-1)}} + W_1$, namely, the variance of the population mean plus the variance of the random effect. Applying Eqs. (11)–(13) gives

$$E(\beta_t|\mathbf{y}_t^*) = \beta_t^{(t)} = (1 - \omega_2)\gamma^{(t-1)} + \omega_2 y_t \tag{23}$$

and

$$V(\beta_t|\mathbf{y}_t^*) = V_{\beta_t^{(t)}} = \omega_2 V_t = (1 - \omega_2)(V_{\gamma^{(t-1)}} + W_1), \tag{24}$$

where $\omega_2 = (V_{\gamma^{(t-1)}} + W_1)/(V_{\gamma^{(t-1)}} + W_1 + V_t)$.

The updated posterior mean for β_t is a compromise between the prior mean, $\gamma^{(t-1)}$, and the new data, y_t. A precisely estimated data value (small V_t), substantial prior between-unit heterogeneity (large W_1), or a poor prior estimate of γ [large $V_{\gamma^{(t-1)}}$] will give more weight to y_t. The last condition follows from the idea that if prior knowledge of β_t, as expressed by the current knowledge $\gamma^{(t-1)}$ of γ, is imprecise, the previous observations will have little information to give about β_t.

The posterior variance is also closely related to the precision of the new data. If y_t is well estimated so that V_t is small, then its associated random effect β_t will also be well estimated. Furthermore, even if V_t is not small, but the prior variance $V_{\gamma^{(t-1)}} + W_1$ is small relative to V_t, then β_t can be precisely estimated using information from previous units.

In general, data on new units have the greatest effect on current posterior estimates if the new data are precisely measured and the new units and current unit are closely related. The filtering equations, therefore, quantify the amount of information provided by new data and the effect of the complete hierarchical structure on the posterior distributions of model parameters. Knowledge about a unit-specific parameter depends not just on the data gathered on that unit, but through the model, on information gathered from other units. As data collection proceeds, the influence of new units on population parameters decreases, but the influence of the population parameters on the new units increases.

F. Simplified Computations for Incorporating Additional Data

The sequential form of these computations also leads to efficient incorporation of new data. To update the posterior densities of β_t and γ when data from several new experimental units $y_{T+1}, \ldots, y_{T+k}$ are collected, we could combine the original and new data into a single sequence and replace T with $T + k$ in the Kalman filter algorithm, but this approach ignores the previous processing of the original data and is therefore not computationally efficient.

A better method uses the stored $\hat{\boldsymbol{\theta}}_{t|t}$ and $\mathbf{S}_{t|t}$ from the original calculations as inputs to a Kalman filter, which proceeds forward from $t = T + 1$ to $T + k$. A backward pass from $t = T$ to $t = 1$ with the fixed-interval smoothing and state-space covariance algorithms then completes updating the posterior density. This process saves the first T steps of the forward filter compared to repeating the full algorithm.

When new studies are added incrementally and the second-stage regression model is of primary interest, as it might be in meta-analysis, further efficiency can be realized by skipping the backward pass. Rather, we can update the second-stage regression estimates by running the forward algorithm one step further as each new unit is added. When the regression model is sufficiently precise, we can make one pass back through the data to update all the individual studies.

Another form of data accumulation occurs when new observations arrive for experimental units y_t for which some data have already been collected. In this case, the structure of the problem is different because the updated β_t are no longer partially exchangeable given $\mathbf{y}^*_t$. Hence, the Kalman filter, fixed-interval smoothing algorithm, and state-space covariance algorithms as stated here cannot be applied. In meta-analysis, at least, this type of accumulating data is uncommon because new data usually represent new studies. Nevertheless, interim analysis of several concurrent studies might contribute new data of this type.

III. Computation

The complete posterior distribution of $\boldsymbol{\gamma}$ and $\boldsymbol{\beta} = (\beta_1, \beta_2, \ldots, \beta_T)$ may be expressed algebraically as

$$(\boldsymbol{\beta}, \boldsymbol{\gamma})|\mathbf{y} \sim N(\mathbf{B}\boldsymbol{\eta}, \mathbf{B})$$

with

$$\mathbf{B}^{-1} = \begin{pmatrix} \mathbf{X}^T\mathbf{V}^{-1}\mathbf{X} + \boldsymbol{\Omega}_1^{-1} & -\boldsymbol{\Omega}_1^{-1}\mathbf{Z} \\ -\mathbf{Z}^T\boldsymbol{\Omega}_1^{-1} & \mathbf{Z}^T\boldsymbol{\Omega}_1^{-1}\mathbf{Z} + \mathbf{W}_2^{-1} \end{pmatrix} \text{ and } \boldsymbol{\eta} = \begin{pmatrix} \mathbf{X}^T\mathbf{V}^{-1}\mathbf{y} \\ \mathbf{W}_2^{-1}\mathbf{G}_0\boldsymbol{\gamma}_0 \end{pmatrix},$$

where $\mathbf{y}$ is the vector of T observed responses,

$$\mathbf{X} = \begin{pmatrix} \mathbf{X}_1 & 0 & \cdot & 0 \\ 0 & \mathbf{X}_2 & \cdot & 0 \\ \cdot & \cdot & \cdot & \cdot \\ 0 & 0 & \cdot & \mathbf{X}_T \end{pmatrix}$$

$$\mathbf{V} = \begin{pmatrix} \mathbf{V}_1 & 0 & \cdot & 0 \\ 0 & \mathbf{V}_2 & \cdot & 0 \\ \cdot & \cdot & \cdot & \cdot \\ 0 & 0 & \cdot & \mathbf{V}_T \end{pmatrix}$$

$$\boldsymbol{\Omega}_1 = \begin{pmatrix} \mathbf{W}_1 & 0 & \cdot & 0 \\ 0 & \mathbf{W}_1 & \cdot & 0 \\ \cdot & \cdot & \cdot & \cdot \\ 0 & 0 & \blacklozenge_c & \mathbf{W}_1 \end{pmatrix}$$

and

$$\mathbf{Z} = \begin{pmatrix} \mathbf{Z}_1 \\ \mathbf{Z}_2 \\ \cdot \\ \mathbf{Z}_T \end{pmatrix}.$$

Inverting $\mathbf{B}$ is the major computational task in evaluating this joint posterior distribution. The Kalman filter algorithm described in Section II is generally much faster than brute force inversion as the number of observations and regression parameters increases, but the SWEEP algorithm as described in Carlin (1990) is even quicker.

To compute by SWEEP, first express the hierarchical model as a multivariate Gaussian density

$$\begin{pmatrix} \mathbf{y} \\ \boldsymbol{\beta} \\ \boldsymbol{\gamma} \end{pmatrix} \sim N \left[\begin{pmatrix} \boldsymbol{\mu}_y \\ \boldsymbol{\mu}_\beta \\ \boldsymbol{\mu}_\gamma \end{pmatrix}, \begin{pmatrix} \boldsymbol{\Sigma}_y & \boldsymbol{\Sigma}_{y,\beta} & \boldsymbol{\Sigma}_{y,\gamma} \\ \boldsymbol{\Sigma}_{\beta,y} & \boldsymbol{\Sigma}_\beta & \boldsymbol{\Sigma}_{\beta,\gamma} \\ \boldsymbol{\Sigma}_{\gamma,y} & \boldsymbol{\Sigma}_{\gamma,\beta} & \boldsymbol{\Sigma}_\gamma \end{pmatrix} \right]$$

and put this in the following symmetric tableau (omitting the lower triangular portion for ease of presentation)

$$\begin{array}{ccc} (\mathbf{y} - \boldsymbol{\mu}_y)^T & -\boldsymbol{\mu}_\beta^T & -\boldsymbol{\mu}_\gamma^T \\ \hline \boldsymbol{\Sigma}_y & \boldsymbol{\Sigma}_{y,\beta} & \boldsymbol{\Sigma}_{y,\gamma} \\ & \boldsymbol{\Sigma}_\beta & \boldsymbol{\Sigma}_{\beta,\gamma} \\ & & \boldsymbol{\Sigma}_\gamma \end{array}$$

For a general block matrix $\begin{pmatrix} A & B \\ C & D \end{pmatrix}$, sweeping on D gives $\begin{pmatrix} A - BD^{-1}C & BD^{-1} \\ D^{-1}C & -D^{-1} \end{pmatrix}$. The reverse operation, called reverse sweeping, that undoes the sweeping operation changes $\begin{pmatrix} A & B \\ C & D \end{pmatrix}$ into $\begin{pmatrix} A - BD^{-1}C & -BD^{-1} \\ -D^{-1}C & -D^{-1} \end{pmatrix}$. Thus sweeping on $\mathbf{y}$ in the tableau above gives

$$\begin{array}{ccc} (\mathbf{y} - \boldsymbol{\mu}_y)^T \boldsymbol{\Sigma}_y^{-1} & -\boldsymbol{\mu}_\beta^T - (\mathbf{y} - \boldsymbol{\mu}_y)^T \boldsymbol{\Sigma}_y^{-1} \boldsymbol{\Sigma}_{y,\beta} & -\boldsymbol{\mu}_\gamma^T - (\mathbf{y} - \boldsymbol{\mu}_y)^T \boldsymbol{\Sigma}_y^{-1} \boldsymbol{\Sigma}_{y,\gamma} \\ \hline -\boldsymbol{\Sigma}_y^{-1} & \boldsymbol{\Sigma}_y^{-1} \boldsymbol{\Sigma}_{y,\beta} & \boldsymbol{\Sigma}_y^{-1} \boldsymbol{\Sigma}_{y,\gamma} \\ & \boldsymbol{\Sigma}_\beta - ** \boldsymbol{\Sigma}_{\beta,y} \boldsymbol{\Sigma}_y^{-1} \boldsymbol{\Sigma}_{y,\beta} & \boldsymbol{\Sigma}_{\beta,\gamma} - \boldsymbol{\Sigma}_{\beta,y} \boldsymbol{\Sigma}_y^{-1} \boldsymbol{\Sigma}_{y,\gamma} \\ & & \boldsymbol{\Sigma}_\gamma - \boldsymbol{\Sigma}_{\gamma,y} \boldsymbol{\Sigma}_y^{-1} \boldsymbol{\Sigma}_{y,\gamma} \end{array}$$

from which the joint posterior probability density $[\boldsymbol{\beta}, \boldsymbol{\gamma}|\mathbf{y}]$ may be read directly. The means of $\boldsymbol{\beta}$ and $\boldsymbol{\gamma}$ are given by the negatives of the entries in the second and third columns, respectively, of the first row. The entries in the last two rows give the posterior covariance matrix of $\boldsymbol{\beta}$ and $\boldsymbol{\gamma}$.

Rather than sweeping $\mathbf{y}$ directly, it is actually more efficient to first sweep $\boldsymbol{\gamma}$ and $\boldsymbol{\beta}$ before sweeping $\mathbf{y}$ and then undo the sweeps on $\boldsymbol{\gamma}$ and $\boldsymbol{\beta}$ by reverse sweeps to reproduce the effect of sweeping on $\mathbf{y}$ alone. Though this modified SWEEP algorithm involves five sets of computations, efficiency is gained because the initial sweeps on γ and β diagonalize $\sum_y$ to the simple form $\sigma_\varepsilon^2 \mathbf{I}$ so that $\mathbf{y}$ may be swept quite simply analytically. This leaves only the smaller dimensional reverse sweeps of γ and β for numerical computation.

While the Kalman filter and SWEEP are each efficient algorithms for computing the posteriors of the mean parameters γ and β in Gaussian models, they assume fixed variance parameters $\mathbf{V}_t$, $\mathbf{W}_1$, and $\mathbf{W}_2$. Typically, variances will be unknown and so these algorithms must be embedded inside a larger algorithm in which the variances are first estimated and then the means are computed conditional on the variances.

One algorithm that can be used for meta-analysis uses the result that the within-study covariance matrices $\mathbf{V}_t$ are usually so well estimated from each study that they may be assumed known (DuMouchel, 1990). To incorporate the uncertainty arising from the between-study variance $\mathbf{W}_1$, average the conditional posterior distribution of β_t and γ given $\mathbf{W}_1$ over the posterior distribution of $\mathbf{W}_1$. Assuming a noninformative prior for $\mathbf{W}_2$ so that $\mathbf{W}_2^{-1} \rightarrow 0$ gives

$$[\gamma|\mathbf{y}_T^*, \mathbf{W}_1][\mathbf{W}_1|\mathbf{y}_T^*] \propto [\mathbf{W}_1][\gamma|\mathbf{W}_1][\mathbf{y}_T^*|\gamma, \mathbf{W}_1],$$

where the notation $[\mathbf{Y}|\mathbf{X}]$ represents the conditional density of the random variable $\mathbf{Y}$ given the random variable $\mathbf{X}$. Letting the prior distribution for $[\gamma, \mathbf{W}_1]$ be constant, suggested in Berger (1985) as an appropriate form of a noninformative prior distribution for hierarchical models, gives

$$[\mathbf{W}_1|\mathbf{y}_T^*] \propto [\mathbf{y}_T^*|\gamma, \mathbf{W}_1]/[\gamma|\mathbf{y}_T^*, \mathbf{W}_1].$$

Combining Eqs. (4) and (5), we have

$$\mathbf{y}_t^*|\gamma, \mathbf{W}_1 \sim N(\mathbf{X}_t\mathbf{Z}_t\gamma, \mathbf{V}_t + \mathbf{X}_t\mathbf{W}_1\mathbf{X}_t^T)$$

and standard calculations show that

$$\gamma|\mathbf{y}_T^*, \mathbf{W}_1 \sim N(\mathbf{m}, \mathbf{\Sigma}),$$

where $\mathbf{m} = (\mathbf{Z}'\mathbf{\Sigma}^{-1}\mathbf{Z})^{-1}\mathbf{Z}'\mathbf{\Sigma}^{-1}\mathbf{y}_T^*$, $\sum$ is the diagonal matrix with diagonal elements $\mathbf{V}_t + \mathbf{W}_1$, and $\mathbf{Z}$ is the matrix defined previously. If V_t and W_1 are scalars, it can then be shown using these last two expressions that the log posterior density for W_1 above is proportional to

$$g = \sum_{t=1}^{T} \log[(V_t + W_1)]^{-1} - \log|\mathbf{Z}'\sum^{-1}\mathbf{Z}|$$

$$- \sum_{t=1}^{T} y_t^2/(V_t + W_1) + (\mathbf{Z}'\sum^{-1}\mathbf{y}_T^*)^T(\mathbf{Z}'\sum^{-1}\mathbf{Z})^{-1}(\mathbf{Z}'\sum^{-1}\mathbf{y}_T^*).$$

An estimate of the posterior for β and γ may then be obtained using Monte Carlo integration over this log posterior by setting up a grid of, say, 100 points that span the range of the distribution of $\mathbf{W}_1$ and then computing the ergodic average $\frac{1}{100}\sum_{k=1}^{100} f_k(\beta, \gamma|W_1) * g_k(W_1)$ where the f_k and g_k are evaluated at each of the 100 points. Another technique for performing this numerical integration uses Markov chain Monte Carlo (MCMC) and is discussed in Section VI.

IV. Example: Meta-Regression

To illustrate application of the Bayesian hierarchical linear model to meta-analysis, we look at the effect of the time from chest pain onset until treatment on the relative risk of mortality following thrombolytic therapy for acute myocardial infarction. Eight large (>1000 patients) randomized control trials of the thrombolytic agents streptokinase (SK), urokinase (UK), recombinant tissue plasminogen activator (tPA), and anistreplase (APSAC) have reported outcomes by subgroups of patients according to time-to-treatment (AIMS Trial Study Group, 1988; ASSET Trial Study Group, 1988; EMERAS Collaborative Group, 1993; GISSI Study Group, 1986; ISAM Study Group, 1986; ISIS-2 Collaborative Group, 1988; LATE Study Group, 1993; USIM Collaborative Group, 1991).

The Bayesian hierarchical model may be used to perform a meta-analysis combining the results from the subgroups in these studies. For the first stage, let y_t be the observed log relative risk in the tth subgroup having a Gaussian distribution centered about the true subgroup relative risk β_t with random error ε_t having variance σ_ε^2. The second stage describes the representation of β_t as a simple linear regression on time-to-treatment Time_t with Gaussian error ω_t having variance W_1, that is, $\beta_t = \delta_0 + \delta_1 * \text{Time}_t + \omega_t$. Thus $\boldsymbol{\gamma} = \begin{pmatrix} \delta_0 \\ \delta_1 \end{pmatrix}$ and $\mathbf{Z}_t = (1\ \text{Time}_t)$. Finally, the third stage describes a Gaussian prior distribution for $\boldsymbol{\gamma}$ with mean $\boldsymbol{\gamma}_0$ and variance $\mathbf{W}_2$. This hierarchical structure may be put into the state-space form of Eqs. (7) and

(8) by setting $\boldsymbol{\theta}_t = \begin{pmatrix} \beta_t \\ \delta_0 \\ \delta_1 \end{pmatrix}$, $\mathbf{F}_t = \begin{pmatrix} 0 & 1 & \text{Time}_t \\ 0 & 1 & 0 \\ 0 & 0 & 1 \end{pmatrix}$, $\boldsymbol{\theta}_{t-1} = \begin{pmatrix} \beta_{t-1} \\ \delta_0 \\ \delta_1 \end{pmatrix}$, $\mathbf{X}_t^* = (100)$,

$\mathbf{W}_t^* = \begin{pmatrix} W_1 & 0 & 0 \\ 0 & 0 & 0 \\ 0 & 0 & 0 \end{pmatrix}$, and $V_t = \sigma_\varepsilon^2$.

Our objective is to compute the joint posterior density of β_t and $\boldsymbol{\gamma}$. This may be accomplished by the Kalman filter algorithms conditional on the variance components $\mathbf{V}_t$, W_1, and $\mathbf{W}_2$ together with the Monte Carlo integration as presented above.

Though averaging over the variance component makes the posterior strictly non-Gaussian, the use of a uniform prior and the substantial amount of data available make an asymptotic Gaussian posterior a good approximation. Figure 1 shows the posterior density of W_1 under both a constant regression model, $\beta_t = \delta_0$, and the model linear in time, $\beta_t = \delta_0 + \delta_1 \text{Time}_t + \omega_t$. The introduction of time-to-treatment reduces the between-study variance so that the posterior mode for W_1 is essentially zero. Nevertheless, the skewness of the posterior indicates that it is quite probable that $W_1 > 0$.

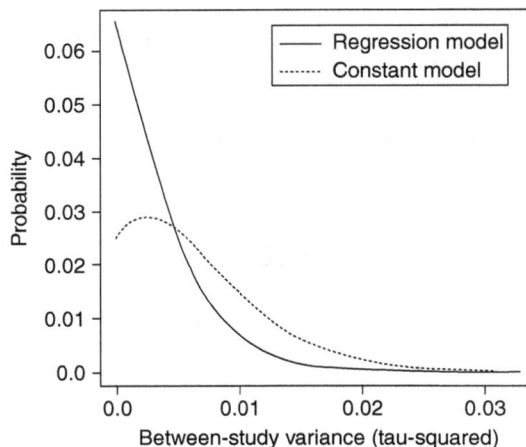

Fig. 1 Unnormalized posterior density of between-study variance (W_1) under models with and without time-to-treatment as a regression variable.

Table I shows the raw data (in terms of the percent risk reduction) for the subgroups together with the point estimates and 95% confidence intervals based on (1) the observed data, y_t; (2) the posteriors for the β_t (averaging over the likelihood for W_1); and (3) the regression line $\delta_0 + \delta_1$ Time$_t$. Figure 2 plots the fitted regression line versus time-to-treatment superimposed on the observed point estimates and Bayes estimates for the subgroups.

Table I shows that the Bayes estimate of each true study effect is a weighted average of the observed effect y_t and the regression estimate, $\delta_0 + \delta_1$ Time$_t$. Figure 2 shows this weighting graphically with the Bayes estimates (represented by + signs) falling between the observed effects (squares) and the regression line. Larger studies (denoted by bigger squares) get more weight; their Bayes estimates are proportionately closer to the observed effect than to the regression line. Conversely, smaller studies carry little weight; their Bayes estimates are pulled almost completely into the regression line. The small amount of between-study variance, however, shrinks all estimates close to the regression line.

Intuitively, if we believe the model, the exchangeability of the study means lets us use information from the other studies to better estimate each treatment effect. Thus, for the first subgroup, the GISSI-1 study patients with mean time-to-treatment of 0.7 h, the Bayes estimate is a 28% reduction, between the 47% observed reduction and the 25% average reduction at 0.7 h from the regression model. Because the evidence from the other studies indicates less benefit, we downweight the estimate from GISSI-1 toward the regression average. If we could discover other factors that explained the additional benefit observed in GISSI-1, we could incorporate them into the regression model, thus increasing the regression estimate (and so the Bayes estimate) toward the observed relative risk for GISSI-1.

Table I
Mortality Data From Nine Large Placebo–Control Studies of Thrombolytic Agents After Myocardial Infarction[a]

Study	Year	Time (h)	Treated Deaths	Treated Total	Control Deaths	Control Total	y_t Lower	y_t Mean	y_t Upper	β_t Lower	β_t Mean	β_t Upper	$\delta_0 + \delta_1 *$ Time Lower	$\delta_0 + \delta_1 *$ Time Mean	$\delta_0 + \delta_1 *$ Time Upper
GISSI-1	1986	0.7	52	635	99	642	27	47	61	17	28	38	16	25	34
ISIS-2	1988	1.0	29	357	48	357	6	40	61	15	26	36	14	25	35
USIM	1991	1.2	45	596	42	538	-45	3	35	12	22	33	14	25	35
ISAM	1986	1.8	25	477	30	463	-37	18	51	13	24	34	11	24	35
ISIS-2	1988	2.0	72	951	111	957	13	35	51	15	26	35	10	24	36
GISSI-1	1986	2.0	226	2381	270	2436	-1	14	28	12	21	30	10	24	36
ASSET	1988	2.1	81	992	107	979	2	25	43	14	24	33	10	24	36
AIMS	1988	2.7	18	334	30	326	-3	41	67	13	24	34	8	23	36
USIM	1991	3.0	48	532	47	535	-51	-3	30	10	21	31	7	23	36
ISIS-2	1988	3.0	106	1243	152	1243	12	30	45	15	24	33	7	23	36
EMERAS	1993	3.2	51	336	56	327	-25	11	37	11	22	31	6	23	37
ISIS-2	1988	4.0	100	1178	147	1181	13	32	46	14	24	32	3	22	37
ASSET	1988	4.1	99	1504	129	1488	2	24	41	13	22	31	3	22	37
GISSI-1	1986	4.5	217	1849	254	1800	2	17	30	12	20	28	1	21	37
ISAM	1986	4.5	25	365	31	405	-49	11	46	10	21	31	1	21	37
AIMS	1988	5.0	14	168	31	176	14	53	74	11	22	32	0	21	38
ISIS-2	1988	5.5	164	1621	190	1622	-5	14	29	10	19	27	-2	20	38
GISSI-1	1986	7.5	87	693	93	659	-17	11	32	7	17	26	-9	18	39
LATE	1993	9.0	93	1047	123	1028	4	26	42	8	18	27	-14	17	39
EMERAS	1993	9.5	133	1046	152	1034	-7	14	30	6	15	24	-16	16	39
ISIS-2	1988	9.5	214	2018	249	2008	-2	14	28	7	16	24	-16	16	39
GISSI-1	1986	10.5	46	292	41	302	-71	-16	21	0	13	24	-20	15	39
LATE	1993	18.0	154	1776	168	1835	-17	5	23	-9	6	18	-49	6	40
ISIS-2	1988	18.5	106	1224	132	1227	-3	20	37	-8	8	21	-51	5	40
EMERAS	1993	18.5	114	875	119	916	-27	0	21	-11	4	17	-51	5	40

[a]The estimated true percent risk reduction (with 95% confidence limits) for each time subgroup is shown for the observed data and results from three regression models: the fixed effects model, the Bayes model, and the estimate that would be predicted if a new study were performed with the same mean time to treatment.

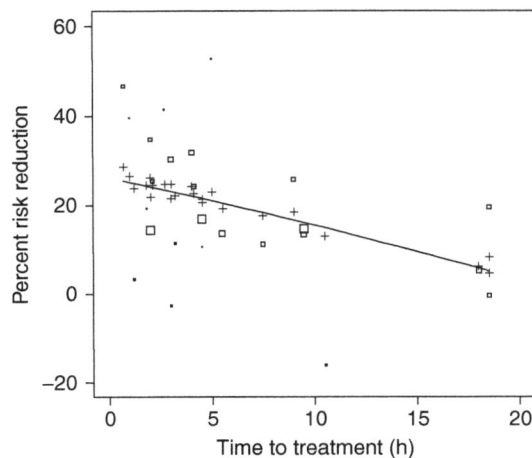

Fig. 2 Estimated percent risk reduction versus time-to-treatment. The regression line is shown together with the Bayes estimates (+ signs) and the observed estimates (squares). The size of the squares is inversely proportional to the study variance.

The Bayes confidence intervals are much narrower than those based solely on the observed data because they have borrowed strength from the other studies to increase the precision available from a single study.

An important clinical objective for cardiologists is determining the latest time at which treatment still shows a benefit. All of the models demonstrate a benefit of treatment in all of the studies, at least up until 18.5 h. The lower bound on the confidence intervals is not as sanguine, of course, but the Bayes model shows some benefit until at least 10.5 h even in the USIM study, which by itself showed little benefit. The lower limits on the intervals about the regression estimates show that we could reasonably expect benefit in any study for which the average time-to-treatment was 5 h or less, but that random variation would make benefit more uncertain if the time were greater than 5 h. Nevertheless, if we are concerned not with the results from a single trial, but rather with pooled results, we can reasonably conclude that thrombolytic treatment is beneficial for at least the first 10.5 h after the onset of chest pain and perhaps for several hours after that also. These conclusions are similar to those drawn by the Fibrinolytic Therapy Trialists' (FTT) Collaborative Group (1994).

To illustrate the updating features of the Kalman filter, Table II shows the change in the estimates of the regression coefficients δ_0 and δ_1 (expressed on the log relative risk scale) from the regression on time-to-treatment as the studies are added in chronological order. Because δ_0 and δ_1 are second-stage parameters common to each study, computing their posterior distribution requires only the Kalman (forward) filter.

Table II
Changing Estimates of Slope (δ_1) and Intercept (δ_0) as New Studies are Added

Study added	Intercept		Slope	
	Estimate	SE	Estimate	SE
GISSI-1	−0.360	0.093	0.0428	0.0192
ISAM	−0.353	0.090	0.0421	0.0190
AIMS	−0.373	0.089	0.0413	0.0189
Wilcox	−0.380	0.082	0.0410	0.0183
ISIS-2	−0.315	0.050	0.0136	0.0072
USIM	−0.291	0.048	0.0114	0.0071
EMERAS	−0.293	0.044	0.0134	0.0056
LATE	−0.296	0.043	0.0130	0.0048

Table III
Three Different GISSI-1 Time Subgroup Estimates (with Standard Errors)[a]

Time subgroup (h)	Subgroup only	All GISSI-1 data	All studies
<1	−0.633 (0.162)	−0.330 (0.082)	−0.287 (0.040)
1–3	−0.155 (0.085)	−0.274 (0.064)	−0.270 (0.036)
3–6	−0.184 (0.086)	−0.167 (0.052)	−0.237 (0.029)
63–9	−0.117 (0.139)	−0.039 (0.084)	−0.198 (0.026)
>9	0.149 (0.198)	0.089 (0.134)	−0.159 (0.031)

[a]The first are the observed estimates from each subgroup. The second are estimates for the subgroup from a hierarchical model fit to all the GISSI-1 data. The third are estimates from a hierarchical model fit to data from all eight studies.

The dominating effect of the two largest studies, GISSI-1 and ISIS-2, is apparent. These two studies reduce the standard errors substantially and also change estimates of the mean and variance. The USIM study also shifts the estimates substantially because its small treatment effects contrast with the larger ones from other short time-to-treatment studies.

Table III shows that an individual study's estimates can also be followed as the algorithm progresses. On the basis of data from the subgroup of patients treated within 1 h in GISSI-1, the estimate of the true effect of treatment for patients treated within 1 h is a reduction in risk of 47% [= exp(−0.633)]. When data from the rest of the GISSI-1 study are incorporated into a regression model, the reduction is estimated to be only 28%. This reduction is shrunk to 25% on the basis of data from all eight studies. Conversely, the estimated effect for patients treated after 9 h changes from a 9% increased risk (16% if only data from that subgroup are used) to a 15% decreased risk with the information from the other eight studies.

This readjustment is not solely of academic interest. The results of the GISSI-1 study were so influential that not only did they establish SK as a standard

treatment for myocardial infarction, but they also convinced physicians that
effectiveness decreased as time since symptom onset increased. On the basis of
this study, treatment was advocated for patients who arrived within 6 h after
symptom onset (GISSI Study Group, 1986). The evaluation of this study by a
Bayesian hierarchical linear model derived from data that also include the other
large clinical trials of thrombolytic therapy suggests now that treatment is also
beneficial for patients arriving much later than 6 h after symptom onset.

V. Example: Saltatory Model of Infant Growth

Lampl *et al.* (1992) reported daily measurements of infants for whom whole-
body length appears to increase at discrete jumps, termed *saltations*, with much
longer intervening periods of no growth, termed *stasis*. Figure 3 shows daily
measurements of total recumbent length (height) of one infant taken over a 4-
month period (from day 321 to day 443 of age) using a specially designed infant
measuring board. The pattern of saltation and stasis is apparent in the figure,
although measurement error does introduce some uncertainty. Sources of this
error include the equipment, measurement technique, and the cooperation of the
individual subjects.

The within-individual variation may involve the random nature of the growth
events or could involve more systematic changes in growth patterns related to the
age of the child or individual genetic, physiologic, or environmental factors
modifying the growth rates. Between-individual variation would describe broad
characteristics of populations and subpopulations of individuals, for instance, the

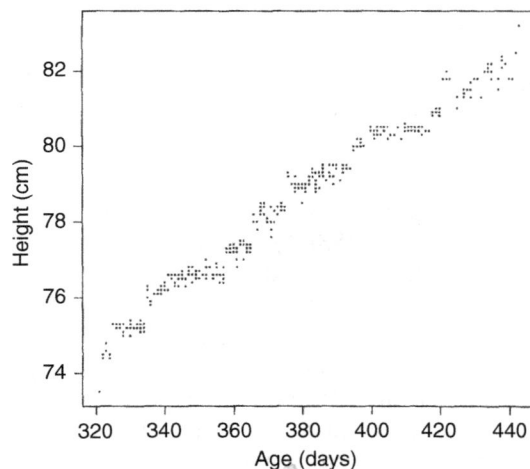

Fig. 3   Height measurements taken between one and six times daily over 4 months on a single infant.

average height at a given age or the distribution of the mean amplitude or mean stasis interval.

These requirements can be described with a three-level Bayesian hierarchical model. The first level represents an individual's height as the sum of the true height and the error of measurement. The second level represents this true height in terms of components that relate to the average height of similar individuals in the population and to the specific growth characteristics of this individual. The third level provides a probabilistic specification of growth in the population. This hierarchical structure permits the growth measurements from one individual to inform the growth distribution of another individual through the population components in the model.

We can represent such saltatory growth as a stochastic process as follows. First, let y_t be the measured height and H_t be the true height at time t for $t = 1, \ldots, T$ time points measured on one individual. Then, take $u = (u_1, u_2, \ldots, u_K)^T$ to be the set of times at which the saltations occur with h_{u_k}, the saltation at time u_k. Then

$$H_t = H_{t-1} + \sum_{t-1 < u_k \leq t} h_{u_k}$$

describes the true growth process and we observe

$$y_t = H_t + \varepsilon_t$$

where ε_t is measurement error distributed as a Gaussian random variable with mean 0 and variance σ_ε^2. We assume that the distribution of each saltation h_{u_k} is Gaussian with mean μ_h and variance σ_h^2. This can be expressed by the equation $h_{u_k} = \mu_h + \varepsilon_{u_k}$ where $\varepsilon_{u_k} \sim N(0, \sigma_h^2)$. Conditional on the set of saltation times, u, the problem may then be set up as a Bayesian hierarchical model as follows.

Let J_t represent the number of saltations that occur during the interval $(t-1, t]$, let the prior distribution for the average amplitude, μ_h, be Gaussian with mean μ_{h_0} and variance $\sigma_{h_0}^2$, and let the prior for the initial height, H_0, be Gaussian with mean μ_0 and variance σ_0^2. We can write this in the state-space form of Eqs. (7) and (8) as

$$\begin{pmatrix} H_t \\ \mu_h \end{pmatrix} = \begin{pmatrix} 1 & J_t \\ 0 & 1 \end{pmatrix} \begin{pmatrix} H_{t-1} \\ \mu_h \end{pmatrix} + \begin{pmatrix} \sum_{t-1 < u_k \leq t} \varepsilon_{u_k} \\ 0 \end{pmatrix}$$

and

$$y_t = (1 \quad 0) \begin{pmatrix} H_t \\ \mu_h \end{pmatrix} + \varepsilon_t$$

so that

$$\boldsymbol{\theta}_t = \begin{pmatrix} H_t \\ \mu_h \end{pmatrix}, \mathbf{F}_t = \begin{pmatrix} 1 & J_t \\ 0 & 1 \end{pmatrix}, \mathbf{X}_t^* = (1 \quad 0), \mathbf{W}_1^* = \begin{pmatrix} J_t \sigma_h^2 & 0 \\ 0 & 0 \end{pmatrix}$$

and $V_t = \sigma_\varepsilon^2$. Given the variances σ_ε^2 and σ_h^2 and the prior distributions for μ_h and H_0, the computations for the Kalman filter are started by setting $\boldsymbol{\theta}_0 = \begin{pmatrix} \mu_0 \\ \mu_{h_0} \end{pmatrix}$ and $\mathbf{S}_0 = \begin{pmatrix} \sigma_0^2 & 0 \\ 0 & \sigma_{h_0}^2 \end{pmatrix}$. For simplicity, we assume that the variance components σ_ε^2 and σ_h^2 are fixed at $\hat{\sigma}_\varepsilon^2 = 0.059$ and $\hat{\sigma}_h^2 = 0.038$ from a prior analysis using the EM algorithm (Little and Rubin, 1987), and that $\boldsymbol{\theta}_0 = \begin{pmatrix} 73 \\ 0.94 \end{pmatrix}$ and $\mathbf{S}_0 = \begin{pmatrix} 4.7 & 0 \\ 0 & 4.9 \end{pmatrix}$, on the basis of values estimated from data from another infant. In fact, the results turn out to be robust even to extreme changes in these prior values. Using Fig. 3, we fix 12 saltations occurring on days 322, 325, 335, 341, 358, 366, 376, 395, 400, 418, 421, and 443. The assumptions of fixed variances and growth times are made for pedagogical reasons in order to illustrate use of the Kalman filter. We will discuss removal of these restrictions presently.

Table IV gives estimates and standard errors for the 12 saltations $\mathbf{h} = (h_1, h_2, \ldots, h_{12})$, the mean saltation μ_h, and the initial height H_0 (Table IV). Figure 4 shows the fitted growth curve overlaid on the data. The infant's initial height was estimated to be 73.6 (± 0.3) cm (mean ± 2 standard errors) and the estimated final height was 83.0 cm, a total growth of 9.4 cm. This corresponds to an average growth of 0.788 cm per growth event with individual growth amplitudes varying between 0.42 and 1.19 cm. These estimates are fairly precise with standard errors ranging from 0.05 to 0.14 cm. All of the amplitudes are significantly different from zero, indicating our choice of saltation times was reasonable.

Table IV
Posterior Means and Standard Errors of Model Parameters for Growth Amplitudes (h_1, h_2, ..., h_{12}), Mean Growth (μ_h), and Initial Height (H_0)

Parameter	Mean	SE
h_1	0.904	0.139
h_2	0.734	0.089
h_3	0.876	0.059
h_4	0.486	0.054
h_5	0.704	0.048
h_6	0.922	0.056
h_7	0.963	0.050
h_8	0.813	0.062
h_9	0.421	0.062
h_{10}	0.550	0.070
h_{11}	0.900	0.071
h_{12}	1.187	0.107
μ_h	0.788	0.058
H_0	73.586	0.137

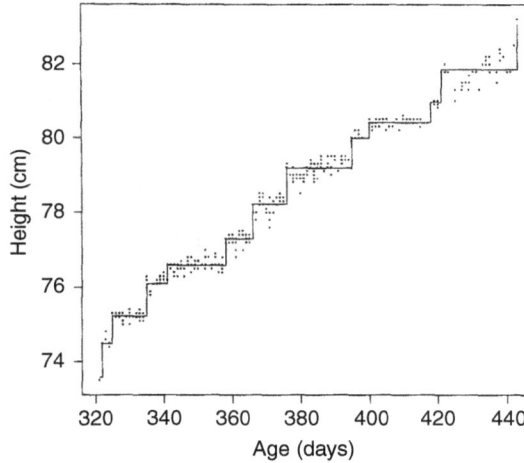

Fig. 4 Fitted curve from the hierarchical growth model overlaid on the data.

VI. Incorporation of Variance Components

Computation of the full model with random variances and growth times requires embedding the Kalman filter inside a larger MCMC algorithm that could incorporate the non-Gaussian structure of these components. A brief outline of MCMC is therefore warranted.

Return to the basic Bayesian formula describing the posterior density of a parameter θ as

$$f(\theta|Y) = \frac{f(\theta)f(Y|\theta)}{f(Y)}.$$

If we wish to calculate the posterior mean (or in fact any posterior moment), we must evaluate the expectation of θ with respect to this density, $E(\theta) = \int \theta f(\theta|Y)d\theta$. This involves an integration which, except in very simple problems, must be performed numerically. In complex problems with many parameters, computing the marginal density requires integrating over all parameters but one. This is simply impossible for many real problems.

MCMC methods work by simulating a series of N draws from the correct posterior distribution for each parameter in the model. With these simulated draws, moments may be calculated simply by Monte Carlo integration. For instance, the mean of θ_i is calculated simply from the average of all the drawn values of the θ_i. Moreover, as draws from the complete joint posterior are available, we can compute the posterior of an arbitrary function g of the parameters by simply averaging the values obtained by applying this function to the simulated parameters, obtaining $E[g(\theta)] \approx \frac{1}{N}\sum_{i=1}^{N} g(\boldsymbol{\theta}_i)$ for the N draws $\tilde{\boldsymbol{\theta}}_i$ from $f(\theta|Y)$.

The theory behind this method is too complex to describe other than briefly here. Interested readers may refer to Gilks *et al.* (1996) and the many papers cited therein for details. Consider starting a chain such that having drawn $\theta^{(t-1)}$ we can draw from $h(\theta^{(t)}|\theta^{(t-1)}, y)$. Because this draw depends only on the previous state, it is a Markov chain. Eventually, the chain will forget its starting value and it can be shown that the chain will converge to its stationary distribution, which is the correct posterior $f(\theta|y)$. Thus, to simulate a Markov chain from the correct posterior, we need only to be able to draw from $h(\theta^{(t)}|\theta^{(t-1)}, y)$. It is often simpler to proceed one parameter at a time, simulating from $h_i(\theta_i^{(t)}|\theta_{[i]}^{(t-1)}, y)$ where the notation $\theta_{[i]}$ represents all members of θ except the ith. If all of these full conditional distributions can be sampled, the MCMC algorithm is described as the Gibbs sampler. Except when conjugate distributions have been used, however, not all of these full conditionals can be represented in terms of known densities that can be simulated.

When the conditional density can be written down but not simulated, the Metropolis-Hastings form of MCMC can be used instead. This consists of drawing $\theta_i^{(t)}$ from a transition density $q_i(\theta_i^{(t)}|\theta_{[i]}^{(t-1)})$ that is constructed to resemble $h_i(\theta_i^{(t)}|\theta_{[i]}^{(t-1)})$ and then accepting this draw with probability

$$R = \min\left(1, \frac{h_i(\theta_i^{(t)}|Y, \theta_{[i]}^{(t-1)})q_i\left(\theta_{[i]}^{(t-1)}|\theta_i^{(t)}\right)}{h_i\left(\theta_i^{(t-1)}|Y, \theta_{[i]}^{(t-1)}\right)q_i(\theta_i^{(t)}|\theta_{[i]}^{(t-1)})}\right).$$

This acceptance ratio can be seen as the ratio of the posterior probability that $\theta_i = \theta_i^{(t)}$ conditional on the current draws of the other parameters to the conditional posterior probability that $\theta_i = \theta_i^{(t-1)}$ weighted by the importance ratio $q_i(\theta_{[i]}^{(t-1)}|\theta_i^{(t)})/q_i(\theta_i^{(t)}|\theta_{[i]}^{(t-1)})$. If the new drawn value has substantially higher posterior probability than the previous value, the acceptance rate will be high. In the Gibbs sampler, the transition density is $q_i(\theta_i^{(t)}|\theta_{[i]}^{(t-1)}) = h_i(\theta_i^{(t)}|Y, \theta_{[i]}^{(t-1)})$ so that the ratio is always one and the draw is always accepted. Gibbs sampling therefore uses the optimal transition density if it can be found and if not uses a Metropolis-Hastings step instead.

In the growth problem, the parameters for which posterior inference are desired include those describing the growth amplitude, $\{\mathbf{h}, H_0, \mu_h, \sigma_h^2\}$, the measurement error variance σ_e^2, and the parameters ϕ involved in describing the saltation times $\mathbf{u}$. The Kalman filter set out above describes the distribution of $\mathbf{h}, H_0, \mu_h|Y, \sigma_e^2, \mathbf{u}, \phi$. MCMC is required to compute the full posterior. This is the subject of current research.

VII. Model Checking

Once models have been fit to data, it is important that they be checked against the data to ensure that they describe the important features. Plots can help. Here, we describe the idea of using posterior predictive checks to evaluate the accuracy of a Bayesian model (Gelman *et al.*, 1995). The basic idea is to simulate draws of the outcomes from the model posterior and then to check these simulated responses against the actual responses.

For a given realization of the parameters from an iteration of the MCMC algorithm, we calculate the test statistic $T(y_k^{rep}, \theta_k)$, a measure of discrepancy where y_k^{rep} is a sample from the predictive distribution $[y|\theta_k]$. $T(y_k^{rep}, \theta_k)$ is then compared with $T(y^{obs}, \theta_k)$ computed using the observed data y^{obs}. A Bayesian p value may then be computed from $\Pr[T(y_k^{rep}, \theta_k) > T(y^{obs}, \theta_k)]$. As in the classical hypothesis test, this p value will be small when the observed discrepancy is most often higher than that would be expected if the data were generated from the model.

A variety of test statistics may be employed to validate the model. Two standard ones are given below:

1. An overall-goodness-of-fit test: $T_1(y, \theta) = \sum_{i=1}^{N} [y_i - E(y_i|\theta)]/[\text{Var}(y_i|\theta)]$
2. The test of maximal deviation: $T_2(y, \theta) = \max|y_i - E(y_i|\theta)|$.

We can also tailor statistics to the problem at hand. For example, to evaluate the growth model, we could examine the following test statistics:

1. Number of sign changes (a rough test of serial correlation):

$$T_3(y, q) = \{\text{sgn}(y_i - E(y_i|q))^1 \, \text{sgn}(y_i - E(y_i|q))\}$$

2. Height at the end of observation: $T_4(y, \theta) = y_N$
3. Largest growth increment: $T_5(y, \theta) = \max|\bar{y}_i - \bar{y}_{i-1}|$
4. Largest stasis interval: $T_6(y, \theta) = \{$longest consecutive run such that $\bar{y}_i - \bar{y}_{i-1} < k \quad \forall i\}$
5. Number of growth events: $T_7(y, \theta) = \sum I\{(\bar{y}_i - \bar{y}_{i-1}) > k\}$ for some constant k where I is the indicator function.

All of these checks use the entire set of data. A second type of posterior check examines how a model predicts on new data. For this purpose, we can split each data series into two parts using the first two-thirds to develop the model and the last third to test the model (Carlin and Louis, 1996). For each realization θ_k, we can compute y_k^{pred} from $[y^{new}|\theta_k, y^{old}]$ where y^{new} represents the last third and y^{old} the first two-thirds of y^{obs}. Bayesian p values are then computed from $\Pr[T(y_k^{pred}, \theta_k) > T(y^{new}, \theta_k)]$.

VIII. Conclusion

Hierarchical models are becoming increasingly popular for representing probability structures with multiple components of variance. Bayesian forms of these models allow a much more flexible description of model features and explicit representation of prior information that may be combined with the data to improve scientific inference. The computation of Bayesian models has been significantly simplified by MCMC techniques, which now allow computation of complex probability models. Representation of a Bayesian hierarchical linear model in state-space form and sequential computation by linear filters provides a useful way to understand the model structure and the interrelationships among model parameters, as well as to appreciate how new data update the posterior densities of the parameters. In problems with an inherent temporal or ordered spatial structure, this representation can also facilitate model construction.

Acknowledgments

This research was supported by grants 19122 and 23397 from the Robert Wood Johnson Foundation, by NAGW 4061 from NASA, and by R01-HS08532 and R01-HS07782 from the Agency for Health Care Policy and Research.

References

AIMS Trial Study Group (1988). *Lancet* **i,** 545.

Ansley, C. F., and Kohn, R. (1982). *Biometrika* **69,** 486.

ASSET Trial Study Group (1988). *Lancet* **ii,** 525.

Berger, J. O. (1985). "Statistical Decision Theory and Bayesian Analysis," 2nd edn Springer-Verlag, New York.

Berry, D. A., and Stang, D. K. (eds.) (1996). "Bayesian Biostatistics," Marcel Dekker, New York.

Brown, E. N., and Schmid, C. H. (1994). *Methods Enzymol.* **240,** 171.

Carlin, J. B. (1990). *Aust. J. Stat.* **32,** 29.

Carlin, J. B. (1992). *Stat. Med.* **11,** 141.

Carlin, B. P., and Louis, T. A. (1996). "Bayes and Empirical Bayes Methods for Data Analysis," Chapman and Hall, New York.

DeJong, P., and Mackinnon, M. J. (1988). *Biometrika* **75,** 601.

DuMouchel, W. (1990). *In* "Statistical Methodology in the Pharmaceutical Sciences," (D. A. Berry, ed.), p. 509. Marcel Dekker, New York.

Duncan, D. B., and Horn, S. D. (1972). *J. Am. Stat. Assoc.* **67,** 815.

EMERAS Collaborative Group (1993). *Lancet* **342,** 767.

Fearn, R. (1975). *Biometrika* **62,** 89.

Fibrinolytic Therapy Trialists' (FTT) Collaborative Group (1994). *Lancet* **343,** 311.

Gelman, A., Carlin, J. B., Stern, H. S., and Rubin, D. B. (1995). "Bayesian Data Analysis," Chapman and Hall, New York.

Gilks, W. R., Richardson, S., and Spiegelhalter, D. J. (eds.) (1996). "Markov Chain Monte Carlo in Practice," Chapman and Hall, New York.

GISSI Study Group (1986). *Lancet* **i,,** 397.

Graver, D., Draper, D., Greenhouse, J., Hedges, L., Morris, C., and Waternaux, C. (1992). "Combining Information: Statistical Issues and Opportunities for Research," National Academy Press, Washington, D.C.

ISAM Study Group (1986). *N. Eng. J. Med.* **314,** 1465.

ISIS-2 Collaborative Group (1988). *Lancet* **ii,** 349.

Lampl, M., Veldhuis, J. D., and Johnson, M. L. (1992). *Science* **258,** 801.

LATE Study Group (1993). *Lancet* **342,** 759.

Lindley, D. V., and Smith, A. F. M. (1972). *J. R. Stat. Soc. B* **34,** 1.

Little, R. J. A., and Rubin, D. B. (1987). "Statistical Analysis with Missing Data," John Wiley and Sons, New York.

Meinhold, R. J., and Singpurwalla, N. D. (1983). *Am. Stat.* **37,** 123.

Morris, C. N., and Normand, S. L. (1992). *In* "Bayesian Statistics 4," (J. M. Bernardo, J. O. Berger, A. P. Dawid, and A. F. M. Smith, eds.), p. 321. Oxford University Press, New York.

Rubin, D. B. (1981). *J. Educ. Stat.* **6,** 377.

Schmid, C. H. (1996). *J. Am. Stat. Assoc.* **91,** 1322.

Shumway, R. H., and Stoffer, D. S. (1982). *J. Time Series Anal.* **3,** 253.

Smith, A. F. M. (1973). *J. R. Stat. Soc. B* **35,** 67.

Smith, T. C., Spiegelhalter, D. J., and Thomas, A. (1995). *Stat. Med.* **14,** 2685.

Strenio, J. F., Weisberg, H. I., and Bryk, A. S. (1983). *Biometrics* **39,** 71.

USIM Collaborative Group (1991). *Am. J. Card.* **68,** 585.

Wecker, W. E., and Ansley, C. F. (1983). *J. Am. Stat. Assoc.* **78,** 381.

CHAPTER 11

Mixed–Model Regression Analysis and Dealing with Interindividual Differences

Hans P. A. Van Dongen,* Erik Olofsen,[†] David F. Dinges,[‡] and Greg Maislin[§]

*Unit for Experimental Psychiatry
University of Pennsylvania School of Medicine
Philadelphia, Pennsylvania, USA

[†]Department of Anesthesiology
P5Q, Leiden University Medical Center
2300 RC, Leiden, The Netherlands

[‡]Unit for Experimental Psychiatry
University of Pennsylvania School of Medicine
Philadelphia, Pennsylvania, USA

[§]Biomedical Statistical Consulting
Wynnewood, Pennsylvania, USA

225

DOI: 10.1016/B978-0-12-384997-7.00011-X

I. Introduction

Repeated-measures study designs are popular in biomedical research because they allow investigation of changes over time within individuals. These temporal changes in subjects may themselves be of interest (e.g., to document the effects of aging), or they may enable statistically powerful comparisons of different conditions within the same individuals (e.g., in cross-over drug studies). Complications arise when analyzing the longitudinal data from repeated-measures studies involving multiple subjects, however, as the data from the different subjects must be combined somehow for efficient analysis of the within-subjects changes. After all, the temporal changes that the various subjects have in common are typically of greater interest than the temporal profiles of each individual subject. The complications stem from the fact that the data collected within a subject are not independent of each other—they are correlated (and the magnitude of this correlation is usually not *a priori* known).[1] The data from different subjects, on the other hand, are typically independent. To properly analyze data from an experiment that combines multiple subjects with multiple data points per subject, two distinct sources of variance in the overall data set must be considered: between-subjects variance and within-subjects variance. Statistical analysis techniques targeted specifically at longitudinal data must keep these sources of variance separated (Burton *et al.*, 1998).

In this chapter, we consider mixed-model regression analysis, which is a specific technique for analyzing longitudinal data that properly deals with within- and between-subjects variance. The term "mixed model" refers to the inclusion of both fixed effects, which are model components used to define systematic relationships such as overall changes over time and/or experimentally induced group differences; and random effects, which account for variability among subjects around the systematic relationships captured by the fixed effects. To illustrate how the mixed-model regression approach can help analyze longitudinal data with large interindividual differences, we consider psychomotor vigilance data from an experiment involving 88 h of total sleep deprivation, during which subjects received either sustained low-dose caffeine or placebo (Dinges *et al.*, 2000; Van Dongen *et al.*, 2001). We first apply traditional repeated-measures analysis of variance (ANOVA), and show that this method is not robust against systematic interindividual variability. The data are then reanalyzed using linear mixed-model regression analysis in order to properly take into account the interindividual differences. We conclude with an application of nonlinear mixed-model regression analysis of the data at hand, to demonstrate the considerable potential of this relatively novel statistical approach. Throughout this chapter, we apply commonly used (scalar) mathematical notation and avoid matrix formulation, so as to provide relatively easy access to the underlying statistical methodology.

---

[1] Counter-intuitively, it is precisely this feature of repeated-measures experimental designs that makes them statistically powerful and efficient (Burton *et al.*, 1998).

II. Experiment and Data

A total of $n = 26$ healthy adult males (moderate caffeine consumers) participated in a 10-day laboratory study. Following a 2-week period in which they refrained from caffeine use, subjects entered the laboratory. After one adaptation night, subjects had two baseline days with bedtimes from 23:30 until 07:30. They then underwent 88 h of total sleep deprivation, during which time they were constantly monitored and kept awake with mild social stimulation. The experiment concluded with three recovery days. Every 2 h of scheduled wakefulness, subjects were tested on a 30-min computerized neurobehavioral assessment battery, which included a 10-min psychomotor vigilance task (PVT). Psychomotor vigilance performance was assessed by counting the number of lapses, defined as reaction times equal to or greater than 500 ms, per 10-min test bout. For the purposes of the present analyses, overall daytime performance was determined by averaging the test bouts at 09:30, 11:30, 13:30, 15:30, 17:30, 19:30, and 21:30 for each day. This served to average out the natural circadian (24-h) rhythm in performance data (Van Dongen and Dinges, 2000). The first daytime period during the 88 h of wakefulness, before any sleep loss was incurred, we call Day 0. The subsequent three daytime periods we refer to as Days 1, 2, and 3.

Subjects were randomized to one of two conditions: $n_1 = 13$ subjects were randomized to receive sustained low-dose caffeine (0.3 mg/kg body weight, or about a quarter cup of coffee, each hour) and $n_2 = 13$ subjects were randomized to receive placebo, in double-blind fashion. Caffeine or placebo pill administration began at 05:30 after 22 h of sustained wakefulness, and continued at hourly intervals for the remaining 66 h of total sleep deprivation. At 1.5-h intervals on average, blood samples were taken via an indwelling intravenous catheter for assessment of blood plasma concentrations of caffeine[2] (these data were available for 10 subjects in the caffeine condition only).

The present investigation focuses on whether caffeine mitigated the psychomotor vigilance performance impairment resulting from total sleep deprivation, and if so, for how long. Fig. 1 shows the psychomotor vigilance data (PVT lapses) for both conditions, as well as the caffeine concentrations in blood plasma for the subjects in the caffeine condition, during the 88 h of total sleep deprivation. The error bars (representing standard deviations) in this figure show large interindividual differences both in the plasma concentrations of caffeine and in the levels of psychomotor vigilance impairment, posing a challenge for the analysis of this data set. In addition, the random assignment to condition resulted in differences between the average performance levels for the two conditions even before the administration of caffeine or placebo. This is also evident in Fig. 2, which shows the daytime averages for psychomotor vigilance performance lapses for Day 0 (before pill administration) and across Days 1–3 (during pill administration). The data points used to construct Fig. 2 are given in Table I.

---

[2] Caffeine concentrations were assessed by EMIT enzyme multiplication immunoassay (Syva, Palo Alto, CA).

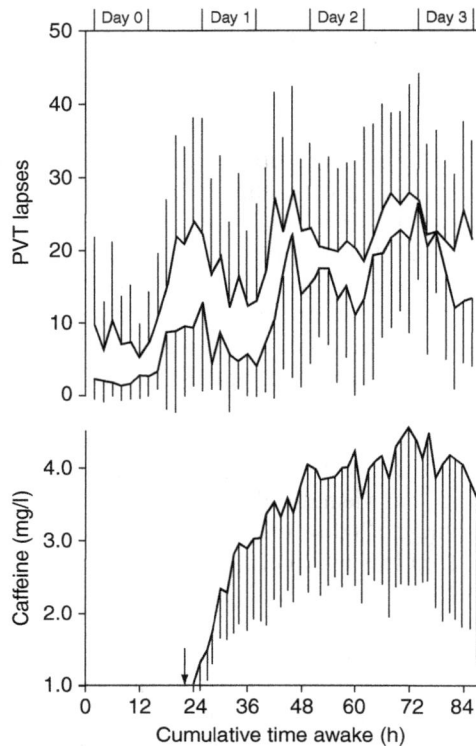

Fig. 1 Psychomotor vigilance task (PVT) lapses for the caffeine condition (downward error bars) and the placebo condition (upward error bars), and caffeine concentrations in blood plasma for the caffeine condition, across 88 h of total sleep deprivation. The curves represent condition averages; the error bars represent the standard deviation over subjects (suggesting large interindividual differences). The arrow indicates the beginning of hourly pill administration (caffeine or placebo).

III. Repeated-Measures ANOVA

As a first analysis of the psychomotor vigilance performance data for the two conditions across Days 0 through 3, we use a commonly applied technique called repeated-measures ANOVA. Before describing repeated-measures ANOVA, we describe the simpler situation in which subjects in N independent groups each contribute only one observation to the overall data set, so that all data points are mutually independent. The data points within each group are assumed to be normally distributed with equal variance but possibly differing mean. If the variance within the groups is relatively small compared to the variance among the group means, then the differences among the means are significant, that is, larger than could have reasonably arisen merely from variability in the data within the groups. This forms the basis of one-way ANOVA, as is illustrated in Fig. 3.

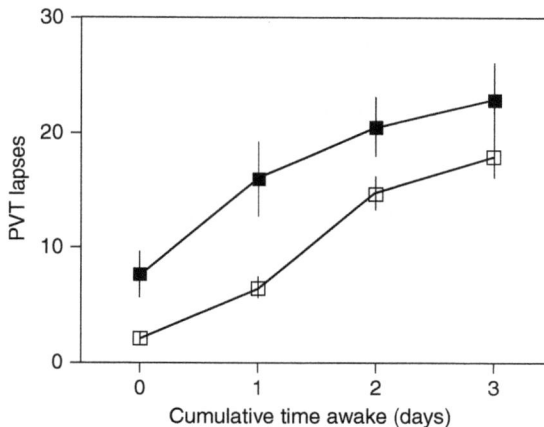

Fig. 2 Average daytime lapses on the psychomotor vigilance task (PVT) for the caffeine condition (open boxes) and the placebo condition (closed boxes) prior to pill administration (Day 0) and during pill administration (Days 1–3). The boxes represent condition means; the error bars represent standard errors of the mean.

We define n as the total number of subjects, n_j as the number of subjects for group j ($j = 1, \ldots, N$), M_j as the observed mean for group j, and s_j^2 as the observed variance for group j:

$$s_j^2 = \{\Sigma_i(y_{ji} - M_j)^2\}/(n_j - 1), \tag{1}$$

where i identifies the different subjects in the group (say, $i = 1, \ldots, n_j$) and y_{ji} are the data points for these subjects. The variance components that form the basis of ANOVA are usually expressed in terms of sums of squares (SS), such as the sum in curly brackets in Eq. (1), and in terms of mean squares (MS), defined as the sum of squares divided by the degrees of freedom (df). For the overall data set, which we indicate with S, the variance is thus described by

$$SS(S) = \sum_j\sum_i(y_{ji} - M)^2, \mathrm{df}(S) = n - 1, MS(S) = SS(S)/\mathrm{df}(S), \tag{2}$$

where M is the grand mean of the data.

The common within-groups variance, indicated here with w, is estimated using the group-specific variances:

$$SS(w) = \sum_j\sum_i(y_{ji} - M_j)^2, \mathrm{df}(w) = \sum_j(n_j - 1) \tag{3}$$

so that

$$MS(w) = SS(w)/\mathrm{df}(w) = \sum_j(n_j - 1)s_j^2/\sum_j(n_j - 1). \tag{4}$$

The between-groups variance (i.e., the variance among the group means) can then be computed from the difference between $SS(S)$ and $SS(w)$:

Table I
Daytime Average Data in Two Experimental Conditions[a]

Id	Cond	Day0	Day1	Day2	Day3
1	1	0.4	2.0	23.0	31.3
2	1	1.0	5.7	19.1	21.9
3	1	0.6	7.4	9.9	25.9
4	1	4.7	11.3	18.0	11.7
5	1	2.6	5.9	13.6	23.7
6	1	1.6	4.0	19.9	24.4
7	1	3.3	5.7	11.4	15.4
8	1	4.4	4.7	19.6	13.3
9	1	0.0	10.1	20.3	12.4
10	1	2.6	10.9	8.0	13.7
11	1	0.3	3.3	9.7	13.6
12	1	3.0	9.3	6.7	13.4
13	1	2.7	3.9	12.0	12.0
14	2	5.0	9.3	31.9	21.7
15	2	5.1	15.6	23.6	43.9
16	2	1.0	4.9	18.3	23.0
17	2	16.6	35.1	31.7	30.1
18	2	15.1	22.0	28.7	32.3
19	2	0.0	0.0	2.4	0.0
20	2	0.3	1.4	9.1	10.6
21	2	1.3	8.1	13.3	22.3
22	2	11.6	23.9	16.9	15.9
23	2	7.1	32.1	27.0	21.9
24	2	12.7	22.0	18.6	20.0
25	2	2.6	11.9	25.7	21.3
26	2	20.7	21.4	19.7	34.1

[a]Daytime averages are given for the number of psychomotor vigilance performance lapses per 10-min test bout, for each of the subjects (column "id") in the caffeine condition (column "cond" value 1) and the placebo condition (column "cond" value 2), across Days 0–3 of total sleep deprivation (columns "day0" through "day3").

$$SS(b) = SS(S) - SS(w), df(b) = N - 1, MS(b) = SS(b)/df(b). \quad (5)$$

To evaluate the statistical significance of the difference between the group means (the "effect of condition"), the one-way ANOVA F statistic [with df(b) and df(w) degrees of freedom] is used[3]:

$$F[df(b), df(w)] = MS(b)/MS(w). \quad (6)$$

For the F test, it is assumed that the data are randomly sampled from normal distributions, although ANOVA results are robust to departures from this assumption. It is also assumed that the underlying distributions from which the data within each group are sampled have equal variances.

[3] When only two groups are compared [i.e., df(b) = 1], the square-root of the F statistic in Eq. (6) yields a t statistic for an equivalent t test with df(w) degrees of freedom.

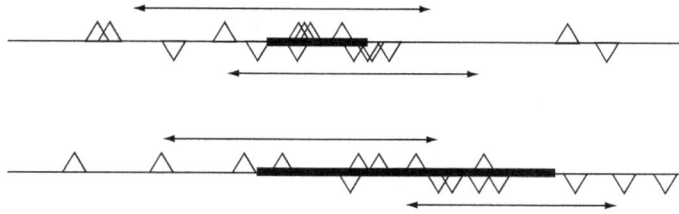

Fig. 3 Illustration of one-way analysis of variance (ANOVA) for two independent groups of eight data points each. On the upper line, data points in the first group (upward triangles) and the second group (downward triangles) show much overlap. The variances of the data in each of the groups, which are illustrated by the arrows (corresponding to the one-standard-deviation intervals around the group means), are relatively large compared to the variance between the group means, which is illustrated by the black bar (corresponding to the one-standard-deviation interval around the mean of the group means). Assuming normal distributions for the data, one-way ANOVA shows that these two groups are not significantly different ($F[1, 14] = 0.03$, $p = 0.96$). On the lower line, data points in the first group (upward triangles) and the second group (downward triangles) show less overlap. The variances of the data in each of these groups are relatively small compared to the variance between the group means. One-way ANOVA shows that these groups are significantly different ($F[1, 14] = 12.01$, $p = 0.004$).

For the analysis of longitudinal data, the ANOVA method is adapted to compare subsets of the data measured in the same subjects at different time points, so as to test whether or not the mean for the subsets is the same at all time points. This is called repeated-measures ANOVA, and it differs from one-way ANOVA in that individual data points can no longer be assumed to be independent. If there is only one experimental condition (i.e., $N = 1$), the variance in the data (in terms of sums of squares) is partitioned into two parts. One variance component is the variance among the means over time, which in repeated-measures ANOVA is a within-subjects factor as each subject is measured at all time points. The other variance component is residual variance, which is the remaining variance within individual subjects. The variance among the means over time is represented by

$$\mathrm{SS}(A) = n\sum\nolimits_t (m_t - M)^2, \mathrm{df}(A) = T - 1, \tag{7}$$

where t identifies the different time points ($t = 0, \ldots, T - 1$), and m_t is the mean at time t. The residual variance can be computed as

$$\mathrm{SS}(R) = \sum\nolimits_i\sum\nolimits_t (t_{it} - m_t)^2 - \mathrm{SS}(A), \mathrm{df}(R) = (n - 1)(T - 1), \tag{8}$$

where y_{it} are the data points for subject i at time t. The following F statistic is used to evaluate the statistical significance of the difference between the means of the different subsets (the "effect of time")[4]:

$$F[\mathrm{df}(A), \mathrm{df}(R)] = \mathrm{MS}(A)/\mathrm{MS}(R). \tag{9}$$

Note that in this test for the effect of time, between-subjects variance (i.e., systematic differences among subjects over time) is automatically filtered out.

---

[4] The test for the effect of time is essentially a generalization of the paired-samples t test.

The principles of one-way ANOVA and repeated-measures ANOVA can be combined in a "mixed design" to compare independent conditions (groups) to each other over time. The full factorial form of the mixed design provides tests for the effect of time (applying to all conditions), for the effect of condition (applying to all time points), and for the interaction of condition with time. The interaction effect concerns any remaining systematic differences for the means among conditions and over time, after those applying to all conditions and those applying to all time points have been taken into account (Rosnow and Rosenthal, 1995). A mixed design involves partitioning the variance into within-subjects and between-subjects variance, with within-subjects variance represented by

$$SS(W) = \sum_j \sum_i \sum_t (y_{jit} - M_{ji})^2, \tag{10}$$

where i identifies different subjects in each condition j ($i = \sum_{j-1} n_j + 1, \ldots, \sum_j n_j$), y_{jit} are the data points for these subjects at time points t, and M_{ji} is the mean over time for these subjects. The between-subjects variance can be shown to be the difference between the overall variance and the within-subjects variance (in terms of sums of squares), as follows:

$$SS(B) = SS(S) - SS(W). \tag{11}$$

For the purpose of testing the effect of condition, the between-subjects variance (in terms of sums of squares) is further partitioned into between-conditions variance (indicated by C) and error variance (indicated by E), as follows:

$$SS(E) = T\sum_j \sum_i (M_{ji} - M_j)^2, df(E) = \sum_j (n_j - 1), \tag{12}$$

where M_j is the mean of all data for condition j, and

$$SS(C) = SS(B) - SS(E), df(C) = N - 1. \tag{13}$$

The following F statistic is then used to evaluate the effect of condition:

$$F[df(C), df(E)] = MS(C)/MS(E) \tag{14}$$

For the purpose of testing the effect of time and the interaction effect, the within-subjects variance is further partitioned into across-times variance (indicated by A) as given by Eq. (7), interaction variance (indicated by I), and residual variance (indicated by R). The latter variance component is represented by

$$SS(R) = \sum_j \sum_i \sum_t [y_{jit} - (m_{jt} + M_{ji} - M_j)]^2, df(R) = \sum_j (n_j - 1)(T - 1), \tag{15}$$

where m_{jt} is the mean for condition j at time t. The interaction is then computed as

$$SS(I) = SS(W) - SS(A) - SS(R), df(I) = (N - 1)(T - 1). \tag{16}$$

The following F statistic is used to evaluate the effect of time[5]:

[5] An adjustment must be made to SS(A) in the case of unequal numbers of subjects in the different conditions. The details of this adjustment for a "proportional sampling model" are omitted here.

$$F[\mathrm{df}(A), \mathrm{df}(R)] = \mathrm{MS}(A)/\mathrm{MS}(R). \qquad (17)$$

Furthermore, the following F statistic is used to evaluate the interaction of condition by time:

$$F[\mathrm{df}(I), \mathrm{df}(R)] = \mathrm{MS}(I)/\mathrm{MS}(R). \qquad (18)$$

We refer to the literature for information about the derivation of these formulas and for more in-depth discourses on repeated-measures ANOVA (Girden, 1992).

A. Repeated-Measures ANOVA Applied to the Data

Let us assume that we have the psychomotor vigilance performance data for Days 0–3 of the sleep deprivation experiment stored in a spreadsheet, organized in columns of $n = n_1 + n_2 = 26$ rows each: a column listing the different subjects by unique identifiers (named "id"), a column indicating in which condition each subject was (named "cond"), and four columns for the subjects' data from Day 0 until Day 3 (named "day0" through "day3," respectively). This conforms to the way the data are organized in Table I. To perform the calculations for repeated-measures ANOVA, we use the following general linear model (GLM) command in the computer software SPSS (2001):

```
GLM
    day0 day1 day2 day3 BY cond
    /WSDESIGN = time
    /WSFACTOR = time 4
    /DESIGN = cond.
```

The second line of this command identifies the data columns for the dependent variables, as well as the between-subjects factor column indicating to which conditions these data belong. The next two lines state that the dependent variables constitute repeated measures of a single underlying variable "time" with four levels (i.e., four different time points). The last line assigns the between-subjects factor (i.e., condition), making this a mixed-design repeated-measures ANOVA.

Table II shows the results of this analysis, revealing a significant effect of time and a significant effect of condition, but no significant interaction effect. We can interpret these results with the help of Fig. 2.[6] According to the analysis, psychomotor vigilance was reduced significantly over days of sleep deprivation (as expected); and the placebo condition performed consistently worse than the

---

[6] The use of figures or tables of the data is crucial for the interpretation of time effects and interactions, because the results of repeated-measures ANOVA do not reveal the direction of changes in the means over time or between conditions. Moreover, repeated-measures ANOVA is insensitive to the order of the time points; if the data points of an upward effect are rearranged (in the same manner for each subject) to form a downward effect, the ANOVA results remain the same. Repeated-measures ANOVA also does not take into account the intervals between the time points; for typical applications, the interval between all adjacent time points would be considered the same.

Table II
Results from Mixed-Design Repeated-Measures ANOVA[a]

Effect	SS	df	MS	F	p
Condition (C)	1082.8	1	1082.8	8.03	0.009
Error (E)	3235.1	24	134.8		
Time (A)	3745.4	3	1248.5	39.88	<0.001
Interaction (I)	83.4	3	27.8	0.89	0.452
Error (R)	2253.9	72	31.3		

[a]Results are shown from repeated-measures ANOVA of psychomotor vigilance performance data over 4 days of sleep deprivation in two different conditions. The sums of squares (SS), the degrees of freedom (df), and the means of squares (MS) for the different variance components are displayed. In addition, the F statistics and p values for the effect of condition, effect of time, and interaction of condition by time are given.

caffeine condition. This latter finding included Day 0, before the beginning of pill administration. Any additional difference between conditions due to the action of caffeine (during Days 1–3, but not Day 0) should have led to an interaction effect in this analysis, but no significant interaction was found. There is relatively little statistical power in interaction effects, (Winer, 1971) however, making it difficult to conclude with any degree of certainty that caffeine was ineffective in this experiment.[7]

If the substantial interindividual variability in psychomotor vigilance performance impairment during total sleep deprivation (Fig. 1) is systematic, as has been previously reported (Van Dongen *et al.*, 2003b), then there is reason to believe that the repeated-measures ANOVA results are inaccurate. Repeated-measures ANOVA does not distinguish variance due to systematic interindividual differences from random error variance (Girden, 1992), lumping these together as a single source of variance. As a consequence, the result for the effect of condition in the current analysis is unreliable, as is evident from the expression for SS(E) in Eq. (12). Depending on the particular design, interindividual variability may lead to overestimation or underestimation of statistical significance (Feldman, 1988). In the present mixed design, the effect of condition is underestimated because of systematic interindividual differences. Therefore, other analyses are warranted to further investigate this data set.

---

[7] By expressing the data on Days 1–3 as relative to those on Day 0, the difference between the two conditions prior to pill administration could be eliminated. Any consistent difference due to caffeine (i.e., present throughout Days 1–3) should then result in a main effect of condition, which is statistically more powerful than an interaction effect. On the other hand, by expressing the data as relative to those on Day 0, the noise in the data for Day 0 would be propagated to the data for Days 1–3. For the present data, it effectively makes no difference in study outcomes (effect of time: $F[2, 48] = 15.64$, $p < 0.001$; effect of condition: $F[1, 24] = 0.35$, $p = 0.559$; interaction effect: $F[2, 48] = 1.04$, $p = 0.362$).

IV. Mixed–Model Regression Analysis

Regression analysis is essentially equivalent to ANOVA; while ANOVA focuses on the variance in the data to assess differences between the means of subsets of the data, however, regression analysis focuses on assessing the parameters of a model (i.e., mathematical function) posited to describe the data set. Depending on the criteria used to determine the optimal parameter values, regression analysis typically involves minimization of the error variance[8] (which, as we shall see, presents a methodological connection between regression analysis and ANOVA[9]. Regression analysis provides a richer framework than ANOVA, in that a wider variety of models for the data can be evaluated.[10] We focus here on mixed-model (or mixed-effects) regression analysis,[11] which means that the model posited to describe the data contains both fixed effects and random effects. Fixed effects are those aspects of the model that (are assumed to) describe systematic features in the data. Fixed effects are used to determine expected or mean values for the subject population (as such, they can be compared to the regression coefficients in a standard regression analysis on pooled data, or to the effects of condition, time, and interaction in repeated-measures ANOVA). Random effects are those aspects of the model that are allowed to vary among subjects (i.e., parameters that take different values depending on which individual subject the data are from). Random effects are variance components that describe the variability (e.g., biological variability) in the observations around the expected values as predicted by the fixed effects. In this chapter, random effects will be indicated by Greek letters, to distinguish them from fixed effects.

To estimate model parameters in a standard regression analysis, the least-squares method, which involves minimization of the error variance, can be used. For mixed-model regression analysis, which requires more complicated computations to be made, an alternative method for parameter estimation is used: maximum likelihood estimation.[8] The basic idea underlying maximum likelihood estimation is that the data reflect the most probable outcome under the conditions

---

[8] In standard regression analysis, as in ANOVA, minimization of the error variance can be achieved by means of the least-squares method (i.e., minimization of the squares of the deviations between the model predictions and the data points). In mixed-model regression analysis, (approximate) minimization of the error variance is rather a by-product of the maximum likelihood approach used to estimate the model parameters.

[9] Note that regression analysis is not contingent upon having a complete data set. In traditional repeated-measures ANOVA, however, a missing data point eliminates the entire subject from the data set.

[10] In contrast with repeated-measures ANOVA, regression analyses usually involve models that are sensitive to the order of the time points, and take into account the intervals between the time points.

[11] There are other regression techniques that could have utility with regard to the present data, such as analysis of covariance (ANCOVA). However, ANCOVA is a fixed-effects method, and it is more restricted than mixed-effects regression analysis (Feldman, 1988).

specified in the model. Thus, given certain assumptions about the statistical distribution(s) of the stochastic process(es) involved in the mechanisms that generated the data (such as random between-subjects variability or noise), maximum likelihood estimation involves finding the model parameter values that maximize the likelihood of observing the data at hand.

To put this in mathematical equations, let us consider a mixed-effects model of the form

$$y_{jit} = f_{jt} + \eta_{ji} + \varepsilon_{jit}, \tag{19}$$

where f_{jt} is a function of t representing the fixed effects posited to describe the data at the group level (i.e., for conditions j overall). The term η_{ji} represents the random effect—more precisely, the η_{ji} are the subject-specific instances of the random effect, for subjects i in conditions j (which are lumped together here), that are usually assumed to arise from some family of parametric probability distributions such as a normal distribution with zero mean and variance ω^2 over subjects (where ω^2 is to be estimated). For the present purpose, the random effect is assumed to be additive to the fixed effects f_{jt}. The ε_{jit} represent independent noise assumed to have a normal distribution with zero mean and variance σ^2 over subjects (where σ^2 is to be estimated) for all times t. Further, the distributions of η and ε are assumed to be independent (i.e., zero covariance). Let us temporarily assume there is only one group j, so that Eq. (19) can be simplified to

$$y_{it} = f_t + \eta_i + \varepsilon_{it}. \tag{20}$$

For each subject i, the likelihood l_i of observing the subject's time series data y_{it} is given by

$$l_i = \Pi_t cN[f_t + \eta_i; \sigma^2](y_{it}), \tag{21}$$

where Π_t denotes multiplication over t (equivalent to Σ_t for summation), and c is a (here irrelevant) normalization factor. $N[f_t + \eta_i; \sigma^2]$ is the density function for a normal distribution with mean $f_t + \eta_i$ and variance σ^2; in Eq. (21) this function is evaluated at the data points y_{it}. Assuming that the η_i are taken from a normal distribution $N[0;\omega^2](\eta_i)$ with zero mean and variance ω^2 over subjects, we can define the marginal likelihood L_i to integrate η_i out:

$$L_i = C \int_{\eta_i=-\infty}^{\infty} l_i \cdot N[0; \omega^2](\eta_i) d\eta_i, \tag{22}$$

where C is a normalization factor. The likelihood L of observing the entire data set is then given by

$$L = \Pi_i L_i. \tag{23}$$

The likelihood L is a function of the parameters that contributed to its derivation: the (currently unspecified) parameters constituting the fixed effects f_t, the (between-subjects) variance ω^2 of the random effect, and the (within-subjects) error variance σ^2. By maximizing the value of L, maximum likelihood estimates of these parameters are obtained.[12] In practice, computer software is used to estimate the parameters, as numerical approximation is needed to find the parameter values that maximize L.

A. Linear Mixed–Model Regression Analysis Applied to the Data: Mixed–Model ANOVA

To illustrate the use of mixed-model regression analysis in practice, we first replicate the repeated-measures ANOVA performed previously, using the data presented in Fig. 2. The only essential difference with the repeated-measures ANOVA is, of course, that we take systematic interindividual differences into account. Because of the equivalence with repeated-measures ANOVA, this application of mixed-model regression analysis is also known as mixed-model ANOVA. The procedure involves explicit estimation of the model parameters, however, by means of the maximum likelihood method outlined previously.

Let us define day indicator variables d_{tu}, which equal 1 if $t = u$ and 0 otherwise, and condition indicator variables c_{jk}, which equal 1 if $j = k$ and 0 otherwise ($j = 1$ corresponds to the caffeine condition). We consider the following linear mixed-effects model for the data:

$$y_{jit} = I_{ji} + d_{t1}Y_1 + d_{t2}Y_2 + d_{t3}Y_3 + c_{j1}(d_{t0}Z_0 + d_{t1}Z_1 + d_{t2}Z_2 + d_{t3}Z_3) + \varepsilon_{jit}, \quad (24)$$

where parameter I_{ji} is the intercept (which represents the mean for Day 0 in the placebo condition), involving a random effect such that

$$I_{ji} = I_0 + \eta_{ji}. \quad (25)$$

The η_{ji} constitute the random effect for subjects i in conditions j (which are lumped together), assumed to arise from a normal distribution with zero mean and variance ω^2 over subjects (where ω^2 is not known in advance). Parameters Y_1, Y_2, and Y_3 represent the means for Days 1, 2, and 3 in the placebo condition, respectively, expressed as differences from the intercept; and parameters Z_0 through Z_3 represent the means for Days 0 through 3 in the caffeine condition, respectively, expressed as differences from their counterparts in the placebo condition. As an example, the data point for subject 5 in the caffeine condition (i.e., condition 1) on Day 2 is modeled as

$$y_{152} = I_0 + \eta_{15} + Y_2 + Z_2 + \varepsilon_{152} \quad (26)$$

as all other terms in Eq. (24) cancel out.

---

[12] Maximum likelihood (ML) parameter estimates tend to be (slightly) biased (Feldman, 1988). An improved methodology called restricted maximum likelihood (REML) is available for linear mixed-model regression analysis (but not for nonlinear mixed-model regression analysis). REML provides unbiased parameter estimates that are preferable to those resulting from conventional ML in virtually all cases (except when comparing models with different fixed-effects structures on the basis of the likelihood ratio c^2 statistic; SAS, 2001b). In this chapter, we report REML parameter estimates for all linear mixed-effect regression analyses, and ML parameter estimates for all nonlinear mixed-model regression analyses (for which REML is not available). For more on this topic, see Diggle et al. (1996).

We assume that we have the psychomotor vigilance performance data for Days 0–3 of the sleep deprivation experiment stored in a spreadsheet, organized in columns of $(n_1 + n_2) T = 26 \times 4 = 104$ rows each: a column listing the different subjects (named "id"), a column equivalent to the indicator variable c_{j1} indicating in which condition each subject was (named "cond"), four columns equivalent to the indicator variables d_{tu} (named "day0" through "day3") marking the day in the experiment (such that column "day$u$" equals 1 if the data in that row are from day u and 0 otherwise), and a column for the data of each subject on each day (named "y"). For (nonessential) technical reasons, the spreadsheet must be ordered by subject (i.e., all the data for a single subject must appear consecutively in the spreadsheet).[13] To perform the calculations for this linear mixed-model regression analysis, we use the following PROC MIXED command in the computer software SAS (2001a):

```
proc mixed;
   class id;
   model y = day1 day2 day3 day0*cond day1*cond day2*cond day3*
   cond / solution;
   random intercept / solution subject = id;
run;
```

The second line of this command specifies a categorical variable by which the data are classified—as repeated measures were obtained for each subject, we classify the data by subject (column "id"). The third line codes for the fixed effects in the model of Eq. (24). In the PROC MIXED command it is unnecessary to explicitly specify the parameters of the model; each term on the second line automatically has a parameter associated with it (e.g., "day1*cond" refers to Z_1, which is automatically multiplied by the value of column "day1" times the value of column "cond," i.e., by $c_{j1} d_{t1}$). Furthermore, unless otherwise specified, an intercept is automatically assumed to be included in the model. The "solution" option requests that the parameter estimates be reported. The fourth line puts a random effect on the intercept, with the random elements being the different subjects (column "id") as specified in the "subject=" option. The "solution" option requests that the empirical best linear unbiased predictors (EBLUPs) of the random effect for the individual subjects (i.e., the estimates for the η_{ji}) be reported. The "run" statement in the last line requests execution of the analysis.

[13] Mixed-model regression analyses are robust against random deviations from a balanced design (i.e., different numbers of observations among subjects) and randomly occurring missing values. However, the PROC MIXED command in SAS expects missing values to be indicated in the data spreadsheet by means of periods, so that the location of the missing values relative to the available observations is clear. This can be circumvented by using the "repeated" statement in the PROC MIXED command (SAS, 2001b).

The previous command represents a complete analysis, but in order to get results equivalent to those obtained with repeated-measures ANOVA, a few statements must be added to the code:

```
proc mixed;
    class id;
    model y = day1 day2 day3 day0*cond day1*cond day2*cond day3*
    cond / solution;
    random intercept /solution subject = id;
    estimate ''effect of condition'' day0*cond 0.25 day1*cond
    0.25
                            day2*cond 0.25 day3*cond 0.25;
    contrast ''effect of time'' day1 1 day1*cond 0.5 day0*cond
    -0.5,
                day2 1 day2*cond 0.5 day0*cond -0.5,
                day3 1 day3*cond 0.5 day0*cond -0.5;
    contrast ''interaction effect'' day1*cond 1 day0*cond -1,
                day2*cond 1 day0*cond -1,
                day3*cond 1 day0*cond -1;
run;
```

The contrast statements, appropriately labeled "effect of time" and "interaction effect," are equivalent to those effects in repeated-measures ANOVA. The contrast for the effect of time again constitutes a test that the means over all subjects (lumping the two conditions) are equal for all time points (i.e., for Day 0 vs. Day 1, for Day 0 vs. Day 2, and for Day 0 vs. Day 3). As the number of subjects in each condition is the same, this translates into the following three-fold null hypothesis:

$$H_0 : \begin{cases} I_0 + Z_0/2 = I_0 + Y_1 + Z_1/2 \\ I_0 + Z_0/2 = I_0 + Y_2 + Z_2/2 \\ I_0 + Z_0/2 = I_0 + Y_3 + Z_3/2 \end{cases} \tag{27}$$

which can be simplified as follows:

$$H_0 : \begin{cases} Y_1 + Z_1/2 - Z_0/2 = 0 \\ Y_2 + Z_2/2 - Z_0/2 = 0 \\ Y_3 + Z_3/2 - Z_0/2 = 0. \end{cases} \tag{28}$$

The latter form of the null hypothesis is coded in the contrast statement for the effect of time shown previously. For details about the formulation of the "contrast" statement in the PROC MIXED command, see the online SAS user's guide (SAS, 2001a).

As with its equivalent in repeated-measures ANOVA, the contrast for the interaction effect is essentially a test that the difference between the condition means is the same for all time points (i.e., for Day 0 vs. Day 1, for Day 0 vs. Day 2, and for Day 0 vs. Day 3). This translates into the following null hypothesis:

$$H_0 : \begin{cases} Z_0 = Z_1 \\ Z_0 = Z_2 \\ Z_0 = Z_3 \end{cases} \tag{29}$$

which can be reformulated as follows:

$$H_0 : \begin{cases} Z_1 - Z_0 = 0 \\ Z_2 - Z_0 = 0 \\ Z_3 - Z_0 = 0. \end{cases} \tag{30}$$

The latter form of the null hypothesis is coded in the contrast statement for the interaction effect shown previously.

In repeated-measures ANOVA, the effect of condition represents a test of whether the grand means are the same for the different conditions. In the context of Eq. (24), this corresponds to a test of the following null hypothesis:

$$H_0 : \begin{matrix} [I_0 + (I_0 + Y_1) + (I_0 + Y_2) + (I_0 + Y_3)]/4 = \\ [(I_0 + Z_0) + (I_0 + Y_1 + Z_1) + (I_0 + Y_2 + Z_2) + (I_0 + Y_3 + Z_3)]/4 \end{matrix} \tag{31}$$

which can be reduced to

$$H_0 : [Z_0 + Z_1 + Z_2 + Z_3]/4 = 0. \tag{32}$$

This leads directly to a simple test for the effect of condition, namely through evaluation of the estimated value for $[Z_0 + Z_1 + Z_2 + Z_3]/4$, the actual difference between the grand means for the two conditions. This test for the effect of condition is different, and more powerful, than the one available in repeated-measures ANOVA. Moreover, it is robust against systematic interindividual differences, as these are absorbed by the random effect for the intercept. In the PROC MIXED command shown previously, the test is implemented by means of the "estimate" statement (SAS, 2001b), labeled "effect of condition" here. It yields the estimated value of the expression at hand and automatically performs a t test against zero, with the following df[14]:

$$df_c = \Sigma_j(n_j - 1)(T - r), \tag{33}$$

where r is the number of random effects (which equals 1 in the present model).

Table III shows the results of the linear mixed-model regression analysis. The effect of time and the interaction effect are identical to those found for repeated-measures ANOVA (cf. Table II). The effect of condition is different, however, as was to be expected. Provided that the assumption of a normal distribution for the

---

[14] The resulting t statistic, when squared, yields the F statistic for an equivalent F test with 1, df_c degrees of freedom.

Table III
Results from Mixed–Model Anova[a]

Effect	Test	Statistic	Df	p
Condition	T	2.83	1	0.006
Time	F	39.88	3, 72	<0.001
Interaction	F	0.89	3, 72	0.452

[a]Results are shown from mixed-model regression analysis, mimicking repeated-measures ANOVA, on psychomotor vigilance performance data over 4 days of sleep deprivation in two different conditions. The type of test (F or t test), the value of the test statistic, the degrees of freedom, and the p value are given for the effect of condition, effect of time, and interaction of condition by time. These effects are defined equivalently to those for repeated-measures ANOVA (cf. Table II).

random effect is correct, in mixed-model regression analysis the effect of condition is not influenced by systematic interindividual differences. Still, even with this improved test for the effect of condition, the overall results are the same as for repeated-measures ANOVA: a significant effect of time and a significant effect of condition, but no significant interaction effect. Thus, psychomotor vigilance was reduced significantly over days of sleep deprivation; and the placebo condition performed consistently worse than the caffeine condition, as was already clear on Day 0 (before pill administration began).

V. An Alternative Linear Mixed–Effects Model

As mentioned earlier, any effect of caffeine (during Days 1–3) on psychomotor vigilance should have led to a significant interaction effect. However, as there is relatively little statistical power in interaction effects (Winer, 1971), the results from this mixed-model ANOVA (Table III) could not be much more helpful than the results from the repeated-measures ANOVA (Table II) given that the interaction effect is identical for the two approaches. Yet, the mixed-effects model in Eq. (24) also yields *a priori* day-by-day comparisons between the conditions (i.e., the parameters Z_0 through Z_3). These same comparisons would require post-hoc tests in the repeated-measures ANOVA approach,[15] which again would yield results confounded by systematic interindividual variability in the data.

The "solution" option for the "model" statement in the PROC MIXED command shown previously produces the parameter estimates for the fixed effects in the model, and automatically performs a t test against zero (with df$_c$ degrees of freedom) for each. The fixed effects "day0*cond" through "day3*cond" are of interest as they correspond to the parameters Z_0 through Z_3. The results are shown in Table IV, revealing a significant difference between the caffeine and placebo conditions on Day 1 only. This finding would suggest that the effect of sustained

[15] In SPSS, these specific tests can be obtained with the "parameter estimates" option for the repeated-measures ANOVA procedure.

Table IV
Fixed Effects in Mixed–Model ANOVA[a]

Day	Mixed-model ANOVA		Repeated-measures ANOVA	
	$t[72]$	p	$t[24]$	p
0	1.87	0.066	2.79	0.010
1	3.20	0.002	2.95	0.007
2	1.96	0.053	2.03	0.054
3	1.67	0.100	1.39	0.177

[a]The table shows statistical tests of the differences in psychomotor vigilance performance between conditions for each of the 4 days of total sleep deprivation. The t statistics (with 72 degrees of freedom) and p values for these differences as resulting from mixed-model ANOVA are given. For comparison, the t statistics (with 24 degrees of freedom) and p values for the equivalent (post-hoc) tests in repeated-measures ANOVA are also shown.[15] These latter results are confounded by the systematic interindividual variability in the data.

low-dose caffeine was limited to the first day of administration. However, we are still faced with the systematic (albeit nonsignificant) difference between conditions on Day 0 (i.e., prior to pill administration; see Fig. 2). This makes it difficult to tell to what extent the difference between conditions on Day 1 might be nonspecific to caffeine.

The substantial flexibility we have in formulating a model for mixed-model regression analysis is helpful to address this problem. As assignment to condition in this double-blind study was random, the difference between conditions before pill administration should be random and unrelated to condition. We may therefore consider a slightly modified (and more parsimonious) mixed-effects model for the data:

$$y_{jit} = I_{ji} + d_{t1} Y_1 + d_{t2} Y_2 + d_{t3} Y_3 + c_{j1}(d_{t1} Z_1 + d_{t2} Z_2 + d_{t3} Z_3) + \varepsilon_{jit} \qquad (34)$$

which is identical to the model in Eq. (24) except that the Z_0 term for the difference between conditions on Day 0 is left out. The estimated means for the two conditions on Day 0 in this model are a function of the EBLUPs for the random effect on the intercept in Eq. (25), as follows:

$$m_{j0} = I_0 + \Sigma_i \eta_{ji}/n_j. \qquad (35)$$

As the mean of all the EBLUPs should be (almost) identical to zero (recall that the η_{ji} are assumed to arise from a normal distribution with zero mean), it follows that $(m_{10} + m_{20})/2 = I_0$.

To perform the calculations of the mixed-model regression analysis for Eq. (34), we use the following **PROC MIXED** command in the computer software SAS:

```
proc mixed;
    class id;
```

```
modely = day1 day2 day3 day1*cond day2*cond day3*cond /
    solution;
random intercept /solution subject = id;
run;
```

The EBLUPs are generated by the "solution" option to the "random" state-ment. They are shown in Table V; substantial interindividual differences are apparent. Using Eq. (35) and the parameter estimates for Eq. (34), we derive the estimated means for each day in each of the two conditions to get a sense of how well the model fits the data. The results are shown in Fig. 4. As expected, this model

Table V
EBLUPs Resulting from Mixed–Model ANOVA[a]

Id	cond	EBLUP
1	1	1.7
2	1	0.0
3	1	−0.8
4	1	−0.4
5	1	−0.4
6	1	0.4
7	1	−2.3
8	1	−1.1
9	1	−1.0
10	1	−2.5
11	1	−4.1
12	1	−3.0
13	1	−3.4
14	2	1.5
15	2	5.4
16	2	−2.6
17	2	10.4
18	2	7.4
19	2	−11.3
20	2	−7.6
21	2	−3.0
22	2	1.5
23	2	5.4
24	2	2.5
25	2	0.2
26	2	6.9

[a]Empirical best linear unbiased predictors (EBLUPs) for the η_{ji} in the mixed-effects regression model of Eq. (34), representing subject-specific deviations in the intercept relative to the overall intercept I_0. The EBLUPs are given in number of psychomotor vigilance performance lapses per 10-min test bout, for each of the subjects (column "id") in the caffeine condition (column "cond" value 1) and the placebo condition (column "cond" value 2).

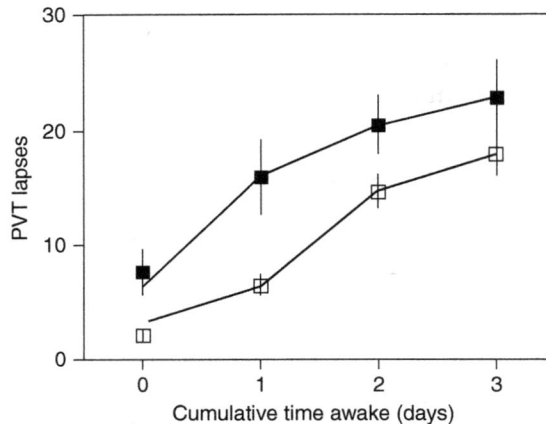

Fig. 4 Estimates for the daily means of lapses on the psychomotor vigilance task (PVT) for the caffeine condition (lower curve) and the placebo condition (upper curve), derived from the model in Eq. (34). The boxes represent condition means (with standard errors of the mean) for the actual data (see Fig. 2).

fits the means as accurately as the model of Eq. (24) or the repeated-measures ANOVA approach (both of which provided a perfect fit to the means; cf. Fig. 2), except on Day 0. This suggests that the distribution of the random elements (i.e., the subjects' performance on Day 0) is not precisely normal.[16]

We make use again of the *a priori* day-by-day comparisons between the conditions (i.e., fixed-effect parameters Z_1 through Z_3) in the model of Eq. (34) to confirm the earlier tentative finding that the effect of sustained low-dose caffeine was limited to the first day of pill administration. The estimated values for Z_1 through Z_3 and the t tests for whether they differ significantly from zero are shown in Table VI. These results confirm the significant difference between the caffeine and placebo conditions on Day 1 only, independently of the systematic interindividual differences in the data (i.e., independently of the random effect). Thus, we have gained more definitive evidence that the effect of sustained low-dose caffeine was limited to the first day of intake, despite the fact that plasma caffeine concentrations were high throughout the 66 h of pill administration (Fig. 1).

[16] The estimate for I_0 is 4.9 ± 1.5 (mean $\pm$ standard error). Inspection of Table V reveals that the estimated subject-specific performance $I_0 + h_{ji}$ on Day 0 is negative (i.e., less than zero lapses per 10-min test bout) for some subjects ($i = 19, 20$). This anomaly again suggests that the distribution of the random elements is not precisely normal. Specification of other types of distributions for the random effect(s), such as the lognormal distribution that always yields positive values, is possible in SAS (2001b). However, the results of mixed-effects modeling are not critically dependent on the assumptions about the distribution of the random effect(s), especially if many repeated measures are available. See Olofsen et al. (2010).

Table VI

Day–by–Day Comparisons in Linear Mixed–Model Regression Analysis[a]

Day	Z	$t[72]$	P
1	6.9	2.58	0.012
2	3.2	1.21	0.231
3	2.4	0.88	0.382

[a]Differences in psychomotor vigilance performance between conditions are shown for each of the 3 days of total sleep deprivation following the beginning of pill administration (i.e., Days 1–3), as assessed with the adjusted mixed-effects regression model of Eq. (34). The differences in the number of lapses Z and the corresponding t statistics (with 72 degrees of freedom) and p values are given.

VI. Nonlinear Mixed–Model Regression Analysis

We now shift our attention from linear mixed-model regression analysis to nonlinear mixed-model regression analysis. Although linear and non-linear mixed-effects models are formulated quite differently in most published literature and computer software (e.g., see the online SAS user's guide; SAS, 2001b, they are actually intimately related, linear mixed-effects modeling being a special case of nonlinear mixed-effects modeling (for this reason, we have standardized the notation throughout this chapter). Nonlinear mixed-model regression is frequently needed to analyze hypothesis-driven models (i.e., models that go beyond describing the data in terms of unspecified changes over time and/or differences among conditions as in ANOVA), as such models tend to include nonlinear combinations of fixed and/or random effects.[17] The extensive numerical calculations required for nonlinear mixed-model regression analysis have become feasible in the last 5 years because of the increasing computational power of standard computer hardware. We can take advantage of this development for the further analysis of our study data.

Considering the evidence we gathered thus far that the attenuation of performance impairment by caffeine dissipated over days of sleep deprivation, we wonder about the precise duration of the efficacy of sustained low-dose caffeine in this experiment. We therefore consider the following mixed-effects model for the study data as a function of days t[18]:

[17] Hybrid models, containing mixed-model ANOVA elements as well as hypothesis-driven components, can be readily constructed from the formulas in this chapter, and typically require nonlinear mixed-model regression analysis as well.

[18] It is generally advisable to centralize the independent (and dependent) variables of a regression analysis (i.e., adding constants to each so that the ranges of values they take center on zero). This practice tends to reduce the covariance among model parameters (especially in linear regression models) and promotes the reliability of model convergence. In Eq. (36), however, centralization is problematic for the independent variable t, since the term $b_0 t^s$ would be undefined for $t < 0$.

$$y_{ji}(t) = I_{ji} + b_0 t^s - [(1 - d_{t0})c_{j1}z(t)] + \varepsilon_{ji}(t), \tag{36}$$

where model parameters I and ε and indicator variables c and d are defined as in the previous sections. The function $b_0 t^s$ has previously been shown to describe the data in the placebo condition (Van Dongen *et al.*, 2003a); it involves a curvature parameter $0 < s < 1$ and a scale factor b_0. The term between square brackets represents the hypothesized temporal change in caffeine's efficacy—for the caffeine condition only and on Days 1–3 only (i.e., during caffeine pill administration). We hypothesize that beginning with the first pill (i.e., on Day 1), the efficacy $z(t)$ of hourly administration of caffeine diminishes exponentially over days t:

$$z(t) = ae^{-(t-1)/T_0}, \tag{37}$$

where a is a scale factor and T_0 is a time constant for the decline of caffeine's efficacy. The model of Eq. (36), though linear in the random effect included in the intercept I_{ji} of Eq. (25), cannot be cast in the form of a linear mixed-model regression model, and must be subjected to nonlinear mixed-model regression analysis.

Let us assume that we have the psychomotor vigilance performance data for Days 0–3 of the sleep deprivation experiment stored in a spreadsheet similar to that described in the previous sections, with a column listing the different subjects (named "id"), a column indicating the days (named "t"), a column equivalent to the indicator factor $[(1 - d_{t0})\,c_{j1}]$ (named "caff"), and a column for the data of each subject on each day (named "y").[19] For technical reasons, the spreadsheet must again be ordered by subject. To perform the calculations for the nonlinear mixed-model regression analysis, we use the following **PROC NLMIXED** command in the computer software SAS:

```
proc nlmixed;
    parms
        i0 = 5.0,
        s2i = 25.0,
        b0 = 10.0,
        a = 20.0,
        s= 0.5,
        t0 = 1.0,
        s2e = 35.0;
        z= a*caff*exp (-(t - 1)/t0);
    if t = 0 then v = 0;
        else v = t**s;
```

[19] For model convergence and reliability of analysis outcomes in PROC NLMIXED, it is desirable that the dependent variable y and the independent variabes of the model (in this case only t) have comparable ranges (i.e., same order of magnitude). Linear transformations should be used as necessary to accomplish this.

```
model y ~ normal (i0 + vari + b0*v - z, s2e);
random vari ~ normal (0, s2i) subject=id out=ebes;
run;
```

The "parms" part of this command introduces the seven parameters of the model, and their initial values[20] (which the computer software requires to begin the calculations). The parameters are "i0" for the intercept I_0, "s2i" for the variance ω^2 of the random effect η for the intercept as in Eq. (25), "b0" and "a" for scale factors b_0 and a in Eqs. (36) and (37), respectively, "s" for the curvature parameter s, "t0" for the time-constant T_0, and "s2e" for the variance σ^2 of the error term ε. These are all the parameters explicitly and implicitly contained in the model of Eq. (36). The "z = ..." line in the command computes the function $z(t)$ of Eq. (37). The "if t=0 ..." part of the command is necessary because SAS does not adopt the convention that $t^s = 0$ for $t = 0$. Thus, we introduce a substitute variable "v" defined by $v = t^s$ for $t > 0$ and $v = 0$ for $t = 0$; v replaces t^s in Eq. (36) without changing the model.

The "model" statement defines the representation for the data y as normally distributed random fluctuations with variance σ^2 (i.e., "s2e") around the model of Eq. (36). That model is described by the following code in the "model" statement:

$$i0 + vari + b0 * v - z$$

which follows directly from Eqs. (36) and (37), except for the term "vari" that represents the random effect for the intercept. This random effect is defined in the "random" statement as a normal distribution with mean zero and variance ω^2 (i.e., "s2i"); the declaration "subject=id" specifies that the random effect pertains to variability among subjects. Finally, the "out=" option stores the empirical Bayes estimates (EBEs) for the random effect η_{ji} (the equivalent of the EBLUPs in linear mixed-model regression analysis) in a spreadsheet called "ebes."

Table VII shows the parameter estimates resulting from the nonlinear mixed-model regression analysis. Of primary interest is the estimate for T_0 (i.e., 1.2336 before rounding), from which we can derive the half-life $T_{0.5}$ of the efficacy of sustained low-dose caffeine using the following expression:

$$e^{-T_{0.5}/T_0} = 0.5. \tag{38}$$

It follows that $T_{0.5} = 0.86$ days. Thus, it appears that sustained low-dose caffeine lost half of its efficacy in less than a day, which is consistent with what we derived using linear mixed-model regression analysis (see previously). This finding could reflect a rapid build-up of tolerance to caffeine. Alternatively, the build-up of sleepiness during the extended sleep deprivation could have simply overwhelmed the stimulating effect of caffeine after about a day.

---

[20] Depending on the complexity of the model, the choice for the initial values can be critical for model convergence and for the success of the analysis. Proper initial values can often be derived from a two-stage analysis of the same data (Burton *et al.*, 1998; Feldman, 1988).

Table VII
Parameter Estimates for Nonlinear Mixed–Effects Regression Model[a]

Parm	Name	Est	SE
I_0	i0	4.8	1.5
ω^2	s2i	27.0	10.1
b_0	b0	10.1	1.9
a	a	7.0	2.6
s	s	0.44	0.15
T_0	t0	1.2	1.1
σ^2	s2e	29.5	4.8

[a]Results are shown from nonlinear mixed-model regression analysis of psychomotor vigilance performance data over 4 days of sleep deprivation, using the hypothesis-driven model in Eq. (36) to assess the duration of the efficacy of sustained low-dose caffeine (relative to placebo). As computed using the PROC NLMIXED command in SAS, the table shows the parameters (Parm), their names in the command (Name), their estimates (Est), and their standard errors (SE).

The relatively large standard error in Table VII for the T_0 estimate would seem to indicate that the effect of sustained low-dose caffeine in this experiment may not be very robust. In fact, the estimate for T_0 is not significantly different from zero ($t[25] = 1.08$, $p = 0.29$). This warrants investigation of whether removal from the model of this parameter, and thereby the entire function $z(t)$, would constitute a significant deterioration in how well the model describes the data. This can be assessed with the likelihood ratio test, which involves calculation of -2 times the natural logarithm of the likelihood (i.e., the value of $-2 \log L$) for the full model (with all parameters included) and for the reduced model (with the parameters that might be unnecessary being removed). By subtracting the $-2 \log L$ value for the full model from the corresponding value for the reduced model, the likelihood ratio is computed. This statistic approximately has a χ^2 distribution, and the difference in the number of free parameters between the full and reduced models determines the df for that χ^2 distribution (see the online SAS user's guide; SAS, 2001b). Using the likelihood ratio test, we find that inclusion of parameter T_0 (and thereby also parameter a) results in a significant improvement over the model without T_0 ($\chi^2[2] = 7.50$, $p = 0.024$).

It is useful also to graphically check how well the nonlinear mixed-effects model of Eq. (36) fits the data. Using Eq. (35), we first estimate the means for the two conditions on Day 0 in this model, which are a function of the EBEs for the random effect on the intercept in Eq. (36). The EBEs are shown in Table VIII; they are similar to those found for the linear mixed-effects regression model of Eq. (34) (cf. Table V). Fig. 5 shows the nonlinear mixed-effects model overlaid on the group mean data. It appears that the model in Eq. (36) fits the means well, except on Day 0. As in the model of Eq. (34), this suggests that the distribution of the random effect for the intercept among subjects is not precisely normal. However, Fig. 5 is reassuring with regard to the approximate validity of the hypothesized dissipation profile in Eq. (37), at least as it pertains to the group means.

Table VIII
EBEs Resulting from Nonlinear Mixed–Model Regression Analysis[a]

Id	Cond	EBE
1	1	1.6
2	1	−0.1
3	1	−0.9
4	1	−0.5
5	1	−0.5
6	1	0.3
7	1	−2.5
8	1	−1.2
9	1	−1.1
10	1	−2.6
11	1	−4.2
12	1	−3.1
13	1	−3.5
14	2	1.6
15	2	5.6
16	2	−2.5
17	2	10.6
18	2	7.5
19	2	−11.3
20	2	−7.5
21	2	−2.9
22	2	1.6
23	2	5.6
24	2	2.6
25	2	0.3
26	2	7.1

[a]Empirical Bayes estimates (EBEs) for the η_{ji} in the nonlinear mixed-effects regression model of Eq. (36), representing subject-specific deviations in the intercept relative to the overall intercept I_0. The EBEs are given in number of psychomotor vigilance performance lapses per 10-min test bout, for each of the subjects (column "id") in the caffeine condition (column "cond" value 1) and placebo condition (column "cond" value 2).

An explanation for the relatively large standard error for the time constant T_0 might be that an additional random effect is needed in the model. Previous analyses have revealed large variability in the scale factor b_0 for the placebo condition (Van Dongen *et al.*, 2003a). Therefore, we add a random effect to the model in Eq. (36) as follows:

$$y_{ji}(t) = I_{ji} + b_{ji}t^s - [(1 - d_{t0})c_{j1}z(t)] + \varepsilon_{ji}(t) \qquad (39)$$

where

$$b_{ji} = b_0 + \beta_{ji}. \qquad (40)$$

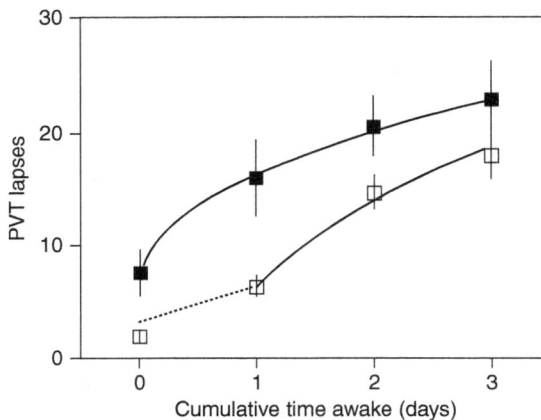

Fig. 5 Hypothesis-driven model for the performance-impairing effect of total sleep deprivation and the transient mitigating effect of caffeine, as measured by the daily means of lapses on the psychomotor vigilance task (PVT). The model for the placebo condition is shown by the upper curve, and the model for the caffeine condition is shown by the lower curve. The dotted part of the lower curve connects the period prior to caffeine administration with the period during which caffeine was administered; the boundary between these two periods involves a discontinuity in the model. The boxes represent condition means (with standard errors of the mean) for the actual data (Fig. 2).

The β_{ji} are assumed to arise from a normal distribution with zero mean and variance λ^2 over subjects (with λ^2 not known in advance). The model now has two random effects (one for the intercept I and one for the scale factor b); we assume that the covariance between these two random effects is zero.[21]

Even though the PROC NLMIXED command in SAS (2001a) can deal with two (but not more than two) random effects, it is now more convenient to use the specialized computer software NONMEM (1998) to perform the calculations for the nonlinear mixed-model regression analysis.[22] Using the same data in the same spreadsheet (named "CAFF.DAT") as for PROC NLMIXED in SAS, ordered by subject, we apply the following NONMEM macro:

```
$PROBLEM CAFFEINE
$DATA CAFF.DAT
$INPUT ID T CAFF DV
$PRED
    I=THETA(1)+ETA(1)
    B=THETA(2)+ETA(2)
```

[21] For small subject populations, the covariance between random effects is usually not well estimable; inappropriately setting it to zero is probably no more problematic than estimating it poorly.

[22] We have found that in NONMEM (1998) the numerical computations for models with two random effects are more likely to converge, over a wider range of initial values for the parameters.

```
A=THETA(3)
S=THETA(4)
TNULL=THETA(5)
Z=A*CAFF*EXP(-(T-1)/TNULL)
IF (T.EQ.0) THEN
    V=0
    ELSE
    V=T**S
    ENDIF
MODEL=I+B*V-Z
Y=MODEL+ERR(1)
$THETA (0.001, 5,100) (0.001, 10, 100) (0.001, 4,100) (0.001,
    1,100) (0.001,1,100)
$OMEGA 10 10
$SIGMA 35
$ESTIMATION METH=1
$COVR
```

The "$PROBLEM" statement introduces the nonlinear mixed-model regression analysis to NONMEM, and gives it a name (arbitrarily set to "CAFFEINE"). The "$DATA" statement tells NONMEM where to find the data. In the "$INPUT" statement, the four columns in the data spreadsheet are assigned to the variables ID, T, CAFF, and DV, where the latter stands for "dependent variable" and corresponds to y.

The "$PRED" statement contains the actual regression model. The five fixed effects in the model are automatically handled as a vector THETA with five elements; the two random effects are represented by a vector ETA with two elements (whose parameter estimates ω^2 and λ^2 are in the corresponding vector OMEGA); and the error term ε is a vector ERR with one element (whose parameter estimate σ^2 is in the corresponding vector SIGMA). The model of Eq. (39) is constructed using these building blocks, with I representing I_{ji}, B representing b_{ji}, A corresponding to a, S corresponding to s, and TNULL standing for T_0. The substitute variable V is defined as in the PROC NLMIXED command shown previously. The last line of the "$PRED" statement (which must begin with "Y=") contains the complete model of Eq. (39).

The "$THETA" statement gives the initial values as well as the boundaries for the fixed effects parameters in the THETA vector, in the format "(lower boundary, initial value, upper boundary)"; the "$OMEGA" and "$SIGMA" statements give the initial values for ω^2 and λ^2, and for σ^2, respectively. The "$ESTIMATION" statement specifies details about the numerical procedures to be used, which are beyond the scope of this chapter. The "$COVR" statement, finally, requests computation of the co-variance matrix, which is needed for estimation of the standard errors of the parameter estimates.

Table IX
Parameter Estimates for Nonlinear Mixed–Effects Regression
Model with Two Random Effects[a]

Parm	Name	Est	SE
I_0	THETA(1)	4.9	1.1
ω^2	OMEGA(1)	20.3	5.6
b_0	THETA(2)	9.6	3.0
λ^2	OMEGA(2)	6.3	5.0
A	THETA(3)	6.5	3.6
s	THETA(4)	0.50	0.24
T_0	THETA(5)	1.3	2.6
σ^2	SIGMA(1)	26.3	5.6

[a]Results are shown from nonlinear mixed-model regression analysis of psychomotor vigilance performance data over 4 days of sleep deprivation, using the hypothesis-driven model in Eq. (39) with two random effects to assess the duration of the efficacy of sustained low-dose caffeine (relative to placebo). As computed using NONMEM, the table shows the parameters (Parm), their names in the NONMEM macro (Name), their estimates (Est), and their standard errors (SE).

Table IX shows the parameter estimates resulting from this nonlinear mixed-model regression analysis. The estimate for the time constant T_0 in the model of Eq. (39) is essentially the same as in the model of Eq. (36) (cf. Table VII). However, the estimated standard error of T_0 is increased rather than decreased,[23] suggesting no improvement in this aspect of the model due to the addition of the random effect for b. Indeed, the improvement in the model is nonsignificant overall, as assessed by comparing the model of Eq. (39) (full model) with that of Eq. (36) (reduced model, without parameter λ^2) by means of the likelihood ratio test ($\chi^2[1] = 2.39$, $p = 0.12$). It follows that among the models we investigated for the data at hand, the preferred model is given by Eq. (36). Further, our best estimate for the half-life of the efficacy of sustained low-dose caffeine in this experiment remains 0.86 days.

[23] A threat to the accuracy of standard error estimates is model misspecification (i.e., when the error variance is not distributed normally as specified). In the PROC NLMIXED command in SAS (2001a), we have observed cases in which the estimated standard errors were 50% smaller than the true standard errors (as assessed with bootstrap simulations). A correction is available for the covariance matrix from which the standard errors are derived, making them more robust against symmetric nonnormality in the error term. It is commonly referred to as quasi-maximum likelihood (QML) estimation of the covariance matrix. The QML estimate of the covariance matrix is the default output of the "$COVR" statement in NONMEM (1998), but it is not available in PROC NLMIXED in SAS (2001a). This may partly explain the increase in the estimated standard error of T_0. See Bollerslev and Wooldridge (1992).

======= ## VII. Correlation Structures in Mixed–Model Regression Analysis

The statistical power and the efficiency of repeated-measures designs like the present study arise from the correlation of data points within individuals, and the associated distinction of within-subjects variance from between-subjects variance (Burton *et al.*, 1998). In most cases, the correlation between the data points within individuals is not *a priori* known. In practice, therefore, a correlation structure is picked during data analysis, in the hopes that it resembles the true correlation structure. There is a rich selection of possible correlation structures (or "covariance structures") that take into account interindividual variability (i.e., random effects) as well as systematic correlations in the residual variance over time (Littell *et al.*, 2000). For linear mixed-model regression analysis, methodology for a variety of correlation structures is readily available and implemented in computer software (e.g., in the PROC MIXED command in SAS (2001b). For nonlinear mixed-model regression analysis, the implementation of covariance structures is less straightforward. The default situation in nonlinear mixed-model regression is known as the compound symmetry correlation structure (Burton *et al.*, 1998), which results from implementing a "variance components" model. A variance components model assumes that the random effects are independent variance components (i.e., having zero covariance).

The variance components model is used for all mixed-model regression analyses in this chapter, including the model of Eq. (36). This model has a random effect on the intercept via the term η of Eq. (25), which has variance ω^2. Although it was not explicitly mentioned in the previous sections, the model actually has a second random effect in the form of the error term ε, which has variance σ^2. These two random effects represent the between-subjects variance and the within-subjects variance, respectively, in this regression model. In a variance components model with a normally distributed random effect on the intercept and normally distributed error variance, the correlation structure is fully determined by the intraclass correlation coefficient (ICC),[24] which is estimated as

$$\text{ICC} = \omega^2/(\omega^2 + \sigma^2). \tag{41}$$

In such variance components models, the correlation between *each pair* of data points of a given subject is assumed to be equal to the ICC. For the nonlinear mixed-effects regression model of Eq. (36), this correlation can be estimated using the results in Table VII:

---

[24] If the random effect is not on the intercept but on another component of the model, computation of the ICC is ambiguous. For instance, suppose that the model in Eq. (39) would not have a random effect on the intercept, leaving only the random effects represented by β and the error term ε. The value and the unit of the variance λ^2 for β would depend on the magnitude (and unit) of the factor t^s in that model (e.g., whether t is expressed in days or in hours). This would cause obvious problems for the computation of the ICC, which do not arise if the random effects are on the intercept and the (additive) error term only (and the distribution of both random effects is assumed to be normal).

$$ICC = 28.6/(28.6 + 29.2) = 0.49 \qquad (42)$$

which means that the correlation between each pair of data points within each subject is (implicitly) assumed to be 49%.[25]

It is noteworthy that the value of the ICC has an additional, complementary interpretation (Van Dongen *et al.*, 2010). By definition of Eq. (41), the ICC expresses the between-subjects variance as a fraction of the total variance not explained by the fixed effects in the model.[26](Snijders and Bosker, 1994) In Eq. (36), therefore, the ICC quantifies the importance of systematic interindividual differences in the intercept with respect to overall variability in the data around the regression model. Using this interpretation of the ICC, studies repeating sleep deprivation in the same individuals have revealed systematic interindividual differences in performance deficits resulting from sleep deprivation, with ICC values greater than 0.5 (Van Dongen *et al.*, 2010). This underlines the importance of taking such interindividual differences into account (e.g., with random effects) when modeling data from sleep deprivation experiments.

VIII. Conclusion

In this chapter, we considered the analysis of longitudinal data in the presence of interindividual differences. We first described repeated-measures ANOVA, a traditional technique for the analysis of longitudinal data tailored to the comparison of the means of subsets of the data. We showed that this technique is not robust to systematic inter-individual differences. We then discussed linear mixed-model regression analysis. We employed this technique to mimic repeated-measures ANOVA while adding robustness against systematic interindividual variability (i.e., mixed-model ANOVA). This application of mixed-effects modeling is especially useful if no *a priori* expectations exist about the shape of the data's temporal profile. For hypothesis-driven analysis of time series data, however, mixed-effects models frequently involve nonlinearity in the parameters. Therefore, we also considered nonlinear mixed-model regression analysis. Our aim was to convey a basic understanding of the mathematical and statistical issues involved in mixed-model regression analysis. For this purpose, we included specific examples for how to implement mixed-effects regression models in computer software (i.e., SAS, 2001a and NONMEM, 1998). In the process, we assessed the duration of the efficacy of sustained low-dose caffeine during an experiment involving 88 h of continuous wakefulness. The data from this study are publicly available

---

[25] Investigating whether or not this correlation structure is realistic given the data at hand is beyond the scope of this chapter. The data can be analyzed repeatedly with different correlation structures and the results compared by means of a statistical information criterion to select a correlation structure that best fits the data. This is often done, for instance, when an autoregressive structure is suspected but the order of autoregression is yet to be determined.

[26] For more about explained variance in mixed-model regression analysis, see Snijders and Bosker (1994).

(see Table I); thus, the findings reported here can be replicated as an exercise for familiarizing oneself with mixed-effects modeling.

As the generic mechanisms of many physiological and pharmacological phenomena are better understood, there will be a growing—and much needed—interest in interindividual differences to explain the diversity in the parameters of these mechanisms within populations. As a consequence, there will be an increasing demand for data analysis techniques capable of dealing with interindividual differences. With the major enhancements of computer power seen in recent years, mixed-model regression analysis has become feasible on common personal computer platforms. We expect, therefore, that mixed-model regression will become a standard by which longitudinal data are analyzed in the twenty-first century. We hope that the present chapter will facilitate this trend.

Acknowledgment

Supported by Grant R01-HS10064 from the Agency for Healthcare Quality and Research of the United States Public Health Service.

References

Bollerslev, T., and Wooldridge, J. M. (1992). *Econom. Rev.* **11**, 143.

Burton, P., Gurrin, L., and Sly, P. (1998). *Stat. Med.* **17**, 1261.

Diggle, P. J., Kiang, K. Y., and Zeger, S. L. (1996). "Analysis of Longitudinal Data", Clarendon Press, Oxford.

Dinges, D. F., Doran, S. M., Mullington, J., Van Dongen, H. P. A., Price, N., Samuel, S., Carlin, M. M., Powell, J. W., Mallis, M. M., Martino, M., Brodnyan, C., Konowal, N., *et al.* (2000). *Sleep* **23**(Suppl. 2), A20.

Feldman, H. A. (1988). *J. Appl. Physiol.* **64**, 1721.

Girden, E. R. (1992). ANOVA: Repeated Measures, Sage, Newbury Park, CA.

Littell, R. C., Pendergast, J., and Natarajan, R. (2000). *Stat. Med.* **19**, 1793.

NONMEM (1998). NONMEM version V level 1.1, GloboMax LLC, Hanover, MD.

Olofsen, E., Dinges, D. F., and Van Dongen, H. P. A. (2010). *Aviat. Space Environ. Med.* (Suppl.), in press.

Rosnow, R. L., and Rosenthal, R. (1995). *Psychol. Sci.* **6**, 3.

SAS (2001a). "SAS System for Windows, Release 8.02", SAS Institute Inc., Cary, NC.

SAS (2001b). "SAS Online Doc SAS/STAT User's Guide, Version 8", SAS Institute Inc., Cary, NC.

Snijders, T. A. B., and Bosker, R. J. (1994). *Sociol. Methods Res.* **22**, 342.

SPSS (2001). SPSS for Windows, release 11.0.1, SPSS Inc., Chicago, IL.

Van Dongen, H. P. A., and Dinges, D. F. (2000). *In* "Principles and Practice of Sleep Medicine," (M. H. Kryger, T. Roth, and W. C. Dement, eds.), 3rd edn., p. 391. W. B. Saunders, Philadelphia, PA.

Van Dongen, H. P. A., Price, N. J., Mullington, J. M., Szuba, M. P., Kapoor, S. C., and Dinges, D. F. (2001). *Sleep* **24**, 813.

Van Dongen, H. P. A., Maislin, G., Mullington, J. M., and Dinges, D. F. (2003a). *Sleep* **26**, 117.

Van Dongen, H. P. A., Rogers, N. L., and Dinges, D. F. (2003b). *Sleep Biol. Rhythms* **1**, 5.

Van Dongen, H. P. A., Maislin, G., and Dinges, D. F. (2010). *Aviat. Space Environ. Med.* (Suppl.), in press.

Winer, B. J. (1971). "Statistical Principles in Experimental Design", 2nd edn. McGraw-Hill, New York, NY.

CHAPTER 12

Distribution Functions from Moments and the Maximum-Entropy Method

Douglas Poland

Department of Chemistry
The Johns Hopkins University
Baltimore, Maryland, USA

I. Introduction

In this chapter, we show how one can use experimental data such as that contained in titration curves and the temperature variation of the heat capacity to obtain distribution functions for proteins and nucleic acids. For example, we show how one can use a titration curve, which gives the average number of ligands bound as a function of ligand concentration, to calculate how many molecules have one ligand bound, two ligands bound, and so on, giving the distribution function for ligand binding. This distribution function gives a detailed picture of the probability of all possible different states of ligand binding. The process of going from the original titration curve, which gives only the average state of ligand binding, to the complete distribution function greatly increases one's knowledge of the different states of binding present in the system. In a similar manner, we show how knowledge of the temperature dependence of the heat capacity of a protein or nucleic acid can be used to obtain a distribution function for the enthalpy content of the molecule.

DOI: 10.1016/B978-0-12-384997-7.00012-1

The enthalpy distribution function tells what fraction of the molecules have a given enthalpy value, and the temperature variation of this distribution gives a detailed picture of the process of denaturation in terms of the probabilities of different enthalpy states of the molecule.

An important aspect of the approach outlined here is that the distribution functions we obtain, ligand binding distributions or enthalpy distributions, are determined solely by experimental data. While models play an important role in understanding biological macromolecules, it is also important to have results that are independent of any assumed model and the distribution functions we obtain are examples of such knowledge.

The starting point for our calculation of distribution functions from experimental data is the realization that information on the average number of ligands bound as a function of ligand concentration or heat capacity as a function of temperature can be used to calculate a set of moments for the appropriate distribution function. The equations that relate titration and heat capacity data to moments of a distribution function are obtained using basic relations from statistical mechanics. In particular, the relations we use require only the basic partition functions of statistical mechanics and are not on the basis of any specific model. In this manner, we convert one set of experimental data (titration curves or heat capacity) into another set (moments of a distribution function).

Given a finite set of moments (say, two to six), the problem then is to calculate the corresponding distribution function. To accomplish the construction of an approximate distribution function from a finite set of moments, we use an algorithm based on the maximum-entropy method due to Mead and Papanicolaou (1984). In this process, one trades experimental knowledge of moments of the distribution function for knowledge of the parameters describing the functional form of the distribution function. An example of the construction of a distribution function from moments is given by the familiar bell-shaped Gaussian distribution function. The parameters required to construct this function are the mean value of the distribution and the standard deviation. The mean value is simply the first moment of the distribution while the square of the standard deviation is the second moment minus the first moment squared. Similarly, application of the maximum-entropy method gives an approximate distribution function using moments as input, but in this approach we are not restricted to the use of just the first two moments; the more the moments used, the better the approximation obtained.

The applications we present in this chapter are on the basis of methodology contained in two publications. The first publication shows how enthalpy distribution functions can be constructed from knowledge of the temperature dependence of the heat capacity (Poland, 2000a) while the second shows how ligand-binding distributions can be constructed using the data contained in titration curves (Poland, 2000b). An outline of the general approach that we use is as follows. One starts with experimental data on the system of interest (a titration curve or heat capacity data) and makes use of basic relations from statistical mechanics to obtain moments for the distribution function as outlined below:

$$\text{Experimental data} \xrightarrow{\substack{\text{Statistical} \\ \text{mechanics}}} \begin{array}{c} \text{Moments of} \\ \text{distribution function.} \end{array} \tag{1}$$
(titration curve, heat capacity)

One then uses the maximum-entropy method to convert the set of moments into parameters of the respective distribution function:

$$\begin{array}{c} \text{Moments of} \\ \text{distribution} \\ \text{function} \end{array} \xrightarrow{\substack{\text{Maximum entropy} \\ \text{method}}} \begin{array}{c} \text{Parameters of} \\ \text{distribution function.} \end{array} \tag{2}$$

The first step, obtaining moments of the distribution function from experimental data, is the more difficult of the two and we devote most of this chapter to describing this process. In Section II, we show how the moments of ligand-binding distribution functions can be obtained from titration curves. Then, in Section III, we outline the maximum-entropy method for determining approximate distribution functions from experimental knowledge of moments. We return to ligand binding in Section IV where, using the binding of protons to the protein lysozyme as an example, we take the moments obtained in Section II and construct distribution functions from them, utilizing the maximum-entropy method outlined in Section III. In Section V, we use this approach to obtain enthalpy distributions for proteins from the temperature dependence of the heat capacity of these molecules. We illustrate how the enthalpy distributions obtained in this manner give a rich picture of the thermal denaturation of proteins. Finally, in Section VI, we use this general method to obtain distribution functions for self-association, using the clustering of ATP as an example.

II. Ligand Binding: Moments

In this section, we show how one can obtain moments of the distribution function for ligand binding from the binding isotherm (or titration curve) that gives the experimental measurement of the average number of ligands bound as a function of the ligand concentration. Our presentation here follows that given in two previous publications (Poland, 2000a, 2001e). We begin by considering the reaction for adding a ligand to a macromolecule. We let P represent any molecule, in particular a biopolymer such as a protein or nucleic acid. We let L represent a general ligand such as Mg^{2+}, H^+, or any small molecule. We begin with the reaction for adding one additional ligand to a molecule with $(n - 1)$ ligands already bound:

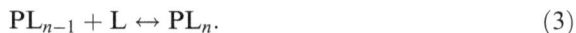

$$PL_{n-1} + L \leftrightarrow PL_n. \tag{3}$$

Taking K_n [with units $(mol/l)^{-1}$] as the equilibrium constant for the reaction in Eq. (3), the standard equilibrium constant expression for this reaction is (using square brackets to indicate concentrations)

$$K_n = \frac{[\text{PL}_n]}{[\text{PL}_{n-1}][\text{L}]}. \tag{4}$$

The development is easier if we relate all states of binding to the same species. This can be accomplished by adding together appropriate reactions of the form given in Eq. (3) as illustrated in the following example:

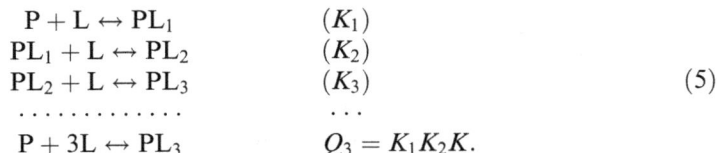

$$\begin{array}{ll}
\text{P} + \text{L} \leftrightarrow \text{PL}_1 & (K_1) \\
\text{PL}_1 + \text{L} \leftrightarrow \text{PL}_2 & (K_2) \\
\text{PL}_2 + \text{L} \leftrightarrow \text{PL}_3 & (K_3) \\
\cdots\cdots\cdots\cdots & \cdots \\
\text{P} + 3\text{L} \leftrightarrow \text{PL}_3 & Q_3 = K_1 K_2 K.
\end{array} \tag{5}$$

Here, we have added together the first three successive steps of binding to give the reaction for binding three ligands directly to the species P. The equilibrium constant for the new reaction, Q_3, is the product of the equilibrium constants for the three reactions used (in general when one adds together two reactions, the equilibrium constant for the new reaction is the product of the constants for the two reactions). The generalization of the process illustrated in Eq. (5) is

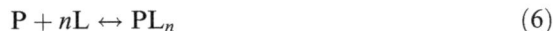

$$\text{P} + n\text{L} \leftrightarrow \text{PL}_n \tag{6}$$

with

$$Q_n = \frac{[\text{PL}_n]}{[\text{P}][\text{L}]^n} \tag{7}$$

where

$$Q_n = \prod_{i=1}^{n} K_i. \tag{8}$$

If the total concentration of the macromolecule is P_T, then the mole fraction of a given state of binding is given by

$$f_n = \frac{[\text{PL}_n]}{\text{P}_\text{T}}. \tag{9}$$

Taking

$$c = [\text{L}] \tag{10}$$

we can rewrite Eq. (7) in terms of the f values and c, giving

$$\frac{f_n}{f_o} = c^n Q_n. \tag{11}$$

Solving Eq. (11) for f_n, we have our fundamental relation for the mole fraction of P molecules having n ligands bound:

$$f_n = f_o c^n Q_n. \tag{12}$$

Now, by definition, the sum of the mole fractions must add up to 1 (conservation of mole fractions),

$$\sum_{n=1}^{N} f_n = 1. \tag{13}$$

Using Eq. (12) in Eq. (13) and solving for f_0 gives

$$f_0 = \frac{1}{1 + cQ_1 + c^2 Q_2 + \ldots + c^N Q_N}. \tag{14}$$

It is useful to define the quantity in the denominator of Eq. (14) as

$$\Gamma = 1 + cQ_1 + c^2 Q_2 + \ldots + c^N Q_N. \tag{15}$$

The quantity Γ in Eq. (15) is known as the binding polynomial and Schellman (1975) has described the use of this quantity to treat ligand binding in biopolymers. The binding polynomial is an example of what is known in statistical mechanics as a partition function. Here, it represents a sum over all possible states of binding. Using the relation for f_0 given in Eq. (14) in Eq. (12), we have the following general relation for the mole fraction of P with n ligands bound:

$$f_n = \frac{c^n Q_n}{\Gamma}. \tag{16}$$

From Eq. (16), one sees that the mole fraction of molecules with n ligands bound (which is equal to the probability of picking a molecule at random that has n ligands bound) is simply given by the term in Γ corresponding to the state representing n ligands bound divided by Γ, that is, the sum over all states of binding.

Equation (16) is a general relation for the mole fractions or probabilities of all different states of binding as a function of the concentration, c, of ligand in solution. As such, this equation is the distribution function for ligand binding and it is this function that we want to construct using moments and the maximum-entropy method. We now show how moments of this function can be obtained from a binding isotherm (titration curve). Notice that in the derivation of Eq. (16) we made no assumptions about the Q_n or the independence of binding sites. We have assumed that the ligand concentration in solution is dilute so that the use of the equilibrium constant expression in Eq. (4) is valid.

Experimentally, one measures the average number of ligands bound as a function of c or $\ln c$ and this set of data gives the binding isotherm or titration curve. From Eq. (16), we have the following relation for average n (average extent of binding),

$$\langle n \rangle = \sum_{n=0}^{N} n f_n. \tag{17}$$

To obtain $\langle n \rangle$ given in Eq. (17) in terms of Γ, we observe that

$$\frac{\partial c^n Q_n}{\partial c} = n c^{n-1} Q_n. \tag{18}$$

We notice that the operation in Eq. (18) "brings down" a factor of n. This type of operation is central to our entire method. Multiplying both sides of Eq. (18) by c gives

$$c \frac{\partial c^n Q_n}{\partial c} = \frac{\partial c^n Q_n}{\partial \ln c} = n c^n Q_n. \tag{19}$$

To obtain the term $n f_n$ appearing in Eq. (17) we need only divide all terms in Eq. (19) by Γ. In this manner, we obtain the basic relation

$$\langle n \rangle = \frac{1}{\Gamma} \frac{\partial \Gamma}{\partial \ln c} = M_1. \tag{20}$$

The quantity $\langle n \rangle$ is the first moment of the distribution that we designate as M_1. It is often convenient to introduce the fraction of binding, which we define as θ:

$$\theta = \langle n \rangle / N \tag{21}$$

or

$$M_1 = N\theta \tag{22}$$

In an experimental study of ligand binding, what one measures is $M_1(c)$ or equivalently $\theta(c)$. We now show how higher moments of the binding distribution can also be obtained from these data. The higher moments (in general the mth) are defined in analogy to the definition of the first moment given in Eq. (17),

$$M_m = \sum_{n=0}^{N} n^m f_n. \tag{23}$$

Letting

$$x = \ln c \tag{24}$$

the analog of the operation given in Eq. (18) for bringing down a single factor of n also works when we want m factors of n,

$$M_m = \frac{1}{\Gamma} \frac{\partial^m \Gamma}{\partial x^m}. \tag{25}$$

If we take the derivative of M_m with respect to x, we obtain the following relation between different moments:

$$\frac{\partial M_m}{\partial x} = \frac{1}{\Gamma} \frac{\partial^{m+1} \Gamma}{\partial x^{m+1}} - \frac{1}{\Gamma^2} \frac{\partial \Gamma}{\partial x} \frac{\partial \Gamma^m}{\partial x^m} = M_{m+1} - M_1 M_m. \tag{26}$$

We define the following symbol for derivatives of M_1 with respect to x as

$$M_1^{(j)} = \frac{\partial^j M_1}{\partial x^j} \tag{27}$$

Then successively using Eq. (26), we obtain

$$M_1^{(1)} = M_2 - M_1^2,$$
$$M_1^{(2)} = M_3 - 3M_1 M_2 + 2M_1^3,$$
$$M_1^{(3)} = M_4 - 4M_1 M_3 - 3M_2^2 + 12M_1^2 M_2 - 6M_1^4.$$
(28)

We can then solve these equations consecutively for M_2, M_3, and M_4 in terms of the derivatives of M_1 with respect to x,

$$M_2 = M_1^{(1)} + M_1^2,$$
$$M_3 = M_1^{(2)} + 3M_1 M_2 - 2M_1^3,$$
$$M_4 = M_1^{(3)} + 4M_1 M_3 + 3M_2^2 - 12M_1^2 M_2 + 6M_1^4.$$
(29)

Thus, knowledge of M_1, or equivalently of $\langle n \rangle$, as a function of $x = \ln c$ (where $c = [L]$) can be used to give higher moments of the binding distribution. In Eq. (29), we give the relations required to obtain the higher moments through M_4. Recall that $M_1 = \langle n \rangle$ is the experimental binding isotherm. Thus, the derivatives of this quantity with respect to x, defined in Eq. (27), can also be obtained from experiment and then used in Eq. (29) to obtain higher moments of the ligand-binding distribution.

In Fig. 1A, we show a typical binding isotherm giving $\langle n \rangle$ as a function of $\ln c \, (= x)$. As an example, we take the case of 20 independent binding sites with a binding constant of $K = 2500 \, (\text{mol/l})^{-1}$. The binding isotherm for this system is shown by the solid curve in Fig. 1A, while we take the solid dots as model experimental data. The derivatives required in Eq. (29) are evaluated at a particular value of x that we will denote as x_0. We are free to pick any value of x to serve as the reference point. We can then expand the function $M_1(x)$ in a Taylor series about $x = x_0$, using the variable

$$\Delta x = x - x_0$$
(30)

giving (through the cubic term in Δx)

$$M_1(x) = a_0 + a_1 \Delta x + a_2 \Delta x^2 + a_3 \Delta x^3 + \ldots$$
(31)

which represents an empirical local fit of the experimental data. The a values in Eq. (31) are determined from experiment and have the following significance:

$$M_1^{(1)} = a_1,$$
$$M_1^{(2)} = 2a_2,$$
$$M_1^{(3)} = 6a_3.$$
(32)

But these are just the quantities required in Eq. (29) to give the first four moments of the binding distribution. Thus, the local expansion of the binding isotherm, $M_1(x) = \langle n \rangle$, given in Eq. (31) allows us to calculate the first four moments of the ligand-binding distribution function.

Another way to obtain the information given in Eq. (32) is to use the derivative of $\langle n \rangle$ with respect to x, as a function of x. A plot of this function is shown in Fig. 1B, where we see that we now have a function that goes through a maximum at the value of x corresponding to the midpoint of the ligand-binding curve.

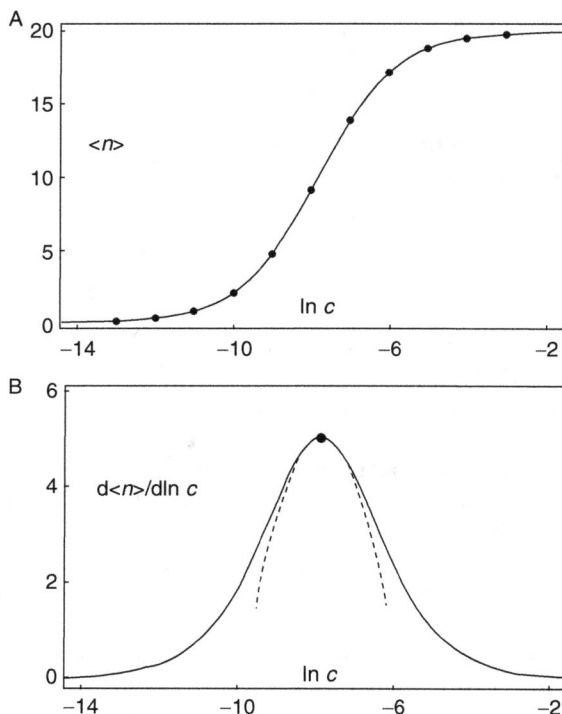

Fig. 1 (A) An example of a typical binding isotherm giving the average number of ligands bound as a function of the logarithm of the ligand concentration in solution. The curve was calculated assuming 20 independent sites with $K = 2500$ (mol/l)$^{-1}$. (B) A plot of the slope of the binding curve given in (A). The dashed curve represents the quadratic expansion given in Eq. (33) about the maximum in the curve, indicated by the solid dot.

A dominant feature of the curve shown in Fig. 1B is the width of the peak in the neighborhood of the maximum in the curve. The local series expansion of this function is given by [using Δx of Eq. (30)]

$$\frac{\partial \langle n \rangle}{\partial x} = M_1^{(1)}(x) = b_0 + b_1 \Delta x + b_2 \Delta x^2 + \dots \tag{33}$$

where now

$$\begin{aligned} M_1^{(1)} &= b_0, \\ M_1^{(2)} &= b_1, \\ M_1^{(3)} &= 2b_2. \end{aligned} \tag{34}$$

Thus, the coefficients in Eq. (33) contain the same information (enough to calculate four moments of the ligand binding distribution) as does the expansion given in Eq. (31), but the coefficients now have simple physical interpretations:

b_0 is the gradient of the binding isotherm at x_0,

b_1 is zero at x_0, (35)

b_2 measures the width of the peak.

The dashed curve in Fig. 1B shows the local quadratic expansion about the maximum in the curve as given by Eq. (33). From Eq. (35), we note that the construction of the quadratic curve shown in Fig. 1B requires only the value of the function at the maximum and the width of the peak. As this is enough experimental information to calculate four moments of the ligand-binding distribution, we see that there is no difficulty in obtaining a set of moments from experimental ligand-binding data. Given a finite set of moments, the next step is to turn information about the moments into parameters of the appropriate distribution function. To this end, we use the maximum-entropy method that we describe in the next section.

III. Maximum–Entropy Distributions

It is an old problem in mathematics to construct an approximation to a distribution function given a finite set of moments of that function. In the preceding section, we have seen how one can easily obtain a set of four to six moments for the distribution function for the binding of ligands to biological macromolecules from the binding isotherm (titration curve). In Section V, we show how one can similarly obtain a finite set of moments for the enthalpy distribution function in proteins and other macromolecules from the temperature dependence of the heat capacity. The problem we address in this section is the use of these moments to construct an approximate distribution function.

The technique we use is the maximum-entropy method as applied to the moment problem by Mead and Papanicolaou (1984) and by Tagliani (1995). To keep our discussion explicit we use a specific example of a distribution function, $f(X)$, which is illustrated in Fig. 2. Note that we are using uppercase X as the independent variable for this distribution; we are saving lowercase x for another upcoming use. The function we have chosen as an example is bimodal with two unequal peaks. The explicit functional form of this distribution is the sum of two unequal Gaussian distributions, as shown:

$$f(X) = (1.6 \exp[-0.25(X - 5)^2] + \exp[-0.35(X + 1)^2])/A \tag{36}$$

where the value of A is chosen to give a normalized function

$$\int_{L_2}^{L_1} f(X) = 1. \tag{37}$$

The bounds L_1 and L_2 in Eq. (37) are taken as practical limits of the extent of the distribution function. From Fig. 2, we take these limits as the points where the value of the distribution function have dropped essentially to zero. Thus, we take

$$L_1 = 12 \text{ and } L_2 = -6. \tag{38}$$

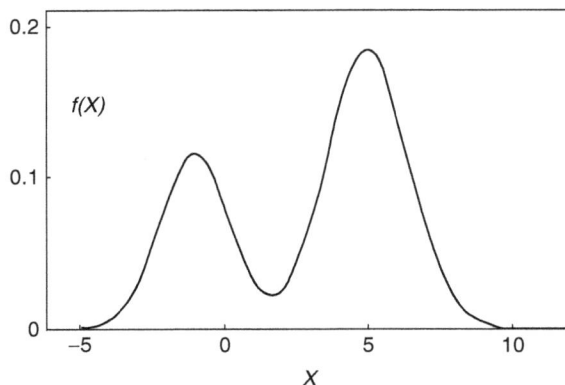

Fig. 2 An example of a bimodal distribution function with peaks of unequal height. The function shown is a sum of two Gaussian distributions as given by Eq. (36).

The choice of reasonable bounds to the range of the distribution function is an important part of the process. For our example given in Eq. (36) with the bounds of Eq. (38), the value of the normalizing constant is $A = 8.6678$.

The input for the calculation of a distribution function using the maximum-entropy method is a finite set of moments of the distribution function. For biological macromolecules, these moments are obtained from experimental data such as titration curves for ligand-binding distributions or heat capacity data for enthalpy distributions. For the example at hand, we know the exact distribution function as given in Eq. (36) and thus for this example we can calculate the moments precisely. The moments of $f(X)$ are given in general by the relation

$$M_m = \int_{L_2}^{L_1} X^m f(X) \mathrm{d}X. \tag{39}$$

For $f(X)$ of Eq. (36), the first four moments are

$$M_1 = 2.92618, M_2 = 18.5069, M_3 = 99.5994, M_4 = 618.546. \tag{40}$$

These are the numbers that, in an actual application of the method, would be determined from experiment.

The next step is technical, but important. To apply the maximum-entropy method, it is most convenient to scale the distribution function in question onto the unit interval so that rather than the limits L_2 and L_1 as given in Eq. (38) one has the limits zero and one. It is best to do this shift in two steps. First, we simply translate the function so that the lower limit is the origin. To do this we simply introduce a new variable,

$$y = X - L_2 \tag{41}$$

so that now when $X = -6$, the new variable has the value $y = 0$. The function now is $f(y)$ with y varying from 0 to $L_1 - L_2$. The moments for the shifted distribution function can easily be obtained form the former moments given in Eq. (40). Taking

$$L = L_1 - L_2 \tag{42}$$

one has

$$M'_m = \int_0^L y^m f(y)\mathrm{d}y = \int_{L_2}^{L_1} (X - L_2)^m f(X)\mathrm{d}X. \tag{43}$$

Defining

$$\alpha = -L_2 \tag{44}$$

the first few moments of the scaled function are given:

$$\begin{aligned} M'_1 &= M_1 + \alpha, \, M'_2 = M_2 + 2\alpha M_1 + \alpha^2, \\ M'_3 &= \mathrm{M}_3 + 3\alpha M_2 + 3\alpha^2 M_1 + \alpha^3, \\ M'_4 &= M_4 + 4\alpha M_3 + 6\alpha^2 M_2 + 4\alpha^3 M_1 + \alpha^4. \end{aligned} \tag{45}$$

For our example, the above moments have the following numerical values (in our example $L = 18$ and $\alpha = 6$):

$$M'_1 = 8.92618, \; M'_2 = 89.621, \; M'_3 = 964.75, \; M'_4 = 10830.6. \tag{46}$$

We have now shifted the distribution so that the lower bound is at the origin, $y = 0$. The next step is to scale the function so that the upper bound is 1. To accomplish this, we introduce yet another new variable,

$$x = y/L \tag{47}$$

so that now when $y = L$ the new variable has the value $x = 1$. The moments for the distribution function expressed in terms of the new variable defined on the interval $x = 0\text{--}1$ are now simply given by

$$\mu_m = M'_m/L^m \tag{48}$$

where L is given by Eq. (42). The final set of moments for the scaled distribution function in our example defined on the unit interval from 0 to 1 is now

$$\mu_1 = 0.495899, \; \mu_2 = 0.276608, \; \mu_3 = 0.165424, \; \mu_4 = 0.103172. \tag{49}$$

Notice that the above moments form a moderate set of monotonically decreasing numbers and that in a sense the scaling process has "tamed" the values of the moments.

The above scaling process is really the most difficult part of the procedure. One problem in scaling experimentally determined moments is that one does not always have a clear sense of what the lower and upper bounds, the L_1 and L_2 of Eq. (38), are. For titration problems, one usually knows the total number of proton-binding sites from the amino acid sequence of the protein, but for other distribution functions, such as the enthalpy distribution, one does not have such knowledge.

If one knows the first and second moments of the distribution, then one can make a first estimate of the width of the distribution, using the standard Gaussian distribution:

$$f(x) = \frac{1}{\sigma\sqrt{2\pi}} \exp[-(M_1 - x)^2/2\sigma^2] \tag{50}$$

where σ is the standard deviation (giving the width of the distribution)

$$\sigma = \sqrt{M_2 - M_1^2}. \tag{51}$$

Of course, for a distribution such as that illustrated in Fig. 2, the above function can give only a crude estimate of the overall range of the distribution function in question.

We now assume that one has a set of moments like those shown in Eq. (49) that have been scaled onto the unit interval from 0 to 1, starting with moments obtained from experiment. We now outline the approach of Mead and Papanicolaou (1984) to obtain approximate distribution functions given a finite set of moments, using the maximum-entropy method.

The scaled distribution function on the unit interval will be referred to as $p(x)$, denoting a probability distribution function in the continuous variable x where x varies from 0 to 1. The probability distribution function $p(x)$ is normalized as follows:

$$\int_0^1 p(x)\mathrm{d}x = 1 \tag{52}$$

We denote the moments of this function as

$$\mu_m = \int_0^1 x^m p(x)\mathrm{d}x = \langle x^m \rangle \quad \text{(for } m = 1 \text{ to } N) \tag{53}$$

where we indicate that a finite set of N moments is known. For the function given in Eq. (36), the first four of these moments are given in Eq. (49).

In the maximum-entropy method, one defines an entropy-like quantity in terms of the function $p(x)$:

$$S = -\int_0^1 [p(x)\ln p(x) - p(x)]\mathrm{d}x + \sum_{m=0}^{N} \lambda_m \left(\int_0^1 x^m p(x)dx - \mu_m \right) \tag{54}$$

where λ_m are Lagrange multipliers. Functional variation of S with respect to $p(x)$, plus the condition that the moments for $m = 0$ to N are given by Eq. (53), gives the result that $p(x)$ has the following functional form

$$p(x) = \exp[-g(x)] \tag{55}$$

where

$$g(x) = \sum_{n=0}^{N} \lambda_n x^n. \tag{56}$$

Thus, $g(x)$ is simply a finite polynomial in x. One sees that the more the moments that are known (the larger the number N), the more are the terms in the polynomial of Eq. (56) one has and the closer will be the approximate distribution function given by Eqs. (55) and (56) to the actual distribution function. Note that the $g(x)$ polynomial given in Eq. (56) when used in Eq. (55) and then in Eq. (53) gives back exactly the first N known moments of the experimental distribution function.

To illustrate the maximum-entropy distribution functions obtained in this manner, we first give the form for the distribution function when one knows only the first moment of the distribution (the average value of x). In this case, the distribution is given by a simple exponential function,

$$\text{(one moment)} \quad p(x) = A \exp(-\lambda_1 x). \tag{57}$$

If the first two moments of the distribution function are known, one obtains a quadratic (or Gaussian) distribution function:

$$\text{(two moments)} \quad p(x) = A \exp[-(\lambda_1 x + \lambda_2 x^2)]. \tag{58}$$

Given the first four moments of the distribution function, the function $g(x)$ of Eq. (56) will be a quartic polynomial in x giving the following approximate distribution function:

$$\text{(four moments)} \quad p(x) = A \exp[-(\lambda_1 x + \lambda_2 x^2 + \lambda_3 x^3 + \lambda_4 x4)]. \tag{59}$$

The remaining technical problem is to determine the values of the parameters λ_n introduced in Eq. (56). Knowledge of the experimentally determined values of the first N moments as given in Eq. (53) gives one N known numbers. The task is to convert these N known quantities into known values of the N parameters λ_n for $n = 1$ to N (note that the parameter λ_0 simply acts as a normalization parameter for the distribution). This process is schematically shown as follows:

$$\begin{array}{ccc} \text{Experimental values} & & \text{Values of} \\ \text{of } \mu_m \text{ for} & \rightarrow & \text{distribution } \lambda_n \text{ for } n = 1 \text{ to } N. \\ m = 1 \text{ to } N & & \end{array} \tag{60}$$

Mead and Papanicolaou (1984) have given a general iterative algorithm using any number of moments to accomplish the process shown in Eq. (60) and that is the scheme that we will outline here. One initiates the iteration procedure by constructing three vectors having N elements each:

$$\begin{aligned} \boldsymbol{\mu} &= (\mu_1, \mu_2, \ldots, \mu_N), \\ \boldsymbol{\lambda}_0 &= (0, 0, \ldots, 0), \\ \mathbf{x} &= (x, x^2, \ldots, x^N). \end{aligned} \tag{61}$$

The $\boldsymbol{\mu}$ vector contains the known experimental values of the first N scaled moments while the $\boldsymbol{\lambda}_0$ vector contains the initial values (all zero) of the unknown $\boldsymbol{\lambda}_n$ parameters. The $\mathbf{x}$ vector contains the first N powers of the independent variable x. The interaction process is primed by setting a general vector $\boldsymbol{\lambda}$ equal to the initial vector $\boldsymbol{\lambda}_0$:

$$\boldsymbol{\lambda} = \boldsymbol{\lambda}_0. \tag{62}$$

One then begins the general iteration loop given.

A. General Iteration Loop

First, one forms the function $f(x)$, where the dot indicates the dot product of the two vectors:

$$f(x) = \exp[-\boldsymbol{\lambda} \cdot \mathbf{x}]. \tag{63}$$

One then normalizes this function on the interval 0–1:

$$A^{-1} = \int_0^1 f(x)\mathrm{d}x. \tag{64}$$

giving the normalized approximation for the probability distribution

$$p(x) = Af(x). \tag{65}$$

Using this approximate distribution function, one then calculates a set of approximate moments,

$$\mu_m^* = \int_0^1 x^m p(x)\mathrm{d}x \ (m = 1, 2N). \tag{66}$$

Note that one calculates $2N$ moments. Using the first N of the preceding approximate moments, one forms the vector

$$\boldsymbol{\mu}^* = (\mu_1^*, \mu_2^*, \ldots, \mu_N^*). \tag{67}$$

The next step is the formation of the following $N \times N$ matrix:

$$\mathbf{W} = (w_{ij}) \tag{68}$$

where the general matrix element is given by

$$w_{ij} = \mu_{i+j}^* - \mu_i^* \mu_j^*. \tag{69}$$

This matrix is then inverted, giving

$$\mathbf{WI} = \mathbf{W}^{-1}. \tag{70}$$

Finally, one forms a vector that gives the difference between the experimental values of the moments and the approximate moments calculated above:

$$\mathbf{v} = \boldsymbol{\mu} - \boldsymbol{\mu}^*. \tag{71}$$

One then forms the vector

$$\mathbf{a} = \mathbf{WI} \cdot \boldsymbol{\nu}. \tag{72}$$

As a last step, the improved estimate of the vector of λ values is given by

$$\boldsymbol{\lambda}_{\text{new}} = \boldsymbol{\lambda} - \mathbf{a}. \tag{73}$$

One then iterates this procedure until there is no difference between $\boldsymbol{\lambda}_{\text{new}}$ and $\boldsymbol{\lambda}$. If there is a difference, then one sets

$$\boldsymbol{\lambda} = \boldsymbol{\lambda}_{\text{new}} \tag{74}$$

and goes back to the beginning of the general iteration loop above Eq. (63) and repeats the whole process.

The whole procedure is actually quite straightforward, and the computer program to carry out this iteration scheme can be written in about a dozen lines on the back of a postcard. The process usually converges to a limiting set of λ values in about 15 iterations. The only numerical problem that sometimes arises is that when one uses eight or more moments, there are sometimes precision problems in the matrix inversion step and one needs to increase the number of significant figures used in the program.

When one has run the iteration scheme outlined above and has found a set of λ values such that the new values are identical with the values obtained in the previous iteration round, one then wants to use the approximate distribution function $p(x)$ as given by Eqs. (55) and (56) to calculate the set of moments, μ_1 to μ_N, to be sure that the first N moments of the approximate distribution function indeed reproduce the experimental input. Mead and Papanicolaou (1984) have proved that if the set of moments, μ_1 to μ_N, vary monotonically [such as the moments given in Eq. (49) do], then there is a unique set of λ values, λ_1 to λ_N, that results from the maximum-entropy method. This does not mean that the approximate maximum-entropy method function given in Eqs. (55) and (56) is the unique distribution function: it is the unique maximum-entropy approximation to the distribution function. As the number of moments used in the construction of the distribution function is increased, the distribution function, through Eq. (56), is described by longer and longer polynomials until finally one has an infinite series (which, of course, is not attainable from experimental data, where the maximum number of moments that can be reasonably obtained is about eight). In the unrealistic limit where one uses an infinite number of moments, one would obtain a unique infinite series that would give the exact distribution function. Practically, the more moments used, the better the approximate distribution will be.

At the conclusion of the iteration process, one has a distribution function, $p(x)$, defined on the interval $x = 0$–1 that reproduces the first N experimental moments that have been scaled using Eq. (43) and Eq. (47) to the unit interval. The only remaining task is to scale the distribution function back to the interval of interest for the given physical system being treated. Again, we use the function given in Eq. (36) and illustrated in Fig. 2 as an example. The upper and lower bounds for this function, L_1 and L_2, are given in Eq. (38) while the parameter L is given in Eq. (42). To scale back to the interval shown in Fig. 2, we simply invert the scaling process we have already used. First, we scale the function from the unit interval $x = 0$–1 to the new interval $y = 0$ to L. This is achieved by the inverse of the relation given in Eq. (47), involving the substitution

$$y = xL. \tag{75}$$

Then, we shift the distribution by an amount L_2 going back to our original variable X (which describes the actual experimental distribution function of interest),

$$X = y + L_2 \tag{76}$$

where now the variable X varies from L_2 to L_1 (this is -6 to 12 for our example in Fig. 2). This double substitution takes the function $p(x)$ found by the maximum-entropy iteration and scales it back to the correct range of the experimental distribution, $f(X)$, which for our example is the function shown in Fig. 2. The moments calculated from the rescaled distribution give one the original set of moments, which in our example are the moments given in Eq. (39).

We now want to illustrate how this procedure works for our sample distribution function shown in Fig. 2 and given in Eq. (36). Notice that the function given in Eq. (36), a sum of two Gaussian distributions, does not have the same functional form as the maximum-entropy function given in Eqs. (55) and (56). Thus, it would take an infinite number of moments for the maximum-entropy method to give the exact distribution function shown in Eq. (36). Nonetheless, the power of the maximum-entropy method is that for a finite set of moments; it can give an excellent approximation to many distribution functions such as that shown in Fig. 2.

In Fig. 3, we show the maximum-entropy approximations to the function shown in Fig. 2 constructed using a variable number of moments. In the upper left-hand panel, we show the result obtained using only two moments. In this case, the

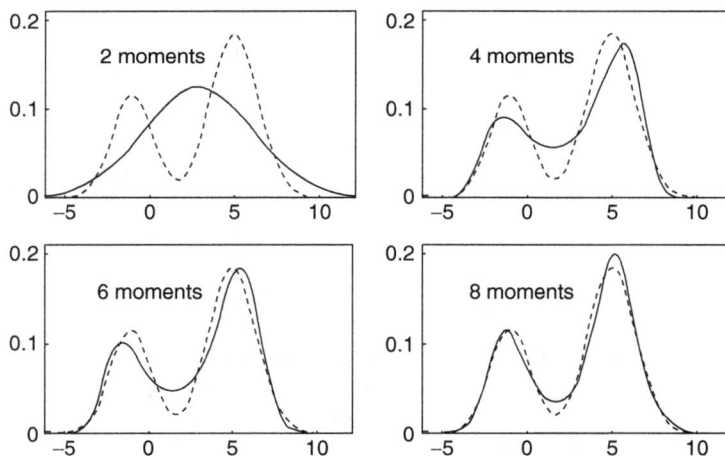

Fig. 3 Maximum-entropy approximations to the bimodal distribution shown in Fig. 2. The solid curves show, respectively, the maximum-entropy approximations obtained using two, four, six, and eight moments. The dashed curve in each box is the exact distribution function given by Eq. (36). In each graph, the approximate distribution function is plotted as a function of X as in Fig. 2.

approximate distribution is a single Gaussian and one obtains a rather poor approximation to the actual function (shown by the dashed curve in all the graphs). In the upper right-hand panel of Fig. 3, we give the result obtained when one uses four moments. Now the maximum-entropy method clearly resolves the bimodal character of the distribution (four is the minimum number of moments that will resolve bi-modal behavior). In the other graphs, we see that as the number of moments used is increased from six to eight, the goodness of fit increases until the function obtained using eight moments gives an excellent fit to the original distribution function.

To test the accuracy of the maximum-entropy method outlined above (Poland, 2002e), we have used exact distribution functions obtained from the two-dimensional Ising model. In that system, we treat a model fluid when it is near the critical point, where it splits into two phases. In that case, the density distribution for the fluid shows marked bimodal character that the maximum-entropy method, using a finite set of moments, reproduces with high accuracy.

IV. Ligand Binding: Distribution Functions

In the previous section, we outlined how an approximation to a molecular distribution function can be constructed from a finite set of the appropriate moments; the more the moments used, the better the approximation, as illustrated in Fig. 3. In Section II, we showed how moments of the ligand-binding distribution function can be obtained from local expansions of the experimental binding isotherm. In particular, we found that the local quadratic fit to the gradient of the binding isotherm, illustrated in Fig. 1B, was sufficient to give the first four moments of the ligand-binding distribution function. In this section, we combine the two results and illustrate the ligand-binding distribution functions obtained in this manner.

As an example, we use the titration curve of the protein lysozyme given by Tanford and Wagner (1954). In this case, the ligand that binds to the macromolecule is a proton, H^+, and the distribution function gives the probability that an arbitrary number of protons are bound to the protein at a given pH. To discuss the binding of protons, we need to briefly review a few facts about acid-base equilibria.

For a weak acid such as acetic acid, the standard dissociation reaction in water (dissociation of a proton, the definition of an acid) is

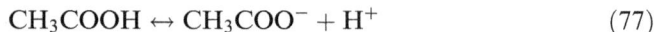

$$CH_3COOH \leftrightarrow CH_3COO^- + H^+ \tag{77}$$

with the standard equilibrium constant expression

$$\frac{[CH_3COO^-][H^+]}{[CH_3COOH]} = K_a = 10^{-pK_a}. \tag{78}$$

We note that Eq. (78) defines the quantity pK_a and that it is defined relative to base 10. For acetic acid at a temperature of 25°C, one has $pK_a = 4.54$, a typical value for a weak acid. We note that pH is also defined relative to base 10:

$$[H^+] = 10^{-pH}. \tag{79}$$

For a weak base, such as methylamine, we have the following reaction when this molecule is placed in water:

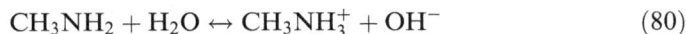

$$CH_3NH_2 + H_2O \leftrightarrow CH_3NH_3^+ + OH^- \tag{80}$$

which gives a slightly basic solution. The equilibrium constant for this reaction is expressed in terms of K_b and pK_b,

$$\frac{[CH_3NH_3^+][OH^-]}{[CH_3NH_2]} = K_b = 10^{-pK_b} \tag{81}$$

where we follow the convention of leaving $[H_2O]$ out of the equilibrium constant expression in Eq. (81) as it is essentially constant. For methylamine at 25°C, one has $pK_b = 3.36$.

To treat the binding of protons to a protein, it is more useful to express the reaction in Eq. (80) as the binding of a proton. We can achieve this by using the reaction for the self-ionization of water reaction,

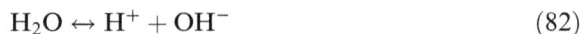

$$H_2O \leftrightarrow H^+ + OH^- \tag{82}$$

with

$$[OH^-][H^+] = K_W = 10^{-14}. \tag{83}$$

If we invert Eq. (80) and add the reaction given in Eq. (82), we obtain the acid dissociation reaction,

$$CH_3NH_3^+ \leftrightarrow CH_3NH_2 + H^+ \tag{84}$$

where now we view $CH_3NH_3^+$ as an acid with the acid dissociation constant,

$$K_a = K_w/K_b = 10^{-pK_a} \tag{85}$$

where $pK_a = 10.64$.

In our treatment of ligand binding, we treated all the reactions involved as binding reactions. Thus we need to turn the reactions given in Eq. (77) and Eq. (84) around and write them as binding reactions. Thus, we have in general for acids,

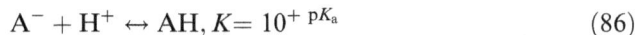

$$A^- + H^+ \leftrightarrow AH, K = 10^{+pK_a} \tag{86}$$

where the binding constant now is 10 raised to the plus pK_a. For weak bases, as written in Eq. (84) we have in general,

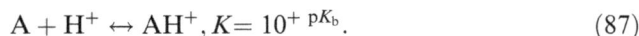

$$A + H^+ \leftrightarrow AH^+, K = 10^{+pK_b}. \tag{87}$$

If we consider each proton-binding site in isolation from the rest of the molecule, then in general we can treat each independent group that can bind a proton as having its own binding polynomial that is a sum over two states, proton not bound and proton bound. We write this function as follows:

$$\gamma = 1 + [\text{H}^+]10^{\text{p}K} \tag{88}$$

so that the mole fractions of the unbound and bound states are given by

$$f_\text{A} = \frac{1}{\gamma} \text{ and } f_\text{AH} = \frac{[\text{H}^+]10^{\text{p}K}}{\gamma}. \tag{89}$$

Nozaki and Tanford (1967) have studied denatured ribonuclease (in 6 M guanidine-HCl) and they find that the proton-binding sites act as if they are independent. In that molecule, they assign the following pK values to the appropriate groups:

$$
\begin{aligned}
&1. \quad \alpha - \text{Carboxyl}(1) \text{ p}K = 3.4\\
&2. \quad \beta - \text{Carboxyl}(5) \text{ p}K = 3.8\\
&3. \quad \gamma - \text{Carboxyl}(5) \text{ p}K = 4.3\\
&4. \quad \text{Imidazole}(4) \text{ p} = 6.5\\
&5. \quad \alpha - \text{Amino}(1) \text{ p}K = 7.6\\
&6. \quad \text{Phenolic}(3) \text{ p}K = 9.75\\
&7. \quad \text{Phenolic}(3) \text{ p}K = 10.15\\
&8. \quad \varepsilon - \text{Amino}(10) \text{ p}K = 10.35.
\end{aligned}
\tag{90}
$$

There are also four guanidyl groups in the molecule with p$K > 12.5$ that are not included in this list. The listing given in Eq. (90) includes $N = 32$ protons that can bind to the protein. The α-carboxyl and α-amino groups are, respectively, the terminal carboxyl and amino groups of the molecule while the β- and γ-carboxyl groups are the carboxyl groups, respectively, on aspartic and glutamic acid (like acetic acid in our example). The imidazole and phenolic pK values refer to proton-binding sites, respectively, on histidine and tyrosine. Finally, the ε-amino groups are the amine groups on lysine (like methyl amine in our example).

For the special case where each binding site is independent (which is not the general case), the binding polynomial for the whole molecule is given by the following product:

$$\Gamma = \gamma_1\gamma_2^5\gamma_3^5\gamma_4^4\gamma_5\gamma_6^3\gamma_7^3\gamma_8^{10}. \tag{91}$$

If we expand this expression, we obtain the polynomial

$$\Gamma = \sum_{n=0}^{N} Q_n[\text{H}^+]^n \tag{92}$$

where in this case $N = 32$. The probability of a given state of binding (number of protons bound) is simply

$$P_n = Q_n[\text{H}^+]^n/\Gamma \tag{93}$$

while the titration curve (average number of protons bound as a function of pH) is given by

$$\langle n \rangle = \sum_{n=0}^{N} nP_n = \frac{\partial \ln \Gamma}{\partial \ln[\text{H}^+]} \tag{94}$$

where we note that

$$\ln[\text{H}^+] = 2.303 \log[\text{H}^+] = -2.303\text{pH}. \tag{95}$$

Note that Q_n represents a sum over all ways that n protons can be bound.

In general, the binding polynomial will not be a product of independent γ values. In the compact form of a native protein, the charged groups will interact with one another and some will be buried, making them less accessible for binding. We now show how one can use moments of the experimental titration curve to obtain the complete binding polynomial for the case where the binding sites are not independent.

In Fig. 4, we show the experimental titration curve for the protein lysozyme on the basis of the data of Tanford and Wagner (1954) This protein contains 22 sites for proton binding. The solid points shown in Fig. 4 were obtained by tracing the curve the authors drew through their experimental points and then taking evenly spaced points on this curve; we will explain the origin of the solid curve shortly. Using the set of experimental points shown in Fig. 4, one can then apply the techniques outlined in Section II to obtain a set of moments at different pH values (Poland, 2001e). First, we take seven different sets of contiguous points, each set

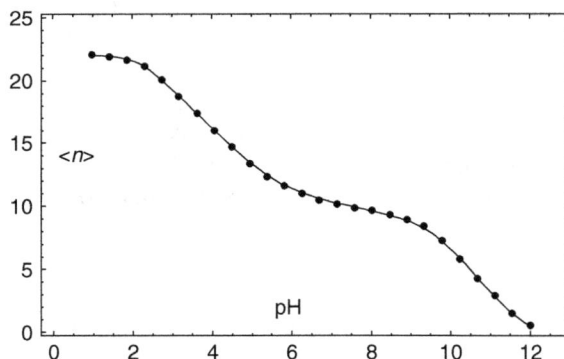

Fig. 4 The experimental titration curve of the protein lysozyme (solid dots) constructed using the original data of Tanford and Wagner (1954). The graph gives the average number of protons bound as a function of the pH of the solution. The maximum number of protons that can bind to this protein is 22. The solid curve represents the theoretical titration curve calculated from the binding polynomial as shown in Eq. (15) using the Q_n given in Fig. 5b. Reprinted from Poland (2001e), with kind permission of Kluwer Academic Press.

containing seven points, and then we fit each set of data to a cubic polynomial centered at the middle point. The pH values at the midpoints of the seven sets of experimental points are as follows: pH 3.19, 4.07, 4.95, 6.28, 8.04, 9.80, and 10.68.

Once we have the set of polynomials, each representing a local expansion of the titration curve about a particular value of the pH, we use the procedure outlined in Section II to obtain the first four moments of the proton-binding distribution from each of these polynomials (each polynomial representing a different pH value). Given these sets of moments, we next use the maximum-entropy method outlined in the previous section to obtain seven different binding distribution functions, one for each of the seven different pH values indicated above. The seven distributions functions so obtained are shown in Fig. 5A. It turns out that for this system the distribution functions calculated using successively two, three, and four moments are virtually superimposable, so the distribution functions shown in Fig. 5A are essentially Gaussian distributions. Note that one needs to obtain four moments in

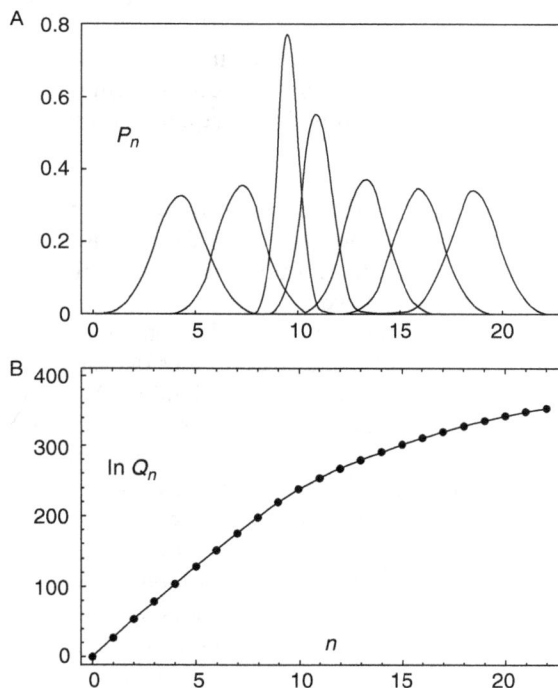

Fig. 5 (A) Maximum-entropy proton-binding distribution functions for the protein lysozyme constructed using moments obtained from local expansions of the titration curve given in Fig. 4. The distribution functions are for the pH values, from left to right: pH 3.19, 4.07, 4.95, 6.28, 8.04, 9.80, 10.68. (B) The values of ln Q_n for $n = 0$–22 (solid dots), giving the complete proton binding polynomial for the protein lysozyme. Reprinted from Poland (2001e), with kind permission of Kluwer Academic Press.

order to confirm that the use of only two moments gives an excellent representation of the proton-binding distribution function.

We now use the proton-binding distributions functions given in Fig. 5A to calculate all of the Q_n in the binding polynomial of Eq. (92). From Eq. (93) the ratio of the probabilities for successive species is given by

$$\frac{P_n}{P_{n-1}} = \frac{Q_n}{Q_{n-1}} [H^+].$$ (96)

Solving this equation for Q_n, we have

$$Q_n = Q_{n-1}\left(\frac{P_n}{P_{n-1}}\right)[H^+].$$ (97)

Using the proton-binding distribution functions obtained from the maximum-entropy method, one can estimate the ratio P_n/P_{n-1} by evaluating the distribution function at integer values of n. The distribution function used refers to a specific value of the pH (one of the seven pH values used and listed previously) and so the value of $[H^+]$ is known. Thus, Eq. (97) represents a recursion relation, giving Q_n in terms of Q_{n-1} and known quantities. To start the recursion process, one must know the first term in the sequence. But from Eq. (15), the zeroth term in the binding polynomial is simply equal to 1 and hence we have

$$Q_0 = 1.$$ (98)

Thus, we start the recursion process with the above value and then successively calculate all of the Q_n values from ratios P_n/P_{n-1} that are obtained from the proton-binding distribution functions, which in turn are obtained from moments of the titration curve.

In this manner, we obtain the complete binding polynomial for the binding of protons to lysozyme. The quantities $\ln Q_n$ thus obtained are shown as a function of n in Fig. 5B. Given the proton-binding polynomial, Γ, one can then calculate the probability of any state of proton binding at any pH value. Thus, all the possible proton-binding information for a given protein is contained in Γ through the coefficients Q_n. Given the complete binding polynomial Γ, one can use Eq. (94) to calculate the titration curve (the average extent of binding as a function of pH). Using the values of $\ln Q_n$ given in Fig. 5B, the calculated titration curve is shown by the solid curve in Fig. 4, which is seen to give an excellent fit to the solid points that are derived from the experimental data.

Thus, the moment/maximum-entropy method can give one the complete proton-binding polynomial for a protein. This function in turn contains all the empirical information possible concerning the binding of protons to the protein. To dissect Q_n into terms that represent different microscopic sets binding constants requires a specific model. Knowledge of Q_n is the most information one can obtain empirically without any specific model.

One can also apply the method outlined in this section to other types of binding. In particular, we have applied this approach to the binding of Mg^{2+} and small molecules to nucleic acids (Poland, 2001a), to the binding of denaturants to proteins (Poland, 2002d), and to the free energy of proton binding in a variety of proteins (Poland, 2004).

V. Enthalpy Distributions

In this section, we apply the use of moments to the calculation of enthalpy distribution functions in proteins. Just as there is a broad distribution of the state of ligand binding in a given biological macromolecule, there also is a broad distribution of enthalpies. In particular, the temperature dependence of the enthalpy distribution gives considerable insight into the process of the thermal unfolding of a protein.

We have already cited the reference (Poland, 2000a) containing the basic outline of the use of the maximum-entropy method to obtain enthalpy distribution functions. In addition, this approach has been applied to the calculation of the density of states for a general substance (Poland, 2000c), to enthalpy distributions in proteins (Poland, 2001b), and to enthalpy distributions in the solid state of high polymers (Poland, 2001d). Most biological systems, including proteins, are usually studied at constant pressure (the system is simply open to the atmosphere). On the other hand, the calculation of the appropriate moments is simpler to describe for a system at constant volume, so we treat that case first and then indicate the minor changes in the formalism required to treat a constant pressure system. In any case, there is little difference between constant volume and constant pressure thermodynamics for condensed matter.

The starting point for the statistical thermodynamics of a constant volume system is the canonical partition function (Hill, 1986; Poland, 1978), which is a sum over all energy levels of the system. Taking ε_i as a general energy level having degeneracy ω_i, the general canonical partition function is given by the following sum:

$$Z_v = \sum_i \omega_i \exp[-\beta \varepsilon_i] \tag{99}$$

where

$$\beta = 1/RT. \tag{100}$$

In Eq. (100), T is the absolute temperature and R is the gas constant. We will measure energies in kilojoules per mole, in which case the gas constant has the value $R = 8.31451 \times 10^{-3}$ kJ mol^{-1} K^{-1}. The connection between the partition function Z_v and thermodynamics is given by the relation

$$Z_v = \exp[-\beta A] \tag{101}$$

where A is the Helmholtz free energy.

Like the binding polynomial, this partition function is a sum over all states. The probability that the system is in a particular state is simply given by the term in the partition function for that state divided by the sum over all terms (the partition function). Thus, our basic equation is

$$P_i = \frac{\omega_i \exp[-\beta \varepsilon_i]}{Z_v}. \tag{102}$$

Given this expression for the probability of a general state, we can use it to obtain expressions for the moments of the energy distribution. The first moment is simply given by

$$\langle E \rangle = \sum_i \varepsilon_i P_i = E_1. \tag{103}$$

One can "bring down" the factor ε_i in Eq. (103) by taking the derivative with respect to $(-\beta)$,

$$\frac{\partial \omega_i \exp[-\beta \varepsilon_i]}{\partial(-\beta)} = (\varepsilon_i)\omega_i \exp[-\beta \varepsilon_i]. \tag{104}$$

The above procedure is analogous to that used in Eq. (19) for the case of ligand binding. The first energy moment, as indicated in Eq. (103), is then given by the relation

$$E_1 = -\frac{1}{Z_v}\frac{\partial Z_v}{\partial \beta}. \tag{105}$$

Higher moments are obtained in analogy with the procedure used in Eq. (25) for ligand binding,

$$E_m = \sum_i \varepsilon_i^m P_i = \frac{(-1)^m}{Z_v}\frac{\partial^m Z_v}{\partial \beta^m}. \tag{106}$$

Finally, we note that the first moment is simply the internal energy of thermodynamics,

$$U = E_1. \tag{107}$$

The next step is to relate the moments given in Eq. (106) to derivatives of E_1 (or U) with respect to β (or, equivalently, with respect to T). One has

$$\begin{aligned} U &= -\frac{Z_v^{(1)}}{Z_v}, \\ \frac{\partial U}{\partial \beta} &= \left(\frac{Z_v^{(1)}}{Z_v}\right)^2 - \left(\frac{Z_v^{(2)}}{Z_v}\right), \\ \frac{\partial^2 U}{\partial \beta^2} &= -2\left(\frac{Z_v^{(1)}}{Z_v}\right) + 3\left(\frac{Z_v^{(1)}}{Z_v}\right)\left(\frac{Z_v^{(2)}}{Z_v}\right) - \left(\frac{Z_v^{(3)}}{Z_v}\right) \end{aligned} \tag{108}$$

and so on, where

$$Z_v^{(j)} = \frac{\partial^j Z_v}{\partial \beta^j}. \tag{109}$$

The first three moments of the energy distribution are then given by

$$E_1 = -\frac{Z_v^{(1)}}{Z_v}, E_2 = \frac{Z_v^{(2)}}{Z_v}, E_3 = -\frac{Z_v^{(3)}}{Z_v}. \tag{110}$$

But as E_1 is the internal energy, U, one also has

$$E_1 = U, E_2 = -U_\beta^{(1)} + E_1, E_3 = U_\beta^{(2)} - 2E_1^3 + 3E_1E_2 \tag{111}$$

where

$$U_\beta^{(j)} = \frac{\partial^j U}{\partial \beta^j}. \tag{112}$$

We recall that $\beta = 1/RT$ and thus we see in Eqs. (111) and (112) that knowledge of the temperature dependence of U allows one to calculate moments of the energy distribution.

One can know the temperature dependence of the internal energy, U, without knowing the value of U itself. That is, one can know the derivatives of U with respect to temperature (such as the heat capacity) without knowing $U = E_1$. In that case, one can construct central moments that are relative to the (unknown) value of E_1. In general, one has

$$M_m = \langle (E - E_1)^m \rangle. \tag{113}$$

We then have the simple results

$$M_1 = 0, M_2 = -U_\beta^{(1)}, M_3 = U_\beta^{(2)}. \tag{114}$$

Given the value of E_1 one can then convert back to regular moments,

$$E_2 = M_2 + E_1^2, E_3 = M_3 + 3E_2E_1 - 2E_1^3. \tag{115}$$

The temperature dependence of U is given by the heat capacity (here at constant volume):

$$C_V = \left(\frac{\partial U}{\partial T} \right)_V. \tag{116}$$

The experimental data concerning the energy moments are thus contained in the temperature dependence of the heat capacity, $C_V(T)$. One can express the temperature dependence of the heat capacity in the neighborhood of a reference temperature T_0 as an empirical Taylor series in ΔT where

$$\Delta T = T - T_0. \tag{117}$$

The general form for this expansion is given below

$$C_V(T) = c_0 + c_1 \Delta T + c_2 \Delta T^2 + \dots. \tag{118}$$

We can also formally write the internal energy as an expansion in ΔT,

$$U(T) = \sum_{m=0}^{\infty} \frac{1}{m!} \left(\frac{\partial^m U}{\partial T^m} \right)_{T_0} \Delta T^m. \tag{119}$$

The coefficients in the $C_V(T)$ series given in Eq. (118) are then related to the derivatives used in Eq. (119) as follows:

$$U_T^{(0)} = U, \ U_T^{(1)} = c_0, \ U_T^{(2)} = c_1, \ U_T^{(3)} = 2c_2 \tag{120}$$

where

$$U_T^{(m)} = \frac{\partial^m U}{\partial T^m}. \tag{121}$$

The equations for the energy moments require derivatives with respect to $\beta \, (= 1/RT)$. To convert from temperature derivatives to derivatives with respect to β, we require the following relations:

$$T_\beta^{(m)} = \frac{\partial^m T}{\partial \beta^m} \tag{122}$$

the first few of which are

$$T_\beta^{(1)} = -\left(\frac{1}{R}\right)(RT)^2, \ T_\beta^{(2)} = \left(\frac{2}{R}\right)(RT)^3, \ T_\beta^{(3)} = -\left(\frac{6}{R}\right)(RT)^4. \tag{123}$$

Then finally we have

$$\begin{aligned}
U_\beta^{(1)} &= T_\beta^{(1)} U_T^{(1)}, \\
U_\beta^{(2)} &= T_\beta^{(2)} U_T^{(1)} + (T_\beta^{(1)})^2 U_T^{(2)}, \\
U_\beta^{(3)} &= T_\beta^{(3)} U_T^{(1)} + 3 T_\beta^{(1)} T_\beta^{(2)} U_T^{(2)} + (T_\beta^{(1)})^3 U_T^{(3)}.
\end{aligned} \tag{124}$$

One sees that if one has the expansion of $C_V(T)$ in Eq. (118) through the c_2 term, [quadratic fit of $C_V(T)$] this then gives $U_T^{(3)}$ from Eq. (120) and, finally, $U_\beta^{(3)}$ from Eq. (124). Thus, knowledge of a quadratic fit of $C_V(T)$ contains enough information to calculate the first four energy moments. If one has the expansion of $C_V(T)$ through c_4 (quartic expansion), this is enough information to give the energy moments through E_6.

An example of a physical system where one knows the energy distribution function exactly (and hence all of the energy moments) is a fluid of hard particles. The only interaction between particles in such a fluid is that of repulsion between the hard cores of the particles. As this fluid includes the effect of excluded volume, it is not an ideal system, but it does not include any attractive interactions. Thus, all of the internal energy U is kinetic energy. The distribution function for this system is most familiar as the Maxwell-Boltzmann velocity distribution. Here, we express it as a distribution of the kinetic energy E,

$$P(E) = \frac{2\beta^{3/2}}{\sqrt{\pi}} \sqrt{E} \exp[-\beta E] \tag{125}$$

with moments

$$E_1 = \frac{3}{2}RT, E_2 = \frac{15}{4}(RT)^2, E_3 = \frac{105}{8}(RT)^3, E_4 = \frac{945}{16}(RT)^4 \qquad (126)$$

where the first moment, $E_1 = 3RT/2$, is the familiar equipartition of energy result for a hard-core fluid. We (Poland, 2000a) have used the moments/maximum-entropy method to approximate the exact distribution function given in Eq. (125). In this case, the use of six moments gives a good fit to the exact distribution function. Note that the functional form of the distribution function given in Eq. (125) is not the same as the maximum-entropy distribution given in Eqs. (55) and (56), because of the preexponential square root of the E term. Thus, it would take an infinite number of moments to reproduce the function given in Eq. (125) exactly. However, the use of a finite number of moments with the maximum-entropy method gives a good approximation to the exact distribution function.

We now indicate how moments and distribution functions are obtained for a system at fixed pressure. For such a system, the appropriate partition function is the isobaric grand partition function (Hill, 1986; Poland, 1978)

$$Z_p = \int_V \sum_i \omega_i \exp[-\beta(\varepsilon_i + pV)] dV \qquad (127)$$

which is related to the Gibbs free energy, G, as follows:

$$Z_p = \exp[-G/RT]. \qquad (128)$$

We can express Eq. (127) as a sum over enthalpy states, using the general definition of enthalpy,

$$H = U + pV. \qquad (129)$$

For a gas, the pV term equals RT per mole, which at room temperature gives a value of approximately 2.5 kJ/mol. As the molar volume of condensed matter is approximately 10^{-3} that of a gas (e.g., 18 cm³/mol for liquid water versus 22.4 l/mol for an ideal gas at STP), there is little difference between H and U for condensed matter.

We can define the enthalpy of a particular state as follows:

$$h_i = \varepsilon_i + pV \qquad (130)$$

giving

$$Z_p = \int_V \sum_i \omega_i \exp[-\beta h_i] dV. \qquad (131)$$

The enthalpy moments are then obtained from Z_p in the same manner as the energy moments were obtained from Z_v of Eq. (99). Thus, in analogy with the result given in Eq. (106), one has

$$H_m = \frac{(-1)^m}{Z_p} \frac{\partial^m Z_p}{\partial \beta^m}. \qquad (132)$$

We note that for the case $m = 1$, Eq. (132) reduces to the familiar equation from thermodynamics giving the average enthalpy (the first moment of the enthalpy distribution) as a temperature derivative of the Gibbs free energy,

$$H = \frac{(-1)}{Z_p} \frac{\partial Z_p}{\partial \beta} = \frac{\partial (G/T)}{\partial (1/T)}. \tag{133}$$

Equation (132) is simply a generalization of this result to higher moments of the enthalpy distribution.

For the constant pressure system, the quantity determined experimentally is the heat capacity at constant pressure, $C_p(T)$. In analogy with Eq. (118), we can express this quantity as a series in ΔT,

$$C_p(T) = c_0 + c_1 \Delta T + c_2 \Delta T^2 + \dots \tag{134}$$

where the c values are empirical parameters determined by the fit of the heat capacity data with respect to temperature. The process of using the c values to calculate enthalpy moments is then exactly the same as the process for calculating energy moments from the expansion of $C_v(T)$ as given in Eq. (118).

We now turn to some actual heat capacity data for proteins. Fig. 6 shows the heat capacity of the protein barnase in units of kJ mol^{-1} K^{-1} on the basis of data of Makhatadze and Privalov (1995). Barnase is a small protein containing 110 amino acids with a molecular mass of 12,365 Da. The dashed curve gives the local quadratic expansion as shown in Eq. (134), where the expansion is taken about the maximum in the heat capacity curve (the point indicated by the solid dot in the graph; Fig. 6). We note that the heat capacity graph shown in Fig. 6 and the graph shown in Fig. 1B for the case of ligand binding are similar: in both cases a local quadratic expansion of the experimental curve gives four moments of the appropriate distribution function. Given the quality of the data shown in Fig. 6, one can

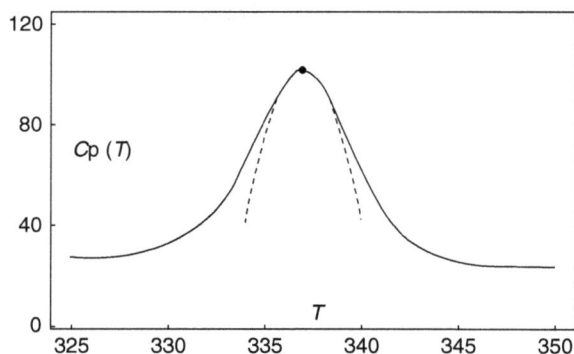

Fig. 6 The heat capacity of the protein barnase (kJ/mol) as constructed from the data of Makhatadze and Privalov (1995) The dashed curve shows a local quadratic expansion of the data about the maximum in the curve (solid dot).

easily obtain six enthalpy moments accurately. In this case, the enthalpy distribution functions calculated using four or more moments are qualitatively different from those calculated using only two moments. This is in marked contrast with the result we found for ligand binding, where the distribution functions based on two to four moments were virtually superimposable.

The enthalpy distribution functions constructed using six enthalpy moments obtained from the experimental heat capacity data for barnase shown in Fig. 6 are given in Fig. 7A for three different values of the temperature [corresponding to expansions as given by Eq. (134) centered around three different temperatures]. The temperatures used were the temperature of the maximum, T_m, and then $T_m - 1$ and $T_m + 1$, where $T_m = 337.1$ K. A natural interpretation of the presence of two peaks in the enthalpy distribution functions is that one peak (the one at lower enthalpy values) represents the native state of the molecule while the other peak (at higher enthalpy) represents the unfolded state of the molecule. It is clear that both species are in fact represented by broad distributions of enthalpy values. At $T_m - 1$, the low-enthalpy species is most probable [higher $P(H)$ peak]. As the temperature is

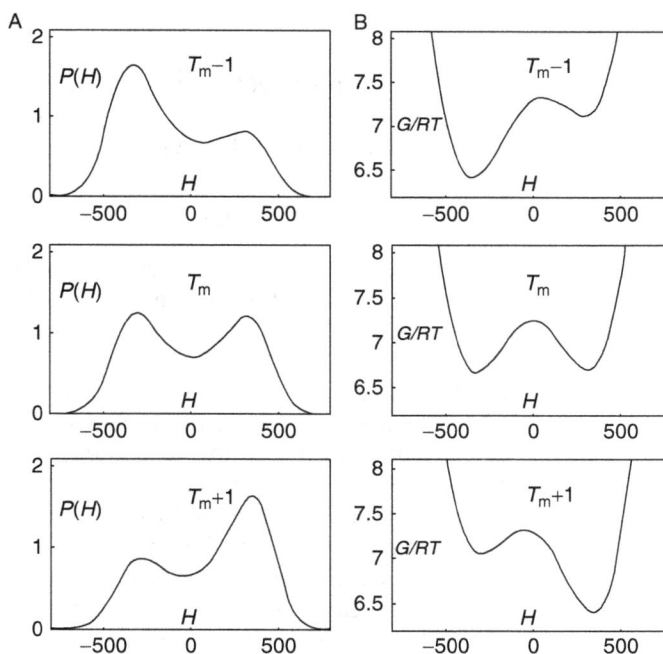

Fig. 7 (A) The enthalpy distribution function, $P(H)$, for the protein barnase near the melting temperature T_m. To simplify the scale, the $P(H)$ functions shown have been multiplied by a factor of 1000. (B) The free energy distribution, G/RT, obtained from $P(H)$ using Eq. (135). A constant value, ln (1000), has been added to these curves reflecting the scaling used for $P(H)$. Reprinted from Poland (2001f), with kind permission of Wiley-Interscience.

increased to T_m, the two peaks become equal in height, indicating that both the native and unfolded species have approximately the same probability at this temperature. Finally, at $T_m + 1$, the high-enthalpy species is more probable, indicating that at this temperature most of the molecules are unfolded. Thus, the temperature dependence of the enthalpy distribution gives a detailed view of the shift in populations between native and unfolded species. Note that in all cases the enthalpy distribution function of both the native and unfolded species is not given by a single enthalpy value (delta function in the distribution), but rather the enthalpy distribution function for each species is represented by a broad peak. The enthalpy distribution function gives one the probability that a molecule picked at random will have a given value of the enthalpy. From our construction of this function, one sees that there is a great variability (even for the native form) in the enthalpy of a protein molecule, this variation being due to, among other causes, vibrations and breathing motions of the protein and variability in solvent structure around the protein.

Given the enthalpy distribution $P(H)$, one can define (Poland, 2001f) a Gibbs free energy that is the potential for this distribution as follows:

$$P(H) = \exp[-G(H)/RT]/\exp[-G/RT] \qquad (135)$$

where

$$\exp[-G/RT] = \int_H \exp[-G(H)/RT]\mathrm{d}H \qquad (136)$$

gives the total Gibbs free energy. From Eq. (135), the quantity $G(H)/RT$ is given by the relation

$$G(H)/RT = -\ln[P(H)] + C \qquad (137)$$

where $C = G/RT$ is a constant that is independent of H but depends on T.

The functions $G(H)/RT$ obtained from the three different enthalpy distribution functions shown in Fig. 7A, using Eq. (137), are illustrated in Fig. 7B. For this function, the most probable species is represented by the lowest valley in the $G(H)/RT$ curve. Thus, at $(T_m - 1)$, the deepest minimum in the free energy curve corresponds to the native (low-enthalpy) species while at T_m there are two minima of equal depth, indicating that at this temperature the native and unfolded species are equal in probability. Then at $(T_m + 1)$, the deepest minimum in the free energy curve shifts to correspond to the unfolded (high-enthalpy) species. Recall that all the functions shown in Fig. 7 were determined from the single set of data shown in Fig. 6, namely, the temperature dependence of C_p. These results clearly illustrate the power of the moments/maximum-entropy method to construct, from standard experimental data, distribution functions that give detailed insight into the behavior of biological macromolecules.

The approach outlined in this section has been applied to the contribution of secondary structure in proteins to the enthalpy distribution (Poland, 2002b), free energy distributions for two different forms of myoglobin (Poland, 2002c), free energy distributions in tRNAs (Poland, 2003), and the enthalpy distribution for the helix-coil transition in a model peptide (Poland, 2001c).

VI. Self-Association Distributions

In this final section, we consider distribution functions for the general clustering, or self-aggregation, of n monomers to give an n-mer or cluster containing n monomers. This process is similar to that of ligand binding treated in Section II except that in this case there is no parent molecule with a fixed set of binding sites. Rather, the monomers simply react with one another to form a cluster. We follow the treatment of self-association using the moments/maximum-entropy method that has been published (Poland, 2002a).

The general reaction for the addition of one monomer to a cluster containing $(n-1)$ monomers with equilibrium constant K_m is

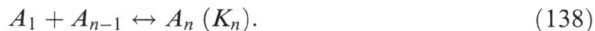

$$A_1 + A_{n-1} \leftrightarrow A_n \ (K_n). \tag{138}$$

As was the case with ligand binding, it is useful to consider the formation of the cluster A_n directly from n monomer molecules,

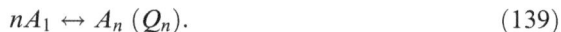

$$nA_1 \leftrightarrow A_n \ (Q_n). \tag{139}$$

The reaction in Eq. (139) is obtained by adding together the stepwise reactions of Eq. (138), thus giving the equilibrium constant Q_n for Eq. (139) as

$$Q_n = \prod_{m=2}^{n} K_m. \tag{140}$$

For completeness, we have the null reaction

$$A_1 \leftrightarrow A_1 \tag{141}$$

where

$$K_1 = Q_1 = 1. \tag{142}$$

From the general conservation of monomer units, one has the relation

$$\sum_{n=1}^{\infty} n[A_n] = c \tag{143}$$

where c is the total original concentration of monomer units.

Using the equilibrium constant expression for the reaction in Eq. (139), one has

$$\frac{[A_n]}{[A_1]^n} = Q_n \tag{144}$$

or, solving for $[A_n]$,

$$[A_n] = [A_1]^n Q_n. \tag{145}$$

We can now define the analog of the binding polynomial that was introduced for the case of ligand binding [see Eq. (15)]. We will call this function the association polynomial,

$$\Gamma = \sum_{n=1}^{\infty} [A_n] = \sum_{n=1}^{\infty} [A_1]^n Q_n. \tag{146}$$

Physically, this is the sum over all cluster concentrations and gives the net concentration of clusters regardless of size. From this polynomial, we obtain a general relation for the probability of a cluster containing n monomers,

$$P_n = [A_1]^n Q_n / \Gamma. \tag{147}$$

This relation follows from the general rule for obtaining probabilities from partition functions: the probability of state n is the term in the partition function representing state n divided by the sum over all states (the partition function or, in this case, the association polynomial). Note that Eq. (147) gives the probability that a cluster picked at random contains n monomer units.

Using the probability of a given cluster size given in Eq. (147), the average cluster size is then given by

$$\langle n \rangle = \sum_{n=1}^{\infty} n P_n = \sum_{n=1}^{\infty} n [A_1]^n Q_n / \sum_{n=1}^{\infty} [A_1]^n Q_n \tag{148}$$

or

$$\langle n \rangle = \frac{[A_1]}{\Gamma} \frac{\partial \Gamma}{\partial [A_1]}. \tag{149}$$

Higher moments are obtained in an analogous fashion,

$$\langle n^m \rangle = \sum_{n=1}^{\infty} n^m P_n \frac{1}{\Gamma} \frac{\partial^m \Gamma}{\partial y^m} \tag{150}$$

where

$$y = \ln[A_1]. \tag{151}$$

One can then take derivatives of $\langle n \rangle$ as given in Eq. (149) with respect to y, giving

$$\frac{\partial \langle n \rangle}{\partial y} = \langle n^2 \rangle - \langle n \rangle^2,$$

$$\frac{\partial^2 \langle n \rangle}{\partial y^2} = \langle n^3 \rangle - 3 \langle n \rangle \langle n^2 \rangle + 2 \langle n \rangle^3. \tag{152}$$

From these relations, one obtains the higher moments, $\langle n^2 \rangle$ and $\langle n^3 \rangle$, in terms of the variation of $\langle n \rangle$ with respect to $[A_1]$ (or $y = \ln[A_1]$). But $[A_1]$ is the concentration of free monomer, not the total original concentration of monomer denoted by c and given in Eq. (143). The total original concentration of monomer, c, is the variable under experimental control and experiment gives $\langle n \rangle$ as a function of c. To obtain expressions for the moments of the distribution in terms of c, one can expand $\langle n \rangle$ about a given value c_0,

$$\langle n(c) \rangle = n_0 + n'(c - c_0) + \frac{1}{2}n''(c - c_0)^2 + \ldots \qquad (153)$$

where

$$n_0 = \langle n(c_0) \rangle, n' = (\partial \langle n \rangle / \partial c)_{c_0}, n'' = (\partial^2 \langle n \rangle / \partial c^2)_{c_0}. \qquad (154)$$

The preceding equations give a local quadratic fit to the experimental data: $\langle n \rangle$ as a function of c. As a result of this empirical fit, one obtains the parameters n_0, n', and n'' evaluated at the point $c = c_0$.

To obtain derivatives with respect to $[A_1]$, as required in Eq. (152), we use the variable y defined in Eq. (151) and introduce the new variable w:

$$y = \ln[A_1] \quad \text{and} \quad w = \ln c. \qquad (155)$$

Note that $[A_1]$ is the concentration of free monomer and c is the total original concentration of monomer. We have the following relations between these variables:

$$\frac{\partial \langle n \rangle}{\partial y} = n^{(1)}\frac{\partial w}{\partial y}, \frac{\partial^2 \langle n \rangle}{\partial y^2} = n^{(2)}\left(\frac{\partial w}{\partial y}\right)^2 + n^{(1)}\frac{\partial^2 w}{\partial y^2}, \qquad (156)$$

where the $n^{(m)}$ are defined as follows:

$$n^{(1)} = \frac{\partial \langle n \rangle}{\partial w} = cn', n^{(2)} = \frac{\partial^2 \langle n \rangle}{\partial w^2} = cn' + c^2 n'' \qquad (157)$$

and are now given in terms of the experimentally determined quantities n' and n'' given in Eq. (154).

We can make the transformation from variable c to variable $[A_1]$, using the conservation relation:

$$c = \sum_{n=1}^{\infty} n[A_n] = \sum_{n=1}^{\infty} n[A_1]^n Q_n. \qquad (158)$$

One then has

$$\frac{\partial w}{\partial y} = \frac{\langle n^2 \rangle}{\langle n \rangle}, \frac{\partial^2 w}{\partial y^2} = \frac{\langle n^3 \rangle}{\langle n \rangle} - \left(\frac{\langle n^2 \rangle}{\langle n \rangle}\right)^2. \qquad (159)$$

Defining the first three moments of the self-association distribution function as

$$M_1 = \langle n \rangle, M_2 = \langle n^2 \rangle, M_3 = \langle n^3 \rangle \qquad (160)$$

we finally have expressions for these moments in terms of experimentally measured quantities:

$$M_1 = n_0,$$
$$M_2 = M_1^2/(1 - n^{(1)}/M_1), \tag{161}$$
$$M_3 = \{(M_2/M_1)^2(n^{(2)} - n^{(1)}) + 3M_1M_2 - 2M_1^3\}/(1 - n^{(1)}/M_1).$$

We will use as an example of this method the association of ATP to form linear clusters. This system has been studied using NMR techniques in the laboratory of H. Sigel (Scheller *et al.*, 1981). The ends of the clusters have a characteristic signal in the NMR and hence one can measure the net concentration of clusters, $\langle n \rangle$, as a function of the total ATP concentration (our variable c). Obtaining the average cluster size as a function of total monomer concentration is the most difficult part of this approach. The curve giving this data for the self-association of ATP on the basis of the work of Scheller *et al.* (1981) is shown in Fig. 8.

Using the data on the average extent of clustering given in Fig. 8, one can then construct the first three moments of the cluster distribution function using Eq. (161). The maximum-entropy distribution function obtained from these moments is shown in Fig. 9A. As was the case for ligand binding, given the cluster probability distribution function, one can calculate the Q_n coefficients in the association polynomial of Eq. (146) and, in turn, the equilibrium constants for the successive binding of monomers as indicated in Eq. (138). The equilibrium constants obtained in this manner are plotted for $n = 2$–5 in Fig. 9B. One sees that for this system there is a subtle decrease in the magnitude of successive binding constants as n increases.

In summary, we have seen that the moments/maximum-entropy method outlined in this chapter is a straightforward way to obtain distribution functions for various molecular variables that characterize biological macromolecules. In this method, one abstracts the appropriate set of moments from experimental data such as the titration curve for lysozyme shown in Fig. 4, the heat capacity curve for

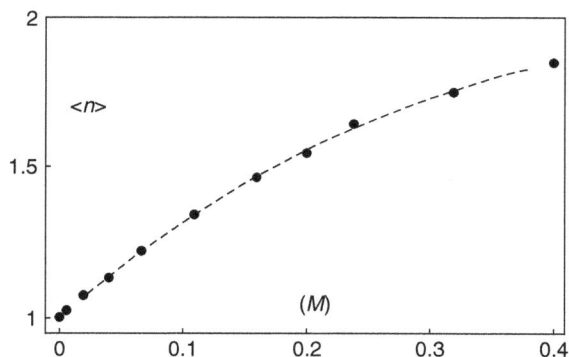

Fig. 8 The self-association of ATP giving the average number of ATP molecules in a cluster as a function of the total amount of ATP. The curve is based on the data of Scheller *et al.* (1981) as constructed by Poland (2002a). Reprinted from Poland (2001g), with kind permission of Elsevier Science.

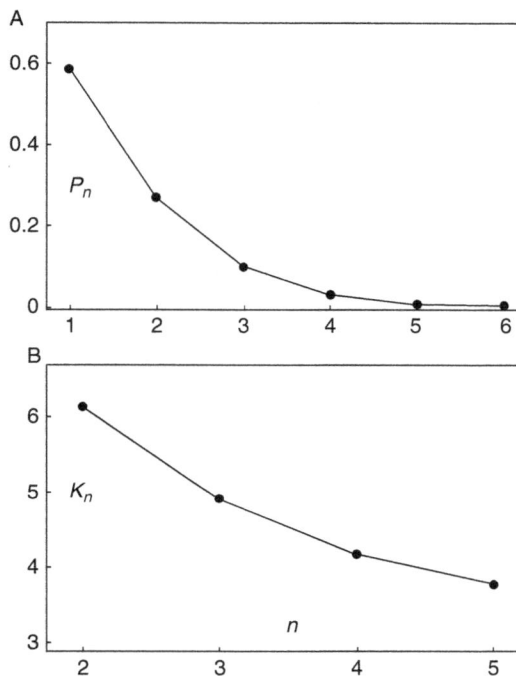

Fig. 9 (A) The maximum-entropy distribution function for aggregates of ATP constructed using moments obtained from local expansions of the self-association curve shown in Fig. 8. (B) The equilibrium constants, K_n, for adding an ATP molecule to a cluster containing $n-1$ units. Reprinted from Poland (2001g), with kind permission of Elsevier Science.

barnase shown in Fig. 6, and the self-association curve for ATP shown in Fig. 8. One then uses the maximum-entropy method to convert knowledge of moments into parameters of the appropriate distribution function, giving, for the examples just cited, the distribution functions for the number of protons bound to lysozyme, the enthalpy of barnase, and the extent of association of ATP. In each case, one gains detailed knowledge about the distribution of the appropriate states in biological macromolecules.

References

Hill, T. L. (1986). *In* "An Introduction to Statistical Thermodynamics". Dover Publications, New York.
Makhatadze, G. I., and Privalov, P. L. (1995). *Adv. Protein Chem.* **47**, 307.
Mead, L. R., and Papanicolaou, N. (1984). *J. Math. Phys.* **25**, 2404.
Nozaki, Y., and Tanford, C. (1967). *J. Am. Chem. Soc.* **89**, 742.
Poland, D. (1978). "CooperativeEquilibriain Physical Biochemistry". Oxford University Press, Oxford.
Poland, D. (2000a). *J. Chem. Phys.* **112**, 6554.
Poland, D. (2000b). *J. Chem. Phys.* **113**, 4774.

Poland, D. (2000c). *J. Chem. Phys.* **113,** 9930.
Poland, D. (2001a). *Biopolymers* **58,** 477.
Poland, D. (2001b). *Biopolymers* **58,** 89.
Poland, D. (2001c). *Biopolymers* **60,** 317.
Poland, D. (2001d). *J. Polym. Sci. B Polym. Phys.* **13,** 1513.
Poland, D. (2001e). *J. Protein Chem.* **20,** 91.
Poland, D. (2001f). *Proteins Struct. Funct. Genet.* **45,** 325.
Poland, D. (2001g). *Biophys. Chem.* **94,** 185.
Poland, D. (2002a). *Biophys. Chem.* **94,** 185.
Poland, D. (2002b). *Biopolymers* **63,** 59.
Poland, D. (2002c). *J. Protein Chem.* **21,** 187.
Poland, D. (2002d). *J. Protein Chem.* **21,** 477.
Poland, D. (2002e). *Physica. A* **309,** 45.
Poland, D. (2003). *Biophys. Chem.* **101,** 485.
Poland, D. (2004). *Biopolymers* in press.
Scheller, K. H., Hofstette, F., Mitchell, P. R., Prijs, B., and Sigel, H. (1981). *J. Am. Chem. Soc.* **103,** 247.
Schellman, J. A. (1975). *Biopolymers* **14,** 999.
Tagliani, A. (1995). *J. Math. Phys.* **34,** 326.
Tanford, C. A., and Wagner, M. L. (1954). *J. Am. Chem. Soc.* **76,** 3331.

CHAPTER 13

The Mathematics of Biological Oscillators

G. Bard Ermentrout

Department of Mathematics
University of Pittsburgh
Pittsburgh, Pennsylvania, USA

I. Introduction

Rhythmic phenomena are ubiquitous in biological processes, ranging from the lunar estrus cycles of certain mammals to the millisecond periodic firing of nerve cells (Rapp, 1987). This repetitive behavior is one of the signatures of living systems and occurs across all levels from local biochemical reactions within a cell to the large-scale periodicity in populations of animals. Rhythmicity is exceedingly important and serves a diversity of functions including communication such as the synchronization of fireflies; locomotor patterns for running, swimming, and chewing; reproduction and menstrual cycles; growth such as the mitotic cycle; secretory processes; and peristaltis and pumping as in the heart, lungs, and bowel.

Rapp (1987) argues that oscillatory behavior in biology is not an artifact or a consequence of a breakdown in regulation. Rather, oscillations can lead to a number of functional advantages. He cites five general categories: temporal organization, entrainment, and synchronization; spatial organization; prediction of repetitive events; efficiency; and frequency encoding of information and precision of control. In spite of the vast number of examples of periodicity, the wide range of frequencies, and the astonishing range of purpose, there is a certain commonality in underlying mechanisms, in how oscillators react to the external world, and how oscillators behave in ensembles. Mathematics provides a tool for understanding this similarity. The objective of this chapter is to describe some of the concepts underlying the formation of biological oscillators and how they interact to create patterns. Obviously, a complete description of oscillatory processes and their analysis is impossible, so a simple model of the glycolytic oscillator will serve as a guide for the techniques described here. In the first section, oscillators and "near oscillators" are described, and some typical mechanisms that lead to their behavior are proposed. In so doing, some simple mathematical techniques for analyzing models are presented. In the next section, the effects of external signals on oscillators are considered. The notions of phase and phase resetting curves are defined and applied to the model biochemical oscillator. In the final section, oscillators are coupled together into networks. Some ways of studying these networks are briefly described.

II. Oscillators and Excitability

Many physical systems exhibit rhythmic activity, that is, some aspect of the systems varies regularly in time. Many of the best known physical examples (e.g., the undamped spring, the pendulum) are poor metaphors for biological oscillators. The pendulum, for example, can oscillate at arbitrary amplitudes, but as soon as the slightest amount of friction is added the oscillations decay to rest. Biological oscillations have a particular robustness; when perturbed, they return to their original magnitude and frequency. Small changes in the environment result in small changes in the rhythm; however, the periodicity remains. For these reasons, this chapter is restricted to a description of the behavior of limit cycle oscillators, that is, oscillators that are locally unique (there exist no other "nearby" oscillations) and stable in the sense that small disturbances quickly die away revealing the original rhythm. Almost all models of biological rhythms have these properties (see Kopell, 1987 for details on this point). This stability has important consequences when external stimuli briefly alter the oscillator. The mechanisms that lead to biological oscillations are intrinsically nonlinear; no linear system has robust limit cycle behavior. Furthermore, all biological oscillators require a constant influx of energy from the environment. In this sense, they are not like the conservative oscillations of classical physics.

Many systems are close to being oscillators but (1) damp slowly; (2) never quite repeat their cycles but maintain a large amplitude; or (3) produce a large response to a stimulus but then return to equilibrium. The first of these, damped oscillations, often precedes the appearance of full rhythmicity as some intrinsic parameter is varied. The second behavior is often erratic and arises in noisy systems and systems that are deterministic but "chaotic" (Glass and Mackey, 1988). The final type of "near oscillation" is called excitable and plays a very important role in many cellular and physiological processes. The existence of excitability often implies that rhythmicity is possible in the presence of a steady applied stimulus.

There are many ways in which biological materials can interact to cause rhythmicity. Many neuronal systems oscillate owing to two separate interactions: (1) positive feedback and (2) delayed negative feedback. The basic idea is that the positive feedback will cause some quantity to self-amplify. This amplification induces a negative feedback which then shuts the original amplification down, and the process is repeated. If the negative feedback is too fast, then it will shut the first process down before it can get started so that no oscillations are possible. Thus, it is necessary for the negative term to be delayed or slow relative to the positive feedback. A biochemical analog is to posit an autocatalytic reaction (e.g., mediated by an enzyme whose efficiency is product dependent) and then a (possibly long) cascade of reactions that eventually inhibit the production of the initial substrate. Systems of this form are called activator-inhibitors and have been suggested as mechanisms for pattern formation in development (Gierer and Meinhardt, 1972) and neurobiology (Edelstein-Keshet, 1988).

A second mechanism for oscillation also involves autocatalysis. Here, however, the autocatalysis of the first substance depends on the presence of the second substance which is itself inhibited by the first. Systems of this type are called positive feedback models. Positive feedback is commonly invoked for biochemical oscillations.

The glycolytic oscillator is an excellent example of a positive feedback oscillator. Segel (1991) provides a set of detailed reactions and complicated equations for modeling this important oscillator. Letting α denote the concentration of ATP and γ denote that of ADP, Segel derives

$$d\gamma/dt = k_s[\lambda\Phi(\alpha,\gamma) - \gamma] \equiv F(\gamma,\alpha), \tag{1}$$

$$d\alpha/dt = \sigma - \sigma_m\Phi(\alpha,\gamma) \equiv G(\gamma,\alpha), \tag{2}$$

where

$$\Phi(\alpha,\gamma) = \frac{\alpha\varepsilon(1+\alpha\varepsilon)(1+\gamma)^2 + L\alpha c\varepsilon'(1+\alpha c\varepsilon')}{L(1+\alpha c\varepsilon')^2 + (1+\alpha\varepsilon)^2(1+\gamma^2)}. \tag{3}$$

For each α, γ fixed, Φ is a monotone increasing sigmoid nonlinearity. The main controllable parameter is σ, which is the normalized rate of substrate infusion. For certain ranges of this parameter, there is a unique equilibrium point, and for σ sufficiently small it is stable. In other words, small perturbations of the system result in a decay to rest. Because Φ increases in both variables, it is clear that both α and γ serve as negative feedback to α. Clearly, α acts positively on γ, and if the slope of Φ with respect to γ is larger than 1 then γ acts positively on itself. Thus, in some parameter ranges, Eqs. (1) and (2) form a positive feedback system.

Enzymatic and biochemical models usually involve many more than two species, but the essential mechanisms that lead to rhythmicity and excitability are easily understood within the restrictions of a two-component model. In fact, many more complicated models can be effectively reduced to two dimensions (Edelstein-Keshet, 1988; Murray, 1989; Segel, 1991).

Once one has devised a model that has the properties of positive feedback, it is easy to analyze its behavior by taking advantage of the special properties of the two-dimensional plane. This approach has been applied to numerous biological systems including nerve membranes (Rinzel and Ermentrout, 1989), population models (Edelstein-Keshet, 1988), and biochemical models (Segel, 1984). To discern the behavior of a positive feedback system, it is helpful to determine several important quantities. First, one seeks the equilibria which satisfy $\dot{\alpha} = \dot{\gamma} = 0$. This usually involves solving systems of nonlinear algebraic equations, which is best done by interactive computer analysis. There are several software packages available for this type of analysis. One of these is described at the end of this section. Equilibria are physically meaningful only if they are also stable, that is, small perturbations from rest damp out and return to the equilibrium state. This is a seemingly difficult calculation; however, it is greatly simplified by taking advantage of a mathematical fact. Having found an equilibrium, one need only compute the matrix of partial derivatives, which for the glycolytic oscillator is

$$M = \begin{bmatrix} \partial F/\partial \gamma & \partial F/\partial \alpha \\ \partial G/\partial \gamma & \partial G/\partial \alpha \end{bmatrix}. \tag{4}$$

If the eigenvalues of the matrix have negative real parts, then the equilibrium is stable; if any of them have positive real parts, it is unstable. Thus, the question of stability is a purely local one. Again, this is an easy numerical calculation and is often done automatically.

With these technical points in mind, it is possible to present a qualitative analysis of the model. The curves defined by setting $F(\gamma, \alpha)$ and $G(\gamma, \alpha)$ to zero are called the nullclines, and their intersections reveal the equilibria of the model. They partition the plane into regions where the variables are increasing or decreasing in time. Thus, by following the trajectories of the differential equations in these regions, it is possible to gain a global picture of the behavior. Suppose that the phase space picture is as shown in Fig. 1. The key feature is that the positive-feedback component (in this model, γ) has a "kinked" nullcline. This is the essential feature for most dynamic phenomena in biological systems. From the direction of the arrows, it is clear that small perturbations from the unique equilibrium point decay back to rest. However, if the perturbation is slightly larger, then there is a large amplification of ADP before everything returns to rest. This is what is meant by an excitable system; small perturbations decay to rest but larger ones result in a big amplification of at least one of the components before ultimately returning to the rest state. The existence of excitability virtually guarantees that one can induce oscillations in the system by applying a constant stimulus. All that is required is to push the equilibrium over into the middle branch of the nullcline of the autocatalytic variable. For the present model, this is implemented by simply increasing the input, σ. In Fig. 2A the oscillation is depicted in the phase plane, and in Fig. 2B the normalized concentrations of ADP and ATP are graphed.

The appearance of this periodic solution is readily understood using techniques from dynamic systems. Consider σ small enough so that there is a unique equilibrium and it is stable. One can numerically follow the equilibrium as the σ is

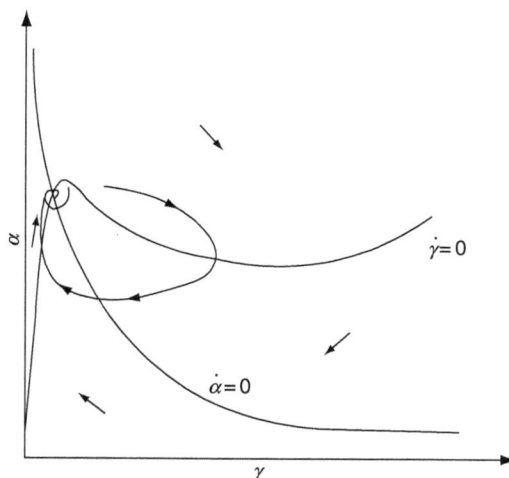

Fig. 1 Phase space diagram of an excitable system.

increased and examine its stability at each point. Stability can be lost in either of two ways: (1) a negative real eigenvalue crosses 0 and becomes a positive real eigenvalue; or (2) a pair of complex conjugate eigenvalues with negative real parts pass the imaginary axis and attain positive real parts. For this model, it is found that the eigenvalues are complex, and at $\sigma = \sigma^*$ they are both imaginary. For $\sigma < \sigma^*$ the equilibrium is stable, and for $\sigma > \sigma^*$ the equilibrium is unstable. The loss of stability by a pair of complex eigenvalues crossing the imaginary axis is the signature for the well-known Hopf bifurcation. When this occurs, a periodic solution to a set of differential equations branches or bifurcates from the branch of equilibrium solutions. It is the main mechanism by which periodic solutions are formed in biological models. Thus, in this simple model, increasing the rate of substrate infusion leads to a destabilization of the equilibrium state and the appearance of regular sustained oscillations. The Hopf bifurcation is not restricted to planar models and has been applied to many higher dimensional systems in order to prove that oscillatory behavior exists. Because the main requirement is obtained by linearizing about a rest state, the Hopf bifurcation is a powerful means by which parameters can be found that lead to oscillation.

It is important to realize that there are other paths to periodicity in models of biology (see, e.g., Rinzel and Ermentrout, 1989) but that the Hopf bifurcation is the most common route. In an experimental system, a hallmark is the existence of very slowly damping oscillations. In the present example, the oscillation appears only for $\sigma > \sigma^*$, that is, only for parameters in which the steady state is not stable. However, there are other models for which there are oscillations on both sides of the critical value of the parameter. Then, as one passes from a stable equilibrium to

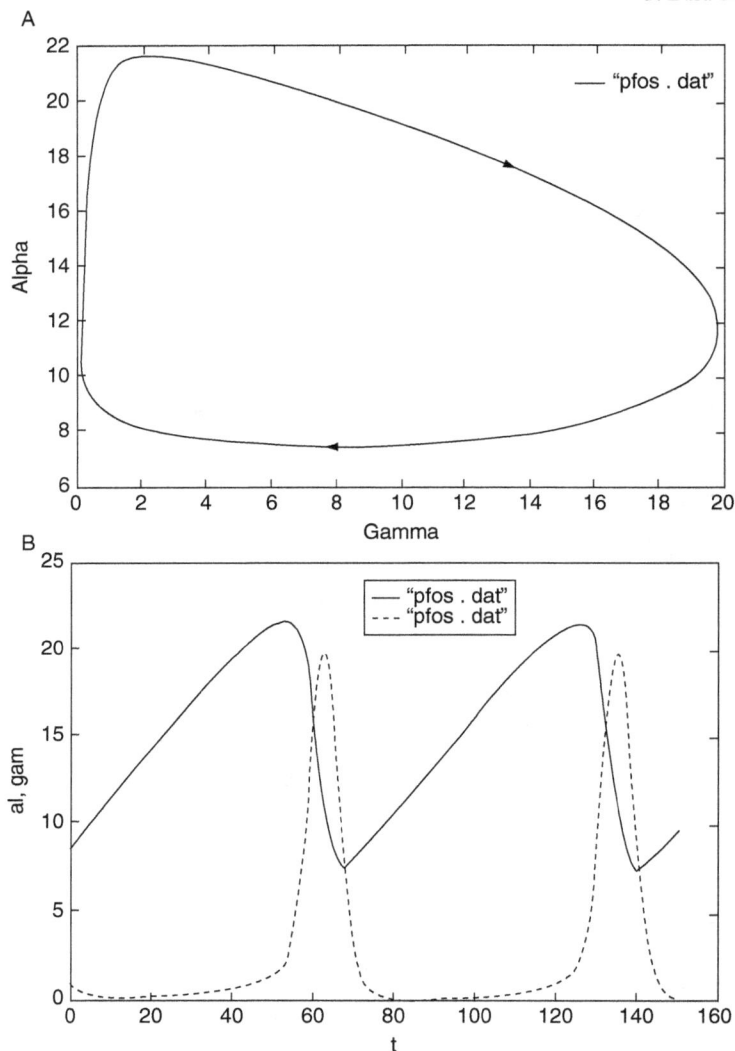

Fig. 2 (A) Phase plane for the glycolytic model during oscillation. (B) Time course of the normalized ATP (solid line) and ADP (dashed line) concentrations.

an unstable equilibrium, there is a jump immediately to large oscillations. Decreasing the parameter at this point maintains the oscillatory behavior well below the critical parameter. This phenomenon is well known in biology and is called hysteresis. In Fig. 3, a diagram for this form of hysteresis is shown. The model for glycolysis does not have this kind of behavior, but there are many examples which do (Rinzel and Ermentrout, 1989).

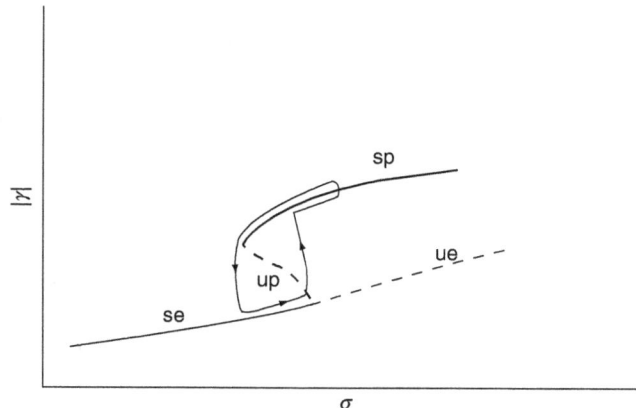

Fig. 3 Diagram illustrating hysteresis with an oscillator, sp, Stable periodic solutions; up, unstable periodic; se, stable equilibrium; ue, unstable equilibrium.

A. Numerical Methods

Numerical and computer methods are an excellent aid to the analysis and understanding of differential equation systems. There are a number of packages that allow the user to input equations and then study their behavior. These packages remove the burden of programming accurate numerical algorithms for the computation of the trajectories, assessment of stability, drawing of nullclines, and other useful techniques. One such program is PhasePlane (Ermentrout, 1989) which runs under MSDOS. A Unix version is available through anonymous ftp at *math.pitt.edu* in the */pub* hardware subdirectory. Both versions compute trajectories, draw nullclines, follow equilibria, and ascertain their stability. All simulations in the rest of this chapter were performed using this tool.

III. Perturbations of Oscillators

A common experiment that can be done with an oscillatory preparation is called phase resetting. This has been exploited clinically as a tool for resetting the circadian pacemaker (Czeisler *et al.*, 1989; Strogatz, 1990). Winfree (1980) describes phase resetting experiments for many biological oscillators, and Glass and Mackey (1988) have used these techniques to explore the behavior of embryonic heart cells. Buck *et al.* (1981) apply this method to characterize the firefly oscillator. The technique exploits a very important property of limit cycle oscillators. When a system is stably oscillating, it is possible to define the phase of the oscillation as a number between 0 and 1 that uniquely defines the current state. Thus, one might set phase 0 to be when ADP peaks. Phase 1/2 then corresponds to a point that is halfway through the oscillatory cycle. Phase 1 is identified with phase 0 as the process is periodic.

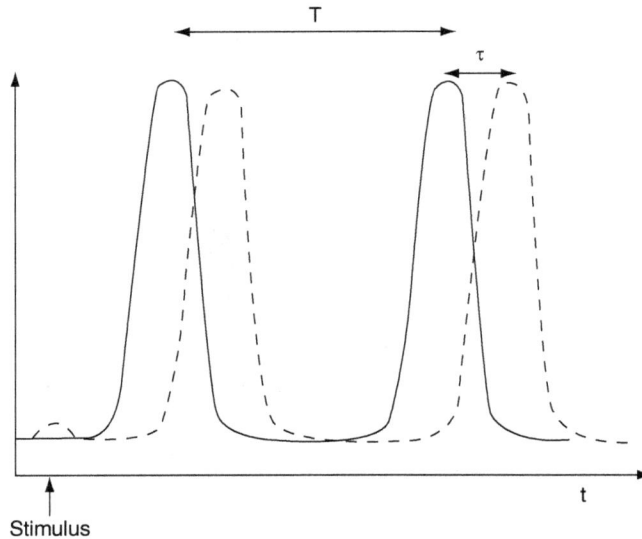

Fig. 4 Effect of a pulse on an oscillator showing the phase shift. Here, T is the unperturbed period of the oscillation, and τ is the phase delay, so that $\Delta = -\tau/T$.

Suppose that one applies a brief stimulus to an oscillator (such as a pulse of ADP) and that this stimulus is not so strong as to stop the oscillation. Then, after some transients, the oscillation will return to its original state but will be phase shifted relative to the unstimulated oscillation (see Fig. 4). The amount of the phase shift depends on two things: (1) the quantitative form of the stimulus (magnitude, duration, chemicals involved) and (2) the timing of the stimulus. In other words, if the stimulus properties are kept constant, then the amount that the phase is shifted depends on the part of the cycle at which the stimulus is presented. For a given stimulus protocol, a function called the phase transition curve (PTC) can be defined which maps the phase at which the stimulus occurs to the new phase after the stimulus. This is easily measured and provides a means by which the response of an oscillator to periodic stimulation and other oscillators can be studied and modeled. If the phase is denoted by θ then the map is given by

$$\theta \mapsto F(\theta, M),$$

where M is the magnitude of the stimulus. Clearly, $F(\theta, 0) = \theta$ as no stimulus results in no phase shift.

There are two qualitatively different PTCs possible: type 0 and type 1. In the latter, the function F obtains all values between 0 and 1 and is thus qualitatively like having no stimulus. More precisely, one requires that $\partial F/\partial\theta > 0$. For type 1 PTCs, one often computes the phase difference, $\Delta(\theta, M) = F(\theta, M) - \theta$ which measures the degree of phase advancement or delay caused by the stimulus.

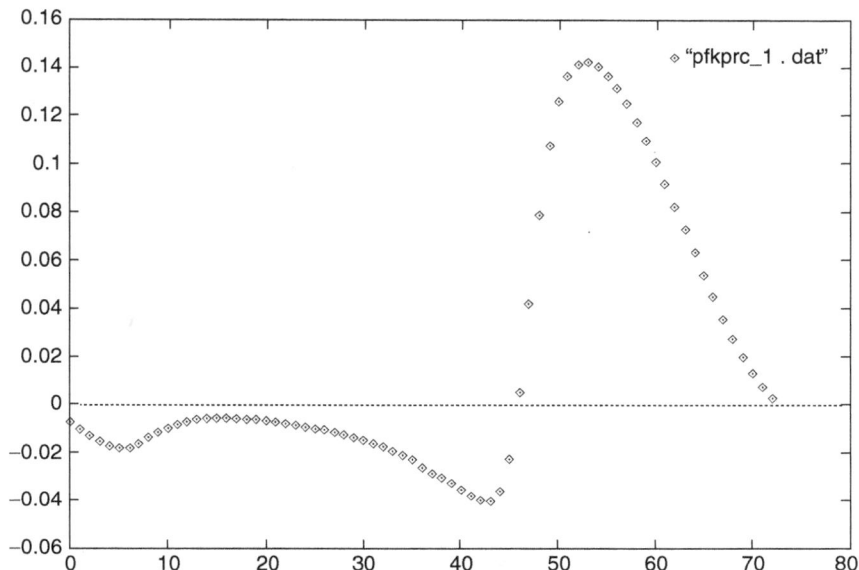

Fig. 5 Phase difference curve for the glycolytic model; Δ for a short pulse showing a weak resetting.

Alternatively, one can view Δ as the change in the period of the oscillator as a function of the timing of the stimulus; phase advances shorten the period, and phase delays lengthen it. Figure 5 shows the phase difference curve for the glycolytic model under a stimulus that consists of a small pulse of ADP. If the stimulus occurs ahead of the ADP maximum, then the oscillator advances in phase, and if it occurs before the ADP maximum, then the oscillator is delayed. The phase difference curve, $\Delta(\theta, M)$ shows that one can advance the phase by a much greater amount than it is possible to delay it. This is typical of biological oscillators.

Type 0 phase resetting requires a much stronger stimulus and is characterized by the fact that the new phase is always in some subset of the full range of phases. In Fig. 6, Δ for the glycolytic oscillator is shown for a large pulse of ADP as the stimulus. Here, Δ is discontinuous, showing that the PTC is type 0. This strong type of phase resetting has been shown to occur in the human circadian pacemaker. For very strong stimuli at critical phases, it is possible to stop an oscillator completely. This is particularly true of the hysteretic oscillators described in Section II. Gutmann *et al.* (1980) use this effect to show how well-timed stimuli could be used to turn on and off the repetitive firing of the squid axon. When such stimuli are presented, the notion of the PTC no longer exists as the oscillation has been stopped. Thus, the computation of the PTC is most useful when it is type 1.

Once the PTC has been computed, it can be used to explore the effects of periodically forcing the oscillator. This leads to a wide range of experiments that

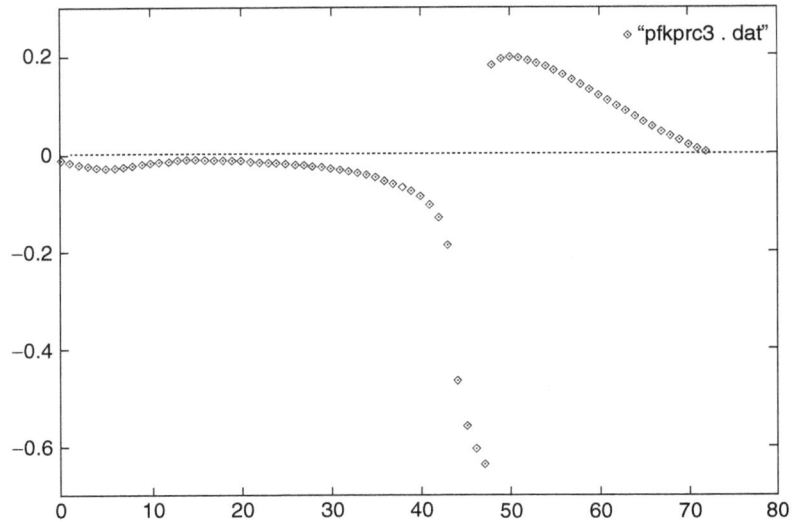

Fig. 6 Phase difference curve; same as Fig. 5, but with a strong pulse showing type 0 resetting.

can be used to study the oscillator. Suppose the PTC is given by $F(\theta)$. The dependence on magnitude has been dropped as the periodic stimulus will always have the same magnitude. Suppose that the stimulus is presented every T seconds and suppose that the period of the unperturbed oscillator is τ. Let r denote the ratio T/τ so that if $r = 1$, the stimulus has the same period as the oscillator. Let θ_n denote the phase of the oscillator right before the nth application of the stimulus. After the stimulus, the phase is $F(\theta_n)$. Then, between stimuli, the oscillator will advance by an amount, r, so that right before the next stimulus,

$$\theta_{n+1} = F(\theta_n) + r \equiv G(\theta_n). \qquad (5)$$

One can ask what possible forms of behavior are possible for Eq. (5). The most common and often the most desirable behavior is called $p{:}q$ phase locking. In $p{:}q$ locking, the oscillator advances p cycles for every q cycles of the stimulus. This implies that $\theta_{n+q} = \theta_n + p$ or, using Eq. (5),

$$\theta + p = G^q(\theta), \qquad (6)$$

where G^q means q iterates of G. For example, 1:1 locking occurs when there is a solution to

$$F(\theta) - \theta \equiv \Delta(\theta) = 1 - r.$$

From Fig. 5, it is clear that it is possible to achieve 1:1 locking for stimuli that are much faster than the oscillator ($r < 1$ and thus $1 - r$ is positive). In contrast, only a very limited range of forcing periods higher than the natural frequency are allowed.

More exotic types of phase locking are possible as well as non-phase-locked behavior. One can define something called the rotation number, ρ, which is

$$\rho = \lim_{n\to\infty} \frac{\theta_n}{n}. \tag{7}$$

For p:q locking, this ratio is exactly p/q. The rotation number as a function of the stimulus period yields a very interesting plot known as the "devil's staircase" (shown in Fig. 7). Each oscillator has a different signature. The flat plateau regions are areas of p:q locking; the largest regimes are 2:1 and 1:1. If one expands one of the sloped regimes, smaller plateaus are found.

The technique of PTCs is not foolproof; there are several ways in which one can be misled. The principal assumption in the use of PTCs is that the perturbation is such that the oscillator very quickly returns to its stable cycle. Otherwise, one must wait for the transients to decay before measuring the phase lag. If the transients are long, however, then the PTC cannot be used to model periodic stimulation unless the period between stimuli is very long.

The method of PTCs is a powerful tool for exploring nonlinear oscillations. It allows the experimenter to quantify the effects of different stimuli and to use these to predict the effects of more complex stimuli from those obtained from a simple protocol. As shown in the next section, the PTC can also be used as a basis for studying many coupled biochemical oscillators.

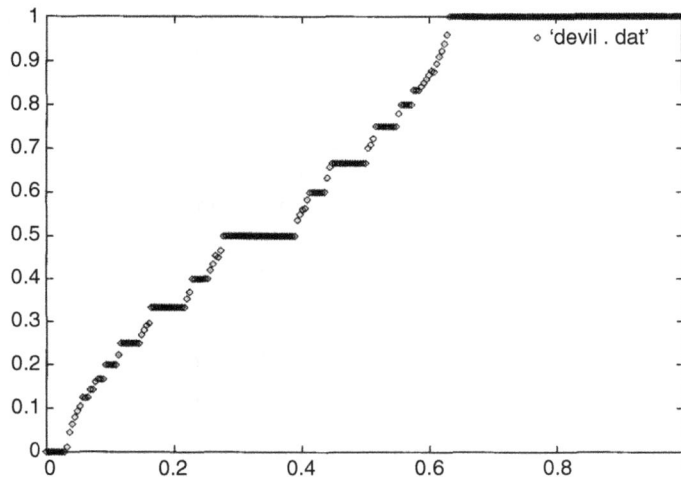

Fig. 7 "Devil's staircase" for the glycolytic oscillator showing rotation number as a function of the stimulus period, r.

IV. Coupled Oscillators

Very few biologically important oscillators occur in isolation. More generally, many are coupled together by a variety of means. The chemical oscillators such as the glycolytic oscillator are coupled via diffusion of the relevant chemical species. In humoral systems such as the β cells of the pancreas, coupling is via gap junctions. The nervous system achieves coupling through synapses. Southeast Asian fireflies, which congregate in trees and flash in synchrony, are coupled via the visual responses to neighboring flashes. Thus, one can justifiably ask whether any general principles can emerge from these diverse phenomena. In many physiologically interesting cases, the answer is a surprising yes. As should be expected, it is very difficult to say anything about a general system of coupled oscillators. However, in some physically interesting circumstances, emergent properties of the network can be understood and classified. If the only "model" available is an experimental one, then one can use the PTC as a tool for simulating a network, but a mathematical analysis is difficult. Such a technique is described below. On the other hand, if a specific mathematical model is known to be a good approximation of a single oscillatory component, then there are several mathematical techniques that can be applied to reduce the coupled system to a new set of equations that is mathematically tractable.

A. Coupling Using PTCs

If one has computed the PTC of an oscillator for a particular stimulus protocol, then it is sometimes possible to use this as a model for the behavior of a coupled system. For oscillators that are mainly quiescent and produce only small pulses of activity, the PTC can be a useful tool for modeling coupling. Consider a pair of oscillators and let θ_1, θ_2 denote their respective phases. Suppose that the only interaction between them is short lasting and pulsatile, and let $P(\theta)$ denote the magnitude of the pulse as a function of the timing of the oscillator (i.e., its phase.) Let $\Delta(\theta)$ denote the phase-difference function as computed from the PTC, that is, the degree of advance or delay of the cycle owing to the pulse. Let ω_1, ω_2 denote the intrinsic frequencies of the two oscillators and assume they are symmetrically coupled. Then a simple model is

$$d\theta_1/dt = \omega_1 + P(\theta_2)\Delta(\theta_1), \qquad (8)$$

$$d\theta_2/dt = \omega_2 + P(\theta_1)\Delta(\theta_2). \qquad (9)$$

In the absence of coupling, each oscillator traverses its cycle with a period of $1/\omega_j$. A pulse from oscillator 1 causes a phase shift in oscillator 2 and vice versa. This type of model has been explored by Winfree (1967), Ermentrout and Kopell (1991), and Strogatz and Mirollo (1990). Typically, one is interested in the existence of $p{:}q$ phase-locked solutions. For a solution to be $p{:}q$ phase-locked, there must exist a $T > 0$ such that

$$\theta_1(t + T) = \theta_1(t) + p$$
$$\theta_2(t + T) = \theta_2(t) + q$$

for all t. A 1:1 phase locking in which each oscillator fires once per cycle is the most common type of locking observed in biological networks. As in the case of forced oscillators, it is possible to define the rotation number and to create plots like the devil's staircase for a pair of coupled oscillators.

More generally, one is concerned with a population of oscillators. The generalization of Eqs. (8) and (9) leads to

$$d\theta_j/dt = \omega_j + \left(\sum_{j=1}^{N} P_{jk}(\theta_k)\right)\Delta_j(\theta_j),$$ (10)

where P_{jk} is the pulse felt by oscillator j from oscillator k, ω_j is the intrinsic frequency, and Δ_j is the phase difference function owing to a pulse. The mathematical analysis of equations such as Eq. (10) is incomplete. Most results have dealt only with specialized connectivities such as "all-to-all" geometry where each oscillator sees all of the other oscillators in the network. Because of the difficulty of analyzing this type of model, it has not been applied to a wide variety of systems. The results of Mirollo (1993) can be applied to Eq. (10) if all of the oscillators are identical and are coupled symmetrically in a ring. For systems of this type, it is possible to show the existence of traveling waves, and so the model may be applicable to systems with circular geometry such as the stems of plants (Lubkin, 1992) or the small bowel (Linkens et al., 1976).

B. Averageable Coupling

The technique of using the PTC to study coupled oscillators applies for pulsatile coupling and is useful as a simulation tool for an experimental system. Naturally, if one has equations such as Eqs. (1) and (2), then the PTC can again be used to model networks of these oscillators. However, as noted above, these systems are difficult to analyze mathematically. Furthermore, if the coupling is through diffusion, then a pulsatile type of interaction is a poor approximation. A better technique would use the fact that interactions are occurring throughout the cycle and are not restricted to a small portion of it. If coupling between the components of a network of oscillators (or only a pair for that matter) is small enough so that the rhythms are not pulled far from their uncoupled behavior, then a powerful technique called the method of averaging is applicable. Averaging is a technique whereby the interactions between two or more oscillators are averaged over a cycle. The result of this averaging is the calculation of an interaction function (one for each input from another oscillator). Unlike PTC coupling, in averaged coupling, the effect of one oscillator on the other depends only on the phase differences between the two oscillators, that is, the relative phase. This is intuitively reasonable as the interaction is averaged over a complete cycle. Because the phase difference between two oscillatory processes is often the only measurable quantity, this approach is very useful in biology (Cohen et al., 1992).

As in the previous section, let θ_1 and θ_2 denote the respective phases of two oscillators. Then, after averaging, the coupled system satisfies an equation of the form

$$d\theta_1/dt = \omega_1 + H_{21}(\theta_2 - \theta_1), \tag{11}$$

$$d\theta_2/dt = \omega_2 + H_{12}(\theta_1 - \theta_2). \tag{12}$$

As before, ω_j is the uncoupled frequency, and H_{jk} is an interaction function that depends only on the difference of the two phases. Equations (11) and (12) result when one seeks 1:1 phase-locked behavior. Other types of $p:q$ locking lead to slightly different equations (Ermentrout, 1981). The computation of H_{jk} is easily done for a given model (Ermentrout and Kopell, 1991), and some software packages (XPP available from the author and described in Section II) will automatically compute the interaction functions given the form of the coupling. As an example, suppose that a pair of glycolytic oscillators is coupled via diffusion of the two principal components. Then, one obtains a four-dimensional system of equations:

$$d\gamma_1/dt = k_s[\lambda\Phi(\alpha_1,\gamma_1) - \gamma_1] + D_\gamma(\gamma_2 - \gamma_1), \tag{13}$$

$$d\alpha_1/dt = \sigma - \sigma_m\Phi(\alpha_1,\gamma_1) + D_\alpha(\alpha_2 - \alpha_1), \tag{14}$$

$$d\gamma_2/dt = k_s[\lambda\Phi(\alpha_2,\gamma_2) - \gamma_2] + D_\gamma(\gamma_1 - \gamma_2), \tag{15}$$

$$d\alpha_2/dt = \sigma - \sigma_m\Phi(\alpha_2,\gamma_2) + D_\alpha(\alpha_1 - \alpha_2). \tag{16}$$

In Fig. 8, the interaction functions $H_{12} = H_{21}$ are drawn for $D_\gamma = 0$, $D_\alpha = 0$, and $D_\gamma = D_\alpha$. If the only knowledge of the intrinsic oscillator comes from the computation of the PTC, then it is still possible to approximate the interaction function as

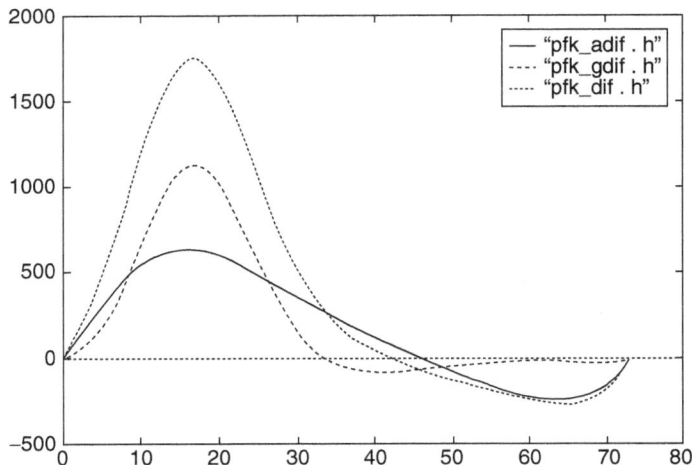

Fig. 8 Averaged interaction functions for diffusively coupled glycolytic oscillators. The solid line shows the interaction if only ADP diffuses, the long dashed line shows only ATP diffusion, and the short dashed line shows the interaction if both species diffuse equally.

$$H(\phi) = \frac{1}{T} \int_0^T P(x + \phi)\Delta(x)\mathrm{d}x, \tag{17}$$

where Δ is the function computed from the PTC and P is an approximation of the pulse interaction.

Equations (11) and (12) have a major advantage over the PTC analogs as the question of phase locking and the ensemble (or coupled) frequency are readily accessible. Let $\phi = \theta_2 - \theta_1$. Then, the phase difference between the two oscillators, ϕ satisfies

$$\mathrm{d}\phi/\mathrm{d}t = \omega_2 - \omega_1 + H_{12}(-\phi) - H_{21}(\phi) \equiv g(\omega_2 - \omega_1, \phi). \tag{18}$$

A rest state for Eq. (18) implies that the oscillators are phase-locked as their phase relationship remains fixed for all time. Thus, 1:1 locking occurs if and only if the scalar equation $g(\delta, \phi) = 0$ has a root where δ is the frequency difference between the two oscillators. Having found such a root, the ensemble frequency is given by

$$\Omega = \omega_1 + H_{21}(\phi) = \omega_2 + H_{12}(\phi).$$

This calculation shows that the ensemble frequency is not necessarily the highest one; rather, it is a consensus of the two frequencies. For example, if one takes $H_{12}(\phi) = H_{21}(\phi) = \sin(\phi)$, then $\Omega = (\omega_1 + \omega_2)/2$, the average of the frequencies.

Equation (18) also shows that if the two oscillators have widely different frequencies, then locking is impossible. In other words, because the interaction functions are bounded and periodic, if the frequency difference gets too large, then no roots to $g(\delta, \phi)$ can be found and the oscillators "drift" apart.

This form of modeling oscillators is powerful and can be extended to many oscillators having the form

$$\mathrm{d}\theta_j/\mathrm{d}t = \omega_j + \sum_{k=1}^{N} H_{kj}(\theta_k - \theta_j). \tag{19}$$

Phase locking reduces to finding roots to a set of nonlinear functions. For many geometries, this can be done explicitly. For example, Kuramoto (1984), Strogatz and Mirollo (1990), and Ermentrout (1985) have explored Eq. (19) when $H_{jk}(\phi) = \sin(\phi)$; Cohen et al. (1982) and Kopell et al. (Kopell and Ermentrout, 1986; Kopell et al., 1990) have looked at chains with nearest neighbor coupling; Ermentrout (1986) considers rings of oscillators; Paullet and Ermentrout (1994) analyze a sheet of oscillators; and Ermentrout (1992) gives conditions for the existence of phase locking of Eq. (19) under very general conditions.

The techniques of this section have been successfully used to suggest and predict properties of the lamprey spinal cord (see Cohen et al., 1982 for a review of this work.) Glass and Mackey (1988) have applied similar models to discern the firing times of heart pacemakers.

The mathematical techniques described in this and the preceding sections are applicable to many physical and biological preparations. The dynamic systems theory of oscillators, including phase plane techniques, bifurcation theory, and averaging, provides a framework for further experimentation and simulation. The methods described here are by no means exhaustive; in particular, no statistical methods such as spectral analysis are considered. Nevertheless, the methods of this chapter are well suited for building a comprehensive mathematical theory of biological rhythms.

References

Buck, J., Buck, E., Hanson, F., Case, J. F., Mets, L., and Atta, G. J. (1981). *J. Comp. Physiol.* **144,** 277.

Cohen, A. H., Holmes, P. J., and Rand, R. H. (1982). *J. Math. Biol.* **13,** 345.

Cohen, A., Ermentrout, B., Kiemel, T., Kopell, N., Sigvardt, K., and Williams, T. (1992). *Trends Neurosci.* **15,** 434.

Czeisler, C. A., Kronauer, R. E., Allan, J. S., Duffy, J. F., Jewett, M. E., Brown, E. N., and Ronda, J. M. (1989). *Science* **244,** 1328.

Edelstein-Keshet, L. (1988). Mathematical Models in Biology. Random House, New York.

Ermentrout, G. B. (1981). *J. Math. Biol.* **12,** 327.

Ermentrout, G. B. (1985). *J. Math. Biol.* **22,** 1.

Ermentrout, G. B. (1986). *J. Math. Biol.* **23,** 55.

Ermentrout, G. B. (1989). PhasePlane: The Dynamical Systems Tool. Brooks/Cole, Pacific Grove, CA.

Ermentrout, G. B. (1992). *SIAM J. Appl. Math.* **52,** 1665.

Ermentrout, G. B., and Kopell, N. (1991). *J. Math. Biol.* **29,** 195.

Gierer, A., and Meinhardt, H. (1972). *Kybernetic* **12,** 30.

Glass, L., and Mackey, M. (1988). From Clocks to Chaos: The Rhythms of Life. Princeton Univ. Press, Princeton, NJ.

Gutmann, R., Lewis, S., and Rinzel, J. M. (1980). *J. Physiol* **305,** 377, (*London*).

Kopell, N. (1987). *In* "Neural Control of Rhythmic Movements in Vertebrates," (A. H. Cohen, S. Grillner, and S. Rossignol, eds.), p. 369. Wiley, New York.

Kopell, N., and Ermentrout, G. B. (1986). *Commun. Pure Appl. Math.* **39,** 623.

Kopell, N., Zhang, W., and Ermentrout, G. B. (1990). *SIAM J. Math. Anal.* **21,** 935.

Kuramoto, Y. (1984). Chemical Oscillations, Waves, and Turbulence. Springer-Verlag, Berlin.

Linkens, D. A., Taylor, I., and Duthie, H. L. (1976). *IEEE Trans. Biomed. Eng* **BME-23,** 101.

Lubkin, S. (1992). Cornell University, Ithaca, NY, Ph.D. Thesis.

Mirollo, R. (1993). *SIAM J. Math. Anal.* **23,** 289.

Murray, J. D. (1989). Mathematical Biology. Springer-Verlag, New York.

Paullet, J. E., and Ermentrout, G. B. (1994). *SIAM J. Appl. Math.* in print.

Rapp, P. E. (1987). *Prog. Neurobiol.* **29,** 261.

Rinzel, J. M., and Ermentrout, G. B. (1989). *In* "Methods of Neuronal Modelling," (C. Koch, and I. Segev, eds.), pp. 135–169. MIT Press, Cambridge, MA.

Segel, L. A. (1984). Modelling Dynamic Phenomena in Molecular and Cellular Biology. Cambridge Univ. Press, Cambridge.

Segel, L. A. (1991). Biological Kinetics. Cambridge Univ. Press, Cambridge.

Strogatz, S. (1990). *J. Biol. Rhythms* **5,** 169.

Strogatz, S., and Mirollo, R. (1990). *SIAM J. Appl. Math.* **50,** 1645.

Winfree, A. T. (1967). *J. Theor. Biol.* **16,** 15.

Winfree, A. T. (1980). The Geometry of Biological Time. Springer-Verlag, New York.

CHAPTER 14

Modeling of Oscillations in Endocrine Networks with Feedback

Leon S. Farhy

Department of Medicine
Center for Biomathematical Technology
University of Virginia
Charlottesville, Virginia, USA

I. Introduction

Important features of different endocrine systems emerge from the interplay between their components. The networks discussed in this chapter (e.g., Figs. 2, 9, 10, 12) are typical in endocrine research and are used to exemplify regulatory hypotheses. Traditionally, the individual components of these networks are studied in isolation from the rest of the system, and therefore, their temporal relationships cannot be assessed. As a result, the mechanism of some key specifics of the system behavior, such as for example its ability to oscillate, which are the result of the time-varying interactions of several components cannot be recovered. Differential equations-based modeling of endocrine networks outlined in this chapter allows for the reconstruction of the dynamic interplay between different hormones and is therefore suitable for the analysis of the structure and behavior of complex endocrine feedback networks.

Since its original publication, the methodology has been utilized in the analysis of different endocrine axes and physiological problems. For example, the growth hormone control network, which is similar to the abstract networks discussed in this chapter (See *"Networks with Multiple Feedback Loops"*), has been studied in Farhy and Veldhuis (2003, 2004) to determine the mechanisms driving its oscillatory behavior. Subsequently, the effects of growth hormone secretagogues in general and the hormone ghrelin in particular have been added to the system and analyzed in detail (Farhy and Veldhuis, 2005; Farhy *et al.*, 2007). More recently (Farhy and McCall, 2009a,b; Farhy *et al.*, 2008; see also Chapter 24 in this volume), the methodology was applied to the glucagon control network in an effort to understand the system-level network control mechanisms that mediate the glucagon counterregulation and their abnormalities in diabetes. These reports exemplify the key role played by the methods presented here within an interdisciplinary approach in which model-based predictions motivate experimental work, the results of which feed back on the modeling effort. In addition, the results in Farhy and McCall (2009b) exemplify the reduction of the number of nodes of a network in an attempt to diminish the model complexity without affecting its performance, and permitting its clinical application as described in *"Networks with Multiple Feedback Loops"*. We have added a new paragraph to this chapter to warn that after any network reduction, the performance of the new simplified model need to be verified (*"Summary and Discussion"*).

The methods described in this Chapter are more appropriate to reconstruct the general "averaged" macroscopic behavior of a given endocrine system rather than to establish its microscopic behavior or the molecular mechanisms that govern this behavior. It is also important to note that these methods are intended to provide means for *in silico* analysis (simulations). Any effort to use the underlying models to fit data and reconstruct/measure individual parameters should be carried out with care because of the nonlinearity of the model equations and the interdependence of the model parameters.

Finally, we have recently published a more comprehensive presentation of the methodology outlined in the current chapter with additional mathematical and biological background and a laboratory manual (Robeva *et al.*, 2008). The text (and in particular, Chapter 10 from Robeva *et al.*, 2008) is intended as a detailed introduction to the methods for modeling of complex endocrine networks with feedback.

II. General Principles in Endocrine Network Modeling

Numerous studies document that the hormone delivery pattern to target organs is crucial to the effectiveness of their action. Hormone release could be altered by pathophysiology, and differences in endocrine output mediate important intraspecies distinctions, for example, some of the sexual dimorphism in body growth and gene expression in humans and rodents. Accordingly, the mechanisms controlling the dynamics of various hormones have lately become the object of extensive biomedical research. Intuitive reconstruction of endocrine axes is challenged by their high complexity, because of multiple intervening time-delayed nonlinear feedback and feedforward inputs from various hormones and/or neuroregulators. Consequently, quantitative methods have been developed to complement qualitative analysis and laboratory experiments and reveal the specifics of hormone release control. The emerging mathematical models interpret endocrine networks as dynamic systems and attempt to simulate and explain their temporal behavior (Chen *et al.*, 1995; Farhy *et al.*, 2001, 2002; Keenan and Veldhuis, 2001a,b; Wagner *et al.*, 1998).

This chapter focuses on the mathematical approximation of endocrine oscillations in the framework of a modeling process structured in three formal phases:

1. Data analysis (examining the available data). We start by studying the available observations and experimental results, by examining the hormone time series, and determining the specifics of the observed profiles. This might include pulse detection, analysis of the variability and orderliness, verifying the baseline secretion and half-life, and detecting the frequency of the oscillations. We identify those phenomena that should be explained by the modeling effort, for example, some specific property of the hormone profiles, combined with selected feedback experiments.

2. Qualitative analysis (designing the formal network). This stage uses the information collected in phase 1 and outlines an intuitive functional scheme of the systems underlying physiology. Qualitative analysis of the available data (Friesen and Block, 1984) identifies the key elements and their interaction, and organizes them as a set of nodes and conduits in a *formal endocrine network*. The *main hypothesis* states that this formal network explains the selected in phase 1 specifics in the experimental data.

3. Quantitative analysis (dynamic modeling). At this phase, the endocrine network is interpreted as a dynamic system and described with a set of coupled ordinary differential equations (ODE). They give the time derivative of each network node and approximate all system positive and negative dose-responsive control loops. The parameters in the ODEs must have a clear physiological meaning and are determined by comparing the model output with the available data (phase 1) as we attempt to address the main hypothesis (phase 2).

The outcome of the modeling effort is a *conditional* answer to the main hypothesis. It formulates necessary physiological assumptions (additional to the main hypothesis) that would allow the formal network to explain the observed data specifics. This further refines the hypothesis and generates new questions to be addressed experimentally.

The general modeling scheme anticipates that the qualitative analysis of the hormone secretion dynamics outlines the formal endocrine network by determining its nodes and conduits. As previously discussed (Friesen and Block, 1984), the main source of oscillations in biology is feedback loops with delay. However, not every network with feedback generates periodic behavior (Thomas *et al.*, 1990). The main goal of this work is to illustrate via a series of abstract examples different conditions under which oscillations can emerge. To this end, we perform quantitative analysis on various abstract endocrine networks, interpreted as dynamic systems. Thus, we will be mainly concerned with phase 3 (previous) and its relations to phases 1 and 2.

We start by describing the approximation of the basic element of an endocrine network: the dynamics of the concentration of a single hormone controlled by one or more other regulators (system nodes). Further, this is used in the simulation and analysis of different feedback networks. The main concepts are illustrated on abstract 2-node/1-feedback reference models. System parameters are introduced on the basis of their physiological meaning and the effect of their modification is examined. Oscillations due to perturbations of systems with damped periodicity are distinguished from oscillations of systems with a true periodic solution (limit cycle). Additionally, we simulate basic laboratory experimental techniques, discuss some of their limitations, and suggest alternatives to reveal more network details.

It should be noted that the theory behind most of the examples in this chapter is not trivial. This is especially valid for those models that include one or more direct delays in the core system. We avoid the abstract mathematical details to make the presentation accessible to a variety of bio-scientists. The simulated networks are abstract and do not correspond to a particular endocrine system. However, the constructs and the modeling techniques can be easily adapted to fit a particular physiology.

III. Simulating the Concentration Dynamics of a Single Hormone

In this section, we describe the quantitative approximation of the concentration dynamics of a single hormone in an abstract pool, where it is secreted (not synthesized). As described elsewhere (Veldhuis and Johnson, 1992), we assume that the hormone concentration rate of change depends on two processes—secretion and ongoing elimination. The quantitative description is given by the ODE

$$\frac{\mathrm{d}C}{\mathrm{d}t} = -\alpha C(t) + S(t). \tag{1}$$

Here, $C(t)$ is the hormone concentration in the corresponding pool, t is the time, $S(t)$ is the rate of secretion, and the elimination is supposed to be proportional to the concentration.

Deconvolution technique, employed to describe hormone pulsatility (Veldhuis and Johnson, 1992), can be used as an alternative approach to introducing Eq. (1). In this context, the observed hormone concentration is described by a convolution integral

$$C(t) = \int_0^t S(\tau)E(t - \tau)\mathrm{d}\tau, \tag{2}$$

where S is a secretion function and E describes the removal of the hormone from the pool. For the purposes of this presentation, E is required to correspond to a model with one half-life. In particular, we assume that the elimination function $E(t)$ satisfies the initial value problem

$$\frac{dE(t)}{dt} = -\alpha E(t),$$

$$E(0) = 1$$

(3)

with some rate of elimination $\alpha > 0$. Consequently, it is easy to see that Eqs. (2) and (3) imply that the right-hand side of Eq. (1) describes the rate of change of $C(t)$. And as the solution of Eq. (3) is the function $E(t) = e^{-\alpha t}$, the hormone concentration [the solution of Eq. (1)] is described as the convolution integral

$$C(t) = \int_0^t S(\tau)e^{-\alpha(t-\tau)}d\tau.$$

Now, suppose that the secretion rate $S = S_A$ (of a hormone A) does not depend explicitly on t and is controlled by some other hormone B. We write $S_A = S_A[C_B(t)]$, where $C_B(t)$ is the concentration of B. In the sequel, S_A is called a control function and its choice, albeit arbitrary to some extent, should conform to a set of general rules.

1. Minimal and maximal endogenous levels: Denote by $C_{A,min}$ and by $C_{A,max}$ the minimal and maximal values (experimentally established or hypothetical) for the concentration of hormone A. Typically (but not always), $C_{A,min}$ is associated with the baseline secretion and $C_{A,max}$ corresponds to the maximal attainable concentration of endogenous A (in a variety of conditions, including responses to external submaximal stimulation). Accordingly, the control function S_A must satisfy the inequalities

$$C_{A,min}/\alpha \leq \min(S_A) \leq \max(S_A) \leq C_{A,max}/\alpha.$$

2. Monotonous and nonnegative: The control function must be nonnegative, as the secretion rate is always nonnegative and monotone (with some rare exceptions briefly mentioned in the sequel). It will be monotone increasing if it represents a positive control. If the control is negative, it will be decreasing.

There are many ways to introduce a control function in an acceptable mathematical form. As many authors do, we use nonlinear, sigmoid functions, known as up- and down-regulatory Hill functions (Thomas $et\ al.$, 1990):

$$F_{up(down)}(G) = \begin{cases} \dfrac{[G/T]^n}{[G/T]^n + 1} & \text{(up)} \\ \quad\quad\text{or} \\ \dfrac{1}{[G/T]^n + 1} & \text{(down)} \end{cases}$$

(4)

where $T > 0$ is called a threshold and $n \geq 1$ is called a Hill coefficient. It should be noted that $F_{up} = 1 - F_{down}$ and $F_{up(down)}(T) = 1/2$. These functions are exemplified in the plots in Fig. 1 (for $n = 5$ and $T = 50$). They are monotone and map $F: [0, \infty) \to [0, 1]$; the Hill coefficient n controls the slope (which also depends on T), and the inflection point I_F is given by

A

B

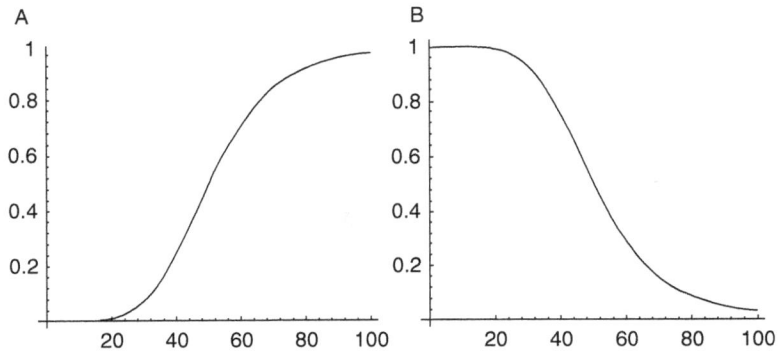

Fig. 1 Exemplary profiles of up-regulatory (A) and down-regulatory (B) Hill functions. In both examples $n = 5$ and $T = 50$.

$$I_F = T\left(\frac{n-1}{n+1}\right)^{\frac{1}{n}} \text{ for } n \geq 2.$$

When $n = 1$ (Michaelis-Menten type equation), the function has no inflection point and its profile is a branch of a hyperbola. If n is large [values, as large as 100, exist in biology (Mikawa *et al.*, 1998; Vrzheshch *et al.*, 1994)] the control function acts almost as an on/off switch.

Using Hill functions, we write the term controlling the secretion of A in the form

$$S_A(C_B) = aF_{\text{up(down)}}(C_B) + S_{A,\text{basal}}, \tag{5}$$

where $S_{A,\text{basal}} \geq 0$ is independent of B and controls the basal secretion of A. The quantities $(a + S_{A,\text{basal}})/\alpha$ and $S_{A,\text{basal}}/\alpha$ represent the previously mentioned $C_{A,\text{max}}$ and $C_{A,\text{min}}$, respectively.

As mentioned earlier, on certain occasions, the monotonousness of the control function may be violated. For example, it might happen that at low to medium concentrations a substance is a stimulator, while at high concentrations it is an inhibitor. Thus, the control function is nonmonotonous and can be written as a combination of Hill functions (Thomas *et al.*, 1990):

$$S_A(G) = a\frac{[G/T_1]^{n_1}}{[G/T_1]^{n_1} + 1}\frac{1}{[G/T_2]^{n_2} + 1}, \quad T_1 < T_2.$$

Next, assume that instead of one, two hormones control the secretion of A. We denote them by B and C with corresponding concentrations $C_B(t)$ and $C_C(t)$. The control function $S_A = S_A(C_B, C_C)$ depends on the specific interaction between A from one side, and B and C from another (Thomas *et al.*, 1990). For example, if both B and C stimulate the secretion of A

$$S_A(C_B, C_C) = a_B F_{\text{up}}(C_B) + a_C F_{\text{up}}(C_C) + S_{A,\text{basal}}, \tag{6}$$

if B and C act independently,

$$S_A(C_B, C_C) = aF_{\text{up}}(C_B)F_{\text{up}}(C_C) + S_{A,\text{basal}} \tag{7}$$

or, if B and C act simultaneously (the secretion of A requires the presence of both). On the other hand, if, for example, the secretion of A is stimulated by B, but suppressed by C, the control function can be introduced as

$$S_A(C_B, C_C) = aF_{\text{up}}(C_B)F_{\text{down}}(C_C) + S_{A,\text{basal}} \tag{8}$$

or

$$S_A(C_B, C_C) = a_B F_{\text{up}}(C_B) + a_C F_{\text{down}}(C_C) + S_{A,\text{basal}} \tag{9}$$

Note, that Eq. (8) simulates a noncompetitive and simultaneous action of B and C. If B and C compete as they control the secretion of A, the secretion term can be described with a modified Hill function:

$$S_A(C_B, C_C) = a\frac{(C_B/T_B)^{n_B}}{(C_B/T_B)^{n_B} + (C_C/T_C)^{n_C} + 1} + S_{A,\text{basal}}. \tag{10}$$

IV. Oscillations Driven by a Single System Feedback Loop

In this section, we discuss in detail networks with a single (delayed) feedback loop that can generate oscillatory behavior. We focus on 2-node/1-feedback networks, in which the concentration of one hormone A regulates the secretion of another hormone B, which in turn controls the release of A. This construct can generate oscillations, even if there is no explicit (direct) delay in the feedback.[1] However, in this case, the oscillations will fade to the steady state of the system. A nonzero delay and a large nonlinearity in the control functions (sufficiently high Hill coefficients) guarantee steady periodic behavior, because of the existence of a nontrivial limit cycle. On the other hand, a network may incorporate a single feedback loop by means of only one or more than two nodes. We comment on some peculiarities of such models in the last section.

A. Formal 2-node/1-Feedback Network

We study the abstract endocrine networks shown in Fig. 2. These particular examples anticipate that two hormones, A and B, are continuously secreted in certain pool(s) (systemic circulation, portal blood, etc.), where they are subject to elimination. The release of hormone B is up-(down-)regulated by hormone A. Hormone B itself modulates negatively (positively) the secretion of A. The A/B interactions are assumed to be dose responsive. The resulting delayed control loop is capable of driving hormone oscillations, if certain conditions (discussed later) are provided.

---

[1] The thresholds in the control functions provide implicit delays in the corresponding conduits.

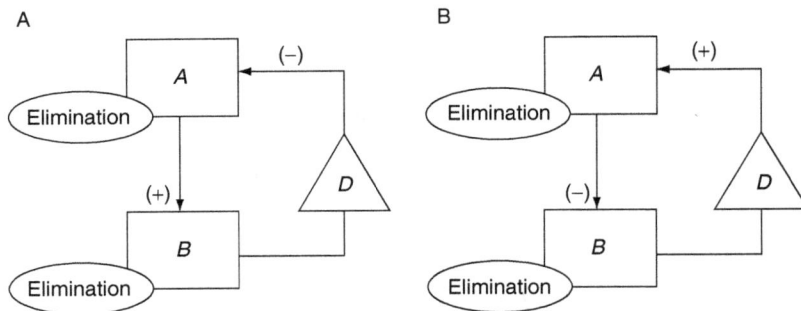

Fig. 2 Formal network of a two-node/one-feedback oscillator. (A) A network in which the main hormone B is stimulated; (B) a model in which B is inhibited. D denotes a delay in the interconnection. In both networks, A and B are subject to elimination.

To formalize the networks depicted in Fig. 2, we denote the concentrations of hormones A and B by $C_A(t)$ and $C_B(t)$, respectively. We assume that the elimination of each hormone is proportional to its concentration with positive constants α and β. The secretion rate S_A of A is supposed to depend on the history of the concentration of B and vice versa. In particular, we assume that $S_A(t) = S_A\{h_1[C_B(t)]\}$ and $S_B(t) = S_B\{h_2[C_A(t)]\}$. The functional h_1 (h_2) incorporates the lag in the action of B on A (A on B). To formally introduce the delays, one can account for the time-averaged effect of the hormone action in a past time interval related to the current moment (Keenan and Veldhuis, 2001a). However, this method requires two parameters for each delayed action—the onset and the termination of the delayed action (Keenan and Veldhuis, 2001a). Here, to keep the model as minimal as possible, we use a "direct" delay (with only one parameter for each delayed control action) and assume that the secretion control functions can be written as

$$S_A(t) = S_A[C_B(t - D_B)] \text{ and } S_B(t) = S_B[C_A(t - D_A)]$$

with some nonnegative delay times D_A and D_B. Then, the system of ordinary (nonlinear) delayed differential equations, which describes a formal two-node/one-feedback endocrine network (Fig. 2), has the form

$$\begin{aligned}
\frac{dC_A}{dt} &= -\alpha C_A(t) + S_A[C_B(t - D_B)], \\
\frac{dC_B}{dt} &= -\beta C_B(t) + S_B[C_A(t - D_A)]
\end{aligned} \tag{11}$$

with some elimination constants α, $\beta > 0$, lag times D_A, $D_B \geq 0$, and secretion rate control functions S_A, $S_B \geq 0$.

B. Reference Systems

To describe the dose-responsive relationships between A and B, corresponding to the network from Fig. 2A, we use the recommendations outlined in "Hormone release approximation" [Eq. (5)]. We write the control functions that appear in (11) as follows:

$$S_A[C_B(t - D_B)] = aF_{\text{down}}[C_B(t - D_B)] + S_{A,\text{basal}},$$
$$S_B[C_A(t - D_A)] = bF_{\text{up}}[C_A(t - D_A)] + S_{B,\text{basal}}.$$

With this special choice, the core system first-order nonlinear differential equations, describing the network from Fig. 2A, have the form

$$\frac{\mathrm{d}C_A}{\mathrm{d}t} = -\alpha C_A(t) + S_{A,\text{basal}} + a\frac{1}{[C_B(t - D_B)/T_B]^{n_B} + 1},$$
$$\frac{\mathrm{d}C_B}{\mathrm{d}t} = -\beta C_B(t) + S_{B,\text{basal}} + b\frac{[C_A(t - D_A)/T_A]^{n_A}}{[C_A(t - D_A)/T_A]^{n_A} + 1}. \tag{12}$$

The units in this model are as follows:

C_A, C_B, T_A, T_B	mass/volume,
$a, b, S_{A,\text{basal}}, S_{B,\text{basal}}$	mass/volume/time,
α, β	time^{-1},
D_A, D_B	time.

However, in the sequel, we avoid specifying the specific unit and the simulated profiles have arbitrary magnitude, which could be rescaled with ease to fit a desired physiology.

In most of the simulations, we assume no basal secretions and a direct action of A on B (no delay). This transforms the core equations [Eq. (12)] into

$$\frac{\mathrm{d}C_A}{\mathrm{d}t} = -\alpha C_A(t) + a\frac{1}{[C_B(t - D_B)/T_B]^{n_B} + 1},$$
$$\frac{\mathrm{d}C_B}{\mathrm{d}t} = -\beta C_B(t) + b\frac{[C_A(t)/T_A]^{n_A}}{[C_A(t)/T_A]^{n_A} + 1}. \tag{13}$$

Note, that solving these equations for $t \geq t_0$ requires the initial condition for C_B to be given on the entire interval $[t_0 - D_B, t_0]$.

From the special form of Eq. (13), we could easily derive that after some time (depending on the initial conditions), the solutions will be bounded away from zero and from above. More formally, for any $\varepsilon > 0$ (and we may choose ε as small as we like), there exists $t_0 > 0$ (depending on ε, the initial conditions, and the system parameters), such that for $t > t_0$ the following inequalities hold and provide upper and lower bounds on the solution of Eq. (13):

$$0 < \frac{a}{\alpha} \frac{1}{(b/[\beta T_B])^{n_B} + 1} - \varepsilon \le C_A(t) \le \frac{a}{\alpha} + \varepsilon,$$

$$0 < \frac{b}{\beta} \frac{1}{(T_A/\min\, C_A)^{n_A} + 1} - \varepsilon \le C_B(t) \le b/\beta + \varepsilon. \tag{14}$$

The upper bounds above are absolute system limits. For example, the model response to exogenous A-bolus cannot exceed the value b/β. However, as $C_A < a/\alpha$, we get from Eq. (14) that the actual endogenous peak concentration of B will never reach b/β. In fact, if there is no external input of energy in the system, it will be less than

$$C_B(t) \le \frac{b}{\beta} \frac{1}{(\alpha T_A/a)^{n_A} + 1} < \frac{b}{\beta}. \tag{15}$$

Hence, changes in four parameters (a, α, n_A, T_A) can model a difference between the maximal amplitude of the internally generated peaks and the eventual response to external stimulation. All estimates may be refined through a recurrent procedure inherent in the core system [Eq. (13)]. For example, one can combine the two inequalities Eq. (14) to get an explicit lower bound for C_B:

$$\frac{b}{\beta} \frac{1}{\left\{ \frac{\alpha T_A [(b/[\beta T_B])^{n_B} + 1]}{a} \right\}^{n_A} + 1} \le C_B(t). \tag{16}$$

Accordingly, we can use this to write an explicit upper bound for C_A:

$$C_A \le \frac{a}{\alpha} \frac{1}{\left(\frac{C_{B,\min}}{T_B} \right)^{n_B} + 1} \le \frac{a}{\alpha} \frac{1}{\left(\frac{M}{T_B} \right)^{n_B} + 1},$$

where

$$M = \frac{b}{\beta} \frac{1}{\left\{ \frac{\alpha T_A \left[\left(\frac{b}{\beta T_B} \right)^{n_B} + 1 \right]}{a} \right\}^{n_A} + 1}.$$

These inequalities can help determine reasonable values for the model parameters.

It is easy to see that (as the control functions are monotonously decreasing and increasing) the system Eq. (13) has a unique fixed point (steady state). It can be shown that if there is no delay ($D_A = D_B = 0$), the fixed point is asymptotically stable (a node or a focus) and attracts all trajectories in the phase space (Fig. 3A). However, even a single nonzero delay [as in Eq. (13)] might change the properties of the steady state. The particular stability analysis is nontrivial, and consists of investigating the real part of eigenvalues, which are roots of equation containing a transcendental term, involving the delay. In the examples that follow, we will encounter one of the two situations depicted in Fig. 3: the steady state will be

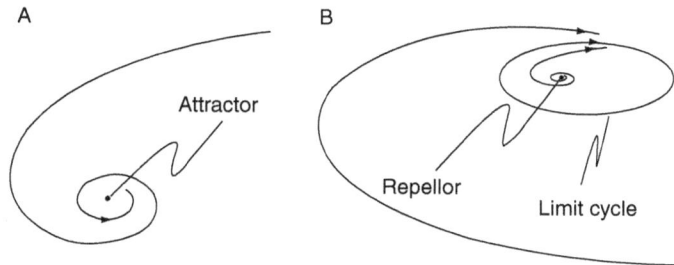

Fig. 3 Illustrative trajectories in the space (C_A, C_B) if the steady state is an attractor (A) or a repellor (B). In the latter case, a unique asymptotically stable periodic solution acts as a limit cycle and attracts all other trajectories (except the fixed point).

either an attracting focus (Fig. 3A) or a repellor (Fig. 3B), and in the latter case there will exist a unique asymptotically stable periodic solution (which encircles the fixed point in the phase space) acting as a global limit cycle by attracting all trajectories (except the one originating from the fixed point).

C. Oscillations Generated by a Periodic Solution

In this section, we present two specific examples describing the networks in Fig. 2. The core system of delayed ODE for the reference models will have unique periodic solution and unique repelling fixed point (Fig. 3B).

Consider a construct, described by the following core equations:

$$\frac{dC_A}{dt} = -1C_A(t) + 5\frac{1}{[C_B(t-3)/20]^2 + 1},$$
$$\frac{dC_B}{dt} = -2C_B(t) + 500\frac{[C_A(t)/5]^2}{[C_A(t)/5]^2 + 1}. \tag{17}$$

These equations simulate the network shown in Fig. 2A (A is a stimulator). The parameters were chosen to guarantee stable oscillations (Fig. 4). Later, we show how the parameter choice affects the periodicity.

Even in this simple example, we have a variety of possibilities to model the specific interactions between A and B. In the previous example, we have surmised the following:

1. The maximal attainable amplitude of C_B is 250.
2. The maximal attainable amplitude of C_A is 5.
3. The threshold T_A is higher than the endogenous levels of C_A.
4. The threshold T_B is approximately 6-fold lower than the highest endogenous levels of C_B.

It follows from 2 and 3 that the response of B to endogenous stimulation is not full. However, a high exogenous bolus of B elicits dose-dependent release of B

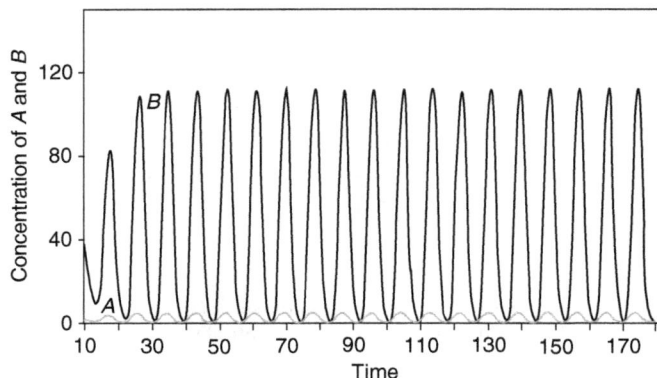

Fig. 4 Dynamics of the concentration of A (the lower profile) and B for the reference model described by Eq. (17).

secretion at levels higher than the typical endogenous B concentration. It is easy to see that because of 2, the maximal endogenous B concentration is less than 125. Because of the choice of T_B (see 4), B almost fully suppresses the release of A between pulses, which in turn results in low intervolley B secretion.

To simulate the network from Fig. 2B (A is an inhibitor), we use the following reference system of delayed ODEs:

$$\begin{aligned}
\frac{dC_A}{dt} &= -1C_A(t) + 50\frac{[C_B(t-3)/20]^2}{[C_B(t-3)/20]^2+1}, \\
\frac{dC_B}{dt} &= -2C_B(t) + 500\frac{1}{[C_A(t)/5]^2+1}.
\end{aligned} \tag{18}$$

The system parameter a in Eq. (17) was increased 10-fold [compared to Eq. (18)] to guarantee the existence of a periodic solution.

D. Simulation of Feedback Experiments

The success of a modeling effort is frequently measured by the capability of the construct to reproduce pivotal experiments. Accordingly, we discuss the correct way of modeling and the system reaction to three common experimental techniques, aimed to disclose the specific linkages within an endocrine system.

1. Antibody Infusion

The introduction of an antibody (Ab) to a certain substance, referred here as S, is generally accompanied by a transformation of S, which results in effectively removing S from the system. The corresponding rate depends on the specific chemical reaction between Ab and S, and increasing the elimination constant of

S (corresponding to the pool where Ab is administered) would model the removal. It remains possible that the reaction specifics change the single half-life pattern into a multiple half-life model. However, the single half-life approximation still might be sufficient in a variety of simulations.

To exemplify the idea, we simulated variable removal of the inhibitor A in the reference model described by Eq. (18). Three simulations were performed, in which the coefficient β was increased 2-fold (left), 6-fold (middle), or 15-fold (right) at time $t = 75$.

The plots in Fig. 5A capture a very interesting phenomenon predicted by the model: a decrease in the peak amplitudes of B, even though an inhibitor is removed from the system. In the current model, this is explained by the actual increase of the rate at which A initiates its rise and reaches its action threshold, which, in turn, promotes an earlier suppression of B secretion.

2. Sensitivity Modification

Modifying the profiles of the control function models alterations in system sensitivity. For example, if the sensitivity of a certain cell group depends on the number of opened receptors, we could simulate receptor blockage/stimulation via

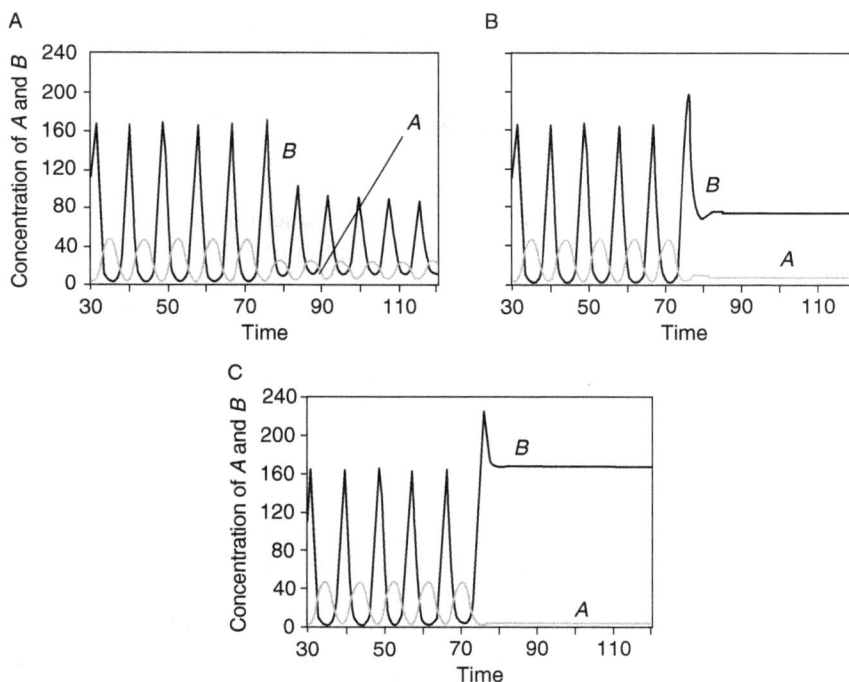

Fig. 5 Simulated variable infusion (starting at $t = 75$) of antibody to the inhibitor A in the reference model outlined in Eq. (18). The plots depict low (A), medium (B), or almost complete (C) removal of A.

changing the parameters of the corresponding control function. In the model described in Eq. (17), this would correspond to changes in the threshold, or in the Hill coefficient. Reducing (increasing) a threshold results in sensitivity increase (decrease). Changes in the Hill coefficient affect the slope of the control function. In general, increasing the Hill coefficient slightly changes the frequency and the amplitude, without affecting the pulsatility character of the profiles. In contrast, a decrease could effectively disturb the oscillations by preventing the system from overshooting the steady state.

We illustrate the effect of changing all thresholds and Hill coefficients in Eq. (17) (Fig. 6). An increase in n_B or n_A (Fig. 6A and C, left panels) produced a slight change in the frequency and amplitude. A decrease in n_B or n_A resulted in pulse shrinking (Fig. 6C, right panel) or in loss of pulsatility (Fig. 6A, right panel) if the control functions can no longer provide the necessary inertia for overshooting the steady-state value. Increasing T_B from 20 to 80 (Fig. 6B, right panel) results in a condition in which B cannot exert the necessary suppression on A. The concentration of B is limited from above and increasing its action threshold gradually obliterates the effect of the delay containing term. Decreasing T_B to 0.2 has no visual effect on the simulated profiles (Fig. 6B, left panel). The pulsatility is not affected because the suppressive action of B on A is not modified. It starts only somewhat earlier, but there is still a 3-h delay in this action, which, in this particular model, is sufficient to maintain oscillations. The analysis of the effect produced by changes in T_A is somewhat different. Both increasing and decreasing might affect the oscillations. When T_A is decreased, even a small amount of A is sufficient to produce a full response, which obliterates the pulsatility (Fig. 6D, left panel). The fact that the concentration of A is bounded from below independently of T_A is crucial [Eq. (14)]. Increasing T_A results in a left shift of the control function S_B, thus, preventing A from stimulating B, which in turn reduces the oscillations (Fig. 6D, right panel).

A more formal approach to explaining the reduction in the range of the oscillations (the "shrinking" of the profile) would consist of (recursive) application of the inequalities [Eq. (14)]. For example, from the right-hand side of Eq. (14), it is evident that if $T_A \to 0$, then $C_B \to b/\beta$ and if $T_B \to \infty$, then $C_A \to a/\alpha$.

3. Exogenous Infusion

The correct way to simulate exogenous infusion of a hormone, which is also a system node, would be to add an infusion term to the right-hand side of the corresponding ODE. This term should correspond to the infusion rate profile in the real experiment. Mathematically, it might be interpreted as a change in the basal secretion. In terms of the specific model described by Eq. (11), if we are simulating infusion of hormone B, the corresponding equation changes as follows:

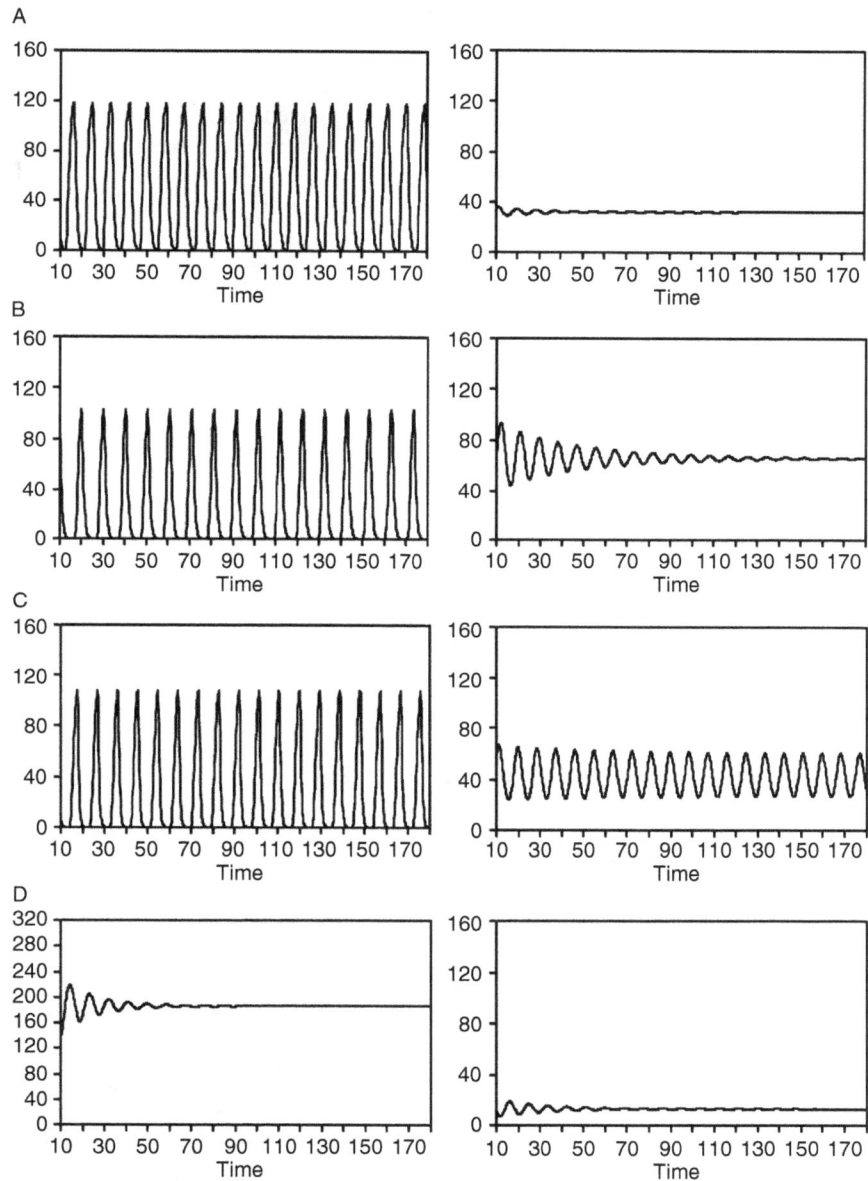

Fig. 6 Model response to alterations in system sensitivity. All profiles depict the dynamics of $C_B(t)$. (A) Changing n_B from 2 to 10 (left) and to 1 (right); (B) changing T_B from 20 to 0.2 (left) and to 80 (right); (C) changing n_A from 2 to 20 (left) and to 2/3 (right); (D) changing T_A from 5 to 1/40 (left) and to 15 (right).

$$\frac{dC_B}{dt} = -\beta C_B(t) + S_B[C_A(t - D_A)] + \inf(t), \tag{19}$$

where $\inf(t)$ is the infusion rate term. The solution of the previous equation is the sum of both endogenous and exogenous concentrations of B. To follow the distinction explicitly, a new equation should be added to the system:

$$\frac{dC_{inf}}{dt} = -\beta C_{inf}(t) + \inf(t)$$

and $C_B(t)$ has to be replaced by $C_B(t) + C_{inf}(t)$ in all model equations, except the one that describes the rate of change of the concentration of B. To sum up, the core equations are

$$\frac{dC_A}{dt} = -\alpha C_A(t) + S_A\{[C_B + C_{inf}](t - D_B)\},$$

$$\frac{dC_B}{dt} = -\beta C_B(t) + S_B[C_A(t - D_A)], \tag{20}$$

$$\frac{dC_{inf}}{dt} = -\beta C_{inf}(t) + \inf(t).$$

The model above [Eq. (20)] is in essence a 3-node/1-feedback construct, where exogenous B is the new node. A particular example, illustrating infusion simulation is shown later in this section (see "Identifying Nodes Controlling the Oscillations").

E. Oscillations Generated by a Perturbation

In the reference models from the previous section, the pulsatility was generated by a system that has a unique periodic solution and a unique fixed repelling point. The purpose of this section is to demonstrate that oscillations may occur as a result of disrupting a system that does not have a periodic solution, and its fixed point is an asymptotically stable focus (Fig. 3A).

We illustrate this concept on an earlier example. Figure 6B (right panel) depicts the profile of the solution to the following delayed ODE:

$$\frac{dC_A}{dt} = -1 C_A(t) + 5\frac{1}{[C_B(t - 3)/80]^2 + 1},$$

$$\frac{dC_B}{dt} = -2 C_B(t) + 500\frac{[C_A(t)/5]^2}{[C_A(t)/5]^2 + 1}. \tag{21}$$

The difference between this model and the reference construct [Eq. (17)] is in the 4-fold increase of the threshold T_B. In this case, there is no periodic solution and the unique fixed point attracts all trajectories in the phase space. Therefore, this system by itself cannot generate stable oscillations. However, if it is externally stimulated, it can be removed from its steady state and oscillations will be detected.

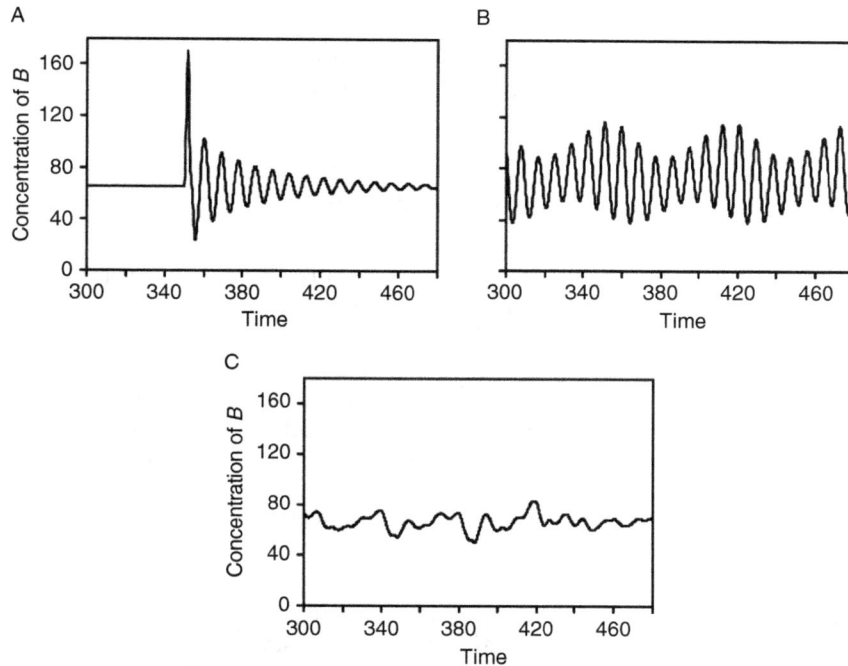

Fig. 7 Oscillations generated by perturbations of the system in Eq. (21). (A) A brief stimulation of the secretion of B at $t = 350$. The rest of the profiles depict external periodic (B) or random (C) control on the coefficient b, which determines the release of B.

For example, assume that at $t = 350$, the secretion of B was briefly stimulated. This removes the trajectory in the phase space away from the fixed point and the system would have enough energy to initiate another waning pulse sequence (Fig. 7A). Moreover, if we allow for some periodic external control on the secretion, the hormone profile displays sustained pulsatility with bursts of variable amplitude (Fig. 7B). The frequency of the pulses is controlled by the coefficients of the core system [Eq. (21)], while the peak amplitudes follow the external stimulus.

If the perturbation is random, it generates pulses of approximately the same frequency as in the previous cases, but with highly variable amplitudes. In the simulation presented in Fig. 7C, we superimposed 40% Gaussian noise on the parameter b. Even though some peaks cannot be detected, an overall pulse periodicity (the same as in Fig. 7A and B) is apparent.

In the previous examples, the perturbation was assumed to be external and independent of the core system. Later on, we show that a delayed system feedback could also provide enough energy and trigger oscillations in submodels with damped periodicity. In the three-node example from "Networks with Multiple Feedback Loops," a 2-node subsystem (with no direct delay in its feedback, and, therefore, without a periodic solution) is perturbed by a delayed system loop via the third node. This removes the whole system from its steady state and drives consecutive pulses during recurrent volleys.

F. Identifying Nodes Controlling the Oscillations

When hormone A cannot be measured directly and is an inhibitor (the network in Fig. 2B), we can test whether it is involved in generating the oscillations of B by neutralizing the action (A-receptor blocker) or by removing (antibody infusion) A from its action pool. On the other hand, if A is a stimulator (Fig. 2A), a large constant infusion of A should remove the oscillations (by exceeding the action threshold, resulting in continuous full response from the target organ). This concept is exemplified in Fig. 8, which depicts two computer-generated predictions for the system response to exogenous infusion of hormone A [assuming that A stimulates B, Eq. (18)]. We simulated constant low (Fig. 8A) and high (Fig. 8B) infusion of A by increasing the basal A-secretion from zero to two different levels, starting at $t = 75$.

The model predicts gradual pulse "shrinking" toward the current steady-state level. If the exogenous administration of A is sufficiently high (Fig. 8B), the pulses wane and the secretion becomes constant. The profiles in Fig. 8 depict the numerical solution (concentration of hormone B) of the system.

$$\frac{dC_A}{dt} = -C_A(t) + \text{Inf}(t) + 5\frac{1}{[C_B(t-3)/20]^2 + 1},$$

$$\frac{dC_B}{dt} = -2C_B(t) + 500\frac{[C_A(t)/5]^2}{[C_A(t)/5]^2 + 1} \tag{22}$$

with two different continuous infusion terms satisfying

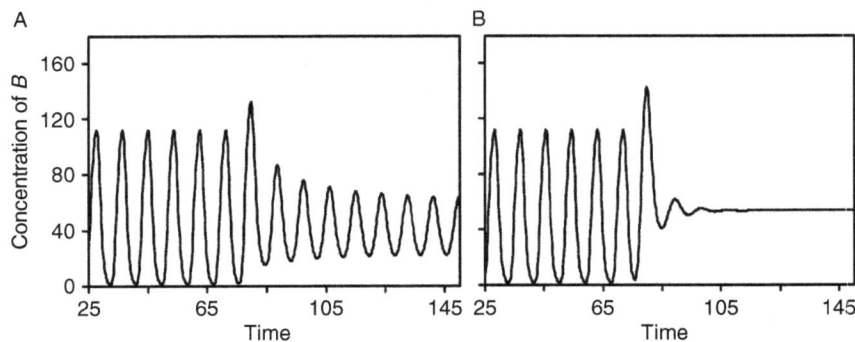

Fig. 8 System response [Eq. (22)] to exogenous infusion of A. The plots show simulation of constant low (A) and high (B) infusion of A starting at $t = 75$.

$$\text{Inf}(t) = \begin{cases} 0 & \text{if } t \le 75 \\ 1 \text{ or } 2 & \text{if } t \ge 76 \end{cases}.$$

The parameters and control functions were chosen arbitrarily to simulate a network like the one in Fig. 2A, which generates stable oscillations.

Almost identical results (Fig. 5) can be achieved by simulating partial or complete removal of A in the case when A is an inhibitor (the network from Fig. 2B). This should be done by increasing the rate of elimination of A to simulate additional removal due to infusion of antibody (see "Simulation of Feedback Experiments" for details).

However, these experiments cannot disclose whether A is actually involved in a feedback with B, or acts merely as a trigger to remove a certain subsystem from its steady state. For example, consider the two networks shown in Fig. 9 and suppose that only the concentrations of hormone B can be measured.

Assume that E stimulates B, and its removal obliterates the secretion of B. As E cannot be measured, we have no direct means to establish whether E is involved in a delayed feedback loop with B. Moreover, in both networks, constant high infusion of E (as proposed previously) removes the pulsatility and elicits constant secretion of B. Therefore, a more sophisticated experiment is required to reveal whether E is indeed involved in a feedback loop with B (Fig. 9A) or acts by perturbing the A-B subsystem (Fig. 9B). A possible approach would include blocking the endogenous E secretion with subsequent introduction of a single exogenous E bolus. The system response would be a single spike of B secretion, if the network were that depicted in Fig. 9A, or a waning train of several B pulses if the network is the one shown in Fig. 9B. Most importantly, the required suppression of endogenous E release must be achieved without affecting the putative A-B relationship.

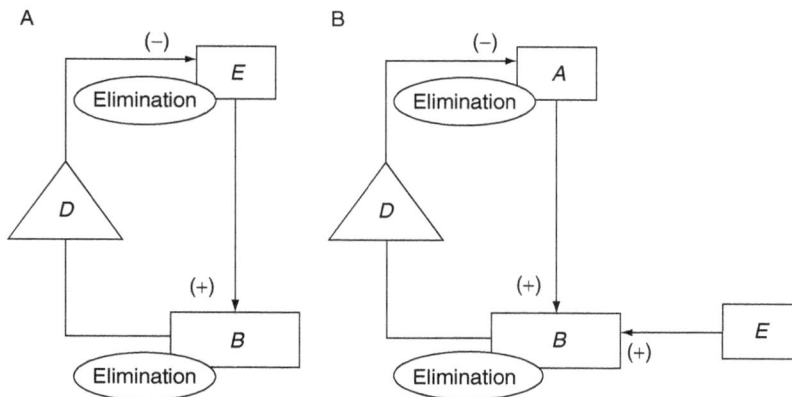

Fig. 9 Two hypothetical networks, in which a hormone E stimulates the secretion of B. E is either involved in a delayed feedback (A) or removes the subsystem A-B (B) from its steady state.

G. Separating Synthesis from Secretion

In certain cases, it would be appropriate to separate on a network level the hormone synthesis from its release. This would be important if a certain compound differently affects these processes. For example, let us consider again the network from Fig. 2A in an attempt to explain a rebound release of B following a withdrawal of continuous infusion of a certain substance C. Assume that during the infusion of C the release of B was suppressed and that we have evidence that C is not affecting the release of A. A possible explanation of the rebound phenomenon would be that C affects the release of B, but not its synthesis. However, as all conduits in the network are affected in this experiment, the intuitive reconstruction of all processes involved is not trivial. The simulation requires introduction of a "storage" pool in which B is synthesized and packed for release and another pool (e.g., circulation) in which B is secreted. This adds a new equation to the model, describing the dynamics of the concentration of B in the storage pool. The following assumptions would be appropriate:

1. The concentration of B in the storage pool (P_B) is positively affected by the synthesis and negatively affected by the release.
2. The concentration P_B exerts a negative feedback on the synthesis of B and cannot exceed a certain limit P_{max}.
3. The rate of release of B from the storage pool is stimulated by the storage pool concentration but might be inhibited by the concentration of B in the exterior.
4. B is subjected to elimination only after it is secreted.

To provide an abstract example, assume that in the network from Fig. 2A we have in addition to A and B a new substance C that inhibits the secretion (competing with A), but does not affect the synthesis of B (Fig. 10).

Using Eq. (10) as a suitable form for the "competitive" control function, we can describe the network by the following system of delayed ODEs:

$$
\begin{aligned}
\frac{dC_A}{dt} &= -\alpha C_A(t) + a\frac{1}{[C_B(t-D_B)/T_B]^{n_B}+1}, \\
\frac{dC_B}{dt} &= -\beta C_B(t) + b\frac{[C_A(t)/T_{A,1}]^{n_{A,1}}}{[C_A(t)/T_{A,1}]^{n_{A,1}}+[C_C(t)/T_C]^{n_C}+1}\frac{[P_B(t)/T_p]^{n_P}}{[P_B(t)/T_P]^{n_P}+1}, \\
\frac{dP_B}{dt} &= c(P_{max}-P_B)\frac{[C_A(t)/T_{A,2}]^{n_{A,2}}}{[C_A(t)/T_{A,2}]^{n_{A,2}}+1} \\
&\quad -b\theta\frac{[C_A(t)/T_{A,1}]^{n_{A,1}}}{[C_A(t)/T_{A,1}]^{n_{A,1}}+[C_C(t)/T_C]^{n_C}+1}\frac{[P_B(t)/T_P]^{n_P}}{[P_B(t)/T_P]^{n_P}+1}.
\end{aligned}
\tag{23}
$$

Here, for simplicity, we assumed that circulating B levels do not feed back on the secretion. This would correspond to a model with a much higher concentration in the storage pool than in the circulation. In the previous presentation, c controls the rate of A-stimulated synthesis of B. The parameter θ represents the ratio between

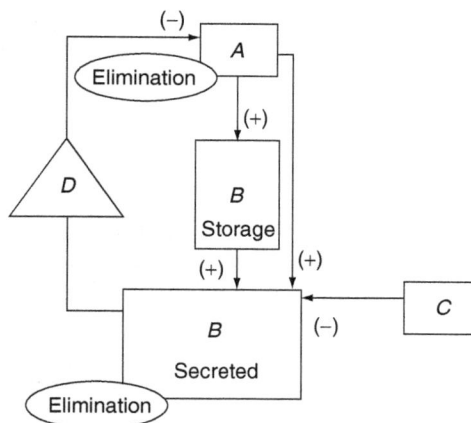

Fig. 10 Formal network depicting the system distinction between synthesis and release. C suppresses the release of B, but not its synthesis.

the volumes of the storage pool and the pool in which B is secreted. Typically, the second pool is larger and $\theta > 1$. We have supposed that the control functions, which correspond to the A-driven synthesis and release, are different with distinct thresholds $T_{A,1}$ and $T_{A,2}$, and corresponding Hill coefficients $n_{A,1}$ and $n_{A,2}$. The control, exerted on the secretion by the current concentration of B in the storage pool, is presented by the up-regulatory function $[P_B(t)/T_P]^{n_P}/\{[P_B(t)/T_P]^{n_P} + 1]\}$. The following values were assigned to the parameters that appear in Eq. (23):

$$\alpha = 1; \beta = 2; \theta = 6; a = 4; b = 4000; c = 2; P_{\max} = 1000;$$
$$T_{A,1} = 4; T_{A,2} = 3; T_B = 40; T_C = 10; T_P = 500;$$
$$n_{A,1} = 2; n_{A,2} = 2; n_B = 2; n_C = 2; n_P = 2.$$

The infusion term $C_C(t)$ is assumed to be a nonzero constant only during the time of infusion:

$$C_C(t) = \begin{cases} 0 & \text{if } t < 55 \\ 500 & \text{if } 56 < t < 95 \\ 0 & \text{if } t > 96. \end{cases}$$

The model output is shown in Fig. 11 and the plots clearly demonstrate a B rebound following the withdrawal of C (Fig. 11A).

During the infusion, the secretion of B is blocked, but not the synthesis and the concentration in the storage pool is elevated (Fig. 11B). The concentration of A increases (Fig. 11C), as low B levels cannot effectively block its release. Thus, the model explains the rebound jointly by the augmented concentration in the storage pool and the increased secretion of A.

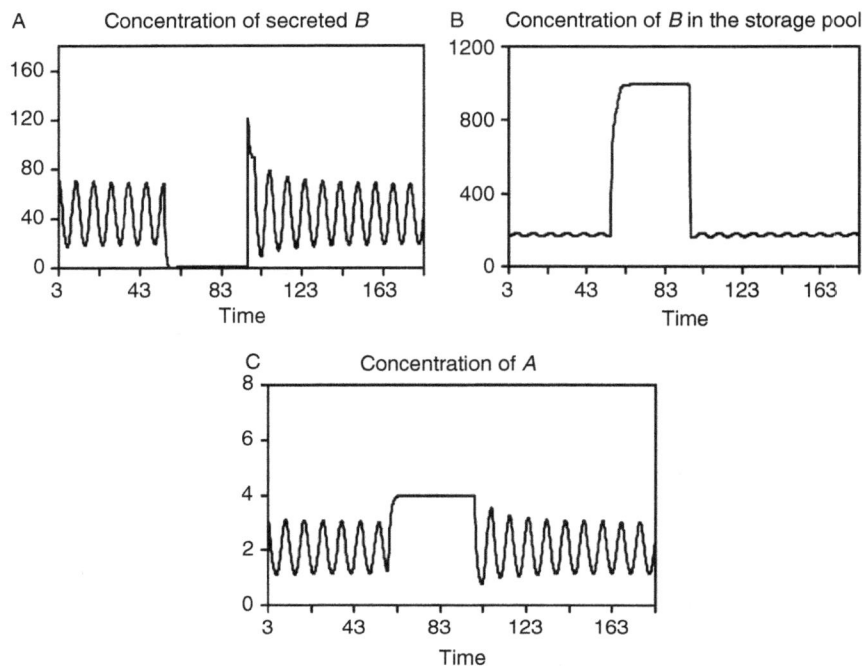

Fig. 11 Simulated rebound response following a withdrawal of continuous *C* infusion (timeline 55–95). (A) Concentration of secreted *B* (in the circulation). (B) Concentration of *B* in the storage pool. (C) *A*-concentration dynamics.

V. Networks with Multiple Feedback Loops

The available experimental data might suggest that the release of a particular hormone *B* is controlled by multiple mechanisms, with different periodicity in the timing of their action. This implies that probably more than one (delayed) feedback loops regulate the secretion of *B* and the formal endocrine network may include more than two nodes. In determining the elements to be included in the core construct, it is important to keep track on the length of the delays in the feedback action of all nodes of interest. For example, if the goal were to explain events recurring every 1–3 h, the natural candidates to be included in the formal network would be nodes, involved in feedback or feedforward relations with *B* with delays shorter than 3 h. Long feedback delays cannot account for high frequency events. In particular, if we hypothesize that a certain delayed feedback is responsible for a train of pulses in the hormone concentration profile, the direct delay must be shorter than the interpulse interval.

In this section, we briefly discuss some features of abstract endocrine networks, incorporating more than one delayed feedback loop. Each loop accounts for its

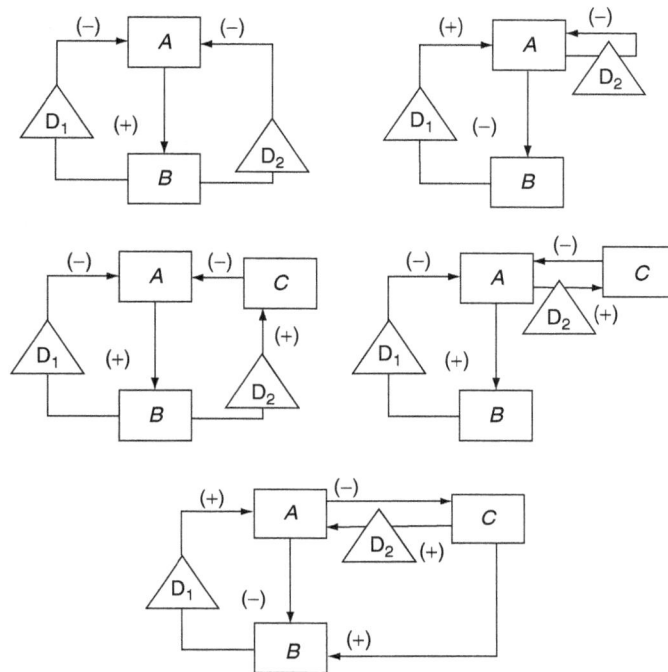

Fig. 12 Examples of hypothetical endocrine networks with more than one delayed feedback loops.

own oscillator mechanism and in what follows, we consider networks with two (delayed) feedback loops. Examples of 2-feedback constructs are shown in Fig. 12.

It should be noted that each of the two 3-node networks, shown in the middle panels of Fig. 12, could be reduced to its corresponding 2-node network from the top panels of Fig. 12. For example, let us consider the 3-node/2-feedback network shown in Fig. 12 (middle left panel). Assuming that both B and C can fully suppress the release of A, we can describe the formal network by the system of delayed ODE:

$$\frac{dC_A}{dt} = -3C_A(t) + 10000 \frac{1}{[C_B(t)/100]^3 + 1} \frac{1}{[C_C(t)/70]^{20} + 1},$$

$$\frac{dC_B}{dt} = -2C_B(t) + 6000 \frac{[C_A(t)/500]^{40}}{[C_A(t)/500]^{40} + 1}, \tag{24}$$

$$\frac{dC_C}{dt} = -3C_C(t) + 180 + 1320 \frac{[C_B(t-1.5)/200]}{[C_B(t-1.5)/200] + 1}.$$

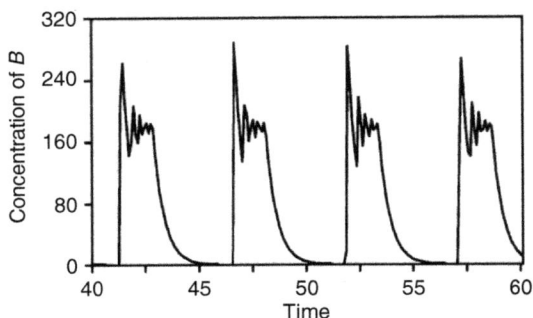

Fig. 13 Computer-generated output (concentration of *B*) of the core system Eq. (24).

Here, for simplicity, we have assumed that there is no delay in the feedback $B \rightarrow$ A. This system is capable of generating recurring multiphase volleys, by the mechanism described in "Oscillations Generated by a Perturbation" (Fig. 13).

However, analogous results can be achieved by reducing the 3-node network to a 2-node model with two feedbacks. In fact, the sequence of nodes and conduits $B \rightarrow$ $C \rightarrow A \rightarrow B$ is, in essence, a negative 2-node delayed feedback loop: $B \rightarrow A \rightarrow B$. Therefore, it can be modeled in the usual way (by simply removing C from the system). The reduced network is the one shown in the upper left panel of Fig. 12.

A corresponding simplified system of delayed ODEs could be

$$\frac{dC_A}{dt} = -3C_A(t) + 10000 \frac{1}{[C_B(t)/100]^3 + 1} \frac{1}{[C_B(t - 1.5)/50]^3 + 1},$$

$$\frac{dC_B}{dt} = -2C_B(t) + 6000 \frac{[C_A(t)/500]^4 0}{[C_A(t)/500]^{400} + 1}$$

and the model output (not shown), even without any special efforts to adjust the system parameters, is almost identical to the profile shown in Fig. 13.

Decreasing the number of equations from three to two reduces the number of parameters to be determined and the time needed for solving the equations numerically. Adding the third node in the formal network can be justified only if the goal is to simulate experiments involving C explicitly. And even then, the initial adjustment of the model would be significantly facilitated if C enters the system after the 2-node construct is validated.

Note that if the network is more complex, a reduction of the number of nodes might be impossible. For example, the network shown in Fig. 12 (lower panel) cannot be transformed into a 2-node model, because of the high system interconnectivity. We comment more on this in the next section.

VI. Summary and Discussion

The mathematical methods presented in this chapter are tailored to quantitatively interpret formal endocrine networks with (delayed) feedbacks. The main goal is to illustrate different conditions, under which oscillations can emerge.

The formal network itself consists of nodes and conduits, and is on the basis of a qualitative analysis of available experimental data (Friesen and Block, 1984). In our presentation, the nodes are hormone concentrations in abstract pools, in which hormones are released or synthesized, under the control of other hormones. The conduits specify how the nodes interact within the network. The quantitative analysis of the formal network is on the basis of approximation of the rate of change of a single system node. This essentially means that the dynamics of the hormone concentration is described with a single (delayed) ODE. To this end, we assume that the rate of change of hormone concentration depends on two processes—secretion and ongoing elimination. We work with a single half-life elimination model and express the control of the secretion as a combination of sigmoid Hill functions, depending on the other system nodes. The derivation of the ODE is demonstrated, along with a brief analysis of the properties of its solution to facilitate the actual determination of all system parameters.

The formal network is then interpreted as a dynamic system by combining all ODEs that describe system nodes dynamics. We exemplify the ideas on a 2-node/1-feedback model—one of the simplest meaningful examples of a network capable of generating and sustaining periodic behavior. In fact, a variety of systems display oscillatory behavior, driven by a single feedback loop. The simplest case is a 1-node/1-feedback network, in which a hormone after being secreted suppresses its own release, immediately or after some lag time. This system can generate periodic behavior only if the delay in the feedback is greater than zero. We do not discuss this case here.

A network may incorporate a single feedback loop in a more complex way, for example, via a combination of two or more nodes. For example, simple stability analysis of the steady state shows that a 3-node/1-feedback network is capable of sustaining periodicity even without a delay in the feedback loop and with relatively low Hill coefficients (Richelle, 1977; Thomas *et al.*, 1990). However, for a variety of practical cases, it is feasible to reduce the 3-node/1-feedback network to a 2-node/1-feedback construct as shown in the previous section.

Some specifics in endocrine network modeling are exemplified on two 2-node/1-feedback networks, in which the concentration of one hormone regulates the secretion of another, which in turn controls the release of the first hormone. This construct could generate oscillations even if there is no explicit delay in the feedback. However, it will be a damped periodicity, as the oscillations will fade and approach the steady state of the system. In contrast, a nonzero delay combined with a sufficiently large non-linearity in the control functions (high Hill coefficients) guarantees steady periodic behavior, as all trajectories approach a nontrivial limit cycle.

We relate all parameters to their physiological meaning and analyze the solutions to our reference systems, which always have only one fixed point (steady state), which is either a repellor or an attractor (Fig. 3). In the first case, the system has a unique limit cycle—a periodic solution, which attracts all trajectories in the phase space and, thereby, generates stable periodic behavior (Fig. 4). In the second case, the steady state is a focus and attracts all trajectories in the phase space. Therefore, the construct displays damped periodic behavior. In particular, if it is in a state close to the fixed point, an external perturbation initiates a waning train of pulses (Fig. 7A). Therefore, oscillations might be generated even by a system that does not have a periodic solution, and its fixed point is an asymptotically stable focus. However, an external source of perturbations must be provided. Note that the frequency of the oscillations is largely independent of the external perturbation (Fig. 7).

We use the two reference systems to illustrate the modeling of three common experimental techniques: infusion of antibody to one of the nodes, sensitivity alterations, and exogenous infusion of one of the system hormones. We comment on the correct way to perform these approximations and examine the corresponding model response. In particular, the simulations illustrate conditions that might disrupt the periodicity.

Increasing the elimination rate of a hormone simulates infusion of antibody and almost a complete removal of one of the nodes, and results in loss of periodicity (Fig. 5). Changes in the profiles of the control functions model alterations in system sensitivity. The analysis shows that if a model has a stable periodic behavior, the increase in one of the Hill coefficients would not change the system performance (Fig. 6A and C, left panels; Glass and Kauffman, 1973). On the other side, a decrease in the same parameter may transform the steady state from a repellor into an attractor and affect the periodic behavior. Changes in the action thresholds may also affect the periodicity (Fig. 6B and D). Exogenous infusion can be simulated by a simple increase in the basal secretion or by introducing a third node, in case we would like to distinguish between exogenous infusion and endogenous secretion of one and the same substance [Eqs. (19) and (20)].

We illustrate how these experiments may be used to disclose whether a certain hormone A is involved in generating the oscillations of another hormone B. The idea is to alter A in such way that the periodic B-profile is transformed into a constant nonzero secretion. When A inhibits B, we can neutralize its action (receptor blocker) or remove (antibody) it from the system. In the later case, the model predicts that the periodicity disappears and is replaced by a stable B secretion (Fig. 8). Alternatively, if A stimulates B, a large continuous A infusion obliterates the oscillations by exceeding the action threshold and eliciting an unvarying full B response from the target organ (Fig. 5). Additionally, the model provides means to disclose whether A is actually involved in a feedback loop with B or generates oscillations by perturbing another subsystem (see "Identifying Nodes Controlling the Oscillations").

To be able to capture a variety of feedback systems, we separate on a network level the hormone synthesis from its release. The proper simulation requires a new "storage" pool in which the hormone is synthesized and stored, in addition to the pool, in which the hormone is secreted. We used this distinction to provide a plausible explanation of a rebound release, following withdrawal of an agent that suppresses the secretion, but not the synthesis.

We would like to emphasize the importance of keeping the model as minimal as possible while performing the initial qualitative analysis of the available experimental data. In general, formal endocrine networks might incorporate multiple feedbacks loops and nodes. However, long feedback delays cannot account for high-frequency events. Therefore, if the model attempts to explain pulses of a hormone that recur every H hours, it might be sufficient to include in the formal network only feedback loops with delay shorter than H. Moreover, if a feedback loop enters the network via a multiple-node subsystem, it might be possible to reduce the number of nodes and simplify the model without affecting its performance. The example provided in the previous section demonstrates a case in which we could safely remove a "passive" node from a feedback loop and still retain the overall periodic behavior. It should be noted, however, that decreasing the complexity of a model typically leads to a reduction in the number of model parameters (degrees of freedom). As a result, the new simplified model may not always have the same behavior as the more complex construct. Therefore, after a node has been removed, the performance of the system has to be reestablished in order to verify that the parameters of the reduced model can be readjusted in such a way that the new construct can explain the same experimental observations already shown to be reconstructed by the older network.

Unfortunately, we cannot always reduce complex networks. The model shown in Fig. 12 (lower panel) is an example in which the system interconnectivity would not allow any simplification. Complex networks with intertwined feedback loops are considered elsewhere (Farhy et al., 2002; Keenan and Veldhuis, 2001b) and their analysis strongly depends on the specific physiology. It should be noted that in this chapter we do not consider more complicated cases, such as networks that have multiple steady states of a different type, which is a significant complication. Such systems can be approached in the early stage of their analysis by Boolean formalization (Thomas, 1973, 1983), which serves as an intermediate between modeling phases 2 and 3 described in the first section. This method describes complex systems in simple terms and allows for preliminary finding of all stable and unstable steady states.

Acknowledgments

I acknowledge support by NIH Grants K25 HD01474R21 AG032555, and R01 DK082805 and would like to thank my mentor, Dr. Johannes Veldhuis, for intellectual support and guidance.

References

Chen, L., Veldhuis, J. D., Johnson, M. L., and Straume, M. (1995). *In* "Methods in Neurosciences," p. 270. Academic Press, New York.

Farhy, L. S., and McCall, A. L. (2009a). System-level control to optimize glucagon counterregulation by switch-off of α-cell suppressing signals in β-cell deficiency. *J. Diabetes Sci. Technol.* **3**(1), 21–33.

Farhy, L., and McCall, A. (2009b). Pancreatic network control of glucagon secretion and counter-regulation. *Methods Enzymol.* **467**, 547–581.

Farhy, L. S., and Veldhuis, J. D. (2003). Joint pituitary-hypothalamic and intrahypothalamic autofeed-back construct of pulsatile growth hormone secretion. *Am. J. Physiol.* **285**, R1240–1249.

Farhy, L. S., and Veldhuis, J. D. (2004). Putative growth hormone (GH) pulse renewal: Periventricular somatostatinergic control of an arcuate-nuclear somatostatin and GH-releasing hormone oscillator. *Am. J. Physiol.* **286**(6), R1030–R1042.

Farhy, L. S., and Veldhuis, J. D. (2005). Deterministic construct of amplifying actions of ghrelin on pulsatile growth hormone secretion. *Am. J. Physiol. Regul. Integr. Comp. Physiol.* **288**, R1649–R1663.

Farhy, L. S., Straume, M., Johnson, M. L., Kovatchev, B. P., and Veldhuis, J. D. (2001). *Am. J. Physiol. Regul. Integr. Comp. Physiol.* **281**, R38.

Farhy, L. S., Straume, M., Johnson, M. L., Kovatchev, B. P., and Veldhuis, J. D. (2002). *Am. J. Physiol. Regul. Integr. Comp. Physiol.* **282**, R753.

Farhy, L. S., Bowers, C. Y., and Veldhuis, J. D. (2007). Model-projected mechanistic bases for sex differences in growth-hormone (GH) regulation in the human. *Am. J. Physiol. Regul. Integr. Comp. Physiol.* **292**, R1577–R1593.

Farhy, L. S., Du, Z., Zeng, Q., Veldhuis, P. P., Johnson, M. L., Brayman, K. L., and McCall, A. L. (2008). Amplification of pulsatile glucagon secretion by switch-off of α-cell suppressing signals in Streptozo-tocin (STZ)-treated rats. *Am. J. Physiol. Endocrinol. Metab.* **295**, E575–E585.

Friesen, O., and Block, G. (1984). *Am. J. Physiol.* **246**, R847.

Glass, L., and Kauffman, S. A. (1973). *J. Theor. Biol.* **39**, 103.

Keenan, D. M., and Veldhuis, J. D. (2001a). *Am. J. Physiol.* **280**, R1755.

Keenan, D. M., and Veldhuis, J. D. (2001b). *Am. J. Physiol.* **281**, R1917.

Mikawa, T., Masui, R., and Kuramitsu, S. (1998). *J. Biochem.* **123**(3), 450.

Richelle, J. (1977). *Bull. Cl. Sci. Acad. R. Belg.* **63**, 534.

Robeva, R. S., Kirkwood, J. R., Davies, R. L., Farhy, L. S., Johnson, M. L., Kovatchev, B. P., and Straume, M. (2008). "An Invitation to Biomathematics" and "Laboratory Manual of Biomathematics", Academic Press, San Diego, CA.

Thomas, R. (1973). *J. Theor. Biol.* **42**, 563.

Thomas, R. (1983). *Adv. Chem. Phys.* **55**, 247.

Thomas, R., D';Ari, R., and Thomas, N. (1990). "Biological Feedback," CRC Press, Boca Raton, FL.

Veldhuis, J. D., and Johnson, M. L. (1992). *Methods Enzymol.* **210**, 539.

Vrzheshch, P. V., Demina, O. V., Shram, S. I., and Varfolomeev, S. D. (1994). *FEBS Lett.* **351**(2), 168.

Wagner, C., Caplan, S. R., and Tannenbaum, G. S. (1998). *Am. J. Physiol.* **275**, E1046.

CHAPTER 15

Boundary Analysis in Sedimentation Velocity Experiments

Walter F. Stafford III[*,†]

[*]Analytical Centrifugation Research Laboratory
Boston Biomedical Research Institute
Boston, MA, USA

[†]Department of Neurology
Harvard Medical School
Boston, MA, USA

337
DOI: 10.1016/B978-0-12-384997-7.00015-7

I. Update

Time derivative sedimentation velocity analysis methods have been used by many investigators since their introduction in 1992. A complete treatment of the theoretical background of the methods was published in 2000 (Stafford, 2000) and represents an extension of this chapter to reversibly interacting systems as well as kinetically limited reversibly interacting systems. A detailed description of laboratory procedures using these methods has been published (Stafford, 2003). The basic concepts of time derivative analysis have been extended to curve fitting methods embodied in a software called SEDANAL that is used to analyze sedimentation velocity data from interacting and noninteracting systems and published by Stafford and Sherwood (2004) as well as in an analysis software called SVEDBERG by Philo (2000). Philo has also published improved methods of time derivative analysis (Philo, 2006). An extensive analysis of the effects of low concentrations of intermediates in multiple step reactions has recently been published by Correia *et al.* (2009) (Meth Enz 467). A general review of sedimentation analysis of interacting ideal and nonideal systems using these methods has been published recently (Stafford, 2009).

II. Introduction

Measurement of the sedimentation velocity of macromolecular particles was the first type of analysis to which the analytical centrifuge was applied (Svedberg and Nichols, 1923; Svedberg and Rinde, 1923, 1924). This chapter describes briefly the earlier methods of analysis of sedimentation velocity data that were used before the advent of digital computers. Then, using this discussion as a point of reference, it proceeds to describe approaches that have become practical because of the availability of computers. It then goes on to describe newer techniques that have become possible through the introduction of modern on-line digital data acquisition systems. Special emphasis is placed on techniques employing the time derivative of the concentration distribution (Stafford,1992, 1992a,b).

Traditionally, sedimentation transport experiments have been observed in two main ways, as either the concentration or concentration gradient as a function of radius depending on the type of optical system employed. The schlieren optical system displays the boundary in terms of refractive index gradient as a function of radius; the Rayleigh optical system, in terms of refractive index as a function of radius; and the absorption optical system, in terms of optical density as a function of radius. This chapter concerns itself mainly with treatment of data obtained with the latter two types of systems as the acquisition and analysis of data from these systems can be nearly completely automated.

Subjects to be discussed are techniques for smoothing and differentiation as well as a new analysis technique that uses the time derivative of the concentration profile. Use of the time derivative results in an automatic baseline elimination with a consequent increase in accuracy. Combination of the time derivative with an averaging procedure has resulted in an increase of 2–3 orders of magnitude in precision. The time derivative technique produces what is referred to as an apparent sedimentation coefficient distribution function, $g(s^*)$, where the symbols are defined below. The function $g(s^*)$ versus s^* is very nearly geometrically similar to the corresponding schlieren pattern that represents dn/dr versus r (where n is the refractive index and r is the radius) and, therefore, can be used in any situation that one would have used a schlieren pattern. The advantage of using the averaged $g(s^*)$ is that it has about 2 orders of magnitude higher signal-to-noise ratio than dn/dr and therefore, can be used to study interacting systems that previously were inaccessible to either the absorbance or Rayleigh optical systems. Extrapolation procedures for minimizing the effects of diffusion on the resolution of boundaries are also discussed briefly.

For convenience, we refer to the data from either system in terms of concentration rather than optical density or refractive index. However, various averages computed from these data will be weighted according to either the extinction coefficients or the refractive index increments of the various components in a mixture depending on the optical system employed.

The concentration data acquired by either optical system can be thought of as composed of two main parts: a time-dependent part that is due to the transport of macromolecule and a time-independent background part due to inhomogeneities in the optical system, detector, and cell windows. If the concentration is sufficiently high, the background part often can be ignored. However, in those cases for which the background contribution is a significant fraction of the signal, a correction for the background must be carried out by some means. The usual ways have been to perform a separate run without protein in the cell or to allow the material to be pelletized completely before taking a background scan. After the sedimentation run has been completed, the background contribution is subtracted from each scan. The first method is usually satisfactory as long as there have been no changes in the background between the main scans and the background scan. The second method requires that there be no slowly sedimenting material. Accumulation of dirt from drive oil, fingerprints, or the effects of moisture condensation, for example, could be sources of variation in the background between the main run and the background run. Ideally, a background run should be performed both before and after the main run to determine whether there have been any changes during the run. The importance of a background correction becomes evident if the concentration data are to be treated by numerical differentiation to produce concentration gradient curves. Variations in the background can often be of the same magnitude as those of the concentration and can obscure the true gradient curves.

III. Methods of Data Acquisition

Traditionally, acquisition of data from the schlieren and Rayleigh optical systems has been done by photography on glass plates or Estar thick based film. Subsequent analysis of the photographs was carried out visually using an optical microcomparator with manual logging of the data. Various schemes for automating this process have been devised. A particularly useful procedure was devised by DeRosier et al. (1972) for analysis of Rayleigh interferograms. Essentially, this procedure, along with its various derivatives, uses a Fourier transform analysis to extract the phase changes, and hence the concentration changes, associated with deflection of fringes in interferograms. A second advance was the introduction of on-line video photography to eliminate the wet photography steps. With the introduction of charge-coupled device (CCD) video cameras, Laue (1981) has devised an on-line video acquisition system that has become the basis for many current video systems. A video acquisition system for the Beckman Instruments (Palo Alto, CA) Model E analytical ultracentrifuge that allows very rapid capture and analysis of the whole cell image of Rayleigh interferograms every 4 s has been devised by Liu and Stafford (1992) and allows the rapid data acquisition necessary for time derivative analysis with multicell rotors. For example, images from five cells in a Model-E AN-G rotor can be acquired, converted to fringe displacement as a function of position, and the results stored on disk every 20 s.

The Spinco Division of Beckman Instruments has introduced the Optima XL-A analytical ultracentrifuge equipped with modern UV scanning optics. Profiles of absorbance as a function of radius are automatically acquired and stored as data files readable by other software.

IV. Measurement of Transport in Analytical Ultracentrifuge

The basis for the analysis of transport in the ultracentrifuge is the continuity equation presented by Lamm (1929)

$$\left[\frac{\partial c(r,t)}{\partial t}\right]_r = \frac{1}{r}\frac{\partial}{\partial r}\left\{Dr\left[\frac{\partial c(r,t)}{\partial r}\right]_t - \omega^2 r^2 sc(r,t)\right\}_t,$$

where c is the concentration as a function of radius and time, r is the radius, t is time, s is the sedimentation coefficient, D is the diffusion coefficient, and ω is the angular velocity of the rotor. Flux, J, because of sedimentation alone is given by the product of the velocity of the particles and their concentration:

$$J_{\text{sed}} = \frac{dr}{dt}c = \omega^2 src.$$

Flux because of diffusion alone is given by Stokes' first law and is proportional to the concentration gradient:

$$J_{\text{diff}} = -D\left(\frac{dc}{dr}\right)_t.$$

V. Traditional Methods of Analysis

A detailed account of the traditional manual methods of data analysis is not given as they are adequately represented in the literature (Chervenka, 1973; Schachman, 1959; Svedberg and Pederson, 1940) and have been supplanted mainly by methods made practical by the introduction of computers. For sedimentation velocity experiments using the schlieren optical system, the main piece of information obtained was the radial position of the peak in the refractive index gradient curve as a function of time. The sedimentation coefficient is obtained from the slope of a plot of the logarithm of the radial position of the peak versus time. Similarly, for the UV photoelectric scanner and the Rayleigh optical systems, the logarithm of the radial position of the midpoint of the sedimenting boundary is plotted versus time. However, neither of these methods will give highly accurate results unless the boundary is symmetrical; therefore, they can be applied rigorously only to a hydrodynamically ideal, homogeneous, monodisperse solution.

The correct method for obtaining accurate estimates of the sedimentation coefficient for cases that exhibit asymmetric or polymodal boundaries is the so-called second moment method in which an equivalent boundary position is computed. It gives the position of a hypothetical boundary that would have been observed in the absence of diffusion and polydispersity (Fig. 1). The rate of movement of the equivalent boundary position gives the weight average sedimentation coefficient for the system and corresponds to the weight average velocity of particles in the plateau region at the plateau concentration. The equivalent boundary position can be computed in several different ways depending on the form of

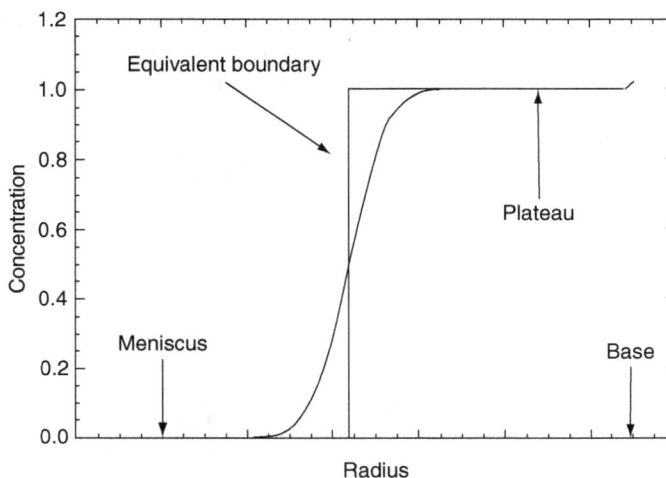

Fig. 1 Hypothetical sedimentation boundary showing the meniscus, equivalent boundary position, plateau, and base of the cell.

the data (Chervenka, 1973; Fujita, 1976; Schachman, 1959). The term second moment is generally applied to data obtained with the schlieren optical system from which the equivalent boundary position is obtained as the second moment of the refractive index gradient curve:

$$\langle r^2 \rangle = \frac{\int_{r_m}^{r_p} \frac{dc}{dr} r^2 dr}{\int_{r_m}^{r_p} \frac{dc}{dr} dr}.$$

However, when the data are obtained as concentration versus radius instead of concentration gradient versus radius, it is convenient to compute the equivalent position from the following relation:

$$\langle r^2 \rangle = \frac{\int_{r_m}^{r_p} r^2 dc}{\int_{r_m}^{r_p} dc}.$$

After obtaining the equivalent boundary position at various times, the weight average sedimentation coefficient is computed from the slope of a plot of the logarithm of the equivalent boundary position versus time. The slope of a plot $\ln(\langle r^2 \rangle)$ versus t is $2\omega^2 s_w$, where s_w is the weight average sedimentation coefficient. It should be noted that this method is valid only if the concentration gradient is zero at r_p. An appropriate baseline correction is required for high accuracy.

VI. Transport Method

The transport method is used in cases for which diffusion does not allow the boundary to move away completely from the meniscus. It is similar in computational complexity to the equivalent boundary method. Essentially, one measures the amount of material between the meniscus and some point in the plateau region that is removed by sedimentation during a given time interval. This amount can be converted to J_{sed}, and, with knowledge of the plateau concentration, one can compute the sedimentation coefficient. The validity of this method also depends on a negligible gradient in the plateau region. The relation may be derived in the following way (for other derivations, see Schachman 1959; Fujita, 1976).

By integrating the Lamm equation,

$$\left[\frac{\partial c(r,t)}{\partial t} \right]_r = \frac{1}{r} \frac{\partial}{\partial r} \left\{ Dr \left[\frac{\partial c(r,t)}{\partial r} \right]_t - \omega^2 r^2 s c(r,t) \right\}_t$$

we have

$$\int_{r_m}^{r_p} \left[\frac{\partial c(r,t)}{\partial t} \right]_r r dr = \left\{ Dr \left[\frac{\partial c(r,t)}{\partial r} \right]_t - \omega^2 r^2 s c(r,t) \right\}_{r_m}^{r_p}$$

and now, after bringing the differential operator outside of the integral, we have

$$\frac{\partial}{\partial t}\left[\int_{r_{\mathrm{m}}}^{r_{\mathrm{p}}} c(r,t)r\mathrm{d}r\right]_r = \left\{Dr\left[\frac{\partial c(r,t)}{\partial r}\right]_t - \omega^2 r^2 sc(r,t)\right\}_{r_{\mathrm{m}}}^{r_{\mathrm{p}}}$$

and as there is no net transport across the meniscus, we have the condition that $\{Dr_{\mathrm{m}}(\partial c(r_{\mathrm{m}},t)/\partial r)_t - \omega^2 r_{\mathrm{m}}^2 sc(r_{\mathrm{m}},t)\} = 0$ at r_{m}. We also have $\partial c/\partial r = 0$ at r_{p}, giving us an expression for mass transport from the region between r_{m} and r_{p} across the cylindrical surface at r_{p}:

$$\frac{\mathrm{d}}{\mathrm{d}t}\left[\int_{r_{\mathrm{m}}}^{r_{\mathrm{p}}} c(r,t)r\mathrm{d}r\right] = -\omega^2 r_{\mathrm{p}}^2 sc(r_{\mathrm{p}},t).$$

If we designate the integral as

$$Q(t) = \int_{r_{\mathrm{m}}}^{r_{\mathrm{p}}} c(r,t)r\mathrm{d}r$$

and note that $c(r_p, t) = c_0 \exp(-2\,\omega^2\,s_w t)$ where c_0 is the initial concentration, then we have

$$\frac{\mathrm{d}Q}{\mathrm{d}t} = -\omega^2 r_{\mathrm{p}}^2 sc_0 \exp(-2\omega^2 st).$$

By rearranging and integrating and as

$$Q(t=0) = c_0 \int_{r_{\mathrm{m}}}^{r_{\mathrm{p}}} r\mathrm{d}r = \frac{c_0}{2}(r_{\mathrm{p}}^2 - r_{\mathrm{m}}^2)$$

we have

$$Q(t) = \frac{c_0 r_{\mathrm{p}}^2}{2} \exp(-2\omega^2 st) - \frac{c_0 r_{\mathrm{m}}^2}{2}.$$

By rearranging and taking the logarithm of both sides, we arrive at

$$\ln\left[\frac{2Q(t)}{c_0 r_{\mathrm{p}}^2} + \left(\frac{r_{\mathrm{m}}}{r_{\mathrm{p}}}\right)^2\right] = -2\omega^2 st.$$

A plot of the left-hand side versus t will have a slope of $-2\omega^2 s$.

For a polydisperse system, the value of s obtained is the weight average sedimentation coefficient, s_w, for the mixture, and c_0 is the total initial concentration. Note again that this method requires that there be no gradient in the plateau region at r_p. It also requires a knowledge of c_0.

A. Transport Method Using Time Derivative

A simple variant of the transport method can be used if the time derivative of the concentration is available (Stafford, 1994). Starting again with the Lamm equation, by rearranging and integrating, but now keeping the differential operator inside the integral, we have

$$\int_{r_m}^{r_p} \left(\frac{\partial c}{\partial t}\right)_r r\,dr = \left[Dr\left(\frac{\partial c}{\partial r}\right)_t - \omega^2 r^2 sc\right]_{r_m}^{r_p}.$$

Noting again that transport across the meniscus is zero and that $\partial c/\partial r = 0$ in the plateau region at r_p, we have after rearranging

$$s = \frac{1}{-\omega^2 r_p^2 c_p} \int_{r_m}^{r_p} \left(\frac{\partial c}{\partial t}\right)_r r\,dr.$$

Again this approach requires that there be no gradient in the plateau region at r_p. The values of $\partial c/\partial t$ as a function of r can be estimated with sufficient accuracy by subtracting concentration profiles closely spaced in time so that $(c_2 - c_1)/(t_2 - t_1)$ at each radial position can be used in place of $\partial c/\partial t$, and $(c_{p,1} + c_{p,2})/2$ in place of c_p, where the numerical subscripts refer to the two scans. For numerical computation, this equation can be recast as

$$s = \frac{2}{-\omega^2 r_p^2 (c_{p,1} + c_{p,2})(t_2 - t_1)} \sum_{r_m}^{r_p} (c_2 - c_1)_i r_i \Delta r.$$

Again, for a polydisperse system, the procedure gives the weight average sedimentation coefficient, s_w. This method is insensitive to noise because the integration averages out the random noise and is unaffected by the baseline contributions except for the determination of c_p.

VII. Smoothing and Differentiating

To obtain a usable derivative curve, dc/dr, from the concentration profile by numerical differentiation, some degree of smoothing usually will be required. This section discusses some techniques used successfully in our laboratory. There are many variations of these techniques that will also work well if applied properly. The particular methods discussed below are presented because they are simple to use and produce reasonably good results.

Although smoothing will always introduce some degree of systematic dispersion and therefore should be carried out with caution, it can be justified in many cases as long as one is aware of the magnitude and type of errors introduced. Several methods of smoothing that produce minimal distortion of the data are discussed along with a Fourier analysis of their frequency response.

As an aside, if one is using least squares fitting for parameter estimation, one should not smooth before fitting as the least squares procedure is itself a smoothing process, and the analysis of the residual noise is necessary for evaluation of the "goodness of fit" and for computation of the confidence limits of the fitting parameters. In the procedure described hereafter, the noise of the original, unsmoothed data is used to compute the error bars for the smoothed $g(s^*)$ plots.

A. Smoothing and Differentiating as Filtering

Smoothing and differentiation can be considered as filtering processes and have an associated frequency response (see Diagram 1) (Williams, 1986).

B. Analysis of Frequency Response of Smoothing Filters

The perfect smoothing filter is one that will remove the unwanted noise and reveal the pure signal without distorting it. In a real situation, this will not be possible if the noise and the signal have overlapping power spectra. If any of the spatial frequency components of the noise are in the same range as those of the true concentration distribution, then those components cannot be removed without also removing them from the data. Therefore, a compromise will have to be reached.

C. Procedure for Determining Frequency Response Given Filter Coefficients

Briefly, the procedure is as follows (see Williams, 1986). First, determine the impulse response of the filtering operation to get the coefficients of the filter. Determining the coefficients in this manner is useful as it would allow one to create a filter that could be applied in one pass instead of several passes if desired. Second, compute the Fourier transform of the filter coefficients to obtain the power spectrum. The power spectrum is the frequency response of the filtering operation. This series of operations is represented in Diagram 2.

For example, if one wanted to know the frequency response for three passes of a simple moving average over the data, one could pass the moving average over the unit impulse three times to obtain the coefficients of the equivalent single-pass filter and then subject those coefficients to a Fourier transform. The power spectrum of that Fourier transform is the frequency response for the combined filtering operation. Figure 2A shows the results of passing a 5-point moving average over the unit impulse four times. Figure 2A shows a plot of the filter coefficients after each pass,

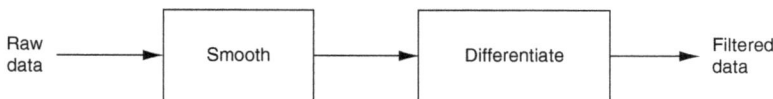

Diagram 1

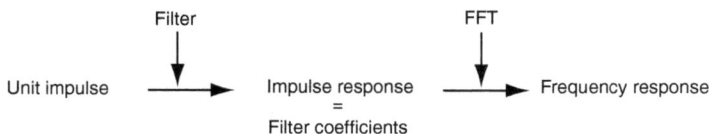

Diagram 2

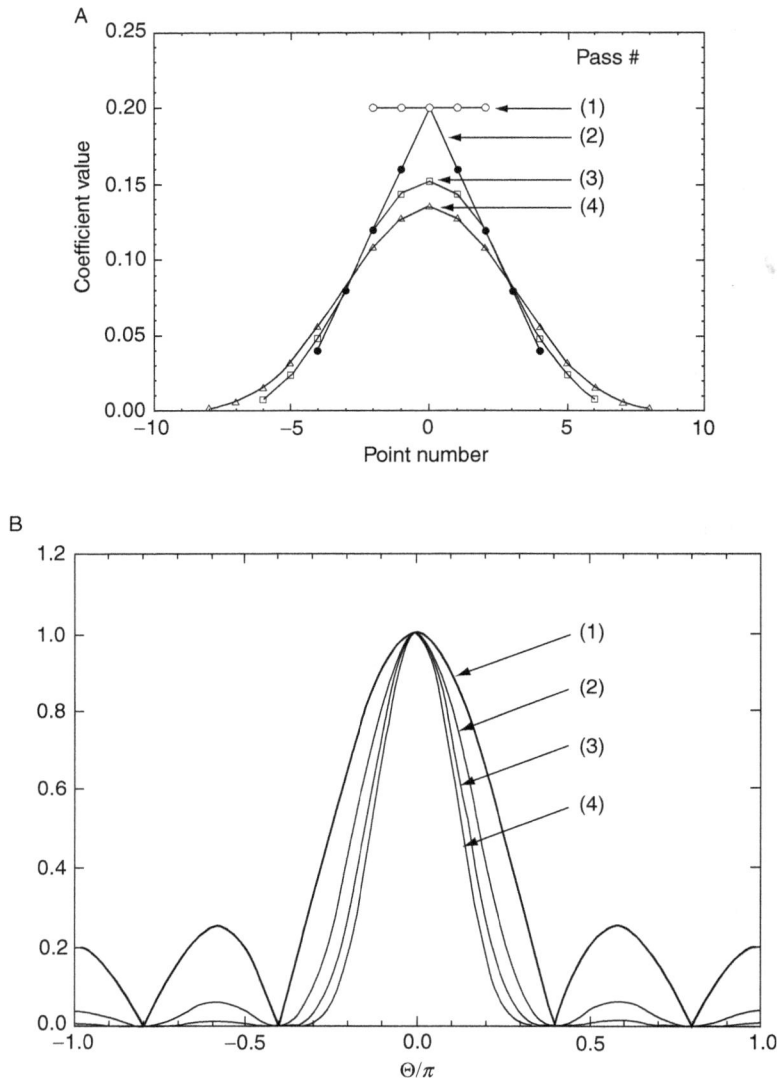

Fig. 2 (A) Plot of the filter coefficients obtained by passing a 5-point moving average over the unit impulse four times. For a 5-point centered moving average filter the coefficients are defined by the following relationship: $c_{-2}x_{i-2} + c_{-1}x_{i-1} + c_0x_i + c_{+1}x_{i+1} + c_{+2}x_{i+2}$, where c_k are the filter coefficients and x_i are a subset of the data to be filtered. For this moving average, the c_k values are all equal to 0.2. On the first pass, one of the $x_i = 1$ (i.e., the unit impulse) and the other x values are set equal to zero. After the first pass, the x_i are replaced by the filter coefficients which now become the data for the second pass of the moving average. Each new set of filter coefficients is generated by repeating this process. (B) Plot of the frequency response of each of the filters represented by the sets of filter coefficients obtained from the impulse response shown in (A). θ is defined by the following equations that were used to compute the Fourier transform of the filter coefficients. The sine transform is given by $S(\theta) = \Sigma \Sigma c_k \sin(k\theta)$; the cosine transform by $C(\theta) = \Sigma \Sigma c_k \cos(k\theta)$; and the power spectrum (i.e., frequency repsonse) by $P(\theta) = [S(\theta)^2 + C(\theta)^2]^{1/2}$, where $\theta = i\pi/N$ and $-N < i < N$, and the sums are over k and i, respectively. N was 300 in this case.

and Fig. 2B shows the frequency response of each of the filters represented in Fig. 2A. A single pass of a moving average results in large side lobes in the frequency response curve allowing some of the high frequency components to pass through. The second and third passes reduce the side lobes to much smaller values, and subsequent passes do not improve the overall frequency response very much. Although each pass lowers the cutoff frequency somewhat, the cutoff frequency is best controlled by varying the number of points in the moving average. Therefore, smoothing and differentiation can be combined into one process, if the appropriate set of coefficients is chosen by the impulse response method.

The rest of this discussion is confined to the repeated application of moving averages (i.e., first-order polynomials) as they can be computed rapidly by recursion, and almost any desired low-pass frequency response can be obtained. For example, it is quicker to pass the 5-point moving average three times than it is to pass the equivalent 13-point filter once as the coefficients of the moving average are all equal and the first point can be removed and the next added without recomputing all the intermediate terms. However, the 13-term equation has unequal coefficients requiring that all 13 terms be recomputed before adding them to compute the value for the smoothed point. The recursion formula for the n-point moving average is

$$\langle y_{i+1}\rangle = \langle y_i\rangle - \frac{y_{i-(n-1)/2}}{n} + \frac{y_{i+1+(n-1)/2}}{n},$$

where the first smoothed value is given by $\langle y_{(n+1)/2}\rangle = 1/n\sum_{j=1}^{j=n}y_j$.

The speed can be increased somewhat more, eliminating all the multiplications on each pass, by casting the process as

$$n\langle y_{i+1}\rangle = n\langle y_i\rangle - y_{i-(n-1)/2} + y_{i+1+(n-1)/2}$$

and then after the pth pass dividing each term by n^p (where p is the number of passes) for a total of N multiplications (where N is the number of data points) and $2pN$ additions. Incidentally, smoothing with formulas for higher order polynomials also requires at least three passes to reduce the side lobes to acceptable levels, and they also have unequal coefficients.

The justification for using repeated applications of a simple moving average, instead of either higher order polynomials or more complicated filtering operations that necessitate convolution, is the speed of application. For example, the three passes of the recursive moving average (this is not a recursive filter, by the way) to a data set of N points require approximately $6N$ additions with N multiplications to rescale the data at the end. Application of a higher order polynomial, for which all the terms would have to be recomputed after shifting to each new point, would require approximately N^2 multiplications and additions.

The filtering process is essentially the convolution of the smoothing polynomial with the data; therefore, application of a sliding polynomial to the data can be treated in terms of convolution. Convolution can be carried out rapidly using the fast Fourier transform (FFT) (Williams, 1986). The convolution theorem states that

$$\mathrm{FFT}(f * g) = \mathrm{FFT}(f)\mathrm{FFT}(g),$$

where g is the data and f is the filter and the asterisk denotes convolution. FFT(f) is just the frequency response of the filter and may be considered to be known as it need be computed only once ahead of time. Convolution by this method would require one FFT, followed by N complex multiplications and followed by an inverse FFT to arrive at the filtered data, $f * g$. Each FFT requires $N \log_2 N$ complex multiplications (4 multiplications and 2 additions) and complex additions (2 additions) for a total of $8N \log_2 N$ multiplications and additions. Because of the savings in computer time afforded by the repeated application of the recursive moving average, in our laboratory it is used almost exclusively. It is interesting to note that the frequency response of the 3-pass moving average filter is nearly identical to that of the popular von Hann window function often used in smoothing by convolution. Figure 3 shows a plot of the window coefficients for a 13-point von Hann window compared to those for the 13-point equivalent 3-pass moving average filter. It is gratifying that the faster recursive moving average procedure can give essentially the same results as the more time-consuming and complex convolution procedure.

In the computer algorithm used in our laboratory, the end points are handled by starting with a 3-point moving average for the first three points and then by expanding the window 2 points at a time until the full window is reached. The full window is used until the end of the data is reached, where the process is reversed by decreasing the window size 2 points at a time. The moving average is passed over the data three times. Variation in the degree of smoothing is controlled by changing the window size. The typical default for smoothing our time derivative data is to use a window spanning about 2% of the distance from the meniscus to the base. If more detail is desired, the smoothing can be turned off. It is sometimes necessary to turn the smoothing off to track down outliers.

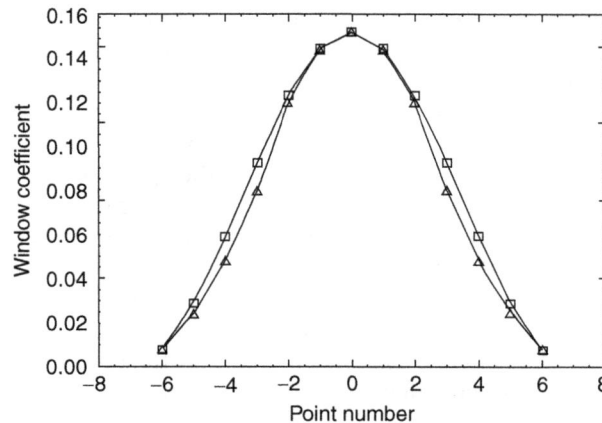

Fig. 3 Comparison of the filter coefficients of the 3-pass moving average (Δ) to those of the corresponding 13-point von Hann window function (κ).

D. Differentiation

The derivative of the concentration profile can be estimated by various polynomial fitting techniques. Differentiation and smoothing often have to be combined to get satisfactorily smooth curves. As mentioned above, a single polynomial fitting equation can be derived that will accomplish both smoothing and differentiation by using the coefficients obtained after application of the smoothing and the differentiating functions to the unit impulse function. The frequency response of the combined process can be computed by taking the Fourier transform of the coefficients. Application of the resulting polynomial can be accomplished by FFT convolution as mentioned above.

A simple and successful procedure used in our laboratory for differentiation has been to use a sliding 4-point cubic polynomial fit to equally spaced data spanning 13 points typically spaced at about 40 μm intervals in the centrifuge cell:

$$\left(\frac{dc}{dr}\right)_i = \frac{1}{12\Delta r}(c_{i-6} - 8c_{i-3} + 8c_{i+3} - c_{i+6}),$$

where Δr is the radial spacing. This differentiating procedure combined with smoothing was used in a study of filament formation of *Acanthamoeba* myosin-II (Sinard *et al.*, 1989).

VIII. Computation of Apparent Sedimentation Coefficient Distribution Functions

The methods to be described for computation of apparent sedimentation coefficient distribution functions are all derived from theoretical relationships predicated on the assumption that the diffusion coefficient of each species is zero. A rigorously correct distribution function will be obtained only for that case. The assumption of zero diffusion coefficient can be made with good approximation for many solutions of very high molecular weight polymers; however, when these relations are used in cases of nonnegligible diffusion, one obtains an *apparent* distribution function that has been broadened by the effects of diffusion. There are several extrapolation methods that can be used to correct for the effects of diffusion (Stafford, 1992; Van Holde and Weischet, 1978).

The techniques described below concentrate mainly on the uses of the uncorrected apparent distribution function as a tool for sedimentation boundary analysis, especially as a technique for revealing details of boundary shape for the analysis of both heterogeneous and reversibly associating systems. As mentioned above, the apparent differential distribution function [$g(s^*)$ vs. s^*] and the refractive index gradient (schlieren) curve (dn/dr vs. r) are very nearly geometrically similar to each other, and, therefore, both show the same details of boundary shape. A derivative pattern is useful because it can reveal features that often are not otherwise obvious by direct inspection of a concentration boundary profile. In addition, the increase in signal-to-noise ratio achieved by using the time derivative combined with averaging to compute $g(s^*)$ has extended the ability to investigate interacting systems to much lower concentrations than previously possible with any given optical system (Stafford, 1992, 1992a,b).

A. Differential Distribution Functions

Before the methods themselves are discussed, it is worth making a point concerning nomenclature. In the past, the apparent distribution function has been designated $g^*(s)$ versus s. However, when these patterns are used for boundary analysis it seems more appropriate to designate them as $g(s^*)$ versus s^* as s^* is a radial coordinate whose value can be interpreted as a sedimentation coefficient only under special circumstances. One special circumstance already mentioned is in the case of negligible diffusion. Another is that the value of s^* at the peak of a symmetrical boundary for a monodisperse solute corresponds to within a very good approximation to the sedimentation coefficient, s, for that species.

B. Method Using Spatial Derivative

In 1942, Bridgman (1942) presented an equation for computing the differential distribution function for systems exhibiting no diffusion:

$$g(s) = \frac{dc}{dr} \left(\frac{\omega^2 rt}{c_0} \right) \left(\frac{r}{r_m} \right)^2,$$

where c_0 is the initial loading concentration and the other symbols have their usual meaning. The radial derivative can be obtained either directly from the schlieren optical system or by numerical differentiation of the concentration profile from either the absorbance or the Rayleigh optical system. Various useful differentiation and smoothing methods have been discussed earlier, and a recent example of the use of the radial derivative, computed by numerical differentiation of absorbance traces from the UV photoelectric scanner of the Beckman Instruments Model-E ultracentrifuge, to compute averaged $g^*(s)$ curves in a study of filament formation of *Acanthamoeba* myosin-II, has appeared Sinard *et al.* (1989) as already mentioned.

C. Method Using Temporal Derivative

Sedimentation patterns in the form of concentration as a function of radius are acquired and stored in digital form. Pairs of curves appropriately spaced in time are subtracted from one another at corresponding radii to produce difference curves. If the time difference between the curves is made sufficiently small, the difference curve will be a good approximation to the time derivative of the original sedimenting boundary at each radial position (Stafford, 1992a). The use of the time derivative results in an automatic baseline elimination. When used in conjunction with rapid data acquisition systems that allow signal averaging, the use of the time derivative can result in several (from 2 to 3) orders of magnitude increase in the signal-to-noise ratio.

After the pairs of concentration curves are subtracted to give a set curves of dc/dt versus r, the x axis of each derivative curve is then converted to s^* by the following relationship, which transforms the dc/dt curves from the stationary reference frame of r to the moving reference frame of s^*:

$$s^* = \frac{1}{\omega^2 t} \ln\left(\frac{r}{r_m}\right). \tag{1}$$

In this moving reference frame, the dc/dt curves obtained over a small time interval will superimpose and can be averaged to reduce the random noise contributions. The value of t to be used in Eq. (1), and in others used for the numerical computation of $g(*s)$, is the harmonic mean of the sedimentation times of the concentration profiles. For data obtained from the XL-A ultracentrifuge, one must get the sedimentation time from the value of $\omega^2 t$ included in each output file.

D. Detailed Procedure

The specific example described here applies to data obtained with the Beckman Instruments Optima XL-A analytical ultracentrifuge. The data are collected in digitized form and stored as text files from which two arrays of data can be extracted; one is the absorbance, and the other is the value of the radius at which the absorbance was measured at a particular sedimentation time, say, t_1:

$$c(1, t_1), c(2, t_1), c(3, t_1), \ldots, c(n, t_1),$$
$$r(1, t_1), r(2, t_1), r(3, t_1), \ldots, r(n, t_1).$$

The value of $(\partial c/\partial t)_r$ at each radial position is estimated by subtracting pairs of sedimentation patterns point by point at corresponding values of r. However, because the data from any two successive scans usually are not acquired at the same radial positions on the XL-A ultracentrifuge, the curves have to be interpolated onto the same radial grid before the subtraction can be performed. Typically, a grid spacing of 0.002 cm is chosen for a grid between 5.8 and 7.2 cm. After a pair of scans has been interpolated onto this grid, they can be subtracted to obtain $\Delta c/\Delta t$ at each value of r. The result is an estimate of $(\partial c/\partial t)_r$ with the automatic elimination of the time-independent optical background. After subtracting and dividing by Δt, we have

$$
\begin{array}{l}
c(r, t_2)_{\mathrm{obs}} = c(r, t_2)_{\mathrm{true}} + c(r)_{\mathrm{background}} + \mathrm{noise}, \\
-[c(r, t_1)_{\mathrm{obs}} = c(r, t_1)_{\mathrm{true}} + c(r)_{\mathrm{background}} + \mathrm{noise}] \\
\hline
\dfrac{\Delta c(r)_{\mathrm{obs}}}{\Delta t} = \dfrac{\Delta c(r)_{\mathrm{true}}}{\Delta t} + 0 + \sqrt{2}\cdot\mathrm{noise}
\end{array}.
$$

If the time interval is sufficiently small that $\Delta c/\Delta t \approx \partial c/\partial t$, then $\Delta c/\Delta t$ can be converted to an apparent sedimentation coefficient distribution function according to the procedure previously presented (Stafford, 1992a):

$$g(s^*) = \left(\frac{\partial c}{\partial t}\right)_{corr}\left(\frac{1}{c_0}\right)\left[\frac{\omega^2 t^2}{\ln(r_m/r)}\right]\left(\frac{r}{r_m}\right)^2, \tag{2}$$

where $(\partial c/\partial t)_{corr}$ is the value of $(\partial c/\partial t)_r$ corrected for the plateau contribution as described (Stafford, 1992a) and repeated in detail in the subsequent text. The product of c_0 and $(r_m/r)^2$ is the plateau concentration at each point, and if these two factors are eliminated from Eq. (2) we have the unnormalized distribution function referred to as $\hat{g}(s)$. This function can be used in much the same way as a schlieren pattern is used.

1. Calculation of $(\partial c/\partial t)_r$

For the purposes of this discussion, we assume that 10 scans have been acquired and will be converted to a single apparent sedimentation coefficient distribution, $g(s^*)$. If we designate the scans by C_1, C_2, C_3, etc., then the difference curves are computed as follows:

$$\left(\frac{\Delta C}{\Delta t}\right)_1 = \frac{C_1 - C_6}{t_6 - t_1},$$

$$\left(\frac{\Delta C}{\Delta t}\right)_2 = \frac{C_2 - C_7}{t_7 - t_2},$$

$$\left(\frac{\Delta C}{\Delta t}\right)_3 = \frac{C_3 - C_8}{t_8 - t_3},$$

$$\left(\frac{\Delta C}{\Delta t}\right)_4 = \frac{C_4 - C_9}{t_9 - t_4},$$

$$\left(\frac{\Delta C}{\Delta t}\right)_5 = \frac{C_5 - C_{10}}{t_{10} - t_5}.$$

The x axis of each curve is converted to s^* using Eq. (1). The later curves are subtracted from the earlier ones to change the sign of $\Delta c/\Delta t$ so that the curves are positive for graphical purposes. Except near the base of the cell, beyond the so-called hinge point, $(\partial c/\partial t)_r$ is negative throughout the cell.

2. Averaging $\partial c/\partial t$

The five curves for $\Delta c/\Delta t$ as a function of s^* are then averaged at constant values of s^* to produce the averaged value, designated as $\langle \Delta c/\Delta t\rangle$. The averaged value is inserted into Eq. (2) for computation of the apparent distribution function.

3. Computation of Error Estimates

The standard deviation, σ_g, for each point on the averaged $g(s^*)$ curves is computed from the standard deviation of the average of $\Delta c/\Delta t$ using the following standard relation for propagation of errors:

$$\sigma_g^2 = \left[\frac{\partial g(s^*)}{\partial(\partial c/\partial t)}\right]^2 \left[\sigma_{(\partial c/\partial t)}^2\right],$$

where $\sigma_{(\partial c/\partial t)}$ is the standard deviation of $\Delta c/\Delta t$ and $\partial g/\partial c$ is obtained by differentiating Eq. (2) and is just

$$\frac{\partial g(s^*)}{\partial(\partial c/\partial t)} = \left(\frac{1}{c_0}\right)\left[\frac{\omega^2 t^2}{\ln(r_m/r)}\right]\left(\frac{r}{r_m}\right)^2 = \frac{t}{c_0 s^*}\exp(2\omega^2 s^* t).$$

The standard error of estimate for each averaged point on the curve is calculated by dividing the standard deviation by $n^{1/2}$, where n is the number of curves used in the average. It is worth pointing out that although the errors in $\delta c/\partial t$ are usually uniformly distributed across the cell, those in $g^*(s)$ are not because of their dependence on $1/s^*$, so that both the noise and the error bars tend to become larger at smaller values of s^*.

E. Correcting for Contribution to $\partial c/\partial t$ in the Plateau Region

Equation (2) gives the normalized differential sedimentation coefficient distribution corrected for radial dilution. This equation was predicated on the assumption that the diffusion coefficient of each species was zero and that each boundary contributing to the curve could therefore be represented by a step function. Under this condition, the contribution from radial dilution to the total value of $\partial c/\partial t$ at the boundary position of any given species is affected only by the more centripetal components (i.e., those having smaller values of s). The value of $(\partial c/\partial t)_p$ at any given value of s is proportional to the product of s_w and c_p at s. Both s_w and c_p can be computed from $\hat{g}(s)$, where $\hat{g}(s)$ is the unnormalized distribution function as already defined. So, now, the plateau value of $\partial c/\partial t$ is given by

$$\left(\frac{\partial c}{\partial t}\right)_p = -2\omega^2 s_w c_p = -2\omega^2 \int s\hat{g}(s)ds$$

and the unnormalized distribution function is

$$\hat{g}(s) = \left(\frac{\partial c}{\partial t}\right)_{corr}\left[\frac{\omega^2 t^2}{\ln(r_m/r)}\right]$$

where

$$\left(\frac{\partial c}{\partial t}\right)_{corr} = \left(\frac{\partial c}{\partial t}\right)_{obs} - \left(\frac{\partial c}{\partial t}\right)_p$$

and "corr" indicates the corrected value and "obs" the observed value of $\partial c/\partial t$.

These three equations imply the following iterative procedure to correct $\partial c/\partial t$. First, compute an approximate value of $\hat{g}(s)$ using the observed value of $\partial c/\partial t$:

$$\hat{g}(s)_{\text{approx}} = \left(\frac{\partial c}{\partial t}\right)_{\text{obs}} \left[\frac{\omega^2 t^2}{\ln(r_{\text{m}}/r)}\right].$$

Then, use this value to compute an approximation to the plateau value of $\partial c/\partial t$ at each point:

$$(\partial c/\partial t)_{\text{p,approx}} = -2\omega^2 s_{\text{w}} c_p = -2\omega^2 \int s\hat{g}(s)_{\text{approx}} \mathrm{d}s.$$

Subtract this from $(\partial c/\partial t)_{\text{obs}}$ to get an approximate value of $(\partial c/\partial t)_{\text{corr,approx}}$. Substitute $(\partial c/\partial t)_{\text{corr,approx}}$ into the above equation to compute a new value for $\hat{g}(s)_{\text{approx}}$. Repeat the cycle until the desired degree of convergence is attained. In practice, three iterations will give satisfactory convergence. The integration is carried out by simply using the trapezoidal rule as follows:

$$\text{Area} = \left(\frac{y_1 + y_n}{2} + \sum_{i=2}^{i=n-1} y_i\right) \Delta s^*,$$

where n is the number of points, Δs^* is the spacing in s^*, and, in this case, $y_i = s_i \hat{g}(s_i)$. In the algorithm used in this laboratory, the $g(s^*)$ curves are interpolated onto an equally spaced grid of s^*, mainly for subsequent extrapolation to correct for diffusion, so that Δs^* is a constant and therefore, can be taken outside of the summation.

IX. Weight Average Sedimentation Coefficient from $g(s^\star)$

The weight average sedimentation coefficient can be estimated from $g(s^*)$, even in the case of significant diffusion, according to the following relationship with quite good accuracy:

$$s_{\text{w}} = \frac{\int_{s^*=0}^{s^*=s_{\text{p}}^*} \hat{g}(s^*)s^* \mathrm{d}s^*}{\int_{s^*=0}^{s^*=s_{\text{p}}^*} \hat{g}(s^*) \mathrm{d}s^*}.$$

This method will give s_{w} to a very good approximation because the $\hat{g}(s^*)$ curves for each component are symmetrical and very nearly Gaussian on the s^* scale. A detailed description of the use of $\hat{g}(s^*)$ for the analysis of self-associating and heteroassociating systems will be presented elsewhere.

A. Experimental Example

As an experimental example of the method, a mixture of bovine serum albumin and aldolase (Sigma, St. Louis, MO) was run at 60,000 rpm in 0.1 M NaCl, 10 mM phosphate, and 1 mM dithiothreitol at pH 7.0 and 20° in the Beckman Instruments

Optimal XL-A analytical ultracentrifuge. Figure 4A shows 10 profiles of optical density ($A_{280 \text{ nm}}$) as a function of radius acquired at approximately 1-min intervals. The raw data were interpolated onto an equally spaced radial grid of 0.002 cm, and pairs of profiles were subtracted at equal radial positions as described above to compute $\Delta c/\Delta t$. Figure 4B shows the resulting $\Delta c/\Delta t$ versus s^* plots. The heavy line shows the average $\Delta c/\Delta t$ plot. Figure 4C shows smoothed and unsmoothed $g(s^*)$ versus s^* plots. Smoothing was performed on the averaged $\Delta c/\Delta t$ data with three passes of a simple moving average using a window spanning 2% of the whole data array from meniscus to base as outlined above. This degree of smoothing has

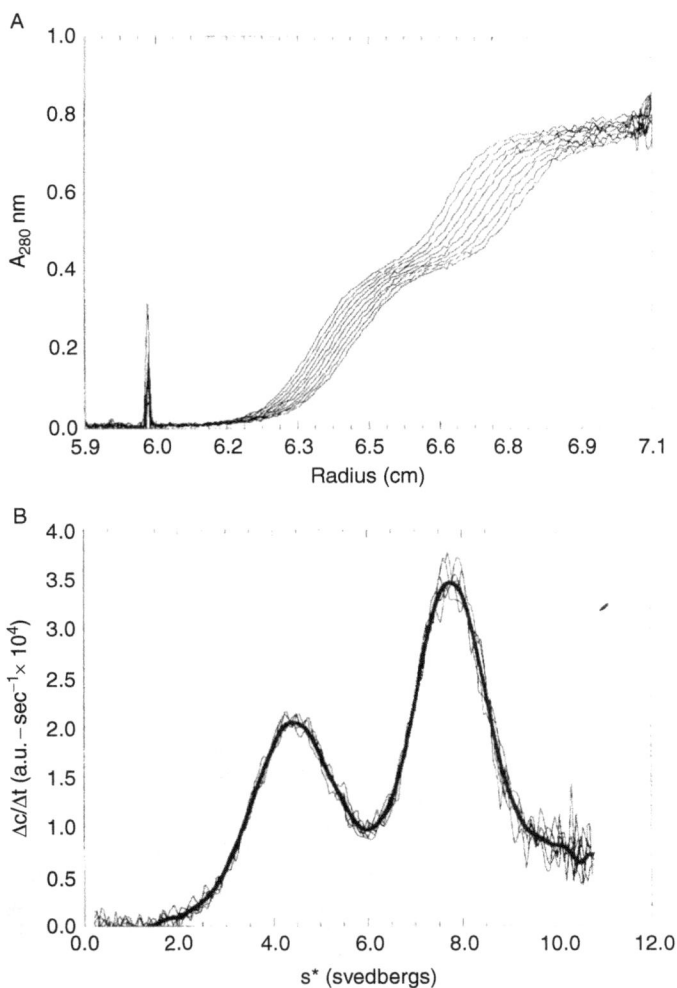

Fig. 4 (Continued)

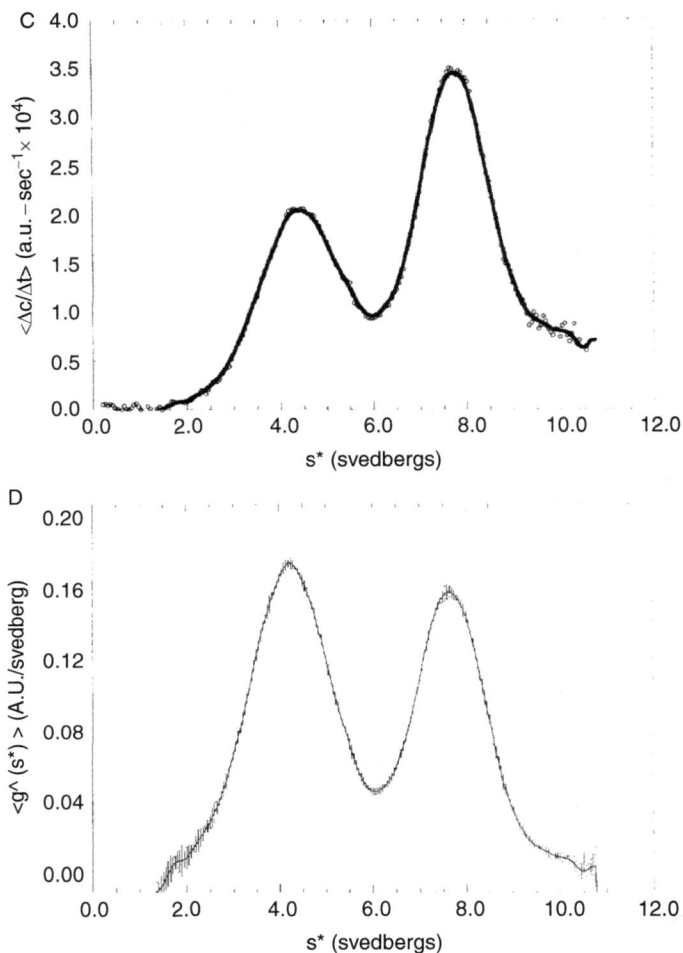

Fig. 4 (A) Data obtained from a sedimentation velocity experiment on the Beckman Instruments Optima XL-A analytical ultracentrifuge. Bovine serum albumin and aldolase (from Sigma, used without further purification) were dialyzed against 0.1 M NaCl, 10 mM phosphate, 1 mM dithiothreitol, pH 7.0, at 2°C overnight. Ten plots of absorbance at 280 nm as a function of radius obtained at approximately 1-min intervals are given. The raw data were interpolated onto an equally spaced radial grid of 0.002 cm. Temperature of the run was 20°C. (B) Plots of $\Delta c/\Delta t$ versus s^* obtained after subtracting pairs of concentration profiles as described in the text. The heavy line shows the average $\Delta c/\Delta t$ plot using a 2% window as outlined in the text. (C) Smoothed and unsmoothed $g(s^*)$ versus plots. Smoothing was performed on the averaged $\Delta c/\Delta t$ data with three passes of a simple moving average using a window spanning 2% of the distance from meniscus to base. (D) Averaged $g(s^*)$ plot for the system. The error bars are the standard error of the mean at each value of $g(s^*)$ propagated from the averaging of $\Delta c/\Delta t$.

produced a visually more pleasing result without introducing significant dispersion. Figure 4D shows the final plot of $g(s^*)$. The error bars are the standard error of the mean propagated from the averaged values of $\Delta c/\Delta t$. It is important to note that the error bars were computed from the original unsmoothed data and reflect the noise in the original curves.

X. Methods of Correcting Distribution Functions for Effects of Diffusion

The rate of boundary spreading because of heterogeneity is roughly proportional to the first power of time, whereas that because of diffusion is roughly proportional to the square root of time. Therefore, if the distribution functions are extrapolated to infinite time, the spreading from diffusion will become negligible compared to that from heterogeneity, and the true distribution function will be recovered. Various forms for extrapolation have been proposed in the literature and are discussed briefly.

A. Differential Distribution Functions

Extrapolation of differential distribution functions to infinite time to remove the effects of diffusion was first described by (Baldwin and Williams, 1950; Williams *et al.*, 1952), who extrapolated the functions at constant values of s^*. The functions of $g^*(s)$ were extrapolated versus $1/t$ to $1/t = 0$. Recently, several new forms that improve the extrapolation have been presented (Stafford, 1992). It was sometimes observed that extrapolations versus $1/t$ resulted in negative values of the distribution function in regions where one expected them to go to zero. If the curves were extrapolated as $\ln[g^*(s)]$ versus a quadratic in $1/t^{1/2}$, better results could be obtained as the extrapolation is constrained to the positive domain of $g(s)$ by taking the logarithm (Stafford, 1992).

B. Integral Distribution Functions

A method for computation of integral sedimentation coefficient distribution functions was first introduced by Gralén and Lagermalm (1952; Fujita, 1976). In that procedure, an apparent sedimentation coefficient was computed at each of several levels of the boundary. The apparent sedimentation coefficient is given by $s^* = \ln(r/r_m)/\omega^2 t$. The values of s^* computed at corresponding levels of successive boundary profiles are then extrapolated to infinite time to remove the contribution from diffusion. In the original method, the values of s^* were extrapolated versus $1/t$ to $1/t = 0$. The method has been improved by Van Holde and Weischet (1978) who noted that a better approximation to the theoretically correct extrapolation could be attained if the values of s^* were extrapolated versus $1/t^{1/2}$.

XI. Discussion

With the introduction of the Beckman Instruments Optima XL-A analytical ultracentrifuge, the processes of data acquisition in sedimentation experiments have now become largely automated. Analysis of these data, from both sedimentation equilibrium and velocity experiments, can be done entirely by digital computer. The speed and efficiency of analysis possible with a computer have allowed not only easier analysis by the older, more common methods but also the implementation of methods, both new and old, that would have been previously impractical.

In this chapter, I have briefly reviewed the older methods and introduced two new methods on the basis of the time derivative of the concentration distribution. The ability to compute the time derivative, especially from data obtained with real-time Rayleigh optical systems (Laue, 1981; Liu and Stafford, 1992), has allowed a significant increase in the sensitivity of sedimentation velocity experiments. The process of computing the time derivative results in an automatic baseline correction. This baseline correction combined with averaging of the data can result in an increase in precision of 1–2 orders of magnitude with the UV photoelectric scanning system and 2–3 orders of magnitude with Rayleigh optics. The new methods are well suited to realtime sedimentation analysis.

References

Baldwin, R. L., and Williams, J. W. (1950). *J. Am. Chem. Soc.* **72,** 4325.

Bridgman, W. B. (1942). *J. Am. Chem. Soc.* **64,** 2349.

Chervenka, C. (1973). "A Manual of Methods". Spinco Division of Beckman Instruments, Palo Alto, CA.

Correia, J. J., Alday, P. H., Sherwood, P., and Stafford, W. F. (2009). Effect of kinetics on sedimentation velocity profiles and the role of intermediates. *Methods Enzymol.* **467,** 135–161.

DeRosier, D. J., Munk, P., and Cox, D. (1972). *Anal. Biochem.* **50,** 139.

Fujita, H. (1976). "Foundations of Ultracentrifugal Analysis", Wiley, New York.

Gralén, N., and Lagermalm, G. (1952). *J. Phys. Chem.* **56,** 514.

Lamm, O. (1929). *Ark. Mat. Astron. Fysik.* **21B**.

Laue, T. M. (1981).Ph.D. Dissertation, University of Connecticut, Storrs.

Liu, S., and Stafford, W. F. (1992). *Biophys. J.* **61,** A476.

Philo, J. S. (2000). A method for directly fitting the time derivative of sedimentation velocity data and an alternative algorithm for calculating sedimentation coefficient distribution functions. *Anal. Biochem.* **279**(2), 151–163, Mar 15.

Philo, J. S. (2006). Improved methods for fitting sedimentation coefficient distributions derived by time-derivative techniques. *Anal. Biochem.* **354,** 238–246.

Schachman, H. K. (1959). "Ultracentrifugation in Biochemistry". Academic Press, New York.

Sinard, J. H., Stafford, W. F., and Pollard, T. D. (1989). *J. Cell Biol.* **109,** 1537.

Stafford, W. F. (1992a). *Anal. Biochem.* **203,** 295.

Stafford, W. F. (1992b). *Biophys. J.* **61,** A476.

Stafford, W. F., III (1994). unpublished.

Stafford, W. F. (2000). Analysis of reversibly interacting macromolecular systems by time derivative sedimentation velocity. *Methods Enzymol.* **323,** 302–325.

Stafford, W. F. (2003). "Analytical Ultracentrifugation. Sedimentation Velocity Analysis". Current Protocols in Protein Science. 20.7.1–20.7.11, John Wiley & Sons.

Stafford, W. F. (2009). Protein-protein and protein-ligand interactions studied by analytical ultracentrifugation in Protein Structure, Stability, and Interactions. *In* "Methods in Molecular Biology Series," (J. W. Shriver, ed.), Vol. 490, pp. 83–113. Humana (Springer), New York.

Stafford, W. F. (1992). *In* "Methods for Obtaining Sedimentation Coefficient Distributions," (S. E. Harding, A. J. Rowe, and J. C. Horton, eds.), p. 359. Royal Society of Chemistry, Cambridge.

Stafford, W. F., and Sherwood, P. J. (2004). Analysis of heterologous interacting systems by sedimentation velocity: Curve fitting algorithms for estimation of sedimentation coefficients, equilibrium and rate constants. *Biophys. Chem.* **108,** 231–243.

Svedberg, T., and Nichols, J. B. (1923). *J. Am. Chem. Soc.* **45,** 2910.

Svedberg, T., and Pederson, K. O. (1940). "The Ultracentrifuge," Oxford University Press, New York.

Svedberg, T., and Rinde, H. (1923). *J. Am. Chem. Soc.* **45,** 943.

Svedberg, T., and Rinde, H. (1924). *J. Am. Chem. Soc.* **46,** 2677.

Van Holde, K. E., and Weischet, W. O. (1978). *Biopolymers* **17,** 1387.

Williams, C. S. (1986). "Designing Digital Filters". Prentice-Hall, Englewood Cliffs, NJ.

Williams, J. W., Baldwin, R. L., Saunders, W., and Squire, P. G. (1952). *J. Am. Chem. Soc.* **74,** 1542.

CHAPTER 16

Statistical Error in Isothermal Titration Calorimetry

Joel Tellinghuisen

Department of Chemistry
Vanderbilt University
Nashville, Tennessee, USA

I. Update

Isothermal titration calorimetry (ITC) is unique among methods of studying binding, in that a single experiment can yield all the key thermodynamic quantities at the temperature of the experiment: $\Delta H°$, the binding constant $K°$ and from it $\Delta G°$,

DOI: 10.1016/B978-0-12-384997-7.00016-9

and hence $\Delta S° = (\Delta H° - \Delta G°)/T$. In all other methods, $\Delta H°$ must be estimated from the T dependence of $K°$, through the van't Hoff relation (Eq. (1) below). Of course, ITC experiments can be run over a range of temperatures, too, making possible the estimation of $\Delta H°$ from the resulting $K°$ values alone. Beginning in the mid-1990s, a number of groups reported that the latter van't Hoff estimates were not in agreement with the directly estimated $\Delta H°$ values, and these observations provided much of the impetus for my work on the role of statistical error in the estimation of thermodynamic quantities from ITC and from van't Hoff analysis of $K°(T)$.

About the time of this work, Mizoue and I conducted ITC experiments on the Ba^{2+} + crown ether complexation reaction (Mizoue and Tellinghuisen, 2004b), and we analyzed the data with the three limiting data error models developed in this ME paper: unweighted, weighted, and correlated-weighted. The second of these was deemed to provide the best results, from which the calorimetric-van't Hoff discrepancy was smaller than in previous studies but still statistically significant. From differences observed for $BaCl_2$ vs. $Ba(NO_3)_2$ as sources of Ba^{2+}, we suggested that some of the discrepancy was due to approximations inherent in the standard procedure of subtracting blanks to correct for heat of dilution. Similar effects were invoked to account for differences observed in later experiments done at very low concentration in a study of variable-volume ITC schemes (Tellinghuisen, 2007b). A full analysis of such effects has not yet been completed, but preliminary results indicate that it cannot account for much of the remaining calorimetric-van't Hoff discrepancy.

The findings in this ME work have led to a number of followup studies:

• An ITC data variance function (VF) was derived from a generalized-least-squares (GLS) global analysis of all data obtained in the work of Mizoue and Tellinghuisen (2004b; Tellinghuisen, 2005a). This study confirmed the presence of both constant and proportional data error, but the titrant volume error was only a minor contributor to the proportional error, contrary to predictions. The derived VF is strictly applicable to only the instrument and working ranges used in the experiments of Mizoue and Tellinghuisen (2004b), but its general properties—constant error dominant at small signal, proportional error at large—are common to many experimental techniques and should hold for all ITC methods.

• Because the titrant volume error was small, it was not possible to tell whether the correlated or uncorrelated model was correct, so I simply adopted the latter. Experiments would need to be conducted with much smaller injection volumes ($<5\ \mu L$) in order to make this error source large enough to draw such a distinction.

• Having a reliable VF, I then conducted a more detailed methods optimization study, leading to two recommendations at odds with procedures then (and unfortunately still now) commonly used to study 1:1 binding processes: (Tellinghuisen, 2005b) (1) Only 10 injections of titrant should be used for optimal precision (and incidentally shortening experiment times). (2) Where possible the range of titration

R_m (ratio of total titrant to total titrate after last injection) should be set according to the empirically derived equation,

$$R_m = \frac{6.4}{c^{0.2}} + \frac{13}{c},$$

where $c = K[M]_0$ (as used in Fig. 3 below). This study also identified a second reduced parameter of importance in ITC—$h \equiv \Delta H°[M]_0$—which determines the amount of heat available. Ordinarily, this quantity should be maximized, by making $[M]_0$ as large as possible. However, for many biophysical processes of interest, adequate precision (better than $\sim$5%) can be achieved for modest h values, avoiding possible systematic errors from working at too-high concentrations.

• A followup optimization study showed how variable injection volumes can significantly enhance precision; it gave simple expressions for setting the volumes for 4- and 5-injection schemes (again just for 1:1 binding) (Tellinghuisen, 2007b). Such methods should be especially useful for systems with both c and h very low.

• For very low c (<0.1) there is another problem: The parameters n and $\Delta H°$ are very highly correlated in the analysis of the data, making it impossible to determine the two independently. However, I have noted that K is remarkably insensitive to uncertainties in n, making it possible to extract reliable values for both n and $\Delta H°$ down to c values as low as 1×10^{-4}, by using temperature dependence to break the correlation between n and $\Delta H°$ (Tellinghuisen, 2008).

• In an attempt to sort out ambiguities in the standard models for treating the perfusion method of ITC (in which each injection of titrant expels an equal volume of reaction mixture from the active region of the cell), I devised a calibration procedure utilizing heat of dilution of NaCl (Tellinghuisen, 2007a). In this work, I checked the absolute T calibration of the instrument used in the earlier experiments on Ba^{2+}/crown ether complexation and discovered an error that amounted to 5% over the typical range 5–45 °C used for T-dependent ITC studies. This error accounted for more than half the apparent calorimetric/van't Hoff discrepancy in the study of Mizoue and Tellinghuisen (2004b), reducing it to $\sim$5%, which is almost within the statistically reasonable range. It seems likely now that most of the discrepancy is due to this and other calibration limitations for the specific instrumentation used in these works.

• Many methods estimate $K°$ values with roughly constant relative standard error. This makes it possible to obtain reliable van't Hoff uncertainties in $\Delta H°$ and $\Delta C_P°$ from knowledge of just the % error in $K°$ and the number of values and T range—an extension and elaboration on results like those shown in Fig. 11 below (Tellinghuisen, 2006).

• Two artifacts of ITC experiments were identified: Backlash in the injection syringe is largely responsible for the long-known "first injection anomaly" (Mizoue and Tellinghuisen, 2004a). Errors in the definition of the cell active volume are responsible for commonly observed values of $n < 1.0$ in cases where solution concentrations are well known and should give $n = 1$ (Tellinghuisen, 2004).

II. Introduction

The method of ITC is widely used to obtain thermodynamics information about binding processes in chemical and biochemical systems (El Harrous *et al.*, 1994a,b; Wiseman *et al.*, 1989). In a typical application of this technique, one of the reactants (M) is contained in a reaction vessel of small volume (0.2-2.0 ml), and the second reactant (the titrant X) is added stepwise to beyond the end point of the reaction. The instrumental responses following each injection of titrant are ana-lyzed to obtain the heat q associated with the chemical changes from that injection, and the experiment thereby produces a titration curve of q versus extent of reaction. Such titration curves are analyzed by means of a nonlinear least-squares (LS) fit to obtain estimates of the enthalpy change $\Delta H°$ and the equilibrium constant $K°$ for the reaction (El Harrous and Parody-Morreale, 1997; El Harrous *et al.*, 1994a,b; Wiseman *et al.*, 1989).

By repeating the experiment over a range of temperatures, one can determine the T dependence of $\Delta H°$ and $K°$. In fact, when $K°$ is known as a function of T, it becomes possible to estimate $\Delta H°$ a second way, from the slope of $\ln K°$ as a function of T^{-1} (the van't Hoff method). Sturtevant's group studied several benchmark reactions by ITC and observed that the level of agreement between the results from the latter method ($\Delta H°_{vH}$) and the directly measured values ($\Delta H°_{cal}$) was not convincing (Liu and Sturtevant, 1995, 1997; Naghibi *et al.*, 1995). These observations prompted a flurry of comments, some of which raised questions about the validity of the van't Hoff relation itself (Holtzer, 1995; Ragone and Colonna, 1995; Ross, 1996; Weber, 1995, 1996). More recent attempts to explain the discrepancies have noted that the van't Hoff estimates are inherently much less precise than the calorimetric values, and have suggested that when this larger error in $\Delta H°_{vH}$ is acknowledged, the two methods are consistent (Chaires, 1997; Horn *et al.*, 2001, 2002). However, when proper error propagation techniques are used to estimate the error in $\Delta H°_{vH}$, inconsistency is still the rule rather than the exception for the available data in the literature (Mizoue and Tellinghuisen, 2004b). There can be nothing wrong with the van't Hoff approach, as the relevant equation,

$$\left(\frac{\partial \ln K°}{\partial T} \right)_P = \frac{\Delta H°}{RT^2} \tag{1}$$

follows directly from the purely mathematical Gibbs-Helmholz equation. Rather, the problems stem from flaws in the procedures for collecting and analyzing ITC data.

One aspect of this problem that has received little attention is the role of statistical error in ITC data and its effect on the determination of $\Delta H°$ and $K°$. In a recent study (Tellinghuisen, 2003), I have noted that most ITC data are probably limited by experimental uncertainties in the delivered titrant volume, which means that the estimates of the heat q are subject to proportional rather than constant error. The LS analyses of such data should employ weighted fits instead of the unweighted fits normally done with the standard software in use. Neglect of weights in such situations leads to significant loss of efficiency in the estimation of $\Delta H°$ and $K°$. For treating the effect of uncertainty in the titrant volume, there are two limiting models that yield radically different results. If the error in the injected volume is assumed to be random, the statistical errors of the derived parameters actually increase with increasing number of titration steps. On the other hand, if one assumes that it is the total accumulated volume after i injections that is subject to random error, the incremental volume in the ith injection is the difference between two such independent quantities and is now correlated with these two volumes. This situation requires the use of weighted, correlated, nonlinear LS for analysis and leads to parameter standard errors that decrease with increasing number of steps.

The key tool for implementing the statistical studies in Mizoue and Tellinghuisen (2004b) and Tellinghuisen (2003) was the use of the LS variance-covariance matrix **V** to assess parameter confidence limits and to properly propagate statistical errors in functions of the LS parameters. As the **V** matrix is heavily underappreciated for this purpose, I have included in the present work some computations designed to illustrate clearly its properties. Chief among these are the following: (1) In linear LS fits with all the usual assumptions (Johnson and Faunt, 1992), especially normally distributed (Gaussian) random error about the true values of the dependent variable (y), the **V** matrix yields exact parameter variances, standard errors, and correlation coefficients when the error structure of the data is known; (2) one can similarly define an "exact" $\mathbf{V}_{nl}$ for nonlinear LS by using exactly fitting data; and (3) although this $\mathbf{V}_{nl}$ does not have the same rigorous validity as in linear LS, previous Monte Carlo (MC) studies have led to a useful "10% rule of thumb": if the predicted standard error for a nonlinear parameter is less than 10% of the parameter's magnitude, the $\mathbf{V}_{nl}$-based prediction is likely to be good within 10% in predicting the confidence limits (Tellinghuisen, 2000d). Note that in doing MC computations on LS models, one must assume an error structure in order to add the random error to the data. Thus, the simple predictions for linear models can be used to verify the MC algorithms. In checking the extent of the error in $\mathbf{V}_{nl}$ for nonlinear models, I have typically used 10^5 simulated data sets.

In the following sections, I briefly review the essential LS and thermodynamics relations relevant to the study by ITC of the simplest binding case, 1:1 complexation ($X + M \rightleftharpoons MX$). The properties of linear and nonlinear fits are illustrated on a simple model for van't Hoff analysis. The $\mathbf{V}_{nl}$ matrix is then used to assess the statistical errors in $\Delta H°$ and $K°$ as estimated from ITC data for various error structures—constant error, proportional error, and both. The use of correlated LS is described in detail for the relevant model of titrant volume delivery. MC calculations are used to check the validity of the $\mathbf{V}_{nl}$ matrix in cases of relatively large parameter error, and also to assess the loss of precision when heteroscedastic and correlated data are analyzed by ordinary unweighted nonlinear LS.

═══════ ## III. Variance–Covariance Matrix in Least Squares

A. Linear Least Squares

The LS equations are obtained by minimizing the sum S,

$$S = \Sigma w_i \delta_i^2 \qquad (2)$$

with respect to a set of adjustable parameters $\boldsymbol{\beta}$, where δ_i is the residual (observed-calculated mismatch) for the ith point and w_i is its weight. In the present matrix notation, $\boldsymbol{\beta}$ is a column vector containing p elements, one for each adjustable parameter. Thus, its transpose is a row vector: $\boldsymbol{\beta}^T = (\beta_1, \beta_2, \ldots, \beta_p)$. The problem is a linear one if the measured values of the dependent variable (y) can be related to those of the independent variable(s) ($x, u, \ldots$) and the adjustable parameters through the matrix equation (Albritton *et al.*, 1976; Hamilton, 1964; Tellinghuisen, 2000d),

$$\mathbf{y} = \mathbf{X}\boldsymbol{\beta} + \boldsymbol{\delta}, \qquad (3)$$

where $\mathbf{y}$ and $\boldsymbol{\delta}$ are column vectors containing n elements (for the n measured values), and the design matrix $\mathbf{X}$ has n rows and p columns, and depends only on the values of the independent variable(s) (assumed to be error free) and not on the parameters $\boldsymbol{\beta}$ or dependent variables $\mathbf{y}$. For example, a fit to $y = ax + b/x^3 + c\,\exp(2u)$ qualifies as a linear fit, with two independent variables (x, u), three adjustable parameters (a, b, c), and $\mathbf{X}$ elements $X_{i1} = x_i, X_{i2} = x_i^{-3}, X_{i3} = \exp(2u_i)$. On the other hand, the fit becomes nonlinear if, for example, the first term is changed to x/a, or the third to $2\exp(cu)$. It also becomes nonlinear if one or more of the "independent" variables are not error free and hence treated (along with y) as dependent variables.

The solution to the minimization problem in the linear case is the set of equations,

$$\mathbf{X}^T\mathbf{W}\mathbf{X}\boldsymbol{\beta} \equiv \mathbf{A}\boldsymbol{\beta} = \mathbf{X}^T\mathbf{W}\mathbf{y}. \qquad (4)$$

When the data are subject to random error only, the square weight matrix $\mathbf{W}$ is diagonal, with n elements $W_{ii} = w_i$; when the data are correlated, $\mathbf{W}$ contains off-diagonal elements. Equations (4) are solved for the parameters $\boldsymbol{\beta}$, for example,

$$\boldsymbol{\beta} = \mathbf{A}^{-1}\mathbf{X}^T\mathbf{W}\mathbf{y}, \qquad (5)$$

where $\mathbf{A}^{-1}$ is the inverse of $\mathbf{A}$. Knowledge of the parameters permits calculation of the residuals $\boldsymbol{\delta}$ from Eq. (3) and thence S, which in matrix form is

$$S = \boldsymbol{\delta}^T\mathbf{W}\boldsymbol{\delta}. \qquad (6)$$

Importantly, the variances in the parameters are the diagonal elements of the variance-covariance matrix $\mathbf{V}$, which is proportional to $\mathbf{A}^{-1}$ (see below).

For these equations to make sense, it is essential that the measurements y_i be drawn from parent distributions of finite variance (Hamilton, 1964). (This, e.g., excludes Lorentzian distributions.) If, in addition, they are unbiased estimates of the true means, then the LS equations will yield unbiased estimates of the parameters $\boldsymbol{\beta}$. If the parent distributions are normal, the parameter estimates will also be normally distributed. For these to be minimum variance estimates as well, it is necessary that the weights be taken as proportional to the inverse variances (Albritton *et al.*, 1976; Hamilton, 1964; Mood and Graybill, 1963),

$$w_i \propto \sigma_{yi}^{-2}. \tag{7}$$

Under these conditions, LS is also a maximum likelihood method. Note that it is possible to have linear LS estimators that are unbiased but not minimum variance, or minimum variance but not unbiased, or even unbiased and minimum variance, but nonnormal.

1. A Priori Weights

At this point, let us assume that the parent distributions for the data are normal (Gaussian) and that we know the σ_{yi} (as we do any time we run a MC LS calculation). With this prior knowledge of the weights, we take the proportionality constant in Eq. (7) to be 1.00. Then, S is distributed as a χ^2 variate for $v = n - p$ degrees of freedom (Albritton *et al.*, 1976; Bevington, 1969; Hamilton, 1964; Mood and Graybill, 1963). Correspondingly, the quantity S/v follows the reduced χ^2 distribution, given by

$$P(z)\mathrm{d}z = Cz^{(v-2)/2}\exp(-vz/2)\mathrm{d}z \tag{8}$$

where $z = \chi_v^2$ and C is a normalization constant. It is useful to note that a χ^2 variate has a mean of v and a variance of $2v$ (Abramowitz and Stegun, 1965), which means that χ_v^2 has a mean of unity and a variance of $2/v$. In the limit of large v, $P(z)$ becomes Gaussian.

With the proportionality constant in Eq. (7) taken as unity, the proportionality constant connecting $\mathbf{V}$ and $\mathbf{A}^{-1}$ is likewise unity, giving

$$\mathbf{V} = \mathbf{A}^{-1}. \tag{9}$$

Since the parent data distributions are normal, the parameter distributions are also normal, as already noted. Then our prior knowledge of the σ_{yi} renders Eq. (9) exact. This is true even when the number of data points equals the number of adjustable parameters, giving an exact solution for the parameters ($v = 0$). For example, the 95% confidence interval on β_1 is $\pm 1.96 V_{11}^{1/2}$, so in MC calculations on linear fit models, 95% of the estimates of β_1 are expected within $\pm 1.96 V_{11}^{1/2}$ of the true value. Conversely, a significant deviation from this prediction indicates a flaw in the MC procedures.

There is much confusion in the literature regarding these matters. In general, the off-diagonal elements in $\mathbf{V}$ (the covariances) are nonzero, for both linear and nonlinear fits. This means that the parameters $\boldsymbol{\beta}$ are correlated. The correlation matrix $\mathbf{C}$ is obtained from $\mathbf{V}$ through

$$C_{ij} = V_{ij}/(V_{ii}V_{jj})^{1/2} \qquad (10)$$

and yields elements that range between -1 and 1. However, each of the parameters in a linear fit is distributed normally about its true value, with $\sigma_{\beta j} = V_{jj}^{1/2}$, irrespective of its correlation with the other parameters. The correlation comes into play only when we ask for joint confidence intervals of two or more parameters, in which case the confidence bands become ellipsoids in two or more dimensions (Press *et al.*, 1986). Then the correlation is also correctly predicted by Eq. (10), obviating MC computations for characterizing such joint confidence ellipsoids for linear LS fits.

It is useful to note from the structure of $\mathbf{A}$ that it scales with σ_y^{-2} and with the number of data points n. Accordingly, $\mathbf{V}$ scales with σ_y^2 and with $1/n$. Thus, the parameter standard errors go as σ_y and as $n^{-1/2}$. For example, if all σ_{yi} are increased by the factor f for a given data structure, all $\sigma_{\beta j}$ increase by the same factor f. To observe the $n^{-1/2}$ dependence exactly, it is necessary to preserve the structure of the data set, for example, by using the same 5 x_i values on going from $n = 5$ to 10, 15, 20, and so on. This means that the $\sigma_{\beta j}$ are to be interpreted in the same manner as the standard deviation in the mean in the case of a simple average. (One can readily verify that for a fit of data to $y = a$, the equations do yield for σ_a the usual expressions for the standard deviation in the mean.)

Of course, all of the foregoing does assume prior knowledge of the statistics of the y_i. Unfortunately, from the experimental side, we never have perfect *a priori* information about σ_{yi}. However, there are cases, especially with extensive computer logging of data, where the *a priori* information may be good enough to make Eq. (9) the proper choice and the resulting $\mathbf{V}$ virtually exact. A good example is data obtained using counting instruments, which often follow Poisson statistics closely, so that the variance in $y_i(\sigma_{yi}^2)$ can be taken as y_i. (For large y_i, Poisson data are also very nearly Gaussian.) An important reason for using prior weighting when it can be justified is that one can then use the χ^2 statistic as an indicator of goodness of fit.

2. A Posteriori Weights

At the other extreme, we have the situation where nothing is known in advance about the statistics of the y_i, except that we believe the parent distributions to be normal and to have the same variance, independent of y_i. In this case, the weights w_i are all the same constant, which without loss of generality we can take to be 1.00. This is the case of unweighted least squares. The variance in y is then estimated from the fit itself, as

$$\sigma_y^2 \approx s_y^2 = \frac{\Sigma \delta_i^2}{n-p} = \frac{S}{\nu} \qquad (11)$$

which is recognized as the usual expression for estimating a variance by sampling. The use of Eq. (11) represents an *a posteriori* assessment of the variance in y_i. (This was designated "external consistency" by Birge (1932) and Deming (1964) as opposed to "internal consistency" for the situation where the σ_{yi} are known *a priori*.) The variance-covariance matrix now becomes

$$\mathbf{V} = \frac{\mathcal{S}}{v} \mathbf{A}^{-1} \tag{12}$$

Under the same conditions as stated before Eq. (7), s_y^2 is distributed as a scaled χ^2 variate. This means, for example, if the s_y^2 values from an MC treatment of unweighted LS are divided by the true value σ_y^2 used to generate the random noise, the resulting ratios are distributed in accord with Eq. (8) for χ_v^2.

In the case of *a posteriori* assessment, the uncertainty in s_y can greatly limit the reliability of the parameter standard error estimates when the data set is small. As the variance in χ_v^2 is $2/v$, the relative standard deviation in s_y^2 is $(2/v)^{1/2}$. From error propagation (see below), the relative standard deviation in s_y is half that in s_y^2, or $[1/(2v)]^{1/2}$. For $v = 200$, this translates into a nominal 5% relative standard deviation in s_y and hence also in all the parameter standard error estimates ($V_{jj}^{1/2}$); but it is a whopping 50% when $v = 2$, as in the test model for van't Hoff analysis explored below (in which case the distribution of s_y^2 is also far from normal, in fact is exponential). It is for this reason that it is highly desirable to characterize the data error independently from a particular experiment in those situations where it may be difficult to obtain a large number of experimental points in each run. Such information then permits use of the *a priori* $\mathbf{V}$ for confidence limits, and the χ^2 test for the data from the experiment in question (Hayashi *et al.*, 1996).

What about the confidence limits on the parameters in the case of *a posteriori* assessment? The need to rely on the fit itself to estimate s_y means the parameter errors are no longer exact but are uncertain estimates. Accordingly, we must employ the t distribution to assess the parameter confidence limits. Under the same conditions that yield a normal distribution for the parameters $\boldsymbol{\beta}$ and scaled χ^2 distributions for s_y^2 and for the V_{jj} from Eq. (12), the quantities $(\beta_j - \beta_{j,\text{true}})/V_{jj}^{1/2}$ belong to the t distribution for v degrees of freedom (Mood and Graybill, 1963), which is given by

$$f(t)\mathrm{d}t = C'(1 + t^2/v)^{-(v+1)/2}\mathrm{d}t \tag{13}$$

with C' another normalizing constant. For small v the t distribution is narrower in the peak than the Gaussian distribution, with more extensive tails. However, the t distribution converges on the unit variance normal distribution in the limit of large v, making the distinction between the two distributions unimportant for large data sets.

3. Intermediate Situations

Sometimes, one has *a priori* information about the relative variation of σ_{yi} with y_i but not a good handle on the absolute σ_{yi}. For example, data might be read from a logarithmic scale, or transformed in some way to simplify the LS analysis. As a specific example of the latter, data might be fitted to $y = ax + bx^2$ by first dividing by x to yield $y' \equiv y/x$, then fitting to $y' = a + bx$. If the original y_i have constant standard deviation σ_y, then simple error propagation shows that the standard deviations in the y'_i values are σ_y/x_i, meaning the weights $w_i \propto x_i^2$. One can readily show that the resulting weighted "straight-line" analysis yields equations [Eq. (4)] that are identical to those for the unweighted fit to $y = ax + bx^2$. This is a general property of linear LS fits to alternative forms relatable by linear variable transformations (which preserve the normal structure of the original data). Also, the results for both $\boldsymbol{\beta}$ and $\mathbf{V}$ [through Eq. (12)] are independent of arbitrary scale factors in the weights. In the present example, if the latter are taken as simply $w_i = x_i^2$, $\mathcal{S}/v$ will be an estimate of σ_y^2.

Another situation is when data come from two or more parent distributions of differing σ_y, but again known in only a relative sense. As before, the results of the calculations are independent of an arbitrary scale factor in the weights. However, to obtain meaningful estimates of the parent variances, it is customary to designate one subset as reference and assign $w_i = 1$ for these data, with all other weights taken as $w_i = s_{ref}^2/s_i^2$ (hence the need for knowledge of the relative precisions). Then the quantity $\mathcal{S}/v(= s_{ref}^2)$ obtained from the fit is more properly referred to as the "estimated variance for data of unit weight," and the estimated variance for a general point in the data set is s_{ref}^2/w_i.

Users of commercial data analysis programs should be aware that those programs that provide estimates of the parameter errors do not always make clear which equation—Eq. (9) or Eq. (12)—is used. For example, recent versions of the program KaleidaGraph (Synergy Software) use Eq. (12) in unweighted fits to user-defined functions, but Eq. (9) in all weighted fits (Tellinghuisen, 2000c). This means that in cases like those just discussed, where the weights are known in only a relative sense, the user must scale the parameter error estimates by the factor $(\mathcal{S}/v)^{1/2}$ to obtain the correct *a posteriori* values. (In the KaleidaGraph program the quantity called "Chisq" in the output box is just the sum of weighted squared residuals $\mathcal{S}$, which is χ^2 only when the input σ_i values are valid in an absolute sense.)

B. Nonlinear Least Squares

In nonlinear fitting, the quantity minimized is again $\mathcal{S}$, and the LS equations take a form similar to Eq. (4) but must be solved iteratively. The search for the minimum in $\mathcal{S}$ can be carried out in a number of different ways (Bevington, 1969; Johnson and Faunt, 1992; Press *et al.*, 1986); but sufficiently near this minimum, the corrections $\Delta\boldsymbol{\beta}$ to the current values $\boldsymbol{\beta}_0$ of the parameters can be evaluated from (Bevington, 1969; Deming, 1964; Press *et al.*, 1986)

$$\mathbf{X}^T\mathbf{W}\mathbf{X}\Delta\boldsymbol{\beta} \equiv \mathbf{A}\Delta\boldsymbol{\beta} = \mathbf{X}^T\mathbf{W}\boldsymbol{\delta} \tag{14}$$

leading to improved values,

$$\boldsymbol{\beta}_1 = \boldsymbol{\beta}_0 + \Delta\boldsymbol{\beta}. \tag{15}$$

The quantities $\mathbf{W}$ and $\boldsymbol{\delta}$ have the same meaning as before; but the elements of $\mathbf{X}$ are $X_{ij} = (\partial F_i/\partial\beta_j)$, evaluated at x_i using the current values $\boldsymbol{\beta}_0$ of the parameters. The resulting matrix $\mathbf{A}$ is now an approximation of the Hessian matrix (Press *et al.*, 1986). The function F expresses the relations among the variables and parameters, and it is convenient to express it in such a way that a perfect fit yields $F_i = 0$. For the commonly occurring case where y can be expressed as an explicit function of x, it can be written

$$F_i = y_{calc}(x_i) - y_i = -\delta_i. \tag{16}$$

In the case of a linear fit, starting with $\boldsymbol{\beta}_0 = 0$, these relations yield for $\boldsymbol{\beta}_1$ equations identical to Eqs. (4) and (5) for $\boldsymbol{\beta}$; and convergence occurs in a single cycle. In the more general case where y cannot be written explicitly in terms of the other variables, these equations still hold, but with $\boldsymbol{\delta}$ in Eq. (14) replaced by $-\mathbf{F}_0$, where the subscript indicates that the F_i values are calculated using the current values $\boldsymbol{\beta}_0$ of the parameters.

Regardless of how convergence is achieved, the variance-covariance matrix is again given by Eq. (9) in the case of *a priori* weighting and Eq. (12) for *a posteriori* weighting, with $\mathbf{X}$ as redefined just below Eq. (15). However, there is an important distinction between $\mathbf{V}$ in the general nonlinear case versus the linear case: the matrix $\mathbf{A}$ now contains a dependence on the parameters. Also, in general, there is no need to distinguish between dependent and independent variables in nonlinear fitting, as all variables can be taken to be uncertain (Deming, 1964). In that case, $\mathbf{A}$ may also depend on the values of all the variables, not just the (previously) independent variables. Thus, even in the case of *a priori* weighting, $\mathbf{V}$ from Eq. (9) will vary from data set to data set. However, one can extract estimates of $\mathbf{V}$ from a perfectly fitting theoretical curve and use this $\mathbf{V}$ in the same fashion as in the case of linear fitting. This is the "exact" $\mathbf{V}_{nl}$ menioned in the Introduction.

Through $\mathbf{V}_{nl}$, one can estimate parameter confidence limits for a particular nonlinear fit model and data structure almost trivially, often with a few minutes of effort using a program like KaleidaGraph. While it is true that the nonlinear LS parameter distributions are generally not normal, often they are close enough thereto to permit estimation of confidence intervals in this *a priori* fashion with a reliability that exceeds that achievable in typical MC calculations. This is because the MC variance estimates are subject to the previously noted statistics of a χ^2 variate, which means for a 1000-set MC calculation a relative standard deviation of about 4.5% in the variances, or half that in the standard errors. And many published studies have employed far fewer than 1000 data sets, with concomitant loss in error precision as $N^{-1/2}$. The 10% rule of thumb for the reliability of $\mathbf{V}_{nl}$, mentioned in the Introduction, actually turns out to be conservative for most of the cases I have examined to date.

C. Statistical Error Propagation

The textbook expression,

$$\sigma_f^2 = \sum \left(\frac{\partial f}{\partial \beta_j} \right)^2 \sigma_{\sigma_j}^2 \tag{17}$$

is normally used to compute the propagated error in a function f of the independent variables $\boldsymbol{\beta}$, where the sum runs over all uncertain variables β_j. However, Eq. (17) assumes that these variables are uncorrelated. This assumption seldom holds for a set of parameters $\boldsymbol{\beta}$ returned by an LS fit, and one must use the more general expression (Tellinghuisen, 2001),

$$\sigma_f^2 = \mathbf{g}^T \mathbf{V} \mathbf{g} \tag{18}$$

in which the jth element of the vector $\mathbf{g}$ is $\partial f / \partial \beta_j$. This expression is rigorously correct for functions f that are linear in variables β_j that are themselves normal variates. For nonlinear functions of normal and nonnormal variates, its validity is limited by the same 10% rule of thumb that applies to parameters estimated by nonlinear LS.

In many cases, the computation of σ_f can be facilitated by simply redefining the fit function so that f is one of the adjustable parameters of the fit (Tellinghuisen, 2001). For example, suppose a set of data is fitted to the quadratic function, $y = a + bz + cz^2/2$, where $z = (x - x_0)$, and it is the errors in this function and its derivatives that are of interest. For $f = y$, $\mathbf{g}^T = (1, z, z^2/2)$, from which it is clear that $\sigma_f = \sigma_a$ for $x = x_0$. Similarly, the statistical errors in the first and second derivatives of f at x_0 are σ_b and σ_c, respectively. Thus, one can bypass Eq. (18) by simply repeating the fit for the several values of x_0 that are of interest.

IV. Monte Carlo Computational Methods

The Monte Carlo LS calculations are done using programs coded in FORTRAN and methods that are detailed elsewhere (Tellinghuisen, 2000d). To minimize post-processing of the very large files that would be produced in a run of 10^5 data sets, the distributional information is obtained by binning "on the fly." The statistical averages and higher moments are similarly computed by running accumulation. For most of the computations, a typical run of 10^5 data sets takes less than 1 min.

The statistics for the various quantities from the MC calculations are calculated by accumulating the relevant sums and then dividing by the number of sets N at the conclusion. For example, the estimated variance in a parameter a is $s_a^2 = \langle a^2 \rangle - \langle a \rangle^2$. For assessing the significance of bias, it is necessary to know the precision of the MC parameter estimates, which (at the 68.3% or 1-σ level, for normal data) is their estimated standard error, $s_a/N^{1/2}$. On the other hand, the sampling estimates of the parameter standard errors are subject to the previously mentioned properties of the χ^2 distribution, for N degrees of freedom in this case. Thus, their relative standard errors are $(2N)^{-1/2} = 0.00224$ for $N = 10^5$.

The histogrammed data are analyzed by fitting to the appropriate models using the user-defined curve-fitting function in KaleidaGraph. The uncertainties in the binned values are taken as their square roots, in keeping with the Poisson nature of the binning process. Bins containing fewer than six counts are normally omitted. For the most part, the values are fitted simply as sampled points. However, technically the bin counts represent integrals over the specified intervals, a distinction that can make a difference if the data are not binned on a fine enough scale. For example, in the present case, proper treatment of narrow χ_v^2 distributions requires breaking each interval into subintervals and integrating.

V. Van't Hoff Analysis of $K^\circ(T)$: Least–Squares Demonstration

With assumption of a functional form for $\Delta H^\circ(T)$, Eq. (1) can be integrated to yield a form suitable for analysis by LS fitting. In the examples discussed below, ΔH° is assumed to be expressible as quadratic in the temperature T over the range encompassed by the data,

$$\Delta H^\circ = a + b(T - T_0) + c(T - T_0)^2 \qquad (19)$$

From this expression, at $T = T_0$, $\Delta H^\circ = a$, $\Delta C_P^\circ = b$, and $\mathrm{d}\Delta C_P^\circ/\mathrm{d}T = 2c$. Integration of Eq. (1) then yields

$$R\ln(K^\circ/K_0^\circ) = A\left(\frac{1}{T_0} - \frac{1}{T}\right) + B\ln\left(\frac{T}{T_0}\right) + c(T - T_0), \qquad (20)$$

where R is the gas constant,

$$A = a - bT_0 + cT_0^2 \text{ and } B = b - 2cT_0. \qquad (21)$$

If we take y as $\ln K^\circ$ and define $\ln K_0^\circ$ as one of the fit parameters, the fit to Eq. (20) becomes linear and should obey all the rules for linear fits described above. Alternatively, the fit of K° to the exponential form of Eq. (20) is nonlinear and can be expected to deviate from these rules. These behaviors are illustrated through a series of MC calculations employing the model spelled out in Table I, which is based on observations for the complexation of Ba^{2+} with 18-crown-6 ether, a reaction that is sometimes used to calibrate ITC instruments (Briggner and Wadsö, 1991) and that has been examined for consistency between ΔH_{vH}° and ΔH_{cal}° (Horn *et al.*, 2001; Liu and Sturtevant, 1995). For simplicity, ΔC_P° is assumed to be constant ($c = 0$); however, the data error is intentionally set large enough to ensure that ΔC_P° is not statistically defined for the five-point data set (as cases of large relative error are more likely sources of significant deviations from linear behavior in nonlinear fits). Note also that the assumption of proportional error in K° means that the error in $\ln K^\circ$ is constant, since $\sigma(\ln y) = \sigma_y/y$.

Results of several MC computations on 10^5 data sets are summarized in Table II and illustrated in Figs. 1 and 2. In cases 1 and 2, the random error is assumed to be normal in the fitted quantity ($\ln K^\circ$), and the MC deviations from the true values for the parameters, their standard errors, and χ_v^2 are reasonable—about equally positive and negative, with 11 of 16 being less than 1σ and none exceeding 2σ. In case 3, $\ln K^\circ$ is still the fitted quantity, but the error is taken to be normal in K° (i.e., the random error is added to K° before the logarithm is taken). As a result of the

Table I
Model Used in Monte Carlo Tests of Van't Hoff Analysis of $K°$ (T)

$\Delta H° = -32,000 \text{ J mol}^{-1}\text{K}^{-1} + 130 \text{ J mol}^{-1}\text{K}^{-2}(T - 298.15) \equiv a + b(T - T_0)$

$\ln(K°) = \ln(K_0°) + \frac{a - bT_0}{R}\left(\frac{1}{T_0} - \frac{1}{T}\right) + \frac{b}{R}\ln\left(\frac{T}{T_0}\right), K_{298}° = 5.5$

$n = 5: t_i(°C) = 5, 15, 25, 35, 45$

$\sigma_{K_i°} = 0.10 K_i°, \sigma_{\ln(K_i°)} = 0.10$

For log fit with $T_0 = 298.15$ K:

$X_{i1} = 1.0, X_{i2} = R^{-1}(298.15^{-1} - T_i^{-1}), X_{i3} = R^{-1}[\ln(T_i/T_0) + T_0/T_i - 1]$

$W_{ii} = 100, W_{ij} = 0, i \neq j$

$\boldsymbol{\beta} = \begin{pmatrix} 1.70475 \\ -32,000 \\ 130 \end{pmatrix}, \mathbf{A} \equiv \mathbf{X}^T\mathbf{W}\mathbf{X} = \begin{pmatrix} 5.00000 \times 10^2 & -4.55537 \times 10^{-4} & 6.80393 \times 10^{-2} \\ -4.55537 \times 10^{-4} & 1.85173 \times 10^{-7} & -2.46001 \times 10^{-7} \\ 6.80393 \times 10^{-2} & -2.46001 \times 10^{-7} & 1.58826 \times 10^{-5} \end{pmatrix}$

$\mathbf{V} = \begin{pmatrix} 4.84137 \times 10^{-3} & -1.59714 \times 10^1 & -2.09873 \times 10^1 \\ -1.59714 \times 10^1 & 5.56650 \times 10^6 & 1.54638 \times 10^5 \\ -2.09873 \times 10^1 & 1.54638 \times 10^5 & 1.55264 \times 10^5 \end{pmatrix}, \mathbf{C} = \begin{pmatrix} 1.00000 & -0.09729 & -0.76548 \\ -0.09729 & 1.00000 & 0.16634 \\ -0.76548 & 0.16634 & 1.00000 \end{pmatrix}$

For log fit with $T_0 = 278.15$ K:

$\boldsymbol{\beta} = \begin{pmatrix} 2.67152 \\ -34,600 \\ 130 \end{pmatrix}; \mathbf{A} = \begin{pmatrix} 5.00000 \times 10^2 & 1.40472 \times 10^{-2} & 2.00600 \times 10^{-1} \\ 1.40472 \times 10^{-2} & 5.79405 \times 10^{-7} & 9.14687 \times 10^{-6} \\ 2.00600 \times 10^{-1} & 9.14687 \times 10^{-6} & 1.53647 \times 10^{-4} \end{pmatrix}$

$\mathbf{V} = \begin{pmatrix} 9.00642 \times 10^{-3} & -5.43633 \times 10^2 & 2.06047 \times 10^1 \\ -5.43633 \times 10^2 & 6.14868 \times 10^7 & -2.95065 \times 10^6 \\ 2.06047 \times 10^1 & -2.95065 \times 10^6 & 1.55264 \times 10^5 \end{pmatrix}, \mathbf{C} = \begin{pmatrix} 1.00000 & -0.73053 & 0.55100 \\ -0.73053 & 1.00000 & -0.95497 \\ 0.55100 & -0.95497 & 1.00000 \end{pmatrix}$

log transformation, the fitted data are no longer normal, and this results in significant bias in χ_ν^2, all parameter standard errors, and $\ln K_0°$. Cases 4–6 involve the nonlinear fitting of $K°$, both with (cases 4 and 5) and without weighting (case 6). As the data error is taken as 10% of $K°$, there are several options for actually assessing the weights: Use the true $K°$ values (theoretical weighting, available only in the MC context), the actual values after the random error has been added (observed weighting), or the adjusted values from the fit itself (not included here) (Tellinghuisen, 2000d). All choices produce bias in most of the tabulated quantities, but the magnitude of the bias varies considerably from case to case.

Even though the biases in cases 3–6 are mathematically significant, they are mostly not practically significant from the standpoint of a single experiment. For example, the disparity of -0.068 in $K_0°$ in case 5 still represents only $\simeq 1/6$ of 1 standard deviation for a single experiment; and remember that this s itself, if estimated a $posteriori$, would be uncertain by 50% ($\nu = 2$). Except for case 6, none of the MC standard error estimates differs by more than 3% from the "exact" predictions, even for the very uncertain $\Delta C_P°$, for which the relative standard error is $\sim$300%. The results for the unweighted fit to the $K°$ values (case 6) demonstrate how neglect of weights for heteroscedastic data leads to standard errors that must exceed the minimum variance values—by 15%, 18%, and

Table II

Monte Carlo Results from 10^5 Data Sets on Model in Table I[a]

Parameter	Value	δ[b]	s[c]	δs[d]
Exact values for $T_0 = 298.15$ and $n = 5$:				
$\ln K_0^\circ$	1.70475		0.06959	
K_0°	5.5		0.3827	
ΔH_0°	−32,000		2359.3	
ΔC_P°	130		394.04	
1. Error normal in $\ln K^\circ$; $\ln K^\circ$ fitted:				
$\ln(K_0^\circ)$	1.70485	0.46	0.06958	−0.09
ΔH_0°	−31,990.7	1.25	2357.5	−0.36
ΔC_P°	129.76	−0.19	395.4	1.49
χ_v^{2d}	0.99709	−0.92	0.99939	−0.27
2. Error normal in $\ln K^\circ$; $\ln K^\circ$ fitted, 20 data points (4 at each T_i):				
$\ln(K_0^\circ)$	1.70476	0.11	0.03484	0.63
ΔH_0°	−31,998.0	0.54	1176.5	−1.21
ΔC_P°	129.66	−0.55	197.8	1.75
χ_v^{2d}	1.00110	1.01	0.34276	−0.31
3. Error normal in K°; $\ln K^\circ$ fitted:				
$\ln(K_0^\circ)$	1.69979	−22.24	0.07049	5.79
ΔH_0°	−31,991.0	1.19	2387.6	5.35
ΔC_P°	129.66	−0.27	400.4	7.25
χ_v^{2d}	1.02276	6.83	1.05416	24.22
4. Error normal in K°; K° fitted using theoretical weighting:				
K_0°	5.50457	3.78	0.38281	0.13
ΔH_0°	−31,995.1	0.65	2378.8	3.68
ΔC_P°	119.93	−8.00	398.1	4.65
χ_v^{2d}	0.99662	−1.07	0.99884	−0.51
5. Error normal in K°; K° fitted using "observed" weighting:				
K_0°	5.43230	−54.50	0.39285	11.86
ΔH_0°	−31,979.8	2.66	2405.5	8.74
ΔC_P°	158.72	22.44	404.8	12.17
χ_v^{2d}	1.01077	3.28	1.03959	17.71
6. Error normal in K°; K° fitted unweighted[e]:				
K_0°	5.50332	2.38	0.44079	67.87
ΔH_0°	−32,185.9	−21.14	2781.0	79.90
ΔC_P°	103.88	−15.40	536.3	161.50

[a]Units: J mol^{-1} for ΔH°, $\text{J mol}^{-1}\,\text{K}^{-1}$ for ΔC_P°. All results for 5-point data sets except where indicated otherwise.

[b]Deviations (observed−true) in units of standard errors, which are $\sigma/(10^5)^{1/2}$ for parameters, $\sigma/(2 \times 10^5)^{1/2}$ for standard deviations.

[c]Ensemble estimates of standard deviations for Monte Carlo data.

[d]Reduced χ^2 has expected value 1.00 and standard error $(2/v)^{1/2}$, $= 1.00$ for 5-point data sets, $(2/17)^{1/2}$ for 20-point data sets.

[e]RMS standard errors from *a posteriori* **V** [Eq. (12)] are 0.3391, 3636, and 408.3, respectively.

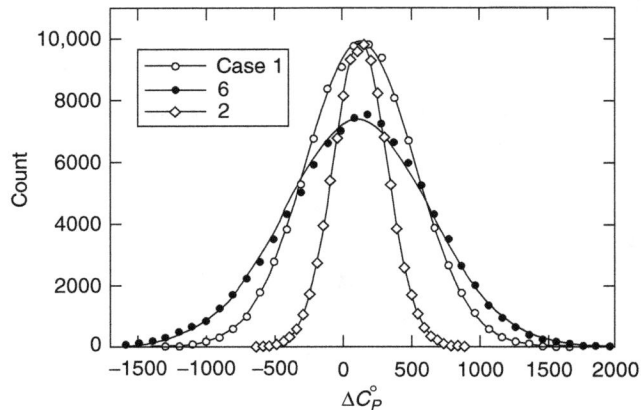

Fig. 1 Histogram data for ΔC_p°, as estimated in the indicated computations summarized in Table II. The results from cases 1 and 2 fit the Gaussian curve with reasonable χ^2 values (32.4 for case 1, 31 points; 36.2 for case 2, 32 points), while the data from the unweighted fit (case 6) are clearly not Gaussian ($\chi^2 =$ 298). Similar data from case 3, although biased in s, do fit the Gaussian distribution adequately at this level of scrutiny ($\chi^2 = 36.2$, not shown). Note that the binning interval for case 2 was half that for the other two, resulting in its factor of 2 smaller area in this display.

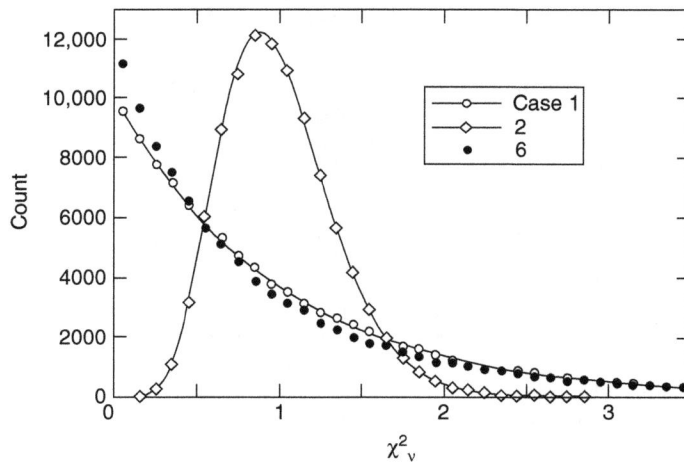

Fig. 2 Histogram data for χ_ν^2 for the indicated computations of Table II. The results from cases 1 and 2 fit the theoretical distribution of Eq. (8), yielding $\nu = 1.997 \pm 0.007$ and 17.085 ± 0.076, respectively. The estimates of S/ν from the unweighted fit (case 6) were scaled to an average value of 1.00 for this plot; they do not follow Eq. (8) for any ν.

36%, respectively. At the same time, the *a posteriori* estimates from Eq. (12) differ from both the exact values (from a properly weighted fit) and the observed (MC) values; relative to the latter they are off by -23, $+31$, and -24%. This demonstrates that the estimated **V** from Eq. (12) is simply wrong when it results from an unweighted fit of data that should properly be analyzed by a weighted fit. Not surprisingly, in this situation, the quantity called χ_ν^2 also fails to follow the true χ_ν^2 distribution (Fig. 2). Interestingly, the unweighted fit does yield the least bias for K_0°.

Case 2 and Fig. 1 illustrate how taking four values at each T_i instead of one drops the parameter standard errors by a factor of 2. We might ask what would happen if we simply averaged each set of four points and fitted only the five average values. Clearly, each single MC data set will yield exactly the same results if we weight each set of four by four times the original weight for individual values. This is the same as taking each of the five averages of K° to have $\sigma_{\bar{y}i} = \sigma_{yi}/2$, that is, the theoretical standard deviation in the mean. Alternatively, if we use the actual statistics of each group of four, we will not get identical results, and the statistics of this process will not follow the usual rules (Tellinghuisen, 1996). In effect, by this procedure, we are granting the data the right to determine their own destiny. This disparity emphasizes the importance of having prior knowledge of the data error and using this prior knowledge to assign the weights. If such prior weighting is used in the present case, all of the results will be identical to those obtained for the original 20-point data sets, except one: χ_ν^2 will now follow the distribution for $\nu = 2$ instead of that for $\nu = 17$ (Fig. 2).

Table I includes the **A**, **V**, and **C** matrices for two different choices of T_0, 298.15 and 278.15 K. Note first that both **V** matrices yield identical V_{33} elements, confirming expectations that the constant ΔC_P° should not have a statistical error that depends on the arbitrary choice of reference temperature. The other two parameters do vary with T, and so do their errors. To assess the error in ΔH° as a function of T, note that $\mathbf{g}^T$ in Eq. (18) is $(0, 1, z)$, where $z = T - T_0$. It is easy to verify that

$$\sigma_{\Delta H^\circ}^2 = V_{22} + 2V_{23}z + V_{33}z^2 \tag{22}$$

and that both **V** matrices thus yield identical estimates of $\sigma_{\Delta H^\circ}^2$ at all T. Of course, the two V values give this quantity directly at 298.15 and 278.15 K, confirming that one can obtain the desired standard error by simply repeating the fit for the several T_0 values of interest.

Finally, it is noteworthy that ΔH° and ΔC_P° are highy correlated at 278.15 K but largely uncorrelated at 298.15 K. This means that the naive error propagation formula of Eq. (17) will work fairly well on the results at the latter T_0 but will be badly in error if used with the results obtained for $T_0 = 278.15$ K.

VI. Isothermal Titration Calorimetry

A. Fit Model for 1:1 Binding

In an ITC experiment, the heat q_i determined for the ith injection of titrant X represents the result of changes in the amount of the complex MX in the active volume V_0 of the reactant vessel. The cells in the most widely used instruments are of the perfusion type, in which a volume v of solution is expelled each time the same volume of titrant is injected. It is assumed that prior to each injection the system is uniform and at equilibrium, and solution of this composition is expelled on injection, after which the injected titrant mixes and reacts to achieve the new equilibrium. Following the ith injection, the total concentrations (free and complexed) of X and M are given by (El Harrous et al., 1994b):

$$[X]_{0,i} = [X]_0(1 - d^i) \text{ and } [M]_{0,i} = [M]_0 d^i, \tag{23}$$

where $[X]_0$ is the concentration of titrant in the syringe, and $[M]_0$ is the starting concentration of M in the reaction vessel. The dilution factor $d = 1 - v/V_0$.

At equilibrium, the concentrations of reactants and product satisfy the equilibrium expression (using the concentration reference state),

$$\frac{[MX]_i}{([X]_{0,i} - [MX]_i)([M]_{0,i} - [MX]_i)} = K \equiv K^\circ \times (1 \, \text{mol}^{-1}). \tag{24}$$

The number of moles of complex produced by the ith injection is thus

$$\Delta n_i = V_0[MX]_i - (V_0 - v)[MX]_{i-1} = V_0\{[MX]_i - d[MX]_{i-1}\} \tag{25}$$

and the associated heat is

$$q_i = \Delta H^\circ \Delta n_i. \tag{26}$$

For notational simplicity, I work with the dimensionless K° below, which is tantamount to taking all activity coefficients to be unity at all times in Eq. (24). I also neglect such experimental complications as the need to estimate heats of dilution for the titrant, and the related concentration dependence of q_i. Within the framework of these assumptions, this model is exact; that is, there is no need for the differential approximation described by Wiseman et al. (1989).

This model has two adjustable parameters, ΔH° and K°, and as many data points as injections. The software in general use for analyzing ITC data includes a third parameter, the "site number" n_s. For 1:1 complexes, this parameter is typically within ~ 0.05 of 1.0 and should usually be viewed as a concentration correction factor, needed to put the concentrations of X and M on a common footing. As inclusion of this factor is important for achieving a good fit of typical ITC data, I have also included it in the present model, where I have taken it as a correction factor to $[M]_0$. (The matter of how this factor should be applied is discussed further below.)

Knowledge of the error structure of ITC data is key to estimating the parameter standard errors, as it is also for realistic MC calculations. There are two clear sources of random error in ITC: (1) the extraction of q_i values from the recorded data, and (2) the delivery of the metered volume v of titrant from the syringe. The first of these is essentially a sensitivity of measurement limitation and is expected to be roughly constant, independent of q_i. The effects of the second depend strongly on the specific assumptions about the volume error. If the incremental volume v is assumed to possess random error, simple error propagation yields a proportional error, $\sigma_{qi} = q_i(\sigma_v/v)$.

On the other hand, if it is the total *accumulated* volume after i steps that is assumed to possess random error, the incremental volume, being the difference between two such independent quantities, possesses correlated error. These three kinds of error affect the precisions differently and are examined individually and in combination in the calculations described below. When simultaneous contributions from the measurement and volume errors are considered, the variances are assumed to be additive, for example, $\sigma_i^2 = \sigma_q^2 + q_i^2(\sigma_v/v)^2$ for the random volume error. Note that random volume error leads to data that are inherently heteroscedastic, requiring a weighted fit for proper analysis. Similarly, correlated volume error requires the use of weighted, correlated LS. For reference, the two uncertainties are estimated as $\sigma_q = 0.28$ μcal and $\sigma_v = 0.015$ μl in Wiseman *et al.* (1989); for the benchmark reaction of 2'CMP with RNase studied there, the volume error dominates over most of the titration curve in the random volume error model (see below).

The point made after Eq. (10), about the dependence of **V** on the data error, is worth revisiting here. If the computations for a given model are repeated after simply scaling σ_q and σ_v by a factor f, the parameter standard errors will scale by the same factor f. As sufficiently small data errors yield adequately Gaussian parameter distributions, this means that the error structure for the model can always be evaluated from $\mathbf{V}_{nl}$. The only question then is the extent to which this structure applies to the actual situation; in previously examined cases, the 10% rule of thumb has proved a reliable guideline for applicability.

In the LS fitting codes, the independent variable is taken as the titration index i, which is rigorously error free. The error in q is taken to be normal at all times, and the solution concentrations are treated as exact. Although uncertainty in the prepared concentrations is significant in most actual experiments, this uncertainty is not manifested as point-to-point random error in a given experiment; and it is anyway partly compensated through the parameter n_s, as was already noted and is discussed further below.

B. Check on 10% Rule of Thumb

For most of the calculations discussed in this and subsequent sections, $[M]_0$ was fixed at 1.00 mM, V_0 was 1.4 ml (as in the instrument of Wiseman *et al.*, 1989), and $\Delta H°$ was 10.0 kcal/mol. The total titrant volume for m injections was typically 0.1 ml; and the precisions in $K°$, n_s, and $\Delta H°$ were investigated as functions of $K°$ (more properly $K[M]_0$), the number of injections m, and the stoichiometry range of the experiment, $R_m = [X]_{0,m}/[M]_{0,m}$.

I address first the circumstances under which $\mathbf{V}_{nl}$ can be trusted to yield reliable estimates of the parameter errors in the analysis of ITC curves. For reference, Fig. 3 illustrates typical titration curves spanning the approximate extremes of analyzable values of the product $K[M]_0$. These curves represent an extension of the example, $K[M]_0 = 1000$, explored in Fig. 3A of El Harrous *et al.* (1994a). As the latter MC analysis also included an assumed 2% error in the concentrations, the present results cannot be compared quantitatively; however, the results for this case are commensurate with

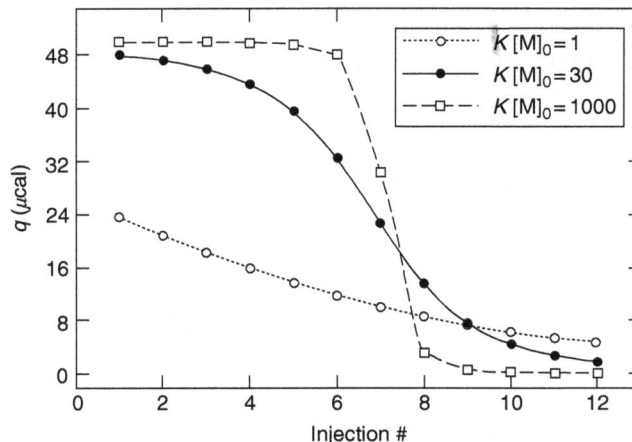

Fig. 3 Computed ITC curves for $K° = 5 \times 10^3$, 1.5×10^5, and 5×10^6 and $[M]_0 = 0.200$ mM. Other conditions: $V_0 = 0.20$ ml, $R_m = 2.08$, $v = 10$ μl, $\Delta H° = 10$ kcal/mol. Neglecting the error in v and taking $\sigma_q = 0.6$ μcal, as in El Harrous et al. (1994a), the predicted standard errors in $K°$ are 1.845×10^3, 1.107×10^4, and 1.008×10^6, respectively, while the corresponding errors in $\Delta H°$ are 2.486, 0.0932, and 0.0533 kcal/mol. The standard errors in the concentration correction parameter n_s (taken to be 1.00) are 0.105, 0.0074, and 0.0030 (same order).

those obtained there. It is interesting that even though $K°$ is uncertain by 20% for $K[M]_0 = 1000$, $\Delta H°$ and n_s are actually quite well defined in this case. On the other hand, all three parameters are much less precise at the other extreme.

Figure 4 illustrates the results of MC calculations for the case $K° = 5 \times 10^6$ in Fig. 3. The results for $K°$ are clearly non-Gaussian, with a bias of +3.4% (even though the peak in the distribution is shifted negatively). However, the MC statistical error in $K°$ exceeds the predicted value by less than 6%, showing that even for this 20% relative error, the predicted value from $\mathbf{V}_{nl}$ would be adequate for many applications. The histograms for $\Delta H°$ and n_s appear to be Gaussian, as expected from their smaller percent standard errors (0.5% and 0.3%, respectively). However, only the former actually fits the Gaussian with an adequately small χ^2—21.6 for 33 points. It is interesting that the reciprocal of $K°$ (or the dissociation constant) is actually much closer to Gaussian than $K°$ itself; similar results were obtained in the large-$K°$ regime in the study of complexation equilibria in Tellinghuisen (2000d).

MC calculations have been carried out for a number of other choices of the ITC parameters, for both constant and proportional error, random and correlated. None of the results has shown any problem with the 10% rule for the analysis of typical ITC data. Its validity extends even to the low limit of three titration increments ($m = 3$), where there are no degrees of freedom and the LS equations yield exact fits at all times. Of course, the $\mathbf{V}_{nl}$ matrix can say nothing about the extent of nonnormality or the bias, so if these are at issue, MC calculations must be employed for the specific cases in question.

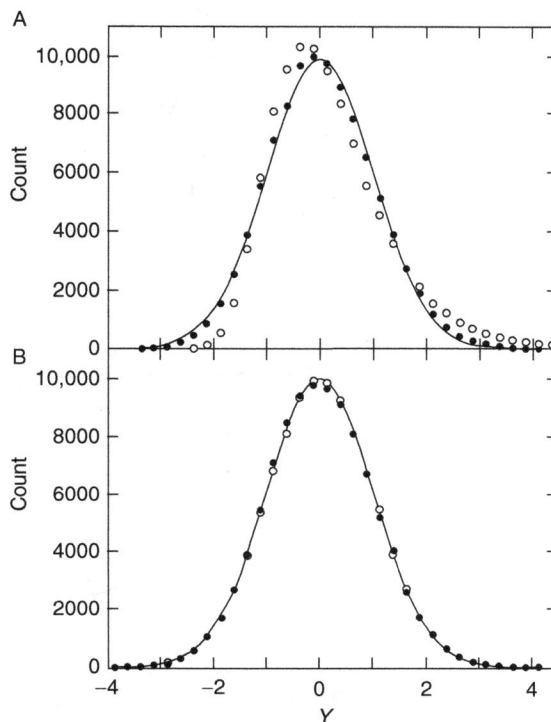

Fig. 4 Histogrammed results for (A) K° (○) and $K^{\circ-1}$ (●), and (B) ΔH° (○) and n_s (●), from 10^5 MC fits of 12-point data sets like that shown for $K^\circ = 5 \times 10^6$ in Fig. 3, with superimposed random error having $\sigma_q = 0.6$ μcal. In each case the histogrammed quantity is $Y = (\beta - \beta_{\text{true}})/\sigma_\beta$, with the values for these quantities given in the caption to Fig. 3. The smooth curves are standard Gaussians, scaled into optimal agreement with the data for $K^{\circ-1}$ in (A) and with ΔH° in (B).

C. Dependence on Stoichiometry Range and Titration Increments

Having demonstrated the approximate validity of the "exact" nonlinear variance-covariance matrix $\mathbf{V}_{\text{nl}}$ under the relatively extreme conditions of Fig. 4, I now use it to investigate the dependence of the ITC parameter standard errors on the other experimental quantities. Figure 5 illustrates the computed standard errors in K° and ΔH° as functions of the range of titration and the number of titration increments, for the midrange $K[M]_0$ value of 36 and constant error. Other parameters were chosen to resemble those for the instrument described in Wiseman et al. (1989) (except the error, which was artificially large). The most interesting result of these calculations is the observation that better precision is achieved with fewer points and a larger titration range than is customarily employed in such work. For small m, the standard errors exhibit structure, showing that they are sensitive to just where the points fall on the titration curve.

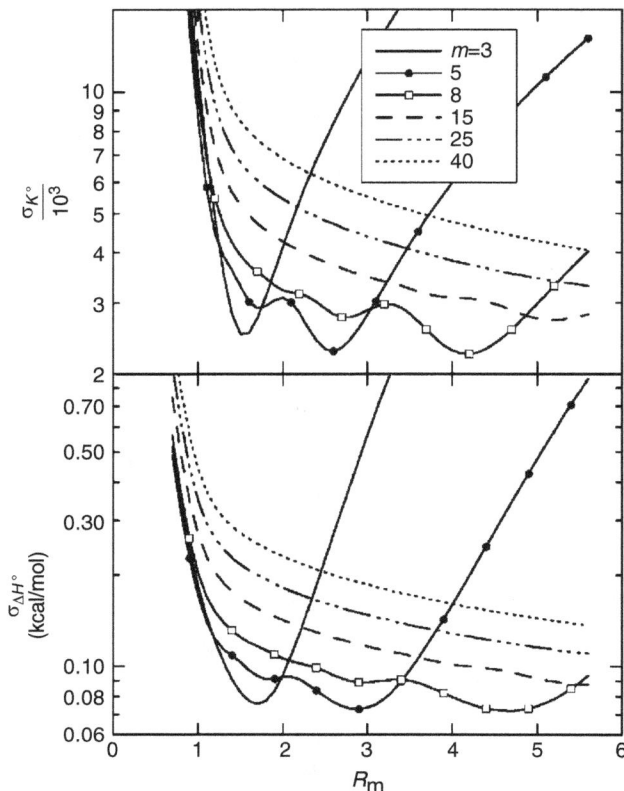

Fig. 5 Calculated standard errors in $K°$ and $\Delta H°$ for $K° = 36{,}000$ and $[M]_0 = 1.0$ mM, as functions of the stoichiometry range and the number of titration steps m, for constant absolute error $\sigma_q = 0.04$ mcal. Other parameters: $V_0 = 1.4$ ml, $\Delta H° = 10$ kcal/mol, and $vm = 0.10$ ml. For reference, the total q for complete reaction is 14 mcal, so for $m = 15$ and $R_m = 5$, the error on q_1 is ~1%.

The loss of precision with increasing number of points seems at odds with expectations, but it can be understood as follows. The total heat q is limited by the fixed amount of M in the reaction vessel, so increasing the number of titration increments m decreases each q_i. At first approximation, if $K°$ and n_s are held constant, one obtains m estimates of $\Delta H°$ from $\Delta H° = q_i/\Delta n_i$ [Eq. (26)]. The relative error in each such estimate of $\Delta H°$ is approximately σ_q/q_i, and as q_i decreases with increasing m, the error in $\Delta H°$ increases concomitantly. This is partially offset by the statistical averaging effect, which goes as $m^{-1/2}$. The net result is a standard error that increases roughly as $m^{1/2}$. Similar observations were made some time ago by Doyle *et al.* in connection with a study of a differential absorption technique for characterizing binding isotherms (Doyle *et al.*, 1990).

The possible role of correlated error in the titrant volumes will be considered in detail below. When the error in the incremental volume v is assumed to be random, there are several cases to consider: (1) v and the titrant concentration $[X]_0$ are fixed, and the operator decides on m, and hence R_m; (2) $[X]_0$ and the stoichiometry range R_m are set in advance, and the operator decides on m; and (3) v and R_m are set, and the operator varies m. In the first case, an increase in m means simply adding additional points at the end of the titration range, which always leads to a decrease in all of the parameter standard errors (i.e., improved precision). In case 2, the total titrant volume is fixed, so increasing m decreases v; this leads to an increase in the relative uncertainty σ_v/v, and hence an m dependence similar to that shown for constant error σ_q in Fig. 5. Case 3 is the least feasible from an experimental standpoint, as it means altering the titrant concentration when either m is changed for fixed R_m, or R_m is changed for fixed m. However, this case does lead to increasing precision with increasing m, as the relative error σ_v/v is now fixed. Results for this case show that minimal error in $\Delta H°$ occurs near $R_m = 1.5$, but a much larger titration range of $R_m \approx 4$ is needed for optimal precision in $K°$ (Tellinghuisen, 2003). Of course, with increasing dilution of titrant, all q_i will decrease as m^{-1}, so that eventually the constant error σ_q in q will dominate.

Figure 6 illustrates the dependence of the relative errors in $K°$ and $\Delta H°$ on $K°$ and R_m for $m = 7$ and constant absolute error. There is a fairly flat minimum in the error surface for $K°$, centered near $K[M]_0 = 10$, and $R_m = 4$. $\Delta H°$ is an order of magnitude more precise, with a large region at large $K°$ where the relative error is less than 1%. For both quantities, there is considerable structure at the error surface for the relatively small m value of 7. Note again that the structure of these contour diagrams is unaffected by the actual value used for σ_q, so, for example, reducing σ_q by a factor of 10 would simply result in a relabeling of the contours by the same factor. Also, in regions where the relative errors in $K°$ and $\Delta H°$ exceed ~ 0.2, the actual statistical distributions may be far from Gaussian, as already noted in the discussion of Fig. 4.

The results illustrated in Figs. 5 and 6 were obtained for fixed $[M]_0 = 1.0$ mM. However, the worker seeking to enhance precision would also consider increasing $[M]_0$, as that increases total reaction heat. This fuller dependence is considered in the work of Tellinghuisen (2005b, 2007b), where also a more reliable data variance function is employed.

Although for efficiency I have used a programming language to generate all the results described in this section, it is worth noting that the "exact" parameter standard errors for a specific data structure can be obtained quite easily using some desktop data presentation and analysis packages. For example, I have used KaleidaGraph to double-check some of the results obtained from the FORTRAN programs. Such calculations are facilitated by defining the key quantities from Eqs. (23)–(26) as library functions, for example,

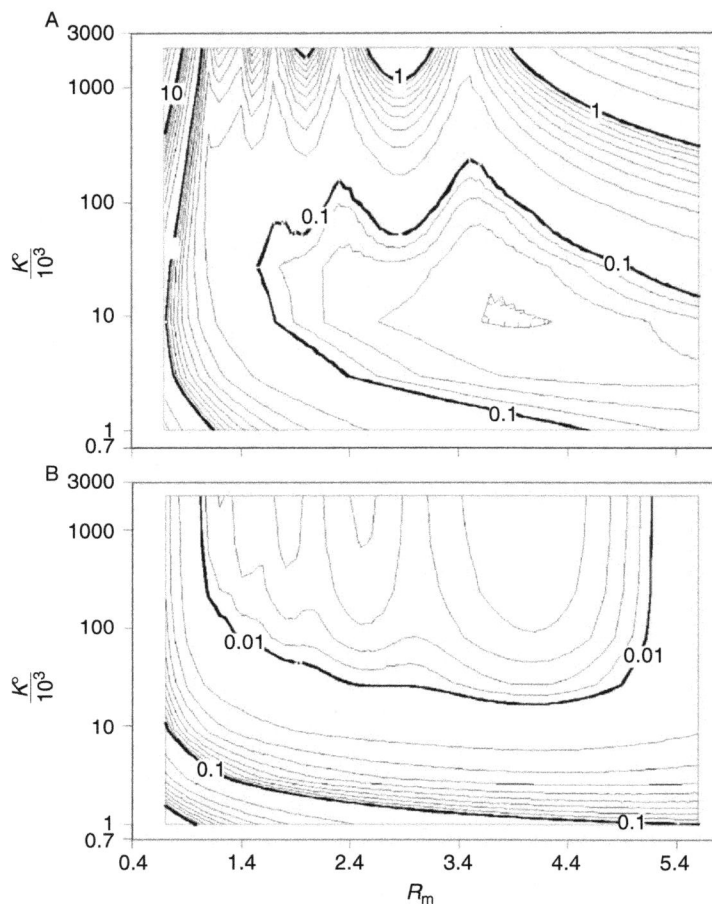

Fig. 6 Contour plots of the relative standard errors σ_β/β as functions of K° and ΔH°, for $m = 7$ and constant error, $\sigma_q = 0.04$ mcal: (A) for K°; (B) for ΔH°. Other parameters as in Fig. 5.

$$
\begin{aligned}
&\mathtt{di(x) = (df\hat{\ }x);}\\
&\mathtt{x0i(x) = (X0*(1.-di(x)));}\\
&\mathtt{y0i(x) = (Y0*c*di(x));}\\
&\mathtt{tsm(x) = (x0i(x)+y0i(x)+1./a);}\\
&\mathtt{tmp(x) = (tsm(x)\hat{\ }2-4.*y0i(x)*x0i(x));}\\
&\mathtt{mxi(x) = ((x>0)?(V0/2*(tsm(x)-sqrt(tmp(x)))*b) : 0);}
\end{aligned}
\tag{27}
$$

where x is the (integer) running index for the titration steps, and `di(x)`, `x0i(x)`, and `y0i(x)` are as defined in Eq. (23), with $\mathtt{c} = n_s$, $\mathtt{a} = K^\circ$, and $\mathtt{b} = \Delta H^\circ$. `tsm` and `tmp` are used in the quadratic solution for $[MX]_i$; and `df`, `X0`, `Y0`, and `V0` are d, $[X]_0$, $[M]_0$, and V_0, respectively, and can be replaced by their numerical values in these expressions or entered through definition statements preceding these in the library. The fit function to be entered in the user-defined function box (General) is then

$$\mathtt{mxi(x) - df * mxi(x - 1)} \tag{28}$$

As the first value of x is 1, the definition of mxi(x) contains a branching statement to set the second term in Eq. (28) = 0 for this first injection of titrant.

D. When Weights Are Neglected

I now take a closer look at the prototype case of Wiseman *et al.* (1989) (Figs. 4A and 5), and ask what happens when such data are analyzed by unweighted LS. As was already noted for case 6 in Table II, when heteroscedastic data are analyzed with neglect of weights, there are two main consequences: (1) the parameter estimators are no longer minimum variance, so the parameter distributions must broaden. (2) The error estimates from the *a posteriori* V of Eq. (12) are not reliable and can be either pessimistic or optimistic. The magnitudes of these effects can be determined only through MC computations.

For the aforementioned experiment in Wiseman *et al.* (1989), the following values of the key parameters apply: $v = 4$ μl, $\Delta H^\circ = -13.7$ kcal/mol, $[M]_0 = 0.651$ mM, $K^\circ = 4.88 \times 10^4$, and $R_m = 2.06$. As was noted earlier, σ_v and σ_q were estimated to be 0.015 μl and 0.28 μcal, respectively. The calculated q_i values range from 1.18 for $i = 1$ to 0.04 mcal for $i = 20$. As the relative error in v is 0.0038, the relative error will exceed the absolute error (σ_q) until $i = 17$, where $q_i = 0.07$ mcal. With data properly weighted for the combined errors, $\mathbf{V}_{\mathrm{nl}}$ yields $\sigma_{K^\circ} = 200$ and $\sigma_{\Delta H^\circ} = 21.0$ cal/mol.

For comparison, MC calculations employing unweighted fits yield the results illustrated in Fig. 7. The neglect of weights has led to increases in both σ values, as anticipated. The corresponding loss of efficiency for the determination of K° is $2.334^2 \approx 5$, meaning one would need to run five equivalent experiments with unweighted analysis to match the precision achievable through proper weighting of a single experiment. On the other hand, the biases in both K° and ΔH° were not statistically significant. Interestingly, the results obtained from the *a posteriori* V (by averaging the appropriate MC V_{jj} elements and taking the square roots) were surprisingly close to correct for σ_{K° (500 versus 472 observed) and only a bit worse for $\sigma_{\Delta H^\circ}$ (17.8 versus 25.7 cal/mol). Still, these results further illustrate how Eq. (12), when applied naively to heteroscedastic data, can both over- and under-shoot the actual values.

E. Correlated Error in Titrant Volume

The motorized syringes used to inject the titrant in ITC apparatuses are programmed to travel a certain distance with each injection. For such devices, it may be more appropriate to consider the end points of travel of the plunger as the

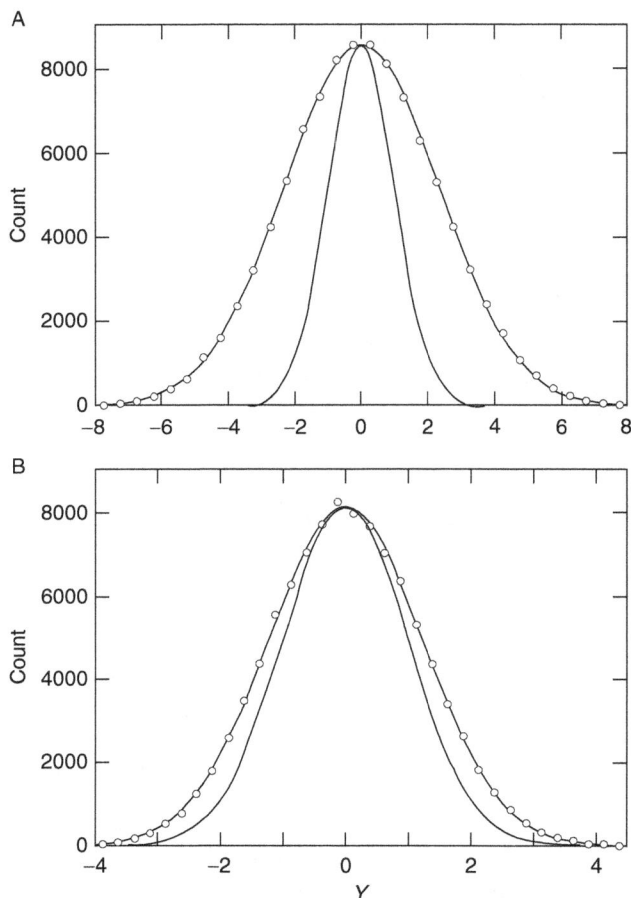

Fig. 7 Histogrammed results from 10^5 unweighted MC fits of 20-point data sets resembling that illustrated in Figs. 4A and 5 of Wiseman *et al.* (1989), for $K°$ (A) and $\Delta H°$ (B). In each case, the histogrammed quantity is $Y = (\beta - \beta_{\text{true}})/\sigma_\beta$, using σ_β values as given in text for a properly weighted fit. The inner curve in each case is the normal curve ($\sigma = 1$) scaled to the peak of the data, while the other curves are fitted Gaussians having $\sigma = 2.334$ (A) and 1.232 (B).

quantities subject to random error, rather than the incremental volume v. Then the latter, being the difference between two independent quantities, possesses correlated error. The effects of this change in assumption about the error can be seen dramatically in MC computations: With increasing number of steps m, the case 2 model above (fixed $[X]_0$ and R_m) now gives decreasing rather than increasing parameter error. This effect, too, was observed previously by Doyle *et al.* (1990) in fitting differences extracted from an absorbance titration curve.

In the correlated LS fit needed for proper analysis of such data, the weight matrix $\mathbf{W}$ of Eqs. (4)–(6) and Eq. (14) is the inverse of the variance-covariance matrix associated with the incremental volumes (Albritton *et al.*, 1976; Hamilton, 1964; Tellinghuisen, 1996). Let us represent the total delivered titrant volume after i injections as u_i. Then, the ith incremental volume is $v_i = u_i - u_{i-1}$. The v_i and u_i are related via a linear transformation,

$$\mathbf{v} = \begin{pmatrix} 1 & 0 & 0 & & 0 & 0 \\ -1 & 1 & 0 & & 0 & 0 \\ 0 & -1 & 1 & \cdots & 0 & 0 \\ & & & \vdots & & \\ 0 & 0 & 0 & & -1 & 1 \end{pmatrix} \mathbf{u} \equiv \mathbf{Lu} \tag{29}$$

and thus $\mathbf{V}_v$ and $\mathbf{V}_u$ are related by (Hamilton, 1964; Tellinghuisen, 2001)

$$\mathbf{V}_v = \mathbf{L}\mathbf{V}_u\mathbf{L}^T \tag{30}$$

Because the u_i are independent, $\mathbf{V}_u$ is diagonal, with elements σ_u^2 (constant by assumption). As it is q_i that is fitted here, these σ_u^2 values must be converted to σ_{qi}^2 by error propagation, using numerical differentiation to assess dq_i/du_i. Calling the resulting matrix $\mathbf{V}'_u$, $\mathbf{V}'_v = \mathbf{L}\mathbf{V}'_u\mathbf{L}^T$ and $\mathbf{W} = \mathbf{V}'^{-1}_v$. It is noteworthy that the matrix $\mathbf{V}'_v$ is tridiagonal, with elements $(i,i) = (\sigma_{qi-1}^2 + \sigma_{qi}^2)$, $(i,i+1) = -\sigma_{qi}^2$, and $(i,i-1) = -\sigma_{qi-1}^2$ (with all indices limited to the range $1-m$ of the data). The diagonal terms are thus seen to be the expected results for subtraction of two uncorrelated quantities.

Use of this $\mathbf{W}$ with the model described and used above yielded formal parameter error estimates (from $\mathbf{V}_{nl}$) in good agreement with the results from the MC computations and well within the framework of the 10% rule of thumb. For example, in computations for $R_m = 3$ on the model described in Fig. 5 but having $\sigma_q = 0$ and $\sigma_v = 0.0015$ ml, the choice $m = 5$ yielded 13% relative standard error in K° and 6.4% in ΔH°, as compared with estimates higher by 5.0% and 4.6%, respectively, from MC computations on 10^5 data sets. However, the reduced χ^2 from the MC computations was 1.034, which deviates more from the expected value of 1.00 than was found in the earlier MC results.

With this demonstrated reliability, the correlated model was used to estimate the parameter errors as a function of m for a more realistic $\sigma_v = 0.015$ μl (Wiseman *et al.*, 1989), while, for comparison, MC computations were carried out on the same model (i.e., random error in the accumulated titrant volume u_i) using ordinary unweighted LS to extract the parameters and their *a posteriori* estimated standard errors. The results (Fig. 8) show that unweighted LS correctly tracks the m dependence predicted by the correlated model $\mathbf{V}_{nl}$, but gives standard errors too large by about a factor of 10 for K° and 2 for ΔH°. On the other hand, the *a posteriori* $\mathbf{V}$ (Eq. 12) gives estimates that happen to be roughly correct only for the range $m \approx 8$–12, deviating sharply to overly optimistic errors for smaller m and pessimistic for larger.

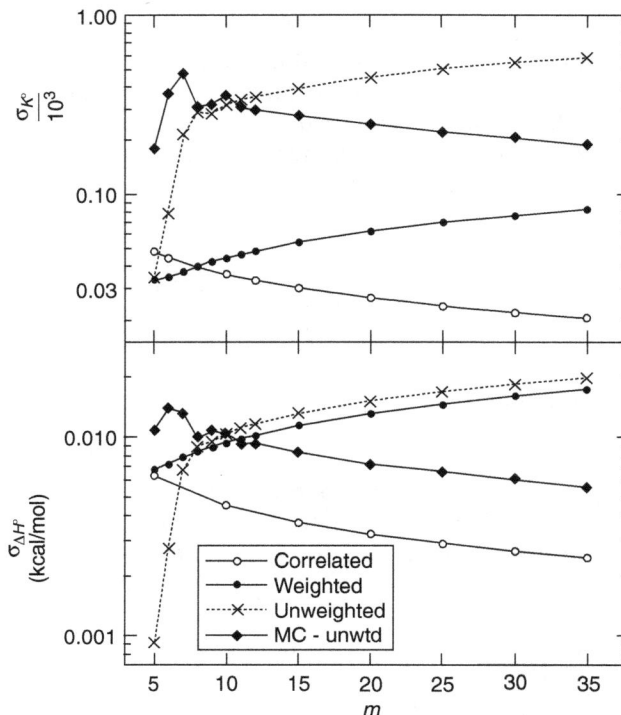

Fig. 8 Statistical errors in $K°$ and $\Delta H°$ as a function of m, for model of Fig. 5, but with $\sigma_q = 0$ and random error $\sigma_v = 0.015 \ \mu l$ in the accumulated titrant volume u_i (hence correlated error in v). The results marked "weighted" are obtained when this same random error is in v. The "MC" results are from the statistics of 10^4 data sets analyzed using unweighted LS, while those marked "unweighted" represent the apparent values (RMS) returned by the *a posteriori* $\mathbf{V}$ of Eq. (12), from the same MC computations. The structure at small m in the "unweighted" and "MC" results is real.

As the error in measuring q_i is presumed to be random, and as this error leads to decreasing precision with increasing m (Fig. 5), we might expect that addition of random error in q_i to this correlated v model should neutralize the increased precision at large m in Fig. 8. That is in fact observed. To generate the data for the MC computations, the effects of the volume error are first computed, as before: the u_i are given random error, v_i is calculated from $v_i = u_i - u_{i-1}$, and the heat q_i is calculated using a variable v version of Eqs. (23) and (25). Then random error is added to each q_i value. This error is correctly accommodated in the correlated fit model by adding σ_q^2 to the diagonal elements of $\mathbf{V}'_v$, as was confirmed in the MC computations. Figure 9 shows that the addition of the random measurement error has rendered $K°$ less precise by a factor of ~ 3, and made $\sigma_K°$ almost independent of m. A smaller reduction of precision occurs for $\Delta H°$, and its standard error still decreases with increasing m.

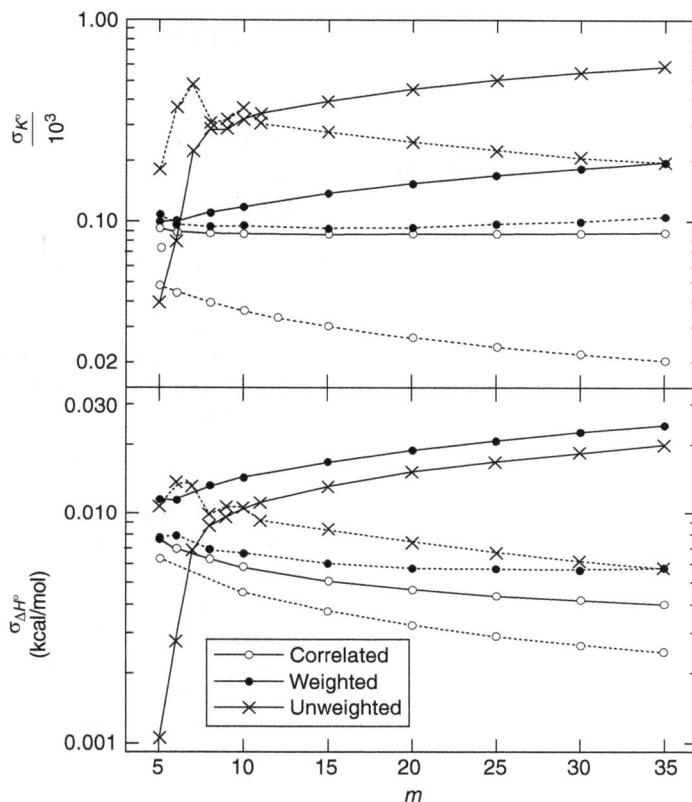

Fig. 9 Results for same model as in Fig. 8, but with the addition of random measurement error $\sigma_q = 0.28$ μcal (as in Wiseman *et al.*, 1989). The dashed "correlated" curves are from Fig. 8 ($\sigma_q = 0$). The other results are obtained by analyzing the same MC data sets with neglect of correlation, using the weighted fit model and unweighted LS. In both cases, the solid curves represent the apparent standard errors, from the *a posteriori* **V**, while the dashed curves represent the actual MC statistics from 10^4 data sets for each point.

Figure 9 also includes results of analyzing these same data with neglect of the correlation, using either unweighted LS or the weighted model that is correct for random error in both v and q_i. The MC statistics show that the weighted model does a good job of extracting the K° values but performs less well for ΔH°. However, without repeating their experiments enough times to obtain ensemble statistics, users of the weighted model would normally rely on the *a posteriori* **V** for error estimates. Accordingly, they would report errors for ΔH° that are significantly too large over the full range of m in Fig. 9, and somewhat less pessimistic errors for K° for most m. Not surprisingly, unweighted LS performs worse almost across the board; the one exception is in the assessment of ΔH°, where the unweighted model actually betters the weighted model for $m > 35$.

VII. Calorimetric Versus Van't Hoff $\Delta H°$ from ITC

A. Test Model

I return now to an issue raised in the Introduction, namely the matter of the precision of $\Delta H°$ as estimated directly from the ITC titration curves versus that determined indirectly from the T dependence of $K°(T)$. To investigate this problem in the ITC framework, I have devised a model that consists of five ITC experiments run on the same thermodynamics model described in Table I. As was already noted, this model was designed to resemble the Ba^{2+}/18-crown-6 ether complexation (except that the actual $K°_{298}$ of 5500 is used in the present ITC model). The model was examined for both constant error and proportional error (random v model only). The reaction volume was taken as 1.4 ml and the initial concentration of the crown ether as 0.020 mol/liter. For each titration curve, 25 incremental 10-μl injections of titrant (Ba^{2+}) were taken, covering a stoichiometry range of 2.0 (Ba^{2+}: ether). The same conditions were assumed for all five temperatures. Under these conditions, the magnitude of q_1 was 70 mJ at 5°, dropping to 59 mJ at 45°. For the constant error model, σ_i was 1.0 mJ for all i. In the proportional error model, the relative σ was set at 2%. This choice made the uncertainty in $\Delta H°_{cal}$ nearly the same in the two models.

The "site parameter" n_s was taken to be 1.00 at all times, and was defined as a correction to $[M]_0$, as in Eq. (27). In these model calculations, $\Delta H°_{cal}$ and $\Delta H°_{vH}$ are identical by definition. In that case, titration data recorded at different temperatures can (and should) be analyzed simultaneously (global analysis) to yield a single reference $K°$ value and a single determination of $\Delta H°(T)$, as defined here by the two parameters a and b. The statistical errors for such a determination were examined both for the assumption of a single n_s value, and for separate n_s values for each titration curve. The latter would be the less presumptuous approach in a set of experiments. In fact, the standard errors in the key thermodynamic parameters (a, b, and $K°_0$) differed very little for these two approaches.

Results for $K°$ are illustrated in Fig. 10. As the largest predicted relative error is only ~8% (for $K°$ in the constant error model), no MC confirmations are deemed necessary, and I present just the "exact" (V_{nl}-based) results here. From the results in Fig. 10, two observations are noteworthy: (1) the constant error model yields results that are less precise by almost an order of magnitude; and (2) in both models the relative error in $K°$ is more nearly constant than the absolute error. The latter observation means that van't Hoff analysis through an unweighted fit to the logarithmic form of Eq. (20) is not a bad approximation, which is reassuring, as the unweighted log fit is the usual approach taken for such analyses.

In keeping with all earlier indications, $\Delta H°_{cal}$ is determined with much better relative precision (<0.8%) than $K°$ in both error models. Results for these directly estimated values are illustrated in Fig. 11, together with their counterpart $\Delta H°_{vH}$ values, as obtained by fitting the five $K°(T)$ values to Eq. (20). Not surprisingly, the poorer precision in $K°$ translates into poorer precision in $\Delta H°_{vH}$. The curves at the bottom of this figure show the reduction of error due to the averaging effect when

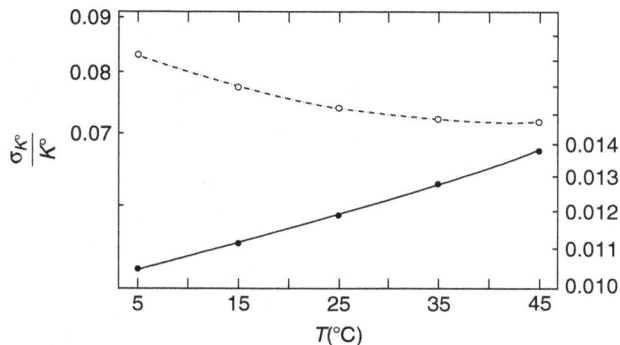

Fig. 10 Relative standard errors in $K°$ from model calculations for van't Hoff analysis, for proportional error (solid, scale to right) and constant error (dashed, scale to left).

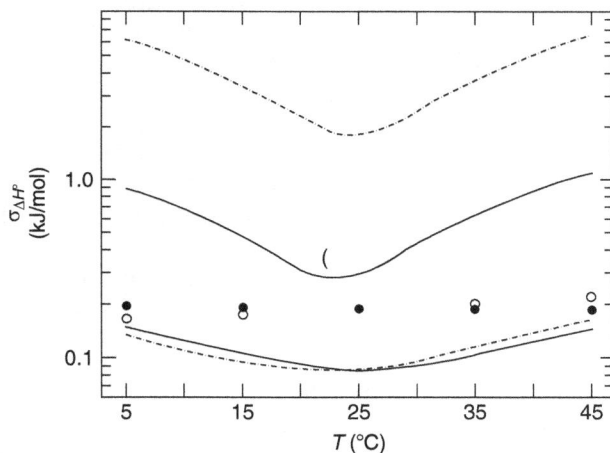

Fig. 11 Standard errors in $\Delta H°$ from the model calculations, as obtained from the individual ITC data sets (points), from weighted fits of these points to a linear relation (curves at bottom), and from weighted fits of $K°(T)$ to the exponential form of Eq. (20) (upper curves). The solid points and curves represent the proportional error model; the open points and dashed curves represent the constant error model.

the five individual $\Delta H°_{cal}$ values are fitted to a straight line; this reduction is not uniform across the T range of the data, rather it is most prominent in the midrange. The van't Hoff results behave similarly, although even more dramatically. In both cases, the reason is as already discussed in connection with Eq. (22). It is noteworthy that the results from the global fits of all five titration curves to the same linear $\Delta H°$ function and a single $K°_0$ reference value yield only slight reduction in the error of the fitted $\Delta H°$ relative to the results in the lower curves in Fig. 11. Further, if $\Delta H°$ is fixed at the fitted results (i.e., if a and b are taken as known), the fits of the five $K°(T)$ values to Eq. (20) return $K°$ with standard errors insignificantly different from those from the global analyses. Thus, while the global analysis of the ITC data is the "proper" statistical procedure, there is little practical gain in this approach over the customary one of just fitting the $\Delta H°_{cal}$ values from the several ITC experiments run at different temperatures to an appropriate function of T. If $\Delta H°_{vH}$ and $\Delta H°_{cal}$ are deemed consistent (as they of course are in the present model), the subsequent fit of the individual $K°(T)$ values with $\Delta H°$ taken as "known" further sharpens the definition of $K°(T)$. Of course, in any effort to determine whether the two estimates of $\Delta H°$ are consistent, the $\Delta H°_{vH}$ values will be limited by the greatly reduced precision shown in the top two curves in Fig. 11.

The fits of the five $K°(T)$ values to Eq. (20) yield parameter errors that are quite large. For example, in the constant error model, this fit yields $\Delta C°_{P,vH} = 130 \pm 299$ J mol^{-1}K^{-1}. While this parameter thus appears to be undefined, it is still necessary to include it in any attempt to evaluate the consistency between $\Delta H°_{vH}$ and $\Delta H°_{cal}$, because it is well defined in the linear fit of the $\Delta H°_{cal}$ values. Omitting it would make the two models incompatible. With it included, the error in $\Delta H°_{vH}$ is also relatively large ($\sim$20%) at the ends of the T range. However, the MC computations summarized in Tables I and II have already shown that this is not a significant source of error in estimating either the parameters or their standard errors in this case.

Although not treated here, the correlated v model of Fig. 8 yields comparative precisions for $K°$ and $\Delta H°$ that resemble those for the present random proportional error model, that is, relative errors a factor of $\sim$2 larger for $K°$ than for $\Delta H°$. With both constant and proportional error present, the comparative precisions fall between the limits of the two error models, for both correlated error (Fig. 9) and random. Thus, the results in Figs. 10 and 11 can be considered to bracket the range of actual observations.

B. Case Study: Ba^{2+} Complexation with Crown Ether

The complexation of Ba^{2+} with 18-crown-6 ether in water (unbuffered) has been studied by at least five groups, but the experimental results hardly show a developing consensus (Fig. 12). The $\Delta H°_{cal}$ values from Liu and Sturtevant (1995) agree well with those from Briggner and Wadsö (1991) in the restricted T region where they overlap, but the former show clear curvature over the full T range. Although statistical errors were not reported for $\Delta H°_{cal}$ in Liu and Sturtevant (1995), the results from the unweighted fit indicate that the errors are comparable to the size of the displayed points. The most recent results, from Horn et al. (2001) are less precise but are still larger in magnitude by a statistically significant 5%. The $K°$ values from Horn et al. (2001) appear to be much less precise than the other results; however, the quoted errors on $K°$ in that study are overly pessimistic, as is discussed further below.

The directly measured $\Delta H°$ values from Liu and Sturtevant are compared with the van't Hoff estimates in Fig. 13. The latter were obtained from weighted fits of their $K°$ values to Eq. (20), taking their tabulated uncertainties as the estimated σ values. Neither the quadratic nor the linear representation of $\Delta H°$ gave a statistically

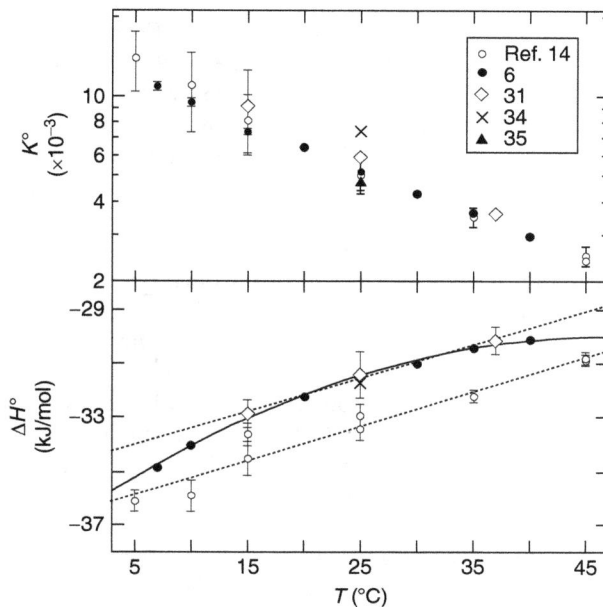

Fig. 12 Literature results for $K°$ and $\Delta H°_{cal}$ for the complexation of Ba^{2+} by 18-crown-6 ether in pure water. Note log scale for $K°$. Error bars represent reported standard errors. Results from Izatt *et al.* (1976) and Hasegawa and Date (1988) were reported only at 25° (no $\Delta H°$ for the latter).

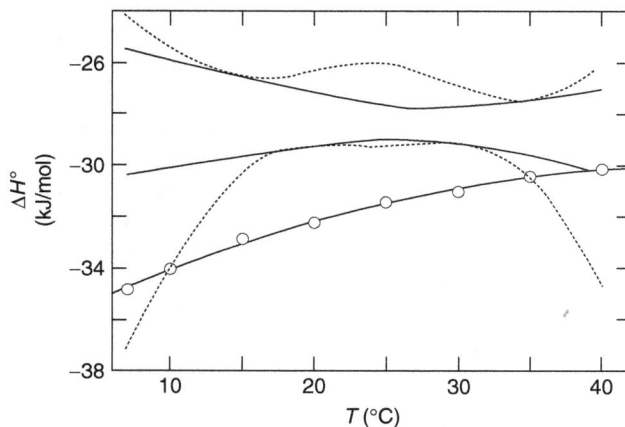

Fig. 13 $\Delta H°_{cal}$ estimates from Liu and Sturtevant (1995) for Ba^{2+}/ether complexation (points and their quadratic least-squares representation), and 1-σ error bands on van't Hoff estimates using quadratic (dashed curves) and linear (solid) representations for $\Delta H°$ in the LS fit function [Eq. (20)].

significant determination of $\Delta C^\circ_{P,\mathrm{vH}}$. Even worse, both fits gave χ^2 values >10 (cf. expected values of 4 and 5, respectively). This suggests that the estimated errors on K° in Liu and Sturtevant (1995) were optimistic. With this interpretation, the conservative approach is to use the *a posteriori* **V** of Eq. (12) to estimate the errors, which is what I have done to generate the error bands in Fig. 13. Even with these enlarged errors, the $\Delta H^\circ_{\mathrm{vH}}$ values are not consistent with the $\Delta H^\circ_{\mathrm{cal}}$ estimates, in agreement with the conclusions of Liu and Sturtevant. If we take the view that only a constant ΔH° is justified by the van't Hoff analysis, we obtain $\Delta H^\circ = -28.4(6)$ kJ mol^{-1}, which, for comparison purposes, should be taken as the estimate in the middle of the T range, or around 23 °C. The inconsistency remains.

Briggner and Wadsö (1991) published values for K° and ΔH° at only three temperatures. Fitting their K° values (weighted) to a constant ΔH° yields -31.7 (1.8) kJ mol^{-1}, in good agreement with their $\Delta H^\circ_{\mathrm{cal}}$ value of -31.42 (20) kJ mol^{-1} at 25 °C. Horn *et al.* (2001) reported that their van't Hoff estimates of ΔH° were statistically consistent with their $\Delta H^\circ_{\mathrm{cal}}$ values. However, if their published K° values are analyzed by weighted LS using their uncertainties, and the error bands on $\Delta H^\circ_{\mathrm{vH}}$ are calculated using Eq. (18) to properly accommodate the interparameter correlation, the case for consistency is less convincing. In this weighted fit, the χ^2 is 0.41, which is much smaller than the expected 6, indicating that the error estimates on K° were pessimistic. Accordingly, use of the *a posteriori* **V** in the error propagation calculation results in a narrowing of the error bands by a factor of ~ 4.

Although the χ^2 from the weighted fit of the K° values in Horn *et al.* (2001) is an order of magnitude too small, a weighted fit of the published $\Delta H^\circ_{\mathrm{cal}}$ values to a linear function of T gives a reasonable value of 9.7 (for $\nu = 7$). This observation is consistent with the use of unweighted LS to analyze ITC data that are actually dominated by proportional error (Mizoue and Tellinghuisen, 2004b). The situation is analogous to that illustrated in Fig. 9, where the actual statistical errors in the parameters from the unweighted analysis are significantly smaller than that indicated by the *a posteriori* **V**. And it is the actual errors in K° and ΔH° that are manifested in subsequent fits of these quantities to functions of T. The proportional error model predicts that the relative error in K° should be a factor of 2–3 larger than that in ΔH°, while the constant error model (implicit in unweighted LS) predicts a factor of ~ 10. The latter is consistent with the published results in Horn *et al.* (2001), while the former is closer to observations from the fits of the extracted values to functions of T.

Beyond the complexation of Ba^{2+} with ether, few other reactions have been examined closely for the consistency of $\Delta H^\circ_{\mathrm{cal}}$ and $\Delta H^\circ_{\mathrm{vH}}$. Those that have been examined also fail to yield convincing agreement (Mizoue and Tellinghuisen, 2004b).

VIII. Conclusion

A computational study of statistical error in the nonlinear LS analysis of ITC data for 1:1 complexation reactions yields the following main results: (1) when the data uncertainty is dominated by constant measurement uncertainty in the heat q, better precision is achieved in both $K°$ and $\Delta H°$ for fewer titrant injections than are customarily used; as few as five may be optimal in many cases; (2) under usual experimental circumstances, the same result holds for data uncertainty dominated by the relative error in the titrant volume v, if this error is assumed to be random; (3) on the other hand, if volume error dominates and it is the error in the integral titrant volume that is random, the precision increases with increasing number of titration steps; (4) for typical conditions, the relative precision for $K°$ is a factor of 2–10 poorer than that for $\Delta H°$; higher precision in $K°$ is generally favored by larger stoichiometry ranges than are customarily used; the same holds for $\Delta H°$ in the constant error model, but $R_m < 2$ is optimal for $\Delta H°$ when the volume error dominates; (5) actual ITC data are typically dominated by the relative error in v for large q (early titrant injections i) and by the constant absolute uncertainty in q for small q (large i); for optimal extraction of $K°$ and $\Delta H°$, such data require analysis by weighted LS, or in the case of random error in the integral titrant volume, by correlated LS; and (6) the larger relative error in $K°$ versus $\Delta H°$ means that $\Delta H°$ estimates obtained from van't Hoff analysis of the T dependence of $K°$ ($\Delta H°_{vH}$) will in turn be inherently less precise than the directly extracted estimates ($\Delta H°_{cal}$).

One of the most intriguing results of this study is the observation that the dependence of the parameter standard errors on the number of titration steps is so completely tied to assumptions about the nature of the error in the titrant volume: If random in the differential volume v, the precision decreases with increasing m; if random in the integrated volume u, it increases with m. The true situation probably lies somewhere between these two extremes, but it can be determined only by properly designed experiments. If the random v error model is found to dominate, experiments should be designed to use much smaller m than is currently the practice.

The present computations have dealt with only the case of 1:1 binding, and even if the random v error is found to dominate, it may be necessary to use $m = 10$ or more to demonstrate that the stoichiometry actually is 1:1. Also, in practice it is wise to work with at least a few degrees of statistical freedom, in order to obtain some indication of the goodness of fit. Other cases, like multiple binding (DiTusa *et al.*, 2001), will have to be examined specifically. However, it is possible that there, too, better precision can be achieved using fewer injections than are commonly employed.

The reduced precision in $\Delta H°_{vH}$ means that $\Delta C°_P$ and its T derivative may not be statistically defined in a van't Hoff analysis, even though they may be well determined by the $\Delta H°_{cal}$ data. However, these parameters may still be necessary as means to an end in the attempt to confirm consistency between $\Delta H°_{cal}$ and $\Delta H°_{vH}$. If they are neglected in the van't Hoff analysis, then the resulting $\Delta H°$ estimate should be reported as the value for the average T of the data set (or an appropriately weighted average, for weighted data). This will eliminate the bias problem noted by Chaires (1997)

With proper attention to error propagation to generate error bands on $\Delta H°_{vH}$, the case for consistency between $\Delta H°_{cal}$ and $\Delta H°_{vH}$ in ITC is not as sanguine as has been suggested in recent work (Horn *et al.*, 2001). A number of factors have been considered in the effort to explain the discrepancies, but one that has largely escaped attention is the role of the ITC site parameter n_s. This parameter is typically determined with greater relative precision than either $K°$ or $\Delta H°$, and it is clearly needed in the analysis of most ITC data to achieve a satisfactory fit. In the 1:1 binding case considered here, n_s is in effect correcting for errors in the stated concentrations of the two solutions. It is normally defined in the stoichiometry sense MX_{ns}, which means that it is serving as a correction factor for M [i.e., $n_s = 1.05$ means the true $[M]_0$ is 5% larger than stated; see Eq. (27)]. As M is often a macromolecule and is typically harder to prepare to known concentrations, this is probably appropriate in most cases. However, when the titrant concentration is less certain, it is proper to redefine the correction. The effect of such redefinition is a correction factor of $1/n_s$ to $[X]_0$, and it results in a change in both $K°$ and $\Delta H°$ by the factor n_s. If experiments are repeated with the same solutions over a range of temperatures, the result will be a systematic shift in both sets of results. A constant proportional error in $K°$ will have no effect on $\Delta H°_{vH}$, so the result will be an error in the apparently more precise $\Delta H°_{cal}$ and possible discord between the two estimates. When the site parameter is covering for a mix of macromolecules (e.g., 2.5% having two sites to yield $n_s = 1.05$), the 1:1 fit model is not really correct. Either way, the deviation of n_s from "chemical" stoichiometry for the process under investigation is an indication of systematic error, and a conservative assessment of the parameter errors in such cases should include its consideration.

Throughout this work I have assumed that equal volume aliquots (v) of titrant are added sequentially to generate the titration curve. This appears to be the only mode used by workers in the field, also. However, from the structure evident in Figs. 5 and 6, it seems likely that for small m and a chosen R_m, some sequence of volumes that vary from step to step might yield smaller parameter errors. Indeed, preliminary results from an examination of this problem show that for a 7-step titration to $R_m = 3$, a variable-v algorithm can reduce the statistical error in $K°$ by 40% from the constant-v approach.

Most of the statistical errors reported, plotted, and discussed in this work have been on the basis of the predictions of the "exact" nonlinear variance-covariance matrix V_{nl}, bolstered by MC calculations in selected cases. To use this approach, it is necessary to know the error structure of the data, which, of course, one always does in an MC calculation. Unfortunately, most experimentalists still take the ignorance approach in their LS analyses, using unweighted fits and the *a posteriori* V from Eq. (12) to estimate parameter errors. Much is to be gained from taking the trouble to assess the experimental statistical error apart from the data for a given run (Hayashi *et al.*, 1996; Tellinghuisen, 2000a,b). The obvious advantage is the narrowed confidence bands that attach to the *a priori* V versus the *a posteriori* V and its concomitant need for the t distribution to assess confidence limits (Hayashi *et al.*, 1996). But in addition, one has the χ^2 statistic to assess goodness of fit—a quantitative answer to the question, "Are my fit residuals commensurate with my data error" (Bevington, 1969; Mood and Graybill, 1963)? In view of the persistent discrepancies between ΔH°_{cal} and ΔH°_{vH}, it is possible that the fit models in current use do not adequately reflect the actual physical situation in an ITC experiment. To address this issue, reliable information about the data error is essential.

There is another "downside" to the naive use of unweighted LS and the *a posteriori* **V** in cases like ITC, where the data are inherently heteroscedastic: Eq. (12) always "lies" in such cases. The extent of the lie can only be determined through MC calculations. If proportional error is assumed to dominate over the entire titration curve, the error in σ_{K° can be as much as a factor of ~ 10 in the case of correlated error (Fig. 8), resulting in a 100-fold loss in efficiency in the estimation of K° (Tellinghuisen, 2003). In linear LS, neglect of weights does not bias the estimation of the parameters. However, nonlinear LS parameters are inherently biased to some degree; and as the bias scales with the variance (Tellinghuisen, 2000d), neglect of weights will exacerbate the bias, possibly converting an insignificant bias into a significant one.

References

Abramowitz, M., and Stegun, I. A. (1965). "Handbook of Mathematical Functions," Dover, New York.

Albritton, D. L., Schmeltekopf, A. L., and Zare, R. N. (1976). *In* "Molecular Spectroscopy: Modern Research II," (K. Narahari Rao, ed.), pp. 1–67. Academic Press, New York.

Bevington, P. R. (1969). "Data Reduction and Error Analysis for the Physical Sciences," McGraw-Hill, New York.

Birge, R. T. (1932). *Phys. Rev.* **40**, 207.

Briggner, L. E., and Wadsö, I. (1991). *J. Biochem. Biophys. Methods* **22**, 101.

Chaires, J. B. (1997). *Biophys. Chem.* **64**, 15.

Deming, W. E. (1964). "Statistical Adjustment of Data," Dover, New York.

DiTusa, C. A., Christensen, T., McCall, K. A., Fierke, C. A., and Toone, E. J. (2001). *Biochemistry* **40**, 5338.

Doyle, M. L., Simmons, J. H., and Gill, S. J. (1990). *Biopolymers* **29**, 1129.

El Harrous, M., and Parody-Morreale, A. (1997). *Anal. Biochem.* **254**, 96.

El Harrous, M., Gill, S. J., and Parody-Morreale, A. (1994a). *Meas. Sci. Technol.* **5**, 1065.

El Harrous, M., Mayorga, O. L., and Parody-Morreale, A. (1994b). *Meas. Sci. Technol.* **5**, 1071.

Hamilton, W. C. (1964). "Statistics in Physical Science: Estimation, Hypothesis Testing, and Least Squares." Ronald Press, New York.

Hasegawa, Y., and Date, H. (1988). *Solvent Extr. Ion Exch.* **6**, 431.

Hayashi, Y., Matsuda, R., and Poe, R. B. (1996). *Analyst* **121**, 591.

Holtzer, A. (1995). *J. Phys. Chem.* **99**, 13048.

Horn, J. R., Russell, D., Lewis, E. A., and Murphy, K. P. (2001). *Biochemistry* **40**, 1774.

Horn, J. R., Brandts, J. F., and Murphy, K. P. (2002). *Biochemistry* **41**, 7501.

Izatt, R. M., Terry, R. E., Haymore, B. L., Hansen, L. D., Dalley, N. K., Avondet, A. G., and Christensen, J. J. (1976). *J. Am. Chem. Soc.* **98**, 7620.

Johnson, M. L., and Faunt, L. M. (1992). *Methods Enzymol.* **210**, 1.

Liu, Y., and Sturtevant, J. M. (1995). *Protein Sci.* **4**, 2559.

Liu, Y., and Sturtevant, J. M. (1997). *Biophys. Chem.* **64**, 121.

Mizoue, L. S., and Tellinghuisen, J. (2004a). *Anal. Biochem.* **326**, 125.

Mizoue, L. S., and Tellinghuisen, J. (2004b). *Biophys. Chem.* **110**, 15.

Mood, A. M., and Graybill, F. A. (1963). "Introduction to the Theory of Statistics," 2nd edn. McGraw-Hill, New York.

Naghibi, H., Tamura, A., and Sturtevant, J. M. (1995). *Proc. Natl. Acad. Sci. USA* **92**, 5597.

Press, W. H., Flannery, B. P., Teukolsky, S. A., and Vetterling, W. T. (1986). "Numerical Recipes,"
    Cambridge University Press, Cambridge.
Ragone, R., and Colonna, G. (1995). *J. Phys. Chem.* **99,** 13050.
Ross, J. (1996). *Proc. Natl. Acad. Sci. USA* **93,** 14314.
Tellinghuisen, J. (1996). *J. Mol. Spectrosc.* **179,** 299.
Tellinghuisen, J. (2000a). *Appl. Spectrosc.* **54,** 1208.
Tellinghuisen, J. (2000b). *Appl. Spectrosc.* **54,** 431.
Tellinghuisen, J. (2000c). *J. Chem. Educ.* **77,** 1233.
Tellinghuisen, J. (2000d). *J. Phys. Chem. A* **104,** 2834.
Tellinghuisen, J. (2001). *J. Phys. Chem. A* **105,** 3917.
Tellinghuisen, J. (2003). *Anal. Biochem.* **321,** 79.
Tellinghuisen, J. (2004). *Anal. Biochem.* **333,** 405.
Tellinghuisen, J. (2005a). *Anal. Biochem.* **343,** 106.
Tellinghuisen, J. (2005b). *J. Phys. Chem. B* **109,** 20027.
Tellinghuisen, J. (2006). *Biophys Chem* **120,** 114.
Tellinghuisen, J. (2007a). *Anal. Biochem.* **360,** 47.
Tellinghuisen, J. (2007b). *J. Phys. Chem. B* **111,** 11531.
Tellinghuisen, J. (2008). *Anal. Biochem.* **373,** 395.
Weber, G. (1995). *J. Phys. Chem.* **99**(1052), 13051.
Weber, G. (1996). *Proc. Natl. Acad. Sci. USA* **93**(7452), 14315.
Wiseman, T., Williston, S., Brandts, J. F., and Lin, L. N. (1989). *Anal. Biochem.* **179,** 131.

CHAPTER 17

Physiological Modeling with Virtual Cell Framework

James C. Schaff, Boris M. Slepchenko, and Leslie M. Loew

R. D. Berlin Center for Cell Analysis and Modeling
University of Connecticut Health Center
Farmington, CT, USA

Abstract

This chapter describes a computational framework for cell biological modeling and simulation that is based on the mapping of experimental biochemical and electrophysiological data onto experimental images. The framework is designed to enable the construction of complex general models that encompass the general class of problems coupling reaction and diffusion.

I. Update

The Virtual Cell computational modeling and simulation software environment has been continuously enhanced and improved over the 10 years since the initial publication of this review. However, the primary philosophy of the project remains the same: to provide a user-friendly, yet comprehensive, software tool for both experimentalists and theoreticians wishing to model complex cell biological events. A major improvement to the software architecture to this end was achieved in 2002 through a multilayered hierarchical interface in the BioModel GUI. The parent layer is called the *Physiology*, which can generate several *Applications*, which, in turn, can spawn multiple *Simulations*. The Physiology houses the details on the chemical species, their locations within volumetric compartments or membranes and kinetic expressions for their reactions and transport processes. Applications provide the model specifications including the kinds of equations to solve (ordinary or partial differential equations or discrete stochastic equations), mappings of the compartments to geometries, domain boundary conditions for spatial models, initial concentrations of all the molecular species and electrophysiology specifications such as voltage or current clamp conditions. The Physiology plus an Application is sufficient to define the mathematics of the system and the software automatically generates a complete mathematical description language for the model. Finally, multiple simulations can be generated from a single Application to permit choice of numerical method, simulation duration, parameter variations and the implementation of sensitivity analysis. Thus, the hierarchical structure of the BioModel enables the creation of multiple scenarios, akin to "virtual experiments," in which the same basic Physiology can be probed to generate many testable hypotheses.

Another major new development since the original review is the VCell database. Models and model components can be searched, reused and updated, as well as privately shared among collaborating groups. When a model is mature and is used as the basis of a published paper, it can be made accessible to any VCell user. Additionally, exchange of models with other tools is possible via import/export of SBML, CellML, and MatLab formats as well as direct linkages to public web-based pathway databases. As a result of this approach, Virtual Cell has been rapidly adopted as a tool of choice for biophysical modeling, in particular by experimental biologists and by researchers interested in spatially resolved simulations. To date, approximately 2000 Virtual Cell users have created tens of thousands of models and simulations, with over 500 publicly available models in the VCell database, and about 100 publications in high-profile journals (see http://vcell.org/ for a current listing).

Other major new features of Virtual Cell include:

- Simulation of diffusion on irregularly shaped three-dimensional (3D) surfaces for full support of reaction/diffusion/transport/flux within a membrane and between the membrane and the adjacent volumes

- The ability to model and simulate stochastic reaction networks, including a choice of Gillespie and hybrid stochastic solvers
- Support for advection (i.e., velocity fields) in spatial reaction-diffusion models and simulations
- Availability of a variety of new solvers, including a stiff-in-time partial differential equation (PDE) solver
- Parameter scans and parameter estimation algorithms
- Support for rule-based-modeling through the BioNetGen modeling language
- "Smart" copy and paste of arrays of parameter and initial concentration values
- Use of image data or results from prior simulations as nonuniform initial conditions for species in spatial models
- Support for "events" in nonspatial models, where a variable or time-dependent expression can trigger a predefined reset or change in the value of other variables
- Integration of convenient two-dimensional (2D) and 3D image segmentation tools to directly convert experimental images into Virtual Cell geometries
- A simulation results viewer that contains a variety of tools for visualization of complex dynamic multivariable spatial simulations, including 3D surface rendering
- Export of simulations results into a wide variety of formats, including Quicktime movies, Excel spreadsheets and a variety of image formats

Additional details on the algorithms within Virtual Cell can be found in several papers and some earlier reviews (Novak *et al.*, 2007; Schaff *et al.*, 2001; Slepchenko *et al.*, 2003). A recent paper (Moraru *et al.*, 2008) provides an updated report on the physics and math supported by Virtual Cell, the numerical methods available for simulating the various classes of supported models, the layered workflow for assembling a BioModel, some of the visualization and analysis features of Virtual Cell, the rationale of our web-deployment strategy, as well as the functionality of the Virtual Cell database. We have also published a comprehensive overview of how Virtual Cell can be used to solve a variety of common cell biological and systems biology problems (Slepchenko and Loew, 2010).

II. Introduction

A general computational framework for modeling cell biological processes, the *Virtual Cell*, is being developed at the National Resource for Cell Analysis and Modeling at the University of Connecticut Health Center. The Virtual Cell is intended to be a tool for experimentalists as well as theorists. Models are constructed from biochemical and electrophysiological data mapped to appropriate subcellular locations in images obtained from a microscope. Chemical kinetics, membrane fluxes, and diffusion are thus coupled and the resultant equations are solved numerically. The results are again mapped to experimental images so that the cell biologist can fully utilize the familiar arsenal of image processing tools to analyze the simulations.

The philosophy driving the Virtual Cell project requires a clear operational definition of the term *model*. The idea is best understood as a restatement of the scientific method. A model, in this language, is simply a collection of hypotheses and facts that are brought together in an attempt to understand a phenomenon. The choices of which hypotheses and facts to collect and the manner in which they are assembled themselves constitute additional hypotheses. A prediction based on the model is, in one sense, most useful if it *does not* match the experimental details of the process—it then unequivocally tells us that the elements of the model are inaccurate or incomplete. Although such negative results are not always publishable, they are a tremendous aid in refining our understanding. If the prediction does match the experiment, it *never* can guarantee the truth of the model, but should suggest other experiments that can test the validity of critical elements; ideally, it should also provide new predictions that can, in turn, be verified experimentally. The Virtual Cell is itself *not* a model. It is intended to be a computational framework and tool for cell biologists to create models and to generate predictions from models via simulations. To ensure the reliability of such a tool, all underlying math, physics, and numerics must be thoroughly validated. To ensure the utility and accessibility of such a tool to cell biologists, the results of such simulations must be presented in a format that may be analyzed using procedures comparable to those used to analyze the results of experiments.

In this chapter, we describe the current status of the mathematics infrastructure, design considerations for model management, and the user interface. This is followed by application to the calcium wave that follows fertilization of a frog egg. Additional details can be found in an earlier publication (Schaff *et al.*, 1997) and on our web site: http://www.nrcam.uchc.edu/.

III. Modeling Abstractions for Cellular Physiology

A. Background

Often, theoreticians develop the simplest model that reproduces the phenomenon under study (Kupferman *et al.*, 1997). These may be quite elegant, but are often not very extensible to other related phenomena. Other modeling efforts characterize single physiological mechanisms (Lit and Rinzel, 1994; Sneyd *et al.*, 1995), but these are often developed *ad hoc* rather than as part of a reusable and consistent framework.

Our approach to modeling concentrates on the mechanisms as well as the phenomena. The goal of this approach is to provide a direct method of evaluating single models of individual mechanisms in the context of several experiments. This approach enables the encapsulation of sufficient complexity that, after independent validation, allows it to be used as a meaningful predictive tool. To include sufficient complexity without over-whelming the user, the models are specified in their most natural form. In the case of chemical reactions, the models are represented by a series of reactants, products, modifiers (e.g., enzymes), their stoichiometry, and their kinetic constants.

One of the obstacles to modeling is the lack of general-purpose simulation and analysis tools. Each potential modeler must have resources in software development and numerical methods at his or her disposal. Each time the model or the computational domain is altered, the program must be changed. And in practice, the modeling of a new phenomenon requires a new simulation program to be written. This is a time-consuming and error-prone exercise, especially when developed without a proper software methodology.

We are developing a general, well-tested framework for modeling and simulation for use by the cell biology community. The application of the underlying equations to our framework with nearly arbitrary models and geometry is rigorously investigated. The numerical approach is then properly evaluated and tuned for performance. This methodology results in a proper basis for a general-purpose framework. Our approach requires no user programming; rather the user specifies models using biologically relevant abstractions such as reactions, compartments, molecular species, and experimental geometry. This allows a very flexible description of the physiological model and arbitrary geometry. The framework accommodates arbitrary geometry and automatically generates code to implement the specified physiological model.

Another problem is the lack of a standard format for expressing those models. Even implementing published models can be a nontrivial exercise. Some of the necessary details required for implementation can be missing or buried in the references. Often the models are obscured by geometrical assumptions used to simplify the problem. A standard modeling format is required to facilitate the evaluation and integration of separate models. This standard format should separately specify physiological models and cellular geometry in an implementation-independent way.

We suggest that the abstract physiological models used with the Virtual Cell framework can form the basis of such a standard.

The current implementation of the cell model description (Schaff *et al.*, 1997) involves the manipulation of abstract modeling objects that reside in the Modeling Framework as Java objects. These modeling objects can be edited, viewed, stored in a remote database, and analyzed using the WWW-based user interface (see User Interface section). These objects are categorized as Models, Geometry, and Simulation Context objects. This corresponds to the naming convention used in the current Modeling Framework software.

There are also mature efforts in Metabolic Pathway modeling such as GEPASI (Mendes, 1993). This package allows a simple and intuitive interface for specifying reaction stoichiometry and kinetics. The kinetics are specified by selecting from a predefined (but extensible) list of kinetic models (such as Michaelis–Menten) describing enzyme-mediated production of metabolites. These packages are focused on biochemical pathways where the spatial aspects of the system are ignored. However, for these simplified descriptions, they provide Metabolic Control Analysis (local sensitivity analysis tools), structural analysis (mass conservation identification), and a local stability analysis.

To provide a simple interface to a general-purpose modeling and simulation capability, the problem must be broken up into manageable pieces. In the case of the Virtual Cell, these pieces consist of abstract physiological models defined within the context of cell structure, experimental cell geometry, and a mapping of the physiological models to the specific geometry including the conditions of that particular problem. For such an interface to be consistent and maintainable, it must map directly to the underlying software architecture.

An intuitive user interface is essential to the usability of a complex application. A prototype user interface was developed to provide an early platform for modeling and simulation, and for investigating user interface requirements. The design goal was to capture the minimum functionality required for practical use of the Virtual Cell.

The Virtual Cell application is built on top of a distributed, component-based software architecture (Fig. 1). The physiological interface is a WWW-accessible Java applet that provides a graphical user interface to the capabilities of the Modeling Framework. The Mathematics Framework is automatically invoked to provide solutions to particular simulations. The architecture is designed such that

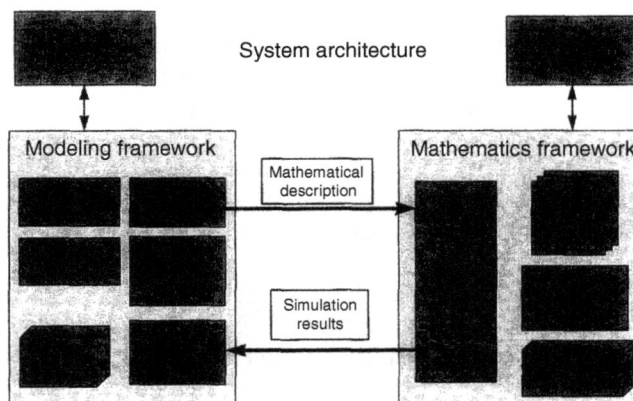

Fig. 1 The System Architecture features two distributed, component-based frameworks that create a stable, extensible, and maintainable research software platform. The Modeling Framework provides the biological abstractions necessary to model and simulate cellular physiology. The Mathematics Framework provides a general-purpose solver for the mathematical problems in the application domain of computational cellular physiology. This framework exposes a very high level system interface that allows a mathematical problem to be posed in a concise and implementation-independent mathematical description language, and for the solution to be made available in an appropriate format.

the location of the user interface and the corresponding back-end services (model storage, simulation, data retrieval) are transparent to the majority of the application. The typical configuration is a Java applet running in a WWW browser, with the Database, Simulation Control, and Simulation Data services executing on a remote machine (WWW server).

B. Specification of Physiological Models

A physiological model of the cell system under study is defined as a collection of Cellular Structures, Molecular Species, Reactions, and Fluxes. These concepts define cellular mechanisms in the context of cell structure, and are sufficient to define nonspatial, compartmental simulations. With the addition of a specific cellular geometry (usually from a microscope image), a spatial simulation is defined. The goal is to capture the physiology independently of the geometric domain (cell shape and scale) and specific experimental context, such that the resulting physiological mechanisms are modular and can be incorporated into new models with minimum modification.

1. Cellular Structures

Cellular structures are abstract representations of hierarchically organized, mutually exclusive regions within cells where the general topology is defined but the specific geometric shapes and sizes are left undefined. These regions are categorized by their intrinsic topology: *compartments* represent 3D volumetric regions (e.g., cytosol), *membranes* represent 2D surfaces (e.g., plasma membrane) separating the *compartments*, and *filaments* represent 1D contours (e.g., microtubules) lying within a single *compartment*. Using this definition, the extracellular region is separated from the cytosol (and hence the ER, nucleus, etc.) by the plasma membrane. All structures can contain *molecular species* and a collection of *reactions* that describe the biochemical behavior of those species within that structure.

These cellular structures are used for bookkeeping purposes, and must be mapped to a specific cellular geometry before any quantitative simulations or analysis can be performed. Although this introduces two parallel representations of anatomical structures (topology and shape), the separation of physiological models from specific simulation geometry allows the same model to be used in various compartmental and 1D, 2D, and 3D geometric contexts without modification.

2. Molecular Species

Species are unique molecular species (e.g., Ca^{2+}, IP_3) or important distinct states of molecular species (e.g., calcium-bound IP_3-Receptor, unbound IP_3-Receptor). These *species* can be separately defined in multiple *compartments, membranes,* or *filaments*. *Species* are described by linear densities when located on *filaments*, by surface densities when located in *membranes*, and by concentrations when located in *compartments*. *Species* diffusion is defined separately for each cellular compartment and is specified by diffusion constants.

Species can participate in reactions, fluxes, and diffusion. The behavior of *species* can be described by

1. Diffusion of *species* within *compartments, membranes,* or *filaments*.
2. Directed motion of *Species* along *filaments*.
3. Flux of *species* between two adjacent *compartments* through the associated *membrane*.
4. Advection (e.g., binding reactions) of *species* between *compartments* and either *membranes* or *filaments*.

3. Reactions and Fluxes

Reactions are objects that represent complete descriptions of the stoichiometry and kinetics of biochemical reactions. *Reactions* are collections of related *reaction steps* (e.g., membrane receptor binding or cytosolic calcium buffering) and *membrane fluxes* (e.g., flux through an ion channel).

Each *reaction step* is associated with a single cellular structure. The stoichiometry of a *reaction step* is expressed in terms of reactants, products, and catalysts, which are related to species in a particular cellular structure (e.g., Ca^{2+} in cytosol). This stoichiometry is depicted in the user interface as a biochemical pathway graph.

Reaction steps located in a compartment involve only those *species* that are present in the interior of that compartment. *Reaction steps* located on a *membrane* can involve *species* that are present on that *membrane* as well as *species* present in the *compartments* on either side of that *membrane*. Reaction steps located on a filament can involve species that are present on that filament as well as species present in the compartment that contains the filament. The kinetics of a *reaction step* can be specified as either mass action kinetics with forward and reverse rate coefficients or, more generally, by an arbitrary rate expression in terms of reactants, products, and modifiers.

Each *flux step* is associated with a single *membrane* and describes the flux of a single flux carrier species through that *membrane*. A flux carrier must be a species that is defined in both neighboring *compartments*. For example, a *flux step* associated with a calcium channel in the endoplasmic reticulum (ER) membrane would have calcium as the flux carrier, and thus calcium must be defined in both the ER and cytosol. A single inward flux is defined by convention ($\mu M \ \mu m \ s^{-1}$) and enforces flux conservation across the *membrane*. The flux is an arbitrary function of flux carrier concentration (on either side of the *membrane*), membrane-bound modifier surface density, and of modifier concentration in the *compartments* on either side of the *membrane*.

C. Specification of Cellular Geometry

A particular experimental context requires a concrete description of the cellular geometry to fully describe the behavior of the cellular system. This geometric description defines the specific morphology of the cell, and any of its spatially resolvable organelles. This geometry is usually taken directly from experimental images that have been segmented into mutually exclusive regions by their pixel intensity. For example, an image may be segmented into only two classes of pixel intensities, white (pixel intensity of 255) and black (pixel intensity of 0), where white regions represent cytosol and black regions represent the extracellular milieu. The actual size of the domain is defined by specifying the scale of a pixel in microns to properly define a simulation domain.

Alternatively, the geometry may be specified analytically as an ordered list of inequalities. Each resolved region can be represented as an inequality in x, y, and z. Each point in the geometric domain can be assigned to the compartment corresponding to the first inequality that is satisfied. For example, if cytosol is represented by a sphere at the origin of radius R, and the rest of the domain is considered extracellular, then the nested compartments can be easily specified as follows.

$$\text{cytosol}: x^2 + y^2 + z^2 > R(\text{true within sphere at origin})$$
$$\text{extracellular}: 1 < 2(\text{always true}). \tag{1}$$

D. Mapping Biology to Mathematical Description

1. Mapping Cellular Structures to Experimental Geometry

The resulting geometry then has to be mapped to the corresponding *cellular structures* defined in the *physiological model*. Each mutually exclusive volumetric region in the geometry is naturally mapped to a single *compartment*, and an interface (surface) separating any two regions maps to the corresponding *membrane*. A *compartment* that is not spatially resolved in the geometry may be considered continuously distributed within the geometric region of its parent *compartment*. For example, if a particular geometry specifies spatial mappings for only extracellular, cytosol, and the plasma membrane, then the ER and its *membrane* (which are interior to the cytosol *compartment*) are continuously distributed within the cytosol region of the geometry. For such continuously distributed *compartments*, the volume fraction (e.g., ER volume to total cytosolic volume) and internal surface to volume ratios (e.g., ER membrane surface to ER volume) are required to reconcile distributed fluxes and membrane binding rates to the spatially resolved *compartments*. Note that unresolved structures, such as the

ER, need not be uniformly distributed within the cytosol *compartment* even if they are described as continuous.

When spatial simulations are not required (compartmental models), specification of volume fractions for all *compartments* and surface to volume ratios for all *membranes* are sufficient to represent the geometric mapping.

2. Mapping of Reactions and Species to Equations and Variables

After the *compartments* and *membranes* have been mapped to concrete geometry, the mapping of *reactions* to systems of equations is well defined. One or more *compartments* and *membranes* are mapped onto a region in the computational domain as discussed in the previous section. All of the *reaction steps* and *membrane fluxes* that are associated with this set of *compartments* and *membranes* are collected.

Within a single computational subdomain, *species* dynamics can be represented as a system of differential equations [Eq. (2)] expressing the time rate of change of each *species* C_i concentration as the sum of their reaction kinetics and their diffusion term (Laplacian scaled by diffusion rate D_i). The reaction kinetics can be expressed in terms of a linear combination of reaction rates v_j (and fluxes through distributed *membranes*) that are scaled by their corresponding stoichiometry c_{ij}:

$$\begin{bmatrix} \dfrac{dC_1}{dt} \\[2mm] \dfrac{dC_2}{dt} \\[2mm] \vdots \\[2mm] \dfrac{dC_n}{dt} \end{bmatrix} = \mathbf{S}\cdot\mathbf{v} = \begin{bmatrix} c_{11} & \cdots & c_{1m} & D_1 & 0 & \cdots & 0 \\ c_{21} & \cdots & c_{2m} & 0 & D_2 & \cdots & 0 \\ \vdots & \ddots & \vdots & \vdots & \vdots & \ddots & \vdots \\ c_{n1} & \cdots & c_{nm} & 0 & 0 & \cdots & D_n \end{bmatrix} * \begin{bmatrix} v_1 \\ \vdots \\ v_m \\ \nabla^2 C_1 \\ \nabla^2 C_2 \\ \vdots \\ \nabla^2 C_n \end{bmatrix} \quad (2)$$

As a result of conservation of mass in biochemical systems, the number of independent differential equations is often less than the number of *species*. The application of a stoichiometry (or structural) analysis (Reder, 1988; Sorensen and Stewart, 1980; Villa and Chapman, 1995) allows the

$$\begin{bmatrix} C_{r+1} \\ \vdots \\ C_n \end{bmatrix} = \begin{bmatrix} L_1 \\ \vdots \\ L_{n-r} \end{bmatrix} \cdot \begin{bmatrix} C_1 \\ C_2 \\ \vdots \\ C_r \end{bmatrix} + \begin{bmatrix} K_1 \\ \vdots \\ K_{n-r} \end{bmatrix} \quad (3)$$

automatic extraction of a minimum set of independent variables ($C_1\cdots C_r$ where r is the rank of $\mathbf{S}$), and a set of dependent variables ($C_{r+1}\cdots C_n$) described by conservation relationships L_i [Eq. (3)] (derived from the left null space of S) that allow expression of all dependent *species* concentrations as linear combinations of independent *species* concentrations. Then, only the equations associated with the independent *species* are used in the mathematical description.

3. Fast Kinetics

Interrelated cellular processes often occur on a wide range of timescales causing a problem that is commonly encountered when computing numerical solutions of reaction/diffusion problems in biological applications. It manifests itself as a set of equations that is said to be stiff in time. The direct approach to solving this type of a system would result in a numerical algorithm taking prohibitively small time steps dictated by the fastest kinetics. The typical techniques used to avoid this are employing stiff solvers (Gear, 1971) or analytic pseudo-steady approximations (Wagner and Keizer, 1994). We have developed a general approach based on separation of fast and slow kinetics (Strang, 1968). We then perform the pseudo-steady approximation on fast processes. Within this approximation, fast reactions are considered to be at rapid equilibrium. This assumption results in a system of nonlinear algebraic equations. To treat it correctly, the set of independent variables within the fast subsystem has to be determined. This can be done by means of the structural analysis described earlier. The difference is that now these identified invariants (mass conservation relationships) are only invariant with respect to the fast kinetics. They therefore have to be updated at each time step via the slow kinetics and diffusion. Our approach does not require any preliminary analytic treatment and can be performed automatically. A user must only specify the subset of reactions that has fast kinetics, and the appropriate "fast system" is automatically generated and integrated into the mathematical description. Currently we are using this capability to investigate the influence of mobile buffers on the properties of calcium waves and on the conditions of their initiation.

4. Stochastic Formulation for Reaction/Diffusion Systems

When the number of particles involved in the cell processes of interest is not sufficiently large and fluctuations become important, the replacement of the continuous description by a stochastic treatment is required. This might be used to simulate the motion and interactions of granules in intracellular trafficking (Ainger et al., 1997). The stochastic approach is also required for the description of spontaneous local increases in the concentration of intracellular calcium from discrete ER calcium channels (calcium sparks or puffs) observed in cardiac myocytes (Cheng et al., 1993). In the case of discrete particles, the problem is formulated in terms of locations (the state variables) and random walks instead of concentration distributions and fluxes in the continuous description, while using the same physical constant—the diffusion coefficient D. The chemical reactions between discrete particles and structures (capture) or between discrete particles and continuously distributed species are described in terms of transition probabilities (reaction rates). In the latter case, we have to incorporate the stochastic formulation into a continuous reaction/diffusion framework. This can be done because the numeric treatment of the corresponding PDEs requires their discretization. In fact, we deal with discrete numbers that characterize each elementary computational volume. Thus for the case of a species treated continuously, the source term of the PDE will contain the random contribution due to chemical reactions with discrete particles.

Monte Carlo techniques have been employed to simulate both the reaction events and Brownian motion of the discrete particles. Since the displacements at time t in unbiased random walks are normally distributed with mean zero and standard deviation $(4Dt)^{1/2}$, one can model the increment of each coordinate within a time step Δt by $(4D\Delta t)^{1/2} r$ where r is a random number described by the standard normal distribution with zero mean and a unity standard deviation (Schmidt *et al.*, 1994). To reproduce the standard normal distribution for the Brownian movement simulation, we use the Box–Muller method (Press *et al.*, 1992). In this method, the variable $\lambda = (-2\ln\xi)^{1/2}\cos 2\pi\eta$ proves to be normally distributed with the standard deviation $\sigma = 1$ provided the random variables ξ and η are uniformly distributed on the interval [0,1]. Methods based on the central limit theorem (Devroye, 1986) might be less computationally intensive, but validation is required. The interaction of particles with the membrane has been described in terms of elastic collisions.

The selection–rejection type method has been used for the stochastic simulation of reaction dynamics. The corresponding transition between particle states is assumed to occur if a generated random number, uniformly distributed on [0,1], is less than the reaction probability. In our case, particles interact with the continuously distributed species. Thus, the reaction probability for small time steps Δt is $k\Delta t[C]$, where k is the reaction on rate, and $[C]$ stands for the concentration of a dispersed species. Clearly, for the reaction to occur, $[C]$ should satisfy the condition $[C]Sd \geq n$, where S is the particle surface, d is the characteristic interaction distance, and n is the stoichiometry number. Hence the reaction is ruled out if $[C] < n/V_0$, provided the control volume is bigger than Sd. This condition is necessary to eliminate the possibility of negative concentrations.

We have implemented a simple model representing the initial step in RNA trafficking: RNA granule assembly, where discrete particles interact with the continuously distributed species (RNA). Thus the two approaches, stochastic and deterministic, have been combined.

E. Compartmental Simulations

For simulation of compartmental models (single-point approximations), the ordinary differential equations (ODEs) representing the reaction kinetics are generated and passed to an interpreted ODE solver (within the client applet).

This system of equations is solved using a simple explicit integration scheme (forward difference) that is first-order accurate in time. A higher order numerical scheme will be integrated in the future.

The Compartmental Simulation (Preview) component executes a compartmental (single-point) simulation based on the defined physiological model and the geometric assumptions entered in the Feature Editor (surface to volume ratios and volume fractions). This results in a set of nonlinear ODEs that typically are solved in seconds. This allows an interactive, though manual, modification of parameters and a quick determination of the effect over time. Once the simulation is complete, each species can be viewed easily.

The Equation Viewer displays the equations generated as a result of mapping the physiological model to either a cellular geometry model (spatial simulation) or a single-point approximation (compartmental model). The parameter values may be substituted (and the expression simplified) or left in their symbolic representation.

1. Model Analysis

It is important to determine the sensitivity of model behavior (i.e., simulation results) to the choice of which physiological mechanisms and their parameter values are incorporated. For a given model structure, the selection of parameter values is constrained, but often not completely determined, by direct empirical measurements and physical limitations, as well as inferred from the steady and dynamic behavior and stability of the composite system.

It is informative to determine the relative change in model behavior due to a relative change in parameter value. For nonspatial, compartmental models, the software computes the sensitivity of any species concentration to any parameter as a function of time evaluated at the nominal solution.

The current implementation lacks a direct steady-state solver. This must either be performed by letting the dynamic system run until it converges to a steady state or by manually doing the analytic calculations (setting rates to zero and solving the simultaneous equations).

F. Spatial Simulations

For the solution of a complete spatial simulation, the PDEs that correspond to diffusive species, and ODEs for nondiffusive species, are generated. These equations are sent to the remote Simulation Server where the corresponding C++ code is automatically generated, compiled, and linked with the Simulation Library. The resulting executable is then run and the results are collected and stored on the server. The Simulation Data Server then coordinates client access to the server-side simulation data for display and analysis.

The system of PDEs is mapped to a rectangular grid using a finite difference scheme based on a finite volume approach (Schaff et al., 1997). The nonlinear source terms representing reaction kinetics are evaluated explicitly (forward difference) and the resulting linearized PDE is solved implicitly (backward difference). Those membranes separating spatially resolved compartments are treated as discontinuities in the solution of the reaction/diffusion equations. These discontinuities are defined by flux jump conditions that incorporate transmembrane fluxes, binding to membrane-bound species, and conservation of mass. Each boundary condition is defined in terms of a known flux (Neumann condition) or a known concentration (Dirichlet condition).

G. Storage

The Database Access Form presents a rudimentary model and simulation storage capability. The current implementation allows whole physiological models, geometric models, and simulation contexts to be stored and retrieved. The simulation context is stored in a way that includes the physiological and geometric models such that it encapsulates all of the information to reproduce and describe a particular spatial simulation. There is, however, currently no ability to query the stored models for specific attributes.

IV. Application to Existing Model

This discussion describes the application of our formalism to a physiological reaction/diffusion system within a spatial context. An existing model of fertilization calcium waves in *Xenopus laevis* eggs (Wagner *et al.*, 1998) is used as an example. This is based on the experimental observation that fertilization results in a wave of calcium that spreads throughout the egg from the point of sperm–egg fusion. The calcium wave depends on the elevation of intracellular inositol 1,4,5-trisphosphate (IP_3) and propagates with a concave shape (Fontanilla and Nuccitelli, 1998). Our discussion is limited to the mechanics of the model representation without an in-depth analysis of the model itself.

A. Background

In the case under consideration, the calcium dynamics is essentially determined by the calcium release from the internal ER stores through IP_3 sensitive channels and the uptake through sarcoplasmic-endoplasmic reticulum calcium ATPase (SERCA) pumps. These processes are modulated by calcium diffusion and binding to calcium buffers that are always present in the cytosol. Additionally, a small constant calcium leak from ER to the cytosol ensures that the flux balance is at the initial steady state. A simplified version (Li and Rinzel, 1994) of the De Young–Keizer model (De Young and Keizer, 1992) for the calcium channel is used. This model implies the independent kinetics of IP_3-binding, calcium activation, and calcium inhibition sites. A Hill equation is employed to describe calcium flux through the pumps.

As has been shown (Wagner *et al.*, 1998), the fertilization calcium wave generation can be explained by the bistability of a matured egg. The system bistability means that the system can maintain two stable steady states with different calcium concentrations, and a wavefront that corresponds to a threshold between the regions with different steady states. Thus, in the bistable system the calcium wave can be initiated even at a steady IP$_3$ concentration, although a spatially heterogeneous IP$_3$ distribution is required to reproduce the details of the wavefront shape. In this chapter, we map the model proposed previously (Wagner *et al.*, 1998) onto the 3D geometry and explicitly introduce buffers as participating species. We treat them as a subsystem with fast kinetics rather than describe buffering with one scaling parameter. For simplicity, we consider only immobile buffers, although our general approach allows us to also treat mobile buffers (for example, fluorescent indicators) as well as any other "fast" kinetics.

B. Extracting Physiological Model

A physiological model first consists of a set of physiological assumptions regarding the mechanisms important to the phenomenon under study. This defines the physiological structure of the model and includes the list of physiological mechanisms as well as the way in which they interact.

The framework application starts with outlining model compartments and participating species (see Fig. 2). In our problem three compartments—extracellular, cytosol, and ER—are separated by the plasma membrane and the ER membrane, respectively. We have no species of interest in extracellular as well as no

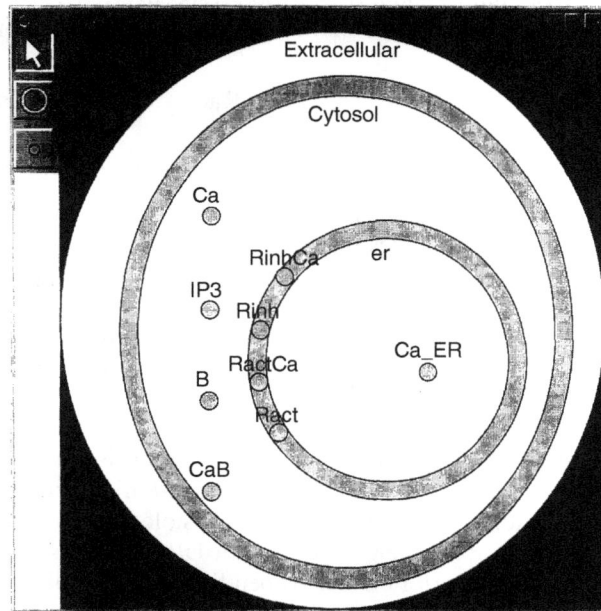

Fig. 2 The Cell Structure Editor allows specification of cellular structure and location of molecular species using a drag-and-drop user interface. In this example, there are three compartments: extracellular, cytosol, and ER, which are separated by the plasma membrane and the ER membrane, respectively. The location of a species within the editor display indicates its association with a cellular structure.

processes that are associated with the plasma membrane in our approximation. Four species are present in the cytosol: calcium (Ca), IP_3, and the free (B) and Ca-bound (CaB) states of fixed calcium buffers. We also have calcium in the ER (Ca_ER). Four species are associated with the ER membrane: the free (Ract, Rinh) and Ca-bound (RactCa, RinhCa) states of the activation and inhibition monomer sites that regulate calcium flux through ER membrane channels.

In our application, the concentration of Ca_ER is considered to be constant, and the constant IP_3 distribution is described by a static, spatial function. Thus, we end up with seven variables, corresponding to the seven identified specified species, of which only cytosolic calcium is diffusive while the other six are treated as spatially fixed.

The initial concentration and diffusion rates are specified for each species present within a compartment. The initial surface densities are specified for each species present within a membrane.

We then turn to the species interaction through the chemical reactions. One can edit the processes associated with the ER membrane in the Biochemical Pathway Editor (see Fig. 3). This Editor permits the user to define reaction models as a series of membrane fluxes and reaction steps associated with either compartments or membranes.

Our model contains three calcium fluxes through the ER membrane due to channels, pumps, and intrinsic leak, respectively. The channel flux is enzymatically regulated by the states of the activation and inhibition sites that participate in two reactions:

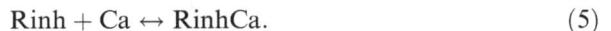

$$Ract + Ca \leftrightarrow RactCa \qquad (4)$$

$$Rinh + Ca \leftrightarrow RinhCa. \qquad (5)$$

Similarly, there are editing windows for compartments. Figure 4 shows the only reaction that takes place in the cytosol:

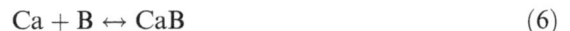

$$Ca + B \leftrightarrow CaB \qquad (6)$$

To complete the reaction description, we have to specify the "fast" subsystem if it exists (as described earlier this subsystem will be treated within the pseudo-steady-state approximation). As in the original published model (17), we assume reactions [Eqs. (4) and (6)] to be in a rapid equilibrium while reaction [Eq. (5)] and fluxes through the ER membrane comprise the slow system dynamics. The complete physiological model is described in Appendix 1.

After specifying the entire physiological model within the user interface, the Modeling Framework then translates the physiological description into a mathematical description (see Appendix 2). While doing this, it automatically creates the system of equations, analyzes it, and determines the minimal set of independent variables and equations that are sufficient for a complete solution. Thus, in our case, we finally end up with four independent variables, Ca, CaB, RactCa, and RinhCa, and the temporally constant IP_3, which was explicitly defined as a variable to facilitate visualization.

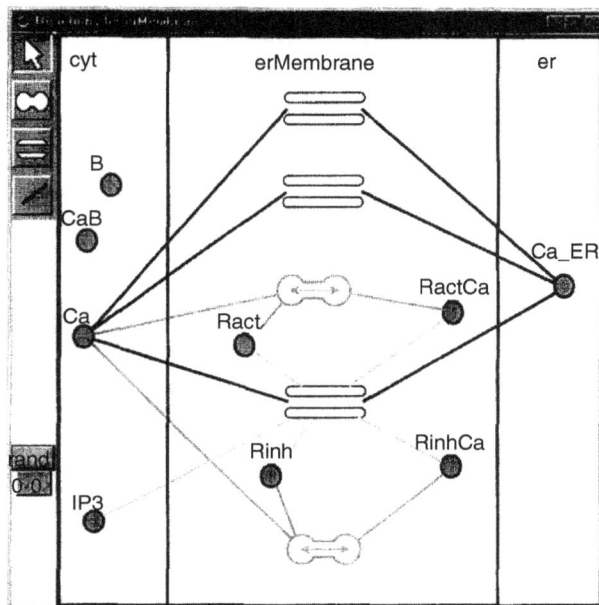

Fig. 3 The Membrane Reaction Editor allows specification of *reaction steps* (represented by dumbbell-shaped objects) and *membrane fluxes* (represented by hollow tube objects). The *reaction steps* are treated as in the Compartment Reaction Editor. The *membrane fluxes* are defined as the transmembrane flux of a single molecular species (in this case Ca^{2+}) from the exterior compartment (cytosol) to the interior compartment (ER lumen). In this example, the three membrane flux shapes represent the behavior of (from top to bottom) the SERCA pump, the membrane leak permeability, and the IP_3-receptor calcium channel. The channel flux is a function of the calcium channel activating binding site states Ract/RactCa and the inhibitory binding site states Rinh/RinhCaRI, which are modeled as independent. As in the reaction steps, flux kinetics are hyperlinked to the corresponding flux kinetics.

It is worth mentioning that modelers have direct access to the Mathematical Framework interface, which allows them to skip the physiological description and create their own mathematical description based on a known system of equations. This Framework is a problem-solving environment (PSE) designed for the class of reaction/diffusion problems involving multiple spatially resolved compartments. Thus, one can easily test published mathematical models by translating them into the Mathematical Description Language (see Appendix 2).

C. Mapping Compartments to Simulation Geometry

Finally, we specify the geometry of the problem and boundary conditions. The egg cell geometry approximates a sphere, so the geometry was defined analytically as a sphere of radius 500 μm. In our case, one can reduce the computational domain by taking into account the rotational symmetry of the problem. Thus, it

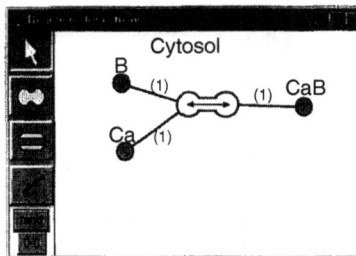

Fig. 4 Compartment Reaction Editor allows specification of *reaction steps* (represented by dumbbell-shaped objects), which are defined within the interior of a particular *compartment* and involve only those *species* that are defined within that compartment. A *species* participating as a reactant or a product has a line connecting itself to the left or right side of the reaction step shape, respectively, where the number in parentheses specifies the stoichiometry. Reaction kinetics is hyperlinked to the corresponding reaction step shape.

is sufficient to simulate the calcium distribution within a quarter of a sphere only. This geometry is specified using the Geometry Builder, a stand-alone Java applet that permits the construction of 2D or 3D cellular geometric models based on a series of image files or analytic geometry.

As in the previous model (Wagner *et al.*, 1998), we do not spatially resolve ER. Instead, we treat it as continuously and uniformly distributed within the cytosol. Correspondingly, species associated with the ER membrane are effectively described as volume variables. The information on the surface-to-volume ratio and the density of ER distribution is incorporated in the parameters LAMBDA1, LAMBDA2, and LAMBDA3 (Appendixes 1 and 2), these factors convert mass flux into rate of change of concentration in cytosol.

D. Spatial Simulation of Generated Equations

Finally, the procedure of automatic code generation is invoked: the mathematical description is read and the corresponding C++ files are automatically created and executed. Figure 5 shows the 3D reconstruction (using VoxelView) of the simulation results for the calcium concentration distribution at time 70 s after the wave initiation. We have also tested the 2D version of the model. Our results are in good agreement with the previous results (Wagner *et al.*, 1998) for the corresponding value of the buffer scaling coefficient.

The Geometry/Mesh Editor allows participation in the choice of spatial resolution, permitting a balance of computational costs and the goodness of geometric representation. This interface directs the binding of regions of the segmented geometry to the corresponding features within the physiological model. An orthogonal mesh is specified and displayed interactively.

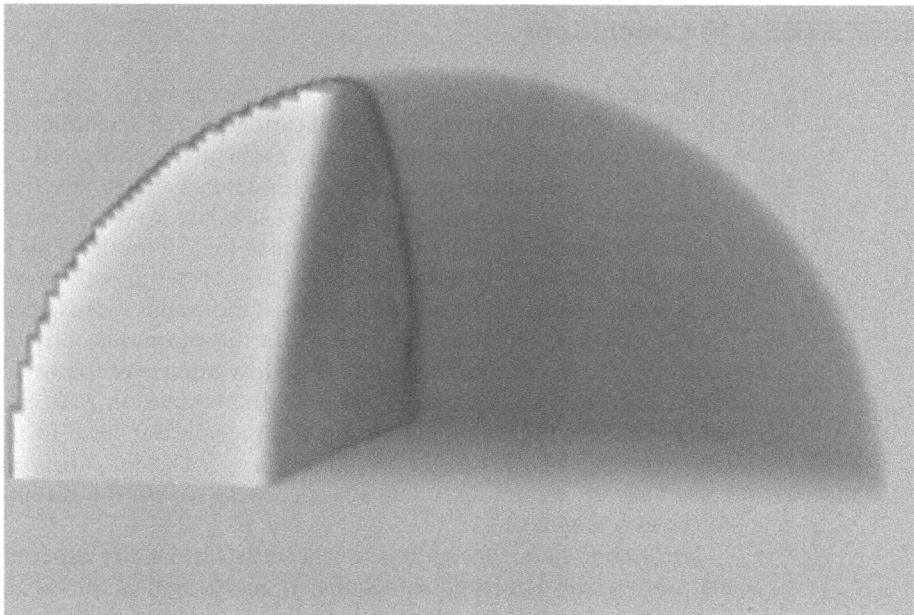

Fig. 5 A three-dimensional simulation of the fertilization calcium wave in a *Xenopus laevis* egg. The original published mathematical model (Wagner *et al.*, 1998) was directly implemented using the math interface in two and three dimensions and achieved essentially identical results. The underlying physiological model was then entered into our physiological framework and the corresponding mathematical model was automatically generated. The simulation results were in good agreement with the previous results (Wagner *et al.*, 1998). This is a quarter of the spherical egg with a 500-μm radius. This is a volume rendering of the simulated intracellular concentration 70 s after wave initiation (using the automatically generated mathematical model).

The Initial/Boundary Condition Editor allows the specification of initial conditions and boundary conditions for each of the species for each feature. To afford maximum flexibility, the boundary conditions for each simulation border may be specified independently for each feature. For example, the concentrations may be specified at simulation boundaries in the extracellular space to indicate a sink. A zero molecular flux may be specified at a simulation boundary belonging to cytosol to ensure the symmetry of function with the missing portion of cytosol (the implied mirror image).

The Simulation Data Viewer displays the results of the current spatial simulation. The species concentrations are displayed superimposed on the mesh. The analysis capability includes graphing the spatial distribution of a species as a line scan and graphing a time series at a single point.

V. Conclusions

The need for the Virtual Cell arises because the very complexity of cell biological processes severely impedes the application of the scientific method. A pair of separate factors that contribute to this problem are addressed by the Virtual Cell.

First, the large number of interdependent chemical reactions and structural components that combine to affect and regulate a typical cell biological process forces one to seek the help of a computer to build a model. This issue is the subject of an eloquent essay by Bray (1997). We are now faced with an overwhelming body of data describing the details of individual molecular events occurring inside cells. As Bray puts it, "What are we to do with the enormous cornucopia of genes and molecules we have found in living cells? How can we see the wood for the trees and understand complex cellular processes?" Brays solution: "Although we poor mortals have difficulty manipulating seven things in our head at the same time, our silicon protégés do not suffer this limitation. . . . The data are accumulating and the computers are humming. What we lack are the words, the grammar and the syntax of the new language."

The second factor recognizes that scientists trained in experimental cell biology are not typically equipped with sufficient mathematical, physical, or computational expertise to generate quantitative predictions from models. Conversely, theoretical biologists are often trained in the physical sciences and have difficulty communicating with experimentalists (bifurcation diagrams, for example, will not serve as a basis for a common language). By maintaining the physical laws and numerical methods in separate modular layers, the Virtual Cell is at the same time accessible to the experimental biologist and a powerful tool for the theorist. Also, by maintaining a direct mapping to experimental biochemical, electrophysiological, and/or image data, it ensures that simulation results will be communicated in a language that can be understood and applied by all biologists.

Acknowledgments

We are pleased to acknowledge the support of NIH (GM35063) and the Critical Technologies Program of the state of Connecticut. The National Resource for Cell Analysis and Modeling is funded through NIH Grant RR13186.

Appendix 1:. Physiological Model Description

Name oocyte3d
```
Species CalciumBufferUnbound
Species CalciumBufferBound
Species IP3ReceptorCalciumActivationSiteUnbound
Species IP3ReceptorCalciumActivationSiteBound
```

```
Species IP3ReceptorCalciumInhibitionSiteUnbound
Species IP3ReceptorCalciumInhibitionSiteBound
Species Calcium
Species IP3

Compartment extracellular {}
Compartment cytosol {
    Context Ca Calcium { InitialCondition 0.1153 DiffusionRate
    300.0 }
    Context IP3 IP3 { InitialCondition 0.12 DiffusionRate 0.0 }
    Context B CalciumBufferUnbound { InitialCondition
    1398.3876590291393 DiffusionRate 0.0 }
    Context CaB CalciumBufferBound { InitialCondition
    1.6123409708605976 DiffusionRate 0.0 }
}
Membrane plasmaMembrane cytosol extracellular {
  SurfaceToVolume 1.0
  VolumeFraction 0.5
}
Compartment er {
  Context Ca_ER Calcium { InitialCondition 10.0 DiffusionRate
  0.0 }
}
Membrane erMembrane er cytosol {
  SurfaceToVolume 1.0
  VolumeFraction 0.5
  Context Ract IP3ReceptorCalciumActivationSiteUnbound {
      InitialCondition 0.009123393902531743 DiffusionRate 0.0 }
  Context RactCa IP3ReceptorCalciumActivationSiteBound {
     InitialCondition 8.766060974682582E-4 DiffusionRate 0.0 }
  Context Rinh IP3ReceptorCalciumInhibitionSiteUnbound {
     InitialCondition 0.009286200705751254 DiffusionRate 0.0 }
  Context RinhCa IP3ReceptorCalciumInhibitionSiteBound {
     InitialCondition 7.137992942487463E-4 DiffusionRate 0.0 }
}
Reaction TestReaction {
  SimpleReaction cytosol {
    Reactant Ca Calcium cytosol 1
    Reactant B CalciumBufferUnbound cytosol 1
```

```
          Product CaB CalciumBufferBound cytosol 1
          Kinetics MassActionKinetics {
            Fast
            Parameter K 100.0;
            ForwardRate 1.0;
            ReverseRate K;
          }
      }
    SimpleReaction erMembrane {
        Reactant Ca Calcium cytosol 1
        Reactant Ract IP3ReceptorCalciumActivationSiteUnbound
        erMembrane 1
         Product RactCa IP3ReceptorCalciumActivationSiteBound
        erMembrane 1
        Kinetics MassActionKinetics {
          Fast
          Parameter dact 1.2;
          ForwardRate 1.0;
          ReverseRate dact;
        }
    }
    SimpleReaction erMembrane {
        Reactant Ca Calcium cytosol 1
        Reactant Rinh IP3ReceptorCalciumInhibitionSiteUnbound
        erMembrane 1
         Product RinhCa IP3ReceptorCalciumInhibitionSiteBound
        erMembrane 1
        Kinetics MassActionKinetics {
          Parameter TAU 4.0;
          Parameter dinh 1.5;
          ForwardRate (1.0 / TAU);
          ReverseRate (dinh / TAU);
        }
    }
    FluxStep erMembrane Calcium {
        Catalyst IP3 IP3 cytosol
        Catalyst RactCa IP3ReceptorCalciumActivationSiteBound
        erMembrane
```

```
          Catalyst Ract IP3ReceptorCalciumActivationSiteUnbound
          erMembrane
          Catalyst Rinh IP3ReceptorCalciumInhibitionSiteUnbound
          erMembrane
          Catalyst RinhCa IP3ReceptorCalciumInhibitionSiteBound
          erMembrane
          Kinetics GeneralKinetics {
            Parameter LAMBDA1 75.0;
            Parameter dI 0.025;
            Rate (-LAMBDA1 * (Ca_ER--Ca) * pow(((IP3 / (IP3 + dI)) *
            (RactCa/(RactCa+Ract)) * (Rinh/(RinhCa+Rinh))),3.0));
          }
        }
        FluxStep erMembrane Calcium {
            Kinetics GeneralKinetics {
              Parameter LAMBDA2 75.0;
              Parameter vP 0.1;
              Parameter kP 0.4;
              Rate (LAMBDA2 * vP * Ca * Ca/((kP * kP) + (Ca * Ca)));
            }
        }
        FluxStep erMembrane Calcium {
            Kinetics GeneralKinetics {
              Parameter LAMBDA3 75.0;
              Parameter vL 5.0E-4;
              Rate (-LAMBDA3 * vL * (Ca_ER--Ca));
            }
          }
        }
```

Appendix 2:. Mathematical Description

```
name oocyte3d_generated
Constant K 100.0; Constant dact 1.2; Constant TAU 4.0;
Constant dinh 1.5; Constant LAMBDA1 75.0; Constant dI 0.025;
Constant LAMBDA2 75.0; Constant vP 0.1; Constant kP 0.4;
Constant LAMBDA3 75.0; Constant vL 5.0E-4; Constant Ca_init
      0.1153;
```

```
Constant B_init 1398.3876590291393; Constant CaB_init
    1.6123409708605976;
Constant Ract_init 0.009123393902531743; Constant RactCa_init
    8.76060974682582E-4;
Constant Rinh_init 0.009286200705751254; Constant RinhCa_init
    7.13799294248746E-4;
Constant Ca_ER 10.0;
VolumeVariable Ca VolumeVariable IP3 VolumeVariable CaB
VolumeVariable RactCa VolumeVariable RinhCa
Function IP3_init 0.12 + (exp(-0.133333 * (500.0 - sqrt(x*x + y*y +
    z*z))) * (0.12 + (0.84 * (x < -170.0) * exp(-pow((0.0025 * sqrt
    (y*y + z*z)), 4.0))))));
Function K_B_total (B_init + CaB_init);
Function K_Ract_total (Ract_init + RactCa_init);
Function K_Rinh_total (Rinh_init + RinhCa_init);
Function B (K_B_total -- CaB);
Function Ract (K_Ract_total -- RactCa);
Function Rinh (K_Rinh_total -- RinhCa);

CartesianDomain {
    Dimension 3
    Size 1050.0 525.0 525.0
    Origin -525.0 0.0 0.0
    Compartment cytosol ((x*x + y*y + z*z) < (500.0*500.0));
    Compartment extracellular 1.0;
}
CompartmentSubDomain cytosol {
    Priority 2
    BoundaryXmDirichlet  BoundaryXp  Dirichlet  BoundaryYm
    Dirichlet
    BoundaryYp  Dirichlet  BoundaryZm  Dirichlet  BoundaryZp
    Dirichlet
    PdeEquation Ca {
        Rate ((vL * (Ca_ER - Ca) * LAMBDA3) - (LAMBDA2*vP*Ca*Ca/
        (kP*kP + Ca*Ca)) -- (Rinh*Ca/TAU) - (RinhCa*dinh/TAU)) +
        ((Ca_ER - Ca) * pow((IP3/(IP3 + dI)*RactCa/(RactCa +
        Ract) * Rinh/(RinhCa + Rinh)), 3.0) * LAMBDA1));
            Diffusion 300.0;
            Initial 0.1153; }
```

```
                    OdeEquation IP3 {
                          Rate 0.0;
                          Initial 0.12 + (exp(-0.13333 * (500.0 - sqrt(x*x +
                             y*y + z*z))) * (0.12 + (0.84 * (x < -170.0) * exp
                             (-pow((0.0025 * sqrt(y*y + z*z)), 4.0)))))); }
                    OdeEquation CaB { Rate 0.0; Initial 1.6123409708605976; }
                    OdeEquation RactCa { Rate 0.0; Initial
                    8.766060974682582E-4; }
                    OdeEquation RinhCa {
                       Rate ((Rinh * Ca/TAU) - (RinhCa * dinh/TAU));
                       Initial 7.137992942487463E-4; }
              FastSystem {
                    FastInvariant (RactCa + Ca + CaB);
                    FastInvariant IP3;
                    FastRate ((Ca * (K_B_total - CaB)) - (K*CaB));
                    FastRate ((Ca * (K_Ract_total - RactCa)) - (dact *
                    RactCa));
                    FastInvariant RinhCa; }
              }
        CompartmentSubDomain extracellular {
              Priority 1
              BoundaryXm Dirichlet BoundaryXp Dirichlet
              BoundaryYm Dirichlet BoundaryYp Dirichlet
              BoundaryZm Dirichlet BoundaryZp Dirichlet
              PdeEquation Ca { Rate 0.0; Diffusion 300.0; Initial 0.0; }
              OdeEquation IP3 { Rate 0.0; Initial 0.0; }
              OdeEquation CaB { Rate 0.0; Initial 0.0; }
              OdeEquation RactCa { Rate 0.0; Initial 0.0; }
              OdeEquation RinhCa { Rate 0.0; Initial 0.0; }
        }
        MembraneSubDomain cytosol extracellular {
              JumpCondition Ca {
                    InFlux 0.0;
                    OutFlux 0.0;
              }
        }
        Mesh { Size 100 50 50 }
        Task { Output 35.0 Unsteady 0.1 0.0 105.0 }
```

References

Ainger, K., Avossa, D., Diana, A. S., Barry, C., Barbarese, E., and Carson, J. H. (1997). *J. Cell Biol.* **138,** 1077.

Bray, D. (1997). *TIBS* **22,** 325.

Cheng, H., Lederer, W. J., and Cannell, M. B. (1993). *Science* **262,** 740.

Devroye, L. (1986). "Non-Uniform Random Variate Generation." Springer-Verlag, New York.

De Young, G., and Keizer, J. (1992). *Proc. Natl. Acad. Sci. USA* **89,** 9895.

Fontanilla, R. A., and Nuccitelli, R. (1998). *Biophys. J.* **75,** 2079.

Gear, C. (1971). "Numerical Initial Value Problems in Ordinary Differential Equations." Prentice-Hall, Englewood Cliffs, New Jersey.

Kupferman, R., Mitra, P. P., Hohenberg, P. C., and Wang, S. S. H. (1997). *Biophys. J.* **72,** 2430.

Lit, Y. X., and Rinzel, J. (1994). *J. Theor. Biol.* **166,** 463.

Li, Y. X., and Rinzel, J. (1994). *J. Theor. Biol.* **166,** 461.

Mendes, P. (1993). *Comput. Applic. Biosci.* **9,** 563.

Moraru, I. I., Schaff, J. C., Slepchenko, B. M., Blinov, M., Morgan, F., Lakshminarayana, A., Gao, F., Li, Y., and Loew, L. M. (2008). *IET Syst. Biol.* **2,** 352–362.

Novak, I. L., Gao, F., Choi, Y. S., Resasco, D., Schaff, J. C., and Slepchenko, B. M. (2007). *J. Comput. Phys.* **226,** 1271–1290.

Press, W. H., Teukolsky, S. A., Vetterling, W. T., and Flannery, B. P. (1992). "Numerical Recipes in C: The Art of Scientific Computing." Cambridge University Press, New York.

Reder, C. (1988). *J. Theor. Biol.* **135,** 175.

Schaff, J., Fink, C. C., Slepchenko, B., Carson, J. H., and Loew, L. M. (1997). *Biophys. J.* **73,** 1135.

Schaff, J. C., Slepchenko, B. M., Choi, Y. S., Wagner, J., Resasco, D., and Loew, L. M. (2001). *Chaos* **11,** 115–131.

Schmidt, C. E., Chen, T., and Lauffenburger, D. A. (1994). *Biophys. J.* **67,** 461.

Slepchenko, B., and Loew, L. M. (2010). *Int. Rev. Cell Mol. Biol.* (in press).

Slepchenko, B. M., Schaff, J. C., Macara, I. G., and Loew, L. M. (2003). *Trends Cell Biol.* **13,** 570–576.

Sneyd, J., Keizer, J., and Sanderson, M. J. (1995). *FASEB J.* **9,** 1463.

Sorensen, J. P., and Stewart, W. E. (1980). *AIChE J.* **26,** 99.

Strang, G. (1968). *SIAM J. Numer. Anal.* **5,** 506.

Villa, C. M., and Chapman, T. W. (1995). *Ind. Eng. Chem. Res.* **34,** 3445.

Wagner, J., and Keizer, J. (1994). *Biophys. J.* **67,** 447.

Wagner, J., Li, Y. X., Pearson, J., and Keizer, J. (1998). *Biophys. J.* **75,** 2088.

CHAPTER 18

Fractal Applications in Biology: Scaling Time in Biochemical Networks

F. Eugene Yates

The John Douglas French Alzheimer's Foundation
11620 Wilshire Blvd, Suite 270
Los Angeles, CA, USA

DOI: 10.1016/B978-0-12-384997-7.00018-2

I. Update

In the 18 years since my essay on *Fractal Applications in Biochemical Networks* appeared, laying out some fundamentals of (then) modern dynamical systems theory, including self-organization, chaos, and fractals, there has been an explosion of developments in these areas of mathematics, physics, and computations, and to a lesser extent in biology. Here, I shall mention only two recent examples from a now vast and rapidly growing literature. However, the fundamentals described in my essay have remained intact and still have usefulness.

In nonequilibrium, highly nonlinear systems with many dynamic degrees of freedom, the phenomenon of self-organized criticality (SOC) appears in their macroscopic behaviors. It involves scale invariances and generates complexity both in nature and in engineered designs and computations, often arising from deceptively "simple" conditions or equations. In them, underlying geometries of space and time express fractals, power laws, and $1/f$ noise. In physics, SOC can be found in the study of phase transitions, and in biology it appears both during development of an organism and in evolution of populations. Neurosciences and genetics have become especially informed by these concepts. I offer two examples below.

(1) A new algorithm explores the effects of long-term evolution on fractal gene regulatory networks (Bentley, 2004). It exhibits a case of "biomimesis" in which the use of genes expressed as fractal proteins enables greater evolvabilty of gene regulatory networks. The investigations led to an algorithm that could be effectively applied to robotic controls and to invention of efficient and fault-tolerant controllers, more like robust natural designs than the brittle designs we commonly propose.

(2) It has proved difficult to characterize the complex neural dynamics of the neocortex that express abilities to learn and adapt, while remaining "plastic." Edelman (2004) suggested that recurrent neural networks must be involved. Recently Lazar *et al.* (2009) introduced SORN: a *s*elf-*o*rganizing *r*ecurrent *n*eural network that illustrates the potential of that approach. It learns to "encode information in the form of trajectories through its high dimensional state space"—an achievement that

mimics the ability of the natural neocortex to learn "representations" of complex spatio-temporal stimuli. Three types of plasticity seem to be required.

In spite of the demonstrated heuristic value of computational modeling, some of the experimental results in systems creating spontaneous fractal networks in the laboratory have been refractory to such modeling. They seem to be "transcomputable"! The vigor of researches on fractal applications in biology promises a rich future for the subject.

II. Introduction

A. Aims of This Chapter

The purpose of this chapter is to present to a readership of biochemists, molecular biologists, and cell physiologists some of the terms and concepts of modern dynamical systems theory, including chaotic dynamics and fractals, with suggested applications. Although chaos and fractals are different concepts that should not be confounded, they intersect in the field of modern nonlinear dynamics. For example, models of chaotic dynamics have demonstrated that complex systems can be globally stable even though locally unstable and that the global stability reveals itself through the confinement of the motion of the system to a "strange attractor" with a microscopic fractal geometry. Some of the technical aspects of chaos, fractals, and complex dynamical systems are sketched in the Glossary at the end of this chapter, where appropriate references to the papers of experts can be found. The details in the Glossary permit me to use a freer style in the body of this chapter, where I shall explore the possible relevance of fractals to the understanding of both structure and function in biology.

The general motivation for biologists to examine chaotic dynamics and fractal geometries arises from recognition that most of physics addresses only simple systems, no matter how elaborate the mathematical apparatus seems to the uninitiated. For the purposes of theorizing, the simple systems examined, both objects and processes, are preferably rather uniform (all electrons are alike; there are only a few quarks), analytically smooth, conservative (nondissipative), time-symmetric (reversible), and linear. Biological systems, in contrast, are notably complex, nonlinear, dissipative, irreversible, and diverse. In the last 20 years, for the first time, we have witnessed the development of several branches of mathematics, both "pure" and applied, that attempt to confront nonlinearity and complexity in a systematic way. That development led to modern dynamical systems theory and vigorous extensions of nonlinear mechanics. These advances have transformed discussions of the shapes of common objects such as trees, clouds, coastlines, and mountains; of the stability of complex systems; of noise and apparent randomness; of the genesis and nature of turbulence and the Red Spot of Jupiter; of the intervals between drops in a dripping faucet; of the music of Bach; and so on. . . . Now we begin to see biological forms and functions examined from the same mathematical perspectives.

To provide a more specific motivation for biochemists to attend to the terms and concepts of modern dynamical systems theory, I ask the reader to accept on trial, as it were, the hypothesis that biological systems depend on a chemical network whose synthetic capabilities lead to many forms with fractal geometries (including bronchial and vascular trees and dendritic branchings of neurons), and many processes that organize time in a fractal manner. We further hypothesize that biological systems are not inherently noisy, but follow state-determined dynamics (motion and change) of the deterministic chaotic class. Their chaotic dynamics produce their marginal stability, many of their structures, and their fractal organization of time. Furthermore, biological systems with their very large numbers of degrees of freedom (i.e., high dimensionality) have biochemical traffic patterns as fluxes, transports, and transformations whose stability can be comprehended formally only through the confinement of their motions to low-dimensional, chaotic, strange attractors. (*Note*: These are all assumptions, for the sake of discussion.)

Because we lack commonplace terms or familiar metaphors to convey the essence of modern dynamical systems theory, I rely on the Glossary in either its alphabetical or indicated logical order to help the reader through the jargonistic jungle. I doubt that deterministic chaos and fractal time will ultimately prove to be the most useful or insightful models of the complexity of biological systems. Fractals and chaos have little in them of theoretical profundity, in spite of their attendant and fashionable mathematical pyrotechnics. But as models of data, they may be the best we now have for complex systems.

B. Discovery of Fractals: Background Reading and Computing

Fractals can pertain to the organization of both space and time. It is astonishing, given that so much of our terrestrial surround has fractal shapes, that the concept can be attributed chiefly to one person: Benoit (1977). A revised edition entitled *The Fractal Geometry of Nature* appeared in 1982–1983. This work is idiosyncratic, imaginative, and difficult for the layman to read. Mandelbrot showed that recursive phenomena can generate fractals, and one kind of basic fractal expression is a statement describing a recursion. For example, for a univariate system, the recursion is

$$x_{n+1} = f(x_n, c). \tag{1}$$

The $(n + 1)$th value of the recursive function is a function of the nth value plus a constant. The function for the recursion may be deterministic, stochastic, or even a combination. The example commonly used for a purely deterministic recursion is $Z_{n+1} = Z_n^2 + C$, where Z and C are complex numbers. Mandelbrot's insight was to examine recursions on the complex plane, instead of on the real number line. His recursion serves as a basis for the definition of both the Mandelbrot and Julia sets (Peterson, 1988). These remarkable sets are portrayed in the complex plane, where they show very irregular and stunning shapes with extremely elaborate boundaries. (Julia sets may be exactly self-similar at all scales of magnification, but a Mandelbrot set, consisting of all values of C that have connected Julia sets, has a fantastic fine structure at all scales, including repetition, and is not exactly self-similar across scales. The details change.)

In the case of generating fractal trees as computer images of this recursion, if the parameters remain constant through various generations of branching, a somewhat regular tree looking like a bracken fern is obtained, but if the parameters for the branchings are random, then very irregular shapes are obtained. In the mathematical generation of fractals by recursion rules, orders of magnitude from 0 to ∞ are permitted. However, in the real, physical world, there are cutoffs both at the lower and at the higher ends. The measurement of the length of a coastline would stop with a grain of sand; most physical and biological systems damp out or filter out very high frequencies.

In book form fractals and the mathematical background (not easy) are available in Peitgen and Richter (1986), Barnsley (1988), Devaney (1989a,b, 1990).

From these sources, the careful reader will find a deeper understanding of fractals than arising from merely supposing that a fractal is a geometric object having a fractional dimension. Some trivial fractal objects actually have integer dimension, but so do some that are not so trivial. In the discussion of biological structure and function in this chapter, I shall be emphasizing both structures that do in fact have fractional dimensions and also functions, revealed as time histories, that produce $1/f^m$ spectra typical of fractal time, where $m \cong 1$ or, more generally, some noninteger value. The key to this fractal view of both structure and function lies in two of the very strong features of fractals: lack of a characteristic scale and some self-similarity across all levels of magnification or minification. In the strongest cases, self-similarity lies in the detailed shapes, or patterns, but in weaker cases it can be found only in statistical characteristics. Because most fractals are not homogeneous (i.e., not identical at every scale), the more closely you examine them, the more details you find, although there will not necessarily be infinite layers of detail.

The creation of fractal structures in the physical/biological world, as opposed to their generation in recursive computer models, may require the operation of chaotic dynamics. Chaotic processes acting on an environment such as the seashore, atmosphere, or lithosphere can leave behind fractal objects such as coastlines, clouds, and rock formations. I wish to emphasize that the mathematics of fractals was developed independently of the mathematics of chaotic dynamics. Although there were antecedants, I think it is fair to say that the general scientific communities became aware of chaotic dynamics through a 1971 paper on turbulence by Ruelle and Takens, and aware of fractals through Mandelbrot's 1977 book already cited. In biology, some of these modern concepts were adumbrated in works by N. Rashevsky.

The common occurrence of fractal geometries in biological morphologies, also found in physical structures, suggests but does not prove that chaotic dynamics are very widespread in nature and should be found in biological morphogenetic processes. That search is a current and advanced topic for biological investigations and should provide, in my opinion, an enrichment of Edelman's "topobiology," (Edelman, 1988) which at present suffers from a lack of detailed, mathematical modeling. J. Lefevre in 1983 attempted to extend Mandelbrot's illustration of a fractal network, space-filling in two dimensions, into three dimensions for the bronchial tree. A fine account of the possible importance of fractal processes in morphogenesis has been provided by West (1987). It is now thought that many aspects of mammalian morphology involve fractal tree branchings, including bronchi, arteriolar networks, cardiac conduction systems, and neuronal dendrites.

III. Fractal Morphology in Mammals: Some Branchings

A. Bronchial Tree and Pulmonary Blood Flow

The bronchial tree meets all of the criteria for a fractal form (see Glossary). The conduits through which gases flow to and from the lungs branch repeatedly from the single trachea to the terminal structures, the alveoli. West *et al.* have recently reanalyzed lung casts of humans and several other mammalian species originally prepared in 1962 by Weibel *et al.* They found the type of scaling characteristic of a fractal geometry. The fractal dimension, D, $2 < D < 3$, may have provided a mechanism for converting a volume of dimension three (blood and air in tubes) into a surface area of dimension two, facilitating gas exchange. Goldberger *et al.* (1990) have discussed these findings. Fractal branching systems greatly amplify surface area available for distribution or collection, or for absorption, or even for information processing.

Glenny and Robertson (1990) have examined the fractal properties of pulmonary blood flow and characterized the spatial heterogeneity. Their data fit a fractal model very well with a fractal dimension (D_s) of 1.09 ± 0.02, where a D_s value of 1.0 reflects homogeneous flow, and 1.5 would indicate a random flow distribution in their model.

B. Vascular Tree

Elsewhere I have discussed the branching rules of the mammalian vascular tree (Yates, 1991). As in the case of the bronchii, casts have been used (by Suwa) to discover the branching rules of the vascular tree, and that tree also turns out to be a fractal structure.

C. Morphology of the Heart

The branching of the coronary arterial network is self-similar, as is that of the fibers (chordae tendineae) anchoring the mitral and triscupid valves to the ventricular muscle. Furthermore, the irregular spatial branching of the conduction system of the heart (His–Purkinje system) sets up a fractal-like conduction network, which has been interpreted as having important functional consequences because it forces a fractal temporal distribution on electrical impulses flowing through (Goldberger and West, 1985).

D. Neurons (Dendrites)

By the criteria for fractals given in the Glossary and the preceding discussion, the branching of the dendritic tree of many neurons is fractal. Goldberger *et al.* (1990) have suggested that this dendritic fractal geometry may be related to chaos in the nervous system, a feature of the nervous system suggested by the work of Freeman (1991). Such conjectures are interesting but not yet proved to be correct. (See the comment of L. Partridge in the Glossary, section on chaos.)

The above examples represent a small sample of the data which seem to establish that branching treelike structures with fractal geometries abound in the biological realm, both for animal and plant life. However, it is less clear that biological processes are organized in fractal time. I consider that question next, cautioning the reader that I am not comfortable with the current fad for casually imposing chaotic dynamics and fractal concepts on models of physiological processes. These models should be examined skeptically, as provocative possibilities yet to be proved.

E. Fractal Time: Prelude to Fractal Function and Temporal Organization

The main purpose of this chapter is to show how fractals might help to describe and explain biochemical and physiological processes. A first step toward that goal is recognition of the general form of fractals; a second step is the examination of fractal structures in space. Now the third step is an examination of temporal organization from the fractal viewpoint. The final step will consist of applications to research in biochemistry and cellular physiology.

F. Time History Analysis, Scaling Noises

Consider a record of the amplitude of a single biochemical or physiological variable over the duration of some experiment. If the variable is continuous, as in the case of blood pressure, it may be recorded continuously or intermittently (discrete sampling). For time history analysis in digital computers, continuous records are ordinarily discretized at constant sampling interval and converted to a string of numbers. If the variable has the nature of an irregular event, such as the beat of the heart, it may be merely counted or treated as a point process. (As a first approximation, these are usually expected to generate a Poisson exponential distribution of interevent intervals, as in radioactive decay of a specific type from a specific source, but recently attention has been directed to more complicated point processes, for example, in the analyses of neuronal firing in parts of the auditory system.)

Time histories of biological variables can be hypothesized to have been generated either by a deterministic process or by a random process. In the field of nonlinear topological dynamics under discussion in this chapter, varying deterministic processes can be periodic, quasi-periodic, or chaotic. The random processes include (1) point processes with Poisson distributions, (2) processes generating intermittent or episodic pulses ("pulsatility"), and (3) processes that generate broad, rather featureless spectra such as white noise ($1/f^0$ spectrum) or Brownian motion noise ($1/f^2$ spectrum; where f is the frequency against which an associated component of variance—roughly amplitude squared—is being plotted). But there is another case: Mandelbrot has called attention to stochastic fractals, an example of which is fractional Gaussian noise. The power spectrum in the low frequency range for such a univariate, real-valued function with fractal dimension D ($1 \leq D \leq 2$) can be shown to be proportional to f^{1-2H}, where $0 \leq H < 1$, and H is a constant related to D. In other words, fractional Gaussian noise corresponds to what is generically termed $1/f^m$ ($m \cong 1$) noise, where the power spectrum reveals concentration of low frequency energy, with a long higher frequency tail. Familiarly, the spectrum of such processes is broadly referred to as "$1/f$." (The more general case where m is not an integer is discussed below.)

Some point processes can be modeled as stochastic fractals, such as the activity of primary discharge patterns of some neurons. However, the enthusiasm for searching for processes with $1/f$ spectra in biology lies not in the direction of stochastic processes, but in the domain of deterministic chaos, whose dynamics can produce both fractal time and fractal spatial structures. (The finding of a $1/f$ spectrum in a biological time history does not prove that the generator was deterministic chaos. It does, however, permit that hypothesis.) The "noise" classification according to the spectral sequence $1/f^0$, $1/f^1$, $1/f^2$ is implemented by a log–log plot in which log (amplitude squared) is plotted against log (frequency or harmonic number). The three kinds of noise in the sequence will produce log–log plots with a slope of 0 (white or Johnson noise), approximately -1 (Mandelbrot noise), or -2 (Brownian noise). Mandelbrot has called these various noises "scaling noises." If an investigator plots the power spectrum of the time history of a finite length record on a single variable, using the log–log transformation, and finds that the background, band-limited but broad, "noise" does not fall on a line with slope 0, -1, or -2, he could either reject the scaling noise model and assume some other basis for the background variations, or he might consider a harmonically modulated fractal noise model (if the data wander back and forth across a straight line with a slope of -1). Such a model has been provided by West (1987) and is of the form

$$y(z) = [A_0 + A_1 \cos(2\pi \ln z / \ln\beta)]/z^\alpha, \qquad (2)$$

where A_0, A_1, α, and β are parameters.

It should be noted that the log–log transformation of the power spectrum does not obliterate any spectral lines identifying periodic processes, which can occur on top of a background of scaling noise. In summary, an investigator believing that the time history of the variable he is looking at might have resulted from chaotic dynamics might well start with the search for the $1/f$ spectrum characteristic of fractal time, remembering that the presence of a $1/f$ spectrum is compatible with, but does not establish, the presence of a deterministic, chaotic generator of the time history. (In practice it is not always easy to decide on the value of a negative slope in a log–log plot of the power spectrum of real biological data.)

Because fractal time, like fractal space, has the features of heterogeneity, self-similarity, and absence of a characteristic scale, it follows that if recorded on a tape, fractal noise always sounds the same when the tape speed is varied (making allowances for changes in loudness). In contrast, the pitch of any periodic process will be a function of tape speed (rising as the tape speed is increased, a familiar phenomenon when fast forward is used).

Schlesinger (1987) offered a technical but clear account of fractal time and $1/f$ noise in complex systems. In what follows, I paraphrase some points of his article, possibly relevant to biochemical and physiological data in which time appears as an independent variable. It should be an embarrassment to biochemists that time rarely appears explicitly in plots of their data, their idea of kinetics too often being merely a relationship between a reaction velocity (time implicit) and a (steady) substrate concentration.

Suppose that a biologist is observing a process that seems to have some kind of "pulsatility" in which there is a variable time between events or peak concentrations. A starting point for the analysis of pulsatility is to suppose that there is some probability for time between events

$$\psi(t)\mathrm{d}t = \mathrm{Prob}[\text{time between events} \in (t, t + \mathrm{d}t)]. \tag{3}$$

There will be a mean time $\langle t \rangle$ and a median time, t_m, between events

$$\langle t \rangle \equiv \int_0^\infty t\psi(t)\mathrm{d}t, \int_0^{t_m} \psi(t)\mathrm{d}t = \frac{1}{2}. \tag{4}$$

Schlesinger gives the references that provide the basis for these and others of his statements that follow. If $\langle t \rangle$ is finite, then we can say that some natural scale exists in which to measure time. For a very long series of data, it will appear that events occur at the constant rate $\langle t \rangle^{-1}$. If $\langle t \rangle$ is very large, then events will occur at a slow rate, but we would not call such events rare. When $\langle t \rangle = \infty$, there is no natural time scale in which to gauge measurements, and events are indeed rare. Note that even under these conditions, t_m is finite, so events still occur.

Schlesinger then considers the case that we have three events in a row at times $t = 0$, $t = \tau$, and $t = T$, where the value of T is known. We want to know the probability of the middle event that occurs at $t = \tau$

$$f(t) = \frac{\psi(\tau)\psi(T - t)}{\int_0^T \psi(s)\psi(T - s)\mathrm{d}s}, \tag{5}$$

where "the denominator insures the proper normalization." For a purely random process $\psi(t) = \lambda \exp(-\lambda t)$, and $f(\tau)$ is a uniform distribution in the interval $(0, T)$. In that case, the most likely time for the middle event is $\tau = T/2$; that is, on the average, events occur at rather regular intervals. However, if $\langle t \rangle$ is infinite, the process cannot resemble a constant rate renewal; so then $\tau = T/2$ is the least likely value of τ, and values of τ closer to $t = 0$ and $t = T$ are more probable. For such rare events (mean renewal rate of zero), the time sequence of events appears in self-similar clusters like points in a Cantor set (one of the best known demonstrations of self-similarity and fractal dimension). For proof of this by no means obvious result, see the references provided by Schlesinger.

One example of a physical mechanism that can generate such a fractal time distribution of events is hopping over a distribution of energy barriers. A small median jump time can exist and be consistent with an infinite mean jump time. In physics, these ideas have become important in describing charge motion in some disordered systems. In that case, $\psi(t)$ can represent the probability density for an electron's not moving, but it can also represent the probability density for remaining in a correlated motion.

It still has to be answered why $1/f$ noise is so prevalent in physical and biological systems. Schlesinger gives a plausible argument which says that this kind of noise is generic in the same sense that the Gaussian distribution arises from the central limit theorem which governs sums of independent random variables with finite second moments. Consider a process that is described by a product of random variables. Then, for an event to occur, several conditions have to be satisfied simultaneously or in sequence. If P is the probability for the event to occur, and if

$$P = p_1 p_2 \cdots p_N, \tag{6}$$

then

$$\log P = \sum_{i=1}^{N} \log p_i \tag{7}$$

has a Gaussian distribution and P has a log-normal distribution:

$$P(\tau) = \frac{1}{\pi\sigma\tau} \exp(-[\log(\tau/\langle\tau\rangle)]^2/2\sigma^2), \tag{8}$$

where $\langle\tau\rangle$ and σ^2 are the mean and variance of the distribution, respectively. As more factors N participate, σ increases (see next section) and $P(\tau) \cong 1/\tau$ over a range of τ values. The greater the value for σ is, the more extensive the range is over which the τ^{-1} behavior persists and thus the larger the range is over which $1/f$ noise is found. Schlesinger concludes, "The underlying product of the random variables idea that leads to a log-normal distribution of relaxation times and naturally to $1/f$ noise provides a generic generation of this phenomenon. ... The real message of $1/f$ noise is that a scale-invariant distribution of relaxation times has been generated." (See the following section for a similar account by Bassingthwaighte.)

An earlier paper by Careri et al. (1975) provides a contrasting background for the analysis of Schlesinger; it examined statistical time events in enzymes. The great progress in understanding spatial aspects of enzyme action has not been matched by an equally important analysis of the temporal aspects. Carerri et al. explored elementary processes and assessed the microscopic mechanisms by comparative studies on representative model systems using the theory of random processes.

They modeled relaxation and concentration fluctuation spectroscopy by use of the fluctuation-dissipation theorem. This theorem relates the average time for the decay of spontaneous microscopic fluctuations to the time course observed after a small perturbation around equilibrium. They focused on the correlation time (especially the autocorrelation time of a statistically stationary variable) as the basic quantity of interest because it gives a direct measure of the time interval over which a variable is behaving more or less regularly and predictably. For longer time intervals, the behavior becomes progressively more random in this model. However, the authors caution that in their kind of analysis of random processes, they must assume that the variables involved are statistically stationary and linearly superimposed, and that such an assumption may not hold for an enzyme, where the different classes of fluctuations may interact in a nonlinear way, merging into a new cooperative with nonstationary effects of great chemical interest— precisely the point of modern, nonlinear dynamics!

The presence of $1/f$ noise in a dynamic system does not imply any particular mechanism for its generation. For that, one needs to draw on phase-space plots, Liapunov exponents, embedding plots, etc., for a more detailed investigation of dynamic behavior in complex biological systems. Recently Sun (1990) considered the general case that a fractal system follows a generalized inverse power law equation of the form $1/f^m$, where m is any fractional number. He reexpresses such a system in fractional power pole form (not shown here) which, as he points out, has a much wider representation of natural and physiological phenomena than does a $1/f^1$ view because its low frequency magnitude is finite instead of infinite (as $f \rightarrow 0$). He then offers a time domain expression of such a fractal system consisting of a set of linear differential equations with time-varying coefficients. If the fractal dimensions m_i approach unity or any other integer numbers, his equations lead to regular time-invariant systems. Most importantly, he shows that certain familiar dynamic systems whose performance criteria are well known become more stable in the fractal domain. Thus, he confirms the view expressed earlier by Bruce West that fractal systems are error tolerant and, in that sense, more stable than their nonfractal counterparts.

A clear and profound explanation of $1/f$ noise has been provided by Bak *et al.* (1987). As they remark, "One of the classical problems in physics is the existence of the ubiquitous '$1/f$' noise which has been detected for transport in systems as diverse as resistors, the hour glass, the flow of the river Nile, and the luminosity of stars. The low-frequency power spectra of such systems display a power-law behavior $f^{-\beta}$ over vastly different time scales. Despite much effort, there is no general theory that explains the widespread occurrence of $1/f$ noise." They then argue and demonstrate numerically that dynamical systems with extended spatial degrees of freedom naturally evolve into self-organized critical structures of states which are barely stable. They propose that this SOC is the common underlying mechanism behind the common occurrence of $1/f$ noise and self-similar fractal structures. The combination of dynamic minimal stability and spatial scaling leads to a power law for temporal fluctuations. It should be emphasized that the criticality in their theory is fundamentally different from the critical point at phase transitions in equilibrium statistical mechanics, which can be reached only by tuning of a parameter. These authors refer to critical points that are attractors reached by starting far from equilibrium, and the scaling properties of the attractor are insensitive to the parameters of the model. In fact, this robustness is essential in supporting their conclusion that no fine tuning is necessary to generate $1/f$ noise and fractal structures in nature. They end their article with the remark, "We believe that the new concept of SOC can be taken much further and might be *the* underlying concept for temporal and spatial scaling in a wide class of dissipative systems with extended degrees of freedom."

The ubiquity of physical and biological processes in which some measure of "intensity" (such as frequency of usage or occurrence, physical power, or probability) varies inversely with f^m (where f is a frequency or a rank order) forces us to speculate about interpretations of different values of m, whether integer (0, 1, 2) or fractional. A rich and varied literature bears on such systems, real and model. The mathematical treatments are advanced, and we have no compact, reduced figure of thought to encompass all $1/f^m$ phenomena. There are, however, some informal inferences that may be drawn, and they are considered next.

G. Fractals and Scatter in Biological Data: Heterogeneity

Bassingthwaighte (1988) has noted that there are spatial variations in concentrations or flows within an organ, as well as temporal variation in reaction rates or flows, which appear to broaden as the scale of observation is made smaller (e.g., smaller lengths, areas, volumes, or times), for the same constant total size or interval, composed of N_i units, where i indexes an observational scale (number of pieces or intervals in the population of samples). He then asks, "How can we characterize heterogeneity independently of scale?" This scale-dependent scatter in physiological and biochemical observations is a property inherent in the biological system and cannot be accounted for entirely by measurement error. Bassingthwaighte considers the example of channel fluctuations: "When the duration of openings of ion channels is measured, the variation is broader when the observations are made over short intervals with high-resolution instrumentation and narrower when made over long intervals with lower fidelity. While this problem seems obvious when considered directly, no standard method for handling it has evolved." Given any arbitrary choice for the size of the domain, one wishes to consider, how could one describe the heterogeneity of the system in a fashion that is independent of the magnitude of the domain or the period of observation? The fractal concept provides the answer.

As an example, consider the measurement of variation in regional flows throughout some organ whose total blood flow is known. The mean flow per gram of tissue is the total flow divided by the mass of the organ. Next, assume that the flows everywhere are steady and that the organ is chopped up into weighed pieces and the flow to each piece is known (from the deposition of indicator or microspheres) so that we have an estimate of the flow per gram for each piece. The relative dispersion of the regional flows is given by the standard deviation divided by the mean (the usual coefficient of variation, CV). Empirically we find that the finer the pieces we chopped the organ into, the greater the relative dispersion or coefficient of variation is. So, what is the true variation? From experimental data Bassingthwaighte obtained the result for relative dispersion (which he designated *RD* and I call CV):

$$CV = CV(N = 1)N^{D-1},\tag{9}$$

where N is the number of pieces and D is the fractal dimension, a measure of "irregularity." In a plot of CV versus N, the real value of $CV(N = 1)$ can only be 0, and so it cannot actually be on the fractal curve. Therefore, the value of $CV(N = 1)$ in these data had to be obtained by extrapolation of the log–log plot to its intercept. It was found that extrapolated $CV(N = 1) = 12.9\%$.

To generalize the expression for relative dispersion,

$$CV(w) = CV(w = 1g)w^{1-D},\tag{10}$$

where w is the mass of the observed pieces of tissues and D is the fractal dimension (which in the case of blood flow in the myocardium had an observed value of 1.18). (The exponent can be $D - 1$ or $1 - D$, depending on whether the measure is directly or inversely proportional to the measuring stick strength.) Thus it is the fractal dimension, not the coefficient of variation for any particular set of data that expresses the "true" heterogeneity.

The same approach can be applied to temporal fluctuations, and the equation has the same form as that for spatial heterogeneity

$$CV(\tau) = CV(\tau = 1 \text{ unit})\tau^{1-D_\tau}.\tag{11}$$

In this case, we examine the standard deviation of flows over a given interval τ, divided by the mean flow determined over a much longer time. The broader the standard deviation is, the shorter the interval τ is over which the flows are measured. Some arbitrary reference interval ($\tau = 1$) must be chosen. Preliminary analyses of capillary flow fluctuations give a D_τ value of approximately 1.3. Similarly, Liebovitch *et al.* (1987) obtained patch-clamp data on ion channel openings or closings in lens epithelial cells. They compared a fractal model with mono- or multiexponential rate constants and obtained better fits with the fractal model.

The important conclusion is that suggested by Bassingthwaighte: fractals link determinism and randomness in structures and functions.

IV. Chaos in Enzyme Reactions

To further illustrate possible applications of concepts of chaos and fractals for the interpretation of biochemical phenomena, I consider the historically important, brief report by Olsen and Degn (1977) of perhaps the first direct experimental demonstration of chaos in a chemical reaction system. They studied the behavior of oxygen concentration as a function of time in a peroxidase-catalyzed oxidation of NADH in a system open to O_2 in a stirred solution, where the O_2 could enter by diffusion from the gas phase. They observed sustained "oscillations" by continuously supplying NADH to the reaction mixture. They discovered that the waveform of the variations depended on the enzyme concentration. At one concentration, sustained, regular, true oscillations were observed; but at other concentrations, the large variations showed no apparent periodicity. This irregular fluctuation was analyzed by plotting each amplitude against the preceding amplitude, and each period against the preceding period, according to the mapping technique introduced by Lorenz. (The study of such iterated maps originally arose from a desire to understand the behavior of solutions of ordinary differential equations, and they can be used to identify random, periodic, or chaotic behaviors.) Such maps draw diagrams of transition functions, and, according to a theorem of Li and Yorke (1975), if the transition function allows the period 3 (which can be tested graphically on the map), then it allows any period and chaos exists. (Devaney credits this idea to an earlier source; see the videotapes already mentioned.) By this means, Olsen and Degn demonstrated graphically that the fluctuations in the concentration of the reactant O_2 at a certain enzyme concentration produced a transition function admitting period 3. By the Li and Yorke criterion (not necessary but sufficient to identify chaos), the reaction kinetics were indeed chaotic. In the absence of chaos theory, the data most likely would have been uninterpretable or regarded merely as showing a contaminating noise of unknown origin. With chaos theory, it could be concluded that the data accurately conveyed the real (deterministic, nonlinear) dynamics of the reaction.

V. Practical Guide to Identification of Chaos and Fractals in Biochemical Reactions

Suppose that a biochemist is monitoring the concentration or thermodynamic activity of a reactant under conditions that the reaction is not at equilibrium. (It may be in a constrained steady state, or it may be tending toward an equilibrium, depending on the conditions.) Traditionally, such kinetic systems are usually thought to have either monotonic or periodic behavior. Examples of nonmonotonic, nonperiodic behavior have long been known, but before the advent of deterministic chaos theory it was usually assumed that the variations had to be attributed to some kind of "noise." It was not easy to imagine where the noise came from during controlled *in vitro* studies. Today, contamination of the reaction kinetics by some random process should not be the first choice for modeling; deterministic chaos deserves to be tried.

The first step in such an analysis consists of converting the concentration data into a string of numbers (usually at equally spaced time intervals). If the data were continuously recorded, as might be the case for hydrogen ion concentration using a pH electrode, then the data must be discretized. Here, the choice of sampling interval depends on the frequency range the experimenter wishes to explore, recognizing restrictions imposed by the frequency response of the recording equipment. Once the string of pairs of numbers (concentration, time of observation) has been obtained, several practical considerations arise that I cannot treat formally here. These concern whether the data should be in any way smoothed or filtered according to what is known about measurement uncertainty. In addition, there must be an *a priori* policy for interpolation across missing points. Choices about the length of the record or the number of successive points to be included in the analysis can bear critically on the results of subsequent computations. Some of the most powerful techniques for exploring chaotic dynamics require more points (more than a thousand) than biochemists may have. If the conditions of the reaction are nonstationary, so that the statistical properties of segments of record are different in different pieces of the record, then it may be desirable to use less than the whole record length for analysis. But there is a caveat: As I have pointed out elsewhere in this chapter, one chaotic dynamic can produce different spectral results depending on the piece of the record used, even when the underlying dynamics are actually the same throughout. This is tricky business, and the investigator must proceed cautiously.

Having chosen a sampling interval and a length of record, the investigator may then apply spectral analysis and examine the power spectrum for evidence of periodicities (peaks) and broad-band, background "noise." If that background has a $1/f^m$ property, as revealed by a log–log plot of power (amplitude^2) versus frequency, yielding a straight line of negative slope $-m$, then the value of m can give some clue as to the nature of the process; m equals 2 is consistent with an assumption that a Brownian motion, diffusion process is dominant; an m of approximately 1 hints that chaotic dynamics may be operating (this is not a critical test).

The next step might be the graphical iteration technique in which each value x_{t+1} is plotted against the preceding value x_t, along the whole time series. The resulting iterated map gives the transition function for the reactant whose concentration was x. If the plot yields a random scattergram, the process is random. However, if the plot reveals regularities, a chaotic "attractor" may be present.

Next, the "correlation dimension" of the data string may be computed using the Grassberger and Procaccia, 1983), not further described here. If the computed dimension has a noninteger value greater than 2, then chaotic dynamics are suggested. (In my experience, if the correlation dimension ranges larger than 10, it is best just to assume that white noise is present.) There are many reservations about the interpretation of the correlation dimension as being the (fractal) dimension of a chaotic attractor (not reviewed here). It is safest just to regard it as some kind of "complexity" estimate on the dynamics of the reactant. As a practical matter, this complexity will range from 2^+ to 10^+; the larger the number is, the greater the complexity of the dynamics will be. (In theory the correlation dimension can range from zero to infinity.)

There is a new measure of the complexity of dynamics that is mathematically robust and makes fewer assumptions than are required for the computation of the correlation dimension. Pincus (1991) has introduced a regularity statistic for biological data analysis that he calls "approximate entropy," abbreviated ApEn. It is a relative, not absolute, measure. Its rigorous use may require as many as 1000 data points, but its robustness permits valuable results from strings of time series data no longer than perhaps 200 points. This measure has been used to detect abnormal hormone pulsatility in physiological systems. Pincus developed ApEn as a corrective to blind applications of certain algorithms to arbitrary time series data. He commented (personal communication, 1991), "These algorithms include estimates of Kolmogorov–Sinai (K–S) entropy and estimates of system dimension. These algorithms were developed for application to deterministic systems, yet are frequently applied to arbitrary data with questionable interpretations and no established statistical results. Sometimes attempts are made to apply these algorithms to very noisy processes, but again, statistical understanding is lacking. Furthermore, correlated stochastic processes are almost never evaluated in this way. There are severe difficulties with such blind applications of these popular algorithms. One cannot establish underlying determinism via the correlation dimension algorithm." At the present time ApEn may be the best we can do to specify "complexity" of a time series, particularly given that biological systems may produce very messy, weakly correlated stochastic processes. However, like all other aggregated measures, ApEn loses information.

VI. Summary, or What Does It All Mean?

I have tried to convey some of the new trends of thought regarding the performance of complex systems that have been provided by mathematical advances, particularly in topology. It would be impossible to try to fill in every step required for a clear understanding of the separate concepts of chaos and fractals that come together in the modern dynamic systems theory addressing the behavior of complex, nonlinear systems. I have provided only a few primary or secondary references out of the hundreds now available in this rapidly moving branch of applied mathematics. The bibliographies of the references I offer do give a foundation for understanding the new developments. Most of the material I have cited, and the references cited in them, are mathematically somewhat advanced and not easily accessible to a nonspecialist reader. Gleick (1987) provide very attractive accounts of some of the excitement in the field, of some of the leading personalities, and of some of the claims. However, in my opinion, a lay article inevitably must fail to convey the substance of chaos and fractals. Here, I have tried to indicate some of that substance, without figures and with only a few equations. This effort, too, must fail, but it should at least avoid misleading the reader about some of the achievements and the perplexities attendant on these new views of dynamics. A scholarly examination of the history of these trends in mathematics would show, as Ralph Abraham has done, that not all the fashionable and even faddish ideas are really new; there are strong antecedants going back to at least the early 1900s.

The study of chaos and fractals in a biological setting provides many valuable lessons, some of which I list below. Complex results do not imply complex causes. Simple systems need not behave in simple ways. Complex systems can have simple behavior. Randomness out does not imply randomness in (or inside). Small transient or distant causes can have large long-term effects. There can be instability at every point in phase space, while trajectories within it may be confined: aperiodic, never repeating, yet staying close together. Erratic behavior can be stable. Global behavior can differ from local behavior and may be predictable even when local behavior may be unpredictable. When a quantity changes, it can change arbitrarily fast. Deterministic, continuum dynamics can be intermittent. Trends can persist through seeming randomness. There can be scaling symmetry (i.e., self-similarity) across all levels of observation in systems having no characteristic scale of space or time. The region of phase space in which trajectories settle down can have noninteger dimension. Transitions from one dynamic mode of behavior to another can occur through bifurcations that can be subtle, catastrophic, or explosive. A nonlinear system, such as may be seen in a hydrodynamic field, may have regions with steady behavioral modes, separated by chaotic boundaries that are stable (e.g., the Red Spot of Jupiter, which may be a vortex stabilized by a chaotic surround, or which may be a soliton). Fluctuations may be enhanced and captured as a new form or function.

The list could go on, but the items above should convey the explanatory richness of the concepts of chaos and fractals and their potential for changing our viewpoints about determinism and randomness, about stability, evolution, speciation, noise. . . . Surely these are nontrivial yields to be had from a study of these subjects.

Perhaps for the first time we have mathematics of complexity mature enough to describe motions and transformations within biochemical and physiological systems. But not all of those competent to understand the mathematical developments in detail are comfortable with the tendency, possibly a rampant distortion, to view more and more of biological processes through the spectacles of chaos and fractals. Are we ready to abandon all the old-fashioned models of deterministic systems contaminated by additive or multiplicative noise of various types? Is there really so much of chaotic dynamics in living systems? How can we tell?

Admittedly, deterministic chaotic dynamics can imitate many different kinds of what we previously thought of as random "noise." Above all, there is a kind of efficiency or compactness about the concepts of chaos and fractals as applied to biological dynamics. They can dispel some of our confusion about properties of data, as I have tried to illustrate, and arguably they justify the claim that underneath many appearances of complexity lies some kind of deeper simplicity—an assumption that physicists have long made.

I have described some crude tests for the presence of fractal time, fractal space, and chaotic dynamics in living systems. There are no standalone, solid tests; however, there are indicators of the operation of chaotic dynamics and the presence of fractal space or time. The most important yield for biologists arising from the study of chaotic dynamics and fractals seems to me to pertain to the fundamental biological concept (originated separately by Sechenov, Bernard, and Cannon) of "homeostasis." Homeostasis says something about the stability, vitality, health, and persistence of a living system and its ability to accommodate perturbations without dying. The very nonlinear world of chaotic dynamics and fractal geometries seems to me to justify substitution of the term "homeodynamics" for homeostasis. Homeodynamics carries with it the potential for a deeper understanding of what it means for a complicated system to be stable and yet show very rich behaviors, including the possibilities of development of individuals and evolution of species. Our thinking itself is evolving as shown in the following diagram.

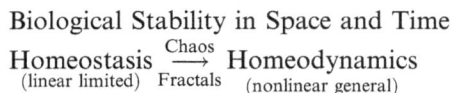

Biological Stability in Space and Time

Homeostasis $\xrightarrow[\text{Fractals}]{\text{Chaos}}$ Homeodynamics

(linear limited)   (nonlinear general)

VII. Glossary

In the setting of this book this glossary must necessarily be only semitechnical. The subject of fractals properly belongs to topology, whose contributions to dynamics cannot be appreciated without knowledge of sets, maps (linear and nonlinear), manifolds, metric spaces, and vectorfields. A brief reading list for some of the relevant mathematics might include the following: Barnsley, 1988; Devaney, 1989a,b; Peitgen and Richter, 1986; Stewart, 1989; Abraham and Shaw, 1982–1988.

A. Instruction to Reader

The following terms are partially defined and presented in alphabetical order: attractor (basin of attraction), bifurcation, chaos, complexity, cycle, dimension, dynamics, dynamical system (and state vector), fractals, limit cycle, linear, manifold, maps (mapping), noise, nonlinear, quasi-periodicity, spectral analysis, stability, state, and vectorfield. Terms in italics in the definitions that follow are themselves defined elsewhere in the Glossary. The alphabetical order is not the logical order. Any reader wishing to use this Glossary as an introduction to modern nonlinear dynamics should read the entries in the following, more logical order: complexity, state, dynamics, dynamical system, manifold, maps, vectorfield, attractor, linear, nonlinear, bifurcation, chaos, fractals, dimension, limit cycle, quasi-periodicity, noise, spectral analysis, and stability.

B. Attractor (Basin of Attraction)

Modern *nonlinear dynamics* provides qualitative predictions of the asymptotic behavior of the system of interest in the long run. These predictions may hold even when quantitative predictions are impossible. In each *vectorfield* of a *dynamical scheme* there are certain asymptotic limit sets reachable sooner or later from a significant set of initial conditions. Each asymptotic set in each vectorfield of a dynamical scheme is an attractor. The set of initial states that tends to a given attractor asymptotically as time goes to infinity comprises the basins of the attractor. Initial conditions from which the system will not reach the attractor belong to a separator. Attractors occur in three types: static (an attractor limit point); periodic (an attractive *limit cycle* or oscillation); and *chaotic* (meaning any other attractive limit set). If the attractor has a topological *dimension* that is not an integer, it is considered "strange," and many *nonlinear* systems have such strange attractors (that are *fractals*). A strange attractor has weird geometry such as fractional dimension or nondifferentiability.

C. Bifurcation

Bifurcation theory asks how the equilibrium solution, or value, of interest for a dynamical system changes as some control parameter or variable is gradually changed. The equilibrium solution (itself a dynamical behavior or dynamical mode) can change in a subtle manner or in a very abrupt, catastrophic, or explosive manner at critical parameter values. Changes in states such as the freezing of liquid water into ice represent a familiar bifurcation. Changes in flows from laminar to turbulent as the velocity is increased (for a given geometry) also represent a familiar bifurcation.

D. Chaos

A readable introduction to the subject of chaos has been provided by Morrison (1991). The simplest view of chaos is that it is an irregular (aperiodic) fluctuation of a variable—unpredictable in the long term—generated by a fully deterministic process without noise. That view is all very well if one knows the deterministic system in advance, but if all one has is irregularly varying data, proof that the dynamics are chaotic is difficult, and may be impossible. It is important to note that even if long-term prediction of chaotic systems is impossible, accurate short-term prediction is possible. The chaotic orbit of the planet Pluto is an example.

Chaotic dynamics are deterministic, not random, although they may imitate various kinds of *noise*. Broad-band power spectra are often associated with chaotic dynamics. They may show the $1/f$ spectrum of Mandelbrot noise. All of the known chaotic attractors have a fractal microstructure that is responsible for their long-term unpredictable behavior. The power *spectrum* of a periodic attractor is a discrete, or line, spectrum, but the power spectra of *chaotic motions on attractors* are usually continuous or "noisy-looking."

Farmer *et al.* computed the spectra for the Rössler chaotic attractor under six different parameter values for the simple equations generating the chaotic dynamics (see discussion by Schaffer, 1987). For each of six parameter values, the motion was indeed chaotic, but the spectra varied from sharp peaks (periodic behavior) to featureless spectra (broad band noise). Similar results from other chaotic systems give a warning: dividing a single chaotic trajectory into subsamples, so that the data sets are relatively short, causes the same chaotic process to reveal itself through very different spectra.

The onset of chaos is often preceded by an apparently infinite cascade of period doublings, so that by the time one reaches the chaotic region all cycles of period 2^n have gone unstable, even though they are still there. (Odd periods also appear; see Devaney videotapes for illustration.) As a result, the chaotic orbit wanders among former basins of attraction. We can conclude that *spectral analysis*, even of rather dense biological data, is not a reliable means to identify chaotic dynamics. However, the spectra are useful in their own right because they do identify such periodicities as may be present, regardless of origin.

Formal definitions of chaos require that systems show, as typical orbits on their attractors, trajectories with a positive Liapunov exponent. Chaos is characterized by the fact that adjacent initial conditions lead to exponentially diverging trajectories (i.e., there is extremely close dependence on initial conditions in the long run), and the exponent describing this divergence is the Liapunov exponent. In addition, the Kolmogorov–Sinai criterion requires that a chaotic system should have nonzero entropy.

There is currently active research under way on the development of practical algorithms that can be used to compute numerically the *dimensions* and Liapunov numbers (which I do not define here), given the values of some variables as a function of time. The algorithms being used have many potential problems. There is debate on two aspects: (1) the requirements for the size of the data set being analyzed and (2) the effects of noise, large derivatives, and geometry of the attractor. Because of these difficulties, unambiguous interpretation of published reports is difficult. Any claim for "chaos" based solely on calculation of dimension or the Liapunov numbers without additional supporting evidence such as well-characterized bifurcations must be viewed with extreme skepticism at present (Glass and Mackey, 1988).

Fractals and chaos are separate concepts and should not be confounded. "Chaos" refers to the dynamics of a system; "strange attractor" characterizes the (often fractal) geometry of an attractor. Chaotic dynamics can have attractors that are not strange; nonchaotic dynamics can display strange attractors, and not all strange attractors are chaotic!

Chaotic dynamics may describe the transient motions of a point in a phase space of high *dimension*, even infinity. The lower dimension region of phase space to which the motions tend asymptotically as time goes to infinity (the chaotic or strange *attractor*) will usually have a *fractal* dimension. It is thought that most strange or chaotic attractors have fractal dimensions, usually greater than 2^+. Even though in principle dimension can reach infinity, as a practical matter biological data rarely support a claim for the dimensionality of an attractor higher than about 6^+-10^+. (White noise, which has no attractor, can generate an apparent dimension of about 10 using the correlation dimension algorithms.)

Some physicists view the world algorithmically, saying that the existence of predictable regularities means that the world is algorithmically compressible. For example, the positions of the planets in the solar system over some interval constitute a compressible data set, because Newton's laws may be used to link these positions at all times to the positions and velocities at some initial time. In this case, Newton's laws supply the necessary algorithm to achieve the compression. However, there is a wide class of theoretical and perhaps actual systems, the chaotic ones, that are not algorithmically compressible.

Whenever one has a mathematical model of a two-variable limit cycle oscillator (a nonlinear system) and adds just one more variable to it (such as a depletable source that regenerates at a finite rate), one is extremely likely to discover a chaotic regime after some fiddling with the parameters (Rössler, 1987). To produce chaos through nonlinear models based on ordinary differential equations, one needs at least three continuous independent variables. However, in the case of nonlinear discrete models (e.g., with finite difference equations), the dynamics of a single variable can have chaotic regimes.

Because random processes are often characterized by their interevent histograms, and because it is well known that in the Poisson process case the interevent histogram is an exponential function, a problem arises: namely, deterministic chaotic systems can also give rise to exponential interevent histograms. Thus, it is not a simple matter to distinguish between random *noise* and deterministic *chaos*, and it is always possible that irregular dynamics in many systems that have been ascribed to chaos may be noise, or vice versa. Glass and Mackey (1988) remark, "The strongest evidence for chaotic behavior comes from situations in which there is a theory for the dynamics that shows both periodic and chaotic dynamics as parameters are varied. Corresponding experimental observation of theoretically predicted dynamics, including irregular dynamics for parameter values that give chaos in the deterministic equations, is strong evidence that the experimentally observed dynamics are chaotic." Other approaches to the identification of chaos include the power *spectrum*, Poincaré map, Liapunov number, and dimension calculations (but see the caveat, above, by Glass and Mackey).

If a chaotic attractor exists inside a basin of attraction in the phase space of a dynamical system, then globally the dynamics will be stable as time goes on, although microscopically, within the attractor, they may be locally unstable. However, if the parameters of the system change in such a way that a chaotic attractor collides with the boundaries of its attraction basin, all the chaotic trajectories will become only transients (metastable chaos), and eventually the system escapes the basin of chaos and evolves toward some other attractor in the phase space. When topological, unstable chaos occurs between two stable attractors, the basins acquire a *fractal* nature with self-similarity at their boundaries, which then possess fine structure at each scale of detail, creating a fuzzy border.

It is an open question whether low-dimension chaos and nonlinear mappings are relevant to the nervous system. Freeman (1991) has argued strongly that they are. Lloyd Partridge (personal communication, 1991) points out that there seems to be no universally accepted definition of what constitutes deterministic chaos, but at least two things seem to appear in all definitions. First, chaos describes the continuously changing pattern of some variable in a system in which future values are determined by the operation of rules of change on the present value. Second, those rules are very sensitive to initial conditions. The generation of a chaotic response is always determined by the internal rules of the system while that system has only constant (including zero) input or uniform periodic input. Outputs that are considered chaotic continue to change in a never exactly repeating manner, yet they stay within specific bounds. Thus, while individual response values cannot be predicted, the bounds of possible responses can only be learned by observations. Partridge closes his presentation as follows: "I conclude that the contributions of nonlinearity, discontinuity, feedback, and dynamics to neural function demands serious examination. At the same time, the division between truly chaotic and nonchaotic effects that may result from these properties falls within the range that these studies should span and does not represent an important division. Thus, while understanding of the results of formal study of chaotic behavior can be a valuable background for neural science, distinguishing sharply between chaos and the variety of related phenomena, in particular neural function, may be relatively useless."

Among some of those who understand chaotic dynamics very deeply, such as Stuart Kauffman, there is a lingering feeling that the emergence of order in self-organizing systems is in some sense "anti-chaotic" (Kauffman, 1991). Using Boolean networks as models, Kauffman tries to show that state cycles, as dynamic attractors of such networks, may arise because Boolean networks have a finite number of states. A system must therefore eventually reenter a state that it has previously encountered. In less formal terms I have tried to advance the same idea under the term "homeodynamics," discussed later. Kauffman concludes that parallel-processing Boolean networks poised between order and chaos can adapt most readily, and therefore they may be the inevitable target of natural selection. According to this view, evolution proceeds to the edge of chaos. Kauffman concludes, "Taken as models of genomic systems, systems poised between order and chaos come close to fitting many features of cellular differentiation during ontogeny—features common to organisms that had been diverging evolutionarily for more than 600 million years. ... If the hypotheses continue to hold up, biologists may have the beginnings of a comprehensive theory of genomic organization, behavior and capacity to evolve." Although I am somewhat dubious that Boolean networks are the best form for modeling the dynamics of biological systems (Boolean networks can have arbitrary dynamics violating physical law), I think Kauffman has captured an important idea.

E. Complexity

There is no general agreement as to what constitutes a complex system. Stein (1989) remarks that complexity implies some kind of nonreducibility: the behavior we are interested in evaporates when we try to reduce the system to a simpler, better understood one. In my opinion the best definition of a complex system is that according to Rosen (1987). He notes that complex systems are counterintuitive, that is, they may do things that are unexpected and apparently unpredictable. Their behavior may show emergent properties, because complex systems do not appear to possess a single, universal, overarching description such as those postulated for simpler physical, mechanical, or thermodynamic systems. For example, an organism admits a multiplicity of partial descriptions, and each partial description, considered by itself, describes a simpler system, that is, one with a prescribed set of *states* and a definite *dynamical law* or state transition rule. Thus, an organism can present itself to different observers in various ways, each of which can be described simply according to a standard Newtonian-like paradigm, but the description will be different for different observers. A complex system is one that cannot be comprehensively described. A complex system is not effectively explainable by a superposition of the simple subsystem descriptions; it does not fit the Newtonian dynamical scheme.

F. Cycle

I use "cycle" and "rhythm" synonymously to describe a time history of a variable (usually referred to as a time series) in which there is a recurrent amplitude variation with statistical regularity (stationarity). The duration of one variation back to its original starting point is the length of one cycle, or the period. The reciprocal of period is frequency. A noise-free, sine-wave generator produces perfect cycles of constant period, amplitude, phase, and mean value. In the presence of certain kinds of *noise*, there may be wobble, that is, there may be variation on the length of the period, as well as on amplitude, but if the dispersion is not so great as to obscure the underlying periodicity around some average period, we may still wish to claim the presence of a cyclic process. The shape of the recurring process can range from a train of spikes (as in nerve impulses, which have height but little width) separated by intervals, on one extreme, to a pure sine wave which is very rounded with no intervals between events on the other extreme. (Fourier-based spectral analysis deals well with the latter, but poorly with the former shapes.)

Time histories that are cyclic can be (1) periodic, (2) nearly periodic (this is an informal term, used when a little unexplained wobble is observed), or (3) *quasiperiodic*. In contrast is the acyclic (usually called aperiodic) time series in which there is no regularity in the occurrence of any amplitude value. Aperiodic time histories can be generated by deterministic chaotic dynamics, by some random processes, or by a mixture of both.

G. Dimension

The familiar three dimensions of Euclidean space give us an intuitive feeling for the meaning of the term. However, the concept of a topological dimension is more elaborate and not easy to explain in lay terms. As Peterson (1988) remarks,

"Experiments with soap films show the tremendous variability in the shapes of surfaces. Computer-generated pictures of 4-dimensional forms reveal unusual geometric features. The crinkly edges of coast lines, the roughness of natural terrain, and the branching patterns of trees point to structures too convoluted to be described as 1-, 2-, or 3-dimensional." Instead mathematicians express the dimensions of these irregular objects as decimal fractions rather than whole numbers. In the case of whole numbers, for example, any set of four numbers, variables, or parameters can be considered as a four-dimensional entity. In the theories of special and general relativity, three-dimensional space and time together make up a four-dimensional continuum. Going beyond the fourth dimension requires only adding more variables.

The *Encyclopedic Dictionary of Mathematics* states (Volume I, Section 121, "Dimension Theory"): "Toward the end of the 19th century, G. Cantor discovered that there exists a 1-to-1 correspondence between the set of points on a line segment and the set of points on a square; and also, G. Peano discovered the existence of a continuous *mapping* from the segment onto the square. Soon, the progress of the theory of point-set topology led to the consideration of sets which are more complicated than familiar sets, such as polygons and polyhedra. Thus it became necessary to give a precise definition to dimension, a concept which had previously been used only vaguely." (The rest of the section gives a highly technical definition of the dimension of metric spaces.)

Glass and Mackey (1988) offer this definition of one of the simplest meanings of (capacity) dimension. Consider a set of points in N-dimensional space. Let $n(\varepsilon)$ be the minimum number of the N-dimensional cubes of side ε needed to cover the set. Then the dimension, d, of the set is

$$d = \lim_{\varepsilon \to 0} \frac{\log \eta(\varepsilon)}{\log(1/\varepsilon)}. \tag{1}$$

For example, to cover the length of a line L, $n(\varepsilon) = L/\varepsilon$, and d is readily computed to be 1. Similarly, for a square of side L, we have $n(\varepsilon) = L^2/\varepsilon^2$ and $d = 2$. Unfortunately, the many different views of dimension touch on mathematical issues much too deep for this chapter. A sense of those intricacies can be found in Mayer-Kress, 1986). It will have to suffice for my purposes to invoke the technical capacity dimension or the intuitive notion that a dynamical system has a hyperspace defined by one dimension for each dynamical degree of freedom. These two views do not always coincide. For example, the capacity dimension of a limit cycle based on a two-dimensional nonlinear differential equation (e.g., the van der Pol equation) is 1 for the limit set of the orbit, but 2 for the basin of attraction feeding asymptotically onto the orbit. The number of degrees of freedom in the van der Pol oscillator is 2 in either the separated x and y form or the acceleration, velocity, and position form for a single variable. (Both forms are presented elsewhere in the Glossary.) In spite of the difficulties, "dimension" can be used as a qualitative term corresponding to the number of independent variables needed to specify the activity of a system at a given instant. It also corresponds to a measure for the number of active modes modulating a physical process. Layne *et al.* (1986) point out that it is therefore a measure of complexity. However, many also make use of the correlation dimension based on the algorithm proposed by Grassberger and Procaccia, 1983). An application can be found in Mayer-Kress *et al.* (1988).

H. Dynamics

The basic concepts of dynamics have been expressed by Abraham and Shaw (1982) as follows: "The key to the geometric theory of dynamical systems created by Poincaré is the phase portrait of a *dynamical system*. The first step in drawing this portrait is the creation of a geometric model for the set of all possible states of the system. This is called the state space. On this geometric model, the dynamics determined a cellular structure of *basins* enclosed by separatrices. Within each cell, or basin, is a nucleus called the *attractor*. The *states* which will actually be observed in this system are the attractors. Thus, the portrait of the dynamical system, showing the basins and attractors, is of primary importance in applications. ... The history of a real system [can be] represented graphically as a trajectory in a geometric state space. Newton added the concept of the instantaneous velocity, or derivative, of vector calculus. The velocity *vectorfield* emerged as one of the basic concepts. Velocities are given by the first time derivative (tangent) of the trajectories. The prescription of a velocity vector at each point in the state space is called a velocity *vectorfield*."

I. Dynamical System and State Vector

A system with *n* degrees of freedom, that is, with *n* different, independent variables, can be thought of as living in an *n*-space. The *n* coordinates of a single point in the *n*-space define all the *n* variables simultaneously. If the motion of the point in *n*-space follows some rule acting on the positions (magnitudes) of the variables and their velocities, then that rule defines a dynamical system; the *n* variables are then more than a mere aggregate. The point in *n*-space that is moving is called a configuration or state vector. The rule defining the dynamical system expresses the law relating the variables and the parameters (Parameters are more or less constant or very slowly changing magnitudes that give the system its particular identity). Influences (controls) may act on the dynamical system either through the variables or the parameters. Controls change the motions of the system. A simple dynamical scheme is a function assigning a smooth *vectorfield* to the *manifold* of instantaneous states for every point in the manifold of control influences. The smooth function is a *mapping*.

J. Fractals

Fractal geometry is based on the idea that the natural world is not made up of the familiar objects of geometry: circles, triangles, and the like. The natural world of clouds, coastlines, and mountains cannot be fully described by the geometry of circles and squares. Fractals are structures that always look the same, either exactly or in a statistical sense, as you endlessly enlarge portions of them. According to a usage becoming standard, fractal forms ordinarily have three features: heterogeneity, self-similarity, and the absence of a well-defined (characteristic) scale of length. The first feature is not an absolute requirement, but the other two are. There must be nontrivial structure on all scales so that small details are reminiscent of the entire object. Fractal structures have both irregularity and redundancy and, as a result, they are able to withstand injury.

Fractals were first conceived by Mandelbrot. They are geometric fragments of varying size and orientation but similar in shape. It is remarkable that the details of a fractal at a certain scale are similar (although not necessarily identical) to those of the structure seen at larger or smaller scales. There is no characteristic scale. All fractals either have this look-alike property called self-similarity, or, alternatively, they may be self-affine. As Goldberger *et al.* (1990) comment, "Because a fractal is composed of similar structures of ever finer detail, its length is not well defined. If one attempts to measure the length of a fractal with a given ruler, some details will always be finer than the ruler can possibly measure. As the resolution of the measuring instrument increases, therefore, the length of a fractal grows. Because length is not a meaningful concept for fractals, mathematicians calculate the '*dimensions*' of a fractal to quantify how it fills space. The familiar concept of dimension applies to the objects of classical, or Euclidean, geometry. Lines have a dimension of one, circles have two dimensions, and spheres have three. Most (but not all!) fractals have noninteger (fractional) dimensions. Whereas a smooth Euclidean line precisely fills a one-dimensional space, a fractal line spills over into a two-dimensional space." A fractal line, a coastline, for example, therefore has a dimension between one and two. Likewise a fractal surface, of a mountain, for instance, has a dimension between two and three. The greater the dimension of a fractal is, the greater the chance is that a given region of space contains a piece of that fractal.

Peterson (1988) points out that for any fractal object of size P, constructed of smaller units of size p, the number, N, of units that fit into the object is the size ratio raised to a power, and that exponent, d, is called the Hausdorff dimension. In mathematical terms this can be written as

$$N = (P/p)^d \tag{2}$$

or

$$d = \log N / \log(P/p). \tag{3}$$

This way of defining *dimension* shows that familiar objects, such as the line, square, and cube, are also fractals, although mathematically they count as trivial cases. The line contains within itself little line segments, the square contains little squares, and the cube little cubes. (This is self-similarity, but without heterogeneity.) But applying the concept of Hausdorff dimension to other objects, such as coastlines, gives a fractional dimension. Fractals in nature are often self-similar only in a statistical sense. The fractal dimension of these shapes can be determined only by taking the average of the fractal dimensions at many different length scales.

The correlation dimension (not defined here, but see Grassberger and Procaccia, 1983)) serves as a lower bound for the fractal dimension. The fractal dimension itself is a number bounded below by the topological dimension (0 for a point, 1 for a line, 2 for a surface, etc.) and above by the Euclidean dimension in which the fractal is located (a point is on a line, Euclidean dimension 1; a line is on a surface, Euclidean dimension 2; etc.) The topological dimension approximately corresponds to the number of independent variables required for the definition of the function. The Euclidean dimension is the dimension of the range of the function.

Mandelbrot (1983) discusses *dimension* in very advanced terms using the concept to define a fractal: "A fractal is by definition a set for which the Hausdorff–Besicovitch dimension strictly exceeds the topological dimension" (p. 15). But he thought this definition incomplete.

Chaotic attractors are all fractals, and usually they are of a dimension greater than 2, but not all systems with self-similar phase space (fractal properties) are necessarily ascribable to a chaotic system. I want to emphasize again that chaos and fractals should not be confused with each other. "Chaos" is about dynamics; "fractals" is about geometry. It happens that chaotic attractors often have a fractal microstructure geometry, but that does not make fractals and chaos synonymous.

In the case of processes, fractals can be stochastic or they can be deterministic. Stochastic fractals are an example of fractional Gaussian *noise* where the power *spectrum* contains large amounts of low frequency energy. Fractal noise ($1/f^n$ spectrum, $m \cong 1$, or, more generally, a noninteger) is very structured and is a long-time scale phenomenon. Measures that deal with short-time scales such as correlation functions and interspike interval histograms cannot assess fractal activity.

To identify fractals in objects, for example, in the case of branching structures common in biological objects, two types of scaling can be compared: one exponential and the other fractal. If a tree structure follows the simple exponential rule: $d(z, a) = d_0 e^{-az}$, where $d(z, a)$ is the average diameter of tubes in the zth generation, d_0 is the diameter of the single parent trunk or vessel, and a is the characteristic scale factor, then a *semilog plot* of ln $d(z, a)$ versus z will give a straight line with negative slope, $-a$. In contrast, a fractal tree has a multiplicity of scales, and each can contribute with a different weighting or probability of occurrence that is revealed by an inverse power law. In that case, $d(z) \propto 1/z^\mu$, where μ is the power law index, $\mu = 1 - D$, and D is the fractal dimension. Bak *et al.* (1987) Now a *log–log plot* of ln $d(z)$ versus ln z gives a straight line of negative slope $-\mu$. (There may be harmonic modulation of the data around the pure power law regression line without overcoming the fractal scaling; West, 1987).

K. Limit Cycle

The limit cycle is a nonlinear cyclic time series creating, when abstracted, a closed orbit (on certain plots described below) of a wide variety of shapes (but not including that of a circle; the circle represents a pure, harmonic, linear oscillation in these plots). In the single-variable picture of a limit cycle, we plot the velocity of a variable against its magnitude. However, in some systems under nonholonomic constraints, velocities and magnitudes or positions are independently specifiable; thus the plot of velocity versus magnitude of a single variable is actually a two-variable plot. Poincaré studied nonlinear differential equations with two variables in which it is possible to have an oscillation that is reestablished following a small perturbation delivered at any phase of the oscillation. He called such oscillations stable limit cycles. One of the most thoroughly examined versions is the simple two-dimensional differential equation proposed by van der Pol to model nonlinear limit cycle oscillations:

$$\frac{d^2\mu}{dt^2} - \varepsilon(1 - \mu^2)\frac{d\mu}{dt} + \mu = \beta\cos(\alpha t). \tag{4}$$

When $\beta = 0$ there is a unique, stable limit cycle oscillation. Alternatively, the van der Pol oscillator is given by the following pair of equations:

$$\frac{dx}{dt} = \frac{1}{\varepsilon}\left(y - \frac{x^3}{3} + x\right), \tag{5a}$$

$$\frac{dy}{dt} = \varepsilon x, \varepsilon > 0. \tag{5b}$$

A one-dimensional, nonlinear finite difference equation (e.g., the logistic equation discussed under the section on maps) at certain parameter values can also generate stable limit cycles.

It has been a challenge to generalize the limit cycle concept beyond the two-dimensional (two degrees of freedom, two variable) case. However, generalization is essential for the understanding of homeodynamic stability of biological systems (Yates, 1982). Complex modes of behavior almost always appear if two nonlinear oscillatory mechanisms are coupled, either in series or in parallel. Any interaction between two limit cycles can produce complex periodic oscillations, or chaos. (In biological systems the term "near-periodic" best describes the motions, and this result is to be expected of homeodynamic systems of all kinds.)

L. Linear

A linear term is one which is first degree in the dependent variables and their derivatives. A linear equation is an equation consisting of a sum of linear terms. If any term of a differential equation contains higher powers, products, or transcendental functions of the dependent variables, it is nonlinear. Such terms include $(dy/dt)^3$, $u(dy/dt)$, and $\sin u$, respectively. $(5/\cos t)(d^2y/dt^2)$ is a term of first degree in the dependent variable y, whereas $2uy^3(dy/dt)$ is a term of fifth degree in the dependent variables u and y.

Any differential equation of the form below is linear, where y is the output and u the input:

$$\sum_{i=0}^{n} a_i(t) \frac{d^i y}{dt^i} = u. \tag{6}$$

If all initial conditions in a system are zero, that is, if the system is completely at rest, then the system is a linear system if it has the following property: (1) if an input $u_1(t)$ produces an output $y_1(t)$, and (2) an input $u_2(t)$ produces an output $y_2(t)$, (3) then input $c_1u_1(t) + c_2u_2(t)$ produces an output $c_1y_1(t) + c_2y_2(t)$ for all pairs of inputs $u_1(t)$ and $u_2(t)$ and all pairs of constants c_1 and c_2.

The principle of superposition follows from the definition above. The response $y(t)$ of a linear system due to several inputs $u_1(t), u_2(t), \ldots, u_n(t)$ acting simultaneously is equal to the sum of the responses of each input acting alone, when all initial conditions in the system are zero. That is, if $y_i(t)$ is the response due to the input $u_i(t)$, then

$$y(t) = \sum_{i=1}^{n} y_i(t). \tag{7}$$

Any system that satisfies the principle of superposition is linear. All others are nonlinear.

M. Manifold

A manifold is a geometrical model for the observed states of a dynamical or experimental situation and is identical to the n-dimensional state space of a model of the situation. Each instantaneous state has a location in n-space, and all those locations achievable by the system following a rule for its motion constitute a manifold.

N. Maps (Mapping)

Functions that determine *dynamical systems* are called mappings or maps. This terminology emphasizes the geometric process of taking one point to another. The basic goal of the theory of dynamical systems is to understand the eventual (i.e., asymptotic) behavior of an iterative or ongoing process. In dynamical systems analysis we ask, where do points go, and what do they do when they get there? The answer is a mapping.

A map is a rule. The rule can be deterministic or statistical, linear or nonlinear, continuous or discrete. In *dynamical systems*, a mapping is the rule or law governing motion and change as a function of state configuration, initial velocities and positions, and constraints. An example of a discrete, nonlinear dynamical law of one dimension is

$$x_{n+1} = f(x_n), f \neq \text{a constant}, \tag{8}$$

where f is a function carrying out the mapping. A notation for this map is $f: x \rightarrow x$, which assigns exactly one point $f(x) \in x$ to each point $x \in x$, when (x, d) is a metric space. A common example is the nonlinear population growth model or logistic map:

$$x_{n+1} = f(x_n) = rx_n(1 - x_n). \tag{9}$$

In this dynamic rule, time appears as a generational step (discrete map). The level of the population, x, as a function of time (or iteration number) goes from some initial condition toward extinction (0), constancy, oscillation, period doubling oscillations, or chaos as r is extended in the range of positive values starting close to zero and increasing past 3. The switching of dynamical outcomes (extinction, constancy, oscillation, period doubling, and chaos) occurs at successive critical values of r, at which the iterated dynamics undergo a bifurcation (change in behavioral mode).

For the logistic equation above as often displayed for illustration, plotting x_{n+1} against x_n, at a given value of parameter r, results in a parabola that opens downward. The parabola sits on the x_n axis between 0 and 1. The plotting of one value of x to the next creates a graph called a return map, which is also a transition function.

Maps can address real or complex numbers. An important example involving complex numbers is

$$z_{n+1} = z_n^2 + c, \tag{10}$$

where c is a complex constant and the mapping rule is

$$f_c : z \rightarrow z^2 + c. \tag{11}$$

This mapping results in quadratic Julia sets and the famous Mandelbrot set, which have led to astonishing computer graphical demonstrations (Douady, 1986).

In summary, if a rule exists that assigns to each element of a set A an element of set B, this rule is said to define a mapping (or simply, map) function or transformation from A into B. (The term transformation is sometimes restricted to the case where $A = B$.) The expression $f: A \rightarrow B$ or $(A \rightarrow B)$ means that f is a function that maps A into B. If $f: A \rightarrow B$ and $A \in A$, then $f(a)$ denotes the element of B, which is assigned to A by f.

O. Noise

Noise is any unwanted (from the point of view of the investigator) variation in data. Noise may be stochastic (random) or deterministic; it is traditionally thought of as being random and pernicious. Curiously, the presence of low levels of noise can actually improve detection of weak signals in systems with stochastic resonance. In that case, a bistable system operates as a detector when a sufficiently strong external force—a signal—provokes it into a change of state. If the force is too weak, the system stays in its original state and detects no signal. The addition of noise injects energy into the variations or fluctuations in each state and changes the probability, if the barrier between states is low, that the system might change state. Then, when a weak signal arrives, combined with the noise, a state change may be accomplished, whereas the signal, without the energizing noise, would not be sufficient to overcome the energy barrier.

One way to describe noise is through its *spectral analysis* or its distributional characteristics. The discrete events making up a Poisson distribution generate an exponential interval histogram. Brownian motion generates a $1/f^2$ *spectrum*; fractal, brown (Mandelbrot, Zipfian) noise generates a $1/f^1$ spectrum; band-limited, "white" (or uniform) noise generates a $1/f^0$ spectrum (where f is frequency). More generally, $1/f^m$ spectra, where m is not an integer, characterizes a fractal system (Sun, 1990).

P. Nonlinear

Because nonlinearity is a more general concept than is linearity, it can be understood from the very restricted definition of linearity already given. There are many different kinds and causes of nonlinearity, some of them "hard," such as sharp thresholds or saturations, and some of them "soft," such as found in some memory elements of a system. For a good discussion of linearity and nonlinearity, see DiStefano *et al.* (1990).

Q. Quasi-Periodicity

An *attractor* that is a torus (i.e., the surface of a doughnut) and that allows a trajectory winding around it an infinite number of times, filling its surface but never intersecting itself, describes quasi-periodicity. The attractor has integral dimensions and is not "strange." It can be created in the case of two rhythms that are completely independent, whose phase relationships we examine. The phase relations between the two rhythms will continually shift but will never repeat if the ratio between the two frequencies is not rational. Glass and Mackey (1988) note that the dynamics are then not periodic, but they are also not *chaotic* because two initial conditions that are close together remain close together in subsequent iterations. If two periodic motions have periods with a common measure, both being integer multiples of the same thing, and are added together, then their result is itself periodic. For example, if one motion has a period of 3 s and the other 5 s, their (linear) combination will repeat every 15 s. But if there is no common measure, then the motion never repeats exactly, even though it does almost repeat—thus the term "quasi-periodic." Quasi-periodicity is often found in theoretical classic, conservative dynamics, although it is not considered to be a motion typical of a general dynamical system in the mathematical world. In the physical/biological world, something that might be called "near-periodicity" is very typical of observational data. Whether this ubiquitous near-periodicity is formal quasi-periodicity is an open question. Some think it is *chaos*, but have not proved it for biological data sets. Elsewhere, I have argued that near-periodicity is neither quasi-periodicity nor chaos, but the temporal organization to be expected of a homeodynamic, complex, system (Yates, 2008).

R. Spectral Analysis

Classic Fourier-based spectral analysis consists of fitting time history data with a linear series of sines and cosines. Any recurrent process can be modeled by a Fourier series, but many of the terms in the series will simply adjust the shape of the fit. The Fourier series is harmonic, and all the terms have commensurate frequencies (the quotient of any two frequencies is a rational number). Spectral analysis consists of partitioning the variance of a variable undergoing a time history into frequency bands or windows or bins. In the absence of noise, a linear, additive mixture of pure sines and cosines will lead to a line spectrum with extremely sharp "peaks" at the relevant frequencies. Roughly, the "power" at each frequency where there is a line is given by the square of the amplitude of the periodic term having that frequency.

A Fourier series model of raw biological time history data usually produces broad "peaks" rather than a line spectrum. One must then have a theory of peak broadening. One possibility is that the underlying process was *quasi-periodic* rather than purely periodic. Another notion would be that the underlying process was purely periodic but contaminated by *noise*. Still another hypothesis would be that the underlying dynamics were *chaotic*. (Chaotic dynamical systems can produce near-periodicity as well as other spectral pictures.) Spectral analysis alone cannot resolve the underlying nature of the generator of a time history. Many theoretical systems can start out quasi-periodic, but as excitation of the system is increased the motion may become "random" or, more exactly, *chaotic*. In the *quasi-periodic* regime, motions can be decomposed into a Fourier series with a few fundamental frequencies and overtones. The spectrum will consist of a small number of possibly sharp lines. In the *chaotic* regime, the trajectories sample much more (perhaps all) of the allowed phase space, and the spectrum becomes broad but may still manifest some frequency bands in which there are hints of "peakedness." But chaotic dynamics can also produce a $1/f$ spectrum of Mandelbrot *noise*.

S. Stability

Stability is as difficult to define as is *dimension*. It has multiple meanings. For linear systems, we have a complete theory of stability, but for nonlinear systems, there can be multiple interpretations. For example, if trajectories from nearby initial conditions stay close to each other and asymptotically approach a fixed point, or an orbit, or else wind themselves around the surface of an invariant 2-torus, we can easily imagine that the dynamical system is stable in the region of those initial conditions. The situation becomes more confusing if from nearby conditions trajectories diverge exponentially but nevertheless asymptotically find a low-dimension, *chaotic attractor* that provides the limit set for all points generated by the trajectories as time approaches infinity. We could then think of the attractor as defining a bounded behavior for the dynamical system and consider the system to be globally stable on the attractor.

The most useful concept of stability for nonlinear systems, in my opinion, is asymptotic orbital stability. It pictures a limit cycle in more than topological one dimension, and takes the view that nearby trajectories converge to an orbital process in which certain states (nearly) recur at (nearly) identical intervals. There may be wobble on amplitudes and wobble on frequencies, but same conditions are seen repetitively, even though the repetition is not precisely periodic. It is near-periodic. (I prefer this term to *quasi-periodic* because the latter has a precise meaning that may be too restricted to account for the observed recurrent behavior of most biological processes.) For the purposes of this chapter, a nonlinear dynamic system will be considered stable if it asymptotically approaches a limit point, a limit cycle, or a chaotic attractor as it evolves from its initial conditions. However, in a broader view, according to the homeodynamic heuristic (Yates, 1982), complex dynamical systems express their stability in a global, limit cycle-like, near-periodic motion. (I have not achieved a formal proof of this conjecture.)

T. State

A state is a set of data that in the deterministic limit (of a state-defined system) gives us all that we need to know to predict future behavior. The present state is an input, some deterministic dynamic rule operates on it, and the future state is the output. The state is a vector whose values allow the estimation of future states. In the absence of noise, reconstruction of a state space can be accomplished even from a single time series as partial information. In other words, we can work with data whose dimension is lower than that of the true dynamics. In the presence of noise, the reconstruction of the state space is not always possible. The concept of state and state-determined systems follows from the Newtonian–causal tradition, and it fails to deal adequately with *complexity* (Rosen, 1987; Yates, 2008).

U. Vectorfield

A vectorfield is a *mapping* that assigns a tangent vector or velocity to each point in some region of interest in a *manifold*. For each control bearing on a dynamical system, the dynamical system rule or mapping creates a particular vectorfield, giving the positions and motions for the state vector point of the system. The vectorfield is a model for the habitual tendencies of the situation to evolve from one state to another and is called the dynamic of the model. The vectorfield may have slow or fast regions in it, according to the velocity of the configurational point.

A vectorfield is a field of bound vectors, one defined at (bound to) each and every point of the state space. The state space, filled with trajectories, is called the phase portrait of the dynamical system. The velocity vectorfield can be derived from the phase portrait by differentiation in state-determined systems. The phrase *dynamical system* specifically denotes this vectorfield. For analytic tractability, we like to hypothesize that the vectorfield of a model of a dynamical system is smooth, meaning that the vectorfield consists of continuous derivatives with no jumps or sharp corners.

Note: Much of dynamical systems theory addresses state-determined systems, whose velocities are specified by the state, for example, $\dot{X} = f(X, \ldots)$. But complex biological systems may not be state-determined systems (Yates, 2008) (This issue depends somewhat on how one defines "state.").

Acknowledgments

I thank Laurel Benton for valuable help with the manuscript, including clarifying discussions. This work was supported by a gift to the University of California, Los Angeles, for medical engineering from the ALZA Corporation, Palo Alto, California.

References

Abraham, R. H., and Shaw, C. D. (1982). "Dynamics: The Geometry of Behavior, Part 1: Periodic Behavior," p. 11. Ariel Press, Santa Cruz, CA.

Abraham, R., and Shaw, C. (1982–1988). "Dynamics: The Geometry of Behavior, Part One: Periodic Behavior; Part Two: Chaotic Behavior; Part Three: Global Behavior; Part Four: Bifurcation Behavior (4-volume series)". The Visual Mathematics Library, Aerial Press, Santa Cruz, CA.

Bak, P., Tang, C., and Wiesenfeld, K. (1987). *Phys. Rev. Lett.* **59**, 381.

Barnsley, M. (1988). Metric spaces, equivalent spaces, classification of subsets, and the space of fractals. "Fractals Everywhere," pp. 6–42. Academic Press, New York.

Bassingthwaighte, J. B. (1988). *NIPS* **3**, 5.

Benoit, B. M. (1977). "Fractals: Form, Chance, and Dimension" W. H. Freeman and Company, New York City.

Bentley, J. (2004). Evolving beyond perfection: An investigation of the effects of long-term evolution on fractal gene regulatory networks. *BioSystems* **76**, 291–301.

Careri, G., Fasella, P., and Gratton, E. (1975). *Crit. Rev. Biochem.* **3**, 141.

Devaney, R. L. (1989a). Dynamics of simple maps. "Chaos and Fractals: The Mathematics Behind the Computer Graphics. *In* "Proceedings of Symposia in Applied Mathematics," (R. L. Devaney, and L. Keen, eds.), Vol. 39, pp. 1–24. American Mathematical Society, Providence, Rhode Island.

Devaney, R. L. (1989b). "An Introduction to Chaotic Dynamical Systems" 2nd edn. Addison-Wesley, New York, (written for the mathematical community).

Devaney, R. L. (1990). "Chaos, Fractals, and Dynamics: Computer Experiments in Mathematics" Addison-Wesley, Menlo Park, CA, (this book is for nonexperts).

DiStefano, J. J., III., Stubberud, A. R., and Williams, I. J. (1990). "Feedback and Control Systems" (Schaum's Outline Series), 2nd edn. McGraw-Hill, New York.

Douady, A. (1986). *In* "The Beauty of Fractals," (H.-O. Peitgen, and P. H. Richter, eds.), p. 161. Springer-Verlag, Berlin.

Edelman, G. M. (1988). "Topolobiology: An Introduction to Molecular Embryology" Basic Books, New York.

Edelman, G. (2004). Wider than the Sky: The Phenomenal Gift of Consciousness Yale. University Press, New Haven.

Freeman, W. F. (1991). *Sci. Am.* **264**, 78.

Glass, L., and Mackey, M. C. (1988). "From Clocks to Chaos: The Rhythms of Life" p. 53. Princeton Univ. Press, Princeton, New Jersey.

Gleick, J. (1987). "Chaos: Making a New Science". Viking Press, New York.

Glenny, R. W., and Robertson, H. T. (1990). *J. Appl. Physiol.* **69**, 532.

Goldberger, A. L., and West, B. J. (1985). *Biophys. J.* **48**, 525.

Goldberger, A. L., Rigney, D. R., and West, B. J. (1990). *Sci. Am.* **262**, 44.

Grassberger, P., and Procaccia, I. (1983). *Physica.* **9D**, 189.

Kauffman, S. A. (1991). *Sci. Am* **265**, 78.

Layne, S. P., Mayer-Kress, G., and Holzufs, J. (1986). *In* "Dimensions and Entropies in Chaotic Systems: Quantification of Complex Behavior," (G. Mayer-Kress, ed.), p. 248. Springer-Verlag, Berlin.

Lazar, A., Pipa, G., and Triesch, J. (2009). SORN: A self-organizing recurrent neural network. Online at http://www.frontiersin.org/neuroscience/computationalneuroscience/paper/10.3389/neuro, (open-access article).

Liebovitch, L. S., Fischbarg, J., Koniarek, J. P., Todorova, I., and Wang, M. (1987). *Biochim. Biophys. Acta* **896**, 173.

Li, T. Y., and Yorke, J. A. (1975). *Am. Math. Monthly* **82**, 985.

Mandelbrot, B. (1983). "The Fractal Geometry of Nature". Revised edn Freeman, New York.

Mayer-Kress, G. (ed.) (1986). *In* "Dimensions and Entropies in Chaotic Systems," Springer-Verlag, New York.

Mayer-Kress, G., Yates, F. E., Benton, L., Keidel, M., Tirsch, W., Poppl, S. J., and Geist, K. (1988). *Math. Biosci.* **90**, 155.

Morrison, F. (1991). "The Art of Modeling Dynamic Systems: Forecasting for Chaos, Randomness, and Determinism". Wiley (Interscience), New York.

Olsen, L. F., and Degn, H. (1977). *Nature (London)* **267**, 177.

Peitgen, H. O., and Richter, P. H. (1986). "The Beauty of Fractals: Images of Complex Dynamical Systems" Springer-Verlag, New York.

Peterson, I. (1988). "The Mathematical Ty34hourist: Snapshots of Modern Mathematics" p. 157. Freeman, New York.

Pincus, S. M. (1991). *Proc. Natl. Acad. Sci. USA* **88**, 2297.

Rosen, R. (1987). *In* "Quantum Implications," (B. J. Hiley, and F. T. Peat, eds.), p. 314. Routledge and Kegan Paul, London.

Rössler, O. E. (1987). *Ann. N.Y. Acad. Sci.* **504**, 229.

Schaffer, W. M. (1987). *In* "Chaos and Biological Systems," (H. Degn, A. V. Holden, and L. F. Olsen, eds.), p. 233. Plenum, New York.

Schlesinger, M. F. (1987). *Ann. N.Y. Acad. Sci.* **504**, 229.

Stein, D. L. (1989). *In* "Lectures in the Sciences of Complexity," (D. L. Stein, ed.), p. xiii. Addison-Wesley, Redwood City, CA.

Stewart, I. (1989). "Does God Play Dice? The Mathematics of Chaos, Basel Blackwell", Cambridge, MA.

Sun, H. H. (1990). *Ann. Biomed. Eng.* **18**, 597.

West, B. J. (1987). *In* "Chaos in Biological Systems," (H. Degn, A. V. Holden, and L. F. Olsen, eds.), p. 305. Plenum, New York.

Yates, F. E. (1982). *Can. J. Physiol. Pharmacol.* **60**, 217.

Yates, F. E. (1991). *In* "The Resistance Vasculature," (J. A. Bevan, W. Halpern, and M. J. Mulvaney, eds.), p. 451. Humana Press, Clifton, NJ.

Yates, F. E. (2008). "Homeokinetics/Homeodynamics: A physical heuristic for life and complexity". *Ecolog. Psycho.* **20**, 148–179.

CHAPTER 19

Analytical Methods for the Retrieval and Interpretation of Continuous Glucose Monitoring Data in Diabetes

Boris Kovatchev, Marc Breton, and William Clarke

University of Virginia Health System
Charlottesville, Virginia, USA

Abstract

Scientific and industrial effort is now increasingly focused on the development of closed-loop control systems (artificial pancreas) to control glucose metabolism of people with diabetes, particularly type 1 diabetes mellitus. The primary prerequisite to a successful artificial pancreas, and to optimal diabetes control in general, is the continuous glucose monitor (CGM), which measures glucose levels frequently (e.g., every 5 min). Thus, a CGM collects detailed glucose time series, which carry significant information about the dynamics of glucose fluctuations. However, a CGM assesses blood glucose (BG) indirectly via subcutaneous determinations. As a result, two types of analytical problems arise for the retrieval and interpretation of CGM data: (1) the order and the timing of CGM readings and (2) sensor errors, time lag, and deviations from BG need to be accounted for. In order to improve the quality of information extracted from CGM data, we suggest several analytical and data visualization methods. These analyses evaluate CGM errors, assess risks associated with glucose variability, quantify glucose system stability, and predict glucose fluctuation. All analyses are illustrated with data collected using MiniMed CGMS (Medtronic, Northridge, CA) and Freestyle Navigator (Abbott Diabetes Care, Alameda, CA). It is important to remember that traditional statistics do not work well with CGM data because consecutive CGM readings are highly interdependent. In conclusion, advanced analysis and visualization of CGM data allow for evaluation of dynamical characteristics of diabetes and reveal clinical information that is inaccessible via standard statistics, which do not take into account the temporal structure of data. The use of such methods has the potential to enable optimal glycemic control in diabetes and, in the future, artificial pancreas systems.

I. 2010 Update of Developments in the Field

The clinical utility of continuous glucose monitoring (CGM) for the optimization of glycemic control in type 1 diabetes (T1DM) has been demonstrated by a landmark study published by the New England Journal of Medicine in 2008, which showed a significant improvement in glycated hemoglobin after 6 months of CGM in adults with T1DM (The Juvenile Diabetes Research Foundation Continuous Glucose Monitoring Study Group, 2008). The methods for analysis of CGM data presented in this chapter have been further elaborated and presented in conjunction with methods for modeling and control of (BG) metabolism in humans in an extensive recent review (Cobelli *et al.*, 2009).

With the rapid development of both CGM devices and analytical methodology in the past 2 years, closed-loop control (artificial pancreas) became possible. The next logical step was therefore the demonstration of the feasibility of subcutaneous closed-loop control. To date, several studies have reported clinical results for closed-loop glucose control using subcutaneous CGM and insulin delivery (Bruttomesso *et al.*, 2009; Clarke *et al.*, 2009; El-Khatib *et al.*, 2010; Hovorka *et al.*, 2010; Kovatchev *et al.*, 2009a; Weinzimer *et al.*, 2008). These studies used one of two algorithmic strategies known as proportional-integral-derivative (PID; Weinzimer *et al.*, 2008) or model-predictive control (MPC; Bruttomesso *et al.*, 2009; Clarke *et al.*, 2009; El-Khatib *et al.*, 2010; Hovorka *et al.*, 2010). In January 2010 the Juvenile Diabetes Research Foundation (JDRF) announced strategic industrial partnerships aiming to bring closed-loop control systems to market within 4 years.

The development of closed-loop control systems is being greatly accelerated by employing computer simulation. Such *in silico* testing provides direction for clinical studies, out-ruling ineffective control scenarios in a cost-effective manner. Comprehensive simulator of the human metabolic system equipped with a "population" of *in silico* images of $N = 300$ "subjects" with type 1 diabetes, separated in three "age" groups, has been recently introduced (Kovatchev *et al.*, 2009b).

The biometric and metabolic parameters of these "subjects" range widely to approximate the interperson variability observed *in vivo*. In addition, the simulation environment includes three continuous glucose sensors, as well as two brands of insulin pumps. With this technology, any meal and insulin delivery scenario can be pilot-tested efficiently *in silico*, prior to its clinical application by placing a virtual CGM and a virtual insulin pump on virtual subjects (Kovatchev *et al.*, 2009b). In 2008, this simulator was accepted by the FDA as a substitute to animal trials in the preclinical testing of closed-loop control algorithms. Since then, a number of algorithm designers have used our simulation environment to test their closed-loop control strategies; animal trials in this area have been almost completely abandoned.

The first international study using closed-loop control algorithm developed entirely *in silico* without prior animal trials has been completed by a collaborative team of researchers from the Universities of Virginia (USA), Padova, (Italy), and Montpellier (France): Twenty adults with type 1 diabetes were recruited. Open- and closed-loop control sessions were scheduled 3–4 weeks apart, continued for 22 h, and used CGM and insulin pump. The only difference between the two sessions was insulin dosing done by the patient under physician's supervision during open loop and by the control algorithm during closed-loop (Bruttomesso *et al.*, 2009; Clarke *et al.*, 2009). Compared to open-loop, closed-loop control reduced significantly nocturnal hypoglycemia and increased the amount of time spent overnight within the overnight target range of 70–140 mg/dl (Kovatchev *et al.*, 2009a).

This study was followed by two clinical trials, which reported positive results from manually controlled artificial pancreas tested in 17 children with T1DM (Hovorka *et al.*, 2010), and from using bihormonal approach employing glucagon to reduce hypoglycemia (El-Khatib *et al.*, 2010).

In conclusion, the methods presented in this chapter have found application in the emerging fields of metabolic simulation and closed-loop control.

II. Introduction

In health, the metabolic network responds to ambient glucose concentration and glucose variability (Hirsch and Brownlee, 2005). The goals of the network are to reduce basal and postprandial glucose elevations and to avoid overdelivery of insulin and hypoglycemia. In both type 1 and type 2 diabetes mellitus (T1DM, T2DM), this internal self-regulation is disrupted, leading to higher average glucose levels and dramatic increases in glucose variability. Recent national data show that nearly 21 million Americans have diabetes, and one in three American children born today will develop the disease. In individuals with T1DM, the immune system destroys the pancreatic β cells. As a result, thousands of insulin shots are needed over a lifetime with T1DM, accompanied by testing of BG levels several times a day. In T2DM, increased insulin resistance is amplified by the failure of the β cell to compensate with adequate insulin delivery. Both T1DM and T2DM are lifelong conditions that affect people of every race and nationality and are leading causes of kidney failure, blindness, and amputations not related to injury. The only treatment of diabetes proven to reduce the risks of serious complications is tight control of BG levels (DCCT 1993; UKPDS, 1998). It is estimated that diabetes and its comorbidities account for more than $132 billion of our nation's annual health care costs and one out of every three Medicare dollars (U.S. Senate hearing, 2006).

Monitoring of BG levels and accurate interpretation of these data are critical to the achievement of tight glycemic control. Since measuring mean BG directly is not practical, the assessment of glycemic control is with a single simple test, namely, hemoglobin A1c (HbA1c; Santiago, 1993). However, the development of CGM technology has changed this conclusion (Klonoff, 2005). It is now feasible to observe temporal glucose fluctuations in real time and to use these data for feedback control of BG levels. Such a control could be patient initiated or actuated by a control algorithm via variable insulin infusion. Increasing industrial and research effort is concentrated on the development of CGM that sample and record frequent (e.g., every 1–5 min) glucose level estimates. Several devices are currently on the U.S. and European markets (Klonoff, 2005, 2007). While the accuracy of these devices in terms of approximating any particular glucose level is still inferior to self-monitoring of blood glucose (SMBG; Clarke and Kovatchev, 2007), CGM yields a wealth of information not only about current glucose levels but also about the BG rate and direction of change (Kovatchev *et al.*, 2005). This is why a recent comprehensive review of this technology's clinical implications, accuracy, and current problems rightfully placed CGM on the roadmap for 21st-century diabetes therapy (Klonoff, 2005, 2007).

While CGM is new and the artificial pancreas is still experimental (Hovorka, 2005;Hovorka *et al.*, 2004; Steil *et al.*, 2006; Weinzimer, 2006), the current gold-standard clinical practice of assessment is SMBG. Contemporary SMBG memory meters store up to several hundred self-monitoring BG readings, calculate various statistics, and visualize some testing results. However, SMBG data are generally one dimensional, registering only the amplitude of BG fluctuations at intermittent points in time. Thus, the corresponding analytical methods are also one dimensional, emphasizing the concept of risk related to BG amplitude, but are incapable of capturing the process of BG fluctuations over time (Kovatchev *et al.*, 2002, 2003). In contrast, CGM can capture the *temporal dimension* of BG fluctuations, enabling detailed tracking of this process. As a result, the statistical methods traditionally applied to SMBG data become unsuitable for the analysis and interpretation of continuous monitoring time series (Kollman *et al.*, 2005; Kovatchev *et al.*, 2005). New analytical tools are needed and are being introduced, ranging from variability analysis (McDonnell *et al.*, 2005) and risk tracking (Kovatchev *et al.*, 2005) to time series and Fourier approaches (Miller and Strange, 2007). Before proceeding with the description of the analytical methods of this chapter, we will first formulate the principal requirements and challenges posed by the specifics of CGM data.

(1) CGM assesses BG fluctuations *indirectly* by measuring the concentration of interstitial glucose (IG), but is calibrated via self-monitoring to approximate BG (King *et al.*, 2007). Because CGM operates in the interstitial compartment, which is presumably related to blood via diffusion across the capillary wall (Boyne *et al.*, 2003; Steil *et al.*, 2005), it faces a number of significant challenges in terms of sensitivity, stability, calibration, and physiological time lag between blood and IG concentration (Cheyne *et al.*, 2002; Kulcu *et al.*, 2003; Stout *et al.*, 2004). Thus, analytical methods are needed to assess and evaluate different types of sensor errors due to calibration, interstitial delay, or random noise.

(2) The CGM data stream has some inherent characteristics that allow for advanced data analysis approaches, but also call for caution if standard statistical methods are used. Most importantly, CGM data represent *time series*, that is, sequential readings that are ordered in time. This leads to two fundamental requirements to their analysis: First, consecutive sensor readings taken from the same subject within a relatively short time are highly interdependent. Second, the order of the CGM data points is essential for clinical decision making. For example, the sequences $90 \rightarrow 82 \rightarrow 72$ mg/dl and $72 \rightarrow 82 \rightarrow 90$ mg/dl are clinically very different. In other words, while a random reshuffling of CGM data in time will not change traditional statistics, such as mean and variance, it will have a profound impact on the temporal interpretation of CGM data. It is therefore imperative to extract CGM information across several dimensions, including risk associated with BG amplitude as well as time.

(3) A most important critical feature of contemporary CGM studies is their limited duration. Because the sensors of the CGM devices are generally short-lived (5–7 days), the initial clinical trials of CGM are bound to be relatively short term (days), and therefore their results cannot be assessed by slow measures, such as HbA1c, which take 2–3 months to react to changes in average glycemia (Santiago, 1993). Thus, it is important to establish an array of clinical and numerical metrics that would allow testing of the effectiveness of CGM over the relatively short-term, few-day life span of the first CGM sensors. Before defining such an array, we would reiterate that the primary goal of CGM and the artificial pancreas is β (and possibly α)-cell replacement. Thus, the effectiveness of CGM methods needs to be judged via assessment of their ability to approximate nondiabetic BG concentration and fluctuation.

These principal requirements are reflected by the analytical methods presented in this chapter, which include (i) decomposition of sensor errors into errors due to calibration and blood-to-interstitial time lag, (ii) analysis of average glycemia and deviations from normoglycemia, (iii) risk and variability analysis of CGM traces that uses a nonlinear data transformation of the BG scale to transfer data into a risk space, (iv) measures of system stability, and (v) prediction of glucose trends and events using time-series-based forecast methods. The proposed analyses are accompanied by graphs, including glucose and risk traces and system dynamics plots. The presented methods are illustrated by CGM data collected during clinical trials using MiniMed CGMS and Freestyle Navigator.

Before proceeding further, it is important to note that the basic unit for most analyses is the glucose trace of an individual, that is, a time-stamped series of CGM or BG data recorded for one person. Summary characteristics and group-level analyses are derived after the individual traces are processed to produce meaningful individual markers of average glycemia, risk, or glucose variation. The analytical methodology is driven by the understanding that BG fluctuations are a continuous process in time, BG(t). Each point of this process is characterized by its value (BG level) and by its rate/direction of BG change. CGM presents the process BG(t) as a discrete time series {BG(t_n), $n = 1, 2, \ldots$} that approximates BG (t) in steps determined by the resolution of the particular device (e.g., a new value displayed every 5 min).

III. Decomposition of Sensor Errors

Figure 1 presents the components of the error of MiniMed CGMS assessed during a hyperinsulinemic hypoglycemic clamp involving 39 subjects with T1DM. In this study reference, BG was sampled every 5 min and then reference data were synchronized with data from the CGMS. The calibration error was estimated as the difference between CGMS readings and computer-simulated recalibration of the raw CGMS current using *all* reference BG points to yield an approximation of the dynamics of IG adjusted for the BG-to-IG gradient (King *et al.*, 2007). The physiologic BG-to-IG time lag was estimated as the difference between reference BG and the "perfectly" recalibrated CGMS signal. The mean absolute deviation (MAD) of sensor data was 20.9 mg/dl during euglycemia and 24.5 mg/dl during

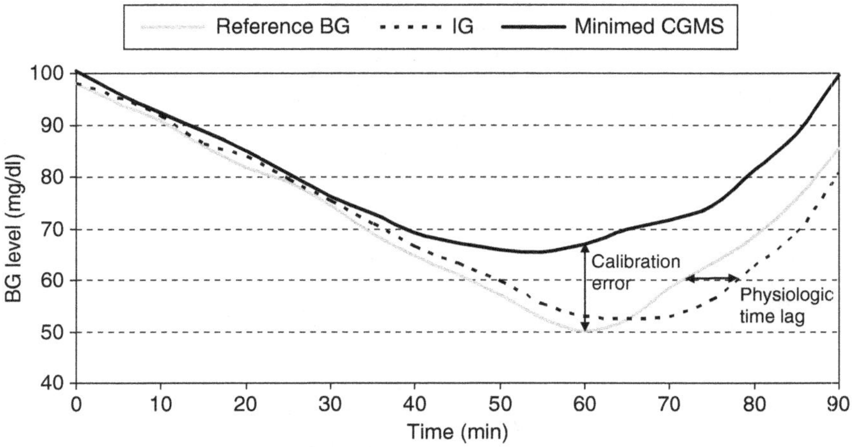

Fig. 1 CGMS error during hypoglycemic clamp decomposed into error of calibration and physiologic deviation caused by blood-to-interstitial time lag.

descent into and recovery from hypoglycemia. Computer-simulated recalibration reduced MAD to 10.6 and 14.6 mg/dl, respectively. Thus, during this experiment, approximately half of the sensor deviation from reference BG was attributed to calibration error; the rest was attributed to BG-to-IG gradient and random sensor deviations.

A diffusion model was fitted for each individual subject's data, as well as globally across all subjects. While the details of this model have been reported previously (King *et al.*, 2007), of particular importance is the finding that "global," across-subjects parameters describe the observed blood-to-interstitial delays reasonably well (King *et al.*, 2007). The availability of global parameters allows glucose concentration in the interstitium to be numerically estimated directly from reference BG data. This in turn allows for (i) setting the accuracy benchmark by simulating a sensor that does not have calibration errors, (ii) tracking and correction of errors due to calibration, and (iii) numerical compensation for BG-to-IG differential through an inverted diffusion model estimating BG from the sensor's IG approximation (in essence, the reference BG line in Fig. 1 could be derived numerically from the IG line, thereby reducing the influence of interstitial time lag).

IV. Measures of Average Glycemia and Deviation from Target

Certain traditional data characteristics are clinically useful for the representation of CGM data. The computation of mean glucose values from CGM data and/or BG data points is straightforward and is generally suggested as a descriptor of overall control. Computing of pre- and postmeal averages and their difference can serve as an indication of the overall effectiveness of meal control. Computing percentage of time spent within, below, or above preset target limits has been proposed as well. The suggested cutoff limits are 70 and 180 mg/dl, which create three commonly accepted glucose ranges: hypoglycemia (BG $\leq$ 70 mg/dl; ADA, 2005); normoglycemia (70 mg/dl < BG $\leq$ 180 mg/dl), and hyperglycemia (BG > 180 mg/dl). Percentage of time within additional bands can be computed as well to emphasize the frequency of extreme glucose excursions. For example, when it is important to distinguish between postprandial and postabsorptive (fasting) conditions, a fasting target range of 70–145 mg/dl is suggested. Further, %time <50 mg/dl would quantify the frequency of severe hypoglycemia, whereas %time >300 mg/dl would quantify the frequency of severe hyperglycemia. Table IA includes the numerical measures of average glycemia and measures of deviation from target.

Plotting glucose traces observed during a set period of time represents the general pattern of a person's BG fluctuation. To illustrate the effect of treatment observed via CGM, we use previously published 72-h glucose data collected pre- and 4 weeks postislet transplantation (Kovatchev *et al.*, 2005). Figure 2 presents glucose traces [of the process BG(t)] pre- and posttransplantation with

Table I
Summary Measures Representing CGM Traces

(A) *Average glycemia and deviations from target*

Mean BG	Computed from CGM or blood glucose data for the entire test
% time spent within target range of 70–180 mg/dl; below 70 and above 180 mg/dl	For CGM, this generally equals to % readings within each of these ranges. For BG measurements that are not equally spaced in time, we suggest calculating the % time within each range via linear interpolation between consecutive glucose readings
% time ≤ 50 mg/dl	Reflects the occurrence of extreme hypoglycemia
% time > 300 mg/dl	Reflects the occurrence of extreme hyperglycemia

(B) *Variability and risk assessment*

BG risk index	= LBGI + HBGI – measure of overall variability in "risk space"
Low BG index (LBGI)	Measure of the frequency and extent of low BG readings
High BG index (HBGI)	Measure of the frequency and extent of high BG readings
SD of BG rate of change	A measure of the stability of closed-loop control over time

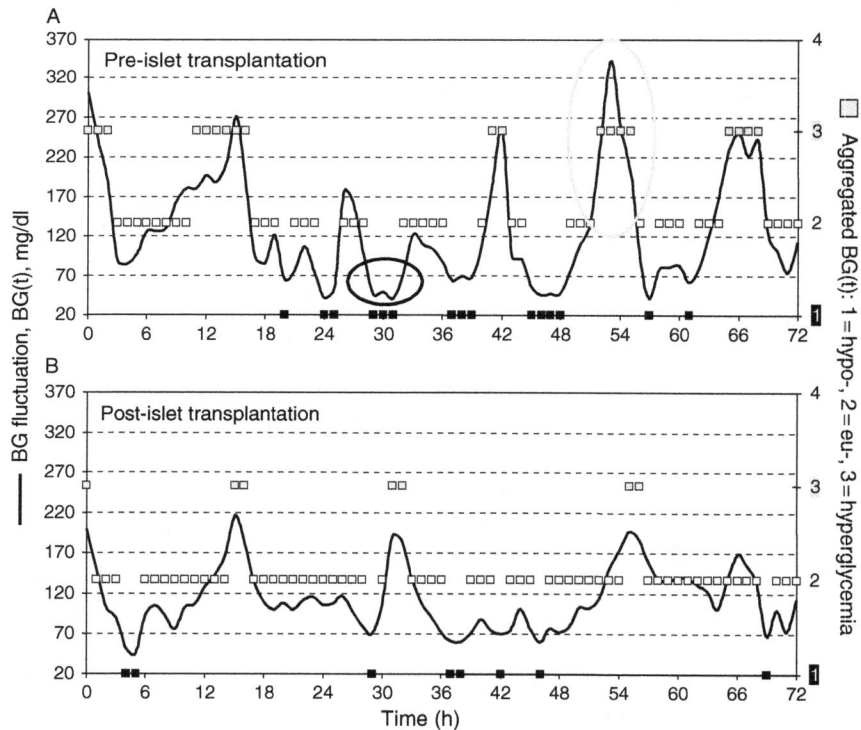

Fig. 2 Glucose traces pre- and postislet transplantation with superimposed aggregated glucose traces. Aggregated traces represent the time spent below/within/above target range.

Table II
Graphs Visualizing CGM Traces and the Effectiveness of Treatment[a]

(A) *Average glycemia and deviations from target*

Glucose trace (Fig. 2)	Traditional plot of frequently sampled glucose data
Aggregated glucose trace (Fig. 2)	Corresponds to time spent below/within/above a preset target range. Visualizes the crossing of glycemic thresholds

(B) *Variability and risk assessment*

Risk trace (Fig. 3)	Corresponds to LBGI, HBGI, and BGRI. Designed to equalize the size of glucose deviations toward hypo- and hyperglycemia, emphasize large glucose excursions, and suppress fluctuation within target range, thereby highlighting essential variance
Histogram of BG rate of change (Fig. 4)	Represents the spread and range of glucose transitions. Related to system stability. Corresponds to SD of BG rate of change
Poincaré plot (Fig. 5)	Represents the spread of the system attractors and can be used for detection of cyclic glucose fluctuations

[a]Each graph corresponds to a numerical measure from Table I.

superimposed aggregated glucose traces. The aggregated traces represent the time spent below/within/above target range.

The premise behind aggregation is as follows: Frequently one is not particularly interested in the exact BG value because close values such as 150 and 156 mg/dl are clinically indistinguishable. It is, however, important whether and when BG crosses certain thresholds, for example, 70 and 180 mg/dl as specified in the previous section. Thus, the entire process BG(t) can be aggregated into a process described only by the crossings of the thresholds of hypoglycemia and hyperglycemia. In Fig. 2A and B, the aggregated process is depicted by squares that are black for hypoglycemia, white for normoglycemia (euglycemia), and gray for hyperglycemia. Each square represents the average of 1 h of CGM data. It is evident that the aggregated process presents a clearer visual interpretation of changes resulting from islet transplantation: posttreatment of most of the BG fluctuations are within target, leading to a higher density of green squares. Possible versions of this plot include adding thresholds, such as 50 and 300 mg/dl, which would increase the levels of the aggregated process to five, and a higher resolution of the plot in the hypoglycemic range where one square of the aggregated process would be the average of 30 min of data. Table IIA includes a summary of the suggested graphs.

V. Risk and Variability Assessment

Computing standard deviation (SD) as a measure of glucose variability is not recommended because the BG measurement scale is highly asymmetric, the hypoglycemic range is numerically narrower than the hyperglycemic range, and the distribution of the glucose values of an individual is typically quite skewed (Kovatchev *et al.*, 1997). As a result from this asymmetry, SD would be predominantly influenced by hyperglycemic excursions and would not be sensitive to hypoglycemia. It is also possible for confidence intervals based on SD to assume unrealistic negative values. Thus, instead of reporting traditional measures of glucose variability, we suggest using risk indices based on a symmetrization transformation of the BG scale into a risk space (Kovatchev *et al.*, 1997, 2001). The symmetrization formulas, published a decade ago, are data independent and have been used successfully in numerous studies. In brief, for any BG reading, we first compute:

$$f(\mathrm{BG}) = 1.509 \times [(\ln(\mathrm{BG}))^{1.084} - 5.381]$$

if BG is measured in mg/dl or

$$f(\mathrm{BG}) = 1.509 \times [(\ln(18 \times \mathrm{BG}))^{1.084} - 5.381]$$

if BG is measured in mmol/liter. Then we compute the BG risk function using

$$r(\mathrm{BG}) = 10 \times f(\mathrm{BG})^2$$

and separate its left and right branches as follows:

$$rl(\mathrm{BG}) = r(\mathrm{BG}) \text{ if } f(\mathrm{BG}) < 0 \text{ and } 0 \text{ otherwise,}$$
$$rh(\mathrm{BG}) = r(\mathrm{BG}) \text{ if } f(\mathrm{BG}) > 0 \text{ and } 0 \text{ otherwise.}$$

Given a series of CGM readings $\mathrm{BG}_1, \mathrm{BG}_2, \ldots \mathrm{BG}_n$, we compute the low and high BG indices (LBGI, HBGI) as the average of $rl(\mathrm{BG})$ and $rh(\mathrm{BG})$, respectively (Kovatchev *et al.*, 2001, 2005):

$$\mathrm{LBGI} = \frac{1}{n} \sum_{i=1}^{n} rl(\mathrm{BG}_i)$$

and

$$\mathrm{HBGI} = \frac{1}{n} \sum_{i=1}^{n} rh(\mathrm{BG}_i)$$

The BG risk index is then defined as $\mathrm{BGRI} = \mathrm{LBGI} + \mathrm{HBGI}$

In essence, the LBGI and the HBGI split the overall glucose variation into two independent sections related to excursions into hypo- and hyperglycemia and, at the same time, equalize the amplitude of these excursions with respect to the risk they carry. For example, in BG space, a transition from 180 to 250 mg/dl would appear threefold larger than a transition from 70 to 50 mg/dl, while in risk space these fluctuations would appear equal. Using the LBGI, HBGI, and their sum BGRI complements the use of thresholds described earlier by adding information about the *extent of* BG *fluctuations*. A simple example would clarify this point. Assume two sets of BG readings: (110,65) and (110,40) mg/dl. In both cases we have 50% of readings below the threshold of 70 mg/dl; thus the percentage readings below target are 50% in both cases. However, the two scenarios are hardly equivalent in terms of risk for hypoglycemia, which is clearly depicted by the difference in their respective LBGI values: 5.1 and 18.2. Table IB includes the suggested measures of glucose variability and associated risks.

Figure 3A and B present 72-h traces of BG dynamics in risk space corresponding to the glucose traces of Fig. 2A and B at baseline and postislet transplantation. Each figure includes fluctuations of the LBGI (lower half) and HBGI (upper half), with both indices computed from 1-h time blocks (Kovatchev *et al.*, 2005). In particular, Figs. 2A and 3A demonstrate the effect of transforming BG fluctuations from glucose to risk values. A hypoglycemic event at hour 30 of study in Fig. 2A is visually expanded and emphasized in Fig. 3A (black circles). In contrast, the magnitude a hyperglycemic event at hour 54 in Fig. 2A is reduced in Fig. 3A to reflect the risk associated with that event (gray circles). Further, comparing Fig. 3A–B, it becomes evident that the magnitude of risk associated with glucose fluctuations decreases as a result of treatment. The average LBGI was 6.72 at baseline and 2.90 posttransplantation. Similarly, the HBGI was reduced from 5.53 at baseline to 1.73 after islet transplantation.

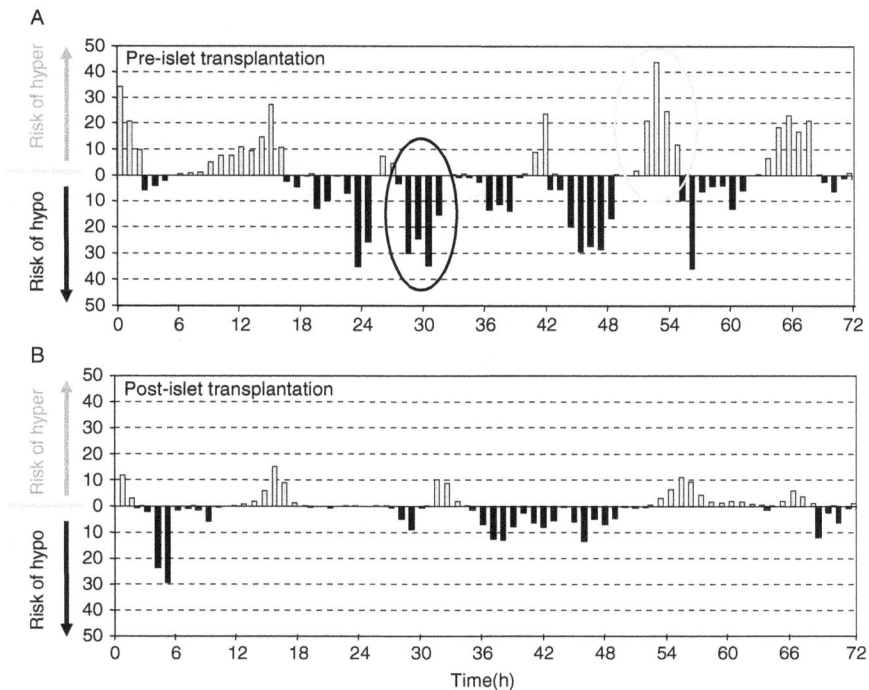

Fig. 3 CGM data in risk space: converting data equalizes numerically the hypoglycemic and hyperglycemic ranges and suppresses the variance in the safe euglycemic range.

Thus, the advantages of a risk plot include the following: (i) the variance carried by hypo- and hyperglycemic readings is equalized; (ii) excursions into extreme hypo- and hyperglycemia get progressively increasing risk values; and (iii) the variance within the safe euglycemic range is attenuated, which reduces noise during data analysis. In essence, Figs. 3A and B link better glycemic control to a narrower pattern of risk fluctuations. Because the LBGI, HBGI, and the combined BGRI can theoretically range from 0 to 100, their values can be interpreted as percentages of maximum possible risk.

VI. Measures and Plots of System Stability

Analysis of BG rate of change (measured in mg/dl/min) is suggested as a way to evaluate the dynamics of BG fluctuations on the timescale of minutes. In mathematical terms, this is an evaluation of the "local" properties of the system as opposed to "global" properties discussed earlier. Being the focus of differential calculus, local functional properties are assessed at a neighborhood of any point in time t_0 by the value $BG(t_0)$ and the derivatives of $BG(t)$ at t_0. The BG *rate of change* at t_i—is computed as the ratio $[BG(t_i) - BG(t_{i-1})]/(t_i - t_{i-1})$, where $BG(t_i)$ and $BG(t_{i-1})$ are CGM or reference BG readings taken at times t_i and t_{i-1} that are close in time. Recent investigations of the frequency of glucose fluctuations show that optimal evaluation of the BG rate of change would be achieved over time periods of 15 min (Miller and Strange, 2007; Shields and Breton, 2007), for example, $\Delta t = t_i - t_{i-1} = 15$. For data points equally spaced in time, this computation provides a sliding approximation of the first derivative (slope) of $BG(t)$. A larger variation of the BG rate of change indicates rapid and more pronounced BG fluctuations and therefore a less stable system. Thus, we use the SD of the BG rate of change as a measure of stability of closed-loop control. Two points are worth noting: (i) as opposed to the distribution of BG levels, distribution of the BG rate of change is *symmetric* and, therefore, using SD is statistically accurate (Fig. 4) and (ii) the SD of BG rate of change has been introduced as a measure of stability computed from CGM data and is known as Continuous Overall Net Glycemic Action (CONGA) of order 1 (McDonnell *et al.*, 2005).

Figure 4A and B present histograms of the distribution of the BG rate of change over 15 min, computed from MiniMed CGMS data of our transplantation case. It is apparent that the baseline distribution is more widespread than the distribution posttransplantation. Numerically, this effect is reflected by 19.3% of BG rates outside of the $[-2, 2]$ mg/dl/min range in Fig. 4A versus only 0.6% BG rates outside that range in Fig. 4B. Thus, pretransplantation the patient experienced rapid BG fluctuations, whereas posttransplantation the rate of fluctuations was reduced dramatically. This effect is also captured by the SD of the BG rate of change, which is reduced from 1.58 to 0.69 mg/dl/min as a result of treatment.

Another look at system stability is provided by the Poincaré plot (lag plot) used in nonlinear dynamics to visualize the attractor of the investigated system

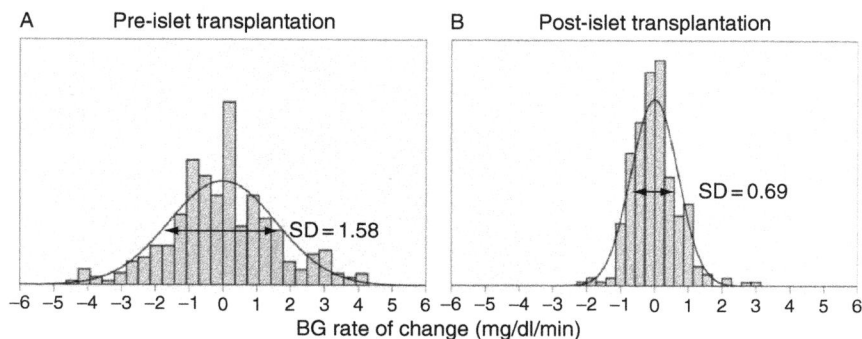

Fig. 4 Histograms of the distribution of the BG rate of change over 15 min, computed from MiniMed CGMS data pre- and postislet transplantation.

(Brennan *et al.*, 2001): a smaller, more concentrated attractor indicates system stability, whereas a more scattered Poincaré plot indicates system irregularity, reflecting poorer glucose control. Each point of the plot has coordinates $BG(t_{i-1})$ on the X axis and $BG(t_i)$ on the Y axis. Thus, the difference $(Y–X)$ coordinates of each data point represents the BG rate of change occurring between times t_{i-1} and t_{i-1}. Figure 5A and B present Poincaré plots of CGM data at baseline and postislet transplantation. It is evident that the spread of the system attractor is substantially larger before treatment compared with posttreatment. Thus, the principal axes of the Poincaré plot can be used as numerical metrics of system stability.

Another use of the Poincaré plot is to scan data for *patterns of oscillation*. Because the plot of an oscillator is an ellipse (see Fig. 5C), elliptical configuration of data points would indicate cyclic glucose fluctuations. Table IIB includes a summary of the suggested graphs.

VII. Time–Series–Based Prediction of Future BG Values

Most contemporary CGM systems include glucose prediction capabilities, in particular hypoglycemia and hyperglycemia alarms. Practically all predictions are currently based on a linear extrapolation of glucose values, for example, *projection* ahead of the current glucose trend. Because glucose fluctuations are generally nonlinear, such projections frequently result in errors and typically have a high false alarm rate. In contrast, a time-series model-based sliding algorithm designed to continually predict glucose levels 30 to 45 min ahead had substantially higher accuracy than typical linear projections (Zanderigo *et al.*, 2007). The sliding algorithm works as follows. For each time series, a linear model is fitted, continually at any sampling time, against past glucose data by weighted least squares. Then, the model is used to predict the glucose level at a preset prediction horizon. In model fitting, data points are "weighted" using a forgetting factor of 0.8 (which was determined to be optimal in numerical experiments), that is, the weight of the

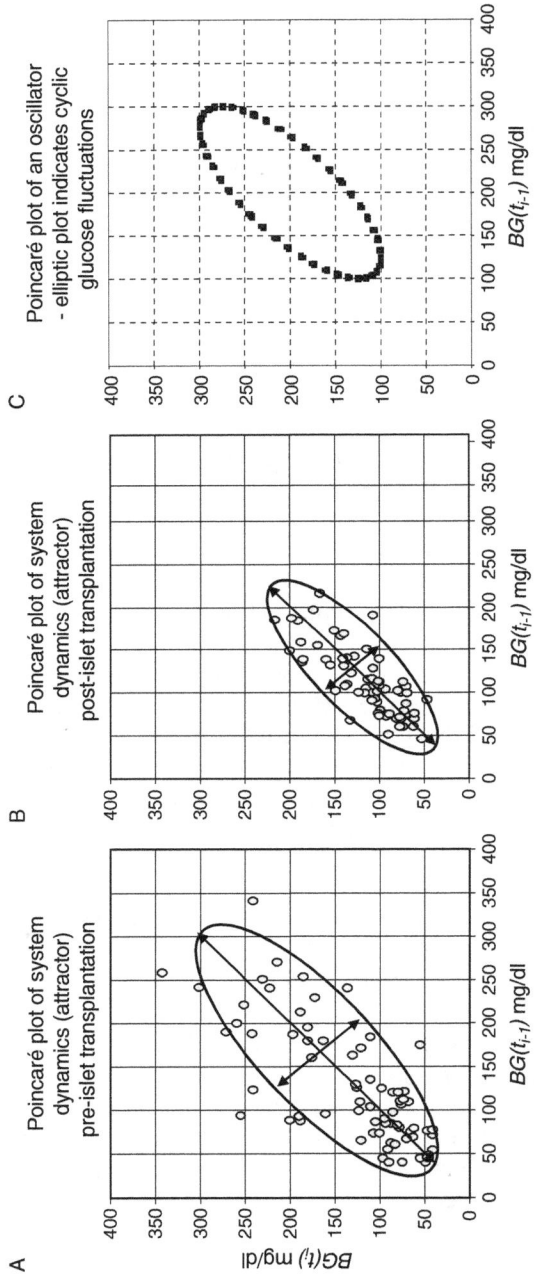

Fig. 5 Poincaré plot (lag plot) of glucose fluctuations: a smaller, more concentrated spread of data indicates system stability, whereas a more scattered Poincaré plot indicates system irregularity, resulting from poorer glucose control.

kth point before the actual sampling time is $(0.8)^k$. Figure 6 presents the action of the prediction algorithm at a prediction horizon of 30 min. While the algorithm tends to exaggerate transient peak and nadir glucose values, its overall predictive capability is very good. Judged by continuous glucose error-grid analysis (CG-EGA; Kovatchev *et al.*, 2004), >80% of predicted versus actual values fall in CG-EGA zone A and >85% fall in zones A + B. As would be expected, the accuracy of the prediction is highest during euglycemia, with 97% of predicted versus actual values falling in CG-EGA zones A + B.

Finally, a statistical disadvantage of the CGM data stream is the high interdependence between data points taken from the same subject within a relatively short time. As a result, standard statistical analyses, such as t tests, while appropriate for independent data points, will produce inaccurate results if applied directly to CGS data. The reason is a severe violation of the statistical assumptions behind the calculation of degrees of freedom, which are essential to compute the p value of any statistical test. In order to clarify the dependence of consecutive CGM data points, we have computed their autocorrelation and have shown that it remains significant for approximately 1 h, after which its significance drops below the level of 0.05 (Kovatchev and Clarke, 2008). Thus, CGM readings separated by more than 1 h in time could be considered *linearly* independent, which is sufficient for some statistical tests. A note of caution is that linear independence does not imply stochastic independence, which might be essential in some cases. Another conclusion from this autocorrelation analysis is that CGM data aggregated in 1-h blocks would be reasonably approximated by a Markov chain, which opens a number of possibilities for the analysis of aggregated data. Finally, the linear dependence between

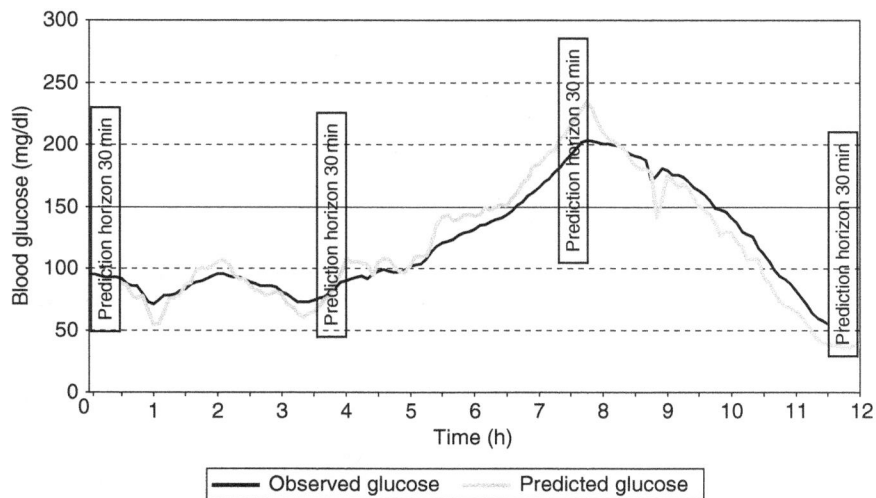

Fig. 6 Thirty-minute real-time prediction of glucose fluctuation using an auto-regression algorithm.

consecutive data points practically disappears at a time lag of ≈30 min. Therefore, a projection of BG levels more than 30 min ahead, which is using linear methods, would be inaccurate. This last point has significant clinical impact on the settings of hypo- or hyperglycemia alarms, many of which are based on linear projections of past CGM data.

VIII. Conclusions

The intent of this chapter was to introduce a set of mathematically rigorous methods that provide statistical and visual interpretation of frequently sampled BG data, which might serve as a basis for evaluation of the effectiveness of closed-loop control algorithms. Because a major purpose of closed-loop control is to reduce BG fluctuations, these methods augment the traditional approaches with understanding and analysis of variability-associated risks and the temporal structure of glucose data. Table I presents a summary of the metrics suggested for the analysis of CGM data, whereas Table II presents corresponding graphs. It is envisioned that this system of methods would be employed both in *in silico* computer simulation trials (Kovatchev *et al.*, 2008) and in clinical trials involving patients.

Acknowledgments

This chapter was prepared with support of the JDRF Artificial Pancreas Consortium. The theoretical development of some of the presented metrics was supported by Grant RO1 DK 51562 from the National Institutes of Health. Data and material support were provided by Abbott Diabetes Care (Alameda, CA).

References

American Diabetes Association (ADA) Workgroup on Hypoglycemia (2005). Defining and reporting hypoglycemia in diabetes: A report from the American Diabetes Association workgroup on hypoglycemia. *Diabetes Care* **28,** 1245–1249.

Boyne, M., Silver, D., Kaplan, J., and Saudek, C. (2003). Timing of changes in interstitial and venous blood glucose measured with a continuous subcutaneous glucose sensor. *Diabetes* **52,** 2790–2794.

Brennan, M., Palaniswami, M., and Kamen, P. (2001). Do existing measures of Poincare plot geometry reflect nonlinear features of heart rate variability? *IEEE Trans. Biomed. Eng.* **48,** 1342–1347.

Bruttomesso, D., Farret, A., Costa, S., Marescotti, M. C., Vettore, M., Avogaro, A., Tiengo, A. C., *et al.* (2009). Closed-loop artificial pancreas using subcutaneous glucose sensing & insulin delivery, and a model predictive control algorithm: Preliminary studies in Padova and Montpellier. *J Diabetes Sci. Technol.* **3,** 1014–1021.

Cheyne, E. H., Cavan, D. A., and Kerr, D. (2002). Performance of continuous glucose monitoring system during controlled hypoglycemia in healthy volunteers. *Diabetes Technol. Ther.* **4,** 607–613.

Clarke, W. L., and Kovatchev, B. P. (2007). Continuous glucose sensors continuing questions about clinical accuracy. *J. Diabetes Sci. Technol.* **1,** 164–170.

Clarke, W. L., Anderson, S. M., Breton, M. D., Patek, S. D., Kashmer, L., and Kovatchev, B. P. (2009). Closed-loop artificial pancreas using subcutaneous glucose sensing and insulin delivery and a model predictive control algorithm: The Virginia experience. *J. Diabetes Sci. Technol.* **3,** 1031–1038.

Cobelli, C., Dalla Man, C., Sparacino, G., Magni, L., Nicolao, G., and Kovatchev, B. P. (2009). Diabetes: Models, signals, and control. *IEEE Rev. Biomed. Eng.* **2**, 54–96.

Diabetes Control and Complications Trial (DCCT) Research Group (1993). The effect of intensive treatment of diabetes on the development and progression of long-term complications of insulin-dependent diabetes mellitus. *N. Engl. J. Med.* **329**, 978–986.

El-Khatib, F. H., Russell, S. J., Nathan, D. M., Sutherlin, R. G., and Damiano, E. R. (2010). A bihormonal closed-loop artificial pancreas for type 1 diabetes. *Sci. Transl. Med.* **2**, 27–35.

Hirsch, I. B., and Brownlee, M. (2005). Should minimal blood glucose variability become the gold standard of glycemic control? *J. Diabetes Complicat.* **19**, 178–181.

Hovorka, R. (2005). Continuous glucose monitoring and closed-loop systems. *Diabet. Med.* **23**, 1–12.

Hovorka, R., Chassin, L. J., Wilinska, M. E., Canonico, V., Akwi, J. A., Federici, M. O., Massi-Benedetti, M., Hutzli, I., Zaugg, C., Kaufmann, H., Both, M., Vering, T., *et al.* (2004). Closing the loop: The adicol experience. *Diabetes Technol. Ther.* **6**, 307–318.

Hovorka, R., Allen, J. M., Elleri, D., *et al.* (2010). Manual closed-loop insulin delivery in children and adolescents with type 1 diabetes: A phase 2 randomised crossover trial. *The Lancet* **375**, 743–751.

King, C. R., Anderson, S. M., Breton, M. D., Clarke, W. L., and Kovatchev, B. P. (2007). Modeling of calibration effectiveness and blood-to-interstitial glucose dynamics as potential confounders of the accuracy of continuous glucose sensors during hyperinsulinemic clamp. *J. Diabetes Sci. Technol.* **1**, 317–322.

Klonoff, D. C. (2005). Continuous glucose monitoring: Roadmap for 21st century diabetes therapy. *Diabetes Care* **28**, 1231–1239.

Klonoff, D. C. (2007). The artificial pancreas: How sweet engineering will solve bitter problems. *J. Diabetes Sci. Technol.* **1**, 72–81.

Kollman, C., Wilson, D. M., Wysocki, T., Tamborlane, W. V., and Beck, R. W. (2005). Diabetes research in children network: Limitation of statistical measures of error in assessing the accuracy of continuous glucose sensors. *Diabetes Technol. Ther.* **7**, 665–672.

Kovatchev, B. P., and Clarke, W. L. (2008). Peculiarities of the continuous glucose monitoring data stream and their impact on developing closed-loop control technology. *J. Diabetes Sci. Technol.* **2**, 158–163.

Kovatchev, B. P., Cox, D. J., Gonder-Frederick, L. A., and Clarke, W. L. (1997). Symmetrization of the blood glucose measurement scale and its applications. *Diabetes Care* **20**, 1655–1658.

Kovatchev, B. P., Straume, M., Cox, D. J., and Farhi, L. S. (2001). Risk analysis of blood glucose data: A quantitative approach to optimizing the control of insulin dependent diabetes. *J. Theor. Med.* **3**, 1–10.

Kovatchev, B. P., Cox, D. J., Gonder-Frederick, L. A., and Clarke, W. L. (2002). Methods for quantifying self-monitoring blood glucose profiles exemplified by an examination of blood glucose patterns in patients with type 1 and type 2 diabetes. *Diabetes Technol. Ther.* **4**, 295–303.

Kovatchev, B. P., Cox, D. J., Kumar, A., Gonder-Frederick, L. A., and Clarke, W. L. (2003). Algorithmic evaluation of metabolic control and risk of severe hypoglycemia in type 1 and type 2 diabetes using self-monitoring blood glucose (SMBG) data. *Diabetes Technol. Ther.* **5**, 817–828.

Kovatchev, B. P., Gonder-Frederick, L. A., Cox, D. J., and Clarke, W. L. (2004). Evaluating the accuracy of continuous glucose-monitoring sensors: Continuous glucose-error grid analysis illustrated by TheraSense Freestyle Navigator data. *Diabetes Care* **27**, 1922–1928.

Kovatchev, B. P., Clarke, W. L., Breton, M., Brayman, K., and McCall, A. (2005). Quantifying temporal glucose variability in diabetes via continuous glucose monitoring: Mathematical methods and clinical applications. *Diabetes Technol. Ther.* **7**, 849–862.

Kovatchev, B. P., *et al.* (2009a). Personalized subcutaneous model-predictive closed-loop control of T1DM: Pilot studies in the USA and Italy. *Diabetes* **58**(Supplement 1), 0228-OR.

Kovatchev, B. P., Breton, M. D., Dalla Man, C., and Cobelli, C. (2009b). In silico preclinical trials: a proof of concept in closed-loop control of type 1 diabetes. *J. Diabetes Sci. Technol.* **3**, 44–55.

Kulcu, E., Tamada, J. A., Reach, G., Potts, R. O., and Lesho, M. J. (2003). Physiological differences between interstitial glucose and blood glucose measured in human subjects. *Diabetes Care* **26**, 2405–2409.

McDonnell, C. M., Donath, S. M., Vidmar, S. I., Werther, G. A., and Cameron, F. J. (2005). A novel approach to continuous glucose analysis utilizing glycemic variation. *Diabetes Technol. Ther.* **7**, 253–263.

Miller, M., and Strange, P. (2007). Use of Fourier models for analysis and interpretation of continuous glucose monitoring glucose profiles. *J. Diabetes Sci. Technol.* **1**, 630–638.

Santiago, J. V. (1993). Lessons from the diabetes control and complications trial. *Diabetes* **42**, 1549–1554.

Shields, D., and Breton, M. D. (2007). Blood vs. interstitial glucose dynamic fluctuations: The Nyquist frequency of continuous glucose monitors. "Proc. 7th Diabetes Technol Mtg" p. A87. San Francisco, CA.

Steil, G. M., Rebrin, K., Hariri, F., Jinagonda, S., Tadros, S., Darwin, C., and Saad, M. F. (2005). Interstitial fluid glucose dynamics during insulin-induced hypoglycaemia. *Diabetologia* **48**, 1833–1840.

Steil, G. M., Rebrin, K., Darwin, C., Hariri, F., and Saad, M. F. (2006). Feasibility of automating insulin delivery for the treatment of type 1 diabetes. *Diabetes* **55**, 3344–3350.

Stout, P. J., Racchini, J. R., and Hilgers, M. E. (2004). A novel approach to mitigating the physiological lag between blood and interstitial fluid glucose measurements. *Diabetes Technol. Ther.* **6**, 635–644.

The Juvenile Diabetes Research Foundation Continuous Glucose Monitoring Study Group (2008). Continuous glucose monitoring and intensive treatment of type 1 diabetes. *N. Engl. J. Med.* **359**, 1464–1476.

U.S. Senate hearing (2006). The Potential of an Artificial Pancreas: Improving Care for People with Diabetes. September 27.

UK Prospective Diabetes Study Group (1998). Intensive blood-glucose control with sulphonylureas or insulin compared with conventional treatment and risk of complications in patients with type 2 diabetes. *Lancet* **352**, 837–853.

Weinzimer, S. (2006). Closed-loop artificial pancreas: Feasibility studies in pediatric patients with type 1 diabetes. "Proc. 6th Diabetes Technology Meeting" p. S55Atlanta, GA.

Weinzimer, S. A., Steil, G. M., Swan, K. L., Dziura, J., Kurtz, N., and Tamborlane, W. V. (2008). Fully automated closed-loop insulin delivery versus semi-automated hybrid control in pediatric patients with type 1 diabetes using an artificial pancreas. *Diabetes Care* **31**, 934–939.

Zanderigo, F., Sparacino, G., Kovatchev, B., and Cobelli, C. (2007). Glucose prediction algorithms from continuous monitoring data: Assessment of accuracy via continuous glucose-error grid analysis. *J. Diabetes Sci. Technol.* **1**, 645–651.

CHAPTER 20

Analyses for Physiological and Behavioral Rhythmicity

Harold B. Dowse*,†

*School of Biology and Ecology
University of Maine
Orono, Maine, USA

†Department of Mathematics and Statistics
University of Maine
Orono, Maine, USA

I. Introduction

Biological systems that evolve in time often do so in a rhythmic manner. Typical examples are heart beating (Bodmer *et al.*, 2004; Dowse *et al.*, 1995), circadian (Dowse, 2007), and ultradian (Dowse, 2008) biological cycles, and acoustic communication, for example, in *Drosophila* mating (Kyriacou and Hall, 1982). Using objective analysis techniques to extract useful information from these time series is central to understanding and working with the systems that produce them. Digital signal analysis techniques originating with astrophysics, geophysics, and electronics have been adapted to biological series and provide critical information on any inherent periodicity, namely its frequency or period as well as its strength and regularity. The latter two may be two separate matters entirely.

The mode of data acquisition is the first concern. Often biological data are records of events as a function of time, or perhaps the number of events during a sequential series of equal time intervals or "bins." Alternatively, output may be a continuous variable, such as the titer of an enzyme or binding protein. Acquisition technique and constraints upon it may affect the outcomes of later analyses and must be taken into consideration. As part of this process, the signal must ultimately be rendered digital for computer analysis. Examples will be considered.

Initial analysis is done in the time domain and may range from something as simple as a plot of the amplitude of the process to powerful statistical techniques such as autocorrelation, which can be used for determining if significant periodicities are present. Analysis in the frequency domain, usually spectral analysis, provides information on the period or frequency of any cycles present. This usually involves one of several variants of Fourier analysis, and recent advances in that area have revealed exceptional detail in biological signals (review: Chatfield, 1989). The mating song of the fruit fly, *Drosophila melanogaster*, is rich with information, but the data stream, as is the case with many other biological systems, is irregular and variable in time. Wavelet analysis is particularly useful in this instance. Digital signals, like their analog counterparts, may be filtered to remove noise or any frequencies in other spectral ranges that can be obscuring those in the range of interest (Hamming, 1983). The strength and regularity of the biological signal are of paramount importance. Spectral analysis algorithms may be altered appropriately to provide an objective measurement of a signal-to-noise (SNR) ratio (Dowse and Ringo, 1987). A related but distinct issue is the regularity of the cycles in the signal. For example, a heart may be beating strongly, but the duration of its pacemaker duty cycle may vary considerably more than normal from beat to beat with occasional skipped beats or, conversely, may be more regular than normal. Either alteration might be a result of pathology (Glass and Mackey, 1988; Lombardi, 2000; Osaka *et al.*, 2003). An index of rhythmic regularity is useful in this regard.

This chapter reviews modern digital techniques used to address each of these problems in turn. It uses the *Drosophila* model cardiac system extensively in this discourse, but the methods are widely applicable and other examples will be used as needed.

II. Types of Biological Data and Their Acquisition

One of the most intensively studied biological signals, of the several we shall consider, is found in the physiological and behavioral records of organisms over-time. These records are commonly found to be rhythmic with periodicity in the ballpark of 24 h, the solar day. In unvarying environmental conditions, such as constant darkness (DD) or low illumination (LL), the periodicity will vary from the astronomical day, hence the term *"circadian,"* or approximately daily rhythms (review: Palmer, 2002). This periodicity is the output of a biological oscillator (or oscillators), and study of this living horologue has been intense in the hopes of finding the mechanism (Dunlap, 1999; Hall, 2003). This field offers the opportunity to discuss generally applicable concepts.

Biological rhythm data take many forms, as clocks may be studied at levels ranging from intracellular fluorescence to running wheel activity. This broaches the topic of sampling. In cases of activity of an enzyme or the fluorescence level of a tag, for example, the variables are continuous and the sampling interval can be chosen arbitrarily. A primary concern here is that it be done rapidly enough to avoid "aliasing" in the periodicity region of interest. Aliasing occurs when the sampling interval is longer than the period being recorded and can be seen in old western movies when the spokes of wagon wheels seem to be going backward, a result of the interaction between the number of still frames/s and the angular velocity of the wheel (Hamming, 1983). Sampling frequency must be at least twice the frequency (half the period) of that of the sampled process. This is the Nyquist or fold-over frequency (Chatfield, 1989). A bit faster is better to be sure detail is not lost, but this is the theoretical tipping point. The tradeoff is an increasing number of data points to store and the commensurate wait for analysis programs to run if the sampling is gratuitously rapid.

The primary event in data acquisition is often an instantaneous reading of an analog signal. This may be transmembrane voltage or current in a *Xenopus* oocyte clamp setup, sound levels picked up by a microphone, light intensity reported out by a photomultiplier tube, or the output of an O_2 electrode; the list is endless. In general, however, whatever is being measured, the transduction process ultimately yields a voltage. This continuous analog voltage signal needs to be converted to a format that the computer can deal with. This process is often now done by analog to digital (A/D) converters within the instruments themselves, which will have a digital computer interface capability as a matter of course. Nonetheless, research equipment must often be built from scratch in-house for a specific purpose, and here analog signals may need to be dealt with by the user. The A/D converter is a unit that assigns numbers to a given input voltage. For example, in my laboratory we monitor fly heartbeat optically (see later; Dowse *et al.*, 1995; Johnson *et al.*, 1998). The output of the system is a voltage between −5 and +5 V, which is monitored on an oscilloscope. The computer has a DAS8, A/D 12-bit interface (Kiethly/Metrabyte) that employs 4096 0.00244 V steps, assigning proportional values between −2048 and +2048. The rate of digitization is programmable up to 1 MHz in this antique but thoroughly serviceable system. We find that for a ≈2- to 3-Hz heartbeat, 100 Hz is more than sufficient to yield excellent time resolution.

For noncontinuous data, there are other considerations. The Nyquist interval must still be factored in as a baseline for maximum sampling interval/minimum frequency (see earlier discussion), but there is a further constraint on how fast sampling may be done that has nothing to do with optimizing the number of data points to grab for computing expedience versus resolution. Common examples of this sort of data are running wheel activity in mammals (DeCoursey, 1960 and the breaking of an infrared light beam by *Drosophila* (Dowse *et al.*, 1987). Here, individual events are being registered and are summed across arbitrary intervals or "bins." The question is how well do these binned time series stand up to the sorts of analyses developed for discretely sampled continuous functions? It has been shown that bin size affects the output of time series analysis and that this effect can be profound when bin size is too small (review: Dowse and Ringo, 1994).

Over an arbitrarily short interval of the day, say a half an hour, the series of occurrences of events, such as a fly breaking a light beam in a chamber, is described by a Poisson process. There is no time structure or pattern and events occur stochastically. The probability, P, of k events occurring during the interval t, $t + 1$ is given by

$$P[N(t + 1) - N(t) = k] = e^{-\lambda} \frac{\lambda^k}{k!}. \tag{1}$$

The mean overall rate in events per unit time (EPUT) is given by λ (Schefler, 1969).

Over the course of a circadian day, for example, EPUT varies in a pattern, notably if the fly is behaving rhythmically. This variation can usefully be thought of as a Poisson process with a time-varying λ. In the case of running wheel data, of course, the events appear regularly spaced with a periodicity dependent on the rate of running in the apparatus, although bouts of running may be stochastically spaced throughout the active period. Nonetheless, the "amplitude" of the process remains EPUT, with the unit of time being the bin length.

On the basis of empirical and practical considerations, bin size much smaller than 10 min may cause artifact, in that perfectly good periodicities may be obscured in the presence of a lot of noise. Half-hour bins are generally small enough for good results in our experience. Longer bin lengths, for example, 1 h or longer, may act as a poorly defined low-pass digital filter, with a reduction in power transferred of about 20% at a periodicity of 2 h. Five-min bins have a flat transfer function (the plot of power transmitted through the filter as a function of period or frequency—see detailed discussion later; Dowse and Ringo, 1994).

III. Analysis in the Time Domain

Time series data may be analyzed in two domains: time and frequency. They may be transformed from one to the other as needed. In the time domain, relatively simple techniques are usually used initially to visualize evolution of the system. There is also a relatively straightforward statistical analysis available, the autocorrelogram, which tests for the presence and significance of any rhythmicity. Frequency or period may be measured crudely from either plots of raw data or the autocorrelogram, and questions of phase and waveform can be addressed directly.

We shall consider the cardiac system of the fly carefully to illustrate the analyses. The heart of the insect is a simple tube that works as a peristaltic pump within an open circulatory system, moving hemolymph from the most posterior region of the abdomen forward to the brain (Curtis *et al.*, 1999; Jones, 1977; Rizki, 1978). As air is carried to the tissues by a tracheal system, there are no pigments for gas transport, so even serious decrements in cardiac function may not necessarily be fatal (e.g., Johnson *et al.*, 1998). Nutrients and wastes are transported by the hemolymph (Jones, 1977). Heartbeat is myogenic, arising in discrete pacemaker cells posteriorly (Dowse *et al.*, 1995; Gu and Singh, 1995; Rizki, 1978), but as heartbeat can be retrograde, there is an alternate pacemaker near the anterior end as well (Dulcis and Levine, 2005; Wasserthal, 2007). The fly heart model is of considerable interest of late, as genes encoding heart structure and ion channels that function in the pacemaker have been shown to have analogous function in the human heart (Bodmer *et al.*, 2004; Wolf *et al.*, 2006).

Figure 1A shows a 60-s sample time series from this system, depicting the heartbeat of a wild-type *D. melanogaster* recorded optically at the P1 pupal stage. At this point in time, the heart can be monitored optically, as the pupal case has not yet begun to tan and remains transparent (Ashburner, 1989). The heart is also transparent, but the nearly opaque fat bodies on either side move as the heart beats, causing a change in the amount of light passing through the animal. This is picked up by a phototransistor (FPT100) affixed in the outlet pupil of one of the eyepieces of a binocular microscope. The signal is preamplified by a 741C op amp and is further amplified by a Grass polygraph. The output voltage is digitized as described previously by a DAS8 AD converter (Kiethley/Metrabyte) at 100 Hz and recorded as a text file in a computer. The temperature of the preparation is controlled by a Sensortek100 unit and, in this instance, is maintained at 25° C (cf. Dowse *et al.*, 1995; Johnson *et al.*, 1997, 1998). The heartbeat is very regular in this animal, although there are several gaps. Recall that this is the plot of voltage as a function of time.

While it is clear that this heart is rhythmic, it is useful to apply an objective statistical test, even in this clear example, to determine the significance of any periodicity. For example, Fig. 1B shows the record of a second wild-type animal's heart that is not nearly so clearly rhythmic, and Fig. 1C may depict a totally arrhythmic organ, also from a wild-type fly. No solid conclusion can be drawn based just on inspection of this erratic plot. An objective statistical method to determine whether a significant rhythm is present is by autocorrelation analysis (Chatfield, 1989; Levine *et al.*, 2002). To conduct this analysis, the time series is paired with itself in register and a standard correlation analysis is done yielding the correlation coefficient, r. Since the correspondence is exactly one to one, the correlation is perfect and the resultant r is 1. The two identical series are then set out of register or "lagged" by one datum. This will cause a corresponding decrement in the correlation coefficient computed. This lagging continues one datum at a time up to about one-third of the length of the entire series. The sequential r values are plotted as a function of the lag, and this is the autocorrelogram or autocorrelation function. If the series is rhythmic, the drop in *r* will continue and will become negative, reaching a nadir as the peaks and valleys in the values become π antiphase. A second positive peak will occur when the peaks and valleys return to phase locking at a full 2π. The general mathematical interpretation of the coefficients in the autocorrelogram is that they are cosines of the angles between

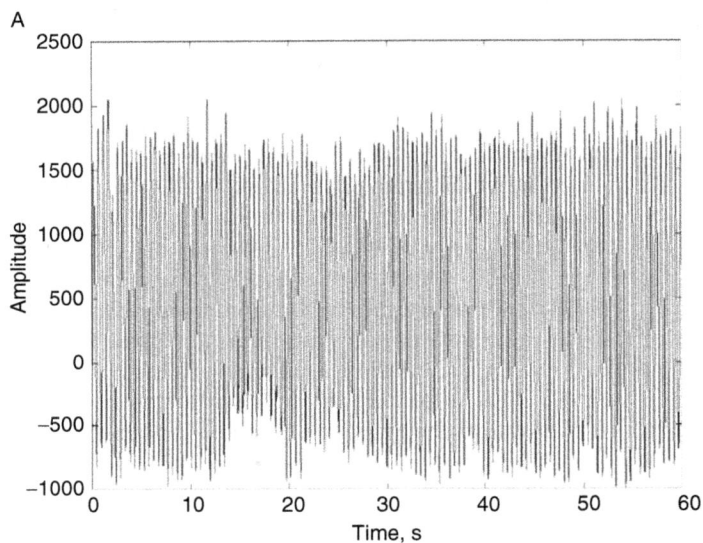

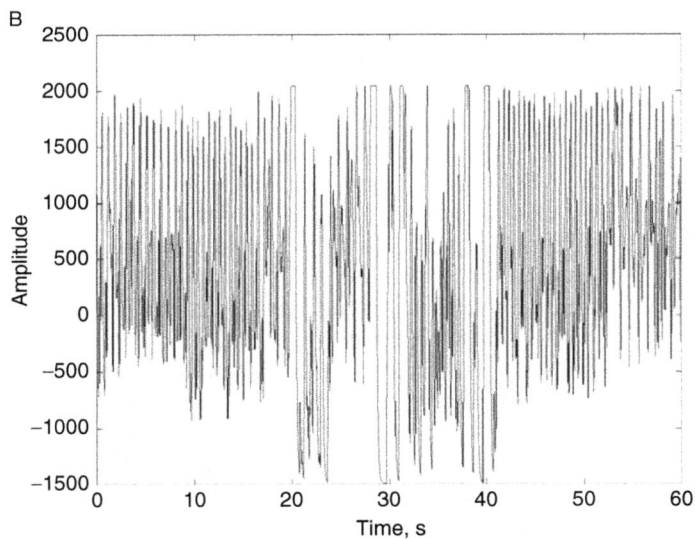

C

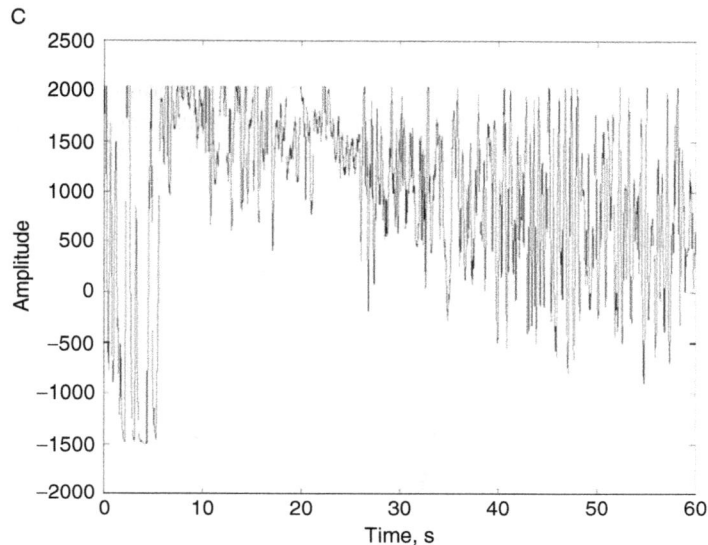

Fig. 1 Optically acquired digital records of wild-type *Drosophila melanogaster* heartbeat. (A)Extremely regular heartbeat with few changes in amplitude or period. (B) This heart is substantially more irregular in function. There are periods during which the beat is fairly erratic interspersed with regular beating. At times, especially from about 53 s on, it can be seen that the beat is bigeminal, with weak beats alternating with the much stronger power beats. (C) Here, the heart is almost arrhythmic. It can be seen to be beating during a few intervals, notably between 45 and 50 s.

two vectors, with the vectors being the original time series and that same series lagged out of register (Wiener, 1949). The process is well approximated by

$$r_k = \left(\sum_{t=1}^{N-k} (x_t - x_m)(x_{t-k} - x_m) \right) \Big/ \sum_{t=1}^{N} (x_t - x_m)^2, \tag{2}$$

where N is the number of samples and x_m is the mean of the series (Chatfield, 1989).

Note that the output as described earlier is normalized at each step by dividing by the variance in the entire data set (the denominator in the aforementioned equation), but need not be. If this is not done, the output is in the form of variance, and this "covariance" can be reported out instead, if this is desired, as the auto-covariance function (Chatfield, 1989). Differences between and relative utilities of these two functions will become apparent when spectral analysis is considered later. Here, the normalization to get an r is useful, as it allows comparisons among experiments and the function will yield yet another useful objective statistic for comparisons, as described later.

Each time the vectors are lagged, the values on the two far ends are no longer paired and must be discarded; hence the power of the test is gradually diminished.

For this reason, the usual limit of the autocorrelation computation is about $N/3$. The 95% confidence interval and hence significance of a given peak is given as $2/\sqrt{N}$, where N is the number of data points (Chatfield, 1989). Plus and minus confidence intervals are plotted as flat lines; the decrement in N as values are discarded is usually ignored. The rule of thumb interpretation of the plot is normally looking for repeated peaks equaling or exceeding the confidence interval,

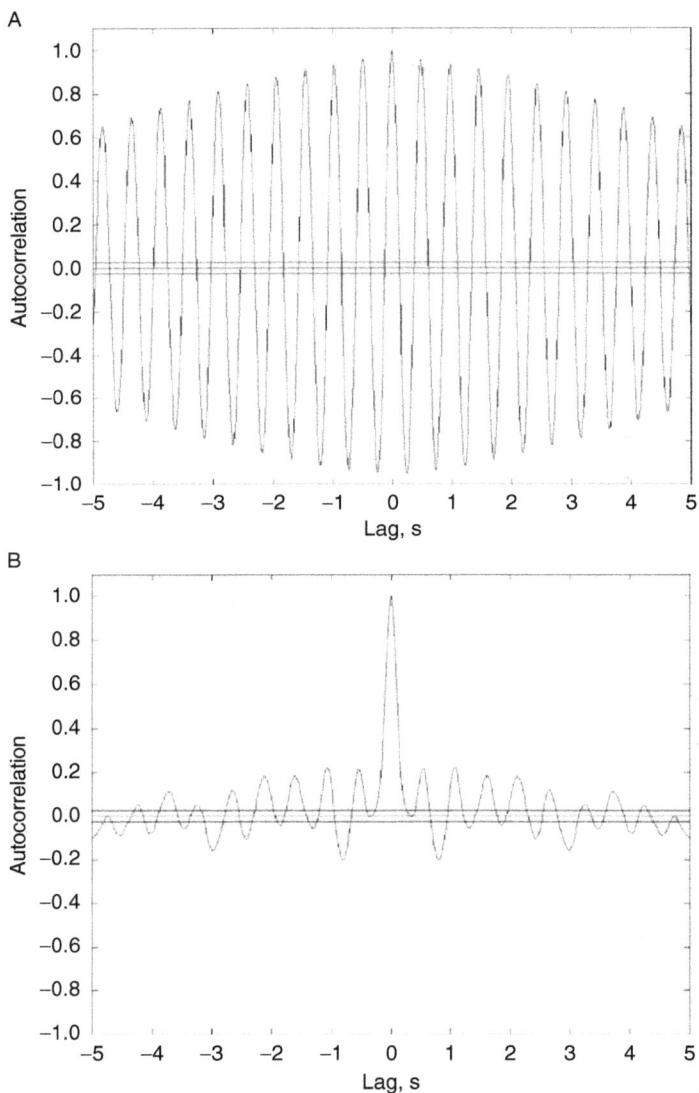

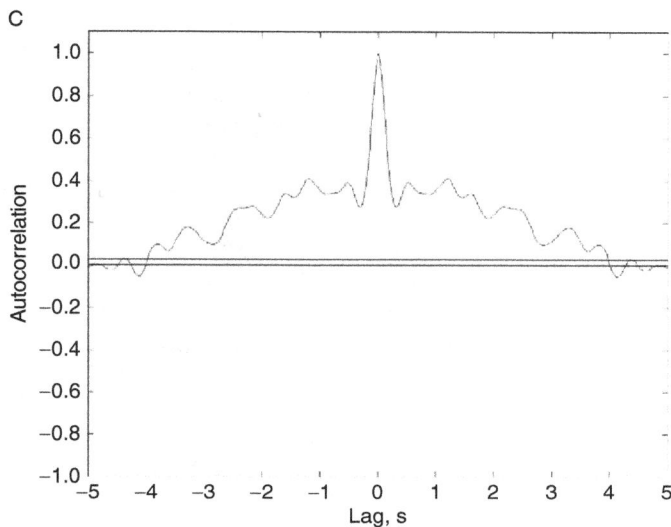

Fig. 2 Autocorrelograms produced from the data in the previous figure, appearing in the same order. The correlogram is a time-domain analysis that allows assessment for the presence or absence of any periodicities in the data as well as their regularity (see text). The autocorrelation values, r, are without units. The horizontal lines above and below the abscissa are ± the 95% confidence interval calculated as $2/\sqrt{N}$, in this case ± 0.0258. (A) Correlogram from the data in Fig. 1A. This heart is exceptionally regular in its rhythmicity. The decay envelope of the function is very shallow indicating long range order and stable frequency. The height of the third peak, counting the peak at lag 0 as #1 is 0.935 and constitutes the Rhythmicity Index (RI; see section below on signal strength and regularity). (B) In keeping with the appearance of reduced regularity in Fig. 1B, the decay envelope is steep. The RI is 0.221. (C) The heart of this animal beats occasionally, but it is erratic in the extreme. Owing to an RI < 0.025, it is considered arrhythmic.

but a long run of peaks not quite reaching this level is usually sufficient if inspection of the raw data plot yields similar results and if the periodicity turns up in this range in the spectral analysis. Use of the autocorrelation function to provide an estimator of regularity in rhythmicity is discussed later. In the examples shown in Fig. 1, the heart of the third pupa is considered arrhythmic, as will be shown. Figure 2 depicts the autocorrelograms of data from the hearts in Fig. 1. As the function is symmetrical and can be lagged in either direction, data from the reverse lagging are plotted here for symmetry and ease in visual interpretation.

In the case of biological rhythm research, another way of displaying data is commonly applied. This is by way of producing a "raster plot" or actogram, in which data are broken up into 24-h segments, which are plotted one below the other sequentially, that is, "modulo" 24 h. In this way, long records may be viewed easily and the relationship between the rhythmicity and the 24-h day can be assessed (e.g., DeCoursey, 1960). However, we shall not consider this technique here. The reader is referred to the following source for a full coverage with examples (Palmer *et al.*, 1994). It is worth noting, however, that such raster plots can be very misleading. Flies bearing mutations in the *period* gene (Konopka and Benzer, 1971), considered central to the biological "clock" (Dunlap, 1999), were reported as arrhythmic based on such raster plotting and the employment of the badly flawed "spectral analysis" program erroneously called the "periodogram" (see later for a discussion; Dowse *et al.*, 1987). By choosing a proper value for the length of the raster based on periodicity revealed by proper spectral analysis, ultradian (faster than 1/day) became clear. Even the relatively insensitive autocorrelograms showed clear, significant rhythmicity in these data (review: Dowse, 2008).

A further use of the autocorrelation algorithm can be done when it is desirable to compare the phase relationship between two time series that have similar frequencies. This may be done by way of "cross correlation." In this case, instead of comparing a time series with itself as it is lagged, a second time series is used. If they are in perfect phase, the peaks in the correlogram will be centered, but insofar as they are out of phase the central peak will be offset one way or the other. This analysis has been covered in detail elsewhere (Levine *et al.*, 2002).

One final technique can be applied in the time domain to enhance the interpretability of data; this is "time averaging." This is done commonly in electrophysiology, but has been used in circadian rhythm research as well (see, e.g., Hamblen-Coyle *et al.*, 1989). In this process, successive cycles are excised from the data stream modulo the period calculated and the peaks within are kept in phase. If this is done for a behavioral rhythm, for example, recorded in a 24-h LD cycle, then the section is simply 24 h. In electrophysiology, data sections containing individual events are excised. These data segments become rows in a matrix in register with one another. The columns produce means, which are plotted to get a composite picture of the signal (Hille, 2001). Figure 3A depicts an artificially produced time series (produced by a program we have written) consisting of a square wave with a period of 25 h and 50% stochastic noise added. We shall use this as an example of circadian periodicity. For comparison, the autocorrelogram of the series is shown in Fig. 3B. Figure 3C shows the result of time averaging the series to produce a single waveform estimate.

IV. Analysis in the Frequency Domain

In frequency analysis, the goal is to determine either the period or the frequency of any cycles ongoing in the process. This is done by looking at signal power as a function of frequency. Power in a signal is the ensemble average of the squared values in the series (Beauchamp and Yuen, 1979). Note that if the mean is zero, this is the same as variance. The power in the signals depicted in Fig. 1 are $1A = 7.732 \times 10^5$; $1B = 8.735 \times 10^5$; and $1C = 6.813 \times 10^5$. It is informative to think of it this way: the power in data is being partitioned by frequency, and the area under the curve of a spectrum, constructed as described later, is the power in the original signal.

To prepare such a spectrum, the workhorse is Fourier analysis. This begins with the remarkable observation that most functions can be approximated by a series of sine and cosine terms in a process called orthogonal decomposition. Start with an arbitrary function $f(t)$ that conforms to the "Dirichlet conditions," namely that it has a finite number of maxima and minima, that it is everywhere defined, and that there is a finite number of discontinuities (Lanczos, 1956, 1966). Biological time series will almost certainly conform.

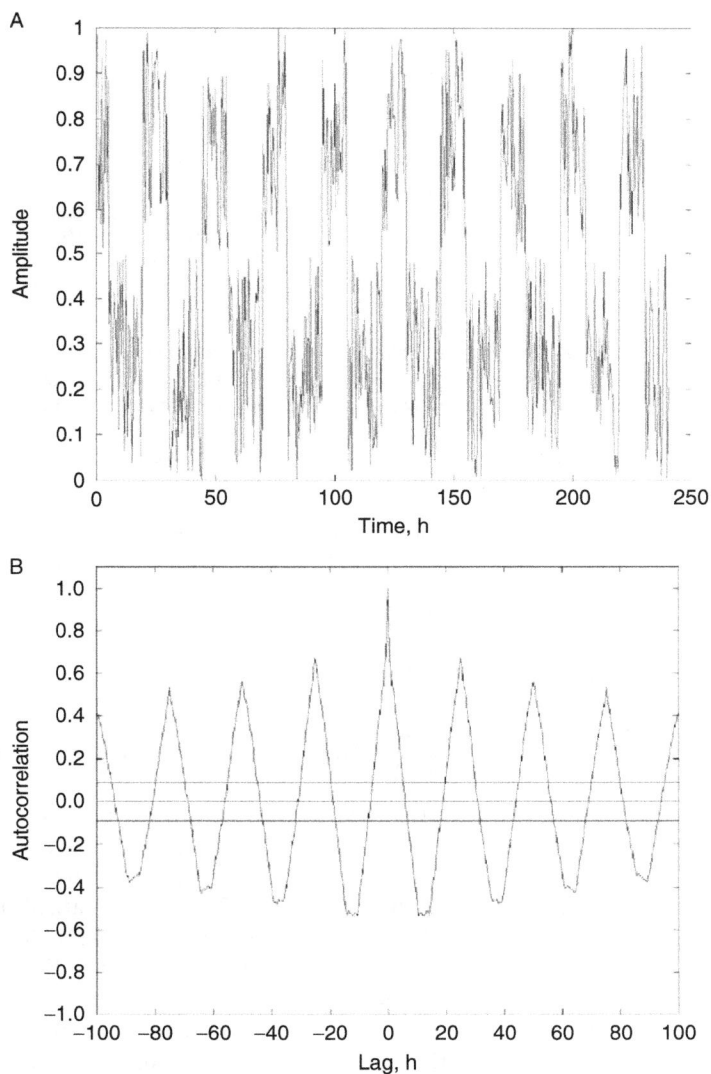

Fig. 3 (Continued)

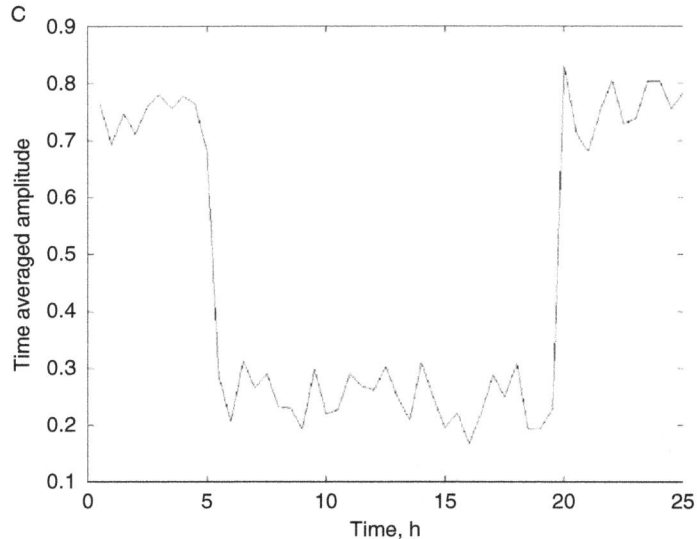

Fig. 3 To simulate a circadian behavioral rhythm, a signal generating program was employed to produce a square wave with a period of 25 h and 50% added white noise. Data acquisition was set at one half hour intervals and 480 data points were produced with a simulated half hour sampling/binning rate. (A) The raw unconditioned signal as it was produced by the program. (B) As with the heartbeat data, the signal was analyzed with the autocorrelogram. Note the strong repeating peaks at lags of 25, 50, and 75 h. Despite the large amount of noise, given the unvarying length of the period, this is to be expected. The decay envelope is not too steep. RI for this signal is 0.561. (C) The signal was broken up into 25-h segments (50 data points each) which were inserted as rows into of a 9 X 50 (Row X Column) Matrix. Extra "odd" points were discarded. The matrix columns were summed and a mean activity was computed. This is the plot of the output of that operation, a time-averaged estimate of the underlying wave form in the presence of high frequency noise.

$$f(t) \cong \frac{a_0}{2} + a_1 \sin t + a_2 \sin 2t + \cdots b_1 \cos t + b_2 \cos 2t + \cdots. \tag{3}$$

The Fourier series used to approximate the function consists of pairs of sine and cosine terms that are orthogonal (Hamming, 1983). An acoustic analogy is good here. Think of the function as a guitar string, with the fundamental vibration first, followed by successive harmonics. The mathematical interpretation is a series of vectors of length R rotating in the complex plane with angular velocity ω that is in radians/s ($\omega = 2\pi f$ or $2\pi/T$, where f is frequency and T is period). Here $R^2 = a^2 + b^2$ for each value of a and b for a given harmonic (Beauchamp and Yuen, 1979). The Fourier transform is a special case of the Fourier series, which in this form can now be used to map the series from the time domain to the frequency domain as $F(\omega)$ with the series of coefficients a and b being extracted from data (Lanczos, 1956):

$$F(\omega) = \int_{-\infty}^{\infty} f(t)e^{-i\omega t}dt, \qquad (4)$$

where the exponential consolidates the sine and cosine terms. A plot of R^2 calculated from the a and b coefficients extracted form the "periodogram" of the series and constitute a representation of the spectrum (Schuster, 1898). Peaks in the periodogram indicate periodicity in data at those given values. The area under the curve, as noted, is the total power in data, and for each value of R, this can be interpreted as the power in the signal at that period or frequency.

This process is not to be confused with another "periodogram" concocted some time later and used extensively in biological rhythm work. Whitaker and Robinson (1924) proposed producing a "Buys-Ballot" table for data using all possible values for frequency. This is much like the rasterizing or signal averaging techniques mentioned earlier. The rows of the series of matrices have varying length, sectioned off modulo each periodicity. The columns of the matrices are then added and means produced. For a matrix with a given set of row lengths, the variance of the column sums becomes the coefficient for the period corresponding to the length of that row. As the length varies, the variance will peak when the row equals the length of the period. The peaks and valleys will all be in register at this point as with the signal-averaged waveform discussed earlier. This method was championed for circadian rhythm studies by Enright (1965, 1990). However, it is not a mathematically sound procedure, as was demonstrated conclusively by Kendall (1946) when he noted that if there is a peak in the output, it does not mean there is any periodicity, as the variance of the column sums is independent of their order. In any event, in practice, this "periodogram" is unable to perform to the standards demanded of modern spectral analysis techniques. Historically, its widespread employment obscured important short-period (ultradian) rhythms in what appeared to be arrhythmic flies for a long time (see earlier discussion; review: Dowse, 2008). Its use is not recommended by this author. At the very least, because of Schuster's (1898) long priority, it cannot legitimately be called a periodogram. Figure 4 shows the Whitaker/Robinson (1924) "periodogram" for the noisy square wave depicted in Fig. 3A and compared further with the corresponding MESA plot in Fig. 6D (full discussion: Dowse and Ringo, 1989, 1991).

Figure 5 shows discrete Fourier analyses of the same three fly heartbeat records shown in Figs. 1 and 2. The algorithm used here is the long "brute force" computational method effected by operating on the original data set. In common usage, the actual transform is done not on original data, but on either autocorrelation or autocovariance functions. If it is the former, then the output is spectral density; if it is the latter, it is the true periodogram$_{SS}$. In the latter case, the area under the curve is the power in the signal, and this can be useful, whereas the normalization inherent in spectral density allows comparisons independent of the amplitude of the process across subjects (Chatfield, 1989). More computationally efficient methods, for example, the fast Fourier transform, are faster (Cooley et al., 1969).

Compromises must be made in standard Fourier spectral analysis. We will consider a brief summary of the concerns here and reference more thorough

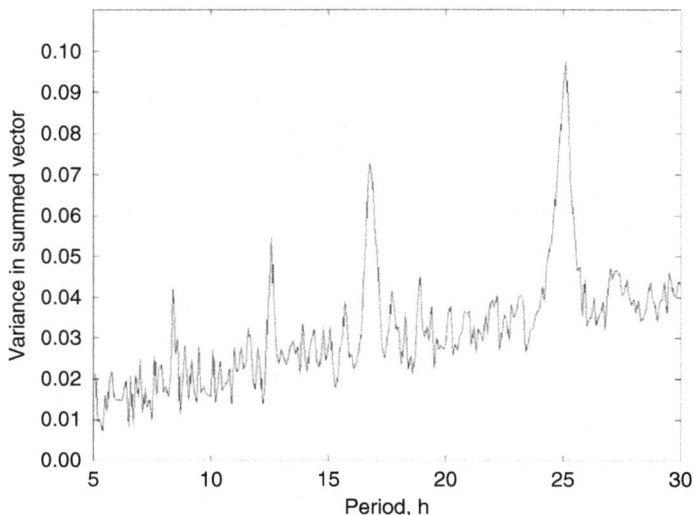

Fig. 4 A Whitaker-Robinson "periodogram" of the data vector shown in Fig. 3A. Note the ragged, noisy output with the monotonically rising background and the multiple "harmonics."

coverage. Recall that as the data vector is lagged in the calculation of the auto-covariance or autocorrelation function, data are lost off the "ends." As noted, this means that confidence intervals widen. Also, this imposes a limit on how long the computed function can be. To achieve the requisite number of samples to do computation of the coefficients, the function is "padded out" with zeros. Also the Fourier transform causes artifactual peaks when there are sharp discontinu-ities, so the step created when the autocorrelation or autocovariance function is terminated is smoothed out by a "window" function. This is a compromise in its own right, as what were perfectly good data points are altered by the smoothing operation. The practical tradeoff is between resolution and what is called side-lobe suppression (reviews: Ables, 1974; Chatfield, 1989; Kay and Marple, 1981).

In recent years, a new method for producing a spectrum that addresses these problems has become popular and, we maintain, is quite a good choice for biological time series. This technique is called "maximum entropy spectral analysis" (MESA; Burg, 1967; 1968; Ulrych and Bishop, 1975). In its most basic sense, it is a way of extending the autocorrelation out to the end in a reasonable manner, which is consistent with maximizing ignorance of the series, that is, entropy in its information sense. In choosing zeros to pad out the AC function, one is making an assumption about the process, creating values arbitrarily. It seems unlikely that if the process had continued, it would have consisted of only zeros. However, orthogonal decomposi-tion creates no model of the system and thus cannot predict. Stochastic modeling of the system is the answer. An autoregressive (AR) function is fitted to data, which describes the evolution of the system in time. The assumption is that the system moves forward in time as a function of previous values and a random noise component. The previous values are weighted by a series of coefficients derived from known data values (Ulrych and Bishop, 1975):

$$X_t = aX_{t-1} + bX_{t-2} + cX_{t-3} + \ldots Z_t, \tag{5}$$

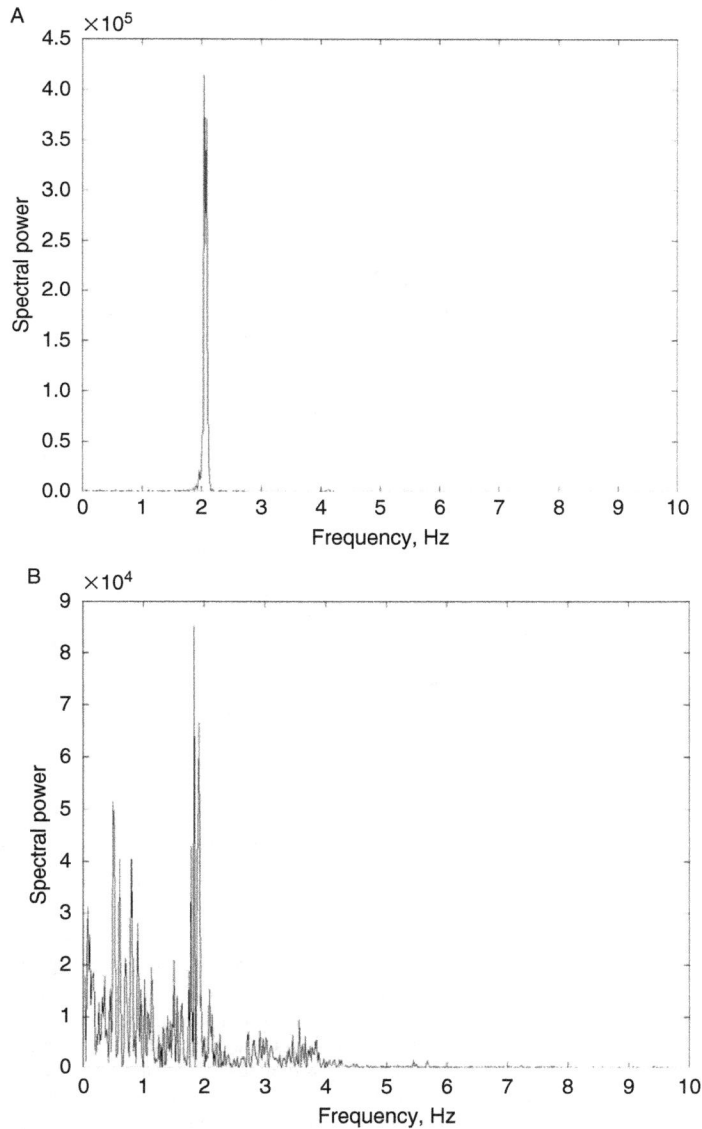

Fig. 5 (Continued)

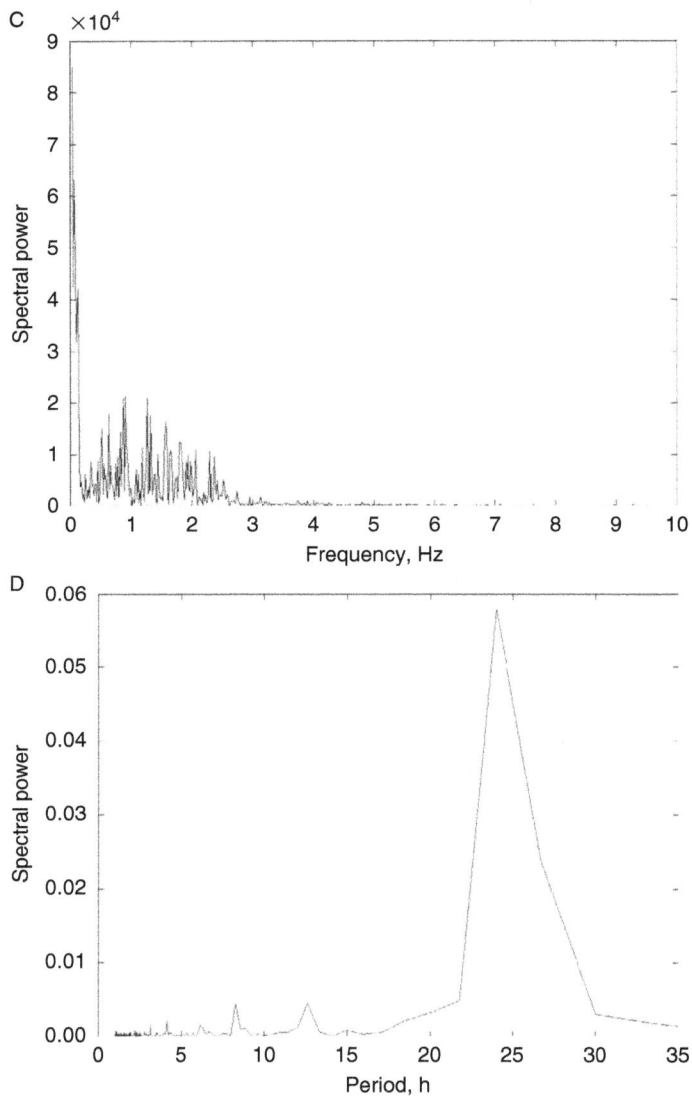

Fig. 5 Discrete Fourier analysis of the heartbeat data shown in Fig. 1. (A) The very regular heart produces a clean spectral output at approximately 2 Hz, which is substantiated by the peaks in the autocorrelogram (Fig. 2A). (B) The Fourier spectrum becomes less regular with this heartbeat which is substantially more erratic. Nonetheless, the output shows a peak at just under 2 Hz. (C) The spectrum appears as just noise in this analysis of a heart that was shown to be arrhythmic by its low RI. Note the relatively large noise component at very high frequency. (D) This analysis, when applied to the artificial square wave shown in Fig. 3A, yields a fairly broad peak at the expected 25 h.

where $a, b, \ldots$ are the model's coefficients and Z_t is random noise. These coefficients constitute the prediction error filter (PEF). It is possible to predict values into the future, in this case functionally taking the autocorrelation function out to the needed number of values. Mathematically, it formally maximizes ignorance of the function, meaning that the values estimated are most likely based on what is known from data in hand. The spectrum is constructed from the coefficients as follows:

$$S(\omega) = \frac{P}{\left| 1 - \sum_{k=1}^{p} a_k e^{-i\omega k} \right|^2}. \tag{6}$$

MESA has proven itself superior to ordinary Fourier analysis, as it does not produce artifacts from the various manipulations, which need to be absent in a model for the function, and both resolution and side-lobe suppression are superior to standard Fourier analysis (Ables, 1974; Kay and Marple, 1981). We employ a computationally efficient algorithm described by Andersen (1974).

The number of coefficients in the PEF is crucial to the output of the analysis. Too few, and resolution and important detail can be lost. If an excessive number is used, the spectrum will contain spurious peaks. In practice, an objective method has been described using the methods of Akaike (Ulrych and Bishop, 1975), based on information theory that chooses a PEF that is consistent with the most amount of real, useful information that can be extracted. This is employed in the MESA software application demonstrated here, but we usually set a minimum filter length of about $N/4$ for biological rhythm analyses to ensure adequate representation of any long period cycles in the presence of considerable noise. This is not usually necessary for the heartbeat analyses.

Figure 6 shows the three heartbeat records shown earlier, subjected here to MESA. Note the relationship between the sharpness of the peaks and the regularity of the rhythms. It should be pointed out that the broadness of the peak in the preceding Fourier spectrum is partially a result of the paucity of coefficients that can be computed. This number may be increased if necessary (Welch, 1967) for greater resolution; however, we did not elect to do this here (Please see discussion in section VIII below). It is substantially easier to increase the number of MESA coefficients to any degree needed, with a concomitant increase in computation time, but for comparison's sake, we left the number at the minimum level. For comparison, the artificially produced circadian rhythm-like signal created to demonstrate signal averaging in Fig. 3 has been analyzed by Fourier analysis [Fig. 5D, MESA (Fig. 6D)]. In the Whittaker–Robinson "periodogram" (Fig. 4), the ragged, weak 25-h peak, along with the multiple subpeaks at resonant harmonics, is striking, as is the inexorably rising background noise level. MESA produces a single peak, much sharper than in the Fourier spectrum (Fig. 5D) and there is little interference by the 50% of the signal that is added noise.

V. Time/Frequency Analysis and the Wavelet Transform

The primary problem with any Fourier-based system is the fundamental assumption that the process goes on unchanged for all time, the definition of a stationary series (Chatfield, 1989). Period, phase, and amplitude are invariant.

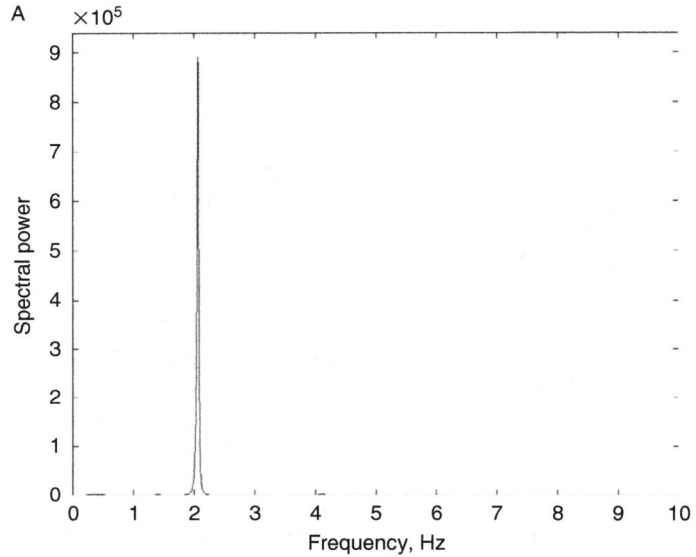

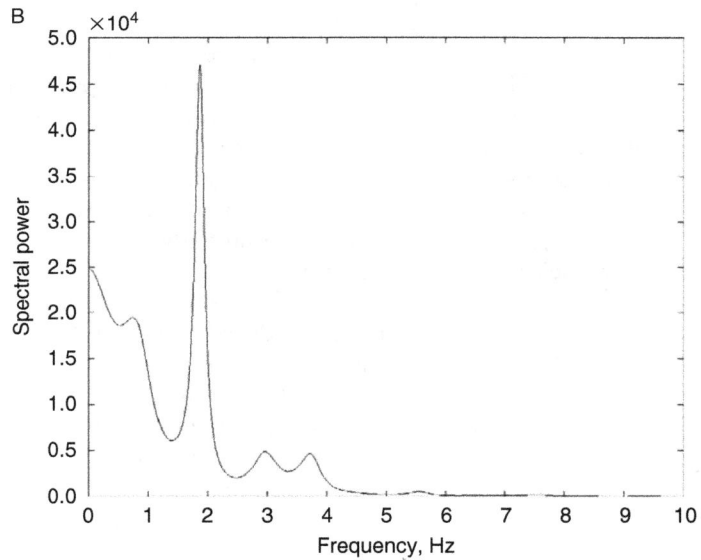

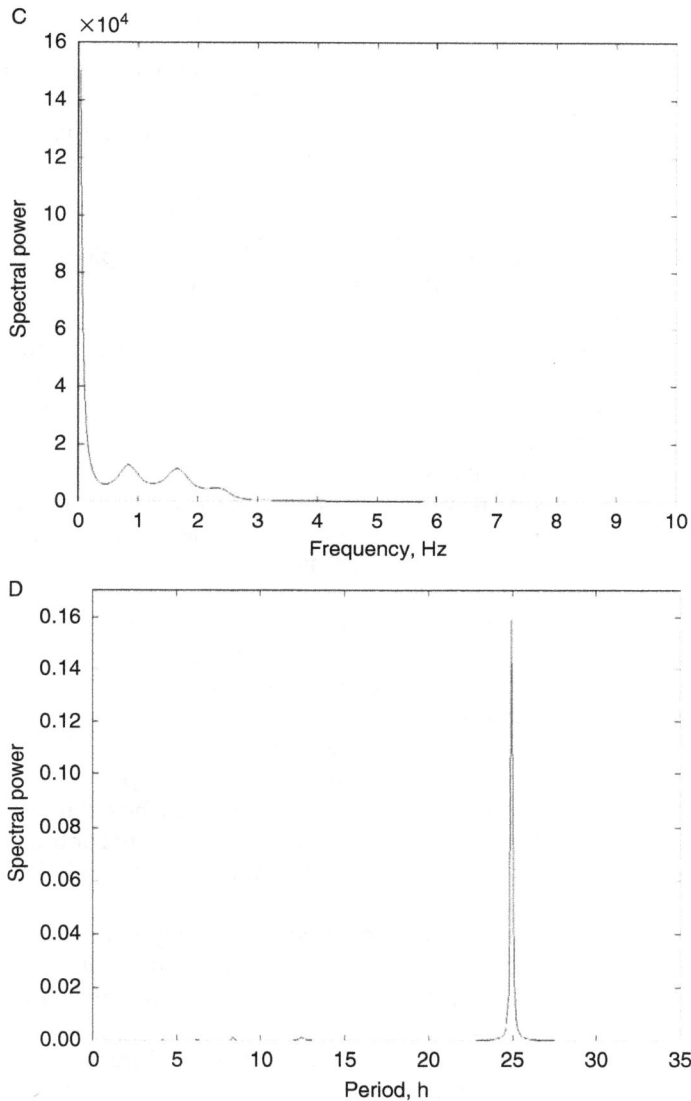

Fig. 6 Maximum Entropy Spectral Analysis (MESA) for the four time series shown in Figs. 1 and 3. (A) The most regular heart once again produces an extremely clean plot with no noise apparent in the spectrum. The peak, taken directly by inspection of the output of the program is 2.07 Hz. (B) While less regular, with a hefty peak of noise in the high frequency range, this heart also produces a relatively clean spectral peak at 1.87 Hz, as taken from the output file as in (A). (C) This is a typical noise spectrum. The few actual beats are lost in the record. This result is common for arrhythmic hearts. (D) For the artificially produced circadian rhythm example, the MESA peak is at exactly 25 h, as would be expected. Compare the sharp, narrow peak here with the Fourier analysis of the same file depicted in Fig. 5D and the sketchy "periodogram" in Fig. 4. There is no evidence of the large amount of noise that was added to the signal when it was produced. This is a typical performance for this advanced signal analysis technique.

The output of the analysis degrades to the extent that the system changes with time. Biological systems are not known for being stationary. To show the effect of changing period, a "chirp" was produced. Here, the signal is generated by the following equation:

$$X_t = \cos(\omega t^{1.3}). \tag{7}$$

The output series is plotted in Fig. 7, along with a plot of a MESA done on the data. Even the redoubtable MESA is incapable of dealing with this continually moving target.

Thus, it would be advantageous to follow a changing system as it evolves in time rather than looking for some consensus peak for the entire record. For example, if the heart slows down at some point, either as a result of treatment or because of alterations in the animal's internal physiology, it would be useful to be able to document this objectively and know when the change occurs. This is the role of "time-frequency analysis" and there are several methodologies (Chui, 1992). One might, for example, do short time fast Fourier transforms, meaning breaking a longer signal down into shorter segments. The loss in power of the analysis is great, however, which is unavoidable. The more you decide you want to know about frequency, the less you know about the time structure and vice versa. This relationship is rooted in quantum theory, literally the uncertainty principle (Chui, 1992; Ruskai et al., 1992). It is useful to think of a plot with the frequency domain of a time series as the ordinate and the time domain as the abscissa. If one draws a rectangular box in that two-dimensional plane delimited along the ordinate by what is known reliably of frequency and on the abscissa by knowledge of time, the uncertainty constraint means that the area of the box can never decrease below a minimum. If you want to know more about time, you contract that interval, but at the expense of widening the interval on the frequency axis, increasing uncertainty about that domain (Chui, 1992).

A very useful way to do time-frequency analysis has turned out to be to use "wavelet decomposition" (Chui, 1992; Ruskai et al., 1992). Wavelets are compactly supported curves, meaning that they are nonzero only within a bounded region. They can be as simple as the Haar wavelet, which is just a single square wave; they may be derived from cardinal spline functions, or they may even be fractal (Chui, 1992; Ruskai et al., 1992). This wavelet is convolved with the time series, meaning that each value of the wavelet is multiplied by a corresponding value of the time series and a sum of the values is computed. The wavelet is then translated (moved systematically) along the series, producing values from the convolution as it goes, which are the wavelet coefficients. The wavelet is then dilated, meaning it retains its form, but is made larger along its x axis. The process of translation is repeated and a second band of values is computed. This continues over a range of dilations. The matrix of the output is a representation of original data, but has substantially fewer points in it; this is one method of data compression. This is the wavelet transform. It is useful for storing signals, but more has been done than just that. For example, if you take a large matrix and waveletize it to produce a compressed "sparse" matrix,

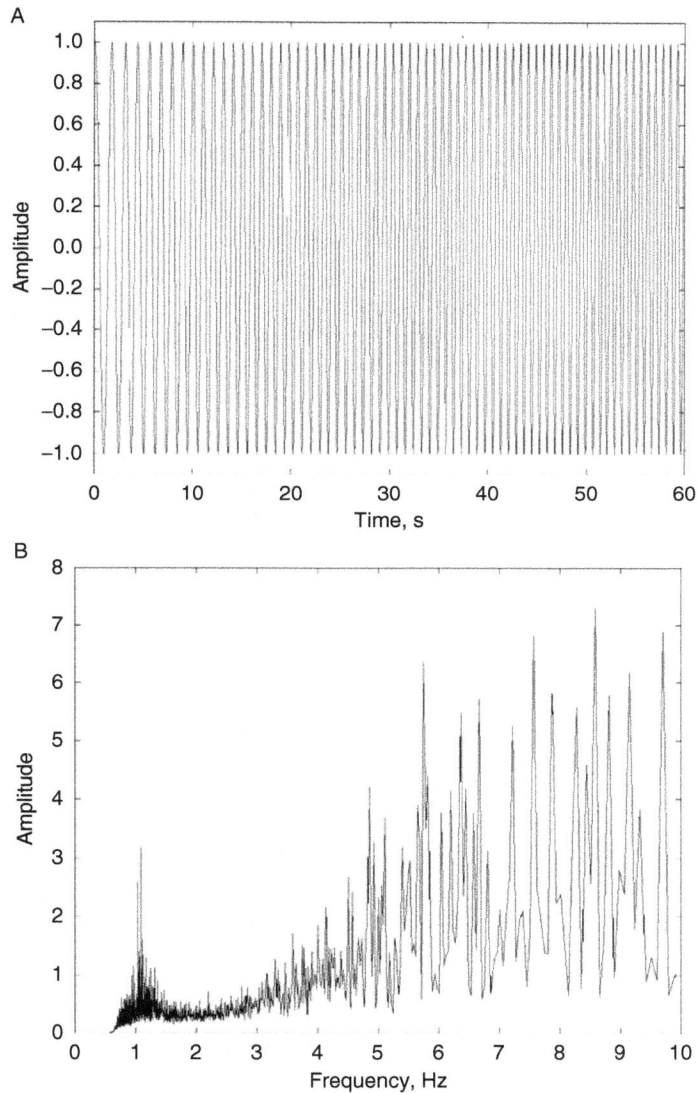

Fig. 7 (A) This is a "chirp," or nonstationary signal artificially produced without any noise. The frequency is rising regularly and monotonically as a power of time (see text). (B) Application of MESA to this signal produces a very erratic output which is virtually uninterpretable.

you can invert it much faster than you can invert the original. Taking the inverse wavelet transform does not restore the original matrix, rather it produces the inverted matrix, which can be useful in many applications (Chui, 1992; Ruskai *et al.*, 1992).

The critical consideration for signal analysis is that as the wavelet is dilated, it has different filtering characteristics. In this so-called "multiresolution analysis," each wavelet band will have a different range of frequencies that have been allowed to pass the filter and, crucially, the information about the TIME when these frequencies occur is retained. Imagine, for example, a trumpet note, which is anything but constant over the course of its evolution. Fourier analysis might simply break down (as did MESA, discussed earlier, when presented with a chirp, a far simpler time series), but when wavelet analysis is done, it would show a useful breakdown of frequency as a function of time.

The utility of wavelet transform analysis for biological data is proving out in the investigation of another fly system. *D. melanogaster* males court females using a number of stylized behavioral gestures (Spieth and Ringo, 1983). Among these is the production of a "mating song." This is produced during courtship by the male extending a single wing and vibrating it, producing either a "hum," also known as a "sine song," or a "buzz," which is a series of short wing flicks called a "pulse song" (Shorey, 1962). Much information of use to the female is inherent in this signaling (Kyriacou and Hall, 1982). The sine portion is of problematic utility (Kyriacou and Hall, 1984; Talyn and Dowse, 2004; von Schilcher, 1976), but the pulse song is species specific and definitely primes the female to mate more rapidly (Kyriacou and Hall, 1982; Talyn and Dowse, 2004). There is a sinusoidal rhythm in the peak-to-peak interval, the IPI, which varies regularly with a periodicity of about a minute. Remarkably, the period of this cyclicity is, to an extent, under the control of the *period* gene, which encodes a molecule of importance in the 24-h clock mechanism (Alt *et al.*, 1998; Kyriacou and Hall, 1980). Given the staccato nature of the signal, wavelet analysis is a natural choice for time-frequency analysis and even potential automation of song analysis.

We illustrate wavelet analysis of this signal as follows (data taken by Dr. Becky Talyn in this laboratory): males and females were housed separately within 10 h after eclosion to ensure adequate mating activity. Individual males and females were aspirated into a 1-cm chamber in an "Insectavox" microphone/amplifier instrument (Gorczyca and Hall, 1987) capable of picking up sounds at this intensity. The amplified signal was collected by a computer with a Sound Blaster card digitizing at 11,025 Hz. The sound card was controlled by and data were further viewed and edited using Goldwave software. Figure 8A depicts a section of sine song from a typical record, while Fig. 8B shows pulses. This section of song was subjected to cardinal spline wavelet decomposition using a program of our devising (program written in collaboration with Dr. William Bray, Department of Mathematics and Statistics and School of Biology and Ecology, University of Maine). Figure 8C shows a 2-s segment of song with both sine and pulse present, along with three bands from a wavelet analysis. The central frequency of the transfer function (see later) increases from top to bottom: 141, 283, and 566 Hz. The pulses appear as sharp peaks in all bands (even the last—they are just not seen readily if the scale of the abscissa is kept the same), meaning that the wavelet sees them as "singularities," while the sine song appears in only one band, appropriate to its commonly accepted species-specific frequency of 155 Hz (Burnet *et al.*, 1977). It is hoped that the differential presence of the two components in these two bands will allow us to automate the song analyses, a project that is ongoing at this time (Dowse and Talyn, unpublished).

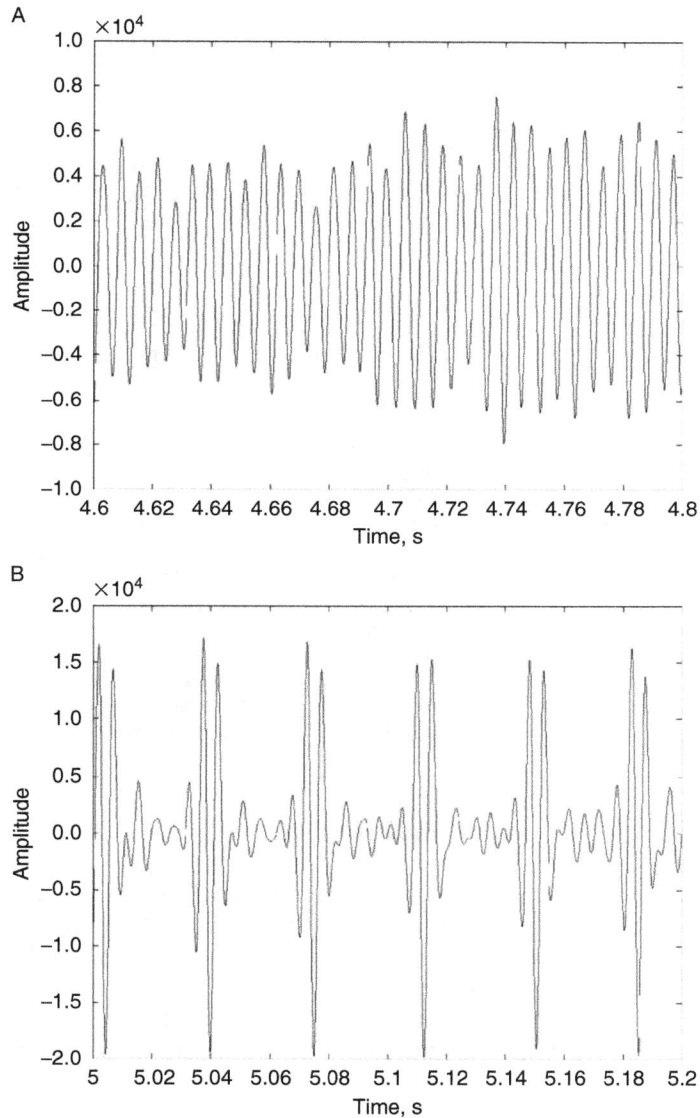

Fig. 8 (Continued)

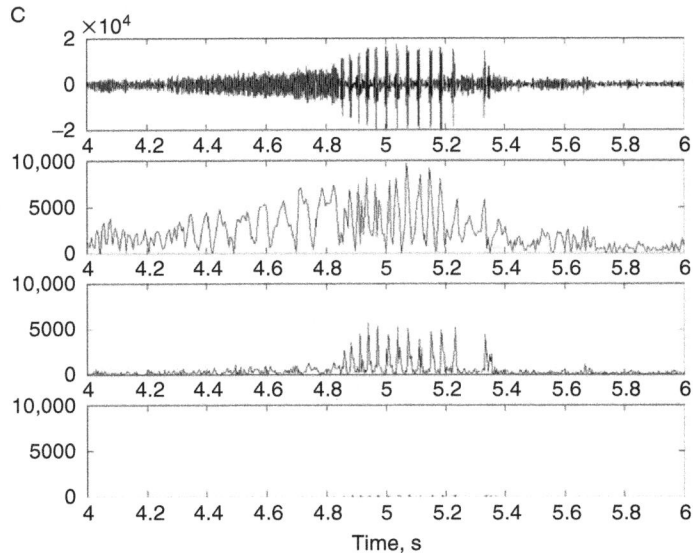

Fig. 8 (A) Male *Drosophila melanogaster* produce mating song by vibration of their wings. The output can come in two forms, either sine song, which sounds like a hum, or pulse song which makes a staccato buzz. (A) short segment of sine song and (B) pulse song. (C) Top panel: two seconds of song from which the above segments were excised as examples. Panel 2: Output of a cardinal spline wavelet analysis with a central frequency pass at 141 Hz, near the commonly reported species norm of ~150 Hz (see text). The ordinates of this panel and those below are power rather than amplitude, so the output can be thought of as a time-frequency analysis, showing how much power is present at a given time rather than a way of looking at all frequencies present across all time. Note that both the pulses and the sine pass power through the operation. Panel 3: Here, the pulses continue to pass power, as they are seen by the wavelet as singularities, while the sine disappears. The central frequency of the band is 283 Hz. Panel 4: There is nothing in the signal that passes through at this central frequency of 566 Hz with the exception of pulses that are barely visible at this scale. Bands higher or lower than the ones shown are similar.

VI. Signal Conditioning

Biological data are seldom "clean." Living systems commonly have noise associated with them, which can be a serious detriment to analysis. If the signal being acquired is analog voltage, electronic filtering can be done to remove at least a portion of this. Sixty-hertz notch filters to remove this omnipresent "hum" in electrophysiological preparations are a common example. However, we deal here with digitized signals in computers and have at hand a satisfying array of digital signal conditioning techniques for improving our analysis output once data have been recorded (Hamming, 1983).

The first sort of problem is very common in the study of biological behavioral rhythms. This is trend. An animal may be consistently rhythmic throughout a month-long running-wheel experiment, for example, but the level of its total activity may increase or decrease through that period of time. There may also be long-period fluctuations as well. This is also commonly seen in experiments where rhythmic enzymatic activity is being recorded and the substrate is depleted throughout the time period, leaving a decay envelope superimposed over the fluctuations of interest. There are numerous examples that could be cited. One exemplary problem is the search for rhythms as amplitude declines monotonically (review: Levine *et al.*, 2002). The way to tackle this is to remove any linear trend. In doing this, it is also sound practice to remove the mean from the series. This is equivalent to removing the "direct current" or DC part of the signal (Chatfield, 1989). There may be a strong oscillation in a parameter superimposed on a large DC offset, which will detract from the analysis. Removing the mean will leave only the "alternating current" or AC portion. The technique is extremely straightforward. One fits a regression line to data and subtracts that regression line from original data point by point (Dowse, 2007).

More problematic is a situation where there are nonlinear trends, for example, if one is looking for ultradian periodicity, and have a strong circadian period upon which it is superimposed. There are several ways to combat this. We will consider two here. First, "Fourier filtering" can be done to remove long-period rhythmicity. The discrete Fourier transform is first taken, and the Fourier coefficients are computed. Recall that these coefficients are orthogonal, meaning that they are totally independent of one another. What one does in one area of the spectrum does not affect actions taken in another. Hence, one simply zeroes out the coefficients for periodicities or frequencies one wishes to eliminate and then does the inverse Fourier transform. The resultant reconstructed time series then no longer has those frequencies and there is no disturbance of the other periodicities of interest (Dowse, 2007; Levine *et al.*, 2002; Lindsley *et al.*, 1999).

This is by way of doing a "high pass filter," meaning that the higher frequencies pass with a removal of the longer. Other filtering techniques are available, and digital filters of many sorts can be applied. We shall consider these in the context of "low pass" filters as the techniques are similar. The function of any filter is characterized by its transfer function, which is a plot of power transmitted as a function of frequency (Hamming, 1983).

The more common problem with biological signals in physiology is high frequency noise. This can arise within the electronics of the data acquisition systems themselves or be part of the actual process. Either way, this "static" can be highly detrimental to analyzing the longer frequency periodicities of interest. The amount and power in the noise portion of the spectrum can be computed for reasons relating to understanding the process itself, which is the subject of the next section. Here, we deal with ways of removing the noise from data to strengthen analysis, for example, to be certain frequency estimates are as accurate as may be obtained from data in hand.

We have already discussed several techniques for minimizing noise. Signal averaging is one done in the time domain, yielding more accurate waveform approximation. Noise is distributed stochastically throughout the spectrum and hence cancels out when multiple cycles are superimposed. The signal, however, reinforces itself continually cycle after cycle. Binning of unary data also removes substantial noise as this process acts as a low pass filter in and of itself (Dowse and Ringo, 1994). However, we now begin work with digital filters per se, beyond the Fourier filtering discussed briefly immediately preceding.

The simplest sort of digital filter is a moving average. This may be nothing more than a three-point process: $Y_t = (X_{t-1} + X_t + X_{t+1})/3$, where $X_t = X_1, X_2, \ldots, X_N$ is the original series, and Y_t is the filtered version. This is surprisingly effective and can be all that is needed. However, far more sophisticated filters are available and there are freeware programs available to compute the coefficients. Chebyshev and Butterworth recursive filters are well-known examples (review: Hamming, 1983). We consider here the Butterworth. This is called recursive because not only are original time series data incorporated into the moving filtering process, but previously filtered values are used as well. Butterworth filters are highly accurate in their frequency stopping characteristics, and the cutoff can be made quite sharp. The cutoff is usually expressed in decibels and is a quantification of the decrement in amplitude at a particular frequency. The cutoff frequency is the value at which the decrement, for example, 3 dB, is specified. In Fig. 9, the artificially produced square wave depicted first in Fig. 3A is shown after filtering with a two-pole low pass Butterworth filter with a $\approx$3-dB amplitude decrement at the cutoff period of 4 h (Hamming, 1983). The filter recursion is

$$Y_t = (X_t + 2X_{t-1} + X_{t-2} + AY_{t-1} + BY_{t-2})/C, \qquad (8)$$

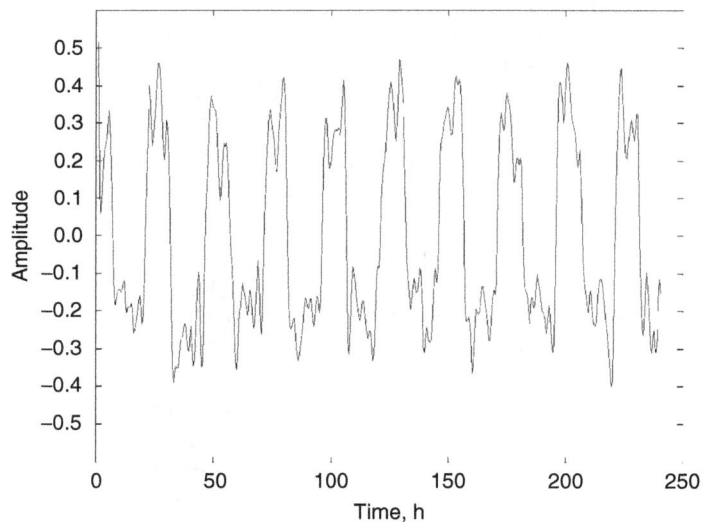

Fig. 9 The signal in Fig. 3A was filtered with a low-pass recursive Butterworth filter (see text) with a 3-dB attenuation at a period of 4 h. The 50% added noise is substantially reduced, and the underlying signal is far easier to see. There is a slight phase lag introduced compared to the original.

where X_t is the original data series and Y_t is the output. A and B are filter coefficients: $A = 9.656$ and $B = -3.4142$. C is the "gain" of the filter, meaning the amplitude change in the filtered output and $C = 10.2426$. Note again that both original data points and points from the output vector are combined. There is a net phase delay in the series produced by the filter, which should be kept in mind if phase is something of interest, as is often the case in biological rhythm work. Note also that the mean has been subtracted out, as was described in an earlier section. One important warning, it is highly inadvisable to run a filter more than once to achieve further smoothing as this will result in multiplication of error in the signal (Hamming, 1983).

VII. Strength and Regularity of a Signal

The amount of noise in a biological signal is more than just an annoyance when it comes to analysis. Assuming that the noise arises in the system being monitored and not in the acquisition hardware, it may be part of the process itself and thus will be of extreme interest. How strong and regular is the signal? Such central questions can reflect on values of parameters in the oscillating system, for example, activity of enzymes, or, in the heart pacemaker, conductivity and kinetics of the component ion channels. Note that strength and regularity are not the same thing. A wild-type fly heart beating at 2 Hz may have a much greater regularity than a mutant heart beating at the same rate and amplitude, all else being equal (Johnson et al., 1998). In the heart, "noise" may be an artifact resulting from interference by unrelated systems, but might also derive from poorly functioning ion channels (Ashcroft, 2000; Dowse et al., 1995). Thus quantification of noise and irregularity can be enlightening and useful (Glass and Mackey, 1988).

One standard way to quantify noise in a system in engineering is by the signal-to-noise ratio (Beauchamp and Yuen, 1979). As noted earlier, power in digital signals is the ensemble average of the squared values of the vector. Note that a noiseless, DC signal would still have power by this definition, while its variance would be zero. If all the noise in the signal is constrained to one region of the spectrum, while the signal is in another, SNR computation from the power spectrum would be simple. This is not usually the case, however, and given the erratic waveforms in biological signals, it is necessary to use alternate strategies.

To characterize such signals, we developed an algorithm based on MESA that allows waveform-independent calculation of SNRs (Dowse and Ringo, 1987, 1989). In this method, we fit the AR function to the vector in the usual manner; however, we use the coefficients calculated *not* in the PEF to compute a spectrum, but plug them into the actual AR model and use this equation, thus fleshed out with numbers derived from data, to predict upcoming values from past ones. A new series is thus generated one datum at a time from the previous values of the original. If, say, there are 30 coefficients, value Y_t in a series of predicted values is generated by the previous 30 values of X ($X_{t-1}, X_{t-2}, \ldots X_{t-30}$), the original series operated on by the filter. This forms a predicted output vector Y in parallel to the original. This process continues, working through the time series one value at a time until the entire predicted series is developed. The power in this series (as defined earlier) is the "signal" as it reflects the output of the model underlying the original series. The generated series Y is subtracted point by point from the values of X, and the difference series is the "noise." The ratio of the power in the generated series to that of the noise series is the SNR. The SNRs of the three heartbeat records are $A = 2134$, $B = 667$, and $C = 488$.

Noise in the system obscuring the output is not the only variation. The output of the oscillator generating the periodicity may also be irregular in another way, namely its period may vary from cycle to cycle. In the heart pacemaker, this may be a result of chaos (Glass and Mackey, 1988). The variation is not stochastic, but rather derives from a deterministic process, which has no perfect repeating orbit in phase space, and is considered "pseudoperiodic." This is considered quite normal and even necessary in a healthy heart, but in excess, chaos can be life-threatening. Fibrillation is the worst-case scenario, but irregular heartbeat plagues millions of patients and may often be fatal; analysis of the regularity of heartbeat may prove to be a useful predictive tool (Lombardi, 2000).

To assess this beat-to-beat variability, one may go to the length of recording these intervals from raw data and looking at the variance. Alternatively, difficult algorithms can be used to compute how chaotic a system might be (Glass and Mackey, 1988). We have chosen a simple way to characterize this phenomenon in physiological oscillators based on the autocorrelation function (review: Levine *et al.*, 2002). We measure the height of the third peak in the autocorrelogram, counting the first peak as the peak at lag zero, which is termed the rhythmicity index (RI). The decay envelope of the autocorrelogram is a function of the long-range regularity in the signal (Chatfield, 1989). If there is a lot of variation between beats, plus possible beat-to-beat decrement in amplitude, the function will decay more rapidly than in a regular series. With a perfect sinusoid, it will not decay at all. We have a program employing a sequential bubble sort algorithm that automatically retrieves this value from the output files of the autocorrelation program. Values of RI for the heights of the third peaks of the autocorrelograms depicted in Fig. 2 are $A = 0.935$, $B = 0.221$, and C = arrhythmic (no significant third peak; in fact, no third peak at all). Note that as in the autocorrelogram itself, from which this statistic is derived, values are normalized to the unit circle. These values are distributed normally and can be compared statistically. While this is not sophisticated enough to be used diagnostically, it is of more than sufficient resolution for us to compare heartbeat among strains with mutations affecting heart function (see, e.g., Sanyal *et al.*, 2005).

VIII. Some Practical Considerations on Statistical Comparisons of Analytical Results

Up to this point, we have considered only the techniques for estimating values, in most cases the period or frequency of a periodic process evolving in time. But once those results are in hand, they may then need to be analyzed using standard statistical tests, for example, to determine if a genetic lesion results in an altered period. The nature of the spacing of the spectral data needs consideration. Comparing Fourier spectral analysis with MESA is instructive. Consider a hypothetical experiment involving monitoring circadian periodicity in a mouse for a week. The spacing of frequencies to compute in these two analyses is substantially different. The time series can be considered a discrete sampling of the process even though the data are likely collected as events per unit time. In Fourier analysis, the frequencies are harmonics of one full cycle and are discrete samples of a continuous spectrum. If we are interested in comparing circadian periodicity, we take the reciprocal of frequency to show period. Alarmingly, independent of our sampling frequency (see above), we get the following periods for which Fourier coefficients can be computed: 18.7, 21.0, 28.0, and 33.6 h. The sampling of the spectrum is sparse and attempting to compare periods between two sample populations would be problematic. This is partially compensated in modern Fourier analysis by padding the data with zeros to get tighter spacing, but as has been argued above, this has its drawbacks (Ables, 1974; Welch, 1967).

It is easier with MESA and there are no compromises. Since the process is based on an AR model, one may arbitrarily decrease the sampling interval. Starting at the base level, the periods (in hours) for which MESA would calculate coefficients in the range near 24 h (rounded off to two decimals) are 22.7, 24.3, and 26.2, not much different from the Fourier analysis. However, by a simple change in the algorithm used to get the periods for which coefficients are calculated, the interval between estimates can be lowered to just a few minutes: 24.00, 24.10, 24.20, 24.31, 24.42, 24.53, 24.64, 24.75, 24.87, and 24.98 h. The interval can be lowered further to an arbitrary level, but little is to be gained beyond this point, as the increased precision of the estimates would be lost in the noise in the system. This is done with no change in the manner in which coefficients are computed.

A simple technique for analysis of spectral data is also worth considering at this point. Often, one may be looking at multiple periodicities in a range. This is common, for example, when ultradian rhythms are of interest (e.g., Dowse, 2008). Here, the use of the discrete Fourier transform becomes useful, not for estimating spectra per se, but for modeling the system in the classical sense (Lanczos, 1956). The resulting Fourier coefficients that result from fitting the sine and cosine series can be compared with standard statistical tests. Dowse *et al.* (2010) compared ultradian rhythms across several genetically distinct strains of mice whose activity was monitored in a 12:12 LD cycle. By doing a standard ANOVA, it was found that there were systematic variations in the ultradian range among the strains, suggesting genetic variation underlying the multiple periodicities we observed.

IX. Conclusions

A complete suite of programs for the analysis of biological time series has been described, capable of dealing with a wide range of signals ranging from behavioral rhythmicity in the range of 24-h periods to fly sine song in the courtship ritual, commonly in the range of 150 Hz in *D. melanogaster*. Frequencies outside this range can be dealt with easily and are limited only by data acquisition systems. It should be possible to pick and choose among the various techniques to assemble a subset for almost any situation. All the programs demonstrated here, other than the proprietary MATLAB used extensively for plotting output and the Goldwave program used in programming the SoundBlaster card, are available from the author free of charge. They may be requested in executable form or in FORTRAN source code from which they may be translated into other programming languages. For the former, a computer capable of running DOS window applications is required, but we have tested all applications in operating systems up to and including Windows XP.

References

Ables, J. G. (1974). Maximum entropy spectral analysis. *Astron. Astrophys. Suppl. Ser.* **15**, 383–393.

Alt, S., Ringo, J., Talyn, B., Bray, W., and Dowse, H. (1998). The period gene controls courtship song cycles in *Drosophila melanogaster*. *Anim. Behav.* **56**, 87–97.

Andersen, N. (1974). On the calculation of filter coefficients for maximum entropy spectral analysis. *Geophysics* **39**, 69–72.

Ashburner, M. (1989). "Drosophila, a Laboratory Manual." Cold Spring Harbor Press, Cold Spring Harbor, NY.

Ashcroft, F. (2000). "Ion Channels and Disease." Academic Press, New York.

Beauchamp, K., and Yuen, C. K. (1979). "Digital Methods for Signal Analysis," Allen & Unwin, London.

Bodmer, R., Wessels, R. J., Johnson, E., and Dowse, H. (2004). Heart development and function. *In* "Comprehensive Molecular Insect Science," (L. I. Gilbert, K. Iatrou, and S. Gill, eds.), Vol 2. Elsevier, New York.

Burg, J. P. (1967). Maximum entropy spectral analysis. *In* "Modern Spectrum Analysis" (1978), (D. G. Childers, ed.), pp. 34–41. Wiley, New York.

Burg, J. P. (1968). A new analysis technique for time series data. *In* "Modern Spectrum Analysis" (1978), (D. G. Childers, ed.), pp. 42–48. Wiley, New York.

Burnet, B., Eastwood, L., and Connolly, K. (1977). The courtship song of male *Drosophila* lacking aristae. *Anim. Behav.* **25**, 460–464.

Chatfield, C. (1989). "The Analysis of Time Series: An Introduction," Chapman and Hall, London.

Chui, C. K. (1992). "An Introduction to Wavelets," Academic Press, New York.

Cooley, J. W., Lewis, P. A. W., and Welch, P. D. (1969). Historical notes on the fast Fourier transform. *IEE Trans. Aud. Elect.* **AU15**, 76–79.

Curtis, N., Ringo, J., and Dowse, H. B. (1999). Morphology of the pupal heart, adult heart, and associated tissues in the fruit fly, *Drosophila melanogaster*. *J. Morphol.* **240**, 225–235.

DeCoursey, P. (1960). Phase control of activity in a rodent. *In* "Cold Spring Harbor Symposia on Quantitative Biology XXV," (A. Chovnik, ed.), pp. 49–55.

Dowse, H. (2007). Statistical analysis of biological rhythm data. *In* "Methods in Molecular Biology: Circadian Rhythms," (E. Rosato, ed.), Vol. 362, pp. 29–45. Humana Press.

Dowse, H. B. (2008). Mid-range ultradian rhythms in *Drosophila* and the circadian clock problem. *In* "Ultradian Rhythms From Molecules to Mind," (D. L. Lloyd, and E. Rossi, eds.), pp. 175–199. Springer Verlag, Berlin.

Dowse, H., and Ringo, J. (1987). Further evidence that the circadian clock in *Drosophila* is a population of coupled ultradian oscillators. *J. Biol. Rhythms* **2**, 65–76.

Dowse, H. B., and Ringo, J. M. (1989). The search for hidden periodicities in biological time series revisited. *J. Theor. Biol.* **139**, 487–515.

Dowse, H. B., and Ringo, J. M. (1991). Comparisons between "periodograms" and spectral analysis: Apples are apples after all. *J. Theor. Biol.* **148**, 139–144.

Dowse, H. B., and Ringo, J. M. (1994). Summing locomotor activity into "bins": How to avoid artifact in spectral analysis. *Biol. Rhythm Res.* **25**, 2–14.

Dowse, H. B., Hall, J. C., and Ringo, J. M. (1987). Circadian and ultradian rhythms in *period* mutants of *Drosophila melanogaster*. *Behav. Genet.* **17**, 19–35.

Dowse, H. B., Ringo, J. M., Power, J., Johnson, E., Kinney, K., and White, L. (1995). A congenital heart defect in *Drosophila* caused by an action potential mutation. *J. Neurogenet.* **10**, 153–168.

Dowse, H., Umemori, J., and Koide, T. (2010). Ultradian components in the locomotor activity rhythms of the genetically normal mouse. *Mus musculus. J. Exp. Biol.* **213**, 1788–1795.

Dulcis, D., and Levine, R. (2005). Innervation of the heart of the adult fruit fly, *Drosophila melanogaster*. *J. Comp. Neurol.* **465**, 560–578.

Dunlap, J. (1999). Molecular bases for circadian clocks. *Cell* **96**, 271–290.

Enright, J. T. (1965). The search for rhythmicity in biological time-series. *J. Theor. Biol.* **8**, 662–666.

Enright, J. T. (1990). A comparison of periodograms and spectral analysis: Don't expect apples to taste like oranges. *J. Theor. Biol.* **143**, 425–430.

Glass, L., and Mackey, M. C. (1988). "From Clocks to Chaos: The Rhythms of Life," Princeton Univ. Press, Princeton, NJ.

Gorczyca, M., and Hall, J. C. (1987). The INSECTAVOX, and integrated device for recording and amplifying courtship songs. *Dros. Inform. Serv.* **66**, 157–160.

Gu, G.-G., and Singh, S. (1995). Pharmacological analysis of heartbeat in *Drosophila. J. Neurobiol.* **28**, 269–280.

Hall, J. (2003). "Genetics and Molecular Biology of Rhythms in Drosophila and Other Insects," Academic Press, New York.

Hamblen-Coyle, M., Konopka, R. R., Zwiebel, L. J., Colot, H. V., Dowse, H. B., Rosbash, M. R., and Hall, J. C. (1989). A new mutation at the *period* locus of *Drosophila melanogaster* with some novel effects on circadian rhythms. *J. Neurogenet.* **5**, 229–256.

Hamming, R. W. (1983). "Digital Filters," Prentice-Hall, New York.

Hille, B. (2001). "Ion Channels of Excitable Membranes," Sinauer, Sunderland, MA.

Johnson, E., Ringo, J., and Dowse, H. (1997). Modulation of *Drosophila* heartbeat by neurotransmitters. *J. Comp. Physiol. B* **167**, 89–97.

Johnson, E., Ringo, J., Bray, N., and Dowse, H. (1998). Genetic and pharmacological identification of ion channels central to *Drosophila's* cardiac pacemaker. *J. Neurogenet.* **12**, 1–24.

Jones, J. (1977). "The Circulatory System of Insects," Charles C. Thomas, Springfield, IL.

Kay, S. M., and Marple, S. G., Jr. (1981). Spectrum analysis, a modern perspective. *IEEE Proc.* **69**, 1380–1419.

Kendall, M. G. (1946). "Contributions to the study of oscillatory time series," Cambridge Univ. Press, Cambridge.

Konopka, R. J., and Benzer, S. (1971). Clock mutants of *Drosophila melanogaster*. *Proc. Natl. Acad. Sci. USA* **68**, 2112–2116.

Kyriacou, C. P., and Hall, J. C. (1980). Circadian rhythm mutation in *Drosophila melanogaster* affects short-term fluctuations in the male's courtship song. *Proc. Natl. Acad. Sci. USA* **77**, 6729–6733.

Kyriacou, C. P., and Hall, J. C. (1982). The function of courtship song rhythms in *Drosophila*. *Anim. Behav.* **30**, 794–801.

Kyriacou, C. P., and Hall, J. C. (1984). Learning and memory mutations impair acoustic priming of mating behaviour in *Drosophila*. *Nature* **308**, 62–64.

Lanczos, C. (1956). "Applied Analysis," Prentice Hall, New York.

Lanczos, C. (1966). "Discourse on Fourier Series," Oliver and Boyd, Edinburgh.

Levine, J., Funes, P., Dowse, H., and Hall, J. (2002). Signal Analysis of Behavioral and Molecular Cycles. *Biomed. Central. Neurosci.* **3**, 1.

Lindsley, G., Dowse, H., Burgoon, P., Kilka, M., and Stephenson, L. (1999). A persistent circhoral ultradian rhythm is identified in human core temperature. *Chronobiol. Int.* **16**, 69–78.

Lombardi, F. (2000). Chaos theory, heart rate variability, and arrhythmic mortality. *Circulation* **101**, 8–10.

Osaka, M., Kumagai, H., Katsufumi, S., Onami, T., Chon, K., and Wantanabe, M. (2003). Low-order chaos in sympatheic nerve activity and scaling of heartbeat intervals. *Phys. Rev. E* **67**, 1–4104915.

Palmer, J. D. (2002). "The Living Clock," Oxford Univ. Press, London.

Palmer, J. D., Williams, B. G., and Dowse, H. B. (1994). The statistical analysis of tidal rhythms: Tests of the relative effectiveness of five methods using model simulations and actual data. *Mar. Behav. Physiol.* **24**, 165–182.

Rizki, T. (1978). The circulatory system and associated cells and tissues. *In* "The Genetics and Biology of *Drosophila*," (M. Ashburner, and T. R. F. Wright, eds.), pp. 1839–1845. Academic Press, New York.

Ruskai, M., Beylkin, G., Coifman, R., Daubechies, I., Mallat, S., Meyer, Y., and Raphael, L. (eds.) (1992). "Wavelets and Their Applications.," Jones & Bartlett, Boston.

Sanyal, S., Jennings, T., Dowse, H. B., and Ramaswami, M. (2005). Conditional mutations in SERCA, the sarco-endoplasmic reticulum Ca^{2+}-ATPase, alter heart rate and rhythmicity in *Drosophila*. *J. Comp. Physiol. B* **176**, 253–263.

Schefler, W. (1969). "Statistics for the Biological Sciences," Addison Wesley, Reading, MA.

Schuster, A. (1898). On the investigation of hidden periodicities with application to a supposed 26-day period of meterological phenomena. *Terrestrial Magn. Atmos. Electr.* **3**, 13–41.

Shorey, H. H. (1962). Nature of sound produced by *Drosophila* during courtship. *Science* **137**, 677–678.

Spieth, H., and Ringo, J. M. (1983). Mating behavior and sexual isolation in *Drosophila*. *In* "The Genetics and Biology of *Drosophila*," (M. Ashburner, H. Carson, and J. Thompson, eds.) Academic Press, New York.

Talyn, B., and Dowse, H. (2004). The role of courtship song in sexual selection and species recognition by female *Drosophila melanogaster*. *Anim. Behav.* **68**, 1165–1180.

Ulrych, T., and Bishop, T. (1975). Maximum entropy spectral analysis and autoregressive decomposition. *Rev. Geophys. Space Phys.* **13**, 183–300.

von Schilcher, F. (1976). The function of pulse song and sine song in the courtship of *Drosophila melanogaster*. *Anim. Behav.* **24**, 622–625.

Wasserthal, L. (2007). *Drosophila* flies combine periodic heartbeat reversal with a circulation in the anterior body mediated by a newly discovered anterior pair of ostial valves and 'venous' channels. *J. Exp. Biol.* **210**, 3703–3719.

Welch, P. (1967). The use of Fast Fourier Transform for the estimation of power spectra: A method based on time averaging over short, modified periodograms. *IEEE Trans. Audio Electroacoust.* Au-**15**, 70–73.

Whitaker, E., and Robinson, G. (1924). The Calculus of Observations, Blackie, Glasgow.

Wiener, N. (1949). "Extrapolation, Interpolation, and Smoothing of Stationary Time-Series," MIT Press, Cambridge, MA.

Wolf, M., Amrein, H., Izatt, J., Choma, M., Reedy, M., and Rockman, H. (2006). *Drosphila* as a model for the identification of genes causing adult human heart disease. *Proc. Nat. Acad. Sci. USA* **103**, 1394–1399.

CHAPTER 21

Evaluation and Comparison of Computational Models

Jay I. Myung, Yun Tang, and Mark A. Pitt

Department of Psychology
Ohio State University
Columbus, Ohio, USA

Abstract

Computational models are powerful tools that can enhance the understanding of scientific phenomena. The enterprise of modeling is most productive when the reasons underlying the adequacy of a model, and possibly its superiority to other models, are understood. This chapter begins with an overview of the main criteria that must be considered in model evaluation and selection, in particular explaining why generalizability is the preferred criterion for model selection. This is followed by a review of measures of generalizability. The final section demonstrates the use of five versatile and easy-to-use selection methods for choosing between two mathematical models of protein folding.

I. Update

Advancement in the field of model evaluation tends to be gradual. This chapter was written sufficiently recently that there have been no developments that would qualify what was written. In place of such an update, we would like to make modelers aware of a growing field that could be of interest in their pursuit of distinguishing computational models. Statistical model selection methods such as Akaike information criterion (AIC) and Bayesian model selection (BMS) are used to compare models *after* data are collected in an experiment. Why not use knowledge of these models and the experimental setting in which they are compared to design better experiments with which to discriminate them? *Design optimization* is a burgeoning field that has the potential to accelerate scientific discovery. Recent advances in Bayesian statistics now make it possible to design experiments that are optimized along various dimensions to distinguish competing computational models. In brief, through a sophisticated search of the design space of the experiment and the parameter spaces of the models, the method identifies the designs that are most likely to discriminate the models if the experiment were conducted. Interested reader should consult the following papers: Muller *et al.* (2004) and Myung and Pitt (2009).

Hints and Tips

(1) A good fit is a necessary, but not a sufficient, condition for judging the adequacy of a model.

(2) When comparing models, one should avoid choosing an unnecessarily complex model that overfits, and instead, should try to identify a model that is sufficiently complex, but not too complex, to capture the regularity in the data.

(3) Model comparison should be based not upon goodness of fit (GOF), which refers to how well a model fits a particular pattern of observed data, but upon generalizability, which refers to how well a model fits not only the observed data at hand but also new, as yet unseen, data samples from the same process that generated the observed data.

(4) If models being compared differ significantly in number of parameters and also the sample size is relatively large, use AIC, AICc, or Bayesian information criterion (BIC).

(5) If the conditions in (4) are not met or when the models have the same number of parameters, start by using CV or accumulative prediction error (APE). Their ease of application makes them a worthwhile first step. Only if they do not provide the desired clarity regarding model choice should BMS or stochastic complexity (SC) be used.

(6) Keep the outcomes of model comparison analyses in perspective. They are only one statistical source of evidence in model evaluation.

II. Introduction

How does one evaluate the quality of a computational model of enzyme kinetics? The answer to this question is important and complicated. It is important because mathematics makes it possible to formalize the reaction, providing a precise description of how the factors affecting it interact. Study of the model can lead to significant understanding of the reaction, so much so that the model can serve not merely as a description of the reaction, but can contribute to explaining its role in metabolism. Model evaluation is complicated because it involves subjectivity, which can be difficult to quantify.

This chapter begins with a conceptual overview of some of the central issues in model evaluation and selection, with an emphasis on those pertinent to the comparison of two or more models. This is followed by a selective survey of model comparison methods and then an application example that demonstrates the use of five simple yet informative model comparison methods.

Criteria on which models are evaluated can be grouped into those that are difficult to quantify and those for which it is easier to do so (Jacobs and Grainger, 1994). Criteria such as *explanatory adequacy* (whether the theoretical account of the model helps make sense of observed data) and *interpretability* (whether the components of the model, especially its parameters, are understandable and are linked to known processes) rely on the knowledge, experience, and preferences of the modeler. Although the use of these criteria may favor one model over another, they do not lend themselves to quantification because of their complexity and qualitative properties. Model evaluation criteria for which there are quantitative measures include *descriptive adequacy* (whether the model fits the observed data), *complexity* or *simplicity* (whether the model's description of observed data is achieved in the simplest possible manner), and *generalizability* (whether the model is a good predictor of future observations). Although each criterion identifies a property of a model that can be evaluated on its own, in practice they are rarely independent of one another. Consideration of all three simultaneously is necessary to assess fully the adequacy of a model.

III. Conceptual Overview of Model Evaluation and Comparison

Before discussing the three quantitative criteria in more depth, we highlight some of the key challenges of modeling. Models are mathematical representations of the phenomenon under study. They are meant to capture patterns or regularities in empirical data by altering parameters that correspond to variables that are thought to affect the phenomenon. Model specification is difficult because our knowledge about the phenomenon being modeled is rarely complete. That is, the empirical data obtained from studying the phenomenon are limited, providing only partial information (i.e., snapshots) about its properties and the variables that influence it. With limited information, it is next to impossible to construct the "true" model. Furthermore, with only partial information, it is likely that multiple models are plausible; more than one model can provide a good account of the data. Given this situation, it is most productive to view models as approximations, which one seeks to improve through repeated testing.

Another reason models can be only approximations is that data are inherently noisy. There is always measurement error, however small, and there may also be other sources of uncontrolled variation introduced during the data collection process that amplifies this error. Error clouds the regularity in the data, increasing the difficulty of modeling. Because noise cannot be removed from the data, the researcher must be careful that the model is capturing the meaningful trends in the data and not error variation. As explained later, one reason why generalizability has become the preferred method of model comparison is how it tackles the problem of noise in data.

The descriptive adequacy of a model is assessed by measuring how well it fits a set of empirical data. A number of GOF measures are in use, including sum of squared errors (SSE), percent variance accounted for, and maximum likelihood (ML; e.g., Myung, 2003). Although their origins differ, they measure the discrepancy between the empirical data and the ability of a model to reproduce those data. GOF measures are popular because they are relatively easy to compute and the measures are versatile, being applicable to many types of models and types of data. Perhaps most of all, a good fit is an almost irresistible piece of evidence in favor of the adequacy of a model. The model appears to do just what one wants it to—mimic the process that generated the data. This reasoning is often taken a step further by suggesting that the better the fit, the more accurate the model. When comparing competing models, then, the one that provides the best fit should be preferred.

GOF would be suitable for model evaluation and comparison if it were not for the fact that data are noisy. As described earlier, a data set contains the regularity that is presumed to reflect the phenomenon of interest plus noise. GOF does not distinguish between the two, providing a single measure of a model's fit to both (i.e., GOF = fit to regularity + fit to noise). As this conceptual equation shows, a good fit can be achieved for the wrong reasons, by fitting noise well instead of the regularity. In fact, the better a model is at fitting noise, the more likely it will provide a superior fit than a competing model, possibly resulting in the selection of a model that in actuality bears little resemblance to the process being modeled. GOF alone is a poor criterion for model selection because of the potential to yield misleading information.

This is not to say that GOF should be abandoned. On the contrary, a model's fit to data is a crucial piece of information. Data are the only link to the process being modeled, and a good fit can indicate that the model mimics the process well. Rather, what is needed is a means of ensuring that a model does not provide a good fit for the wrong reason.

What allows a model to fit noisy data better than its competitors is that it is the most complex. *Complexity* refers to the inherent flexibility of a model that allows it to fit diverse data patterns (Myung and Pitt, 1997). By varying the values of its parameters, a model will produce different data patterns. What distinguishes a simple model from a complex one is the sensitivity of the model to parameter variation. For a simple model, parameter variation will produce small and gradual changes in model performance. For a complex model, small parameter changes can result in dramatically different data patterns. It is this flexibility in producing a wide range of data patterns that makes a model complex. For example, the cubic model $y = ax^2 + bx + c$ is more complex than the linear model $y = ax + b$. As shown in the next section, model selection methods such as AIC and BIC include terms that penalize model complexity, thereby neutralizing complexity differences among models.

Underlying the introduction of these more sophisticated methods is an important conceptual shift in the goal of model selection. Instead of choosing the model that provides the best fit to a single set of data, choose the model that, *with its parameters held constant*, provides the best fit to the data if the experiment were repeated again and again. That is, choose the model that generalizes best to replications of the same experiment. Across replications, the noise in the data will change, but the regularity of interest should not. The more noise that the model captures when fit to the first data set, the poorer its measure of fit will be when fitting the data in replications of that experiment because the noise will have changed. If a model captures mostly the regularity, then its fits will be consistently good across replications. The problem of distinguishing regularity from noise is solved by focusing on generalizability. A model is of questionable worth if it does not have good predictive accuracy in the same experimental setting. Generalizability evaluates exactly this, and it is why many consider generalizability to be the best criterion on which models should be compared (Grunwald *et al.*, 2005).

The graphs in Fig. 1 summarize the relationship among the three quantitative criteria of model evaluation and selection: GOF, complexity, and generalizability. Model complexity is along the x axis and model fit along the y axis. GOF and generalizability are represented as curves whose performance can be compared as a function of complexity. The three smaller graphs contain the same data set (dots) and the fits to these data by increasingly more complex models (lines). The left-most model in Fig. 1 underfits the data. Data are curvilinear, whereas the model is linear. In this case, GOF and generalizability produce similar outcomes because the model is not complex enough to capture the bowed shape of the data. The model in the middle graph of Fig. 1 is a bit more complex and does a good job of fitting only the regularity in the data. Because of this, the GOF and generalizability measures are higher and also similar. Where the two functions diverge is when the model is more complex than is necessary to capture the main trend. The model in the right-most graph of Fig. 1 captures the experiment-specific noise, fitting every data point perfectly. GOF rewards this behavior by yielding an even higher fit score, whereas generalizability does just the opposite, penalizing the model for its excess complexity.

The problem of overfitting is the scourge of GOF. It is easy to see when overfitting occurs in Fig. 1, but in practice it is difficult to know when and by how much

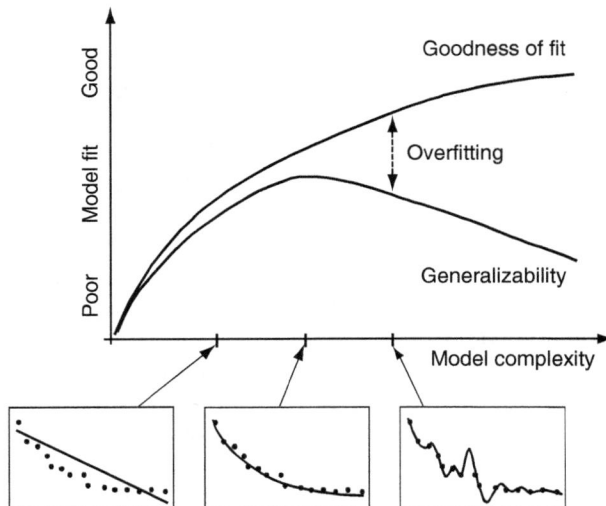

Fig. 1 An illustration of the relationship between GOF and generalizability as a function of model complexity. The *y* axis represents any fit index, where a larger value indicates a better fit (e.g., maximum likelihood). The three smaller graphs provide a concrete example of how fit improves as complexity increases. In the left graph, the model (line) is not complex enough to match the complexity of data (dots). The two are well matched in complexity in the middle graph, which is why this occurs at the peak of the generalizability function. In the right graph, the model is more complex than data, capturing microvariation due to random error. Reprinted from Pitt and Myung (2002).

a model overfits a data set, which is why generalizability is the preferred means of model evaluation and comparison. By using generalizability, we evaluate a model based on how well it predicts the statistics of future samples from the same underlying processes that generated an observed data sample.

IV. Model Comparison Methods

This section reviews measures of generalizability currently in use, touching on their theoretical foundations and discussing the advantages and disadvantages of their implementation. Readers interested in more detailed presentations are directed to Myung *et al.* (2000) and Wagenmakers and Waldorp (2006).

A. Akaike Information Criterion and Bayesian Information Criterion

As illustrated in Fig. 1, good generalizability is achieved by trading off GOF with model complexity. This idea can be formalized to derive model comparison criteria. That is, one way of estimating the generalizability of a model is by appropriately discounting the model's GOF relative to its complexity. In so doing, the aim is to identify the model that is sufficiently complex to capture the underlying regularities in the data but not unnecessarily complex to capitalize on random noise in the data, thereby formalizing the principle of Occam's razor.

The AIC (Akaike, 1973; Bozdogan, 2000), its variation called the second-order AIC (AICc; Burnham and Anderson, 2002; Sugiura, 1978), and the Bayesian information criterion (BIC; Schwarz, 1978) exemplify this approach and are defined as

$$
\begin{aligned}
\text{AIC} &= -2\ln f(y|w^*) + 2k, \\
\text{AICc} &= -2\ln f(y|w^*) + 2k + \frac{2k(k+1)}{n-k-1}, \\
\text{BIC} &= -2\ln f(y|w^*) + k\ln(n),
\end{aligned}
\tag{1}
$$

where y denotes the observed data vector, $\ln f(y \mid w^*)$ is the natural logarithm of the model's maximized likelihood calculated at the parameter vector w^*, k is the number of parameters of the model, and n is the sample size. The first term of each comparison criterion represents a model's lack of fit measure (i.e., inverse GOF), with the remaining terms representing the model's complexity measure. Combined, they estimate the model's generalizability such that the lower the criterion value, the better the model is expected to generalize.

The AIC is derived as an asymptotic (i.e., large sample size) approximation to an information-theoretic distance between two probability distributions, one representing the model under consideration and the other representing the "true" model (i.e., data-generating model). As such, the smaller the AIC value, the closer the model is to the "truth." AICc represents a small sample size version of AIC and is recommended for data with relatively small n with respect to k, say $n/k < 40$ (Burnham and Anderson, 2002). BIC, which is a Bayesian criterion, as the name implies, is derived as an asymptotic expression of the minus two log marginal likelihood, which is described later in this chapter.

The three aforementioned criteria differ from one another in the way model complexity is conceptualized and measured. The complexity term in AIC depends on only the number of parameters, k, whereas both AICc and BIC consider the sample size (n) as well, although in different ways. These two dimensions of a model are not the only ones relevant to complexity, however. Functional form, which refers to the way the parameters are entered in a model's equation, is another dimension of complexity that can also affect the fitting capability of a model (Myung and Pitt, 1997). For example, two models, $y = ax^b + e$ and $y = ax + b + e$, with a normal error e of constant variance, are likely to differ in complexity, despite the fact that they both assume the same number of parameters. For models such as these, the aforementioned criteria are not recommended because they are insensitive to the functional form dimension of complexity. Instead, we recommend the use of the comparison methods, described next, which are sensitive to all three dimensions of complexity.

B. Cross-Validation and Accumulative Prediction Error

Cross-validation (CV; Browne, 2000; Stone, 1974) and the APE (Dawid, 1984; Wagenmakers *et al.*, 2006) are sampling-based methods for estimating generalizability from the data, without relying on explicit, complexity-based penalty terms as in AIC and BIC. This is done by simulating the data collection and prediction steps artificially using the observed data in the experiment.

CV and APE are applied by following a three-step procedure: (1) divide the observed data into two subsamples, the calibration sample, y_{cal}, simulating the "current" observations and the validation sample, y_{val}, simulating "future" observations; (2) fit the model to y_{cal} and obtain the best-fitting parameter values, denoted by $w^*(y_{cal})$; and (3) with the parameter values fixed, the model is fitted to y_{val}. The resulting prediction error is taken as the model's generalizability estimate.

The two comparison methods differ from each other in how the data are divided into calibration and validation samples. In CV, each set of $n-1$ observations in a data set serve as the calibration sample, with the remaining observation treated as the validation sample on which the prediction error is calculated. Generalizability is estimated as the average of n such prediction errors, each calculated according to the aforementioned three-step procedure. This particular method of splitting the data into calibration and validation samples is known as leave-one-out CV in statistics. Other methods of splitting data into two subsamples can also be used. For example, the data can be split into two equal halves or into two subsamples of different sizes. In the remainder of this chapter, CV refers to the leave-one-out CV procedure.

In contrast to CV, in APE the size of the calibration sample increases successively by one observation at a time for each calculation of prediction error. To illustrate, consider a model with k parameters. We would use the first $k+1$ observations as the calibration sample so as to make the model identifiable, and the $(k+2)$-th observation as the validation sample, with the remaining observations not being used. The prediction error for the validation sample is then calculated following the three-step procedure. This process is then repeated by expanding the calibration sample to include the $(k+2)$-th observation, with the validation sample now being the $(k+3)$-th observation, and so on. Generalizability is estimated as the average prediction error over the $(n-k-1)$ validation samples. Time series data are naturally arranged in an ordered list, but for data that have no natural order, APE can be estimated as the mean over all orders (in theory), or over a few randomly selected orders (in practice). Figure 2 illustrates how CV and APE are estimated.

Formally, CV and APE are defined as

$$
\begin{aligned}
\mathrm{CV} &= -\sum_{i=1}^{n} \ln f(y_i | w^*(y_{\neq i})), \\
\mathrm{APE} &= -\sum_{i=k+2}^{n} \ln f(y_i | w^*(y_{1,2,\dots,i-1})).
\end{aligned}
\tag{2}
$$

In the aforementioned equation for CV, $-\ln f(y_i | w^* (y_{\neq i}))$, is the minus log likelihood for the validation sample y_i evaluated at the best-fitting parameter values $w^*(y_{\neq i})$, obtained from the calibration sample $y_{\neq i}$. The subscript signifies

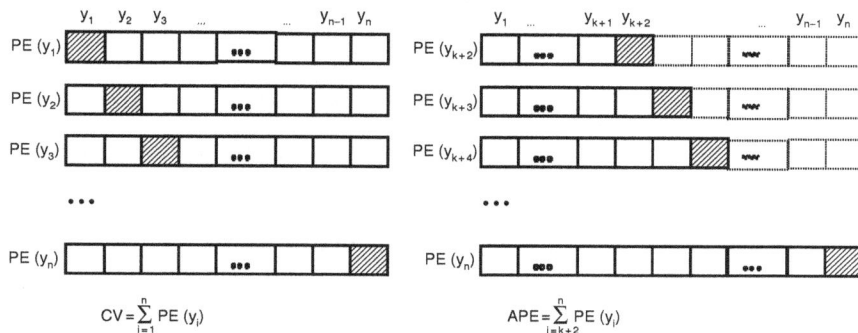

Fig. 2 The difference between the two sampling-based methods of model comparison, cross-validation (CV) and accumulative prediction error (APE), is illustrated. Each chain of boxes represents a data set with each data point represented by a box. The slant-lined box is a validation sample, and plain boxes with the bold outline represent the calibration sample. Plain boxes with the dotted outline in the right panel are not being used as part of the calibration or validation sample. The symbol $PE(y_i)$, $i = 1$, $2, \ldots n$, stands for the prediction error for the ith validation data point. k represents the number of parameters, and n the sample size.

"all observations except for the ith observation." APE is defined similarly. Both methods prescribe that the model with the smallest value of the given criterion should be preferred.

The attractions of CV and APE are the intuitive appeal of the procedures and the computational ease of their implementation. Further, unlike AIC and BIC, both methods consider, albeit implicitly, all three factors that affect model complexity: functional form, number of parameters, and sample size. Accordingly, CV and APE should perform better than AIC and BIC, in particular when comparing models with the same number of parameters. Interestingly, theoretical connections exit between AIC and CV, and BIC and APE. Stone (1977) showed that under certain regularity conditions, model choice under CV is asymptotically equivalent to that under AIC. Likewise, Barron *et al.* (1998) showed that APE is asymptotically equivalent to BIC.

C. Bayesian Model Selection and Stochastic Complexity

BMS (Kass and Raftery, 1995; Wasserman, 2000) and SC (Grunwald *et al.*, 2005; Myung *et al.*, 2006; Rissanen, 1996, 2001) are the current state-of-the-art methods of model comparison. Both methods are rooted in firm theoretical foundations; are nonasymptotic in that they can be used for data of all sample sizes, small or large; and, finally, are sensitive to all dimensions of complexity. The price to pay for this generality is computational cost. Implementation of the methods can be nontrivial because they usually involve evaluating high-dimensional integrals numerically.

BMS and SC are defined as

$$\text{BMS} = -\ln \int f(y|w)\pi(w)dw,$$
$$\text{SC} = -\ln f(y|w^*) + \ln \int f(z|w^*(z))dz. \tag{3}$$

BMS is defined as the minus logarithm of the marginal likelihood, which is nothing but the mean likelihood of the data averaged across parameters and weighted by the parameter prior $\pi(w)$. The first term of SC is the minus log maximized likelihood of the observed data y. It is a lack-of-fit measure, as in AIC. The second term is a complexity measure, with the symbol z denoting the *potential data* that could be observed in an experiment, not the actually observed data. Both methods prescribe that the model that minimizes the given criterion value is to be chosen.

BMS is related to the Bayes factor, the gold standard of model comparison in Bayesian statistics, such that the Bayes factor is a ratio of two marginal likelihoods between a pair of models. BMS does not yield an explicit measure of complexity but complexity is taken into account implicitly through the integral and thus avoids overfitting. To see this, an asymptotic expansion of BMS under Jeffrey's prior for $\pi(w)$ yields the following large sample approximation (Balasubramanian, 1997)

$$\text{BMS} \approx -\ln f(y|w^*) + \frac{k}{2}\ln\left(\frac{n}{2\pi}\right) + \ln \int \sqrt{\det(I(w))}dw, \tag{4}$$

where $I(w)$ is the Fisher information matrix of sample size 1 (e.g., Schervish, 1995). The second and third terms on the right-hand side of the expression represent a complexity measure. It is through the Fisher information in the third term that BMS reflects the functional form dimension of model complexity. For instance, the two models mentioned earlier, $y = ax^b + e$ and $y = ax + b + e$, would have different values of the Fisher information, although they both have the same number of parameters. The Fisher information term is independent of sample size n, with its relative contribution to that of the second term becoming negligible for large n. Under this condition, the aforementioned expression reduces to another asymptotic expression, which is essentially one-half of BIC in Eq. (1).

SC is a formal implementation of the principle of minimum description length that is rooted in algorithmic coding theory in computer science. According to the principle, a model is viewed as a code with which data can be compressed, and the best model is the one that provides maximal compression of the data. The idea behind this principle is that regularities in data necessarily imply the presence of statistical redundancy. The model that is best designed to capture the redundancy will compress the data most efficiently. That is, the data are reexpressed, with the help of the model, in a coded format that provides a shorter description than when the data are expressed in an uncompressed format. The SC criterion value in Eq. (3) represents the overall description length in bits of the maximally compressed data and the model itself, derived for parametric model classes under certain statistical regularity conditions (Rissanen, 2001).

The second (complexity) term of SC deserves special attention because it provides a unique conceptualization of model complexity. In this formulation, complexity is defined as the logarithm of the sum of maximized likelihoods that the model yields collectively for all *potential* data sets that could be observed in an experiment. This formalization captures nicely our intuitive notion of complexity. A model that fits a wide range of data patterns well, actual or hypothetical, should be more complex than a model that fits only a few data patterns well, but does poorly otherwise. A serious drawback of this complexity measure is that it can be highly nontrivial to compute the quantity because it entails numerically integrating the maximized likelihood over the entire data space. This integration in SC is even more difficult than in BMS because the data space is generally of much higher dimension than the parameter space.

Interestingly, a large-sample approximation of SC yields Eq. (4) (Rissanen, 1996), which itself is an approximation of BMS. More specifically, under Jeffrey's prior, SC and BMS become asymptotically equivalent. Obviously, this equivalence does not extend to other priors and does not hold if the sample size is not large enough to justify the asymptotic expression.

V. Model Comparison at Work: Choosing Between Protein Folding Models

This section applies five model comparison methods to discriminating two protein-folding models.

In the modern theory of protein folding, the biochemical processes responsible for the unfolding of helical peptides are of interest to researchers. The Zimm–Bragg theory provides a general framework under which one can quantify the helix–coil transition behavior of polymer chains (Zimm and Bragg, 1959). Scholtz and colleagues (1995) applied the theory "to examine how the α-helix to random coil transition depends on urea molarity for a homologous series of peptides" (p. 185). The theory predicts that the observed mean residue ellipticity q as a function of the length of a peptide chain and the urea molarity is given by

$$q = f_H(g_H - g_C) + g_C. \tag{5}$$

In Eq. (5), f_H is the fractional helicity and g_H and g_C are the mean residue ellipticities for helix and coil, respectively, defined as

$$f_H = \frac{rs}{(s-1)^3} \left(\frac{ns^{n+2} - (n+2)s^{n+1} + (n+2)s - n}{n(1 + [rs/(s-1)^2][s^{n+1} + n - (n+1)s])} \right),$$

$$g_H = H_0 \left(1 - \frac{2.5}{n} \right) + H_U[\text{urea}],$$

$$g_C = C_0 + C_U[\text{urea}], \tag{6}$$

where r is the helix nucleation parameter, s is the propagation parameter, n is the number of amide groups in the peptide, H_0 and C_0 are the ellipticities of the helix and coil, respectively, at $0°$ in the absence of urea, and finally, H_U and C_U are the coefficients that represent the urea dependency of the ellipticities of the helix and coil (Greenfield, 2004; Scholtz *et al.*, 1995).

We consider two mathematical models for urea-induced protein denaturation that determine the urea dependency of the propagation parameter s. One is the linear extrapolation method model (LEM; Pace and Vanderburg, 1979) and the other is called the binding-site model (BIND; Pace, 1986). Each expresses the propagation parameter s in the following form

$$\text{LEM} : \ln s = \ln s_0 - \frac{m[\text{urea}]}{RT},$$

$$\text{BIND} : \ln s = \ln s_0 - d \ln(1 + k(0.9815[\text{urea}] - 0.02978[\text{urea}]^2 + 0.00308[\text{urea}]^3),$$

(7)

where s_0 is the s value for the homopolymer in the absence of urea, m is the change in the Gibbs energy of helix propagation per residue, $R = 1.987 \text{ cal mol}^{-1} \text{ K}^{-1}$, T is the absolute temperature, d is the parameter characterizing the difference in the number of binding sites between the coil and helix forms of a residue, and k is the binding constant for urea.

Both models share four parameters: H_0, C_0, H_U, and C_U. LEM has two parameters of its own (s_0, m), yielding a total of six parameters to be estimated from the data. BIND has three unique parameters ($s_0, d,$ and k). Both models are designed to predict the mean residue ellipticity denoted q in terms of the chain length n and the urea molarity [urea]. The helix nucleation parameter r is assumed to be fixed to the previously determined value of 0.0030 (Scholtz *et al.*, 1991).

Figure 3 shows simulated data (symbols) and best-fit curves for the two models, LEM (in solid lines) and BIND (in dotted lines). Data were generated from LEM for a set of parameter values with normal random noise of zero mean and 1 standard deviation added to the ellipticity prediction in Eq. (5)(see the figure legend for details). Note how closely both models fit the data. By visual inspection, one cannot tell which of the two models generated the data. As a matter of fact, BIND, with one extra parameter than LEM, provides a better fit to the data than LEM (SSE = 12.59 vs. 14.83), even though LEM generated the data. This outcome is an example of the overfitting that can emerge with complex models, as depicted in Fig. 1. To filter out the noise-capturing effect of overly complex models appropriately, thereby putting both models on an equal footing, we need the help of statistical model comparison methods that neutralize complexity differences.

We conducted a model recovery simulation to demonstrate the relative performance of five model comparison methods (AIC, AICc, BIC, CV, and APE) in choosing between the two models. BMS and SC were not included because of the difficulty in computing them for these models. A thousand data sets of 27 observations each were generated from each of the two models, using the same nine points

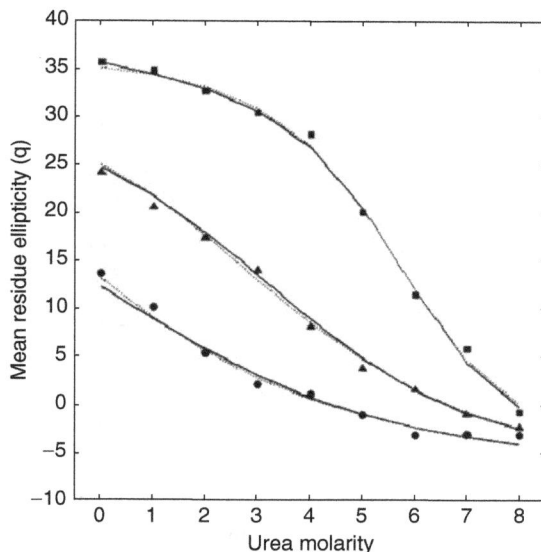

Fig. 3 Best fits of LEM (solid lines) and BIND (dotted lines) models to data generated from LEM using the nine points of urea molarity $(0,1,2,\ldots,8)$ for three different chain lengths of $n = 13$ (●), 20 (▲), and 50 (■). Data fitting was done first by deriving model predictions using Eqs. (5)–(7) based on the parameter values of $H_0 = -44{,}000$, $C_0 = 4400$, $H_U = 320$, $C_U = 340$, $s_0 = 1.34$, and $m = 23.0$ reported in Scholtz *et al.* (1995). See text for further details.

of urea molarity $(0, 1, 2, \ldots, 8)$ for three different chain lengths of $n = 13$, 20, and 50. The parameter values used to generate the simulated data were taken from Scholtz *et al.* (1995) and were as follows: $H_0 = -44{,}000$, $C_0 = 4400$, $H_U = 320$, $C_U = 340$, $s_0 = 1.34$, $m = 23.0$ and temperature $T = 273.15$ for LEM and $H_0 = -42{,}500$, $C_0 = 5090$, $H_U = -620$, $C_U = 280$, $s_0 = 1.39$, $d = 0.52$, $k = 0.14$ for BIND. Normal random errors of zero mean and standard deviation of 1 were added to the ellipticity prediction in Eq. (5).

The five model comparison methods were compared on their ability to recover the model that generated the data. A good method should be able to identify the true model (i.e., the one that generated the data) 100% of the time. Deviations from perfect recovery reveal a bias in the selection method. (The MatLab code that implements the simulations can be obtained from the first author.)

The simulation results are reported in Table I. Values in the cells represent the percentage of samples in which a particular model (e.g., LEM) fitted best data sets generated by one of the models (LEM or BIND). A perfect selection method would yield values of 100% along the diagonal. The top 2×2 matrix shows model recovery performance under ML, a purely GOF measure. It is included as a reference against which to compare performance when measures of model complexity are included in the selection method. How much does model recovery

Table I
Model Recovery Performance of Five Model Comparison Methods

		Data generated from	
Model comparison method	Model fitted	LEM	BIND
ML	LEM	47	4
	BIND	53	96
AIC	LEM	81	16
	BIND	19	84
AICc	LEM	93	32
	BIND	7	68
BIC	LEM	91	28
	BIND	9	72
CV	LEM	77	26
	BIND	23	74
APE	LEM	75	45
	BIND	25	55

Note: The two models, LEM and BIND, are defined in Eq. (7). APE was estimated after randomly ordering the 27 data points of each data set.

improve when the number of parameters, sample size, and functional form are taken into account?

With ML, there is a strong bias toward BIND. The result in the first column of the matrix shows that BIND was chosen more often than the true data-generating model, LEM (53% vs. 47%). This bias is not surprising given that BIND, with one more parameter than LEM, can capture random noise better than LEM. Consequently, BIND tends to be selected more often than LEM under a GOF selection method such as ML, which ignores complexity differences. Results from using AIC show that when the difference in complexity due to the number of parameters is taken into account, the bias is largely corrected (19% vs. 81%), and even more so under AICc and BIC, both of which consider sample size as well (7% vs. 93% and 9% vs. 91%, respectively). When CV and APE were used, which are supposed to be sensitive to all dimensions of complexity, the results show that the bias was also corrected, although the recovery rates under these criteria were about equal to or slightly lower than that under AIC. When the data were generated from BIND (right column of values), the data-generating model was selected more often than the competing model under all selection methods, including ML.

To summarize, the aforementioned simulation results demonstrate the importance of considering model complexity in model comparison. All five model selection methods performed reasonably well by compensating for differences in complexity between models, thus identifying the data-generating model. It is interesting to note that Scholtz and colleagues (1995) evaluated the viability of the same two models plus a third, seven-parameter model, using GOF, and found that all three models provided nearly identical fits to their empirical data. Had they compared the models using one of the selection methods discussed in this chapter, it might have been possible to obtain a more definitive answer.

We conclude this section with the following cautionary note regarding the performance of the five selection methods in Table I: The better model recovery performance of AIC, AICc, and BIC over CV and APE should not be taken as indicative of how the methods will generally perform in other settings (Myung and Pitt, 2004). There are very likely other model comparison situations in which the relative performance of the selection methods reverses.

VI. Conclusions

This chapter began by discussing several issues a modeler should be aware of when evaluating computational models. They include the notion of model complexity, the triangular relationship among GOF, complexity and generalizability, and generalizability as the ultimate yardstick of model comparison. It then introduced several model comparison methods that can be used to determine the "best-generalizing" model among a set of competing models, discussing the advantages and disadvantages of each method. Finally, the chapter demonstrated the application of some of the comparison methods using simulated data for the problem of choosing between biochemical models of protein folding.

Measures of generalizability are not without their own drawbacks, however. One is that they can be applied only to statistical models defined as a parametric family of probability distributions. This restriction leaves one with few options when wanting to compare nonstatistical models, such as verbal models and computer simulation models. Often times, researchers are interested in testing qualitative (e.g., ordinal) relations in data (e.g., condition A < condition B) and comparing models on their ability to predict qualitative patterns of data, but not quantitative ones.

Another limitation of measures of generalizability is that they summarize the potentially intricate relationships between model and data into a single real number. After applying CV or BMS, the results can sometimes raise more questions than answers. For example, what aspects of a model's formulation make it superior to its competitors? How representative is a particular data pattern of a model's performance? If it is typical, the model provides a much more satisfying account of the process than if the pattern is generated by the model using a small range of unusual parameter settings. Answers to these questions also contribute to the evaluation of model quality.

We have begun developing methods to address questions such as these. The most well-developed method thus far is a global qualitative model analysis technique dubbed *parameter space partitioning* (PSP; Pitt *et al.*, 2006, 2007). In PSP, a model's parameter space is partitioned into disjoint regions, each of which corresponds to a qualitatively different data pattern. Among other things, one can use

PSP to identify all data patterns a model can generate by varying its parameter values. With information such as this in hand, one can learn a great deal about the relationship between the model and its behavior, including understanding the reason for the ability or inability of the model to account for empirical data.

In closing, statistical techniques, when applied with discretion, can be useful for identifying sensible models for further consideration, thereby aiding the scientific inference process (Myung and Pitt, 1997). We cannot overemphasize the importance of using nonstatistical criteria such as explanatory adequacy, interpretability, and plausibility of the models under consideration, although they have yet to be formalized in quantitative terms and subsequently incorporated into the model evaluation and comparison methods. Blind reliance on statistical means is a mistake. On this point we agree with Browne and Cudeck (1992), who said "Fit indices [statistical model evaluation criteria] should not be regarded as a measure of usefulness of a model...they should not be used in a mechanical decision process for selecting a model. Model selection has to be a subjective process involving the use of judgement" (p. 253).

Acknowledgments

This work was supported by Research Grant R01-MH57472 from the National Institute of Health to JIM and MAP. This chapter is an updated version of Myung and Pitt (2004). There is some overlap in content.

References

Akaike, H. (1973). Information theory and an extension of the maximum likelihood principle. *In* "Second International Symposium on Information Theory," (B. N. Petrox, and F. Caski, eds.), pp. 267–281. Akademia Kiado, Budapest.

Balasubramanian, V. (1997). Statistical inference, Occam's razor and statistical mechanics on the space of probability distributions. *Neural Comput.* **9,** 349–368.

Barron, A., Rissanen, J., and Yu, B. (1998). The minimum description length principle in coding and modeling. *IEEE Trans. Inform. Theory* **44,** 2743–2760.

Bozdogan, H. (2000). Akaike information criterion and recent developments in information complexity. *J. Math. Psychol.* **44,** 62–91.

Browne, M. W. (2000). Cross-validation methods. *J. Math. Psychol.* **44,** 108–132.

Browne, M. W., and Cudeck, R. C. (1992). Alternative ways of assessing model fit. *Sociol. Methods Res.* **21,** 230–258.

Burnham, L. S., and Anderson, D. R. (2002). "Model Selection and Inference: A Practical Information-Theoretic Approach," 2nd edn. Springer-Verlag, New York.

Dawid, A. P. (1984). Statistical theory: The prequential approach. *J. Roy. Stat. Soc. Ser. A* **147,** 278–292.

Greenfield, N. J. (2004). Analysis of circular dichroism data. *Methods Enzymol.* **383,** 282–317.

Grunwald, P., Myung, I. J., and Pitt, M. A. (2005). "Advances in Minimum Description Length: Theory and Application," MIT Press, Cambridge, MA.

Jacobs, A. M., and Grainger, J. (1994). Models of visual word recognition: Sampling the state of the art. *J. Exp. Psychol. Hum. Percept. Perform* **29,** 1311–1334.

Kass, R. E., and Raftery, A. E. (1995). Bayes factors. *J. Am. Stat. Assoc.* **90,** 773–795.

Muller, P., Sanso, B., and De Iorio, M. (2004). Optimal Bayesian design by inhomogeneous Markov chain simulation. *J. Am. Stat. Assoc.* **99,** 788–798.

Myung, I. J. (2003). Tutorial on maximum likelihood estimation. *J. Math. Psychol.* **44,** 190–204.

Myung, I. J., and Pitt, M. A. (1997). Applying Occam's razor in modeling cognition: A Bayesian approach. *Psychon. Bull. Rev.* **4,** 79–95.

Myung, I. J., and Pitt, M. A. (2004). Model comparison methods. *In* "Numerical Computer Methods, Part D," (L. Brand, and M. L. Johnson, eds.), Vol. 383, pp. 351–366.

Myung, J. I., and Pitt, M. A. (2009). Optimal experimental design for model discrimination. *Psychol. Rev.* **116,** 499–518.

Myung, I. J., Forster, M., and Browne, M. W. (eds.) (2000). Special issue on model selection. *J. Math. Psychol.* **44,** 1–2.

Myung, I. J., Navarro, D. J., and Pitt, M. A. (2006). Model selection by normalized maximum likelihood. *J. Math. Psychol.* **50,** 167–179.

Pace, C. N. (1986). Determination and analysis of urea and guanidine hydrochloride denaturation curves. *Methods Enzymol.* **131,** 266–280.

Pace, C. N., and Vanderburg, K. E. (1979). Determining globular protein stability: Guanidine hydrochloride denaturation of myoglobin. *Biochemistry* **18,** 288–292.

Pitt, M. A., and Myung, I. J. (2002). When a good fit can be bad. *Trends Cogn. Sci.* **6,** 421–425.

Pitt, M. A., Kim, W., Navarro, D. J., and Myung, J. I. (2006). Global model analysis by parameter space partitioning. *Psychol. Rev.* **113,** 57–83.

Pitt, M. A., Myung, I. J., and Altieri, N. (2007). Modeling the word recognition data of Vitevitch and Luce (1998): Is it ARTful? *Psychon. Bull. Rev.* **14,** 442–448.

Rissanen, J. (1996). Fisher information and stochastic complexity. *IEEE Trans. Inform. Theory* **42,** 40–47.

Rissanen, J. (2001). Strong optimality of the normalized ML models as universal codes and information in data. *IEEE Trans. Inform. Theory* **47,** 1712–1717.

Schervish, M. J. (1995). "The Theory of Statistics," Springer-Verlag, New York.

Scholtz, J. M., Qian, H., York, E. J., Stewart, J. M., and Balding, R. L. (1991). Parameters of helix-coil transition theory for alanine-based peptides of varying chain lengths in water. *Biopolymers* **31,** 1463–1470.

Scholtz, J. M., Barrick, D., York, E. J., Stewart, J. M., and Balding, R. L. (1995). Urea unfolding of peptide helices as a model for interpreting protein unfolding. *Proc. Natl. Acad. Sci. USA* **92,** 185–189.

Schwarz, G. (1978). Estimating the dimension of a model. *Ann. Stat.* **6,** 461–464.

Stone, M. (1974). Cross-validatory choice and assessment of statistical predictions. *J. Roy. Stat. Soc. Ser. B* **36,** 111–147.

Stone, M. (1977). An asymptotic equivalence of choice of model by cross-validation and Akaike's criterion. *J. Roy. Stat. Soc. Ser. B* **39,** 44–47.

Sugiura, N. (1978). Further analysis of the data by Akaike's information criterion and the finite corrections. *Commun. Stat. Theory Methods* **A7,** 13–26.

Wagenmakers, E.-J., and Waldorp, L. (2006). Editors' introduction. *J. Math. Psychol.* **50,** 99–100.

Wagenmakers, E.-J., Grunwald, P., and Steyvers, M. (2006). Accumulative prediction error and the selection of time series models. *J. Math. Psychol.* **50,** 149–166.

Wasserman, L. (2000). Bayesian model selection and model averaging. *J. Math. Psychol.* **44,** 92–107.

Zimm, B. H., and Bragg, J. K. (1959). Theory of the phase transition between helix and random coil. *J. Chem. Phys.* **34,** 1963–1974.

CHAPTER 22

Algebraic Models of Biochemical Networks

Reinhard Laubenbacher★ and Abdul Salam Jarrah★,†

★Virginia Bioinformatics Institute at Virginia Tech
Blacksburg, Virginia, USA

†Department of Mathematics and Statistics
American University of Sharjah
Sharjah, United Arab Emirates

Abstract

With the rise of systems biology as an important paradigm in the life sciences and the availability and increasingly good quality of high-throughput molecular data, the role of mathematical models has become central in the understanding of the relationship between structure and function of organisms. This chapter focuses on particular type of models, the so-called algebraic models, which are generalizations of Boolean networks. It provides examples of such models and discusses several available methods to construct such models from high-throughput time course data. One such method, *Polynome*, is discussed in detail.

DOI: 10.1016/B978-0-12-384997-7.00022-4

I. Introduction

The advent of functional genomics has enabled the molecular biosciences to come a long way towards characterizing the molecular constituents of life. Yet, the challenge for biology overall is to understand how organisms function. By discovering how function arises in dynamic interactions, systems biology addresses the missing links between molecules and physiology.

Bruggemann and Westerhoff (2006)

With the rise of systems biology as an important paradigm in the life sciences and the availability of increasingly good quality of high-throughput molecular data, the role of mathematical models has become central in the understanding of the relationship between structure and function of organisms. It is by now well understood that fundamental intracellular processes such as metabolism, signaling, or various stress responses, can be conceptualized as complex dynamic networks of interlinked molecular species that interact in nonlinear ways with each other and with the extracellular environment. Available data, such as transcriptional data obtained from DNA microarrays or high-throughput sequencing machines, complemented by single-cell measurements, are approaching a quality and quantity that makes it feasible to construct detailed mechanistic models of these networks.

There are two fundamental approaches one can take to the construction of mathematical or statistical network models. For each approach the first step is to choose an appropriate model type, which might be, for example, a dynamic Bayesian network (DBN) or a system of ordinary differential equations (ODE). The traditional *bottom-up* approach begins with a "parts list," that is, a list of the molecular species to be included in the model, together with information from the literature about how the species interact. For each model type, there will then likely be some model parameters that are unknown. These will then either be estimated or, if available, fitted to experimental data. The result is typically a detailed mechanistic model of the network. That is, the selected parts are assembled to a larger system.

Systems biology has provided another, so-called *top-down* approach, which attempts to obtain an unbiased view of the underlying network from high-throughput experimental data alone, using statistical or mathematical network inference tools. The advantage of this approach is that the results are not biased by a perception or presumed knowledge of which parts are important and how they work together. The disadvantage is that the resulting model is most likely phenomenological, without a detailed mechanistic structure. It is also the case that at this time the available data sets for this approach are rather small, compared to the number of network nodes, so that the resulting network inference problem is underdetermined. The most beneficial approach, therefore, is to combine both methods by using the available information about the network as prior information for an appropriate network inference method. Thus, network inference becomes an extreme case of parameter estimation, in which no parameters are specified. The development of appropriate network inference methods has become an important research field within computational systems biology.

The goal of this chapter is to illustrate parameter estimation and network inference using a particular type of model, which we will call *algebraic model*. Boolean networks, which are being used increasingly as models for biological networks, represent the simplest instance. First, in Section II, we will provide an introduction to computational systems biology and give some detailed examples of algebraic models. In the following section, we provide an introduction to network inference within several different model paradigms and provide details of several methods. Finally, in the last section, we talk about network inference as it pertains specifically to algebraic models. Few inference methods provide readily available software. We describe one of those, *Polynome*, in enough detail so that the reader can explore the software via the available Web interface.

II. Computational Systems Biology

To understand how molecular networks function, it is imperative that we understand the dynamic interactions between their parts. Gene regulatory networks give an important example. They are commonly represented mathematically as so-called directed graphs, whose nodes are genes and sometimes proteins. There is an arrow from gene A to gene B if A contributes to the regulation of B in some way. The arrow typically indicates whether this contribution is an activation or an inhibition. It is important to understand that such a network provides an abstract representation of gene regulation in the sense that the actual regulation is not direct but could involve a fairly long chain of biochemical reactions, involving mRNA from gene A and/or a protein A codes for.

As an example, we consider the *lac* operon in *Escherichia coli*, one of the earliest and best understood examples of gene regulation. We will use this gene network as a running example for the entire chapter. *E. coli* prefers glucose as a growth medium, and the operon genes allow *E. coli* to metabolize lactose in the absence of glucose. When glucose is present, it is observed that the enzymes involved in lactose metabolism have very low activity, even if intracellular lactose is present. In the absence of glucose, lactose metabolism is induced through expression of the *lac* operon genes. Figure 1 shows a representation of the basic biological mechanisms of the network, commonly referred to as a "cartoon" representation.

While this representation is very intuitive, to understand the biology, a network representation of the system is a first step toward the construction of a dynamic mathematical model. Such a representation is given in Fig. 2.

The dynamic properties of this network are determined by two control mechanisms, one positive, leading to induction of the operon, and another one is negative, leading to the repression of the operon mechanism. The negative control is initiated by glucose, through two modes of action. In the absence of intracellular glucose, the catabolite activator protein CAP forms a complex with another protein, cAMP, which binds to a site upstream of the *lac* promoter region, enhancing transcription of the *lac* genes. Intracellular glucose inhibits cAMP

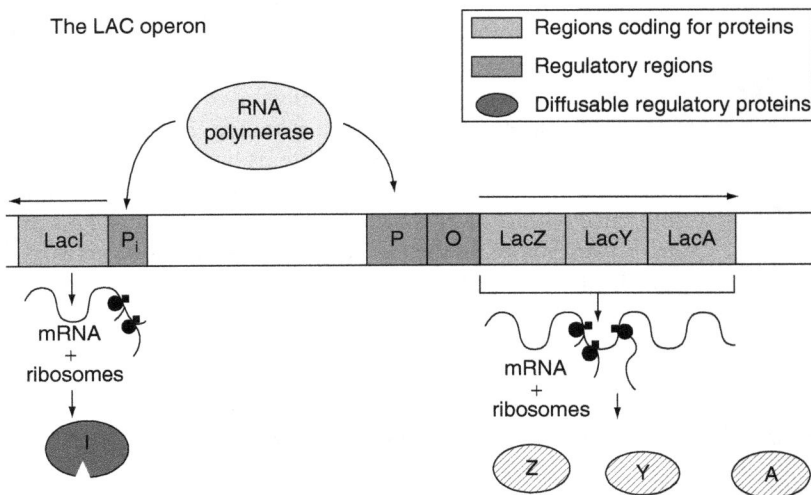

Fig. 1   The *lac* operon (from http://www.uic.edu/classes/bios/bios100/lecturesf04am/lect15.htm).

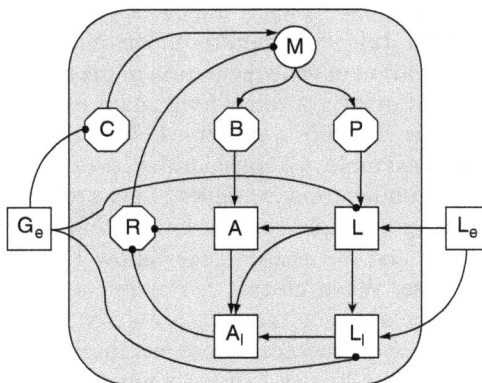

Fig. 2   The *lac* operon network (from Stigler and Veliz-Cuba, 2009).

synthesis and thereby gene transcription. Furthermore, extracellular glucose inhibits the uptake of lactose into the cell via lactose permease, one of the proteins in the *lac* operon. The positive control operates via the action of lactose permease, increasing intracellular lactose, and by disabling of the *lac* repressor protein via increased production of allolactose. To understand how these two feedback loops, one positive and the other negative, work together to determine the dynamic properties of this network it is necessary to construct a mathematical model that captures this synergistic interplay of positive and negative feedback.

Several different mathematical modeling frameworks are available for this purpose. The most commonly used type of model for molecular networks is a system of ODE, one equation for each of the nodes in the network. Each equation describes the rate of change of the concentration of the corresponding molecular species over time, as a function of other network nodes involved in its regulation. As an example, we present a very simplified differential equations model of the *lac* operon taken from Section 5.2 of deBoer (2008). This is an example of a model which was referred to in the introduction as "bottom-up." The model includes only the repressor R, the *lac* operon mRNA M, and allolactose A. The three equations are given below:

$$R = 1/(1 + A^n),$$
$$dM/dt = c_0 + c(1 - R) - \gamma M,$$
$$dA/dt = ML - \delta A - vMA/(h + A).$$

Here, c_0, c, γ, v, δ, h, and L are certain model parameters, n is a fixed positive integer, and the concentrations R, M, and A are functions of time t. The model does not distinguish between intracellular and extracellular lactose, both denoted by L. It is assumed further that the enzyme β-galactosidase is proportional to the operon activity M and is not represented explicitly. The repressor concentration R is represented by a so-called Hill function, which has a sigmoid-shaped graph: the larger the Hill coefficient n, the steeper the shape of the sigmoid function. The constant c_0 represents the baseline activity of the operon transcript M, and the term γM represents degradation. The concentration of allolactose A grows with M, assuming that lactose L is present. Its degradation term is represented by a Michaelis–Menten type enzyme substrate reaction composed of two terms. The various model parameters can be estimated to fit experimental data, using parameter estimation algorithms.

This model is not detailed enough to incorporate the two different feedback loops discussed earlier, but will serve as an illustration of the kind of information a dynamic model can provide. The most important information typically obtained from a model is about the steady states of the network, that is, network states at which all derivatives are equal to 0, so that the system remains in the steady state, once it reaches it. A detailed analysis of this model can be found in deBoer (2008) and also in Laubenbacher and Sturmfels (2009). Such an analysis shows that the model has three steady states, two stable and one unstable. This is what one would expect from the biology. Depending on the lactose concentration, the operon is either "on" or "off," resulting in two stable steady states. For bacteria growing in an environment with an intermediate lactose concentration, it has been shown that the operon can be in either one of these two states, and which one is attained depends on the environmental history a particular bacterium has experienced, a form of hysteresis. This behavior corresponds to the unstable steady state. Thus, the model dynamics agrees with experimental observations.

Another modeling framework that has gained increasing prominence is that of Boolean networks, initially introduced to biology by Kauffman (1969) as a model for genetic control of cell differentiation. Since then, a wide array of such models has been published, as discussed in more detail below. They represent the simplest examples of what is referred to in the title of this chapter as *algebraic models*. They are particularly useful in cases when the quantity or quality of available experimental data is not sufficient to build a meaningful differential equations model. Furthermore, algebraic models are just one step removed from the way a biologist would describe the mechanisms of a molecular network, so that they are quite intuitive and accessible to researchers without mathematical background. In particular, this makes them a useful teaching tool for students in the life sciences (Robeva and Laubenbacher, 2009). To illustrate the concept, we present here a Boolean model of the *lac* operon, taken from Stigler and Veliz-Cuba (2009).

A Boolean network consists of a collection of nodes or variables, each of which can take on two states, commonly represented as ON/OFF or 1/0. Each node has attached to it a Boolean function that describes how the node depends on some or all of the other nodes. Time progresses in discrete steps. For a given state of the network at time $t = 0$, the state at time $t = 1$ is determined by evaluating all the Boolean functions at this state.

Example 1

We provide a simple example to illustrate the concept. Consider a network with four nodes, $x_1, \ldots, x_4$. Let the corresponding Boolean functions be

$$f_1 = x_1 \text{ AND } x_3, f_2 = x_2 \text{ OR}(\text{NOT } x_4), f_3 = x_1, f_4 = x_1 \text{ OR } x_2.$$

Then this Boolean network can be described by the function

$$f = (f_1, \ldots, f_4) : \{0, 1\}^4 \rightarrow \{0, 1\}^4,$$

where

$$f(x_1, \ldots, x_4) = (x_1 \text{ AND } x_3, x_2 \text{ OR } (\text{NOT } x_4), x_1, x_1 \text{ OR } x_2). \tag{1}$$

As a concrete example, $f(0, 1, 1, 0) = (0, 1, 0, 1)$. Here, $\{0, 1\}^4$ represents the set of all binary 4-tuples, of which there are 16. The dependencies among the variables can be represented by the directed graph in Fig. 3, representing the wiring diagram of the network.

The dynamics of the network is represented by another directed graph, the discrete analog of the phase space of a system of differential equations, given in Fig. 4.

The nodes of the graph represent the 16 possible states of the network. There is a directed arrow from state a to state b if $f(a) = b$. The network has three steady states, $(0, 1, 0, 1)$, $(1, 1, 1, 1)$, and $(1, 0, 1, 1)$. (In general, there might be periodic points as well.)

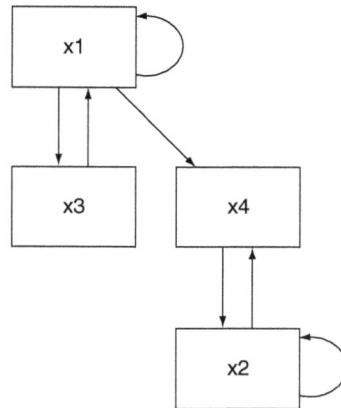

Fig. 3 Wiring diagram for Boolean network in Example 1 (constructed using the software tool DVD; http://dvd.vbi.vt.edu).

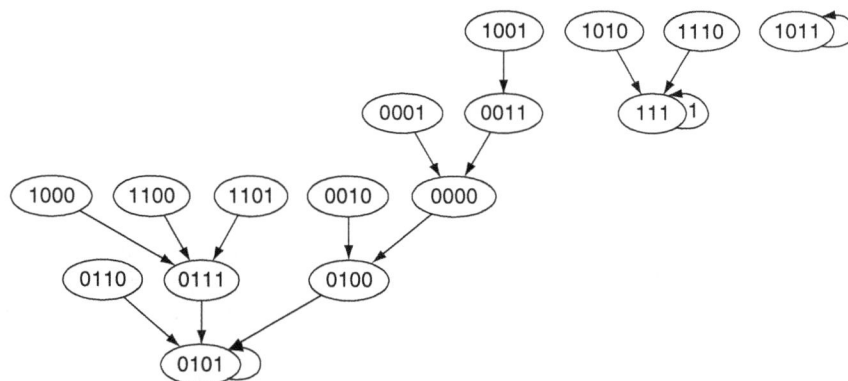

Fig. 4 Dynamics of the Boolean network in Example 1 (constructed using the software tool DVD; http://dvd.vbi.vt.edu).

A Boolean model of the lac operon: the following model is presented in Stigler and Veliz-Cuba (2009). In addition to the three variables M (mRNA for the 3 *lac* genes), R (the repressor protein), and A (allolactose) in the previous ODE model, we need to include these additional variables:

- *Lac* permease (P)
- *β-Galactosidase (B)*
- Catabolite activator protein CAP (C)
- Lactose (L)
- Low concentrations of lactose (L_{low}) and allolactose (A_{low})

The last two variables are needed for the model to be accurate because we need to allow for three, rather than two, possible concentration levels of lactose and allolactose: absent, low, and high. Introducing additional binary variables to account for the three states avoids the introduction of variables with more than two possible states. (Models with multistate variables will be discussed below.) The model depends on two external parameters: a, representing the concentration of external lactose and g, representing the concentration of external glucose. They can both be set to the values 0 and 1, providing four different choices. The interactions between these different molecular species are described in Stigler and Veliz-Cuba (2009) by the following Boolean functions:

$$
\begin{aligned}
4f_M &= (\text{NOT } R) \text{ AND } C, \\
f_P &= M, \\
f_B &= M, \\
f_C &= \text{NOT } g, \\
f_R &= (\text{NOT } A) \text{ AND } (\text{NOT } A_{\text{low}}), \\
f_A &= L \text{ AND } B, \\
f_A \text{ low} &= A \text{ OR } L \text{ OR } L_{\text{low}}, \\
f_L &= (\text{NOT } g) \text{ AND } P \text{ AND } a, \\
f_{L\text{low}} &= (\text{NOT } g) \text{ AND } (L \text{ OR } a).
\end{aligned}
\tag{2}
$$

To understand these Boolean statements and how they assemble to a mathematical network model, consider the first one. It represents a rule that describes how the concentration of *lac* mRNA evolves over time. To make the time dependence explicit, one could write $f_M(t+1) = (\text{NOT } R(t)) \text{ AND } C(t)$ (similarly for the other functions). This is to be interpreted as saying that the *lac* genes are transcribed at time $t+1$ if the repressor protein R is absent at time t and the catabolite activator protein C is present at time t. The interpretation of the other functions is similar. One of the model assumptions is that transcription and translation of a gene happens in one time step, and so does degradation of mRNA and proteins.

Thus, choosing the above ordering of these nine variables, a state of the *lac* operon network is given by a binary 9-tuple, such as $(0, 1, 1, 0, 0, 1, 0, 1, 1)$. The above Boolean functions can be used as the coordinate functions of a time-discrete dynamical system on $\{0, 1\}^9$, that is, a function

$$
f = (f_M, f_P, f_B, f_C, f_R, f_A, f_{A\text{low}}, f_L, f_{L\text{low}}) : \{0, 1\}^9 \to \{0, 1\}^9.
$$

For a given network state $\boldsymbol{a}$ in $\{0, 1\}^9$, the function f is evaluated as

$$
f(a) = (f_M(a), f_P(a), f_B(a), f_C(a), f_R(a), f_A(a), f_{A\text{low}}(a), f_L(a), f_{L\text{low}}(a))
$$

to give the next network state. For the exemplary network state above, and the parameter setting $a = 1$, $g = 0$, we obtain

$$
f(0, 1, 1, 0, 0, 1, 0, 1, 1) = (0, 0, 0, 1, 0, 1, 1, 1, 1).
$$

It is shown in Stigler and Veliz-Cuba (2009) that this model captures all essential features of the *lac* operon, demonstrated through several other published ODE models. One of these is its bistability, which has already been discussed in the simple model above. It is shown that for each of the four possible parameter settings, the models attain a steady state, corresponding to the switch-like nature of the network. It is in principle possible to compute the entire phase space of the model, using a software package such as DVD (http://dvd.vbi.vt.edu).

We have discussed this particular model in some detail, for three reasons. It represents a model of an interesting network that continues to be the object of ongoing research, yet is simple enough to fit the space constraints of this chapter. For the reader unfamiliar with algebraic models of this type, it provides a detailed realistic example to explore. Finally, we will use this particular model subsequently to demonstrate a particular network inference method.

Boolean network models of biological systems are the most common type of discrete model used, including gene regulatory networks such as the cell cycle in mammalian cells (Faure *et al.*, 2006), in budding yeast (Li *et al.*, 2004) and fusion yeast (Davidich and Bornholdt, 2007), and the metabolic networks in *E. coli* (Samal and Jain, 2008) and in *Saccharomyces cerevisiae* (Herrgard *et al.*, 2006).

Also, Boolean network models of signaling networks have recently been used to provide insights into different mechanisms such as the molecular neurotransmitter signaling pathway (Gupta *et al.*, 2007), the T cell receptor signaling pathways (Saez-Rodriguez *et al.*, 2007), the signaling network for the long-term survival of cytotoxic T lymphocytes in humans (Zhang *et al.*, 2008), and the abscisic acid signaling pathway (Li *et al.*, 2006a).

A more general type of discrete model, the so-called *logical model*, was introduced by the geneticist (Renee Thomas and D'Ari, 1989) for the study of gene regulatory networks. Since then they have been developed further, with published models of the cell-fate determination in *Arabidopsis thaliana* (Espinosa-Soto *et al.*, 2004), the root hair regulatory network (Mendoza and Alvarez-Buylla, 2000), the Hh signaling pathway (Gonzalez *et al.*, 2008), the gap gene network in Drosophila (Sanchez and Thieffry, 2001), and the differentiation process in T helper cells (Mendoza, 2006), to name a few.

A logical model consists of a collection of variables $x_1, \ldots, x_n$, representing molecular species such as mRNA, where variable x_j takes values in a finite set S_j. The number of elements in S_j corresponds to the number of different concentrations of x_j that trigger different modes of action. For instance, when transcribed at a low level, a gene might perform a different role than when it is expressed at a high level, resulting in three states: absent, low, and high. The variables are linked in a graph, the *dependency graph* or *wiring diagram* of the model, as in the Boolean case, by directed edges, indicating a regulatory action of the source of the edge on the target. Each edge is equipped with a sign +/- (indicating activation or inhibition) and a weight. The weight indicates the level of transcription of the source node required to activate the regulatory action. The state transitions of each node are given by a table involving a list of so-called logical parameters. The dynamics of the system is encoded by the *state space graph*, again as in the Boolean case, whose edges indicate state transitions of the system. An additional structural feature of this model type is that the variables can be updated sequentially, rather than in parallel, as in the previously discussed Boolean model. This feature allows the inclusion of different time scales and stochastic features that lead to asynchronous updating of variables. The choice of different update orders at any given update step can result in a different transition. To illustrate the logical model framework, we briefly describe the T cell differentiation model in Mendoza (2006).

As they differentiated from a common precursor called T0, two distinct functional subsets T1 and T2 of T helper cells were identified in the late 1980s. T1 cells secrete IFNγ, which promotes more T1 differentiation while inhibiting that of T2. On the other hand, T2 cells secret IL-4, a cytokine which promotes T2 differentiation and inhibits that of T1. The most general logical model is presented in Mendoza (2006), where the gene regulatory network of T1/T2 differentiation is synthesized from published experimental data. The multilevel network includes 19 genes and four stimuli and interactions at the inter- and intracellular levels.

Some of the nodes are assumed to be Boolean while a few others are multistates (*Low, Medium,* or *High*). Based on the value at the source of an edge being above or below some threshold, that edge is considered active or not, in the sense, that it will contribute to changing the value at the sink of that edge. When there is more than one incoming edge, the combinations of different active incoming edges and their thresholds are assembled into a logical function.

It is worth mentioning briefly that the algebraic model framework is well suited for the study of the important relationship between the structure and the dynamics of a biological network. In systems biology, the work of Uri Alon has drawn a lot of attention to this topic, summarized in his book (Alon, 2006). An important focus of Alon's work has been the effect of so-called network motifs, such as feed-forward loops, on dynamics. Another topic of study in systems biology has been the logical structure of gene regulatory and other networks. In the context of Boolean networks, significant work has been devoted to identifying features of Boolean functions that make them particularly suitable for the modeling of gene regulation and metabolism. As an example, we present here a summary of the work on so-called *nested canalyzing functions* (NCFs), a particular class of Boolean functions that appear frequently in systems biology models.

Biological systems in general and biochemical networks in particular are robust against noise and perturbations and, at the same time, can evolve and adapt to different environments (Balleza *et al.*, 2008). Therefore, a realistic model of any such system must possess these properties, and hence the update functions for the network nodes cannot be arbitrary. Different classes of Boolean functions have been suggested as biologically relevant models of regulatory mechanisms: biologically meaningful functions (Raeymaekers, 2002), postfunctions (Shmulevich *et al.*, 2003b), and chain functions (Gat-Viks and Shamir, 2003). However, the class that has received the most attention is that of NCFs, introduced by Kauffman *et al.* (2004) for gene regulatory networks. We first give some precise definitions.

Definition 1

A Boolean function $g(x_1, \ldots, x_n)$: $\{0, 1\}^n \to \{0, 1\}^n$ is called *canalyzing* in the variable x_i with the input value a_i and output value b_i if x_i appears in g and

$$g(x_1, \ldots, x_{i-1}, a_i, x_{i+1}, \ldots, x_n) = b_i$$

for all inputs of all variables x_j and $j \neq i$.

The definition is reminiscent of the concept of "canalization" introduced by the geneticist (Waddington, 1942) to represent the ability of a genotype to produce the same phenotype regardless of environmental variability.

Definition 2

Let f be a Boolean function in n variables.

Let σ be a permutation of the set $\{1, \ldots, n\}$. The function f is a *NCF in the variable order* $x_{\sigma(1)}, \ldots, x_{\sigma(n)}$ with canalizing input values $a_1, \ldots, a_n$ and canalized output values $b_1, \ldots, b_n$, if

$$
\begin{aligned}
f(x_1, \ldots, x_n) &= b_1 \text{ if } x_{\sigma(1)} = a_1, \\
&= b_2 \text{ if } x_{\sigma(1)} \neq a_1 \text{ and } x_{\sigma(2)} = a_2, \\
&\qquad \cdots \\
&= b_{n-1} \text{ if } x_{\sigma(1)} \neq a_1, \ldots, x_{\sigma(n-1)} \neq a_{n-1} \text{ and } x_{\sigma(n)} = a_n, \\
&= b_n \text{ if } x_{\sigma(1)} \neq a_1, \ldots, x_{\sigma(n-1)} \neq a_{n-1} \text{ and } x_{\sigma(n)} \neq a_n.
\end{aligned}
$$

The function f is nested canalyzing if it is nested canalyzing for some variable ordering σ.

As an example, the function $f(x,y,z) = x$ AND (NOT y) AND z is nested canalyzing in the variable order x,y,z with canalizing values 0,1,0 and canalized values 0,0,0, respectively. However, the function $f(x,y,z,w) = x$ AND y AND (z OR w) is not nested canalyzing because, if $x = 1$ and $y = 1$, then the value of the function is not constant for any input values for either z or w.

One important characteristic of NCFs is that they exhibit a stabilizing effect on the dynamics of a system. That is, small perturbations of an initial state should not grow in time and must eventually end up in the same attractor of the initial state. The stability is typically measured using so-called Derrida plots which monitor the Hamming distance between a random initial state and its perturbed state as both evolve over time. If the Hamming distance decreases over time, the system is considered stable. The slope of the Derrida curve is used as a numerical measure of stability. Roughly speaking, the phase space of a stable system has few components and the limit cycle of each component is short.

Example 2

Consider the Boolean networks.

$$f = (x_4, x_3 \text{ XOR } x_4, x_2 \text{ XOR } x_4, x_1 \text{ XOR } x_2 \text{ XOR } x_3) : \{0,1\}^4 \to \{0,1\}^4,$$
$$g = (x_4, x_3 \text{ AND } x_4, x_2 \text{ AND } x_4, x_1 \text{ AND } x_2 \text{ AND } x_3) : \{0,1\}^4 \to \{0,1\}^4.$$

Notice that f and g have the same dependency graph and that g is a Boolean network constructed with NCFs while f is not. It is clear that the phase space of g in Fig. 5 has fewer components and much shorter limit cycles compared to the phase space of f in Fig. 6, and therefore g should be considered more stable than f.

In Kauffman *et al.* (2004), the authors studied the dynamics of nested canalyzing Boolean networks over a variety of dependency graphs. That is, for a given random graph on n nodes, where the in-degree of each node is chosen at random

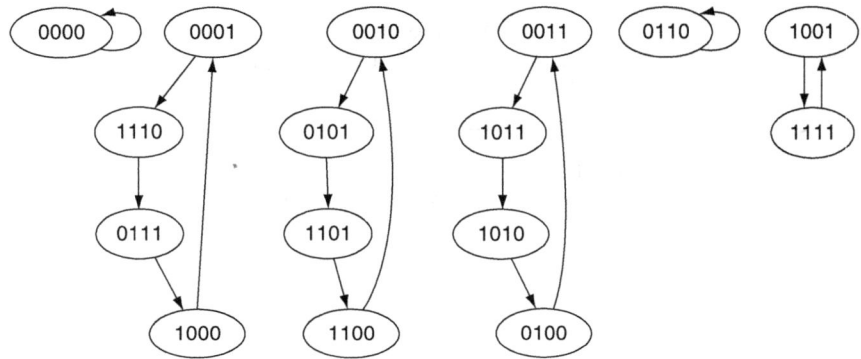

Fig. 5   Phase space of the network f in Example 2.

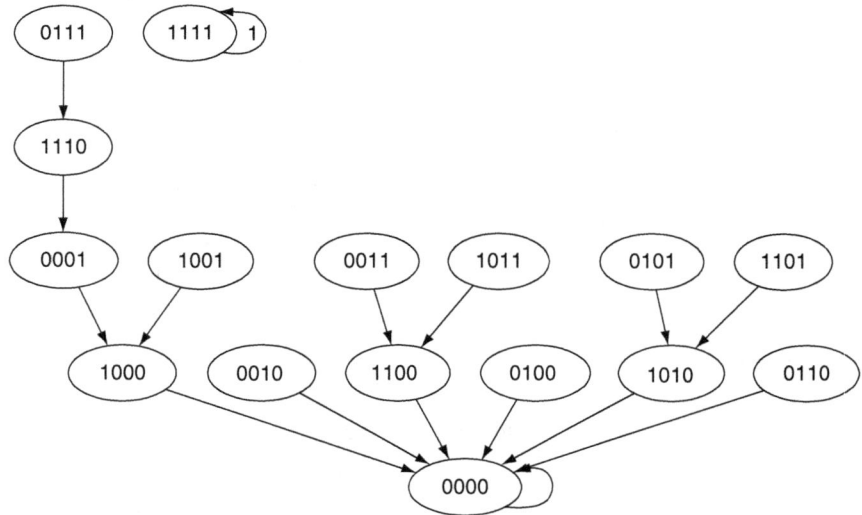

Fig. 6   Phase space of network g in Example 2.

between 0 and k, where $k < n + 1$, a NCF is assigned to each node in terms of the in-degree variables of that node. The dynamics of these networks were then analyzed and the stability measured using Derrida plots. It is shown that nested canalyzing networks are remarkably stable regardless of the in-degree distribution and that the stability increases as the average number of inputs of each node increases.

An extensive analysis of available biological data on gene regulations (about 150 genes) showed that 139 of them are regulated by canalyzing functions (Harris *et al.*, 2002; Nikolayewaa *et al.*, 2007). In Kauffman *et al.* (2004) and Nikolayewaa *et al.* (2007) it was shown that 133 of the 139 are, in fact, nested canalyzing.

Most published molecular networks are given in the form of a wiring diagram, or dependency graph, constructed from experiments and prior published knowledge. However, for most of the molecular species in the network, little knowledge, if any, could be deduced about their regulatory mechanisms, for instance in the gene transcription networks in yeast (Herrgard *et al.*, 2006) and *E. coli* (Barrett *et al.*, 2005). Each one of these networks contains more than 1000 genes. Kauffman *et al.* (2003) investigated the effect of the topology of a subnetwork of the yeast transcriptional network where many of the transcriptional rules are not known. They generated ensembles of different models where all models have the same dependency graph. Their heuristic results imply that the dynamics of those models that used only NCFs were far more stable than the randomly generated models. Since it is already established that the yeast transcriptional network is stable, this suggests that the unknown interaction rules are very likely NCFs.

In Balleza *et al.* (2008), the whole transcriptional network of yeast, which has 3459 genes as well as the transcriptional networks of *E. coli* (1481 genes) and *B. subtillis* (840 genes) have been analyzed in a similar fashion, with similar findings. These heuristic and statistical results show that the class of NCFs is very important in systems biology. We showed in Jarrah *et al.* (2007a,b) that this class is identical to the class of so-called unate cascade Boolean functions, which has been studied extensively in engineering and computer science. It was shown in Butler *et al.* (2005) that this class produces the binary decision diagrams with shortest average path length.

In this section, we have shown that algebraic models, in particular Boolean network models, play an important role in systems biology, as models for a variety of molecular networks. They also are very useful in studying more theoretical questions, such as design principles for molecular networks. In Section III, we will show how to construct such models from experimental data, in the top-down fashion discussed earlier.

III. Network Inference

In 2006 the first "Dialogue on Reverse-Engineering Assessment and Methods (DREAM)" workshop was held, supported in part by the NIH Roadmap Initiative. The rationale for this workshop is captured in the words of the organizers:

"The endless complexities of biological systems are orchestrated by intricate networks comprising thousands of interacting molecular species, including DNA, RNA, proteins, and smaller molecules. The goal of systems biology is to map these networks in ways that provide both fundamental understanding and new possibilities for therapy. However, although modern tools can provide rich data sets by simultaneously monitoring thousands of different types of molecules, discerning the nature of the underlying network from these observations—reverse engineering—remains a daunting challenge."

Traditionally, models of molecular regulatory systems in cells have been created *bottom-up*, where the model is constructed piece by piece by adding new components and characterizing their interactions with other molecules in the model. This process requires that the molecular interactions have been well characterized, usually through quantitative numerical values for kinetic parameters. Note that the construction of such models is biased toward molecular components that have already been associated with the phenomenon. Still, modeling can be of great help in this bottom-up process, by revealing whether the current knowledge about the system is able to replicate its *in vivo* behavior. This modeling approach is well suited to complement experimental approaches in biochemistry and molecular biology, since models thus created can serve to validate the mechanisms determined *in vitro* by attempting to simulate the behaviors of intact cells. While this approach has been dominant in cellular modeling, it does not scale very well to genome-wide studies, since it requires that proteins be purified and studied in isolation. This is not a practical endeavor due to its large scale, but especially because a large number of proteins act on small molecules that are not available in purified form, as would be required for *in vitro* studies.

With the completion of the human genome sequence and the accumulation of other fully sequenced genomes, research is moving away from the molecular biology paradigm to an approach characterized by large-scale molecular profiling and *in vivo* experiments (or, if not truly *in vivo*, at least, *in situ*, where experiments are carried out with intact cells). Technologies such as transcript profiling with microarrays, protein profiling with 2D gels and mass spectrometry, and metabolite profiling with chromatography and mass spectrometry, produce measurements that are large-scale characterizations of the state of the biological material probed. Other new large-scale technologies are also able to uncover groups of interacting molecules, delineating interaction networks. All these experimental methods are data rich, and it has been recognized (e.g., Brenner, 1997; Kell, 2004; Loomis and Sternberg, 1995) that modeling is necessary to transform such data into knowledge. A new modeling approach is needed for large-scale profiling experiments. Such a *top-down* approach starts with little knowledge about the system, capturing at first only a coarse-grained image of the system with only a few variables. Then, through iterations of simulation and experiment, the number of variables in the model is increased. At each iteration, novel experiments will be suggested by simulations of the model that provide data to improve it further, leading to a higher resolution in terms of mechanisms. While the processes of bottom-up and top-down modeling are distinct, both have as an objective the identification of molecular mechanisms responsible for cell behavior. Their main difference is that the construction of top-down models is biased by the data of the large-scale profiles, while bottom-up models are biased by the preexisting knowledge of particular molecules and mechanisms.

While top-down modeling makes use of genome-wide profiling data, it is conceptually very different from other genome-wide data analysis approaches. Top-down modeling needs data produced by experiments suitable for the approach. One should not expect that a random combination of arbitrary molecular snapshots would be of much use for the top-down modeling process. Sometimes they may serve some purpose (e.g., variable selection) but overall, top-down modeling requires perturbation experiments carried out with appropriate controls. In the face of modern experimental methods, the development of an effective top-down modeling strategy is crucial. Furthermore, we believe that a combination of top-down and bottom-up approaches will eventually have to be used. An example of a first step in this direction is the apoptosis model in Bentele *et al.* (2004).

A variety of different network inference methods have been proposed in recent years, using different modeling frameworks, requiring different types and quantities of input data, and providing varying amounts of information about the system to be modeled. There are fundamentally three pieces of information one wants to know about a molecular network: (i) its wiring diagram, that is, the causal dependencies among the network nodes, for example, gene activation or repression; (ii) the "logical" structure of the interactions between the nodes, for example, multiplicative or additive interaction of transcription factors; and (iii) the dynamics of the network, for example, the number of steady states. At one end of the model spectrum are statistical models that capture correlations among network variables. These models might be called *high level* (Ideker and Lauffenburger, 2003). The output of methods at the other end of the spectrum is a system of ODE which models network dynamics, provides a wiring diagram of variable dependencies as well as a mechanistic description of node interactions. In-between is a range of model types such as information-theory-based models, difference equations, Boolean networks, and multistate discrete models.

The literature on top-down modeling, or network inference, has grown considerably in the last few years, and we provide here a brief, but essentially incomplete review. The majority of new methods that have appeared utilize statistical tools. At the high-level end of the spectrum recent work has focused on the inference of relevance networks, first introduced in Butte *et al.* (2000). Using pairwise correlations of gene expression profiles and appropriate threshold choices, an undirected network of connections is inferred. Partial correlations are considered in de la Fuente *et al.* (2004) for the same purpose. In Rice *et al.*, (2005) conditional correlations using gene perturbations are used to assign directionality and functionality to the edges in the network and to reduce the number of indirect connections. Another modification is the use of time-delayed correlations in Li *et al.* (2006b) to improve inference. Using mutual information instead of correlation, together with information-theoretic tools, the ARACNE algorithm in Margolin *et al.* (2006) reports an improvement in eliminating indirect edges in the network.

Probably the largest part of the recent literature on statistical models is focused on the use of DBNs, a type of statistical model that gives as output a directed graph depicting causal dependency relations between the variables. These dependency relations are computed in terms of a time evolution of joint probability distributions on the variables, viewed as discrete random variables for reverse-engineering. Originally proposed in Friedman *et al.* (2000), the use of causal Bayesian network methods has evolved to focus on DBNs, to avoid the limitation of not capturing feedback loops in the network. These can be thought of as a sequence in time of Bayesian networks that can represent feedback loops over time, despite the fact that each of the Bayesian networks is an acyclic directed graph. A variety of DBN algorithms and software packages have been published, see, for example, Beal *et al.* (2005), Dojer *et al.* (2006), Friedman, (2004), Nariai *et al.* (2005), Pournara and Wernisch (2004), Yu *et al.* (2004), and Zou and Conzen (2005). Probably the largest challenge to DBN methods, as to all other methods, is the typically small sample sizes available for microarray data. One proposed way to meet this challenge is by bootstrapping, that is, the generation of synthetic data with a similar distribution as the experimental data; see, for example, Pe'er *et al.* (2001).

These methods all provide as output a wiring diagram, in which each edge represents a statistically significant relationship between the two network nodes it connects. Other approaches that result in a wiring diagram includes (Tringe *et al.*, 2004), building on prior work by Wagner (2001, 2004). If time course data are available it is useful to obtain a dynamic model of the network. There have been some recent results in this direction using Boolean network models, first introduced in Kauffman (1969). Each of the methods in Mehra *et al.* (2004), Kim *et al.* (2007), and Martin *et al.* (2007) either modifies or provides an alternative to the original Boolean inference methods in Liang *et al.* (1998), Akutsu *et al.* (1999, 2000a,b), and Ideker *et al.* (2000). Moving farther toward mechanistic models, an interesting network inference method resulting in a Markov chain model can be found in Ernst *et al.* (2007). Finally, methods using systems of differential equations include (Andrec *et al.*, 2005; Bansal *et al.*, 2006, 2007; Chang *et al.*, 2005; Deng *et al.*, 2005; Gadkar *et al.*, 2005; Gardner *et al.*, 2003; Kim *et al.*, 2007; Yeung *et al.*, 2002). Reverse-engineering methods have also been developed for the S-system formalism of Savageau (1991), which is a special case of ODE-based modeling, such as Kimura *et al.* (2005), Marino and Voit, (2006), and Thomas *et al.*, (2004).

Validating reverse-engineering methods and comparing their performance is very difficult at this time. Typically, each method is validated using data from simulated networks of different sizes and more or less realistic architecture. This is a crucial first step for any method, since it is important to measure its accuracy against a known network. The most common organism used for validation is yeast, with a wide collection of published data sets (very few of which are time course data). One of the key problems in comparing different methods is that they typically have different data requirements, ranging from time course data to steady-state measurements.

Some methods require very specific perturbation experiments, whereas others need only a collection of single measurements. Some methods use continuous data, others, for instance most Bayesian network methods, require discretized data. Some methods take into account information such as binding site motifs. At present, there is no agreed-upon suite of test networks that might provide a more objective comparison. Nonetheless, comparisons are beginning to be made (Bansal *et al.*, 2007; Kremling *et al.*, 2004; Werhli *et al.*, 2006), but even there it is hard to interpret the results correctly. Most methods show some success with specialized data sets and particular organisms, but in all the cases many theoretical as well as practical challenges remain. The stated goal of the DREAM effort mentioned in the beginning is to develop precisely such a set of benchmark data sets that can be used as a guide for method developers and a way to carry out more systematic comparisons.

We briefly describe two such methods here, one using parameter estimation for systems of differential equations as the principal tool, the other using statistical methods. A comparison of different methods, including these two, was done in Camacho *et al.* (2007). In the Section IV, we describe in detail a method that has as output either a wiring diagram or a Boolean network, using the interpolation of data by a Boolean network as the main tool. First, we describe a reverse-engineering method that uses multiple regression, proposed by Gardner *et al.* (2003), which is similar to de la Fuente and Mendes (2002) and Yeung *et al.* (2002). The method uses linear regression and requires data that are obtained by perturbing the variables of the network around a reference steady state. A crucial assumption of the method is that molecular networks are sparse, that is, each variable is regulated only by a few others. This method assumes no more than three regulatory inputs per node. The network is then recovered using multiple regressions of the data. It estimates the coefficients in the Jacobian matrix of a generic system of linear differential equations representing the rates of change of the different variables. (Recall that the Jacobian matrix of a linear system of differential equations has as entry in position (i, j) the coefficient of the variable x_i in the equation for variable x_j.) The assumption that the wiring diagram of the network is sparse translates into the assumption that only a few of the entries of this matrix are different from 0.

The second method, originally published in Hartemink *et al.* (2002), uses the framework of *DBNs*. The data required for this reverse-engineering method are time courses representing temporal responses to perturbations of the system from a steady state. The method has been implemented in the software package BANJO, described in Bernard and Hartemink (2005).

IV. Reverse-Engineering of Discrete Models: An Example

A. Boolean Networks: Deterministic and Stochastic

As mentioned in the introduction, Boolean networks were first proposed as models for gene regulatory networks by Kauffman (1969), where a gene is considered either expressed (1) or not expressed (0), and the state of a gene is determined by the states of its immediate neighbors in the network. One interpretation of the different attractors (steady states, in particular) could be as the different phenotypes into which a cell will differentiate, starting from an arbitrary initialization. As there is evidence that gene regulation as well as metabolism could be stochastic, Boolean networks, which are deterministic, have been generalized to account for stochasticity. *Boolean networks with perturbations* (BNps) and *Probabilistic Boolean networks* (PBNs) have been developed for modeling noisy gene regulatory networks, see, for example, Akutsu *et al.* (2000a,b), Shmulevich *et al.* (2002, p. 225), and Yu *et al.* (2004).

Definition 3

A *Boolean network with perturbations* (BNp) is a Boolean network where, at each time step, the state of a randomly chosen node is flipped with some probability p. That is, if the current state of the network is $x = (x_1, \ldots, x_n)$, the next state y is determined as follows. A unit vector $e = (e_1, \ldots, e_n)$ is chosen at random where e_i is zero for all but one coordinate j for which $e_j = 1$. Then $y = f(x) + e$ with probability p and $y = f(x)$ with probability $1 - p$. In particular, in a BNp, one and only one node could be perturbed with probability p at each time step. That is, the phase space of a BNp is a label-directed graph, where the vertices are all possible states of the network, and the label of an edge (x,y) is the probability that $y = f(x) + e$ for some unit vector e. It is clear that the out-degree of each node x is n, where the edge (x,y) is labeled with $1 - p$ if $y = f(x)$, and p if $y = f(x) + e$ for some unit vector e.

Definition 4

A PBN is a Boolean network in which each node in the network could possibly have more than one update function, in the form of a family of Boolean functions, together with a probability distribution. When it is time to update the state of a node, a function is chosen at random from the family of that node and is used to decide its new state; namely, for each node x_i in the network, let $\{f, g, \ldots\}$ be the set of local functions, where the probability of choosing f is p, that of choosing g is q, etc. and $p + q + \ldots = 1$. Then the phase space of the network is a labeled directed graph with a directed edge (x,y), if for all i, we have $y_i = f(x)$, for some local update function f for node i. The label on the edge is the product of the probabilities of each of the coordinates, where the probability of a coordinate is the sum of all probabilities p such that $y_i = f(x)$.

Example 3

Consider the network on three nodes in Fig. 7 below, and, let $f_{11} = x_1$ OR x_2, $f_{12} = x_2$ AND x_3, with probabilities 0.7, and 0.3, respectively. Suppose node 2 has only one local function, $f_2 = x_2$ OR x_3, and suppose node 3 has two functions,

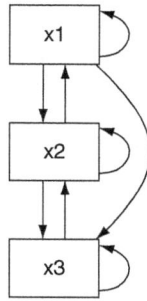

Fig. 7   The wiring diagram of the probabilistic Boolean network in Example 3.

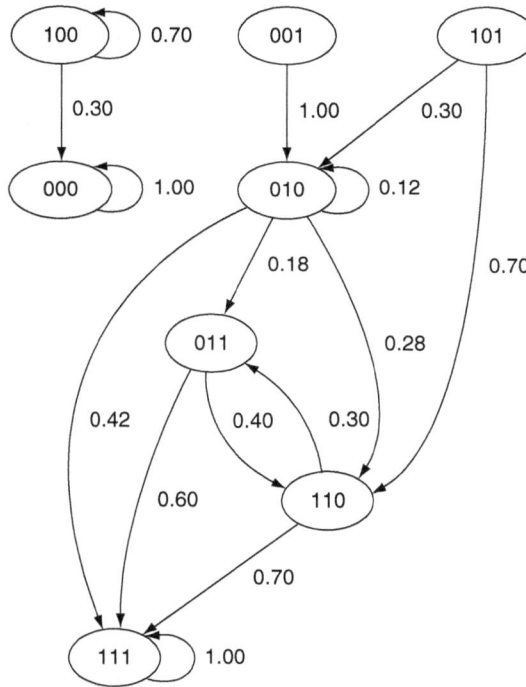

Fig. 8   The phase space of the probabilistic Boolean network in Example 3. Notice that there are three fixed points: 000 and 111 with probability 1, while the state 010 is a fixed point with probability 0.12. Furthermore, the two states 011 and 110 form a limit cycle with probability 0.12.

$f_{31} = x_1$ AND x_2 with probability 0.4, and $f_{32} = x_2$ with probability 0.6. The phase space of this network is depicted in Fig. 8.

Another way to introduce stochasticity into a deterministic model is by updating the network nodes asynchronously, and, at each time step, the order at which they are updated is chosen at random. These networks are clearly biologically relevant as they capture the order at which different events and processes take place and hence could change the outcome. In Chaves *et al.* (2005), the authors present a stochastic update-order model of the *segment polarity network* in Drosophila. The model captures aspects of this biological process that the original model in Albert and Othmer (2003) did not account for. One way to accomplish update-stochastic simulation of deterministic Boolean networks is to represent them as PBNs, where each node has two local update functions; the original one and the identity function, with appropriate choice of probabilities. This approach is implemented in the parameter estimation package *Polynome*, which we discuss below. Next, we briefly review some of the known network inference methods for BNp and PBNs.

It goes without saying that, to infer a Boolean network from experimental data sets, one has to start by assuming that each node in the network can only be in one of two states at any given time. In particular, the data used for inferring the network must also be binary, and hence the experimental data first have to be discretized into two qualitative states. There are several different methods for discretizing continuous data, see, for example, Dimitrova *et al.* (2010). However, for the Boolean case, all methods come down to deciding the proper threshold that should be used to decide if a given molecular species is present (1) or absent (0). For DNA microarray data, this may be done by, for example, choosing a fold change above which a gene is considered upregulated compared to a control value. Or it may be done by inspection of a time course.

B. Inferring Boolean Networks

As we described above, Boolean networks have emerged as a powerful framework for modeling and simulating gene regulatory networks. Therefore, it is natural to infer these networks from experimental data, and different methods have been proposed. Liang *et al.* (1998) pioneered this approach with the algorithm REVEAL, where information-theoretic principles are applied to reduce the search space. In Akutsu *et al.* (1999), the authors proposed a simple algorithm for identifying a Boolean network from a data set assuming that the in-degree of each node is relatively small. They discussed requirement on the data for such networks to exist. Recently, Martin *et al.* (2007) presented an algorithm for identifying all activation–inhibition Boolean networks (here each edge is either an activator or a strong inhibitor) from a given data set. Here, too, a small upper bound for the in-degree of each node is assumed. The dynamics of the identified Boolean networks are then used to shed light on the biological system.

Using the Boolean framework, Ideker *et al.* (2002) presented a method that identifies a minimal wiring diagram of the network from time course data. The network is minimal in the sense that each edge in the network is essential to reproduce the time course data.

C. Inferring Stochastic Boolean Networks

Deterministic Boolean network models seem inadequate for modeling some biological systems, as uncertainty is a prominent feature of many known systems. This is due either to hidden variables, intrinsic or extrinsic noise, or measurement noise. Different algorithms have been proposed for inferring stochastic Boolean networks within the framework of PBNs (Ching *et al.*, 2005; Shmulevich *et al.*, 2002, 2003a) or BNps (Akutsu *et al.*, 2000a,b; Yu *et al.*, 2004).

Shmulevich and his collaborators developed an inference method that identifies a set of local functions of a given node using either time course data or a set of steady states (Shmulevich *et al.*, 2002). For each node x_i in the network, a set of local Boolean functions $X_i = \{f, g, \ldots\}$ is assigned with probabilities $P_i = \{p, q, \ldots\}$. The set X_i and P_i correspond to the highest coefficients of determination (CoD) of the node x_i relative to randomly chosen subsets of variables that could be possible input sets for node x_i. On the other hand, Yu *et al.* (2004) presented an algorithm for inferring a Boolean network with perturbations from steady-state data. Based on certain assumptions about the size of the basin of attraction for each observed state and lengths of transients, a matrix describing the transition between different attractors is computed.

Some of the algorithms mentioned above have been implemented as either C++ code, such as the algorithm of Akutsu *et al.*, or within other software packages, such as the algorithms of Shmulevich *et al.* (2002), which require the commercial software package Matlab. Furthermore, experimental data need to be booleanized ahead of time before applying these algorithms. In Section IV.D, we describe the software package *Polynome* (Dimitrova *et al.*, 2009) that incorporates several different algorithms using tools from computational algebra and algebraic geometry. The software is capable of inferring wiring diagrams as well as deterministic ones and PBNs. Furthermore, the software can be used to simulate and explore the dynamics of the inferred network.

D. Polynome: Parameter Estimation for Boolean Models of Biological Networks

As described earlier, the goal of parameter estimation is to use experimental time course data to determine missing information in the description of a Boolean network model for the biological system from which the data were generated. This can be done with either partial or no prior information about the wiring diagram and dynamics of the system. *Polynome* will infer either a static wiring diagram alone or a dynamical model, with both deterministic and stochastic model choices. The software is available via a Web interface at http://polymath.vbi.vt. edu/polynome. Figure 9 shows a screenshot of the interface of *Polynome*.

The main idea behind the algebraic approach underlying *Polynome* is that any Boolean function can be written uniquely as a polynomial where the exponent of any variable is either 0 or 1 (hence the name). The dictionary is constructed from the basic correspondence:

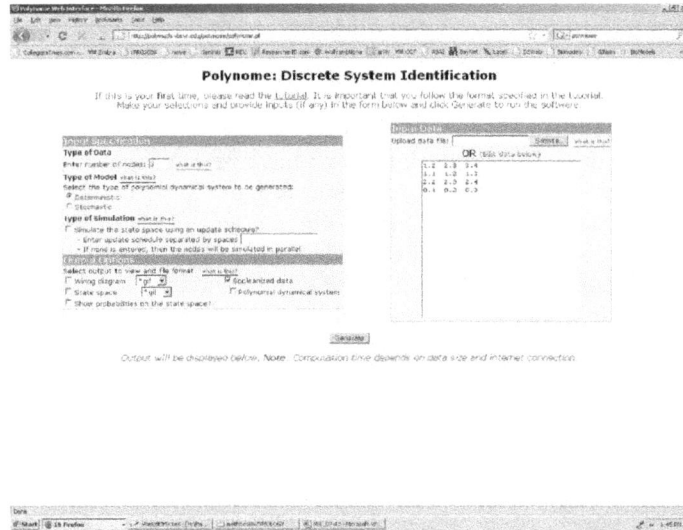

Fig. 9   A screenshot of POLYNOME at http://polymath.vbi.vt.edu/polynome.

$$x \text{ AND } y = xy, x \text{ OR } y = x + y + xy, \text{ NOT } x = x + 1.$$

Therefore, any Boolean network f can be written as a polynomial dynamical system

$$f(x_1, \ldots, x_n) = f(x) = (f_1(x), \ldots, f_n(x)) : \{0,1\}^n \to \{0,1\}^n,$$

where the polynomial function f_i is used to compute the next state of node i in the network. For example, the Boolean network in Eq. (1) has the following polynomial form:

$$f(x_1, \ldots, x_4) = (x_1 x_3, 1 + x_4 + x_2 x_4, x_1, x_1 + x_2 + x_1 x_2).$$

The Boolean network from Eq. (2) has the polynomial form:

$$f(x_1, \ldots, x_9) = ((1 + x_4)x_5, x_1, x_1, 1, (1 + x_6)(1 + x_7), x_3 x_8, x_6 + x_8 + x_9 + x_6 x_8$$
$$+ x_6 x_9 + x_8 x_9, x_2(x_8 + 1)),$$

where $g = 1$ and $a = 1$, and $(x_1, \ldots, x_9) = (M, P, B, C, R, A, A_{\text{low}}, L, L_{\text{low}})$.

Studying Boolean networks as polynomial, dynamical systems has many advantages, primarily that within the polynomial framework, a wide variety of algorithmic and theoretical tools from computer algebra and algebraic geometry can be applied. The remainder of the section will describe the parameter estimation algorithms implemented in *Polynome* and illustrate the software using the *lac* operon example described in detail above. This example is also used in Dimitrova *et al.* (2009) to validate the software.

Input: The user can input two kinds of information. The first kind consists of one or more time courses of experimental data. While several different data types are possible, we will focus here on DNA microarray data, for the sake of simplicity. If the network model to be estimated has nodes $x_1, \ldots, x_n$, then a data point consists of a vector $(a_1, \ldots, a_n)$ of measurements, one for each gene in the network. Since the model is Boolean, the first step is to discretize the input data into two states, 0 and 1. The user can provide a threshold to discriminate between the two states. As default algorithm, *Polynome* uses the algorithm in Dimitrova *et al.* (2010), which incorporates an information-theoretic criterion and is designed to preserve dynamic features of the continuous time series and to be robust to noise in the data.

The second type of input consists of biological information. This can take the form of known edges in the wiring diagram or known Boolean functions for some of the network nodes. Recall that an edge from node x_i to node x_j in the wiring diagram indicates that node x_i exerts causal influence on the regulation of node x_j. In other words, the variable x_i appears in the local update function f_j for x_j. In the absence of this type of information, the problem is equal to what is often called *reverse-engineering* of the network, that is, network inference using exclusively system-level data for the network.

To understand the algorithms, it is necessary to clarify the relationship between the input data and the networks produced by the software. The software produces networks that fit the given experimental data in the following sense. Suppose that the input consists of a time course $s_1, \ldots, s_t$, which each $s_i \in \{0, 1\}^n$. Then, we say that a Boolean network f fits the given time course if $f(s_i) = s_{i+1}$ for all i.

Software output: There are five types of output the user can request. We briefly describe these and the algorithms used to obtain them.

A static wiring diagram of the networ: It, is a directed graph with vertices in the nodes of the network and edges indicating causal regulatory relationships. Since there is generally more than one such diagram for the given information (unless a complete wiring diagram is already provided as input), the user can request either a diagram with weights on the edges, indicating the probability of a particular edge being present, or a collection of topscoring diagrams. The algorithm used for this purpose has been published in Jarrah *et al.* (2007b). It computes all possible wiring diagrams of Boolean networks that fit the given data. The algorithm outputs only *minimal wiring diagrams*. Here, a wiring diagram is minimal if it is not possible to remove an edge and still obtain a wiring diagram of a model that fits the given data. In this sense, the output is similar to that in Ideker *et al.* (2002). However, the approach in Jarrah *et al.* (2009) is to encode the family of all wiring diagrams as an algebraic object, a certain monomial ideal, which has the advantage that ALL minimal wiring diagrams can be calculated, in contrast to a diagram produced by a heuristic search.

A deterministic dynamic model in the form of a Boolean network: This model fits the given data exactly and satisfies the constraints imposed by the input on the wiring diagram and the Boolean functions, using the algorithm described in Laubenbacher and Stigler (2004). This is done by first computing the set of all

Boolean networks that fit the given data and the constraints. Using tools from computational algebra, this can be done by describing the entire set of models, that is, the entire parameter space, in a way similar to the description of the set of all solutions to a system of nonhomogeneous linear equations. As in the case of nonhomogeneous linear equations, if f and g are two Boolean networks that fit the given data set, that is, $f(s_t) = s_{t+1} = g(s_t)$, then $(f-g)(s_t) = 0$ for all t. Hence, all networks that fit the data can be found by finding one particular model f and adding to it any Boolean network g such $g(s_t) = 0$ for all t. The space of all such g can be described by a type of basis that is similar to a vector space basis for the null space of a homogeneous system of linear equations.

A PBN that fits the given data: That is, the network has a family of update functions for each node, together with a probability distribution on the functions, as described earlier. This network has the property that for any choice of function at any update the network fits exactly the given data. The network is constructed using an algorithm that builds on the one described in Dimitrova *et al.* (2007).

A Boolean network that optimizes data fit and model complexity. In contrast to the previous two choices of output, this network does not necessarily fit the given data exactly but gives a network that is optimized with respect to both data fit and model complexity. This is a good model choice if the data are assumed to contain significant noise, since it reduces the tendency to overfit the data with a complex model. This option uses an evolutionary algorithm (Vera-Licona *et al.*, 2009) that is computationally intensive and is only feasible for small networks at this time.

A deterministic model that is simulated stochastically: This model is constructed by estimating Boolean functions that fit the data exactly, when simulated with synchronous update. But the network is then simulated using a stochastic update order. That is, the simulated network may not fit the given data exactly, but will have the same steady states as the synchronous model. The stochastic update order is obtained by representing the deterministic system as a PBN by adding the identify function to each node. At a given update, if the identity function is chosen, this represents a delay of the corresponding variable. Choosing an appropriate probability distribution, one can in this way simulate a stochastic sequential update order. The resulting phase space is a complete graph, with transition probabilities on the edges. This approach is also computationally very intensive, so this option is only feasible for small networks.

E. Example: Inferring the *lac* Operon

In this section, we demonstrate some of the features of *Polynome* by applying it to data generated from the Boolean *lac* operon model in Eq. (1) above. That is, we take the approach that this model represents the biological system we want to construct a model of, based on "experimental" data generated directly from the model. This approach has the advantage that it is straightforward to evaluate the performance of the estimation algorithm in this case.

Table I
A Set of Time Courses from the *lac* Operon Model, Generated Using Eq. (2)

All are *high*	*R* is *high*	*M* is *high*	*L* and L_{low} are *high*
1 1 1 1 1 1 1 1 1	0 0 0 0 1 0 0 0 0	1 0 0 0 0 0 0 0 0	0 0 0 0 0 0 0 1 1
0 1 1 1 0 1 1 1 1	0 0 0 1 1 0 0 0 1	0 1 1 1 1 0 0 0 1	0 0 0 1 1 0 1 0 1
1 0 0 1 0 1 1 1 1	0 0 0 1 1 0 1 0 1	0 0 0 1 1 0 1 1 1	0 0 0 1 0 0 1 0 1
1 1 1 1 0 0 1 0 1	0 0 0 1 0 0 1 0 1	0 0 0 1 0 0 1 0 1	1 0 0 1 0 0 1 0 1
1 1 1 1 0 0 1 1 1	1 0 0 1 0 0 1 0 1	1 0 0 1 0 0 1 0 1	1 1 1 1 0 0 1 0 1
1 1 1 1 0 1 1 1 1	1 1 1 1 0 0 1 0 1	1 1 1 1 0 0 1 0 1	1 1 1 1 0 0 1 1 1
1 1 1 1 0 1 1 1 1	1 1 1 1 0 0 1 1 1	1 1 1 1 0 0 1 1 1	1 1 1 1 0 1 1 1 1
	1 1 1 1 0 1 1 1 1	1 1 1 1 0 1 1 1 1	1 1 1 1 0 1 1 1 1
	1 1 1 1 0 1 1 1 1	1 1 1 1 0 1 1 1 1	

The data in Table I include four time courses: all molecules are *high*, only *R* is *high*, only *M* is *high*, and only *L* and L_{low} are *high*.

Table II shows a PBN (in polynomial form) inferred from the data in Table I, using *Polynome*. Here, for each node, a list of update functions and their probabilities is given. The bold functions are the ones with probability higher than 0.1. (This threshold is provided by the user.) Notice that the true function $(1 + x_4)x_5$ for x_1 is in the list of inferred functions for x_1 with the second highest probability, the same as for the true function x_3x_8 for x_6. The inferred functions with highest probability for nodes 2–4 are the correct ones. In the case of node 7, the only inferred polynomial x_9 is clearly not the "true" function, which is $x_6 + x_8 + x_9$. However, it is important to remember that we are using four time courses involving only 26 states from the phase space of 512 states. Parameter estimation methods cannot recover information about the network that is missing from the data.

The phase space of this system has 512 states and many edges connecting them and so a visual inspection of the phase space graph is not possible. *Polynome* in this case provides a summary of the dynamics that includes the number of components, the number of limit cycles of each possible length as well as the stability of these cycles. Here, the stability of a cycle is the probability of its remaining in that cycle. Table III shows that our inferred system in Table II has only one component which has the steady state (111101111), and its stability is 0.33. Note that the original Boolean *lac* operon model in Eq. (2) has only one component and the same steady state as the inferred model. The wiring diagram of the inferred network is shown in Fig. 10.

V. Discussion

Mathematical models have become important tools in the repertoire of systems biologists who want to understand the structure and dynamics of complex biological networks. Our focus has been on algebraic models and methods for constructing them from experimental time course data. For differential equations-based models, the standard approach to dealing with unknown model parameters

Table II
A Probabilistic Boolean Model Inferred from the Data in Table I Using *Polynome*

f1 = {
$x5*x8+x1*x5+x5+x2*x6+x2*x8+x6+x1*x7+x8+x1+1$ #.0222222
$x5+x7*x8+x1*x9+x8+x1+1$ #.0222222
$x5+x4+x1*x9+x9+x1+1$ #.133333
$x5+x1*x4+x4+x9+x1+1$ #.0666667
$x4*x5+x4$ #.2
$x5+x7*x8+x1*x7+x8+x1+1$ #.0666667
$x5*x7+x7$ #.244444
$x5*x9+x4$ #.0444444
$x2*x5+x5*x8+x5+x1*x4+x1*x2+x2+x1*x8+x8+x1+1$ #.0222222
$x5+x4*x8+x1*x4+x8+x1+1$ #.0666667
$x5*x9+x7*x8+x8+x9$ #.0222222
$x5+x4+x1*x7+x9+x1+1$ #.0444444
$x5*x6+x5*x8+x5*x9+x2*x6+x2*x8+x6+x8+x9$ #.0222222
$x5*x6+x5*x8+x5+x1*x4+x1*x6+x6+x1*x8+x8+x1+1$ #.0222222
}
f2 = $x1$
f3 = $x1$
f4 = 1
f5 = $x7 + 1$
f6 = {
$x2*x5+x1*x5+x2*x6+x1*x2+x6+x2+x1*x8$ #.0222222
$x5*x6+x5*x8+x3*x6+x4+x6+x8+x9$ #.0222222
$x3*x6+x1*x6+x1*x8$ #.0444444
$x5*x8+x5*x7+x1*x5+x5+x3*x6+x6+x1*x7+x8+x7+x1+1$ #.0444444
$x2*x8$ #.377778
$x3*x8$ #.355556
$x2*x6+x1*x6+x1*x8$ #.0888889
$x3*x6+x6+x1*x8+x3*x7+x1*x3$ #.0222222
$x5*x6+x5*x8+x5*x7+x5+x2*x6+x6+x1*x7+x8+x7+x1+1$ #.0222222
}
f7 = {
$x5*x8+x1*x5+x2*x6+x2*x8+x4+x6+x8$ #.0222222
$x5*x6+x5*x8+x4+x1*x6+x6+x1*x8+x8$ #.0222222
$x4*x8+x4+x8$ #.111111
$x5*x7+x5+x4+x1*x7+x7+x1+1$ #.0444444
$x9$ #.666667
$x4*x5+x5+x1*x4+x1+1$ #.0666667
$x4+x7*x8+x8$ #.0444444
$x2*x5+x5*x8+x4+x1*x2+x2+x1*x8+x8$ #.0222222
}
f8 = {
$x2$ #.511111
$x3$ #.488889
}
f9 = 1

Table III
The Analysis of the Phase Space of the Probabilistic Boolean Network Using Local Functions with Probability More than 0.1 (The Bold Functions in Table II)

Analysis of the phase space [m = 2, n = 9]
Number of components 1
Number of fixed points 1
Fixed point, component size, stability
(1 1 1 1 0 1 1 1 1), 512, 0.33

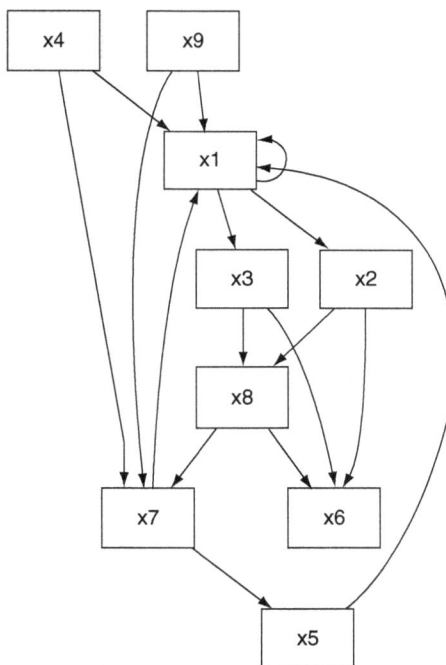

Fig. 10 The wiring diagram of the inferred network in Table II.

is to estimate them by fitting the model to experimental data. The same approach is taken here to the estimate of unknown model parameters in an algebraic model. We have described several approaches to this problem in the literature. For one of these approaches, implemented in the software package *Polynome*, we have provided a detailed guide to how the software can be used with experimental data via a Web interface.

The extreme case of parameter estimation is the lack of any prior biological information, so the network is to be inferred from experimental data alone. This is typically referred to as *reverse-engineering* or *network inference*. Many different approaches to this "top-down" approach to modeling have been published. There are still significant challenges ahead, arising primarily due to the lack of sufficiently large, appropriately collected time course data sets. Nonetheless, the field has advanced to the point where there are some first successes. It is our hope that this chapter will encourage the reader to try this approach to data modeling, using algebraic models, or others based on differential equations or statistics.

References

Akutsu, T., Miyano, S., *et al.* (1999). Identification of genetic networks from a small number of gene expression patterns under the Boolean network model. *Pac. Symp. Biocomput.* 17–28.

Akutsu, T., Miyano, S., *et al.* (2000a). Algorithms for inferring qualitative models of biological networks. *Pac. Symp. Biocomput.* 293–304.

Akutsu, T., Miyano, S., *et al.* (2000b). Inferring qualitative relations in genetic networks and metabolic pathways. *Bioinformatics* **16**(8), 727–734.

Albert, R., and Othmer, H. G. (2003). The topology of the regulatory interactions predicts the expression pattern of the segment polarity genes in *Drosophila melanogaster*. *J. Theor. Biol.* **223**(1), 1–18.

Alon, U. (2006). "An Introduction to Systems Biology: Design Principles of Biological Circuits." CRC Press, Boca Raton, FL.

Andrec, M., Kholodenko, B. N., *et al.* (2005). Inference of signaling and gene regulatory networks by steady-state perturbation experiments: Structure and accuracy. *J. Theor. Biol.* **232**(3), 427.

Balleza, E., Alvarez-Buylla, E. R., *et al.* (2008). Critical dynamics in gene regulatory networks: Examples from four kingdoms. *PLoS One* **3**(6), e2456.

Bansal, M., Gatta, G. D., *et al.* (2006). Inference of gene regulatory networks and compound mode of action from time course gene expression profiles. *Bioinformatics* **22**(7), 815–822.

Bansal, M., Belcastro, V., *et al.* (2007). How to infer gene networks from expression profiles. *Mol. Syst. Biol.* **3**, 78, doi:10.1038/msb4100120.

Barrett, C. B., Herring, C. D., *et al.* (2005). The global transcriptional regulatory network for metabolism in *Escherichia coli* exhibits few dominant functional states. *Proc. Natl. Acad. Sci. USA* **102**(52), 19103–19108.

Beal, M. J., Falciani, F., *et al.* (2005). A Bayesian approach to reconstructing genetic regulatory networks with hidden factors. *Bioinformatics* **21**(3), 349–356.

Bentele, M., Lavrik, I., *et al.* (2004). Mathematical modeling reveals threshold mechanism in CD95-induced apoptosis. *J. Cell Biol.* **166**(6), 839–851.

Bernard, A., and Hartemink, A. (2005). Informative structure priors: Joint learning of dynamic regulatory networks from multiple types of data. *Pac. Symp. Biocompu. Conf. Proc.* 459–470.

Brenner, S. (1997). Loose ends. *Curr. Biol.* 73.

Bruggemann, F. J., and Westerhoff, H. (2006). The nature of systems biology. *Trends Microbiol.* **15**(1), 45–50.

Butler, J. T., Tsutomu, S., *et al.* (2005). Average path length of binary decision diagrams. *IEEE Trans. Comput.* **54**(9), 1041–1053.

Butte, A. J., Tamayo, P., *et al.* (2000). Discovering functional relationships between RNA expression and chemotherapeutic susceptibility using relevance networks. *PNAS* **97**(22), 12182–12186.

Camacho, D., vera-Licona, P., *et al.* (2007). Comparison of reverse-engineering method using an in silico network. *Ann. NY Acad. Sci.* **1115**, 73–89.

Chang, W.-C., Li, C.-W., *et al.* (2005). Quantitative inference of dynamic regulatory pathways via microarray data. *BMC Bioinform.* **6**(1), 44.

Chaves, M., Albert, R., *et al.* (2005). Robustness and fragility of Boolean models for genetic regulatory networks. *J. Theor. Biol.* **235**, 431–449.

Ching, W. K., Ng, M. M., Fung, E. S., and Akutsu, T. (2005). On construction of stochastic genetic networks based on gene expression sequences. *Int. J. Neural Syst.* **15**(4), 297–310.

Davidich, M. I., and Bornholdt, S. (2007). Boolean network model predicts cell cycle sequence of fission yeast. *PLoS One* **3**(2), e1672.

deBoer, R. J. (2008). Theoretical biology. Undergraduate Course at Utrecht University, available at http://theory.bio.uu.nl/rdb/books/.

de la Fuente, A., and Mendes, P. (2002). Quantifying gene networks with regulatory strengths. *Mol. Biol. Rep.* **29**(1–2), 73–77.

de la Fuente, A., Bing, N., *et al.* (2004). Discovery of meaningful associations in genomic data using partial correlation coefficients. *Bioinformatics* **20**(18), 3565–3574.

Deng, X., Geng, H., *et al.* (2005). EXAMINE: A computational approach to reconstructing gene regulatory networks. *Biosystems* **81**(2), 125.

Dimitrova, A., Jarrah, A., *et al.*, (2007). A Groebner-Fan-based method for biochemical network modeling, Proceedings of the International Symposium on Symbolic and Algebraic Computation, Assoc. Comp. Mach. Waterloo, CA.

Dimitrova, E., Garcia-Puente, L., *et al.* (2009). Parameter estimation for Boolean models of biological networks. *Theor. Comp. Sci.* (in press).

Dimitrova, E., Vera-Licona, P., *et al.* (2010). Data discretization for reverse-engineering: A comparative study. *J. Comp. Biol.* **17** (6), 853–868.

Dojer, N., Gambin, A., *et al.* (2006). Applying dynamic Bayesian networks to perturbed gene expression data. *BMC Bioinform.* **7**(1), 249.

Ernst, J., Vainas, O., *et al.* (2007). Reconstructing dynamic regulatory maps. *Mol. Syst. Biol.* **3**, 74.

Espinosa-Soto, C., Padilla-Longoria, P., *et al.* (2004). A gene regulatory network model for cell-fate determination during *Arabidopsis thaliana* flower development that is robust and recovers experimental gene expression profiles. *Plant Cell* **16**(11), 1923–1939.

Faure, A., Naldi, A., *et al.* (2006). Dynamical analysis of a generic Boolean model for the control of the mammalian cell cycle. *Bioinformatics* **22**(14), 124–131.

Friedman, N. (2004). Inferring cellular networks using probabilistic graphical models. *Science* **303** (5659), 799–805.

Friedman, N., Linial, M., *et al.* (2000). Using Bayesian networks to analyze expression data. *J. Comput. Biol.* **7**(3–4), 601–620.

Gadkar, K., Gunawan, R., *et al.* (2005). Iterative approach to model identification of biological networks. *BMC Bioinform.* **6**(1), 155.

Gardner, T. S., di Bernardo, D., *et al.* (2003). Inferring genetic networks and identifying compound mode of action via expression profiling. *Science* **301**(5629), 102–105.

Gat-Viks, I., and Shamir, R. (2003). Chain functions and scoring functions in genetic networks. *Bioinformatics* **19**, 108–117.

Gonzalez, A., Chaouiya, C., *et al.* (2008). Logical modelling of the role of the Hh pathway in the patterning of the *Drosophila* wing disc. *Bioinformatics* **24**(234–240), 16.

Gupta, S., Bisht, S. S., *et al.* (2007). Boolean network analysis of a neurotransmitter signaling pathway. *J. Theor. Biol.* **244**(3), 463–469.

Harris, S. E., Sawhill, B. K., *et al.* (2002). A model of transcriptional regulatory networks based on biases in the observed regulation rules. *Complex Syst.* **7**(4), 23–40.

Hartemink, A., Gifford, D., *et al.* (2002). Bayesian methods for elucidating genetic regulatory networks. *IEEE Intel. Syst.* **17**, 37–43.

Herrgard, M. J., Lee, B. S., *et al.* (2006). Integrated analysis of regulatory and metabolic networks reveals novel regulatory mechanisms in *Saccharomyces cerevisiae*. *Genome Res.* **16**, 627–635.

Ideker, T. E., and Lauffenburger, D. (2003). Building with a scaffold: Emerging strategies for high- to low-level cellular modeling. *Trends Biotechnol.* **21**(6), 256–262.

Ideker, T. E., Thorsson, V., *et al.* (2000). Discovery of regulatory interactions through perturbation: Inference and experimental design. *Pac. Symp. Biocomput.* **5**, 305–316.

Ideker, T. E., Ozier, O., *et al.* (2002). Discovering regulatory and signalling circuits in molecular interaction networks. *Bioinformatics* **18**(Suppl. 1), S233–S240.

Jarrah, A., Raposa, B., *et al.* (2007a). Nested canalyzing, unate cascade, and polynomial functions. *Physica D* **233**(2), 167–174.

Jarrah, A., Laubenbacher, R., Stigler, B., and Stillman, M. (2007b). Reverse-engineering polynomial dynamical systems. *Adv. Appl. Math.* **39,** 477–489.

Kauffman, S. A. (1969). Metabolic stability and epigenesis in randomly constructed genetic nets. *J. Theor. Biol.* **22**(3), 437–467.

Kauffman, S. A., Peterson, C., *et al.* (2003). Random Boolean network models and the yeast transcriptional network. *Proc. Natl. Acad. Sci. USA* **100**(25), 14796–14799.

Kauffman, S. A., Peterson, C., *et al.* (2004). Genetic networks with canalyzing Boolean rules are always stable. *Proc. Natl. Acad. Sci. USA* **101**(49), 17102–17107.

Kell, D. B. (2004). Metabolomics and systems biology: Making sense of the soup. *Curr. Opin. Microbiol.* **7**(3), 296–307.

Kim, J., Bates, D., *et al.* (2007). Least-squares methods for identifying biochemical regulatory networks from noisy measurements. *BMC Bioinform.* **8**(1), 8.

Kimura, S., Ide, K., *et al.* (2005). Inference of S-system models of genetic networks using a cooperative coevolutionary algorithm. *Bioinformatics* **21**(7), 1154–1163.

Kremling, A., Fischer, S., *et al.* (2004). A benchmark for methods in reverse engineering and model discrimination: Problem formulation and solutions. *Genome Res* **14**(9), 1773–1785.

Laubenbacher, R., and Stigler, B. (2004). A computational algebra approach to the reverse engineering of gene regulatory networks. *J. Theor. Biol.* **229**, 523–537.

Laubenbacher, R., and Sturmfels, B. (2009). Computer algebra in systems biology. *Am. Math. Mon.* **116**, 882–891.

Liang, S., Fuhrman, S., *et al.* (1998). REVEAL, a general reverse engineering algorithm for inference of genetic network architectures. *Pac. Symp Biocomput.* **3,** 18–29.

Li, F., Long, T., *et al.* (2004). The yeast cell-cycle network is robustly designed. *Proc. Natl. Acad. Sci. USA* **101**(14), 4781–4786.

Li, S., Assman, S. M., *et al.* (2006a). Predicting essential components of signal transduction networks: A dynamic model of guard cell abscisic acid signaling. *PLoS Biol.* **4**(10), e312.

Li, X., Rao, S., *et al.* (2006b). Discovery of time-delayed gene regulatory networks based on temporal gene expression profiling. *BMC Bioinform.* **7**(1), 26.

Loomis, W. F., and Sternberg, P. W. (1995). Genetic networks. *Science* **269**(5224), 649.

Margolin, A. A., Nemenman, I., *et al.* (2006). ARACNE: An algorithm for the reconstruction of gene regulatory networks in a mammalian cellular context. *BMC Bioinform.* **7**(Suppl. 1), S7.

Marino, S., and Voit, E. (2006). An automated procedure for the extraction of metabolic network information from time series data. *J. Bioinform. Comp. Biol.* **4**(3), 665–691.

Martin, S., Zhang, Z., *et al.* (2007). Boolean dynamics of genetic regulatory networks inferred from microarray time series data. *Bioinformatics* **23**(7), 866–874.

Mehra, S., Hu, W.-S., *et al.* (2004). A Boolean algorithm for reconstructing the structure of regulatory networks. *Metab. Eng.* **6**(4), 326.

Mendoza, L. (2006). A network model for the control of the differentiation process in Th cells. *Biosystems* **84,** 101–114.

Mendoza, L., and Alvarez-Buylla, E. R. (2000). Genetic regulation of root hair development in Arabidopsis thaliana: A network model. *J. Theor. Biol.* **204**, 311–326.

Nariai, N., Tamada, Y., *et al.* (2005). Estimating gene regulatory networks and protein–protein interactions of *Saccharomyces cerevisiae* from multiple genome-wide data. *Bioinformatics* **21**(Suppl. 2), ii206–ii212.

Nikolayewaa, S., Friedela, M., *et al.* (2007). Boolean networks with biologically relevant rules show ordered behavior. *Biosystems* **90**(1), 40–47.

Pe'er, D., Regev, A., *et al.* (2001). Inferring subnetworks from perturbed expression profiles. *Bioinformatics* **17**(Suppl. 1), S215–S224.

Pournara, I., and Wernisch, L. (2004). Reconstruction of gene networks using Bayesian learning and manipulation experiments. *Bioinformatics* **20**(17), 2934–2942.

Raeymaekers, L. (2002). Dynamics of Boolean networks controlled by biologically meaningful functions. *J. Theor. Biol.* **218**(3), 331–341.

Rice, J. J., Tu, Y., *et al.* (2005). Reconstructing biological networks using conditional correlation analysis. *Bioinformatics* **21**(6), 765–773.

Robeva, R., and Laubenbacher, R. (2009). Mathematical biology education: Beyond calculus. *Science* **325**(5940), 542–543.

Saez-Rodriguez, J., Simeoni, L., *et al.* (2007). A logical model provides insights into T cell receptor signaling. *PLoS Comp. Biol.* **3**(8), e163.

Samal, A., and Jain, S. (2008). The regulatory network of *E. coli* metabolism as a Boolean dynamical system exhibits both homeostasis and flexibility of response. *BMC Syst. Biol.* **2**, 21.

Sanchez, L., and Thieffry, D. (2001). A logical analysis of the *Drosophila* gap–gene system. *J. Theor. Biol.* **211**, 115–141.

Savageau, M. A. (1991). Biochemical systems theory: Operational differences among variant representations and their significance. *J. Theor. Biol.* **151**(4), 509.

Shmulevich, I., Dougherty, E. R., *et al.* (2002). Probabilistic Boolean networks: A rule-based uncertainty model for gene regulatory networks. *Bioinformatics* **18**(2), 261–274.

Shmulevich, I., Gluhovsky, I., *et al.* (2003a). Steady-state analysis of genetic regulatory networks modelled by probabilistic Boolean networks. *Comp. Funct. Genomics* **4**(6), 601–608.

Shmulevich, I., Lahdesmaki, H., *et al.* (2003b). The role of certain post classes of Boolean network models of genetic networks. *Proc. Natl. Acad. Sci. USA* **100**(19), 10734–10739.

Stigler, B., and Veliz-Cuba, A. (2009). Network topology as a driver of bistability in the lac operon. http://arxiv.org/abs/0807.3995.

Thomas, R., and D'Ari, R. (1989). "Biological Feedback." CRC Press, Boca Raton, FL.

Thomas, R., Mehrotra, S., *et al.* (2004). A model-based optimization framework for the inference on gene regulatory networks from DNA array data. *Bioinformatics* **20**(17), 3221–3235.

Tringe, S., Wagner, A., *et al.* (2004). Enriching for direct regulatory targets in perturbed gene-expression profiles. *Genome Biol* **5**(4), R29.

Vera-Licona, P., Jarrah, A., *et al.* (2009). "An optimization algorithm for the inference of biological networks" (in preparation).

Waddington, C. H. (1942). Canalisation of development and the inheritance of acquired characters. *Nature* **150**, 563–564.

Wagner, A. (2001). How to reconstruct a large genetic network from *n* gene perturbations in fewer than *n*(2) easy steps. *Bioinformatics* **17**(12), 1183–1197.

Wagner, A. (2004). Reconstructing pathways in large genetic networks from genetic perturbations. *J. Comput. Biol.* **11**(1), 53–60.

Werhli, A. V., Grzegorczyk, M., *et al.* (2006). Comparative evaluation of reverse engineering gene regulatory networks with relevance networks, graphical Gaussian models and Bayesian networks. *Bioinformatics* **22**(20), 2523–2531.

Yeung, M. K., Tegner, J., *et al.* (2002). Reverse engineering gene networks using singular value decomposition and robust regression. *Proc. Natl. Acad. Sci. USA* **99**(9), 6163–6168.

Yu, J., Smith, V. A., *et al.* (2004). Advances to Bayesian network inference for generating causal networks from observational biological data. *Bioinformatics* **20**(18), 3594–3603.

Zhang, R., Shah, M. V., *et al.* (2008). Network model of survival signaling in large granular lymphocyte leukemia. *Proc. Natl. Acad. Sci. USA* **105**(42), 16308–16313.

Zou, M., and Conzen, S. D. (2005). A new dynamic Bayesian network (DBN) approach for identifying gene regulatory networks from time course microarray data. *Bioinformatics* **21**(1), 71–79.

CHAPTER 23

Monte Carlo Simulation in Establishing Analytical Quality Requirements for Clinical Laboratory Tests: Meeting Clinical Needs

James C. Boyd and David E. Bruns
Department of Pathology
University of Virginia Health System
Charlottesville
Virginia, USA

Abstract

Introduction. Patient outcomes, such as morbidity and mortality, depend on accurate laboratory test results. Computer simulation of the effects of alterations in the test performance parameters, on outcome measures, may represent a valuable approach to define the quality of assay performance required to provide optimal outcomes.

DOI: 10.1016/B978-0-12-384997-7.00023-6

Methods. We carried out computer simulations using data from patients on intensive insulin treatment to determine the effects of glucose meter imprecision and bias on (1) the frequencies of glucose concentrations >160 mg/dL; (2) the frequencies of hypoglycemia (<60 mg/dL); (3) the mean glucose; and (4) glucose variability. For each patient, starting with a randomly selected initial glucose concentration and individualized responsiveness to insulin, hourly glucose concentrations were simulated to reflect the effects of (1) IV glucose administration, (2) gluconeogenesis, (3) insulin doses as determined using regimens from the University of Washington and Yale University, and (4) errors in glucose measurements by the meter. For each of the 45 sets of glucose meter bias and imprecision conditions, 100 patients were simulated, and each patient was followed for 100 h.

Results. For both insulin regimens, mean glucose was inversely related to assay bias, glucose variability increased with negative assay bias and assay imprecision, the frequency of glucose concentrations >160 mg/dL increased with negative assay bias and assay imprecision, and the frequency of hypoglycemia increased with positive assay bias and assay imprecision. Nevertheless, each regimen displayed unique sensitivity to variations in meter imprecision and bias.

Conclusions. Errors in glucose measurement exert important regimen-dependent effects on glucose control during intensive IV insulin administration. The results of this proof-of-principle study suggest that such simulation of the clinical effects of measurement error is an attractive approach to assess assay performance and formulate performance requirements.

I. Update

In the time since we prepared our article for *Methods in Enzymology*, there has been an increasing interest in the area of simulation modeling of medical outcomes associated with use of clinical laboratory testing, and there has been particularly intense interest in medical outcomes associated with use of glucose meters.

From the time our paper appeared, the leading journal in clinical chemistry and laboratory medicine, *Clinical Chemistry*, has published a major Review (Klee, 2010) on the broad topic of how to determine the requisite analytical performance of medical tests. In the same journal, we published a critique of the devices used for measurement of hemoglobin A_{1c} at the point of care, and pointed out that few of the devices had analytical characteristics of sufficient quality to meet clinical needs (Bruns and Boyd, 2010).

With regard to glucose meters, specifically, there still is virtually no direct clinical data on which to base decisions about the requirements for the analytical quality of meters. Simulation modeling, thus, is playing an important role. Since the publication of our paper in *Methods in Enzymology,* a variety of related activities have taken place, three of which we summarize here:

In March 2010, the FDA convened a conference to address the analytical quality that will be required prior to the clearance of such devices for sale in the U.S.

One of us (DEB) was invited to discuss our work presented in the *Methods in Enzymology* paper, but could not attend because of illness. At that conference, Marc Breton presented new simulation modeling work that described the impact of meter error on medical decision making, where the simulated outcomes were determined using the well-validated physiological model of glucose metabolism developed by Kovachev and Breton and cited in our *Methods* paper. A manuscript describing that work is in preparation.

In April 2010, the Clinical Laboratory Standards Institute Subcommittee on Point-of-Care Blood Glucose Testing in Acute and Chronic Care Facilities completed a draft of the new recommendations on glucose meters. The simulation modeling studies in the *Methods in Enzymology* paper and the results of another simulation modeling study that used the actual distribution of glucose results derived from a tight glucose control program at Mayo Clinic (Karon *et al.*, 2010) were the key pieces of information considered by the subcommittee.

Finally, in the same time period, a committee of the International Standards Organization also addressed the topic of the requirements for analytical quality of glucose meters. They too carefully considered the simulation modeling studies. Although pressures of time prevented us from serving on the committee, we were consulted by the chair of the committee and provided input.

It is hoped that these efforts will do two things: (1) Lead to tighter standards of performance requirements for glucose meters to better meet the clinical needs of patients and (2) serve as models of how simulation modeling can be used for assessing the analytical quality requirments of medical tests, to better ensure that clinical needs will be met.

We offer the following suggestions, based on our experience with simulation modeling:

- Work with others who understand the relevant physiology, pathophysiology, relevant treatments and relationship of outcomes to type and timing of therapy.
- Compare your results with the results of others in the field doing similar work: Each model has strengths and potential weaknesses.
- Communicate the findings to clinicians and others (even the FDA), who may have vested interest: They will raise valuable points that need to be considered.

II. Introduction

Quantitative laboratory measurements are now playing an increasingly important role in medicine. Well-known examples include (a) quantitative assays for cardiac troponins for diagnosing acute coronary syndromes (heart attacks) (Morrow *et al.*, 2007) and (b) measurements of LDL cholesterol to guide decisions on use of statin drugs (Expert Panel on Detection, Evaluation, and Treatment of High Blood Cholesterol in Adults, 2001). Errors in these assays have been recognized to lead to misdiagnoses and to inappropriate treatment or lack of appropriate treatment.

A growing problem is to define the degree of accuracy of laboratory measurements. Various approaches have been used to define quality specifications (or analytical goals) for clinical assays. A hierarchy of these approaches has been proposed (Fraser and Petersen, 1999; Petersen, 1999). At the low end of the hierarchy lies the approach of comparing the performance of an assay with the "state of the art" or with the performance criteria set by regulatory agencies or with the opinions of practicing clinicians or patients. A step higher, biology can be used as a guide to analytical quality by considering the average inherent biological variation within people; for substances whose concentration in plasma varies dramatically from day to day, there is less pressure to have assays that are highly precise because the analytical variation constitutes a small portion of the total variation. However, none of these approaches directly examine the relationship between the quality of test performance and the clinical outcomes. The collected opinions of physicians are likely to be anecdotal and reflect wide variation in opinion, whereas criteria based on biological variation or the analytical state of the art, or even the criteria of regulatory agencies, may have no relation to clinical outcomes.

Few studies have examined instrument analytical quality requirements based on the highest criterion, that is, patient outcomes. Clinical trials to determine the effects of analytical error on patient outcomes (such as mortality) are extremely difficult to devise and are expensive (Price et al., 2006). Unlike trials of new drugs, the costs of such studies are large in relation to their potential for profit. For ethical reasons, it may be impossible to conduct prospective randomized clinical trials in which patients are randomized to different groups defined by the use of high- or low-quality analytical testing methods. The lack of common standardization of methods used in different studies usually undermines the efforts to draw useful general conclusions on this question, based on systematic reviews of published and unpublished clinical studies. In contrast to these approaches, computer simulation studies allow a systematic examination of many levels of assay performance.

There are many common clinical situations in which patient outcomes are almost certainly connected with the analytical performance of a laboratory test. These situations represent ideal models for the use of simulation studies. One such situation occurs when a laboratory test result is used to guide the administration of a drug: a measured concentration of drug or of a drug target determines the dose of drug. Errors in measurement lead to the selection of an inappropriate dose of the drug. A common example is the use of measured concentrations of glucose to guide the administration of insulin. In this situation, higher glucose concentrations are a signal that indicate the need for a higher dose of insulin, whereas, a low glucose concentration is a signal to decrease or omit the next dose of insulin.

Several years ago, we carried out a simulation modeling of the use of home glucose meters by patients to adjust their insulin doses (Boyd and Bruns, 2001). In clinical practice, the insulin dose is determined from a table that relates the measured glucose concentration and the necessary dose of insulin. We examined the relationships between errors in glucose measurement and the resulting errors in selection of the insulin dose that is appropriate for the true glucose concentration. The simulation model addressed glucose meters with a specified bias (average error) and imprecision (variability of repeated measurements of a sample, expressed as coefficient of variation—CV). We found that to select the intended insulin dosage, 95% of the time, required that both the bias and the CV of the glucose meter be <2%, which is considerably less than the error seen in commonly used meters (Boyd and Bruns, 2001). Based on these results, we concluded that simulation modeling studies could be used to provide a clinically relevant basis for setting quality specifications for home glucose meters which are used to adjust insulin doses.

Recently, several randomized controlled trials have found that tight control of patients' glucose in surgical, medical, and neonatal intensive care units improved clinical outcomes, including rates of mortality and morbidity (see, e.g., Van den Berghe et al., 2001, 2006; Vlasselaers et al., 2009). Although some subsequent studies also showed improved patient outcomes with tight glucose control (TGC) and others did not, such that meta-analyses of all available studies showed no improvement in rates of mortality or morbidity (Griesdale et al., 2009; Wiener et al., 2008). The three studies cited above measured glucose with devices known to have good accuracy and precision, but most other studies, many of which have reported disappointing results, used devices with lower accuracy and precision (Scott et al., 2009). Apart from this suggestive observation, however, little is known regarding the quality parameters of glucose assays that are required to achieve optimum results in TGC programs.

We set out to use simulation modeling, as an alternative to clinical trials in patients, to address the quality requirements for measurements of glucose in TGC programs. Any simulation model used for evaluating the clinical success of TGC regimens requires the selection of clinical measures of success. The currently used, popular measure for assessing the success of TGC regimens is mean blood glucose. Additional measures for assessing the tightness of blood glucose control include the frequencies of hypo- or hyperglycemia, the percent of time during which the patients' blood glucose concentrations are within the target interval, and the relative variability in blood glucose concentrations over time.

We have developed a modeling approach to assess the impact of analytical imprecision and inaccuracy in glucose testing on clinical measures of outcome in TGC regimens. Although the data we present are preliminary and have some weaknesses that we will point out, we believe that such a modeling approach may represent a generally useful method to help answer the question of the degree of analytical testing quality required in a variety of clinical circumstances to meet medical needs.

III. Modeling Approach

A. Simulation of Assay Imprecision and Inaccuracy

Laboratory tests generally display both inaccuracy and imprecision. Systematic assay inaccuracy is reflected in assay bias—the mean deviation of test results from the true concentrations. Assay bias is assessed by comparing test measurements on samples with measurements made by a reference measurement system that is known to have very low bias.

Assay imprecision is assessed by repetitive measurements of quality control materials that simulate patient samples. The data are used to determine the assay imprecision (standard deviation) at several concentrations of the analyte. The average assay imprecision is a reasonably good estimate of the imprecision that might be seen in the analysis of patient samples. Assay imprecision is usually expressed as a relative imprecision, or CV, in percent, and is obtained by dividing the observed assay S.D. by the mean and multiplying the result by 100.

Quality control data, obtained upon repeated analyses of the same samples over months or years, are statistically well behaved and follow a Gaussian distribution. Thus, assuming that the imprecision of a laboratory test for patient samples is similar to that observed on quality control samples, the simulation of assay imprecision can easily be accomplished using a random number generator that yields normally distributed values with mean $= 0$ and S.D. $= 1$, such as the RANNOR function in SAS.

To generate a series of simulated test results that has a bias of B% and a relative imprecision of S%, the following equation is used:

$$\text{Test}_{(\text{simulated})} = \text{Conc}_{(\text{true})} + (B/100) \times \text{Conc}_{(\text{true})} + \text{RANNOR}(\text{seed}) \\ \times (S/100) \times \text{Conc}_{(\text{true})} \tag{1}$$

where $\text{Test}_{(\text{simulated})}$ is the test result with $B\%$ bias and $S\%$ relative imprecision, $\text{Conc}_{(\text{true})}$ is the true concentration of analyte in the sample, and RANNOR(seed) is the RANNOR function output at a given "seed" value.

Alternative equations can also be easily generated to simulate a constant, rather than a proportional, bias or standard deviations expressed in concentration units rather than relative standard deviations (CVs) can be used. Combinations of these approaches allow simulation of test results that, for instance, have a constant concentration-based S.D. at low assay values, but above a threshold concentration, have a constant relative S.D.

B. Modeling Physiologic Response to Changing Conditions

Where laboratory measurements are used to guide patient treatment, simulation modeling requires a good simulation model of the physiological response to drug administration. For the example shown in this chapter, we have modeled the physiological response of glucose to insulin administration. Although the model we have developed is very simplistic (and, therefore, may not reflect the true physiological response very accurately), it is sufficient for the purpose of demonstration. Sophisticated models that give highly accurate representations of the true physiological response of glucose to insulin have been developed for both Type 1 and Type 2 diabetes, and have received FDA approval for their use in preclinical trials of closed-loop control of glucose (Kovatchev *et al.*, 2009). These models would be the ideal models to apply in the simulation studies we present below, but due to their complexity and cost, we have used our simple physiological model to demonstrate the concept being presented here.

IV. Methods for Simulation Study

We utilized simulation modeling studies to evaluate the effect of analytical errors in glucose measurements on the relevant outcomes in simulated intensive care unit patients on TCG regimens. Our simulation models of the TCG regimens were designed to determine the effects of assay imprecision and bias on four measures of success: (1) the frequency of plasma glucose concentrations above goal range (>160 mg/dL); (2) the frequency of plasma glucose concentrations in the hypoglycemic range (defined as <60 mg/dL for this study, but easily redefined if desired); (3) the mean blood glucose; and (4) the variability of plasma glucose (expressed as the standard deviation of repeated measurements of glucose in the same individual).

We modeled two published TGC regimens—the Yale University protocol (referred to herein as Yale) and the University of Washington (UW) protocol. The authors of the Yale protocol describe it as a "safe and effective insulin infusion protocol in a medical intensive care unit" (Goldberg *et al.*, 2004). The UW protocol was developed in the context of diabetic patients undergoing surgery, but also can be applied to patients without diabetes (Trence *et al.*, 2003). These protocols differ in their underlying approach, and each has a different goal for the range of glucose concentrations in patients. We evaluated the effects of glucose assay imprecision and bias on the four measures of clinical success (mentioned above) and whether the two regimens differed in their sensitivity to errors in glucose measurement.

The physiological release of insulin from the pancreatic beta cell is known to be linearly related to the prevailing glucose concentration (Toffolo *et al.*, 1980), and this relationship forms the basis for selection of the pharmacological dose of insulin to be given to patients who lack adequate endogenous insulin. In our computer model, true glucose concentrations, after administration of insulin, were generated based on this model of the relationship between glucose and insulin. For each patient modeled, the *initial* glucose concentration and the patient's individual responsiveness to insulin were randomly selected: the starting glucose was selected from a range of 40 to 600 mg/dL and the responsiveness to insulin was selected from a range of 10 to 54 mg/dL decrease in glucose per unit of insulin per hour. To simplify calculations, the insulin responsiveness in a given patient was assumed to remain constant, but it can be programmed to change predictably or randomly. The glucose concentration was modified each hour to reflect IV glucose administration and normal physiological glucose generation via gluconeogenesis. We chose a mean increment (±S.D.) from IV and from endogenous sources to be 50 (±10) mg/dL. Each hour, the glucose concentration was decremented according to the patient's underlying insulin responsiveness and the insulin dose determined by the regimen. Combinations of analytical bias ranging from −20% to +20% in 5% increments, and analytical imprecision (expressed as percent coefficients of variation) ranging from 0% to +20% in 5% increments were modeled. For each set of analytical error conditions (paired values of % bias and % relative imprecision expressed as CV), 100 patients were simulated, and each patient was followed for 100 h. This gave 10,000 glucose measurements for each of the 45 sets of analytical error conditions simulated, and a total of 450,000 glucose measurements for each simulation.

We performed a side-by-side comparison of the simulated glucose concentration, as measured by a perfect glucose assay, and an assay with inherent analytical error. These glucose concentrations were used to determine the rate of insulin administration, according to the regimens outlined earlier. Based on the insulin responsiveness for a given patient, the decrement in true glucose concentration resulting from an insulin dose can be calculated. An iterative application of this approach, on consecutive glucose measurements, will generate a series of true glucose values in each patient, from which the frequencies of hypoglycemia and above-target glucose concentrations can be determined, along with the variability in plasma glucose concentrations.

All of the computer modeling was performed using SAS software (SAS Institute, Cary, NC). The SAS code used for modeling the Yale and the UW regimens is included in Appendices 1 and 2, respectively.

V. Results

A. Yale Regimen

Figures 1–4 show examples of the true and measured glucose concentrations in simulated patients and the related insulin infusion rates during the 100 h of simulation. The left panel in each figure shows representative simulated patients in whom the Yale regimen was used to determine insulin infusion rates. The right panels show similar patients treated using the UW regimen. We will first describe the key features of these example patients treated according to the Yale regimen (left panels).

The upper and lower left panels of Fig. 1 show a comparison of glucose control achieved by the Yale regimen using a perfect glucose assay (top panel) versus the glucose control achieved in the same patient using an imperfect glucose assay

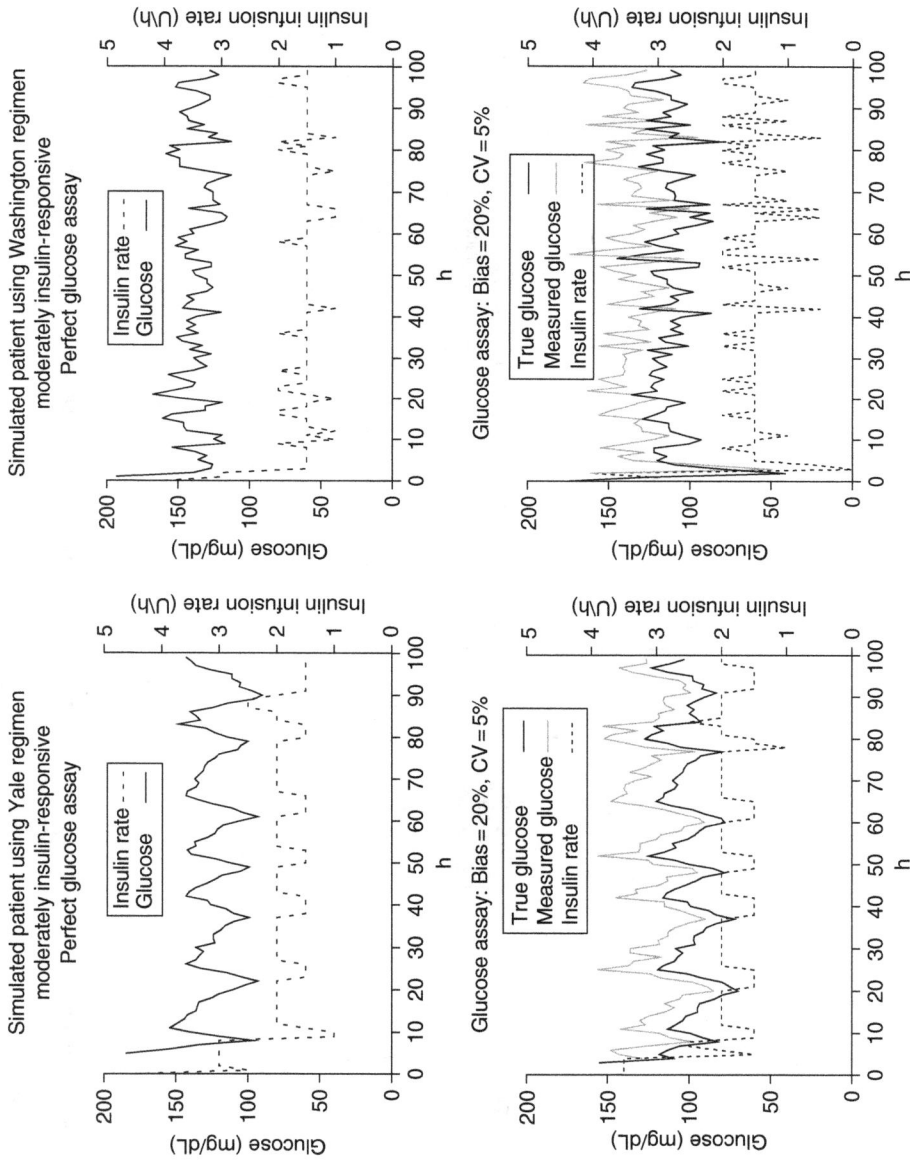

Fig. 1 Glucose concentrations and insulin infusion rates in two simulated, moderately insulin-responsive patients. *Upper panels*: Results with a perfect glucose assay. *Lower panels*: Results with a glucose assay with 20% bias and imprecision (expressed as CV) of 5%. Insulin infusion rates for patients were determined by the Yale regimen (*left panels*) or the University of Washington regimen (*right panels*).

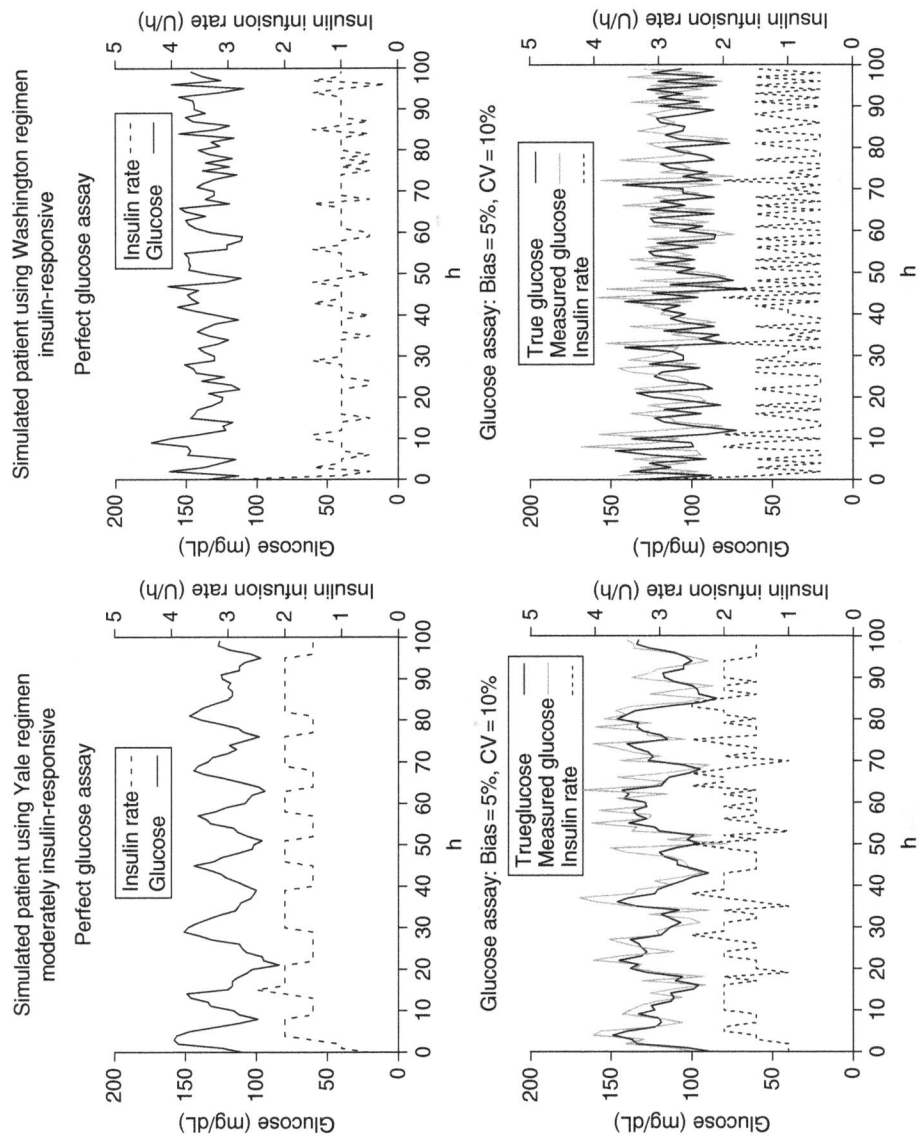

Fig. 2 Glucose concentrations and insulin infusion rates in two simulated patients. *Upper panels*: Results with a perfect glucose assay. *Lower panels*: Results with a glucose assay with 5% bias and imprecision (expressed as CV) of 10%. Insulin infusion rates for patients were determined by the Yale regimen (*left panels*) or the University of Washington regimen (*right panels*).

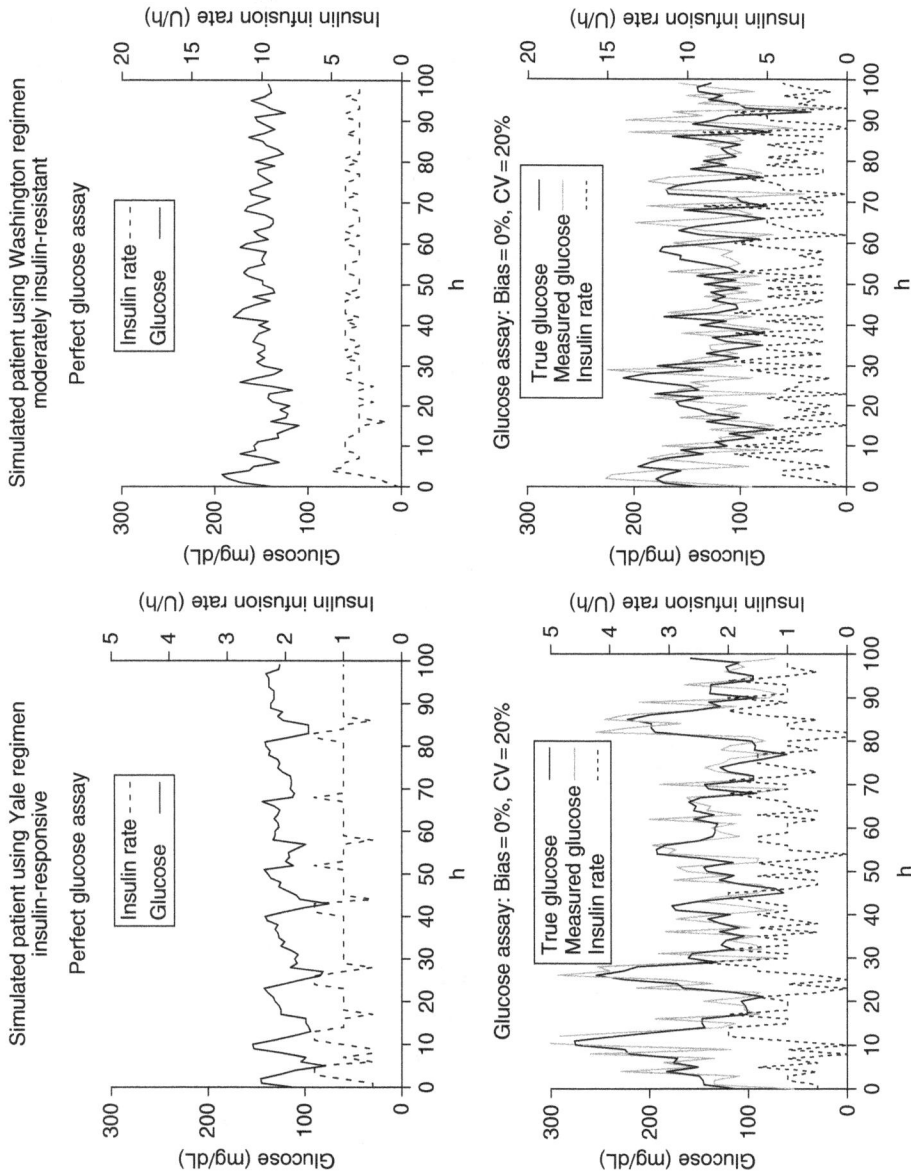

Fig. 3 Glucose concentrations and insulin infusion rates in two simulated patients. *Upper panels*: Results with a perfect glucose assay. *Lower panels*: Results with a glucose assay with 0% bias and imprecision (expressed as CV) of 20%. Insulin infusion rates for patients were determined by the Yale regimen (*left panels*) or the University of Washington regimen (*right panels*).

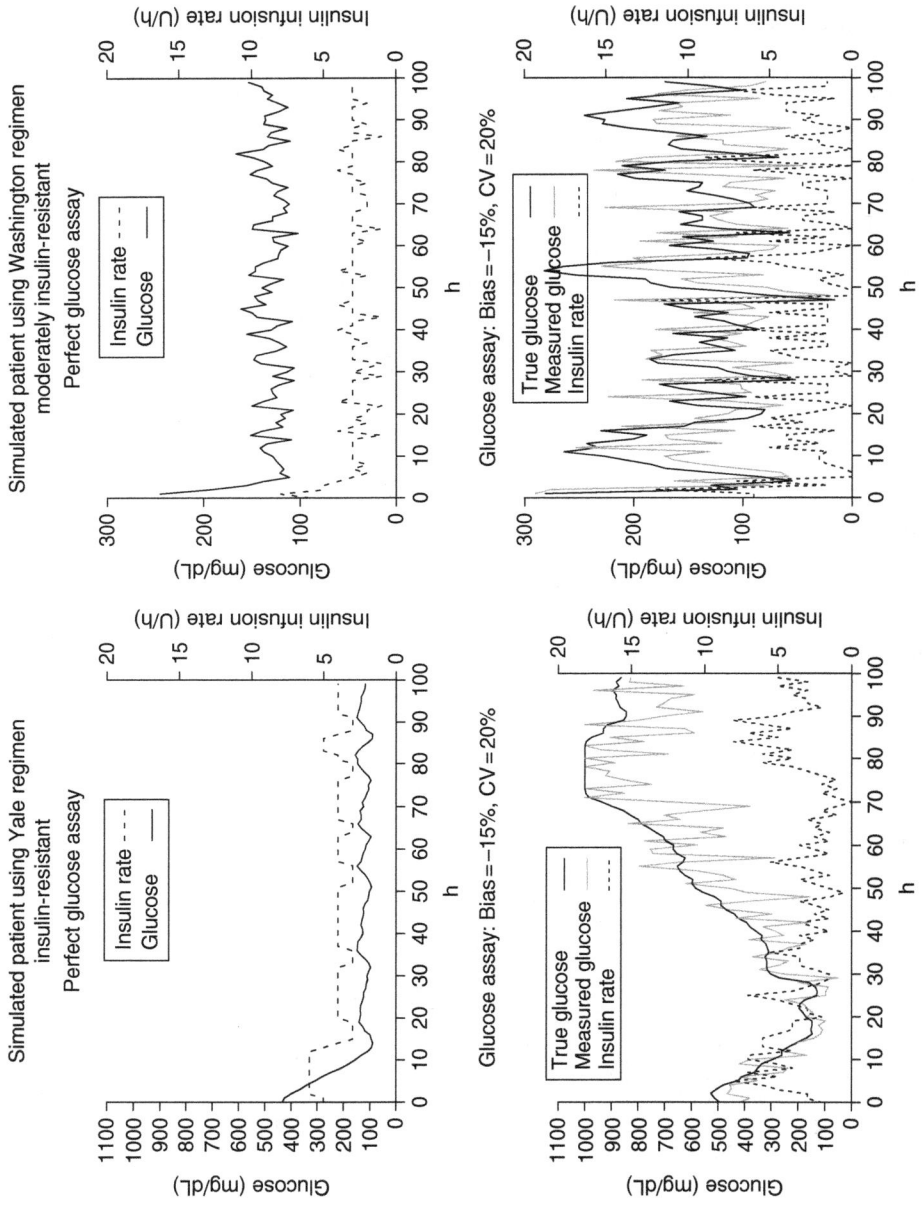

Fig. 4 Glucose concentrations and insulin infusion rates in two simulated patients. *Upper panels:* Results with a perfect glucose assay. *Lower panels:* Results with a glucose assay with −15% bias and imprecision (expressed as CV) of 20%. Insulin infusion rates for patients were determined by the Yale regimen (*left panels*) or the University of Washington regimen (*right panels*).

(lower panel) with a positive 20% bias and a 5% CV. With the perfect assay, in the top panel, periodic oscillation of glucose concentrations and insulin administration rates occur due to insulin dosage adjustments specified by the Yale regimen. In the lower panel, with a positively biased glucose assay, similar oscillations can be seen, but now, as expected, the true glucose concentration is displaced downward, compared to the perfect assay in the top panel which oscillates between approximately 70 and 120 mg/dL rather than between 100 and 150 mg/dL. The biased assay gives measured glucose concentrations that are higher than the true glucose concentrations, and thus higher rates of insulin infusion are selected to hold measured glucose near the target range, all measurements as measured by the meter. True glucose is displaced downwards.

Figure 2, left shows a different simulated patient (with moderate insulin responsiveness) on the Yale regimen, in the bottom panel is a less-severely biased glucose assay (5%) but one with an imprecision that is higher (CV = 10%); these are well within the range of bias and impression reported for glucose meters. It is immediately apparent that the control of the glucose concentrations and insulin rates in the bottom panel is "noisier" than in the top panel. Thus, an increase in the imprecision of the glucose assay results in an increase in the variability of the simulated patient's glucose control.

Figure 3, left shows another simulated patient, on the Yale regimen, with low insulin resistance, as can be judged by the relatively low hourly insulin administration rates. In the bottom panel, a highly imprecise glucose assay (as judged by the assay CV which is 20%) that has a zero bias has been used to make the glucose measurements. It is easy to appreciate a serious degradation in the precision of glucose control in this simulated patient when a more-imprecise assay is used. The true glucose concentrations range from approximately 70 to 280 mg/dL in this panel. A major motivation in TGC protocols is to avoid such high concentrations of glucose. Note that these high concentrations are seen despite the absence of a bias in the measurements, there is only an increase in variability (imprecision) of measurement.

Figure 4, left, presents a simulated patient who is insulin resistant (i.e., requires higher doses of insulin to control glucose). With the perfect assay, the patient's glucose is well controlled, although much higher insulin doses are required to control it. In the bottom panel, with a highly imprecise (CV = 20%) and strongly negatively biased (bias = −15%) assay, we again see wide fluctuations in glucose control, and the eventual loss of glucose control. The imprecise assay results appear to have totally fooled the insulin regimen, such that it is giving inappropriately low doses of insulin even in the face of ever-increasing glucose concentrations that eventually reach 1000 mg/dL. Although such escape from control can be detected by caregivers who could intervene, the example points out that it is possible to fool an insulin regimen when using glucose measurements of very poor quality.

Each of the contour plots in Figs. 5–8 shows a summary of the 450,000 glucose measurements made in 4500 simulated patients (100 patients for each combination

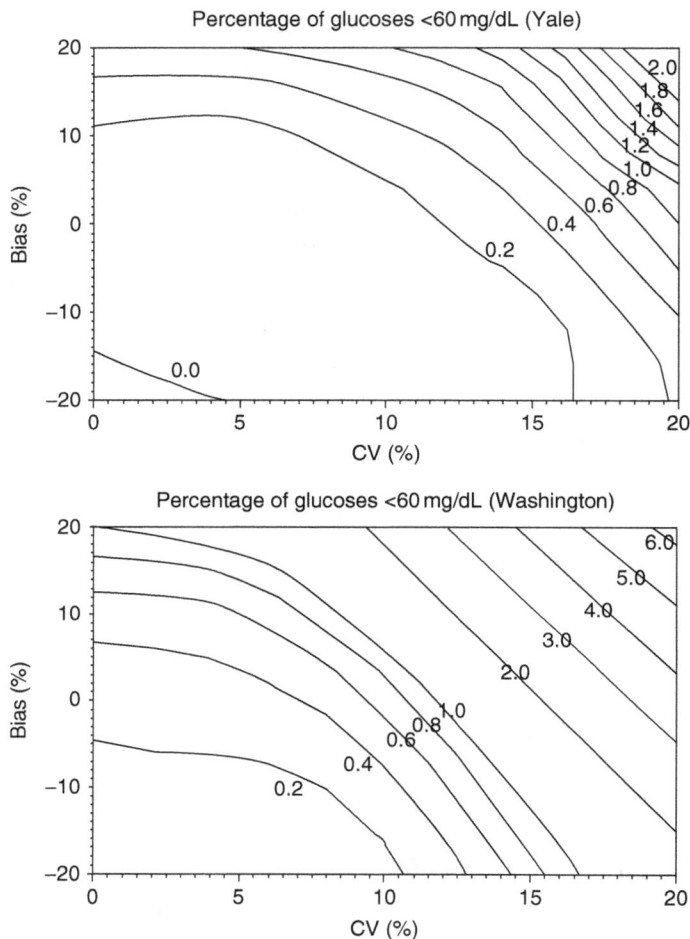

Fig. 5 Contour plots showing percentage of glucose results that were <60 mg/dL when using meters that have the indicated bias and imprecision (CV). Insulin infusion rates were determined according to the Yale regimen (*top panel*) or the University of Washington regimen (*bottom panel*).

of measurement bias and imprecision), as described in Section 3. The input for each patient is based on simulated measurements of glucose, adjustments of insulin infusion rate based on the measured glucose, and a calculation of the change in glucose concentration; this is repeated for 100 h, as shown for the example patients in Figs. 1–4. The upper panels in Figs. 5–8 show results for the Yale regimen and the lower panels for the UW regimen. As before, we will first describe the findings in the Yale regimen (upper panels).

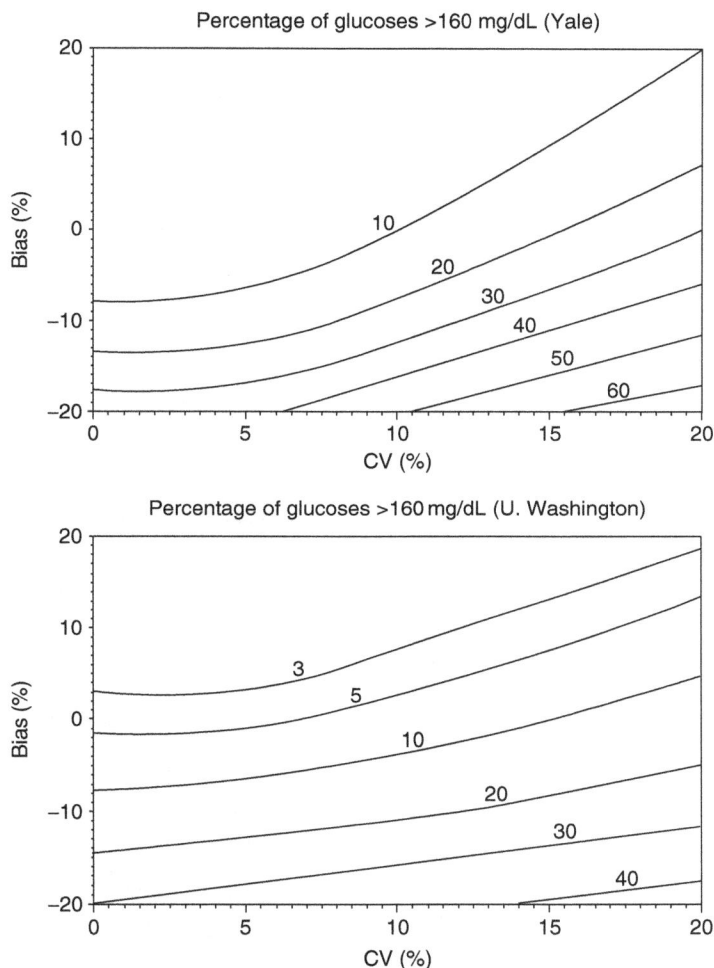

Fig. 6 Contour plots showing percentage of glucose results that were >160 mg/dL when using meters that have the indicated bias and imprecision (CV). Insulin infusion rates were determined according to the Yale regimen (*top panel*) or the University of Washington regimen (*bottom panel*).

Figure 5, top panel, shows the relationship between assay quality measures and the frequency of hypoglycemia observed with the Yale insulin regimen. To use this plot, a particular set of bias and imprecision conditions that match a given glucose assay were chosen. Suppose that an assay has a bias of +7% and an imprecision (CV) of 5%. Reading from the isocontours that represent the rate of hypoglycemia (in percent of readings), an assay with these performance characteristics would lead to a rate of hypoglycemia between 0.0% and 0.2%. Thus, for bias and

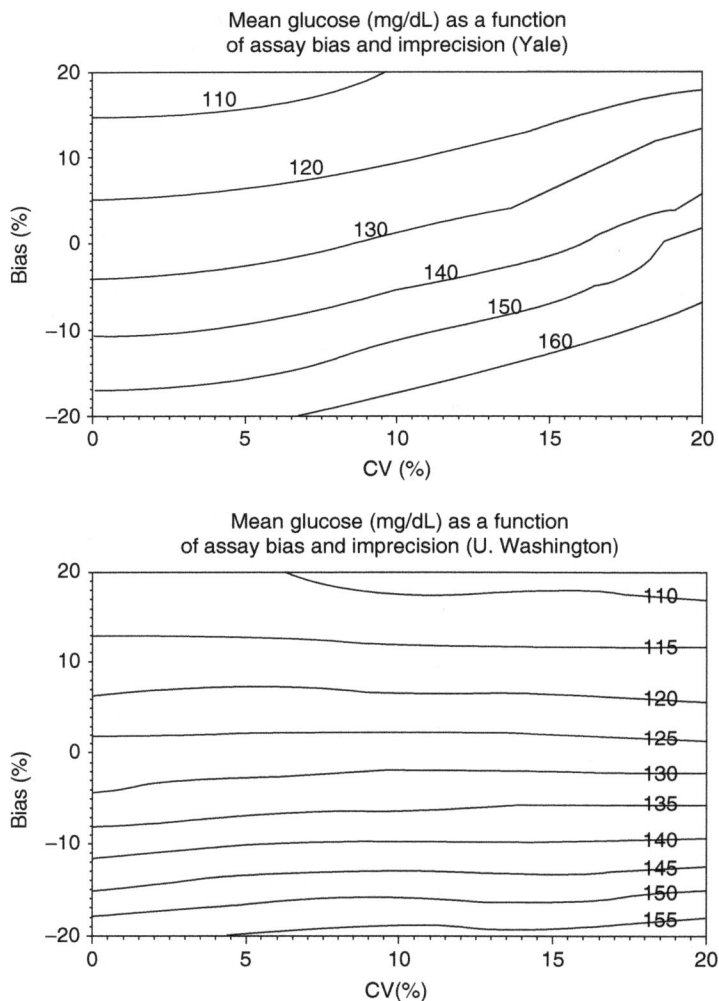

Fig. 7 Contour plots showing mean glucose concentrations when using meters that have the indicated bias and imprecision (CV). Insulin infusion rates were determined according to the Yale regimen (*top panel*) or the University of Washington regimen (*bottom panel*).

imprecision conditions that fall below the 0.2% isocontour, hypoglycemia would be predicted, by the simulation, to occur no more frequently than 0.2% of the time. The frequency of hypoglycemia is increased by positive assay bias and by an increase in imprecision. As the assay bias increases toward +20% and the imprecision increases toward a CV of 20%, the simulation results suggest that the observed rates of hypoglycemia could exceed 2% of observations.

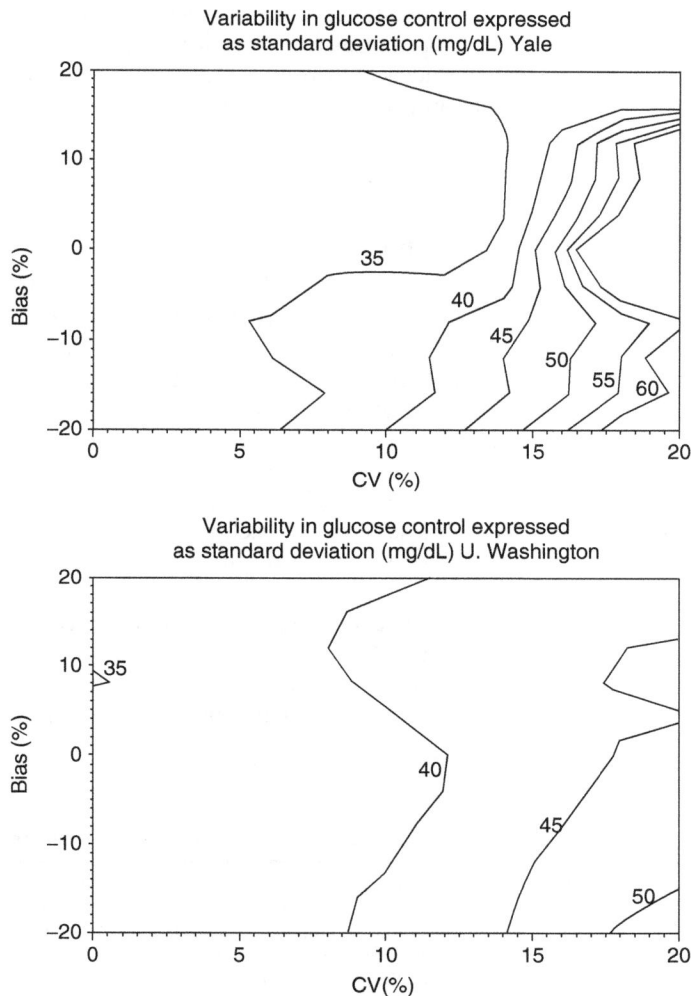

Fig. 8 Contour plots showing imprecision of glucose control (expressed as S.D.) when using meters that have the indicated bias and imprecision (CV). Insulin infusion rates were determined according to the Yale regimen (*top panel*) or the University of Washington regimen (*bottom panel*).

Figure 6 shows similar plots for the percentage of true glucose measurements that exceed 160 mg/dL—a rough measure of the control of hyperglycemia. As the assay becomes more negatively biased (toward −20%), and assay imprecision (as CV) increases toward 20%, the simulation suggests that more than 60% of true glucose concentrations could exceed 160 mg/dL.

The mean glucose concentration was inversely related to assay bias (Fig. 7)—the higher the bias in the positive direction, the lower the mean true glucose (Fig. 7). Variability in glucose control (as measured by the average standard deviation of glucose results in a patient, Fig. 8) increased rapidly when the assay imprecision exceeded 10%.

B. University of Washington Regimen

We will now turn to the UW regimens, which require some additional description. Table I shows the four regimens that dictate the insulin infusion rate (in units per hour) for a given glucose concentration. The regimens have been designed for patients with differing insulin resistance. Thus, regimen 1 is for insulin-sensitive patients (respond readily to low doses of insulin), whereas regimens 2, 3, and 4 are used in increasingly insulin-resistant patients. Higher insulin doses are administered for a given glucose concentration as the regimens progress from regimen 1 to regimen 4. To apply the UW approach, the correct insulin regimen has to be selected for each patient. Most patients start on regimen 1, and are moved from one regimen to another. Separate criteria are defined for moving up one regimen and moving down a regimen. For deciding to move up to the next higher regimen the current regimen should be deemed a failure; this occurs when the measured blood glucose is above the target range (80–180 mg/dL) and does not change by at least 60 mg/dL within 1 h after administration of insulin on the current regimen. When this happens, the decision is made to move to the next

Table I
The University of Washington Standardized Insulin Administration Regimen[a]

Regimen 1		Regimen 2		Regimen 3		Regimen 4	
Glucose (mg/dL)	Units/h	Glucose (mg/dL)	Units/h	Glucose (mg/dL)	Units/h	Glucose (mg/dL)	Units/h
<60 = Hypoglycemia (admin 50 mL $D_{50}W$. Notify MD if unresolved in 20 min)							
<70	Off	<70	Off	<70	Off	<70	Off
70–109	0.2	70–109	0.5	70–109	1	70–109	1.5
110–119	0.5	110–119	1	110–119	2	110–119	3
120–149	1	120–149	1.5	120–149	3	120–149	5
150–179	1.5	150–179	2	150–179	4	150–179	7
180–209	2	180–209	3	180–209	5	180–209	9
210–239	2	210–239	4	210–239	6	210–239	12
240–269	3	240–269	5	240–269	8	240–269	16
270–299	3	270–299	6	270–299	10	270–299	20
300–329	4	300–329	7	300–329	12	300–329	24
330–359	4	330–359	8	330–359	14	330–359	29
>360	6	>360	12	>360	16		

[a]Modified from Clement et al. (2004), as adapted from Trence et al. (2003).

higher regimen. The next lower regimen is used when the measured blood glucose is <70 mg/dL for two consecutive measurements.

Returning to Fig. 1, we can compare results in the example patients for the UW regimen in the right two panels with the Yale University regimen in the left two panels. The upper panel on each side presents the results when glucose is measured using a perfect assay, and the lower panel presents results when an imperfect method is used for glucose measurement. With a perfect glucose assay (top), each insulin regimen adequately controls blood glucose concentrations (although with a different pattern of oscillation of the values). When glucose is measured using an assay that is strongly positively biased and has a 5% imprecision (CV), which is shown in the lower panels, both regimens appear to control glucose, eventhough true glucose has been displaced to lower values. Note that an episode of hypoglycemia in the lower right panel at about 3 h is obscured by the high bias of the meter and would have gone unrecognized.

Figure 2 shows, for the example patients, that the UW and Yale regimens appear to control glucose within similar bounds when the glucose assay has a bias of 5% and an imprecision of 10%. The UW regimen appears, for this example patient, to show much more variability in glucose and requires more-frequent changes of insulin infusion rate compared to the Yale algorithm.

Figure 3 shows a comparison of the regimens in the bottom panels for a glucose assay with no bias, but with an imprecision of 20%. With such a highly imprecise glucose assay, both insulin regimens allow extreme variability in glucose measurements, including, for the UW regimen, an episode of marked hypoglycemia (at about 94 h, lower right panel). Here, it should be noted that the particular patient chosen for the UW simulation on the right is much more highly insulin resistant than the patient chosen for the Yale simulation on the left. Nevertheless, the effects of high assay imprecision seem to be similar.

Figure 4 shows the representative results of a highly inaccurate and imprecise glucose assay. Whereas the Yale regimen allows extreme hyperglycemia and appears to have totally lost control of glucose (lower left), the UW regimen shows extreme fluctuations of glucose and again allows at least one episode of hypoglycemia. In common between the two regimens is the fact that worsening assay performance has detrimental effects on glucose control.

Returning to the contour plots that show the results of 4500 simulated patients (Figs. 5–8), the lower panel of Fig. 5, shows the effect of simulated assay bias and imprecision on the percentage of glucose measurements <60 mg/dL, with the UW regimen. Compared to the Yale regimen, the UW regimen appeared to give a higher frequency of hypoglycemia with an increase in glucose assay imprecision. This effect suggests that it is particularly important to maintain glucose assay imprecision at low levels when using the UW regimen.

Figure 6, bottom panel, shows the effect of simulated assay bias and imprecision on the percentage of glucose measurements >160 mg/dL, with the UW regimen. The rate of above-target glucose concentrations appears to increase more slowly with an increase in negative assay bias while using the UW regimen compared to the Yale regimen.

As with the Yale regimen, the mean glucose concentration obtained using the UW regimen was inversely related to assay bias—the higher the bias in the positive direction, the lower the mean glucose (Fig. 7, bottom panel). Interestingly, an increase in assay imprecision is associated with an increase in mean glucose when using the Yale regimen, but this effect is not seen with the UW regimen.

Variability in glucose control (as measured by the mean standard deviation of glucose results in a patient) also rapidly increased when glucose assay imprecision exceeded 10% (Fig. 8, bottom panel). As noted earlier, the frequency of glucose concentrations >160 mg/dL was directly related to negative assay bias and an increase in imprecision.

VI. Discussion

In this study, we have modeled the effect of errors in glucose measurement on the ability to achieve TGC in patients. The model predicts that measurement error degrades glucose control, and that the effects of measurement error on glucose control depend on the regimen chosen for selecting the rate of intravenous infusion of insulin.

Current approaches to the measurement of glucose in TGC programs vary widely in accuracy (freedom from bias) and imprecision. An early, large study that showed a decrease in mortality due to TGC used a precise and accurate analyzer (Radiometer ABL Blood Gas Analyzer) with an imprecision (expressed as CV) of <2.8% and <3.5% at concentrations of 92 and 220 mg/dL, respectively, including all results such as rare outliers (personal communication to DEB, Roger Bouillon, 16 March 2002). By contrast, many subsequent studies used "glucose meters," often from unspecified manufacturers. Numerous studies have demonstrated that glucose meters have greater imprecision and biases than central laboratories or blood gas analyzers . One study from the CDC (Kimberly et al., 2006) of five common glucose meters showed mean differences versus a central laboratory method to be as high as 32% and an imprecision (CVs) of 6–11% when performed by a single trained medical technologist. Several studies have documented that glucose results produced from glucose meters by healthcare workers and patients have worse imprecision (higher CVs) than those generated by laboratory technologists. The College of American Pathologists proficiency testing shows that the CVs of 17 glucose meter methods (19,597 sites) is 12–14% and that bias between any two methods as high as 41%. Our results suggest that use of such meters will severely degrade the control of blood glucose with either the Yale regimen or the UW regimen. We do not envision a protocol that can overcome these limitations, short of using very frequent sampling.

We emphasize that the simulation model implemented in these studies does not account for the variations in test results, which can be due to patient factors (drugs, interferents, matrix effects), sample collection artifacts (such as drawing blood from IV lines, or collecting skin-puncture blood in the presence of hypoperfusion of skin capillaries as is often seen in critically ill patients), or the occurrence of random spurious test results. All of these factors are important considerations and will serve to only inflate the observed estimates of assay bias and imprecision. Thus, merely establishing that an assay operates in an apparently acceptable range of imprecision and bias does not mean that these other factors cannot degrade the ability of any regimen to achieve glucose concentrations in the target range. Finally, it appears from these studies that the effects of measurement inaccuracy and imprecision should be carefully weighed while making decisions to implement intensive IV insulin regimens.

Our study has several limitations. Any simulation model, applied to the evaluation of analytical quality required for TGC regimens, would need to be adaptable to the wide variety of regimens that have been proposed. Each regimen may have been designed to meet the needs of differing patient populations, may require different levels of nursing attention, and may have a different goal range for glucose and a greater or lesser emphasis on avoiding hypoglycemia. It is left to future studies to investigate the effects of assay errors on other regimens beyond the two modeled here.

As mentioned, earlier our model of physiological control of plasma glucose accounts for only some of the many characteristics of the patient and does not address sample collection, matrix effects, variations in patient activity, variations in nutritional intake, and many other potentially relevant variables. Thus, this work represents a proof-of-concept approach to the use of simulation modeling.

Despite these limitations, the work described here points to the performance quality of glucose measurements as a critical but overlooked factor in the success of TGC programs. The landmark, large study of Van den Berghe *et al.* (2001) used a precise and accurate glucose method (mentioned earlier) and demonstrated a marked decrease in mortality with TGC. Subsequent reports have, most often, used relatively imprecise and inaccurate glucose meters, and a meta-analysis of the studies on TGC found no decrease in overall mortality with TGC when data from all studies (including Van den Berghe *et al.*, 2001) were analyzed. Moreover, a recent multinational study showed an *increase in* mortality with TGC (The NICE-SUGAR Study Investigators, 2009). This latter finding is not surprising given that meters from various manufacturers were used with a single regimen for adjusting the insulin infusion rates. A single regimen cannot be appropriate for the variety of available glucose meters, some of which have positive biases and others have negative biases. The single regimen will lead to administration of too much insulin to patients monitored by glucose meters that give false high results and too little insulin to patients monitored with meters that report false low results. Finally, it is simply unrealistic to aim to keep glucose within an interval that represents a range of plus or minus a few percent when the glucose measuring device has a CV of more than 10%. Future studies of TGC must concentrate on use of the better methods for the measurement of glucose or risk losing the benefits of TGC or even risk harm to patients. It is anticipated that simulation modeling can be a valuable tool in design of future studies and in understanding the effect of measurement accuracy and precision on desired outcomes for patients being treated using TGC protocols.

References

Boyd, J. C., and Bruns, D. E. (2001). Quality specifications for glucose meters: Assessment by simulation modeling of errors in insulin dose. *Clin. Chem.* **47**, 209–214.

Bruns, D. E., and Boyd, J. C. (2010). Few point-of-care hemoglobin A1c assay methods meet clinical needs. *Clin. Chem.* **56**, 4–6.

Clement, S., Braithwaite, S. S., Magee, M. F., Ahmann, A., Smith, E. P., Schafer, R. G., and Hirsch, I. B. (2004). American Diabetes Association Diabetes in Hospitals Writing Committee Management of diabetes and hyperglycemia in hospitals. *Diabetes Care* **27**, 553–597.

Expert Panel on Detection, Evaluation, and Treatment of High Blood Cholesterol in Adults (2001). Executive summary of the third report of the National Cholesterol Education Program (NCEP) expert panel on detection, evaluation, and treatment of high blood cholesterol in adults (adult treatment panel III). *J. Am. Med. Assoc.* **285**, 2486–2497.

Fraser, C. G., and Petersen, P. H. (1999). Analytical performance characteristics should be judged against objective quality specifications. *Clin. Chem.* **45**, 321–323.

Goldberg, P. A., Siegel, M. D., Sherwin, R. S., Halickman, J. I., Lee, M., Bailey, V. A., Lee, S. L., Dziura, J. D., and Inzucchi, S. E. (2004). Implementation of a safe and effective insulin infusion protocol in a medical intensive care unit. *Diabetes Care* **27**, 461–467.

Griesdale, D. E. G., de Souza, R. J., van Dam, R. M., Heyland, D. K., Cook, D. J., Malhotra, A., Dhaliwal, R., Henderson, W. R., Chittock, D. R., Finfer, S., and Talmo, D. (2009). Intensive insulin therapy and mortality among critically ill patients: A meta-analysis including NICE-SUGAR study data. *Can. Med. Assoc. J.* **180**, 821–827.

Karon, B. S., Boyd, J. C., and Klee, G. G. (2010). Glucose meter performance criteria for tight glycemic control estimated by simulation modeling. *Clin. Chem.* **56**, 1091–1097.

Kimberly, M. M., Vesper, H. W., Caudill, S. P., Ethridge, S. F., Archibold, E., Porter, K. H., and Myers, G. L. (2006). Variability among five over-the-counter blood glucose monitors. *Clin. Chim. Acta* **364**, 292–297.

Klee, G. G. (2010). Establishment of outcome-related analytic performance goals. *Clin. Chem.* 10.1373/clinchem.2009.133660.

Kovatchev, B. P., Breton, M., Man, C. D., and Cobelli, C. (2009). In silico preclinical trials: A proof of concept in closed-loop control of type 1 diabetes. *J. Diabetes Sci. Technol.* **3**, 44–55.

Morrow, D. A., Cannon, C. P., Jesse, R. L., Newby, L. K., Ravkilde, J., Storrow, A. B., Wu, A. H., and Christenson, R. H. (2007). National Academy of Clinical Biochemistry, National Academy of Clinical Biochemistry laboratory medicine practice guidelines: Clinical characteristics and utilization of biochemical markers in acute coronary syndromes. *Circulation* **115**, e356–e375.

Petersen, P. H. (1999). Quality specifications based on analysis of effects of performance on clinical decision-making. *Scand. J. Clin. Lab. Invest.* **59**, 517–521.

Price, C. P., Bossuyt, P. M. M., and Bruns, D. E. (2006). Introduction to laboratory medicine and evidence-based laboratory medicine. *In* "Tietz Textbook of Clinical Chemistry and Molecular Diagnostics," (C. A. Burtis, E. R. Ashwood, and D. E. Bruns, eds.), 4th edn., pp. 323–351. Elsevier, Philadelphia, PA.

Scott, M. G., Bruns, D. E., Boyd, J. C., and Sacks, D. B. (2009). Tight glucose control in the intensive care unit: Are glucose meters up to the task? *Clin. Chem.* **55**, 18–20.

The NICE-SUGAR Study Investigators (2009). Intensive versus conventional glucose control in critically ill patients. *N. Engl. J. Med.* **360**, 1283–1297.

Toffolo, G., Bergman, R. N., Finegood, D. T., Bowden, C. R., and Cobelli, C. (1980). Quantitative estimation of beta cell sensitivity to glucose in the intact organism: A minimal model of insulin kinetics in the dog. *Diabetes* **29,** 979–990.

Trence, D. L., Kelly, J. L., and Hirsch, I. B. (2003). The rationale and management of hyperglycemia for in-patients with cardiovascular disease: Time for change. *J. Clin. Endocrinol. Metab.* **88,** 2430–2437.

Van den Berghe, G., Wouters, P., Weekers, F., Verwaest, C., Bruyninckx, F., Schetz, M., Vlasselaers, D., Ferdinande, P., Lauwers, P., and Bouillon, R. (2001). Intensive insulin therapy in critically ill patients. *N. Engl. J. Med.* **345,** 1359–1367.

Van den Berghe, G., Wilmer, A., Hermans, G., Meersseman, W., Wouters, P. J., Milants, I., Van Wijngaerden, E., Bobbaers, H., and Bouillon, R. (2006). Intensive insulin therapy of medical intensive care patients. *N. Engl. J. Med.* **354,** 449–461.

Vlasselaers, D., Milants, I., Desmet, L., Wouters, P. J., Vanhorebeek, I., van den Heuvel, I., Mesotten, D., Casaer, M. P., Meyfroidt, G., Ingels, C., Muller, J., Van Cromphaut, S., *et al.* (2009). Intensive insulin therapy for patients in paediatric intensive care: A prospective, randomised controlled study. *Lancet* **373,** 547–556.

Wiener, R., Wiener, D. C., and Larson, R. J. (2008). Benefits and risks of tight glucose control in critically ill adults: A meta-analysis. *J. Am. Med. Assoc.* **300,** 933–944.

CHAPTER 24

Pancreatic Network Control of Glucagon Secretion and Counterregulation

Leon S. Farhy and Anthony L. McCall

Department of Medicine
Center for Biomathematical Technology
University of Virginia
Charlottesville, Virginia, USA

585
DOI: 10.1016/B978-0-12-384997-7.00024-8

Abstract

Glucagon counterregulation (GCR) is a key mechanism of protection against hypoglycemia, which is compromised in insulinopenic diabetes by an unknown mechanism. In this work, we present an interdisciplinary approach to the analysis of GCR control mechanisms. Our results indicate that a pancreatic network, which unifies a few explicit interactions between the major islet peptides and blood glucose (BG), can replicate the normal GCR axis and explain its impairment in diabetes. A key and novel component of this network is an α-cell autofeedback mechanism, which drives glucagon pulsatility and mediates the triggering of a pulsatile GCR due to hypoglycemia by switching off β-cell suppression of the α-cells. We have performed simulations, based on our models of the endocrine pancreas, which explain the GCR response, *in vivo,* to hypoglycemia of the normal pancreas and the enhancement of the defective pulsatile GCR during β-cell deficiency; the mechanism involves switching off the intrapancreatic α-cell suppressing signals. The models also predicted that reduced insulin secretion would decrease and delay the GCR. In conclusion, based on experimental data, we have developed and validated a model of the normal GCR, its control mechanisms and their dysregulation in insulin-deficient diabetes. One advantage of this construct is that all model components are clinically measurable, thereby permitting its transfer, validation, and application to the study of the GCR abnormalities of the human endocrine pancreas *in vivo.*

I. Update

Largely overshadowed by insulin in the past, the importance of glucagon in maintaining glucose homeostasis is now being increasingly recognized, perhaps because therapies for diabetes that work partly through α-cell inhibition have been developed . This chapter presents a model-based, dynamic, network approach for the analysis of the glucagon counterregulation (GCR). We have proposed control networks that unify a few explicit interactions between the major pancreatic peptides and blood glucose (BG) to mathematically approximate the GCR control axis and explain its impairment in diabetes. Since its publication, there have been several experimental findings that may require future extension or modification of the postulated pancreatic models. First, the importance of somatostatin in the regulation of glucagon secretion, suggested by our model-based predictions (Section V.B), has been supported by a recent publication which showed that δ-cell somatostatin likely exerts a tonic inhibition on α-cell glucagon and also on β-cell insulin (Hauge-Evans et al., 2009). Our reconstruction of the GCR control mechanisms outlined in this chapter assumes that somatostatin suppresses α-cell glucagon release (see Fig. 1); this is based both on experimental data and model-based predictions. However, the impact of somatostatin on β-cells was not considered, which will be of little

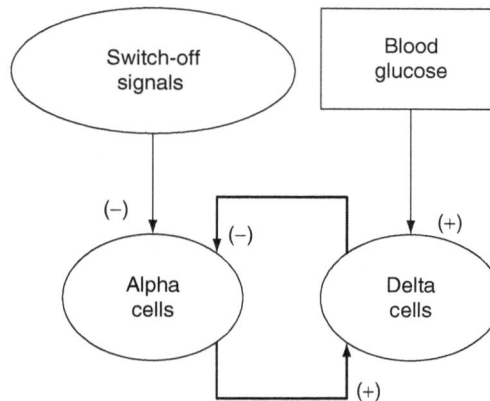

Fig. 1 Schematic presentation of a network model of the GCR control mechanisms in STZ-treated rats.

consequence if the goal is to approximate the GCR control mechanisms and their dysregulation in insulinopenic diabetes. However, future efforts to extend the GCR control network to approximate the normal pancreas may need to take this new pathway into account. For example, in Fig. 3, which shows the proposed interactions between BG and the α- and β-cells that are assumed to regulate the GCR in the normal pancreas, capabilities of the α-cells to amplify their own secretion may have to be included. Such self-modulation could occur as follows: α-cell glucagon stimulates δ-cell somatostatin, which in turn suppresses β-cell insulin, thereby repressing one of the inhibitory inputs to the α-cells. Second, various central nervous system influences continue to be identified and these may modulate glucagon secretion and GCR (Fioramonti et al., 2010; Haywood et al., 2009; Leu et al., 2009). We have briefly noted their importance in Section VIII of this chapter. However, the growing support for their importance in GCR regulation may require these influences to be either separately studied or be incorporated into the model.

Finally, we would like to warn against certain applications of our methodology which may lead to misleading interpretations. As briefly mentioned in Section VII. B, the approximation of the GCR control mechanisms reconstructs portal, rather than circulating, pancreatic hormone concentrations. Therefore, they are suited for the analysis of experimental data in which the pancreatic hormones are sampled in the portal vein. However, collecting blood samples from the portal vein may be unfeasible or unethical. In such cases, a modification of the model (e.g., following Dalla Man et al., 2005) will be required to relate the concentration of a hormone in one compartment (portal vein) to the concentration of the same hormone in another compartment (general circulation), assuming that the hormone is transported (transferred) from one of the compartments to the other. Such a model should take into account the delay, spread, and the partial clearance of glucagon and insulin by the liver and circulation.

II. Introduction

BG homeostasis is maintained by a complex, ensemble control system which is characterized by a highly coordinated interplay between and among various hormone and metabolite signals. One of its key components, the endocrine pancreas, responds dynamically to changes in BG, nutrients, neural, and other signals by releasing insulin and glucagon, in a pulsatile manner, to regulate glucose production and metabolism. Abnormalities in the secretion and interaction of these two hormones mark the progression of many diseases, including diabetes, the metabolic syndrome, the polycystic ovary syndrome, and others. Diminished or complete loss of endogenous insulin secretion in diabetes is closely associated with the failure of the pancreas to respond by appropriate changes in glucagon under not only to hyperglycemia, but also to hypoglycemia. The latter is not caused by loss of glucagon secreting α-cells, but is instead, due to defects in GCR signaling through an unknown mechanism; this is generally recognized as a major barrier to safe treatment of diabetes (Cryer and Gerich, 1983; Gerich, 1988) since unopposed hypoglycemia can cause coma, seizures, or even death (Cryer, 1999, 2002; Cryer et al., 2003).

Our recent experimental (Farhy et al., 2009) and mathematical modeling (Farhy and McCall, 2009; Farhy et al., 2009) results show that a better understanding of the defects in the GCR control mechanisms can be gained if these are viewed as abnormalities of a *network* of intrapancreatic interactions that control glucagon secretion, rather than as defects in an isolated molecular interaction or pathway.

In particular, we have demonstrated that, in a β-cell-deficient rat model, the GCR control mechanisms can be approximated by a simple, feedback network (construct) of dose–response interactions between BG and the islet peptides. Within the framework of this construct, the defects in GCR response to hypoglycemia can be explained by the loss of rapid switching off of β-cell signaling during hypoglycemia, which results in failure to trigger an immediate GCR response. These results support the "switch-off" hypothesis which posits that α-cell activation during hypoglycemia requires both the availability and rapid decline of intraislet insulin (Banarer et al., 2002). Our findings also support an extension of this hypothesis by refocusing from the lack of endogenous insulin signaling to the α-cells as the sole mechanistic explanation to the possible abnormalities in the general manner in which the β-cells regulate the α-cells. In addition, the experimental and theoretical modeling data collected so far indicate that the GCR control network must have two key features: a (direct or indirect) feedback of glucagon secreting α-cells on themselves (autofeedback) and a (direct or indirect) negative regulation of glucagon by BG. In our published model, these two properties are mediated by δ-cell somatostatin and we have shown that such connectivity adequately explains our [and that of others (Zhou et al., 2004)] experimental data (Farhy and McCall, 2009; Farhy et al., 2009).

The construct that we have recently proposed (Farhy and McCall, 2009; Farhy *et al.*, 2009) is suitable for the study and analysis of rodent physiology, but the explicit involvement of somatostatin limits its applicability to clinical studies since, in humans, pancreatic somatostatin cannot be reliably measured and therefore, the ability of the model to adequately describe human physiology, and its potential differences from rodent physiology, cannot be verified. In this work, we review our existing models and show that a control network in which somatostatin is not explicitly involved (but implicitly incorporated) can also adequately approximate GCR control mechanisms. We confirm that the (new) construct can substitute the older, more complex construct by verifying that it adequately explains the same experimental observations previously shown to be reconstructed by the older network (Farhy and McCall, 2009; Farhy *et al.*, 2009). We also demonstrate that the new network can explain the regulation of the normal pancreas by BG and the gradual reduction in the GCR response to hypoglycemia during the transition from a normal to an insulin-deficient state. As a result, a more precise description of the components that are the most critical for the system is provided by a model of GCR regulation. This model can be used to study abnormalities in glucagon secretion and counterregulation and to identify hypothetical ways to repair these abnormalities, not only in rodents but also in humans.

III. Mechanisms of Glucagon Counterregulation (GCR) Dysregulation in Diabetes

Studies of tight BG control in type 1 and type 2 diabetes, to prevent chronic hyperglycemia-related complications, have found a threefold excess of severe hypoglycemia (The Action to Control Cardiovascular Risk in Diabetes Study Group, 2008; The Diabetes Control and Complications Trial Research Group, 1993; The UK Prospective Diabetes Study Group, 1998). Hypoglycemia impairs quality of life and increases the risk of coma, seizures, accidents, brain injury, and death. Severe hypoglycemia is usually due to overtreatment against a background of delayed and deficient hormonal counterregulation. In health, GCR curbs dangerously low BG nadirs and stimulates quick recovery from hypoglycemia (Cryer and Gerich, 1983; Gerich, 1988). However, in type 1 (Fukuda *et al.*, 1988; Gerich *et al.*, 1973; Hoffman *et al.*, 1994) and type 2 diabetes (Segel *et al.*, 2002), the GCR is impaired by unknown mechanisms, and if it is accompanied by a loss of epinephrine counterregulation, it leads to severe hypoglycemia and thus presents a major barrier to safe treatment of diabetes (Cryer, 1999, 2002). Understanding the mechanisms that mediate GCR, its dysregulation and how it can be repaired, are therefore major challenges in the struggle for safer treatment of diabetes.

Despite more than 30 years of research, the mechanism by which hypoglycemia stimulates GCR and how it is impaired in diabetes have yet to be elucidated (Gromada *et al.*, 2007). First described by Gerich *et al.* (1973), defective GCR is common after about 10 years of T1DM. The loss of GCR appears to be more rapid with younger ages of onset and may occur within a few years after onset of T1DM. Although unproven, the appearance of defective GCR seems to parallel secretory loss of insulin in these patients. The defect appears to be stimulus specific, since α-cells retain their ability to secrete glucagon in response to other stimuli, such as arginine (Gerich *et al.*, 1973). Three mechanisms have been proposed, as potential sources that impair the GCR. Those that account for the stimulus specificity of the defect include impaired BG-sensing in α-cells (Gerich *et al.*, 1973) and/or autonomic dysfunction (Hirsch and Shamoon, 1987; Taborsky *et al.*, 1998). The "switch-off" hypothesis envisions that α-cell activation by hypoglycemia requires both the availability and rapid decline of intraislet insulin and it attributes the defect in the GCR, in insulin deficiency, to the loss of a (insulin) "switch-off" signal from the β-cells (Banarer *et al.*, 2002).

These theories are not mutually exclusive, but they all could be challenged. For example, α-cells do not express GLUT2 transporters (Heimberg *et al.*, 1996) and it is unclear whether the α-cell GLUT1 transporters can account for the rapid α-cell response to variations in BG (Heimberg *et al.*, 1995). In addition, proglucagon mRNA levels are not altered by BG (Dumonteil *et al.*, 2000) and whether BG variations in the physiological range can affect the α-cells is debatable (Pipeleers *et al.*, 1985). The switch-off hypothesis can also be disputed, since in α-cell-specific insulin receptor knockout mice, the GCR response to hypoglycemia is preserved (Kawamori *et al.*, 2009). Finally, the hypothesis for autonomic control contradicts the evidence which shows that a blockade of the actions of both epinephrine and acetylcholine did not reduce the GCR in humans (Hilsted *et al.*, 1991), and that a denervated human pancreas could still release glucagon in response to hypoglycemia (Diem *et al.*, 1990).

Recent *in vivo* experiments by Zhou *et al.* support the "switch-off" hypothesis. They have shown that, in STZ-treated rats GCR is impaired, but that it can be restored if their deficiency in intraislet insulin is reestablished and decreased (switched off) during hypoglycemia (Zhou *et al.*, 2004). Additional *in vitro* and *in vivo* evidence to support the switch-off hypothesis has also been reported (Hope *et al.*, 2004; Zhou *et al.*, 2007a). The supposition that insulin is the trigger of GCR in the studies by Zhou *et al.* (2004, 2007a) has been challenged by the results of the experiments conducted by the same group, in which zinc ions, not the insulin molecule itself, provided the switch-off signal to initiate glucagon secretion during hypoglycemia (Zhou *et al.*, 2007b).

In view of the above background, it is clear that the mechanisms that control the secretion of glucagon, and its dysregulation in diabetes, are not well understood. This lack of understanding prevents the restoration of GCR in patients with diabetes and the development of treatments that can effectively repair the defective GCR, which in turn will allow a safer control of hyperglycemia. No such treatment currently exists.

IV. Interdisciplinary Approach to Investigating the Defects in the GCR

The network that underlies the GCR response to hypoglycemia includes hundreds of components from numerous pathways and targets in various pools and compartments. It would, therefore, not be feasible to collect and relate experimental data pertaining to all components of this network. Nevertheless, understanding the glucagon secretion control network is vital for furthering knowledge concerning the control of GCR and its compromise in diabetes, and for developing treatment strategies. To address this problem, we have used a minimal model approach in which the system is simplified by clustering all known and unknown factors into a small number of explicit components. Initially, these components were chosen with the goal of testing whether the recognized physiological relationships can explain the key experimental findings. In our case, the initial reports that described the *in vivo* enhancement of GCR by a "switch-off" of insulin (Zhou *et al.*, 2004) prompted us to propose a parsimonious model of the complex GCR control mechanisms, and this included relationships between the α- and δ-cells, BG and the switch-off signals (below). According to these initial efforts (Farhy *et al.*, 2009), the postulated network explains the switch-off phenomenon by interpreting the GCR as a rebound. This model further predicts that: (i) in β-cell deficiency, multiple α-cell suppressing signals should enhance GCR if they are terminated during hypoglycemia, and (ii) that the "switch-off"-triggered GCR must be pulsatile. The model-based predictions motivated a series of *in vivo* experiments, which showed that, indeed, in STZ-treated male Wistar rats, an intrapancreatic infusion of insulin and somatostatin, followed by their switch-off during hypoglycemia, enhanced the pulsatile GCR response (Farhy *et al.*, 2009). These experimental results confirmed that the proposed network could be a good candidate for a model of the GCR control axis.

In addition to confirming the initial model predictions, our experiments also suggested some new features of the GCR control network, including indications that different α-cell suppressing switch-off signals can not only enhance the GCR during β-cell deficiency but can also do so via different mechanisms. For example, the results suggest a higher response to insulin switch-off and a more substantial suppression of glucagon by somatostatin (Farhy *et al.*, 2009). To show that these observations are consistent with our network model, we had to extend it to reflect the assumption that the α-cell activity can be regulated differently by different α-cell suppressing signals. We have shown that this assumption can explain the differences in the GCR-enhancing action of two α-cell-suppressing signals (Farhy and McCall, 2009).

These simulations have suggested strategies to repair defective GCR by manipulating the system using α-cell inhibitors. However, they also indicate that not all α-cell inhibitors may be suitable for this purpose, and that the infusion rate, of the ones that are, should be carefully selected. In this regard, a clinically verified and tested model of the GCR control axis can greatly enhance our ability to precisely and credibly simulate changes that result from certain interventions, which ultimately will assist us in defining the best strategy to manipulate the system in humans, *in vivo*. However, the explicit involvement of somatostatin and the δ-cells in our initial network and model limits the potential for clinical applications, as pancreatic somatostatin cannot be reliably measured in humans *in vivo*, and therefore the ability of the model to describe the human glucagon axis cannot be verified. To address this limitation, we have recently reduced our initial network into a Minimal Control Network (MCN) of the GCR control axis, in which somatostatin and the δ-cells are no longer explicitly involved, but their effects are implicitly incorporated into the model. Our analysis (presented below) shows that the new MCN is an excellent model of the GCR axis and that it can substitute the older, more complex structure. Thus, we have developed a model that can be verified clinically and also be used to assist in the analysis of the GCR axis, *in vivo*, in humans. Importantly, the new model is not limited to β-cell deficiency and hypoglycemia only. In fact, it describes the transition of the pancreas from a normal state to a β-cell-deficient state, and can explain the failure of suppression of basal glucagon secretion in response to the increase in BG observed in this state. If it is experimentally confirmed that this MCN model can successfully describe both the normal and β-cell-deficient pancreas, future studies can focus on the defects of the pancreatic network not only in type 1 but also in type 2 diabetes, or more generally, in any pathophysiological condition that is accompanied by metabolic abnormalities in the endocrine pancreas.

V. Initial Qualitative Analysis of the GCR Control Axis

To understand the mechanisms of GCR and its dysregulation, pancreatic peptides have been extensively studied and much of the evidence suggests that a complex network of interacting pathways modulates glucagon secretion and the GCR. Some of the well-documented relationships between the different types of islet cell signals are summarized in the following subsections.

A. β-Cell Inhibition of α-Cells

Pancreatic perfusions with antibodies to insulin, somatostatin, and glucagon have suggested that the blood within the islets flows from β- to α- to δ-cells in dogs, rats, and humans (Samols and Stagner, 1988, 1990; Stagner *et al.*, 1988, 1989). It was then proposed that insulin regulates glucagon, which in turn regulates somatostatin. Various β-cell signals provide an inhibitory stimulus to the α-cells and thus suppress glucagon secretion. These include cosecreted insulin, zinc, GABA, and amylin (Gedulin *et al.*, 1997; Gromada *et al.*, 2007; Ishihara *et al.*, 2003; Ito *et al.*, 1995; Maruyama *et al.*, 1984; Rorsman and Hellman, 1988; Rorsman *et al.*, 1989; Samols and Stagner, 1988; Wendt *et al.*, 2004; Xu *et al.*, 2006). In particular, β-cells store and secrete GABA, which can diffuse to neighboring cells and bind to receptors, localized within the islets, on only the α-cells (Rorsman and Hellman, 1988; Wendt *et al.*, 2004). Insulin can directly suppress glucagon by binding to its own receptors (Kawamori *et al.*, 2009) or to IGF-1 receptors on the α-cells (Van Schravendijk *et al.*, 1987). Insulin also translocates to and activates GABA$_A$ receptors on the α-cells, which leads to membrane hyperpolarization and, ultimately, the suppression of glucagon. Hence, insulin may directly inhibit the α-cells, and indirectly potentiate the effects of GABA (Xu *et al.*, 2006). Infusion of amylin, in rats, inhibits arginine-stimulated glucagon (Gedulin *et al.*, 1997), but not the hypoglycemia stimulated GCR (Silvestre *et al.*, 2001). Similar results were obtained with the synthetic amylin analog, pramlintide (Heise *et al.*, 2004), even though in some studies the degree of hypoglycemia was increased; but it is unclear if this is a GCR effect or if it is related to the failure in reducing meal insulin adequately (McCall *et al.*, 2006). Finally, a negative effect of zinc on glucagon has been proposed (Ishihara *et al.*, 2003), including a role in the control of GCR (Zhou *et al.*, 2007b). The role of zinc is unclear as zinc ions do not suppress glucagon in the mouse (Ravier and Rutter, 2005).

B. δ-Cell Inhibition of α-Cells

Exogenous somatostatin inhibits insulin and glucagon; however, the role of the endogenous hormone is controversial (Brunicardi *et al.*, 2001, 2003; Cejvan *et al.*, 2003; Gopel *et al.*, 2000a; Klaff and Taborsky, 1987; Kleinman *et al.*, 1994; Ludvigsen *et al.*, 2004; Portela-Gomes *et al.*, 2000; Schuit *et al.*, 1989; Strowski *et al.*, 2000; Sumida *et al.*, 1994; Tirone *et al.*, 2003). The concept that δ-cells are downstream of α- and δ-cells favors the perception that, *in vivo*, intraislet somatostatin cannot directly suppress the α- or the β-cell through the islet microcirculation (Samols and Stagner, 1988, 1990; Stagner *et al.*, 1988, 1989). On the other hand, the pancreatic α- and β-cells express at least one of the somatostatin receptors (SSTR1–5) (Ludvigsen *et al.*, 2004; Portela-Gomes *et al.*, 2000; Strowski *et al.*, 2000), and recent *in vitro* studies involving somatostatin immunoneutralization (Brunicardi *et al.*, 2001) or the application of selective antagonists to different somatostatin receptors have suggested that Δ-cell somatostatin inhibits the release of glucagon (Cejvan *et al.*, 2003; Strowski *et al.*, 2000). In addition, δ-cells are in close proximity to α-cells in both rat and human islets, and δ-cell processes were observed to extend into α-cell clusters in rat islets (Kleinman *et al.*, 1994, 1995). Therefore, somatostatin may act via existing common gap junctions or by diffusion through the islet interstitium.

C. α-Cell Stimulation of δ-Cells

The ability of endogenous glucagon to stimulate δ-cell somatostatin is supported by a study in which the administration of glucagon antibodies into a perfused human pancreas resulted in the inhibition of somatostatin release (Brunicardi *et al.*, 2001). Earlier studies on immunoneutralization, by perfusion of the rat or dog pancreas, also showed that glucagon stimulates somatostatin (Stagner *et al.*, 1988, 1989). These studies showed that the glucagon receptor colocalized with 11% of the immunoreactive somatostatin cells (Kieffer *et al.*, 1996), suggesting that the α-cells may directly regulate some of the δ-cells. Exogenous glucagon can also stimulate somatostatin (Brunicardi *et al.*, 2001; Epstein *et al.*, 1980; Kleinman *et al.*, 1995; Utsumi *et al.*, 1979). Finally, glutamate, which is cosecreted with glucagon under low-glucose conditions, stimulates somatostatin release from diencephalic neurons in primary culture (Tapia-Arancibia and Astier, 1988), and it is possible that a similar relation exists in the islets of the pancreas.

D. Glucose Stimulation of β- and δ-Cells

It is well established that hyperglycemia directly stimulates β-cells, which react instantaneously to changes in BG (Ashcroft *et al.*, 1994; Bell *et al.*, 1996; Dunne *et al.*, 1994; Schuit *et al.*, 2001). Additionally, it has been proposed that δ-cells have a glucose-sensing mechanism similar to that in β-cells (Fujitani *et al.*, 1996; Gopel *et al.*, 2000a) and consequently, that somatostatin release is increased in response to glucose stimulation (Efendic *et al.*, 1978; Hermansen *et al.*, 1979), possibly via a Ca^{2+}-dependent mechanism (Hermansen *et al.*, 1979).

E. Glucose Inhibition of α-Cells

Hyperglycemia has been proposed to inhibit glucagon even though hypoglycemia alone appears insufficient to stimulate a high amplitude GCR (Gopel *et al.*, 2000b; Heimberg *et al.*, 1995, 1996; Reaven *et al.*, 1987; Rorsman and Hellman, 1988; Schuit *et al.*, 1997; Unger, 1985).

In addition to the above, mostly consensus, findings which show that α-cell activity is controlled by multiple intervening pathways, there are other indirect evidences that suggest that the dynamic relationship between the islet signals are important for the regulation of glucagon secretion and the GCR. For example, this concept is supported by the *pulsatility* of the pancreatic hormones (Genter *et al.*, 1998; Grapengiesser *et al.*, 2006; Grimmichova *et al.*, 2008), which implies feedback control (Farhy, 2004), and by results which suggest that insulin and somatostatin pulses are in phase (Jaspan *et al.*, 1986; Matthews *et al.*, 1987), pulses of insulin and glucagon recur with a phase shift (Grapengiesser *et al.*, 2006), pulses of somatostatin and glucagon appear in antisynchronous fashion (Grapengiesser *et al.*, 2006), and that insulin pulses entrain α- and δ-cell oscillations (Salehi *et al.*, 2007).

A pancreatic network consistent with these findings is shown in Fig. 1. It summarizes interactions (mostly consensus) between BG, β-, α-, and δ-cells: somatostatin (or more generally the δ-cells) is stimulated by glucagon (α-cells) and BG, glucagon (α-cells) is inhibited by the δ-cells (by somatostatin) and by β-cell signals, and BG stimulates the β-cells. This network could easily explain the GCR response to hypoglycemia. Indeed, hypoglycemia would decrease both β-and δ-cell activity, which would entail an increase in the release of glucagon from α-cells, after the suppressive signals from the neighboring β- and δ-cells are removed. However, it is not apparent whether this network can explain the defect in GCR observed in β-cell deficiency or the above mentioned restoration of the defective GCR by a "switch-off." This dampens the appeal of the network as a simple unifying hypothesis for the regulation of GCR, and for the compromise of this regulation in diabetes. The difficulties encountered during the intuitive reconstruction of the properties of the network emerge from the surprisingly complex behavior of this system due to the α–δ-cell feedback loop.

Shortly after the first reports that described the *in vivo* repair of GCR by intrapancreatic infusion and a "switch-off" of insulin (Zhou *et al.*, 2004) were published, we used mathematical modeling to analyze and reconstruct the GCR control network. These considerations demonstrated that the network in Fig. 1 could explain the switch-off effect (Farhy and McCall, 2009; Farhy *et al.*, 2009). We have also presented experimental evidence to support the predictions of this model (Farhy *et al.*, 2009). These efforts are described in the following section.

VI. Mathematical Models of the GCR Control Mechanisms in STZ–Treated Rats

We have developed and validated (Farhy and McCall, 2009; Farhy *et al.*, 2009) a mathematical model of the GCR control mechanisms in the β-cell-deficient rat pancreas which explains two key experimental observations: (a) that in STZ-treated rats, rebound GCR, which is triggered by a switch-off signal (a signal that is intrapancreatically infused and terminated during hypoglycemia), is pulsatile; and (b) that the switch-off of either somatostatin or insulin enhances the pulsatile GCR. The basis of this mathematical model is the network outlined in Fig. 1, which summarizes the major interactive mechanisms of glucagon secretion during β-cell deficiency, by selected consensus interactions between plasma glucose, α-cell suppressing switch-off signals, α-cells, and δ-cells. We should note that the β-cells were part of the network proposed in Farhy *et al.* (2009), but are not part of the corresponding mathematical model, which was designed to approximate the insulin-deficient pancreas. In addition to explaining glucagon pulsatility during hypoglycemia and the switch-off responses mentioned above, this construct predicts each of the following experimental findings in diabetic STZ-treated rats:

(i) Glucagon pulsatility during hypoglycemia after a switch-off, with pulses recurring at 15–20 min as suggested by the results of the pulsatility deconvolution analysis we have previously performed (Farhy *et al.*, 2009);

(ii) Pronounced (almost fourfold increase over baseline) pulsatile glucagon response following a switch-off of either insulin or somatosatin during hypoglycemia (Farhy *et al.*, 2009);

(iii) Restriction of the GCR enhancement by insulin switch-off under high BG conditions (Zhou *et al.*, 2004);

(iv) Lack of a GCR response to hypoglycemia when there is no switch-off signal (Farhy *et al.*, 2009);

(v) Suppression of GCR when insulin is infused into the pancreas but is not switched off during hypoglycemia (Zhou *et al.*, 2004);

(vi) More than 30% higher GCR response to insulin, compared to somatostatin, switch-off (Farhy *et al.*, 2009);

(vii) Better glucagon suppression by somatostatin than by insulin before a switch-off (Farhy *et al.*, 2009).

We note that, in our prior study (Farhy *et al.*, 2009), the comparisons between insulin and somatostatin switch-off in (vi) and (vii) were not significant. However, the difference in (vii) was close to being significant at $p = 0.07$. Therefore, one of the goals of the latter study (Farhy and McCall, 2009) was to test, *in silico*, whether the differences (vi) between a higher GCR response to insulin switch-off and (vii) a better glucagon suppression by somatostatin switch-off were likely and whether these could be predicted by the proposed model of the insulin-deficient pancreas (Fig. 1).

To demonstrate the above predictions, we used dynamic network modeling and formalized the relationships shown in Fig. 1 using a system of nonlinear, ordinary differential equations. Thereby, we approximate to approximate the rates of change of glucagon and somatostatin concentration assuming that the secretion of these hormones is; under the control of switch-off signals and BG. Then, we were able to adjust the model parameters to reconstruct the experimental findings listed above in (i)–(vii), which validates the model based on the network shown in Fig. 1.

The model equations are:

$$ GL' = -k_{GL}GL + r_{basal}\frac{1}{1 + I_1(t)} + r_{GL}\frac{1}{1 + [SS(t - D_{SS})/t_{SS}]^{n_{SS}}}\frac{1}{1 + I_2(t)} \quad (1) $$

$$ SS' = -k_{SS}SS + r_{SS}\frac{[GL(t - D_{GL})/t_{GL}]^{n_{GL}}}{1 + [GL(t - D_{GL})/t_{GL}]^{n_{GL}}} + b_{SS}\frac{[BG(t)/t_{BG}]^{n_{BG}}}{1 + [BG(t)/t_{BG}]^{n_{BG}}} \quad (2) $$

Here, $GL(t)$, $SS(t)$, $BG(t)$, $I_1(t)$, and $I_2(t)$ denote the concentrations of glucagon, somatostatin, BG, and exogenous switch-off signal(s) [acting on the pulsatile or/and the basal glucagon secretion], respectively; the derivative is with respect to time, t. The meaning of the remaining parameters is explained in the following section (see also Farhy and McCall, 2009). We note that the presence of two terms, $I_1(t)$ and $I_2(t)$, which represent the switch-off signal in Eq. (1), reflects the assumption that different switch-off signals may have a different impact on glucagon secretion and may differently suppress the basal and/or δ-cell-regulated glucagon release.

We have used the above model (Farhy and McCall, 2009) to show that the glucagon control axis, postulated in Fig. 1, is consistent with the experimental findings listed in (i)–(vii) [above] and that insulin and somatostatin differently affect the basal and the system-regulated α-cell activity. After the model was validated, we used it to predict the outcome of different switch-off strategies and explored its potential to improve the GCR during β-cell deficiency, as shown in Fig. 2 (Farhy and McCall, 2009). The figure summarizes results from the *in silico* experiments, tracking the dynamics of glucagon from time $t = 0$ h (start) to $t = 4$ h (end). In some simulations, intrapancreatic infusion of insulin or somatostatin started at $t = 0.5$ h and was either continued to the end or was switched off at $t = 2.5$ h. When hypoglycemia was simulated, BG = 110 mg/dL from $t = 0$ h to $t = 2$ h, glucose decline started at $t = 2$ h, BG = 60 mg/dL at $t = 2.5$ h (switch-off point), at the end of the simulations ($t = 4$ h) BG = 43 mg/dL. At the top of the bar graph (a), we show baseline results, without the switch-off signals. The black bar illustrates the glucagon level before $t = 2$ h, which is the time when BG = 110 mg/dL and

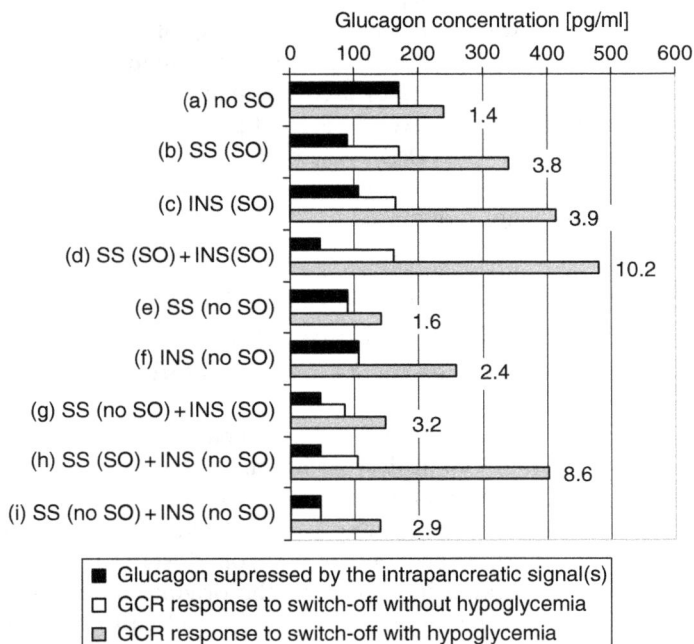

Fig. 2 Summary of the model-predicted GCR responses to different switch-off signals with or without simulated hypoglycemia (see text for more detail). SO, switch-off; no SO, the signal was not switched off; SS, somatostatin; INS, insulin. Modified from Farhy and McCall (2009).

glucagon would be maximally suppressed if a switch-off signal were present. The white and the gray bars illustrate the maximal glucagon response in a 1-h interval from $t = 2.5$ h to $t = 3.5$ h without (white) and with (gray) a hypoglycemia stimulus. This interval corresponds to the 1-h interval after a switch-off in all other simulations. The black and white bars are the same since glucagon levels will remain unchanged if there is no hypoglycemia. Each subsequent set of three bars indicates the effects with single switch-off [(b) and (c)], combined switch-off (d), and no switch-off of a single signal [(e) and (f)], a mixture of switch-off and no switch-off for the two signals [(g) and (h)], and no switch-off for a combination of the two signals (i). Thus, the bar graph gives the following results: glucagon suppression by the intrapancreatic signal (black bars: the glucagon concentration immediately before the onset of BG decline at $t = 2$ h: at that time glucagon is maximally suppressed by the intrapancreatic infusion and is not affected by the decline in glucose); GCR response to a switch-off if hypoglycemia was not induced (white bars: the maximal glucagon concentrations achieved within a 1-h interval after the switch-off); and GCR response if hypoglycemia was induced (gray bars: the maximal glucagon concentrations achieved within a 1-h interval after the switch-off). The graph also includes (at the right of the grey bars) the maximal fold increase in glucagon in response to a switch-off during hypoglycemia which is relative to the glucagon levels before the onset of a decline in BG.

Thus, we concluded that the impact of an α-cell inhibitor on the GCR depends on the nature of the signal and the mode of its delivery. These comparisons between the strategies used for manipulating the network, to enhance the GCR by a switch-off, revealed a good potential for the combined switch-off to amplify the benefits provided by each of the individual signals (Farhy and McCall, 2009), and even a potential to explore certain scenarios in which the α-cell suppressing signal is not terminated.

VII. Approximation of the Normal Endocrine Pancreas by a Minimal Control Network (MCN) and Analysis of the GCR Abnormalities in the Insulin-Deficient Pancreas

The explicit involvement of somatostatin in the model described above limits its potential clinical application, as pancreatic somatostatin cannot be reliably measured in humans *in vivo,* and therefore, the ability of the model to describe the human glucagon axis cannot be verified. It is, however, possible to simplify the network in such a way that somatostatin is no longer explicitly involved, but is incorporated implicitly. In the original model shown in Fig. 1, somatostatin appears in the following two compound pathways, the "α-cell → δ-cell → α-cell" feedback loop and in the "BG → δ-cell → α-cell" pathway. By virtue of its interactions in the "α-cell → δ-cell → α-cell" pathway, the α-cells effectively control their own activity and therefore this pathway can be replaced by a delayed "α-cell → α-cell" autofeedback loop. Such regulation is also consistent with reports, which suggest that glucagon may directly suppress its own release (Kawai and Unger, 1982), possibly by binding to glucagon receptors located on a subpopulation of the α-cells (Kieffer *et al.*, 1996) or by other autocrine mechanisms. Through the "BG → δ-cell → α-cell" pathway, BG downregulates the release of glucagon and this action is mediated by somatostatin. Therefore, this pathway can be simplified and substituted by a BG → α-cell interaction. The outcome of the described procedure of network reduction is a new MCN of GCR control, in which somatostatin and the δ-cells are no longer explicitly involved (Fig. 3). As was originally proposed in our prior work (Farhy *et al.*, 2009), the β-cells of the normal pancreas are now part of the MCN (and thus, of the mathematical model). This feature also extends the physiological relevance of the model. The β-cells are assumed to be stimulated by hyperglycemia and assumed to suppress the activity of the α-cells. The latter action is based on extensive data which show that the β-cells (co)release a variety of signals, including insulin, GABA, zinc, and amylin; all of which are known to suppress α-cell activity (Gedulin *et al.*, 1997; Ishihara *et al.*, 2003; Ito *et al.*, 1995; Reaven *et al.*, 1987; Rorsman and Hellman, 1988; Samols and Stagner, 1988; Van Schravendijk *et al.*, 1987; Wendt *et al.*, 2004; Xu *et al.*, 2006). In addition, it has been reported that the pulses of insulin and glucagon recur with a phase shift (Grapengiesser *et al.*, 2006); this is consistent with the postulated negative regulation of the α-cells by the β-cells. An extensive background that justifies all the postulated MCN relationships has been presented in Section V.

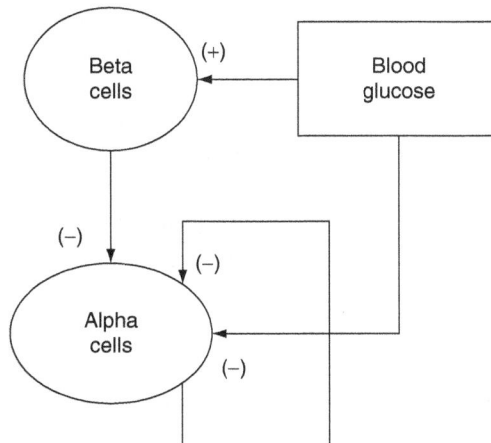

Fig. 3 A Minimal Control Network (MCN) of the interactions between BG and the α- and β-cells postulated to regulate the GCR in the normal pancreas. In this network the δ-cells are not represented explicitly.

A. Dynamic Network Approximation of the MCN

Similar to the analysis presented with the old network, dynamic network modeling methods were used to study the properties of the MCN, as shown in Fig. 3. In particular, two differential equations approximate the rate of change of glucagon and insulin concentration:

$$GL' = -k_{GL}GL + r_{GL,basal}\frac{t_{INS}}{t_{INS} + INS} + r_{GL}\frac{1}{1 + (BG/t_{BG})^{n_{BG}}}$$
$$\times \frac{1}{1 + [GL(t - D_{GL})/t_{GL}]^{n_{GL}}}\frac{t_{INS}}{t_{INS} + INS} \tag{3}$$

$$INS' = -k_{INS}INS + \left[r_{INS}\frac{(BG/t_{BG,2})^{n_{BG,2}}}{1 + (BG/t_{BG,2})^{n_{BG,2}}} + r_{INS,basal}\right] \times Pulse \tag{4}$$

Here, $GL(t)$, $BG(t)$, and $INS(t)$ denote time-dependent concentrations of glucagon, BG, and insulin (or the exogenous switch-off signal in the β-cell-deficient model) respectively; the derivative is the rate of change with respect to the time t. The term *Pulse* in Eq. (4) denotes a pulse generator specific to the β-cells, superimposed to guarantee the physiological relevance of the simulations. The meaning of the parameters is defined as follows:

k_{GL} and k_{INS} are rates of elimination for glucagon and insulin, respectively;

r_{GL} is the BG- and autofeedback-regulated rate of release of glucagon;

$r_{GL,basal}$ is glucagon basal rate of release;

r_{INS} is BG-regulated rate of release of insulin;

$r_{INS,basal}$ is insulin basal rate of release;

t_{INS} is half-maximal inhibitory dose for the negative action of insulin on glucagon(ID_{50});

t_{BG} and $t_{BG,2}$ are half-maximal inhibitory doses for BG (ID_{50});

t_{GL} is half-maximal inhibitory dose for glucagon (ID_{50});

n_{BG}, $n_{BG,2}$, and n_{GL} are Hill coefficients describing the slope of the corresponding dose–response interactions;

D_{GL} is delay in the autofeedback.

B. Determination of the Model Parameters

The half-life ($t_{1/2}$) of glucagon was assumed to be ~2 min, to match the results of our pulsatility analysis (Farhy *et al.*, 2009) and other published data. Therefore, we fixed the parameter $k_{GL} = 22$ h^{-1}. The half-life of insulin was assumed to be ~3 min, as suggested in the literature (Grimmichova *et al.*, 2008). Therefore, to approximate the insulin $t_{1/2}$, we fixed the parameter $k_{INS} = 14$ h^{-1}. The remaining parameters used in the simulations were determined functionally and some of the concentrations presented below are in arbitrary units (specifically, those related to insulin). These units, however, can be easily rescaled to match real concentrations. The delay in the autofeedback, $D_{GL} = 7.2$ min, was functionally determined, together with the potencies, $t_{BG} = 50$ mg/dL, $t_{GL} = 85$ pg/mL and sensitivities $n_{BG} = 5$, $n_{GL} = 5$ in the autofeedback control function, to ensure that glucagon pulses during GCR recur at intervals of 15–20 min, so as to correspond to the number of pulses after a switch-off point which was detected in the pulsatility analysis (Farhy *et al.*, 2009). The parameters $r_{INS} = 80,000$ and $r_{INS,basal} = 270$, together with the amplitude of the pulses of the pulse generator and the parameters $t_{BG,2} = 400$ mg/dL and $n_{BG,2} = 3$ were functionally determined to ensure that BG is capable of stimulating more than a ninefold increase in insulin over baseline in response to a glucose bolus. The ID_{50}, $t_{INS} = 20$, was functionally determined based on the insulin concentrations, to guarantee that insulin withdrawal, during hypoglycemia, will trigger the GCR. The glucagon release rate ($r_{GL} = 42,570$ pg/ mL/h) and basal secretion rate ($r_{GL,basal} = 2128$ pg/mL/h) were functionally determined, so that a strong hypoglycemic stimulus can trigger a more than 10-fold increase in glucagon from the normal pancreas. The parameters of the pulse generator, *Pulse*, were chosen to generate, every 6 min, a square wave of height = 10 over a period of 36 s, based on published reports on insulin pulsatility, which report that insulin pulses recur every 4–12 min (Pørksen, 2002). We note that insulin pulsatility was modeled to mimic the variation of insulin in the portal vein, rather than in systemic circulation. This explains the deep nadirs between the pulses, which is evident in the simulations. The parameter values of the model are summarized in Table I.

C. *In Silico* Experiments

The simulations were performed as follows:
Simulation of glucose input to the system. We performed two different simulations to mimic hypoglycemia: (a) BG decline from 110 to 60 mg/dL in 1 h and (b) a

Table I
Summary of Core Interactive Constants in the Autofeedback MCN

	Rate constant		Dose–response control functions		
	Elimination (1 h^{-1})	Release (concentration/h)	ID_{50} (concentration)	Slope	Delay (min)
Glucagon	$k_{GL} = 22$ h^{-1}	$r_{GL} = 42,570$ pg/mL/h $r_{GL,basal} = 2128$ pg/mL/h	$t_{GL} = 85$ pg/mL	$n_{GL} = 5$	$D_{GL} = 7.2$ min
BG			$t_{BG} = 50$ mg/dL $t_{BG,2} = 400$ mg/dL	$n_{BG} = 5$ $n_{BG,2} = 3$	
Insulin	$k_{INS} = 14$ h^{-1}	$r_{INS} = 80,000$ $r_{INS,basal} = 270$	$t_{INS} = 20$		
Pulse	Periodic function: a square wave of height = 10 over a period of 36 s recurring every 6 min				

stepwise (1 h steps) decline in BG from 110 to ~60 (same as in (a)), then to ~45, and then to ~42 mg/dL. This stepwise decline into hypoglycemia was used to investigate a possible distinction between the responses of the model to 60 mg/dL (a) and to a stronger hypoglycemic stimulus (b); it also mimics a commonly employed human experimental condition (staircase hypoglycemic clamp). To generate glucose profiles that satisfy (a) and (b), we used the equation $BG' = -3BG + 3 \times \text{step} + 330$, where the function *step* changes from 110 to 60, 45, and 42 mg/dL at 1-h steps. Then, we used the solution to the above equation in Eqs. (3) and (4). Similarly, an increase of glucose was simulated by using the above equation and a step function which increased the BG levels from 110 to 240 mg/dL, to mimic an acute hyperglycemia.

Transition from a normal state to an insulin-deficient state. This simulation was performed by gradually reducing, to zero, the amplitude of the pulses generated by the pulse generator, *Pulse*.

Simulation of an intrapancreatic infusion of different α-cell suppressing signals. These simulations were performed in an insulin-deficient model. Equation (4) is replaced by an equation which describes the dynamics of the infused signal:

$$SO' = -k_{SO}SO + \text{Infusion}(t)$$

Here, SO represents the concentration of the switch-off signal, and is an abrupt termination of an α-cell suppressing signal. The function *Infusion* describes the rate of its intrapancreatic infusion (equal to Height if the signal is infused or is 0, otherwise) and k_{SO} is its (functional) rate of elimination. Then, the terms $(1 + m_1 \times SO)$ and $(1 + m_2 \times SO)$ are used in Eq. (3) to divide the parameters r_{GL} and $r_{GL,basal}$, respectively, to simulate suppression of the α-cell activity by the signal. Differences in the parameters m_1 and m_2, model the unequal action of the infused signal on the basal and BG/autofeedback-regulated glucagon secretion. In particular, to simulate an insulin switch-off, we used parameters $k_{SO} = 3$, Height $= 55$, $m_1 = 0.08$, and $m_2 = 0.5$; to simulate somatostatin switch-off, we used $k_{SO} = 3.5$, Height $= 10$, $m_1 = 1$, and $m_2 = 1.4$. These parameters were functionally determined to explain our experimental observations (below) and the possible differences in the glucagon response to the two types of switch-offs (Farhy *et al.*, 2009). In particular, the effect of exogenous insulin on BG/autofeedback-regulated and basal glucagon secretion resembles a 1:6.3 ratio. Similar to our previous work (Farhy and McCall, 2009), exogenous insulin suppressed the basal, more than the pulsatile, glucagon release; for somatostatin, the suppressive effect was more uniform, with a 1:1.4 ratio.

D. Validation of the MCN

To validate the new network we performed an *in silico* study in three steps:

- *Demonstrate that the (new) MCN (Fig. 3) is compatible with the mechanism of GCR and its response to "switch-off" signals during insulin deficiency.* We have already shown that our original network, which includes somatostatin as an explicit node, is consistent with key experimental data. To confirm that the new MCN can substitute the older, more complex, construct, we tested the hypothesis that it can approximate the same key experimental observations [(i)–(vii) listed in the beginning of Section VI], which were predicted by the old network (Fig. 1).

• *Show that the mechanisms underlying the dysregulation of GCR in insulin deficiency can be explained by the MCN.* To this end, we demonstrated that the BG-regulated MCN can explain (i) a high GCR response if the β-cells are intact and can provide a potent switch-off signal to the α-cells; and (ii) a reduction in GCR following a simulated gradual decrease in insulin secretion to mimic the transition from normal physiology to an insulinopenic state.

• *Verify that the proposed MCN approximates the basic properties of the normal endocrine pancreas.* Even though our primary goal is to explain GCR control mechanisms and their dysregulation, we have demonstrated that the postulated MCN can explain the increase in insulin secretion and the decrease in glucagon release in response to BG stimulation.

The goal of this *in silico* study was to validate the MCN by demonstrating that the parameters of the mathematical model (Eqs. (3) and (4)), that approximate the MCN (Fig. 3), can be determined in such a way that the output of the model can predict certain general features of the *in vivo* system. Therefore, the simulated profiles are expected to reproduce the overall behavior of the system, rather than exactly match the experimentally observed, individual hormone, dynamics.

To integrate the equations, we used a Runge–Kutta 4 algorithm and its specific implementation within the software package, Berkeley-Madonna.

E. *In Silico* Experiments with Simulated Complete Insulin Deficiency

We demonstrate that the proposed MCN model, which has changed significantly since it was initially introduced (Farhy and McCall, 2009; Farhy *et al.*, 2009), is consistent with the experimental observations, reported by us and others, in STZ-treated rats (Farhy *et al.*, 2009; Zhou *et al.*, 2004).

1. Defective GCR Response to Hypoglycemia with the Absence of a Switch–Off Signal in the Insulin-Deficient Model

The plot in Fig. 4 (bottom left panel) shows the predicted lack of a glucagon response to hypoglycemia if a switch-off signal is missing—a key observation reported in our experimental study (Farhy *et al.*, 2009) and by others (Zhou *et al.*, 2004, 2007a,b). The system responds with only about 30% increase in the pulse amplitude of glucagon in the 45-min interval after BG reaches 60 mg/dL; this agrees with our experimental observations (Fig. 4, top panels) and shows that the model satisfies condition (iv) from Section VI (no GCR response to hypoglycemia without a switch-off signal).

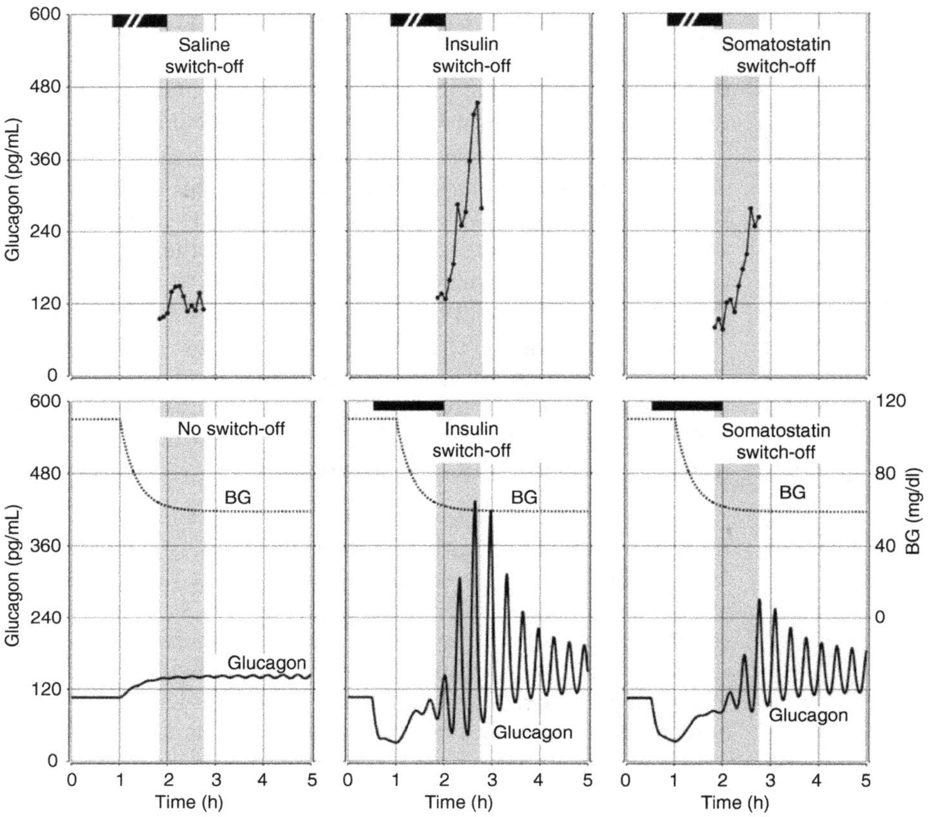

Fig. 4 The mean observed (top) and model-predicted (bottom) glucagon response to hypoglycemia and saline switch-off or no switch-off (left), insulin switch-off (middle), and somatostatin switch-off (right). The shaded area marks the period monitored in our experimental study. The simulations were performed with a complete insulin deficiency.

2. GCR Response to Switch–Off Signals in Insulin Deficiency

The response of the model to a 1.5-h intrapancreatic infusion of insulin or somatostatin, switched off at hypoglycemia (BG = 60 mg/dL), is shown in the bottom middle and right panels of Fig. 4. The infusion was initiated at time $t = 0.5$ h (arbitrary time units) and switched off at $t = 2$ h. A simulated gradual decline in BG started at $t = 1$ h and BG = 60 mg/dL was attained at the switch-off point. This response illustrates a pulsatile, rebound, glucagon secretion after the switch-off with an almost fourfold increase in glucagon in the 45-min period after the switch-off compared to the preswitch-off levels; and this is similar to experimental observations: Fig. 4 (top middle and right panels).

Therefore, the model satisfies the following conditions (i) the pulsatility timing and (ii) the pulsatility amplitude increase from Section VI, in regard to insulin and somatostatin switch-off.

In addition, the bottom middle and right panels of Fig. 4 demonstrate that the model satisfies conditions (vi) >30% higher GCR response to insulin switch-off compared to somatostatin switch-off and (vii) a better glucagon suppression by somatostatin before a switch-off compared to suppression by insulin, from Section VI. Of interest is the prediction that an insulin switch-off signal more potently suppresses the basal, rather than the pulsatile, glucagon release and that it is similar to the predictions of the previous model (Farhy and McCall, 2009).It is necessary to explain the differences between the insulin switch-off and somatostatin switch-off: Fig. 4, middle versus left panels.

Note that the pulsatility of glucagon is not apparent in the plots presented in the top panels of Fig. 4 since they reflect averaged experimental data ($n = 7$ in the saline group and $n = 6$ in the insulin and somatostatin switch-off groups). In Farhy et al. (2009), glucagon pulsatility was confirmed on the individual profiles of glucagon measured in circulation by deconvolution analysis. The current simulations, which approximate the dynamics of glucagon in the portal circulation, agree well with these results.

3. Reduction of the GCR Response by High Glucose Conditions During the Switch-Off Or by Failure to Terminate the Intrapancreatic Signal

For purposes of comparison, Fig. 5 depicts the GCR response when an insulin signal was infused and then switched off but without hypoglycemia (top panel) or if the intrapancreatic insulin was infused, but not switched off during hypoglycemia (bottom panel). In the first simulation, glucagon increases by only 60 pg/mL, relative to the concentration at the switch-off point, but in the second simulation the GCR response is reduced by approximately twofold, compared to the response depicted in Fig. 4 (middle bottom panel). This result agrees with the observations reported by Zhou et al. (2004) who demonstrate a lack of significant increase in glucagon in this 1 h interval if insulin is not switched off. In an additional analysis (results not shown), we increased the simulated rate of insulin infusion switch-off signal by fourfold by increasing the parameter Height from 55 to 220 (see Section VII.C) and using a stronger hypoglycemic stimulus ($\sim$40 mg/dL). The model responded with an increase in glucagon after the switch-off, which reached concentrations above 800 pg/mL in the 1-h interval after the switch-off point. When the same signal was not terminated in this simulation, the response was restricted to a rise till about 180 pg/mL only. This outcome reproduces, more closely, the observations by Zhou et al. (2004).

Thus, the model satisfies the following conditions: (iii) restriction of the response to an insulin switch-off under high BG conditions and (v) the absence of a pronounced GCR when no insulin switch-off is performed, as detailed in Section VI.

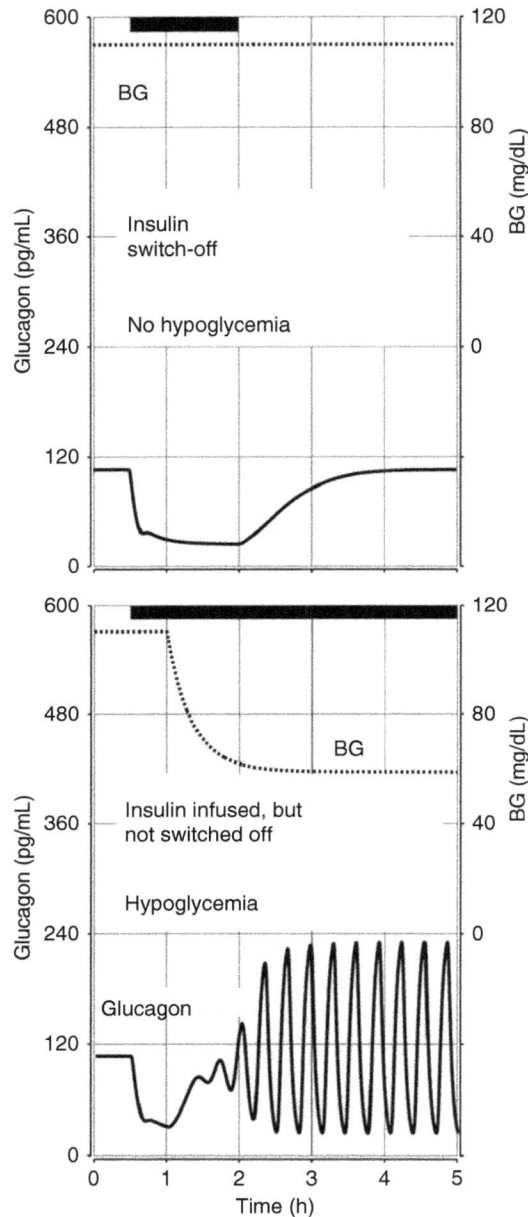

Fig. 5 Model-predicted minimal absolute glucagon response to insulin switch-off if the intrapancrea-tic signal (black bar) is terminated during euglycemia (top panel: glucagon increases minimally with only 60 pg/mL greater than the concentration at the switch-off point) and to insulin intrapancreatic insulin infusion if the signal is not switched off (bottom panel: glucagon increases only 85 pg/mL greater than to the concentration at the time when BG = 60 mg/dL and only about twofold relative to baseline)—these values are by contrast increased more than 3.5-fold when the switch-off occurs—see Fig. 4, the bottom middle panel. All of these simulations were performed with a complete insulin deficiency.

F. Simulated Transition from a Normal Physiology to an Insulinopenic State

One set of simulations was performed to evaluate the model-generated glucagon response to a stepwise decline in BG towards hypoglycemia, with both a normal and an insulin-deficient pancreas. The response of the normal model, shown in Fig. 6 (top panel), illustrates a pronounced glucagon response to hypoglycemia (about fourfold increase when BG = 60 mg/dL and about 14-fold increase over baseline when BG approaches 42.5 mg/dL). Of interest, the model predicts that when BG starts to fall, the high-frequency glucagon pulsatility during the basal period, entrained by the insulin pulses, will be replaced by low-frequency oscillations maintained by the α-cell autofeedback response.

The model also predicts that a complete absence of BG-stimulated and basal insulin release will result in the following abnormalities in glucagon secretion and the response to hypoglycemia (Fig. 6, bottom panel):

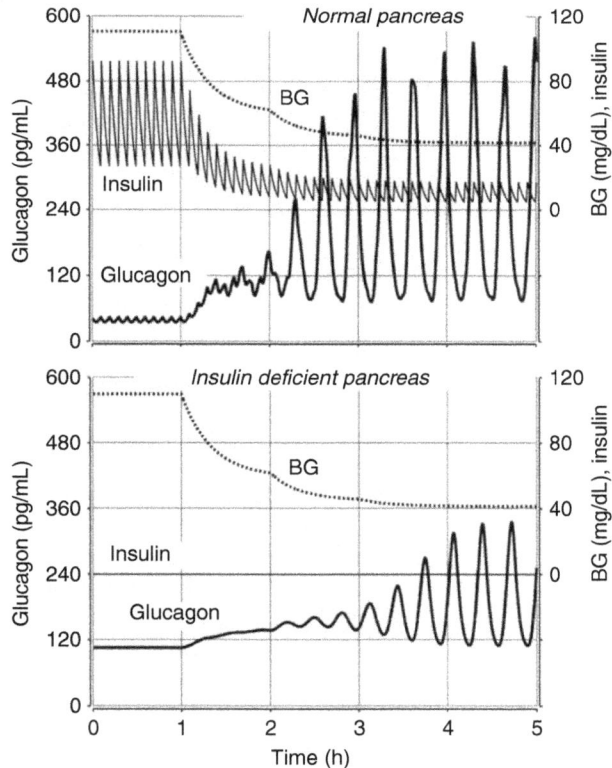

Fig. 6 Model-derived glucagon response to hypoglycemia (stepwise BG decline) in normal physiology with intact insulin release (top) and predicted decrease and delay in GCR following a simulated removal of insulin secretion to mimic a transition from a normal to an insulin-deficient state.

- A significant reduction in glucagon response to hypoglycemia relative to baseline (only about 1.3-fold increase when BG = 60 mg/dL and only about threefold increase when BG approaches 42 mg/dL).
- A reduction in the absolute glucagon response to hypoglycemia (15% lower response when BG = 60 mg/dL and 42% lower response when BG approaches 42.5 mg/dL).
- A delay in the GCR response (BG remains below 60 mg/dL for more than 1 h without any sizable change in glucagon).
- A 2.5-fold increase in basal glucagon.
- Disappearance of the insulin-driven high-frequency glucagon pulsatility.

A comparison between the response of the model to hypoglycemia when BG remained at 60 mg/dL (Fig. 4, lower left panel) and when BG fell further to about 42.5 mg/dL, in the staircase hypoglycemic clamp (Fig. 6, bottom panel), reveals an interesting prediction that a sufficiently strong hypoglycemic stimulus may still evoke some delayed glucagon release. However, additional analysis (results not shown) disclosed that if the basal glucagon release (model parameter $r_{GL,basal}$) is 15–20% higher, this response will be completely suppressed. Therefore, the model predicts that GCR abnormalities may be due to both, the lack of an appropriate switch-off signal and significant basal hyperglucagonemia. The same simulations were also performed under the assumption that BG declines only to 60 mg/dL and remains at that level, similar to the experiments depicted in the lower panels of Fig. 4 (results not shown). We detected that the glucagon pulses released by the normal pancreas were about 47% lower, which stresses the importance of the strength of the hypoglycemic stimulus in eliciting a GCR response. Under conditions of a complete absence of insulin, the weaker hypoglycemic stimulus evokes practically no response (this outcome has already been shown in Fig. 4, lower left panel) and the concentration of glucagon was 57% lower compared to the response stimulated by the stepwise decline (Fig. 6, bottom panel).

A second set of simulations was designed to test the hypothesis that this model of the MCN can correctly predict a typical increase in insulin secretion and a decrease in glucagon following an increase in BG. We also monitored how these two system responses change during the transition from normal physiology to an insulinopenic state. To this end, an increase in BG was simulated (see Section VII.C) with an elevation in the BG concentration from 110 to about 240 mg/dL in 1 h followed by a return to normal levels in the next 1.5 h. The predicted response of the normal pancreas is shown on the top panel of Fig. 7. In this simulation, the BG-driven release of insulin increased by almost ninefold which caused a significant suppression of glucagon release. The bottom plot in Fig. 7 illustrates the effect on the response of the system to a 100% reduction in BG-stimulated insulin release. As expected, insulin deficiency results in an increase in glucagon and a limited ability of hyperglycemia to suppress glucagon (Meier *et al.*, 2006).

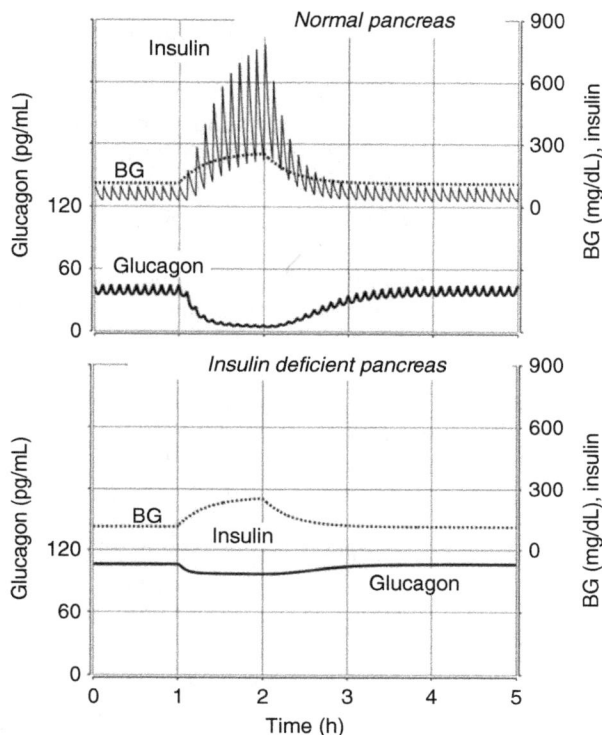

Fig. 7 Simulated progressive decline of the ability of glucose to suppress glucagon resulting from a gradual transition (same as in Fig. 6) from a normal physiology (top) to an insulinopenic state (bottom).

VIII. Advantages and Limitations of the Interdisciplinary Approach

A key conclusion of our model-based simulations is that some of the observed behaviors of the system (like the system response to a switch-off) emerge from an interplay between multiple components. Models, like the networks in Figs. 1 and 3, are certainly not uncommon in endocrine research, and typically exemplify regulatory hypotheses. Traditionally, such models are studied using methods that probe individual components or interactions, in isolation from the rest of the system. This approach has been used in the majority of the published studies that have investigated GCR regulation (see Section V). The limitation of this approach is that the temporal relationships between the components of the system and the relative contribution of each interaction to the overall behavior of the system cannot be properly assessed. Therefore, especially when the model contains feedbacks, the individual approach cannot answer the question of whether the model explains systemic control mechanisms. The main reason for this limitation is that some key specifics of system behavior, like its capability to oscillate and respond with a rebound to a switch-off, both require, and are the result of inter-actions of several components that vary over a time-period. If these are studied in isolation, little information will be gained about the dynamic behavior of this network-like mechanism. Numerous reports have documented that the glucagon control axis is indeed a complex network-like structure, and therefore it lends itself to complex, dynamic behavior analysis. This highlights both the significance and necessity of the mathematical methods that we have used to analyze the experimental data. Using differential equations-based modeling is perhaps the only way to estimate the dynamic interplay of the pancreatic hormones and their importance in GCR control.

Mathematical models have not been applied to study the GCR control mechanisms, but have been used to explore other aspects of BG homeostasis and its control (Guyton et al., 1978; Insel et al., 1975; Steele et al., 1974; Yamasaki et al., 1984). For example, the minimal model of Bergman and colleagues, proposed in 1979 for estimating insulin sensitivity (Bergman et al., 1979), has received considerable attention and been developed further (Bergman et al., 1987; Breda et al., 2001; Cobelli et al., 1986, 1990; Mari, 1997; Quon et al., 1994; Toffolo et al., 1995, 2001). We have previously used modeling methods to successfully estimate and predict the onset of the counterregulation in T1DM patients (Kovatchev et al., 1999, 2000) and to study other complex endocrine axes (Farhy, 2004; Farhy and Veldhuis 2003, 2004, 2005; Farhy et al., 2001, 2002, 2007). However, despite the proven utility of this methodology, our recent efforts were the first to apply a combination of network modeling and in vivo studies to dissect the GCR control axis (Farhy and McCall, 2009; Farhy et al., 2009).

The few selected MCN components cannot exhaustively recreate all the signals that control the GCR. Indeed, in the normal pancreas, glucagon may control its own secretion via α/β-cell interactions. For example, the human β-cells express glucagon receptors (Huypens et al., 2000; Kieffer et al., 1996) and exogenous glucagon stimulates insulin by glucagon- and GLP-1-receptors (Huypens et al., 2000). One immunoneutralization study suggests that endogenous glucagon stimulates insulin (Brunicardi et al., 2001), while other results imply that α-cell glutamate may bind to receptors on β-cells to stimulate insulin and GABA (Bertrand et al., 1992; Inagaki et al., 1995; Uehara et al., 2004).

It has been recently reported that, in human islets, α-cell glutamate serves as a positive autocrine signal for glucagon release by acting on the ionotropic glutamate receptors (iGluRs) on α-cells (Cabrera et al., 2008). Thus, the absence of functional β-cells may cause glutamate hypersecretion, followed by desensitization of the α-cell iGluRs, and ultimately, the defects in GCR, as conjectured (Cabrera et al., 2008). Interestingly, a similar hypothesis to explain the defective GCR in diabetes, by an increase in chronic α-cell activity due to lack of β-cell signaling, can be formulated based on our results. However, in our case hyperglucagonemia is the main reason for the defects in GCR. These two hypotheses are not mutually exclusive, but ours can also explain the *in vivo* GCR pulsatility during hypoglycemia observed by us (Farhy *et al.*, 2009) and others (Genter *et al.*, 1998). Most importantly, the positive autoregulation by α-cells is consistent with the proposed negative and delayed α-cell autofeedback, which could be mediated in part by desensitization of the iGluRs, as suggested (Cabrera *et al.*, 2008). This autocrine regulation is implicitly incorporated in our model equations in the parameter r_{GL}.

The β-cells may control the δ-cells, which are downstream from β-cells in the order of intraislet vascular perfusion. However, in one study, anterograde infusion of insulin antibody in a perfused rat pancreas stimulated both glucagon and somatostatin (Samols and Stagner, 1988), while another immunoneutralization study documented a decrease in somatostatin at high glucose concentrations (Brunicardi *et al.*, 2001). Suppression of the α-cells by insulin (as proposed here) could explain this apparent contradiction. It is also possible that the δ-cells inhibit the β-cells (Brunicardi *et al.*, 2003; Huypens *et al.*, 2000; Schuit *et al.*, 1989; Strowski *et al.*, 2000).

Finally, the MCN components can be influenced by numerous extrapancreatic factors, some of which have important impacts on glucagon secretion and GCR, including autonomic input, catecholamines, growth hormone, ghrelin, and incretins (Gromada *et al.*, 2007; Havel and Ahren, 1997; Havel and Taborsky, 1989; Heise *et al.*, 2004). For example, the incretin, GLP-1, inhibits glucagon, though the mechanism of this inhibition is still controversial (Gromada *et al.*, 2007). Also, there are three major autonomic influences on the α-cell: sympathetic nerves, parasympathetic nerves, and circulating epinephrine, all of which are activated by hypoglycemia and are capable of stimulating glucagon and suppressing insulin (Bolli and Fanelli, 1999; Brelje *et al.*, 1989; Taborsky *et al.*, 1998). We cannot track all signals that control the GCR and most of them have no explicit terms in our model. However, they have not been omitted nor have they been considered unimportant. In fact, when we mathematically describe the MCN, we are including the impact of the nervous system and other factors, even though they have no individual terms in the equations. Thus, the MCN unifies all factors that control glucagon release, based on the assumption that the primary physiological relationships, that are explicit in the MCN, are influenced by these factors.

The model-based simulations suggest that the postulated MCN model of the regulation of GCR is consistent with the experimental data. However, at this stage we cannot estimate how good this model is, and it is therefore hard to assess the validity of its predictions. The simulations can only reconstruct the general "averaged" behavior of the *in vivo* system, and new experimental data are required to support a very important property of the model—that it can explain the GCR response in individual animals. These studies should involve interventional studies which can manipulate the vascular input to the pancreas and analyze the corresponding changes in the output by simultaneously collecting frequently sampled portal vein data for multiple hormones. These must be analyzed by the mathematical model to estimate whether the MCN provides an objectively good description of the actions of the complex GCR control mechanism. Note that, with this approach, we cannot establish the model-based inferences in "micro" detail, since they imply molecular mechanisms that are out of reach of *in vivo* methodology. The approach cannot, nor is it intended to, address the microscopic behavior of the α-cells or the molecular mechanisms that govern this behavior. In this regard, insulin and glucagon (and somatostatin) should be viewed only as (macroscopic) surrogates for the activity of the different cell types, under a variety of other intra- and extrapancreatic influences.

Even though it is usually not stated explicitly, simple models are always used in experimental studies, and, especially in *in vivo* experiments, many factors are ignored or postulated to have no impact on the outcome. Using constructs, like the ones described in this work, to analyze hormone concentration data has the advantage that the underlying model is very explicit, incorporates multiple relationships and uses well-established mathematical and statistical techniques to show its validity and reconstruct the involved signals and pathways.

IX. Conclusions

In the current work, we present our interdisciplinary efforts to investigate the system-level, networking control mechanisms that mediate the GCR and their abnormalities in diabetes—a concept as yet almost completely unexplored for GCR. The results confirm the hypothesis that a streamlined model, which omits an explicit (but not implicit) somatostatin (δ-cell) node entirely, reproduces the results of our original more complex models. Our new findings more precisely define the components that are most critical for the system and strongly suggest that a delayed α-cell autofeedback mechanism plays a key role in GCR regulation. The results demonstrate that such a regulation is consistent not only with most of the *in vivo* system behavior typical for the insulin-deficient pancreas, but also explains key features that are characteristic during the transition from a normal state to an insulin-deficient state. A major advantage of the current model is that its only explicit components are BG, insulin, and glucagon. These are clinically measurable, which would allow the new construct to be used for the study of the control, function, and abnormalities of the human glucagon axis.

Acknowledgment

The study was supported by NIH/NIDDK grants R21 DK072095 and R01 DK082805.

References

Ashcroft, F. M., Proks, P., Smith, P. A., Ammala, C., Bokvist, K., and Rorsman, P. (1994). Stimulus-secretion coupling in pancreatic beta cells. *J. Cell. Biochem.* **55**(Suppl.), 54–65.
Banarer, S., McGregor, V. P., and Cryer, P. E. (2002). Intraislet hyperinsulinemia prevents the glucagon response to hypoglycemia despite an intact autonomic response. *Diabetes* **51**(4), 958–965.

Bell, G. I., Pilkis, S. J., Weber, I. T., and Polonsky, K. S. (1996). Glucokinase mutations, insulin secretion, and diabetes mellitus. *Annu. Rev. Physiol.* **58,** 171–186.

Bergman, R. N., Ider, Y. Z., Boeden, C. R., and Cobelli, C. (1979). Quantitative estimation of insulin sensitivity. *Am. J. Physiol.* **236,** E667–E677.

Bergman, R. N., Prager, R., Volund, A., and Olefsky, J. M. (1987). Equivalence of the insulin sensitivity index in man derived by the minimal model method and the euglycemic glucose clamp. *J. Clin. Invest.* **79,** 790–800.

Bertrand, G., Gross, R., Puech, R., Loubatieres–Mariani, M. M., and Bockaert, J. (1992). Evidence for a glutamate receptor of the AMPA subtype which mediates insulin release from rat perfused pancreas. *Br. J. Pharmacol.* **106**(2), 354–359.

Bolli, G. B., and Fanelli, C. G. (1999). Physiology of glucose counterregulation to hypoglycemia. *Endocrinol. Metab. Clin. North Am.* **28,** 467–493.

Breda, E., Cavaghan, M. K., Toffolo, G., Polonsky, K. S., and Cobelli, C. (2001). Oral glucose tolerance test minimal model indexes of beta-cell function and insulin sensitivity. *Diabetes* **50**(1), 150–158.

Brelje, T. C., Scharp, D. W., and Sorenson, R. L. (1989). Three-dimensional imaging of intact isolated islets of Langerhans with confocal microscopy. *Diabetes* **38**(6), 808–814.

Brunicardi, F. C., Kleinman, R., Moldovan, S., Nguyen, T. H., Watt, P. C., Walsh, J., and Gingerich, R. (2001). Immunoneutralization of somatostatin, insulin, and glucagon causes alterations in islet cell secretion in the isolated per fused human pancreas. *Pancreas* **23**(3), 302–308.

Brunicardi, F. C., Atiya, A., Moldovan, S., Lee, T. C., Fagan, S. P., Kleinman, R. M., Adrian, T. E., Coy, D. H., Walsh, J. H., and Fisher, W. E. (2003). Activation of somatostatin receptor subtype 2 inhibits insulin secretion in the isolated perfused human pancreas. *Pancreas* **27**(4), e84–e89.

Cabrera, O., Jacques–Silva, M. C., Speier, S., Yang, S. N., Köhler, M., Fachado, A., Vieira, E., Zierath, J. R., Kibbey, R., Berman, D. M., Kenyon, N. S., Ricordi, C., *et al.* (2008). Glutamate is a positive autocrine signal for glucagon release. *Cell Metab.* **7**(6), 545–554.

Cejvan, K., Coy, D. H., and Efendic, S. (2003). Intra-islet somatostatin regulates glucagon release via type 2 somatostatin receptors in rats. *Diabetes* **52**(5), 1176–1181.

Cobelli, C., Pacini, G., Toffolo, G., and Sacca, L. (1986). Estimation of insulin sensitivity and glucose clearance from minimal model: New insights from labeled IVGTT. *Am. J. Physiol.* **250,** E591–E598.

Cobelli, C., Brier, D. M., and Ferrannini, E. (1990). Modeling glucose metabolism in man: Theory and practice. *Horm. Metab. Res. Suppl.* **24,** 1–10.

Cryer, P. E. (1999). Hypoglycemia is the limiting factor in the management of diabetes. *Diabetes Metab. Res. Rev.* **15**(1), 42–46.

Cryer, P. E. (2002). Hypoglycemia the limiting factor in the glycaemic management of type I and type II diabetes. *Diabetologia* **45**(7), 937–948.

Cryer, P. E., and Gerich, J. E. (1983). Relevance of glucose counterregulatory systems to patients with diabetes: Critical roles of glucagon and epinephrine. *Diabetes Care* **6**(1), 95–99.

Cryer, P. E., Davis, S. N., and Shamoon, H. (2003). Hypoglycemia in diabetes. *Diabetes Care* **26,** 1902–1912.

Dalla Man, C., *et al.* (2005). Measurement of selective effect of insulin on glucose disposal from labeled glucose oral test minimal model. *Am. J. Physiol.* **289,** E909–E914.

Diem, P., Redmon, J. B., Abid, M., Moran, A., Sutherland, D. E., Halter, J. B., and Robertson, R. P. (1990). Glucagon, catecholamine and pancreatic polypeptide secretion in type I diabetic recipients of pancreas allografts. *J. Clin. Invest.* **86**(6), 2008–2013.

Dumonteil, E., Magnan, C., Ritz–Laser, B., Ktorza, A., Meda, P., and Philippe, J. (2000). Glucose regulates proinsulin and prosomatostatin but not proglucagon messenger ribonucleic acid levels in rat pancreatic islets. *Endocrinology* **141**(1), 174–180.

Dunne, M. J., Harding, E. A., Jaggar, J. H., and Squires, P. E. (1994). Ion channels and the molecular control of insulin secretion. *Biochem. Soc. Trans.* **22**(1), 6–12.

Efendic, S., Nylen, A., Roovete, A., and Uvnas-Wallenstein, K. (1978). Effects of glucose and arginine on the release of immunoreactive somatostatin from the isolated perfused rat pancreas. *FEBS Lett.* **92**(1), 33–35.

Epstein, S., Berelowitz, M., and Bell, N. H. (1980). Pentagastrin and glucagon stimulate serum somatostatin-like immunoreactivity in man. *J. Clin. Endocrinol. Metab.* **51,** 1227–1231.

Farhy, L. S. (2004). Modeling of oscillations in endocrine networks with feedback. *Methods Enzymol.* **384,** 54–81.

Farhy, L. S., and McCall, A. L. (2009). System-level control to optimize glucagon counter-regulation by switch-off of α-cell suppressing signals in β-cell deficiency. *J. Diabetes Sci. Technol.* **3**(1), 21–33.

Farhy, L. S., and Veldhuis, J. D. (2003). Joint pituitary-hypothalamic and intrahypothalamic autofeedback construct of pulsatile growth hormone secretion. *Am. J. Physiol. Regul. Integr. Comp. Physiol.* **285**(5), R1240–R1249.

Farhy, L. S., and Veldhuis, J. D. (2004). Putative GH pulse renewal: Periventricular somatostatinergic control of an arcuate-nuclear somatostatin and GH-releasing hormone oscillator. *Am. J. Physiol. Regul. Integr. Comp. Physiol.* **286**(6), R1030–R1042.

Farhy, L. S., and Veldhuis, J. D. (2005). Deterministic construct of amplifying actions of ghrelin on pulsatile growth hormone secretion. *Am. J. Physiol. Regul. Integr. Comp. Physiol.* **288,** R1649–R1663.

Farhy, L. S., Straume, M., Johnson, M. L., Kovatchev, B., and Veldhuis, J. D. (2001). A construct of interactive feedback control of the GH axis in the male. *Am. J. Physiol. Regul. Integr. Comp. Physiol.* **281**(1), R38–R51.

Farhy, L. S., Straume, M., Johnson, M. L., Kovatchev, B., and Veldhuis, J. D. (2002). Unequal autonegative feedback by GH models the sexual dimorphism in GH secretory dynamics. *Am. J. Physiol. Regul. Integr. Comp. Physiol.* **282**(3), R753–R764.

Farhy, L. S., Bowers, C. Y., and Veldhuis, J. D. (2007). Model-projected mechanistic bases for sex differences in growth-hormone (GH) regulation in the human. *Am. J. Physiol. Regul. Integr. Comp. Physiol.* **292,** R1577–R1593.

Farhy, L. S., Du, Z., Zeng, Q., Veldhuis, P. P., Johnson, M. L., Brayman, K. L., and McCall, A. L. (2009). Amplification of pulsatile glucagon secretion by switch-off of α-cell suppressing signals in streptozotocin treated rats. *Am. J. Physiol. Endocrinol. Metab.* **295,** E575–E585.

Fioramonti, X., *et al.* (2010). Ventromedial hypothalamic nitric oxide production is necessary for hypoglycemia detection and counterregulation. *Diabetes* **59**(2), 519–528.

Fujitani, S., Ikenoue, T., Akiyoshi, M., Maki, T., and Yada, T. (1996). Somatostatin and insulin secretion due to common mechanisms by a new hypoglycemic agent, A-4166, in perfused rat pancreas. *Metab. Clin. Exp.* **45**(2), 184–189.

Fukuda, M., Tanaka, A., Tahara, Y., Ikegami, H., Yamamoto, Y., Kumahara, Y., and Shima, K. (1988). Correlation between minimal secretory capacity of pancreatic beta-cells and stability of diabetic control. *Diabetes* **37**(1), 81–88.

Gedulin, B. R., Rink, T. J., and Young, A. A. (1997). Dose–response for glucagonostatic effect of amylin in rats. *Metabolism* **46,** 67–70.

Genter, P., Berman, N., Jacob, M., and Ipp, E. (1998). Counterregulatory hormones oscillate during steady-state hypoglycemia. *Am. J. Physiol.* **275**(5), E821–E829.

Gerich, J. E. (1988). Lilly lecture: Glucose counterregulation and its impact on diabetes mellitus. *Diabetes* **37**(12), 1608–1617.

Gerich, J. E., Langlois, M., Noacco, C., Karam, J. H., and Forsham, P. H. (1973). Lack of glucagon response to hypoglycemia in diabetes: Evidence for an intrinsic pancreatic alpha cell defect. *Science* **182**(108), 171–173.

Gopel, S. O., Kanno, T., Barg, S., and Rorsman, P. (2000a). Patch-clamp characterisation of somatostatin-secreting-cells in intact mouse pancreatic islets. *J. Physiol.* **528**(3), 497–507.

Gopel, S. O., Kanno, T., Barg, S., Weng, X. G., Gromada, J., and Rorsman, P. (2000b). Regulation of glucagon release in mouse-cells by KATP channels and inactivation of TTX-sensitive Na$^+$ channels. *J. Physiol.* **528,** 509–520.

Grapengiesser, E., Salehi, A., Quader, S. S., and Hellman, B. (2006). Glucose induces glucagon release pulses antisynchronous with insulin and sensitive to purinoceptors inhibition. *Endocrinology* **147,** 3472–3477.

Grimmichova, R., Vrbikova, J., Matucha, P., Vondra, K., Veldhuis, P., and Johnson, M. (2008). Fasting insulin pulsatile secretion in lean women with polycystic ovary syndrome. *Physiol. Res.* **57**, 1–8.

Gromada, J., Franklin, I., and Wollheim, C. B. (2007). α-Cells of the endocrine pancreas: 35 Years of research but the enigma remains. *Endocr. Rev.* **28**(1), 84–116.

Guyton, J. R., Foster, R. O., Soeldner, J. S., Tan, M. H., Kahn, C. B., Koncz, L., and Gleason, R. E. (1978). A model of glucose-insulin homeostasis in man that incorporates the heterogeneous fast pool theory of pancreatic insulin release. *Diabetes* **27**, 1027–1042.

Hauge-Evans, A. C., *et al.* (2009). Somatostatin secreted by islet delta-cells fulfills multiple roles as a paracrine regulator of islet function. *Diabetes* **58**(2), 403–411.

Havel, P. J., and Ahren, B. (1997). Activation of autonomic nerves and the adrenal medulla contributes to increased glucagon secretion during moderate insulin-induced hypoglycemia in women. *Diabetes* **46**, 801–807.

Havel, P. J., and Taborsky, G. J. Jr. (1989). The contribution of the autonomic nervous system to changes of glucagon and insulin secretion during hypoglycemic stress. *Endocr. Rev.* **10**(3), 332–350.

Haywood, S. C., *et al.* (2009). Central but not systemic lipid infusion augments the counterregulatory response to hypoglycemia. *Am. J. Physiol. Endocrinol. Metab.* **297**(1), E50–E56.

Heimberg, H., De Vos, A., Pipeleers, D., Thorens, B., and Schuit, F. (1995). Differences in glucose transporter gene expression between rat pancreatic alpha- and beta-cells are correlated to differences in glucose transport but not in glucose utilization. *J. Biol. Chem.* **270**(15), 8971–8975.

Heimberg, H., De Vos, A., Moens, K., Quartier, E., Bouwens, L., Pipeleers, D., Van Schaftingen, E., Madsen, O., and Schuit, F. (1996). The glucose sensor protein glucokinase is expressed in glucagon-producing alpha-cells. *Proc. Natl. Aca. Sci. USA* **93**(14), 7036–7041.

Heise, T., Heinemann, T., Heller, S., Weyer, C., Wang, Y., Strobel, S., Kolterman, O., and Maggs, D. (2004). Effect of pramlintide on symptom, catecholamine, and glucagon responses to hypoglycemia in healthy subjects. *Metabolism* **53**(9), 1227–1232.

Hermansen, K., Christensen, S. E., and Orskov, H. (1979). Characterization of somatostatin release from the pancreas: The role of potassium. *Scand. J. Clin. Lab. Invest.* **39**(8), 717–722.

Hilsted, J., Frandsen, H., Holst, J. J., Christensen, N. J., and Nielsen, S. L. (1991). Plasma glucagon and glucose recovery after hypoglycemia: The effect of total autonomic blockade. *Acta Endocrinol.* **125**(5), 466–469.

Hirsch, B. R., and Shamoon, H. (1987). Defective epinephrine and growth hormone responses in type I diabetes are stimulus specific. *Diabetes* **36**(1), 20–26.

Hoffman, R. P., Arslanian, S., Drash, A. L., and Becker, D. J. (1994). Impaired counter-regulatory hormone responses to hypoglycemia in children and adolescents with new onset IDDM. *J. Pediatr. Endocrinol.* **7**(3), 235–244.

Hope, K. M., Tran, P. O., Zhou, H., Oseid, E., Leroy, E., and Robertson, R. P. (2004). Regulation of alpha-cell function by the beta-cell in isolated human and rat islets deprived of glucose: The "switch-off" hypothesis. *Diabetes* **53**(6), 1488–1495.

Huypens, P., Ling, Z., Pipeleers, D., and Schuit, F. (2000). Glucagon receptors on human islet cells contribute to glucose competence of insulin release. *Diabetologia* **43**(8), 1012–1019.

Inagaki, N., Kuromi, H., Gonoi, T., Okamoto, Y., Ishida, H., Seino, Y., Kaneko, T., Iwanaga, T., and Seino, S. (1995). Expression and role of ionotropic glutamate receptors in pancreatic islet cells. *FASEB J.* **9**(8), 686–691.

Insel, P. A., Liljenquist, J. E., Tobin, J. D., Sherwin, R. S., Watkins, P., Andres, R., and Berman, M. (1975). Insulin control of glucose metabolism in man. A new kinetic analysis. *J. Clin. Invest.* **55**, 1057–1066.

Ishihara, H., Maechler, P., Gjinovci, A., Herrera, P. L., and Wollheim, C. B. (2003). Islet β-cell secretion determines glucagon release from neighboring α-cells. *Nat. Cell Biol.* **5**, 330–335.

Ito, K., Maruyama, H., Hirose, H., Kido, K., Koyama, K., Kataoka, K., and Saruta, T. (1995). Exogenous insulin dose-dependently suppresses glucopenia-induced glucagon secretion from perfused rat pancreas. *Metab. Clin. Exp.* **44**(3), 358–362.

Jaspan, J. B., Lever, E., Polonsky, K. S., and Van Cauter, E. (1986). *In vivo* pulsatility of pancreatic islet peptides. *Am. J. Physiol.* **251**(2 Pt 1), E215–E226.

Kawai, K., and Unger, R. H. (1982). Inhibition of glucagon secretion by exogenous glucagon in the isolated, perfused dog pancreas. *Diabetes* **31**(6), 512–515.

Kawamori, D., Kurpad, A. J., Hu, J., Liew, C. W., Shih, J. L., Ford, E. L., Herrera, P. L., Polonsky, K. S., McGuinness, O. P., and Kulkarni, R. N. (2009). Insulin signaling in alpha cells modulates glucagon secretion *in vivo*. *Cell Metab.* **9**(4), 350–361.

Kieffer, T. J., Heller, R. S., Unson, C. G., Weir, G. C., and Habener, J. F. (1996). Distribution of glucagon receptors on hormone-specific endocrine cells of rat pancreatic islets. *Endocrinology* **137**(11), 5119–5125.

Klaff, L. J., and Taborsky, G. J. Jr. (1987). Pancreatic somatostatin is a mediator of glucagon inhibition by hyperglycemia. *Diabetes* **36**(5), 592–596.

Kleinman, R., Gingerich, R., Wong, H., Walsh, J., Lloyd, K., Ohning, G., De Giorgio, R., Sternini, C., and Brunicardi, F. C. (1994). Use of the Fab fragment for immunoneutralization of somatostatin in the isolated perfused human pancreas. *Am. J. Surg.* **167**(1), 114–119.

Kleinman, R., Gingerich, R., Ohning, G., Wong, H., Olthoff, K., Walsh, J., and Brunicardi, F. C. (1995). The influence of somatostatin on glucagon and pancreatic polypeptide secretion in the isolated perfused human pancreas. *Int. J. Pancreatol.* **18**(1), 51–57.

Kovatchev, B. P., Farhy, L. S., Cox, D. J., Straume, M., Yankov, V. I., Gonder-Frederick, L. A., and Clarke, W. L. (1999). Modeling insulin–glucose dynamics during insulin induced hypoglycemia. Evaluation of glucose counterregulation. *J. Theor. Med.* **1**, 313–323.

Kovatchev, B. P., Straume, M., Farhy, L. S., and Cox, D. J. (2000). Dynamic network model of glucose counterregulation in subjects with insulin-requiring diabetes. *Methods Enzymol.* **321**, 396–410.

Leu, J., *et al.* (2009). Hypoglycemia-associated autonomic failure is prevented by opioid receptor blockade. *J. Clin. Endocrinol. Metab.* **94**(9), 3372–3380.

Ludvigsen, E., Olsson, R., Stridsberg, M., Janson, E. T., and Sandler, S. (2004). Expression and distribution of somatostatin receptor subtypes in the pancreatic islets of mice and rats. *J. Histochem. Cytochem.* **52**(3), 391–400.

Mari, A. (1997). Assessment of insulin sensitivity with minimal model: Role of model assumptions. *Am. J. Physiol.* **272**, E925–E934.

Maruyama, H., Hisatomi, A., Orci, L., Grodsky, G. M., and Unger, R. H. (1984). Insulin within islets is a physiologic glucagon release inhibitor. *J. Clin. Invest.* **74**(6), 2296–2299.

Matthews, D. R., Hermansen, K., Connolly, A. A., Gray, D., Schmitz, O., Clark, A., Orskov, H., and Turner, R. C. (1987). Greater *in vivo* than *in vitro* pulsatility of insulin secretion with synchronized insulin and somatostatin secretory pulses. *Endocrinology* **120**(6), 2272–2278.

McCall, A. L., Cox, D. J., Crean, J., Gloster, M., and Kovatchev, B. P. (2006). A novel analytical method for assessing glucose variability: Using CGMS in type 1 diabetes mellitus. *Diabetes Technol. Ther.* **8**(6), 644–653.

Meier, J. J., Kjems, L. L., Veldhuis, J. D., Lefebvre, P., and Butler, P. C. (2006). Postprandial suppression of glucagon secretion depends on intact pulsatile insulin secretion: Further evidence for the intraislet insulin hypothesis. *Diabetes* **55**(4), 1051–1056.

Pipeleers, D. G., Schuit, F. C., Van Schravendijk, C. F., and Van de Winkel, M. (1985). Interplay of nutrients and hormones in the regulation of glucagon release. *Endocrinology* **117**(3), 817–823.

Pørksen, N. (2002). The *in vivo* regulation of pulsatile insulin secretion. *Diabetologia* **45**(1), 3–20.

Portela-Gomes, G. M., Stridsberg, M., Grimelius, L., Oberg, K., and Janson, E. T. (2000). Expression of the five different somatostatin receptor subtypes in endocrine cells of the pancreas. *Appl. Immunohistochem. Mol. Morphol.* **8**(2), 126–132.

Quon, M. J., Cochran, C., Taylor, S. I., and Eastman, R. C. (1994). Non-insulin mediated glucose disappearance in subjects with IDDM. Discordance between experimental results and minimal model analysis. *Diabetes* **43**, 890–896.

Ravier, M. A., and Rutter, G. A. (2005). Glucose or insulin, but not zinc ions, inhibit glucagon secretion from mouse pancreatic alpha cells. *Diabetes* **54**, 1789–1797.

Reaven, G. M., Chen, Y. D., Golay, A., Swislocki, A. L., and Jaspan, J. B. (1987). Documentation of hyperglucagonemia throughout the day in nonobese and obese patients with noninsulin-dependent diabetes mellitus. *J. Clin. Endocrinol. Metab.* **64**(1), 106–110.

Rorsman, P., and Hellman, B. (1988). Voltage-activated currents in guinea pig pancreatic alpha 2 cells. Evidence for Ca^{2+}-dependent action potentials. *J. Gen. Physiol.* **91**(2), 223–242.

Rorsman, P., Berggren, P. O., Bokvist, K., Ericson, H., Mohler, H., Ostenson, C. G., and Smith, P. A. (1989). Glucose-inhibition of glucagon secretion involves activation of GABAA-receptor chloride channels. *Nature* **341**(6239), 233–236.

Salehi, A., Quader, S. S., Grapengiesser, E., and Hellman, B. (2007). Pulses of somatostatin release are slightly delayed compared with insulin and antisynchronous to glucagon. *Regul. Pept.* **144,** 43–49.

Samols, E., and Stagner, J. I. (1988). Intra-islet regulation. *Am. J. Med.* **85**(5A), 31–35.

Samols, E., and Stagner, J. I. (1990). Islet somatostatin–microvascular, paracrine, and pulsatile regulation. *Metab. Clin. Exp.* **39**(9 Suppl 2), 55–60.

Schuit, F. C., Derde, M. P., and Pipeleers, D. G. (1989). Sensitivity of rat pancreatic A and B cells to somatostatin. *Diabetologia* **32**(3), 207–212.

Schuit, F. C., Huypens, P., Heimberg, H., and Pipeleers, D. G. (2001). Glucose sensing in pancreatic beta-cells: A model for the study of other glucose-regulated cells in gut, pancreas, and hypothalamus. *Diabetes* **50**(1), 1–11.

Schuit, F., De Vos, A., Farfari, S., Moens, K., Pipeleers, D., Brun, T., and Prentki, M. (1997). Metabolic fate of glucose in purified islet cells. Glucose-regulated anaplerosis in beta cells. *J. Biol. Chem.* **272**(30), 18572–18579.

Segel, S. A., Paramore, D. S., and Cryer, P. E. (2002). Hypoglycemia-associated autonomic failure in advanced type 2 diabetes. *Diabetes* **51**(3), 724–733.

Silvestre, R. A., Rodríguez-Gallardo, J., Jodka, C., Parkes, D. G., Pittner, R. A., Young, A. A., and Marco, J. (2001). Selective amylin inhibition of the glucagon response to arginine is extrinsic to the pancreas. *Am. J. Physiol. Endocrinol. Metab.* **280**, E443–E449.

Stagner, J. I., Samols, E., and Bonner-Weir, S. (1988). Beta-alpha-delta pancreatic islet cellular perfusion in dogs. *Diabetes* **37**(12), 1715–1721.

Stagner, J. I., Samols, E., and Marks, V. (1989). The anterograde and retrograde infusion of glucagon antibodies suggests that A cells are vascularly perfused before D cells within the rat islet. *Diabetologia* **32**(3), 203–206.

Steele, R., Rostami, H., and Altszuler, N. (1974). A two-compartment calculator for the dog glucose pool in the nonsteady state. *Fed. Proc.* **33**, 1869–1876.

Strowski, M. Z., Parmar, R. M., Blake, A. D., and Schaeffer, J. M. (2000). Somatostatin inhibits insulin and glucagon secretion via two receptors subtypes: An *in vitro* study of pancreatic islets from somatostatin receptor 2 knockout mice. *Endocrinology* **141**(1), 111–117.

Sumida, Y., Shima, T., Shirayama, K., Misaki, M., and Miyaji, K. (1994). Effects of hexoses and their derivatives on glucagon secretion from isolated perfused rat pancreas. *Horm. Metab. Res.* **26**(5), 222–225.

Taborsky, G. J. Jr., Ahren, B., and Havel, P. J. (1998). Autonomic mediation of glucagon secretion during hypoglycemia: Implications for impaired alpha-cell responses in type 1 diabetes. *Diabetes* **47**(7), 995–1005.

Tapia-Arancibia, L., and Astier, H. (1988). Glutamate stimulates somatostatin release from diencephalic neurons in primary culture. *Endocrinology* **123**, 2360–2366.

The Action to Control Cardiovascular Risk in Diabetes Study Group (2008). Effects of intensive glucose lowering in type 2 diabetes. *N. Engl. J. Med.* **358**, 2545–2559.

The Diabetes Control and Complications Trial Research Group (1993). The effect of intensive treatment of diabetes on the development and progression of long-term complications in insulin-dependent diabetes mellitus. *N. Engl. J. Med.* **329**, 977–986.

Tirone, T. A., Norman, M. A., Moldovan, S., DeMayo, F. J., Wang, X. P., and Brunicardi, F. C. (2003). Pancreatic somatostatin inhibits insulin secretion via SSTR-5 in the isolated perfused mouse pancreas model. *Pancreas* **26**(3), e67–e73.

Toffolo, G., De Grandi, F., and Cobelli, C. (1995). Estimation of beta-cell sensitivity from intravenous glucose tolerance test C-peptide data. Knowledge of the kinetics avoids errors in modeling the secretion. *Diabetes* **44,** 845–854.

Toffolo, G., Breda, E., Cavaghan, M. K., Ehrmann, D. A., Polonsky, K. S., and Cobelli, C. (2001). Quantitative indices of β-cell function during graded up&down glucose infusion from C-peptide minimal models. *Am. J. Physiol.* **280,** 2–10.

Uehara, S., Muroyama, A., Echigo, N., Morimoto, R., Otsuka, M., Yatsushiro, S., and Moriyama, Y. (2004). Metabotropic glutamate receptor type 4 is involved in autoinhibitory cascade for glucagon secretion by alpha-cells of islet of Langerhans. *Diabetes* **53**(4), 998–1006.

UK Prospective Diabetes Study Group (1998). Intensive blood-glucose control with sulphonylureas or insulin compared with conventional treatment and risk of complications in patients with type 2 diabetes. *Lancet* **352,** 837–853.

Unger, R. H. (1985). Glucagon physiology and pathophysiology in the light of new advances. *Diabetologia* **28,** 574–578.

Utsumi, M., Makimura, H., Ishihara, K., Morita, S., and Baba, S. (1979). Determination of immunoreactive somatostatin in rat plasma and responses to arginine, glucose and glucagon infusion. *Diabetologia* **17,** 319–323.

Van Schravendijk, C. F., Foriers, A., Van den Brande, J. L., and Pipeleers, D. G. (1987). Evidence for the presence of type I insulin-like growth factor receptors on rat pancreatic A and B cells. *Endocrinology* **121**(5), 1784–1788.

Wendt, A., Birnir, B., Buschard, K., Gromada, J., Salehi, A., Sewing, S., Rorsman, P., and Braun, M. (2004). Glucose inhibition of glucagon secretion from rat alpha-cells is mediated by GABA released from neighboring beta-cells. *Diabetes* **53**(4), 1038–1045.

Xu, E., Kumar, M., Zhang, Y., Ju, W., Obata, T., Zhang, N., Liu, S., Wendt, A., Deng, S., Ebina, Y., Wheeler, M. B., Braun, M., *et al.* (2006). Intraislet insulin suppresses glucagon release via GABA-GABAA receptor system. *Cell Metab.* **3,** 47–58.

Yamasaki, Y., Tiran, J., and Albisser, A. M. (1984). Modeling glucose disposal in diabetic dogs fed mixed meals. *Am. J. Physiol.* **246,** E52–E61.

Zhou, H., Tran, P. O., Yang, S., Zhang, T., LeRoy, E., Oseid, E., and Robertson, R. P. (2004). Regulation of alpha-cell function by the beta-cell during hypoglycemia in Wistar rats: The "switch-off" hypothesis. *Diabetes* **53**(6), 1482–1487.

Zhou, H., Zhang, T., Oseid, E., Harmon, J., Tonooka, N., and Robertson, R. P. (2007a). Reversal of defective glucagon responses to hypoglycemia in insulin-dependent autoimmune diabetic BB rats. *Endocrinology* **148,** 2863–2869.

Zhou, H., Zhang, T., Harmon, J. S., Bryan, J., and Robertson, R. P. (2007b). Zinc, not insulin, regulates the rat α-cell response to hypoglycemia in vivo. *Diabetes* **56,** 1107–1112.

INDEX

www.ingramcontent.com/pod-product-compliance
Lightning Source LLC
Chambersburg PA
CBHW080348220326
41598CB00030B/4638